W9-COZ-786

Name Reactions
for Functional Group Transformations

Edited by

Jie Jack Li
Pfizer Global Research & Development

E. J. Corey
Harvard University

WILEY-INTERSCIENCE
A John Wiley & Sons, Inc., Publication

Published by John Wiley & Sons, Inc., Hoboken, New Jersey.
Published simultaneously in Canada.

For general information on our other products and services or for technical support, please contact our Customer Care Department within the United States at (800) 762-2974, outside the United States at (317) 572-3993 or fax (317) 572-4002.

Wiley also publishes its books in a variety of electronic formats. Some content that appears in print may not be available in electronic format. For information about Wiley products, visit our web site at www.wiley.com.

Wiley Bicentennial Logo: Richard J. Pacifico

Library of Congress Cataloging-in-Publication Data:

Name reactions for functional group transformations / edited by Jie Jack Li,
E. J. Corey.
p. cm.
Includes index.
ISBN 978-0-471-74868-7
1. Organic compounds—Synthesis. 2. Chemical reactions. I. Li, Jie Jack.
II. Corey, E. J.
QD262.N36 2007
547'.2—dc22 2007010254

Printed in the United States of America.

10 9 8 7 6 5 4 3 2 1

Dedicated To

Li Wen-Liang and Chen Xiao-Ying

Foreword

Part of the charm of synthetic organic chemistry derives from the vastness of the intellectual landscape along several dimensions. First, there is the almost infinite variety and number of possible target structures that lurk in the darkness waiting to be made. Then, there is the vast body of organic reactions that serve to transform one substance into another, now so large in number as to be beyond credibility to a non-chemist. There is the staggering range of reagents, reaction conditions, catalysts, elements, and techniques that must be mobilized in order to tame these reactions for synthetic purposes. Finally, it seems that new information is being added to that landscape at a rate that exceeds the ability of a normal person to keep up with it. In such a troubled setting any author, or group of authors, must be regarded as heroic if through their efforts, the task of the synthetic chemist is eased.

Modern synthetic chemistry is a multifaceted discipline that greatly benefits from the development of unifying concepts. One of the most useful of these is the idea of the "functional group," generally considered to be a specific collection of connected atoms that occur frequently in organic structures and that exhibit well defined and characteristic chemical behavior. The simplest and most common functional groups (e.g., C=C, CHO, OH, COOH, NH_2) dominate the organization of entry-level organic chemistry textbooks and provide a framework for understanding the fundamentals of the subject. The more complex functional groups, formed using additional elements or by concatenation of simpler groups, play a similar unifying role. This volume, *Name Reactions for Functional Group Transformations*, provides a survey of important transformations that are characteristic of the whole range of functional groups and also serve to interconnect them. In the more than six hundred pages that follow, a highly qualified team of nineteen authors from academia and industry has provided an up-to-date account of forty-seven major classes of functional group transformations. The reviews are clear, concise, and well-referenced. This book serves as a fine companion to the first volume of this series, *Name Reactions in Heterocyclic Chemistry*.

E. J. Corey

E. J. Corey

November 13, 2006

Preface

This book is the second volume of the series *Comprehensive Name Reactions*, an ambitious project conceived by Prof. E. J. Corey of Harvard University in the summer of 2002. Volume 1, *Name Reactions in Heterocyclic Chemistry*, was published in 2005 and was warmly received by the organic chemistry community. After publication of the current Volume 2 in 2007, we plan to roll out Volume 3, *Name Reactions on Homologation,* in 2009; Volume 4, *Name Reactions on Ring Formation* in 2011; and Volume 5, *Name Reactions in Heterocyclic Chemistry-2*, in 2013, respectively.

Continuing the traditions of Volume 1, each name reaction in Volume 2 is also reviewed in seven sections:

1. Description;
2. Historical Perspective;
3. Mechanism;
4. Variations and Improvements;
5. Synthetic Utility;
6. Experimental; and
7. References.

I also have introduced a symbol [R] to highlight review articles, book chapters, and books dedicated to the respective name reactions.

I have incurred many debts of gratitude to Prof. E. J. Corey. What he once told me — "*The desire to learn is the greatest gift from God*" — has been a true inspiration. Furthermore, it has been my great privilege and a pleasure to work with a collection of stellar contributing authors from both academia and industry. Some of them are world-renowned scholars in the field; some of them have worked intimately with the name reactions that they have reviewed; some of them even discovered the name reactions that they authored in this book. As a consequence, this book truly represents the state-of-the-art for *Name Reactions for Functional Group Transformations*.

I welcome your critique.

Jack Li

October 24, 2006

Contributing authors:

Dr. Nadia M. Ahmad
Institute of Cancer Research
Haddow Laboratories
15 Cotswold Road
Sutton, Surrey
SM2 5NG, UK

Dr. Marudai Balasubramanian
Research Informatics
Pfizer Global Research & Development
2800 Plymouth Road
Ann Arbor, MI 48105

Dr. Alice R. E. Brewer
Novartis Horsham Research Centre
Wimblehurst Road
Horsham, West Sussex
RH12 5AB, UK

Dr. Julia M. Clay
Chemistry Department
Princeton University
Princeton, NJ, 08544-1009

Dr. Timothy T. Curran
Department of Chemical R&D
Pfizer Global Research & Development
2800 Plymouth Road
Ann Arbor, MI 48105

Dr. Matthew J. Fuchter
Department of Chemistry
Imperial College London
Exhibition Road, London
SW7 2AZ, UK

Dr. Paul Galatsis
Department of Chemistry
Pfizer Global Research & Development
2800 Plymouth Road
Ann Arbor, MI 48105

Prof. Gordon W. Gribble
Department of Chemistry
6128 Burke Laboratory
Dartmouth College
Hanover, NH 03755

Dr. Timothy J. Hagen
Department of Chemistry
Pfizer Global Research & Development
2800 Plymouth Road
Ann Arbor, MI 48105

Dr. Daniel D. Holsworth
Department of Chemistry
Pfizer Global Research & Development
2800 Plymouth Road
Ann Arbor, MI 48105

Dr. Donna M. Iula
Department of Chemistry
Pfizer Global Research & Development
2800 Plymouth Road
Ann Arbor, MI 48105

Dr. Jacob M. Janey
Process Research
Merck Research Laboratories
P. O. Box 2000 RY800-B363
Rahway, NJ 07065-0900

Dr. Manjinder S. Lall
Department of Chemistry
Pfizer Global Research &
Development
2800 Plymouth Road
Ann Arbor, MI 48105

Dr. Jie Jack Li
Department of Chemistry
Pfizer Global Research & Development
2800 Plymouth Road
Ann Arbor, MI 48105

Dr. Jin Li
Medicinal Chemistry
BioDuro
No. 5, KaiTuo Road
Beijing, PRC 100085

Dr. Dustin J. Mergott
Lilly Research Laboratories
Eli Lilly and Company
Indianapolis, IN 46285

Prof. Richard J. Mullins
Department of Chemistry
Xavier University
3800 Victory Parkway
Cincinnati, OH 45207-4221

Prof. Kevin M. Shea
Department of Chemistry
Clark Science Center
Smith College
Northampton, MA 01063

Prof. John P. Wolfe
Department of Chemistry
University of Michigan
930 N. University Avenue
Ann Arbor, MI 48109

Table of Contents

Chapter 1 Asymmetric Synthesis 1

1.1 CBS Reduction

1.1.1 Description

The Corey–Bakshi–Shibata (CBS) reduction[1] employs the use of borane in conjunction with a chiral oxazaborolidine catalyst to conduct enantioselective reductions of ketones.

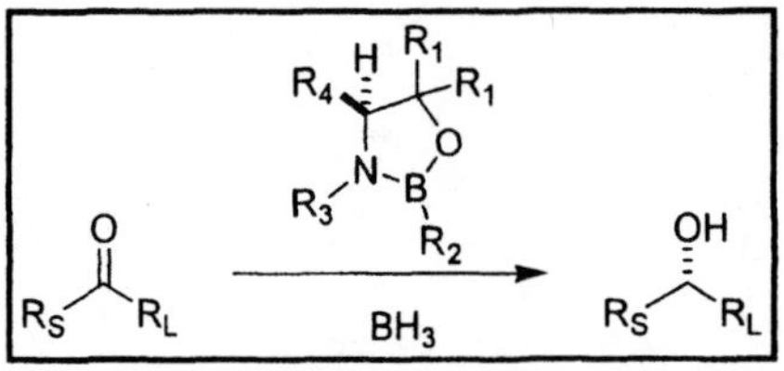

This reduction method has a number of advantages that include wide scope, predictable absolute stereochemistry, ready availability of the chiral catalyst in both enantiomeric forms, high yields, experimental ease, recovery of the catalyst (as the amino alcohol), and low cost of goods. The most common form of the chiral oxazaborolidine is derived from prolinol and has a methyl substituent on the boron atom (*B*-Me-CBS) **1**. When one conducts a reduction on a novel system for the first time, this catalyst provides a good compromise of cost, enantioselectivity, and experimental ease. If sufficient control is not observed with this reagent, one can then systematically evaluate the numerous variations of this framework.

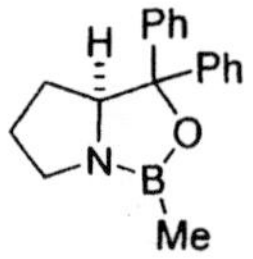

1.1.2 Historical Perspective

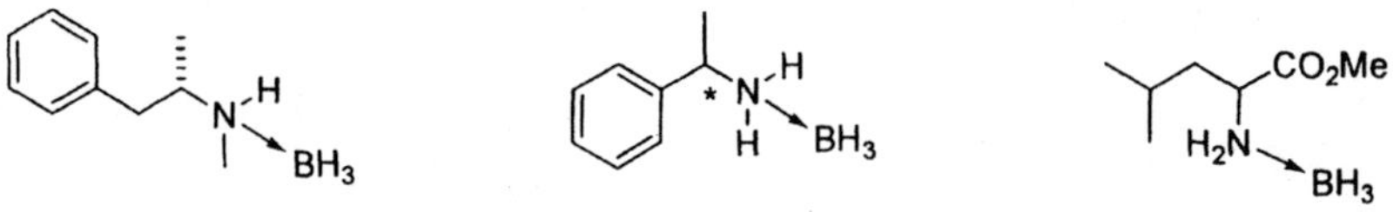

The use of optically active borane reagents for asymmetric reductions was first reported by Fiaud and Kagan in 1969.[2] These workers used the desoxyephedrine–boron complex **2** as a reductant. However, the asymmetric induction was very poor and no greater than 5% *ee* in the reduction of acetophenone was observed. Borch observed similar results employing *R*-(+)- and *S*-(−)-α-phenethylamine–borane complexes **3**[3] as the chiral reagent with a variety of ketones.

Continuing on this tack, Grundon and co-workers[4] were able to obtain optical purities in the range of 14–22% *ee*. They achieved this improvement by employing 1:1 complexes of leucine methyl ester and diborane **4** in THF. Furthermore, their results were facilitated by the addition of one equivalent of BF_3–etherate. Other chiral auxiliaries used include *L*-valine methyl ester and β-phenylalanine methyl ester.

A major advance in the evolution of chiral boron reagents was reported initially by Itsuno and co-workers in 1981.[5] Stereoselectivities up to 73% *ee* were observed using the 1,3,2-oxazaborolidine derived from β-amino alcohols. Thus (*S*)-valinol **5** in reaction with borane afforded **6**.

OH
NH_2
2 BH_3
H
H N O B
H_3B H
+ 2 H_2↑
5 6

This result sat dormant in the literature until a thorough review of B- and Al-based reductants with chiral auxiliaries was conducted by the Corey group. They were intrigued by the work of Itsuno and began detailed studies of the reaction to understand the mechanistic and stereochemical underpinnings of this reduction reaction. Their efforts resulted in the CBS reduction[6] in which improved chiral auxiliaries (**7** → **8**) were developed and a model was formulated to rationalize the stereochemical outcome of this reaction.

H Ph Ph OH
N H
BH_3–THF
H Ph Ph
N B O
H
7 8

1.1.3 Mechanism

The great utility of this asymmetric reduction system is a result of the detailed and systematic analysis of its mechanism by the Corey group at Harvard and others.[1f, 6, 7] Using the Itsuno conditions as a starting point, the Corey laboratories obtained pure (after sublimation) oxazaborolidine **10** from the reaction of amino alcohol **9** with two equivalents of BH_3–THF at 35 °C. The structure of this intermediate was confirmed by FT–IR, ^{1}H and ^{11}B NMR, and mass spectroscopy.

Ph Ph H_2N OH — 2 eq BH_3–THF → Ph Ph H–N B O R + 2 H_2↑

9 10

A solution of **10** in THF with acetophenone did not effect reduction even after several hours at 23 °C. Rapid reduction (less than one minute) was only observed after the addition of BH_3–THF (0.6 equivalents) to afford (*R*)-1-phenylethanol in 94.7% *ee*. This stands in contrast to the reduction in the absence of **10** which required much longer reaction times at 23 °C.

Follow-up studies indicated one could reduce the number of equivalents of the oxazaborolidine species to make the process catalytic. With the establishment of this mechanistic foundation, it became possible to rationalize the outcome of this reaction knowing the structure of the catalyst. ^{11}B NMR confirmed the formation of a 1:1 complex between **10** and BH_3–THF for R = H (**11**), while for the species R = Me (**11**) a single crystal X-ray structure was obtained.[8] The *cis*-fused nature of this complex is a result of the concave shape of this bicyclo[3.3.0]octane framework.

Ph H Ph N O B–R — BH_3–THF ⇌ Ph H Ph N O B–R BH_3

10 11

Figure 1 illustrates the 3-dimensional nature of **11**. The oxazaborolidine ring forms the horizontal core to this scaffold with the proline-derived five-membered ring forming the β-face back wall. The *gem*-diphenyl substituents create an additional aspect to the back wall on the β-face and a blocking group on the α-face. The borane moiety complexes to the nitrogen of the heterocyclic ring on the α-face due to the steric interactions it would encounter on the β-face. The only site "open" for a ligand to complex with this catalyst in on the α-face adjacent to the borane group.

Figure 1

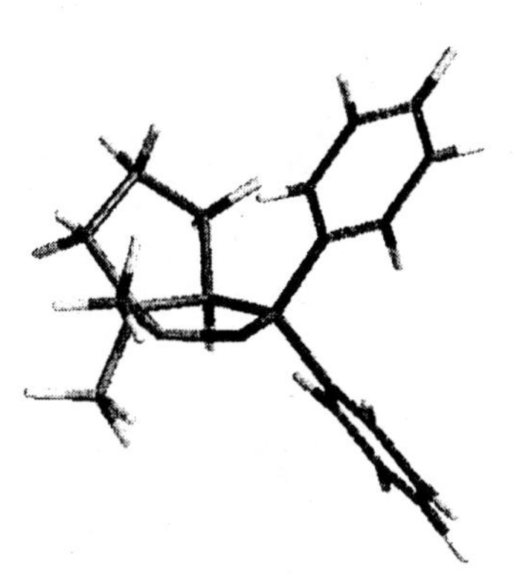

The formation of this complex between the pre-catalyst and borane, sets up this system for interaction with the carbonyl group by activating borane as a hydride donor and increasing the Lewis acidity of the endocyclic boron atom. The later effect serves as the point of coordination to the carbonyl oxygen atom. Once this complexation occurs to form **12**, the chirality of the scaffold restricts the orientation that the substituents on the carbonyl can adopt. In order to minimize steric interactions with the catalyst, the coordination must occur from the α-face (*vide supra*) and the small substituent (R_S) must be oriented in the β-face direction to minimize steric interactions with the substituent on the endocyclic boron atom, compared to the large substituent (R_L). The consequence of these interactions is to place the hydride equivalent in an optimal position for delivery to the carbonyl carbon atom *via* a six-membered transition-state.[9] The result of this hydride transfer is **13**, in which the carbonyl has undergone a net reduction in an enantiocontrolled fashion. This orientation of the reduction can be predicted based on the relative sizes of the carbonyl substituents and the orientation they must adopt in the transition-state **12**. The limited Hammett linear free energy analysis conducted and a measured kinetic isotope effect ($k_H/k_D = 1.7$) indicate that both the coordination of the carbonyl compound and the transfer of hydride are probably fast and comparably rate-determining.

11 **12** **13**

The decomposition of intermediate **13** to the isolated alcohol **14** can occur by one of two possible pathways. The first is a net cyclo-elimination that regenerates the catalyst and forms boronate **10**.

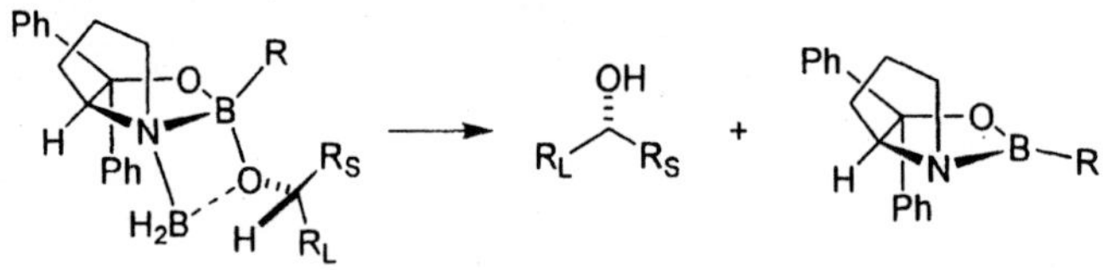

The alternate pathway occurs by the addition of borane to **13** to form the six-membered BH_3-bridged complex **15**. This species then decomposes to regenerate the active catalyst **11**.

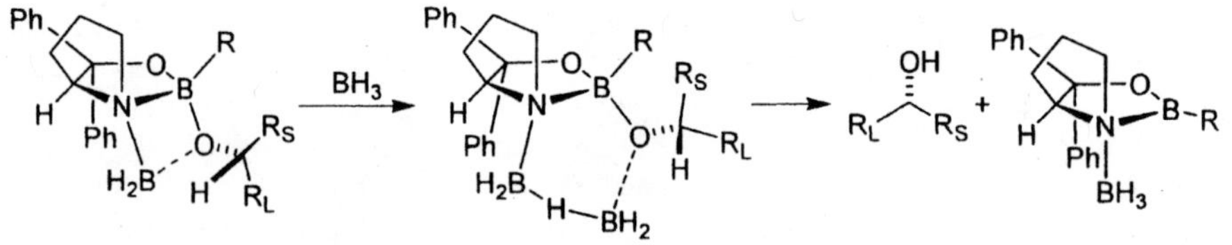

1.1.4 Variations and Improvements or Modifications

Figure 2

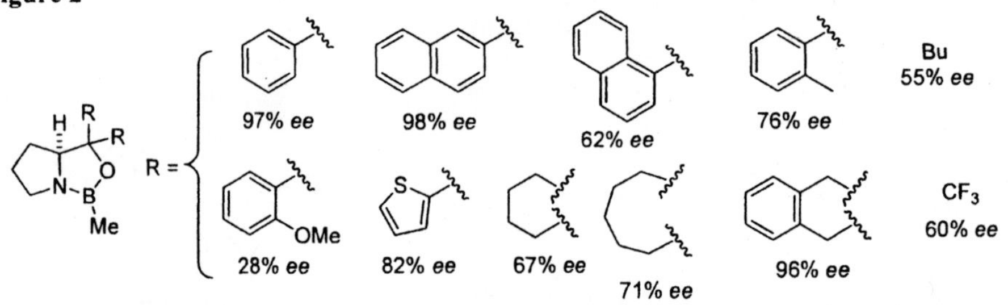

Various laboratories, in an effort to improve reaction yield and stereoselectivity, have made targeted modifications on the core structure of the oxazaborolidine catalyst.[1h] Figure 2 illustrates the level of stereocontrol in the CBS reduction of acetophenone as the R-group was systematically investigated to assess the varying degrees of enantiocontrol. The best compromise of stereocontrol and synthetic complexity was observed to be the phenyl substituent.

Figure 3

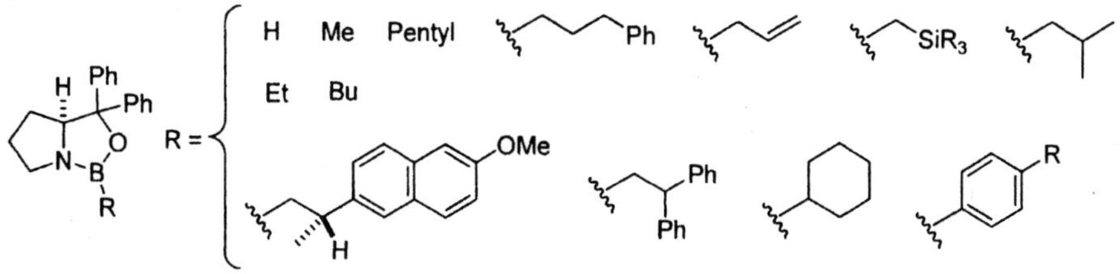

With the heterocyclic substituent optimized, a similar investigation of the B-group was carried out. Examples of the substituents studied are shown in Figure 3.

Additional modifications of the framework have included ring size **16**, ring fusion **17**, and ring substitution **18**.

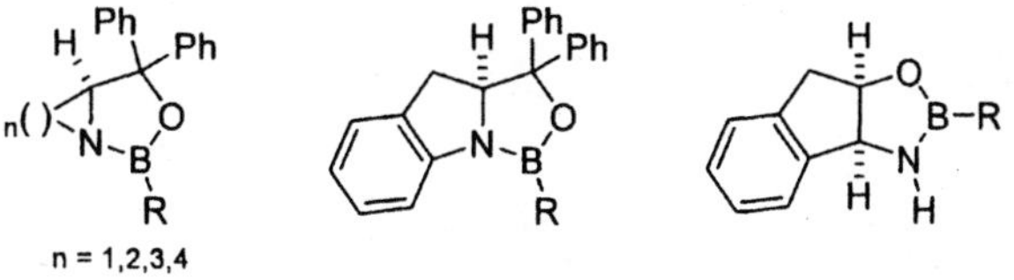

For example, (–)-β-pinene **19** has been used to construct such a modified catalyst.[10] Oxazaborolidine **21** could be prepare in three steps from the monoterpene and was found to be an efficient catalyst for the reduction of ketones. Thus **22** could be reduced with pre-catalyst **20** and trimethoxyboron to alcohol **23**. The chirality of **23** could be rationalized based on the transition-state structure **24**.

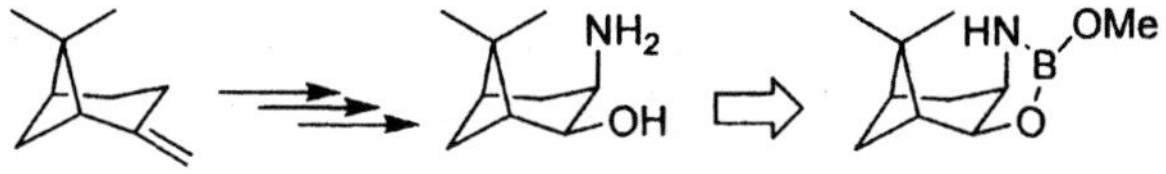

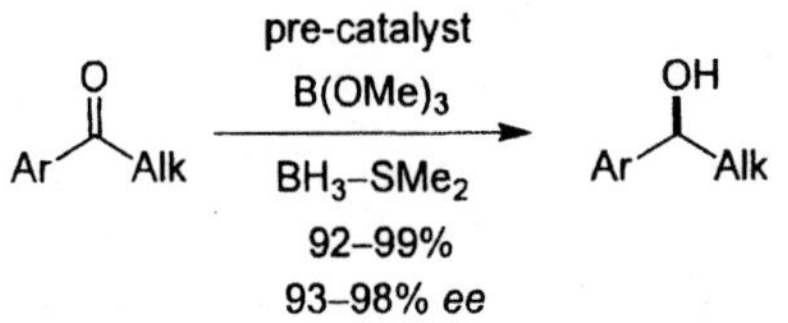

There are reports that extend the nature of the catalyst beyond an oxazaborolidine framework. One such example made use of a chiral guanidine catalyst.[11] Proline-derived **25** was converted to guanidine **26** in good yield. This species was capable of reducing ketones **27** to alcohols **28** by the addition of BH_3–SMe_2.

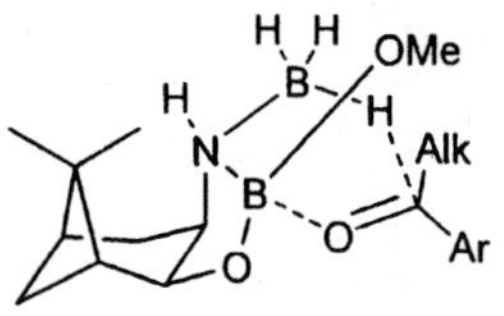

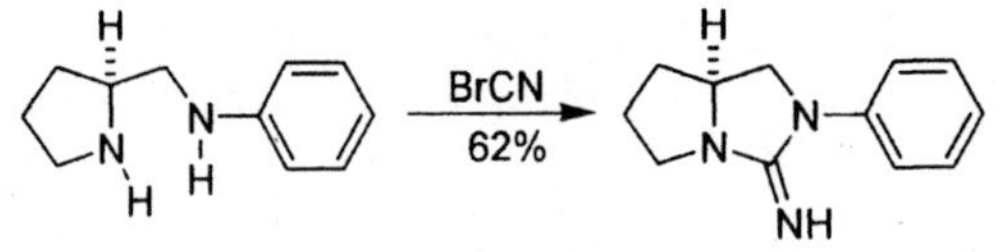

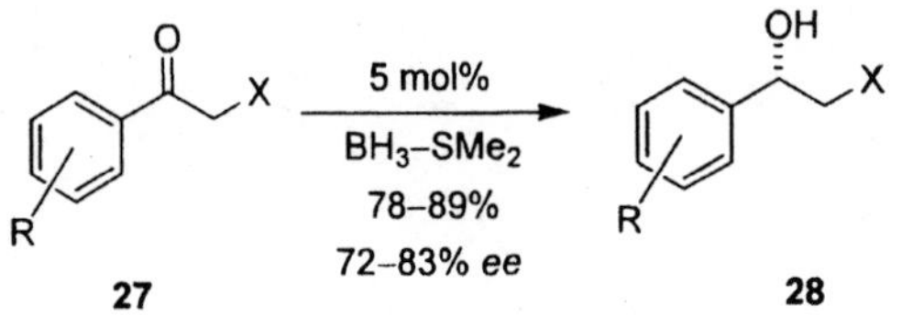

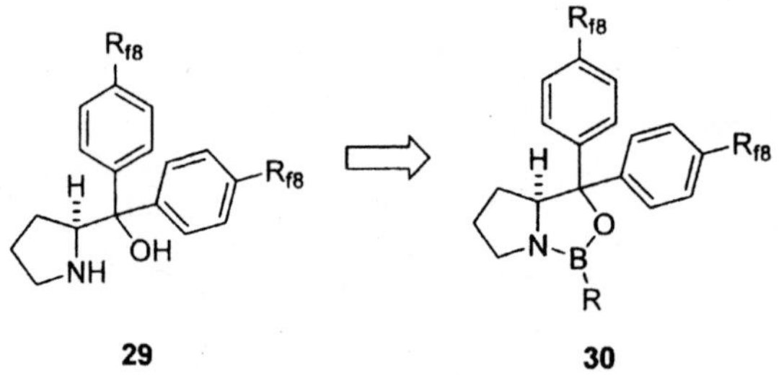

In an attempt to improve the ability to recycle the catalyst, fluorous versions of the oxazaborolidine have been constructed.[12] Pre-catalyst **29** could be prepared in five steps. This species was able to form the requisite chiral catalyst **30** *in situ*. Ketones **31** could be reduced to alcohols **32** in good to excellent chemical and optical yields. It was noted that aryl ketones were observed to be more efficient in the reduction process. After the reaction was carried out, the catalyst could be recovered in 99% yield.

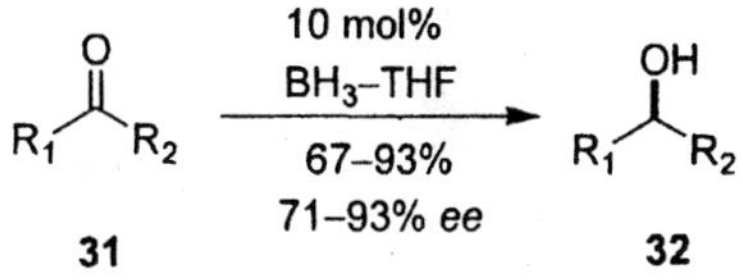

10 mol%
BH_3–THF
67–93%
71–93% *ee*

31 **32**

Polymer-bound versions of the oxazaborolidine catalysts have been constructed.[13] The linkage to the polymer has been reported on the phenyls of the heterocycle **34** or through a substituent on the nitrogen **36**. These polymeric catalysts are recyclable and reuseable without significant loss of activity or selectivity. Placing the linker on the nitrogen appears to create steric interactions that weaken complex formation thus giving rise to diminished enantiocontrol in the reduction. Moving the point of attachment for the resin to the phenyl substituent provided a superior reagent. The reduction of aryl ketones **37** proceeded in good to excellent yields with poor to excellent optical purities in the formation of alcohols **38**.

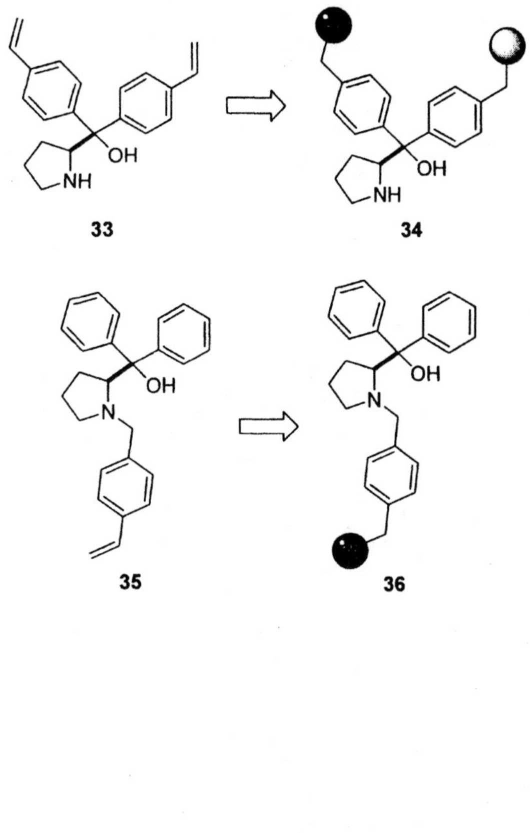

33 **34**

35 **36**

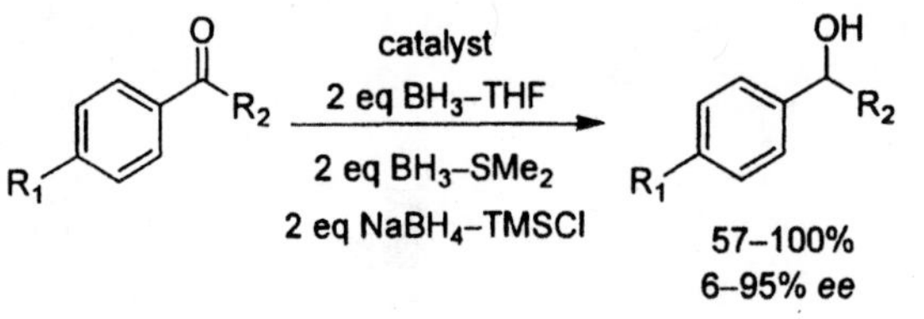

catalyst
2 eq BH_3–THF
2 eq BH_3–SMe_2
2 eq $NaBH_4$–TMSCl

57–100%
6–95% *ee*

37 38

In addition to the reduction of ketones (e.g., aromatic and aliphatic ketones, α-halo ketones, hydroxyketones, enones, and ketoesters), oximes can be reduced to the corresponding amine with this reagent. In general, ketone oxime ethers, such as **39**, can give rise to amines **40** in excellent chemical yield with good to excellent optical purity.[5d]

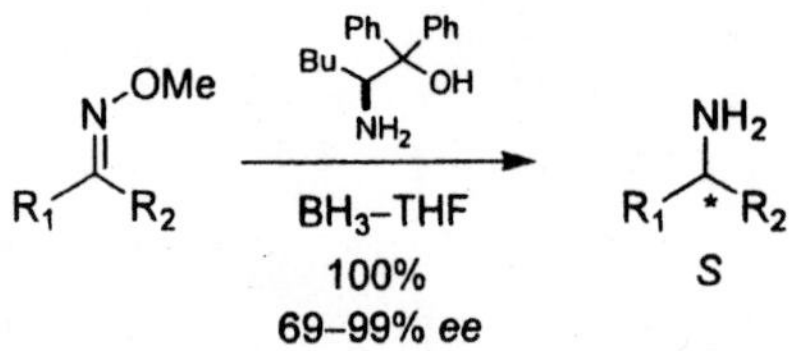

BH_3–THF
100%
69–99% *ee*

39 40

This method was used in the preparation of conformationally constricted analogs of the neurotransmitter glutamate **41**, such as (carboxycyclopropyl)glycines (*L*-CCG I) **42**, that could act as metabatropic glutamate receptor (mGluR) antagonists.[14] Reduction of oxime **43** using the oxazaborolidine derived from prolinol afforded amine **44**. Conversion of the furan rings to carboxylic acids afforded the requisite target **42**.

41 42

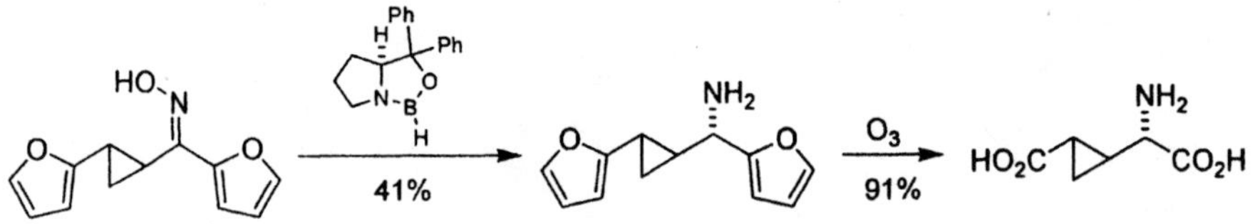

41% O_3 91%

43 44 42

Proline-derived oxazaborolidines **45** have shown to be effective pre-catalysts with triflic acid as an activator to generate cationic Lewis acids.[1g,15] The optimal proportions of **45** and triflic acid was found to be 1.2:1. Protonation of **45** produced a 1.5:1 mixture of **46** and **47** as determined by low temperature ^{1}H NMR. Their interconversion at low temperature (−80 °C) is slow on the NMR timescale. However, this interconversion increases as the temperature rises and at 0 °C this becomes rapid (T_c). Phenyl or *o*-tolyl were determined to be the best substituents for the R group in **45**. For the Ar group of **45**, phenyl and 3,5-dimethylphenyl were determined to be optimal.

H Ar Ar O N B R **45** —TfOH→ H Ar Ar ⊕ O N B TfO⊖ H R **46** ⇌ H Ar Ar ⊕ ⊖ O N B H R OTf **47**

This *in situ* formed cationic Lewis acid catalyst coordinates enals in a highly organized fashion (**48**) that allows for the execution of asymmetric Diels–Alder reactions. Thus for the initially disclosed acrolein examples, the Diels–Alder adducts **51** produced from enals **49** and dienes **50** could be isolated in good to excellent yields with very high optical purities.

46 ⇅ **47** → **48** ≡

49 + **50** —catalyst, 1–2 h, −78 to −95 °C, 58–99%, 91–97% *ee*→ **51** (CHO, R_1, R_2)

The stereochemical outcome from these reactions could be readily rationalized by examining the interactions present in **48**. The [3.3.0]bicyclic system provided a rigid convex scaffold that only allowed the enal to coordinate from the more exposed face of this molecule. The carboxaldehyde hydrogen forms an H-bonding interaction with the endocyclic oxygen atom of the heterocyclic scaffold thus only allowing the diene to approach from the periphery of the complex.

An extension to enones has been accomplished but opposite face selectivities were observed. To rationalize this result, an alternate transition-state structure **52** was formulated. Single crystal X-ray structure analysis examining the coordination of BF_3-etherate with enones and enoates was used to provide support for this novel mode of complexation.

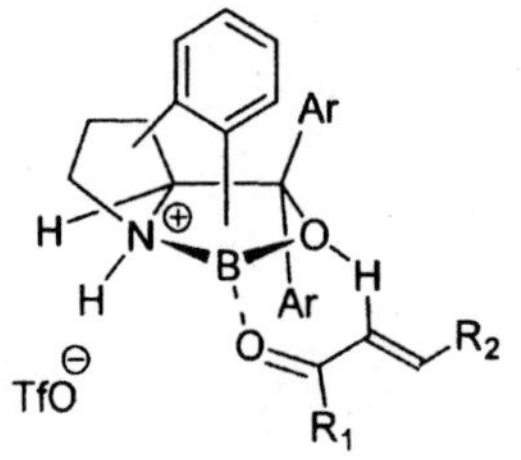

Ar
H N⊕ B O H
H Ar
⊖ O R2
TfO
R1

52

Catalysts derived from triflic acid decompose at appreciable rates at or above 0 °C, which limits the utility of these reagents in Diels–Alder reactions. Switching to triflimide as the acid source resulted in protonation of **45** to produce **53**.[16] ^{1}H NMR from –80 to 23 °C showed the formation of three species including **53** and two diastereomeric tetracoordinated boron species in a ratio of 1:1.2. Additionally, **53** was found to have greater thermal stability and superior catalytic efficiency compared to **46/47** for less reactive dienes. This was illustrated in the Diels–Alder reaction of **54** with **55** to produce adduct **56**.

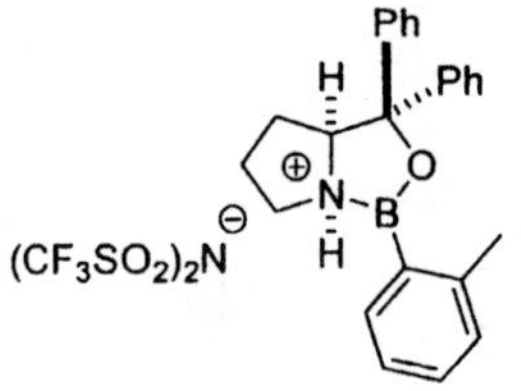

Ph
H Ph
⊕ O
⊖ N B
$(CF_3SO_2)_2N$ H

53

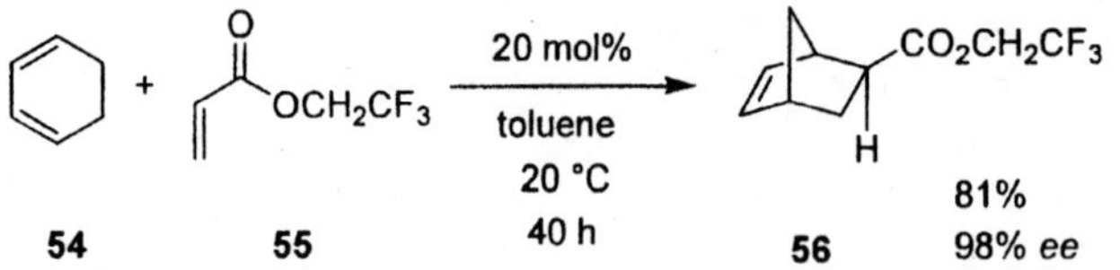

O
+ OCH_2CF_3 —20 mol% / toluene / 20 °C / 40 h→ $CO_2CH_2CF_3$
H

54 **55** **56** 81% 98% *ee*

This last observation was capitalized on by the Corey labs in their efficient synthesis of Tamiflu® (oseltamivir) **59**. The emergence of the virulent strain of influenza, H5N1, coupled with the lack of supply due to the current synthetically challenging route, highlighted the need for such reagents that allow for the rapid and efficient construction of difficult targets. The reaction could be conducted on a multigram scale of **57** and **55** to generate sufficient quantities of the Diels–Alder adduct **58** to complete the target **59**.[17]

+ OCH_2CF_3 —10 mol%, neat, 23 °C, 30 h→ $CO_2CH_2CF_3$ 97% >97% *ee*

Ac–NH, $H_3N^{\oplus}$, $H_2PO_4^{\ominus}$, CO_2Et

57 **55** **58** **59**

Follow-up work by other groups examined alternate sources of generating Lewis acids from various oxazaborolidines (**60** → **61**).[18] One report scanned the commonly used metal halides and found tin tetrachloride to be the best when coupled to valinol-based oxazaboroline. Thus, cyclopentadiene **61** reacting with methacrolein **63** using such a catalyst afforded the Diels–Alder adduct **64** in excellent yield and excellent optical purity.

Ph, Ph, O, R–N–B —LA→ Ph, Ph, O, R–N–B, L–M(L)(L)

R = *n*-octyl
$-CH_2$(1-naphthyl)

60 **61**

+ CHO —1 mol%, $SnCl_4$ 1 mol%, −78 °C, 2 h→ $CO_2CH_2CF_3$ 99% 95% *ee*

62 **63** **64**

Rather than accessing the chiral pool *via* amino acid precursors for CBS catalysts, the (*R*,*R*)- and the (*S*,*S*)-sulfonamide derivatives of 1,2-diphenyl-1,2-diaminoethane (stilbenediamine, stien) **66** in complex with boron **65** or aluminum **67** have also been applied to the Diels–Alder reaction.[19]

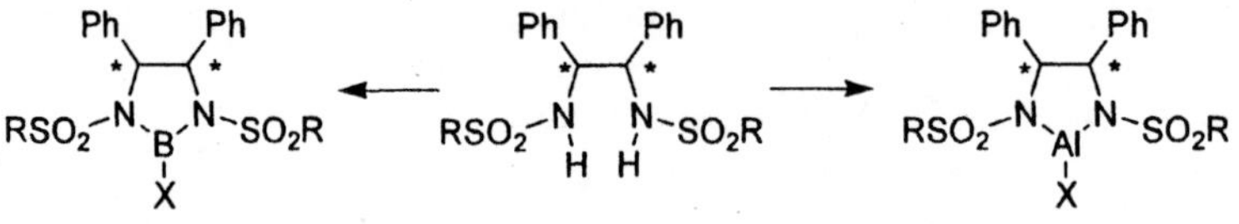

65 **66** **67**

The use of these reagents was exemplified in the preparation of an advanced intermediate in the synthesis of prostaglandins **71**. Diene **68** and dienophile **69** were allowed to undergo the Diels–Alder reaction catalyzed by a derivative of **67** to afford adduct **70**.[19] This intermediate was subsequently transformed into **71**, a well-known precursor in the synthetic preparations of prostaglandins.

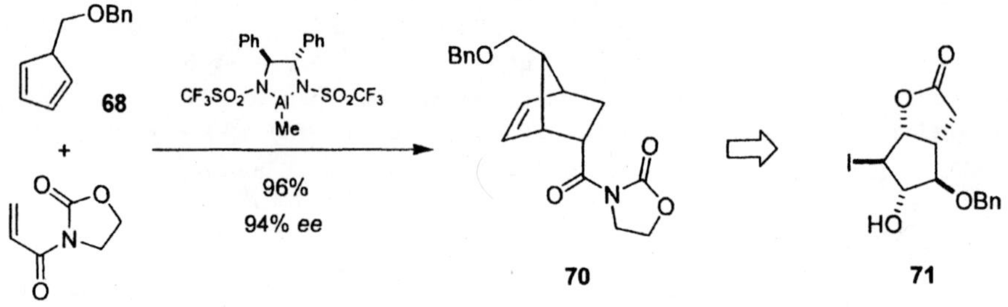

68 **69** **70** **71**

The Lewis acidic nature of these catalysts has permitted their extended use in the Mukaiyama aldol reaction. In this application of CBS reagents, one such example involved the condensation of ketene acetals **72** with aldehydes **73** to produce adducts **74**.[20]

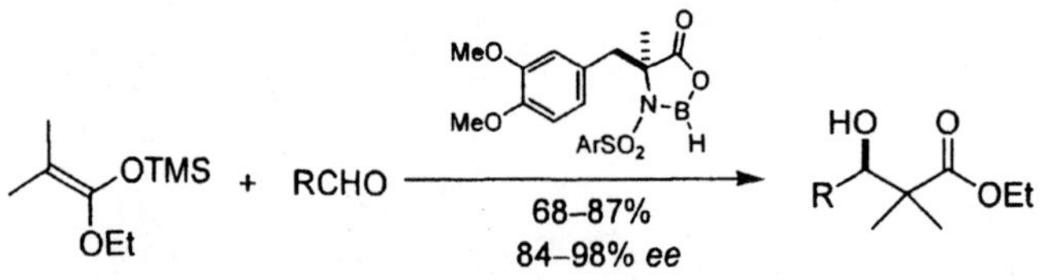

72 **73** **74**

In a similar manner, the insect attractant *endo*-1,3-dimethyl-2,9-dioxabicyclo[3.3.1]nonane **78** could be prepared.[21] To this end, masked keto aldehyde **75** and silyl enol ether **76** underwent the CBS catalyzed Mukaiyama aldol reaction with a high degree of optical purity to generate **77**. Deprotection and acid-catalyzed spiroketal formation afforded the desired product **78**.

75 + 76 → (Ts-N-B(H) oxazaborolidinone; 36%, 94% *ee*) → 77 → (1. TBAF; 2. TsOH) → 78

Using the boron complex of the stien system, enantioselective aldol reactions were also possible. Ester **79** could be converted to the corresponding boron enolate using a derivative of **65**, which could then undergo an aldol reaction with aldehydes **73** to afford **80**.[22]

79 → (1. bis(3,5-bis(trifluoromethyl)phenylsulfonyl) diazaborolidine B–Br, Et_3N; 2. RCHO (**73**)) → 80 (93–97% *ee*)

1.1.5 Synthetic Utility

The literature documents a myriad of examples of the synthetic utility of this reduction methodology. The following examples provide a limited glimpse into the scope of substrates that can undergo this type of reduction.

Prostaglandin PGE_2 **81** elicits bone growth, so EP4 receptor antagonists, such as **82**, would have utility in the pharmacological treatment of osteoporosis.[23] Standard reduction conditions investigated (Luche, *L*-selectride, or (*R*)-CBS) for the reduction of **83** to afford **84** gave the wrong stereochemistry. Alcohol **84** with the desired stereochemistry could only be obtained upon use of the reagent (*S*)-CBS.

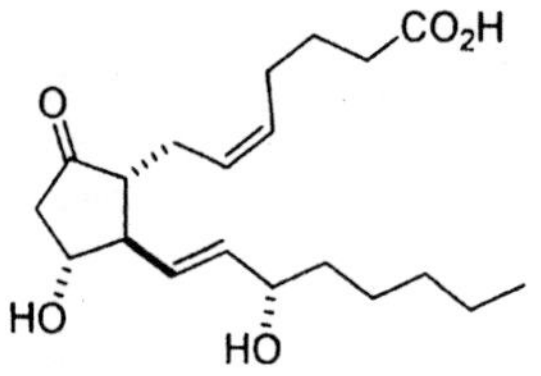

81

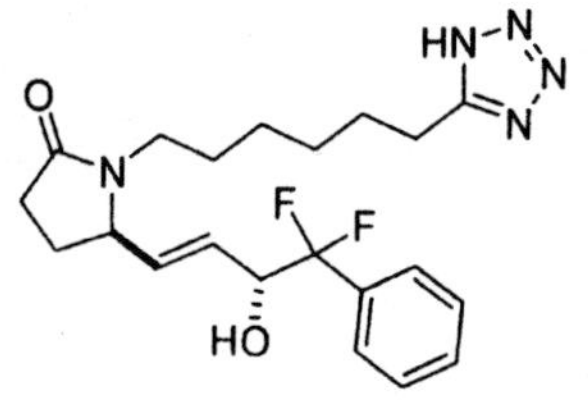

82

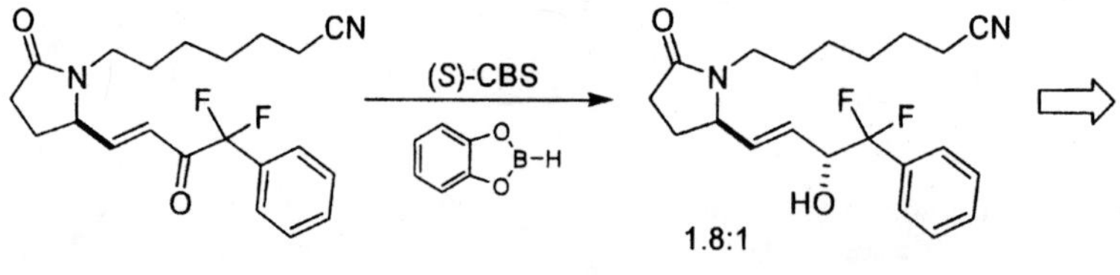

(S)-CBS

1.8:1

83

84

82

Epothilone **85** possesses antimitotic properties similar to those of paclitaxel and could therefore have utility in the treatment of cancer. A key intermediate in the synthesis of this natural product is **86**, a C_{12}–C_{21} subunit of **85**.[24] One approach to this compound employed the CBS reduction of ketone **87** to generate **88**, an alcohol related to **86**.

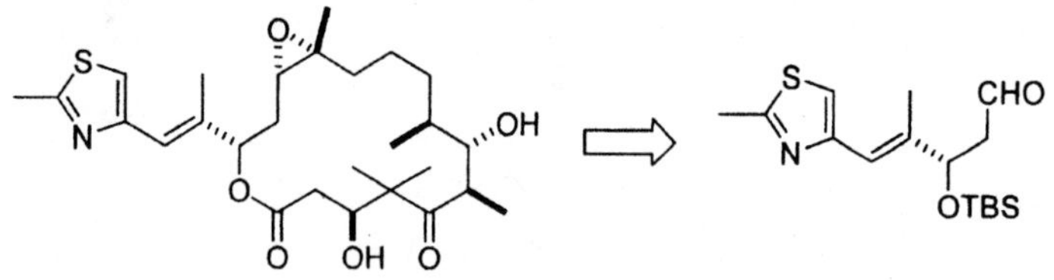

85

86

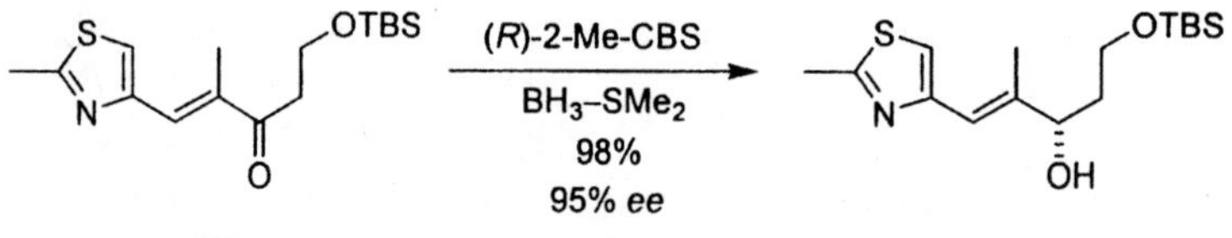

(R)-2-Me-CBS

BH_3–SMe_2

98%

95% *ee*

87

88

The mis-processing of amyloid precursor protein (APP) by α-secretase, β-secretase, and γ-secretase is proposed to be one pathway leading to formation of Aβ protein. One theory for the cause of Alzheimer's disease (AD) is based on the idea that Aβ protein aggregates to form amyloid plaques (a well-known hallmark of AD) that eventually precipitate in the brain. This deposition results in neuronal cell death. Inhibition of any one of these enzymes could be a treatment for AD, thus **89**, a γ-secretase inhibitor, is of pharmacological interest.[25] Epoxy amine **90** is a key synthetic intermediate on the way to **89**. Reduction of ketone **91** using a CBS reagent afforded alcohol **92**. This alcohol is a single transformation away from **90**.

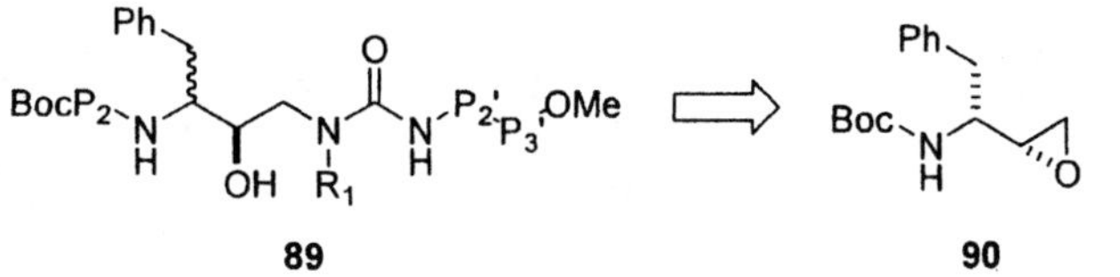

BocP2 Ph OH R1 P2' P3 OMe **89** ⟹ Boc Ph **90**

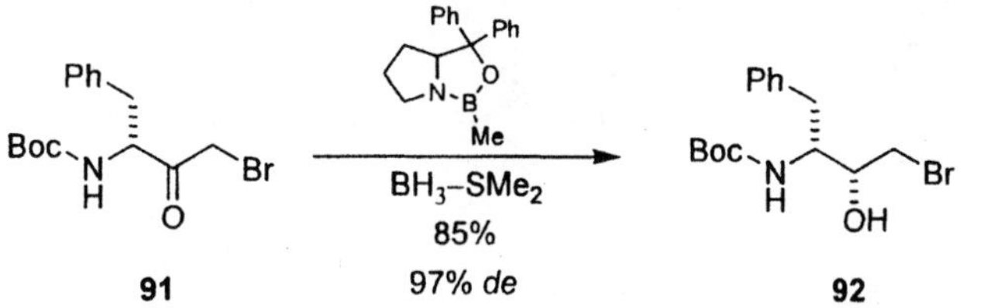

91 → BH_3–SMe_2, 85%, 97% *de* → **92**

Norepinephrine reuptake antagonists (NETs) have antidepressant activity and have been found to be effective in the treatment of attention deficit hyperactivity disorder (ADHD). (*R*)-tomoxetine (Strattera®) **95** is a marketed drug with a label indication for ADHD. Control of the absolute stereochemistry of these compounds is critical in that the (*S*)-enantiomer is nine times less potent. The CBS reduction could be used to set the desired chirality as demonstrated in the reduction of **93** to produce alcohol **94**.[26] This intermediate could then be readily converted into **95**.

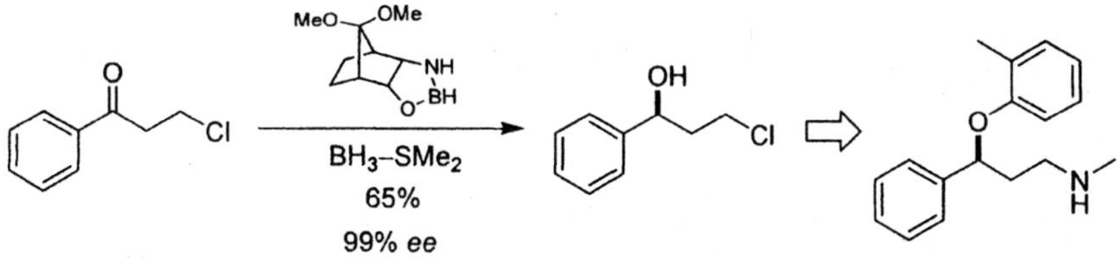

93 → BH_3–SMe_2, 65%, 99% *ee* → **94** ⟹ **95**

Malfunction of hexosaminidase enzymes can lead to such diseases as Tay-Sachs or Sandhoff. The study of *N*-acetylhexosaminidase inhibitors, such as XylNAc-Isofagomine **98**, could provide greater understanding of the method of malfunction.[27] Enone **96** could be reduced by a CBS reagent to afford allylic alcohol **97**. The bromine was required to provide differentiation of the faces of ketone to improve % *ee*.

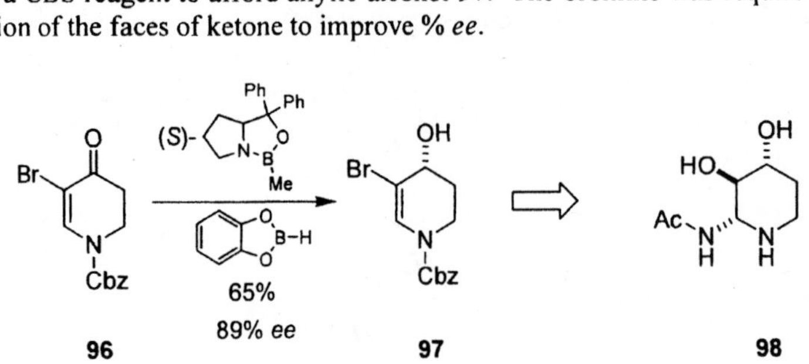

Endothelin and its related peptides play an important role in the biology of vascular smooth muscle. Antagonists of the endothelin receptor would be of great interest in the treatment of cardiovascular and pulmonary disease. One such compound is SB-209670 **101** and could be prepared from alcohol **100**.[28] This alcohol could be accessed *via* the CBS reduction of indenone **99** in excellent yield and excellent optical purity.

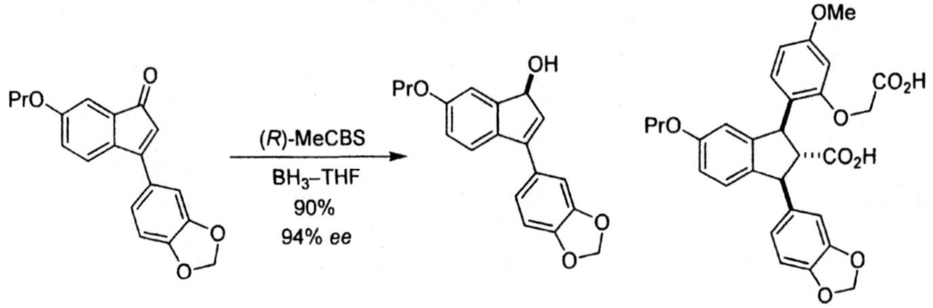

1.1.6 Experimental

The two examples presented exemplify the utility of this reaction. One can access either enantiomer of the product through the proper choice of chirality contained in the CBS reagent. This reagent possesses great scope in that simple, as well as complex, substrates can be reduced with high efficiency, chemical and optical yields. Additionally, these examples illustrate the relatively simplistic experimental conditions required to conduct these reduction reactions.

102 → 103: (*S*)-CBS, BH_3–SMe_2, CH_2Cl_2, 96%, 98.6% *ee*

(*R*)-1-(3′-Bromophenyl)-ethanol (103)[29]

Ketone **102** (10.8 g, 64.6 mmol) in 50 mL of CH_2Cl_2 at –20 °C was added dropwise over 2 h to a solution of (*S*)-4,5,6,7-tetrahydro-1-methyl-3,3-diphenyl-1*H*,3*H*-pyrrolo-[1,2-*c*]-[1,3,2]-oxazaborolidine-borane (0.9 g, 3.2 mmol) and BH_3–SMe_2 (5.1 mL, 51 mmol) in 30 mL of CH_2Cl_2. The reaction mixture was stirred for a total 8 h before slowly adding to 50 mL of methanol at –20 °C. Once gas evolution ceased, the solvent was removed *in vacuo* and the residue purified by bulb-to-bulb distillation (94 °C, 0.5 mmHg) to afford **103** as a colorless oil.

104 → 105: (*R*)-CBS, catecholborane (B–H), –78 °C, 65%, >98:2 *dr*

Allylic alcohol (105)[30]

To ketone **104** (196 mg, 0.395 mmol) in 7 mL of toluene was added (*R*)-CBS (0.43 mL, 0.395 mmol, 1.0 M toluene). The mixture was cooled to –78 °C and catecholborane (0.79 mL, 0.79 mmol, 1.0 M THF) was added dropwise. The reaction was stirred for 12 h before quenching with 0.8 mL of methanol and warmed to room temperature. The mixture was diluted with ether and washed with NaOH (1 N), saturated with $NaHCO_3$, until the aqueous washings were colorless. The combined aqueous phases were back extracted 3 times with ether and the combined organic phases were washed with brine, dried (Na_2SO_4), and concentrated *in vacuo*. Purification by chromatography gave **105** (160 mg).

1.1.7 References

1. (a) [R] Corey, E. J. *Pure Appl. Chem.* **1990**, *62*, 1209–1216. (b) [R] Wallbaum, S.; Martens, J. *Tetrahedron: Asymm.* **1992**, *3*, 1475–1504. (c) [R] Singh, V. K. *Synthesis* **1992**, 605–617. (d) [R] Deloux, L.; Srebnik, M. *Chem. Rev.* **1993**, *93*, 763–784. (e) [R] Taraba, M.; Palecek, J. *Chem. Listy* **1997**, *91*, 9–22. (f) [R] Corey, E. J.; Helal, C. J. *Angew. Chem. Int. Ed.* **1998**, *37*, 1986–2012. g) [R] Corey, E. J. *Angew. Chem. Int. Ed.* **2002**, *41*, 1650–1667. (h) [R] Itsuno, S. *Org. Reac.* **1998**, *52*, 395–576. (i) [R] Cho, B. T. *Aldrichimica Acta* **2002**, *35*, 3–16. (j) [R] Glushkov, V. A.; Tolstikov, A. G. *Russ. Chem. Rev.* **2004**, *73*, 581–608. (k) [R] Cho, B .T. *Tetrahedron* **2006**, *62*, 7621–7643.
2. Fiaud, J.-C.; Kagan, H.-B. *Bull. Soc. Chim. Fr.* **1969**, 2742–2743.
3. Borch, R. F.; Levitan, S. R. *J. Org. Chem.* **1972**, *37*, 2347–2349.
4. Grundon, M. F.; McCleery, D. G.; Wilson, J. W. *Tetrahedron Lett.* **1976**, *17*, 295–296.
5. (a) Hirao, A.; Itsuno, S.; Nakahama, S.; Yamazaki, N. *Chem. Commun.* **1981**, 315–317. (b) Itsuno, S.; Ito, K.; Hirao, A.; Nakahama, S. *Chem. Commun.* **1983**, 469–470. (c) Itsuno, S.; Hirao, A.; Nakahama, S.; Yamazaki, N. *Perkin Trans. I* **1983**, 1673–1676. (d) Itsuno, S.; Nakano, M.; Miyazaki, K.; Masuda, H.; Ito, K.; Hirao, A.; Nakahama, S. *Perkin Trans. I* **1985**, 2039–2044.
6. (a) Corey, E. J.; Bakshi, R. K.; Shibata, S. *J. Am. Chem. Soc.* **1987**, *109*, 5551–5553. (b) Ganem, B. *Chemtracts-Org. Chem.* **1988**, *1*, 40–42.
7. (a) Corey, E. J.; Link, J. O.; Bakshi, R. K. *Tetrahedron Lett.* **1992**, *33*, 7107–7110. (b) Mathre, D. J.; Thompson, A. S.; Douglas, A. W.; Hoogsteen, K.; Carroll, J. D.; Corley, E. G.; Grabowski, E. J. J. *J. Org. Chem.* **1993**, *58*, 2880–2888. (c) Douglas, A. W.; Tschaen, D. M.; Reamer, R. A.; Shi, Y.-J. *Tetrahedron: Asymm.* **1996**, *7*, 1303–1308. For papers related to computational chemistry studies see: (d) Nevalainen, V.; Uggla, R.; Sundberg, M. R. *Tetrahedron: Asymm.* **1995**, *6*, 1431–1440. e) Nevaleinen, V. *Tetrahedron: Asymm.* **1994**, *5*, 289–296. (f) Li, M.; Tian, A. *J. Mol. Struct.* **2001**, *544*, 37–47. (g) Harb, W.; Ruiz-Lopez, M. F.; Coutrot, F.; Grison, C.; Coutrot, P. *J. Am. Chem. Soc.* **2004**, *126*, 6996–7008. (h) Alagona, G.; Ghio, C.; Tomasi, S. *Theor. Chem. Acc.* **2004**, *111*, 287–302.
8. Corey, E. J.; Azimoiara, M.; Sarhar, S. *Tetrahedron Lett.* **1992**, *33*, 3429–2430.
9. (a) Evans, D. A.; *Science* **1988**, *240*, 420–426. (b) Jones, D. K.; Liotta, D. C.; Shinkai, I.; Mathre, D. J. *J. Org. Chem.* **1993**, *58*, 799–801. (c) Quallich, G. J.; Blake, J. F.; Woodall, T. M. *J. Am. Chem. Soc.* **1994**, *116*, 8516–8525.
10. Krzeminski, M. P.; Wojtczak, A. *Tetrahedron Lett.* **2005**, *46*, 8299-8302.
11. (a) Basavaiah, D.; Rao, K. V.; Reddy, B. S. *Tetrahedron: Asymm.* **2006**, *17*, 1036–1040. b) Basavaiah, D.; Rao, K. V.; Reddy, B. S. *Tetrahedron: Asymm.* **2006**, *17*, 1041–1044.
12. (a) Dlicsek, Z.; Pollreisz, F.; Gomory, A.; Soos, T. *Org. Lett.* **2005**, *7*, 3243–3246. b) Park, J. K.; Lee, H. G.; Bolm, C.; Kim, B. M. *Chem. Eur. J.* **2005**, *11*, 945.
13. Degni, S.; Wilen, C.-E.; Rosling, A. *Tetrahedron: Asymm.* **2004**, *15*, 1495–1499.
14. Demir, A. S.; Tanyeli, C.; Cagir, A.; Tahir, M. N.; Ulku, D. *Tetrahedron: Asymm.* **1998**, *9*, 1035–1042.
15. (a) Corey, E. J.; Shibata, T.; Lee, T. W. *J. Am. Chem. Soc.* **2002**, *124*, 3808–3809. (b) Ryu, D. H.; Lee, T. W.; Corey, E. J. *J. Am. Chem. Soc.* **2002**, *124*, 9992–9993.
16. Ryu, D. H.; Corey, E. J. *J. Am. Chem. Soc.* **2003**, *125*, 6388–6390.
17. Yeung, Y.-Y.; Hong, S.; Corey, E. J. *J. Am. Chem. Soc.* **2006**, *128*, 6310–6311.
18. (a) Futatsugi, K.; Yamamoto, H. *Angew. Chem. Int. Ed.* **2005**, *44*, 1484–1487. (b) Harada, T.; Inui, C. *J. Org. Chem.* **2006**, *71*, 1277–1279.
19. Corey, E. J.; Imwinkelried, R.; Pikul, S.; Xiang, Y. B. *J. Am. Chem. Soc.* **1989**, *111*, 5493–5495.
20. (a) Parmee, E. R.; Tempkin, O.; Masamune, S. *J. Am. Chem. Soc.* **1991**, *113*, 9365-9366. (b) Fujiyama, R.; Goh, K.; Kiyooka, S.-I. *Tetrahedron Lett.* **2005**, *46*, 1211–1215.
21. Kiyooka, S.-I.; Kaneko, Y.; Harada, Y.; Matsuo, T. *Tetrahedron Lett.* **1995**, *36*, 2821–2822.
22. Corey, E. J.; Kim, S. S. *J. Am. Chem. Soc.* **1990**, *112*, 4976–4977.
23. Young, R. N.; Billot, X.; Han, Y.; Slipetz, D. A.; Chauret, N.; Belley, M.; Metters, K.; Mathieu, M.-C.; Greig, G. M.; Denis, D.; Girard, M. *Heterocycles*, **2004**, *64*, 437–446.
24. Reiff, E. A.; Nair, S. K.; Reddy, B. S. N.; Inagaki, J.; Henri, J. T.; Greiner, J. F.; Georg, G. I. *Tetrahedron Lett.* **2004**, *45*, 5845–5847.
25. Bakshi, P.; Wolfe, M. S. *J. Med. Chem.* **2004**, *47*, 6485–6489.

26. Lapis, A. A. M.; de Fatima, A.; Martins, J. E. D.; Costa, V. E. U.; Pilli, R. A. *Tetrahedron Lett.* **2005**, *46*, 495–498.
27. Knapp, S.; Yang, C.; Pabbaraja, S.; Rempel, B.; Reid, S.; Withers, S. G. *J. Org. Chem.* **2005**, *70*, 7715–7720.
28. Clark, W. N.; Tickner-Eldridge, A. M.; Huang, G. K.; Pridgen, L. N.; Olsen, M. A.; Mills, R. J.; Lantos, I.; Baine, N. H. *J. Am. Chem. Soc.* **1998**, *120*, 4550–4551.
29. Powell, M. T.; Porte, A. M.; Reibenspies, J.; Burgess, K. *Tetrahedron* **2001**, *57*, 5027–5038.
30. Crimmins, M. T.; DeBaillie, A. C. *J. Am. Chem. Soc.* **2006**, *128*, 4936–4937.

Paul Galatsis

1.2 Davis Chiral Oxaziridine Reagents

1.2.1 Description

Davis oxaziridine reagents such as **1** have exhibited ample synthetic utility as oxidizing agents for the hydroxylation of enolates to provide α-hydroxy carbonyl compounds, such as **2** with superb yield. When the oxaziridine is chiral and nonracemic, the hydroxylation has been shown to proceed with high stereoselectivity.[1]

1. KHMDS
2. Ph–SO₂–N(oxaziridine)–Ph **1**

OH

O

2

1.2.2 Historical Perspective

Prior to Davis and co-workers' introduction of *trans*-(±)-2-(phenylsulfonyl)-3-phenyloxaziridine (**1**) for direct enolate oxidation in 1984, there were several nonoxidative procedures for the formation of optically active α-hydroxy carbonyl compounds, but only one actively practiced oxidative method for the synthesis of such compounds.[2]

[HO]⁻ (1)

[R"]⁺ (2)

[R"]⁻ (3)

The most commonly used nonoxidative method used by chemists prior to 1958 is the substitution reaction using either chiral α-amino acids (L = NH_3) or α-halo amides (L = F, Cl,

Br, or I) (equation 1). Another method relies on alkylation of α-hydroxy or α-alkoxy carbonyl compounds. The induction of diastereoselectivty, in these cases, is achieved through the use of chiral auxiliaries and other stereodirecting groups (equation 2). The third method frequently utilizes the nucleophilic addition of hydride or other carbanions to α-dicarbonyl compounds (equation 3). In addition to being laborious, nonoxidative methods are limited to the synthesis of acyclic compounds, which greatly reduces the magnitude of their synthetic practicality.

In 1958, Rubottom and co-workers introduced an oxidative methodology, which first requires the conversion of a carbonyl compound to either an enol silane or silyl enol ether. As generated, the enol silane **3**, for example, is then treated with *m*-chloroperbenzoic acid (*m*-CPBA) to form the α-hydroxy carbonyl compound **4** following epoxidation and desilylation.[3]

$OSiMe_3$ CO_2R'' R R' **3** —*m*-CPBA→ O R CO_2R'' HO R'' **4**

Despite high yield, the Rubottom oxidation is limited by the necessity for synthesis of the requisite silane ethers. The direct oxidation of enolates has thus emerged as the preferred method for the stereoselective formation of α-hydroxy carbonyl compounds because of the method's effectiveness for both acyclic and cyclic substrates. Davis's oxaziridine reagents have proved to be ideally suited for the one-step enolate hydroxylation process. The following chiral oxaziridine reagents have been utilized effectively in this protocol and will be showcased throughout the chapter.

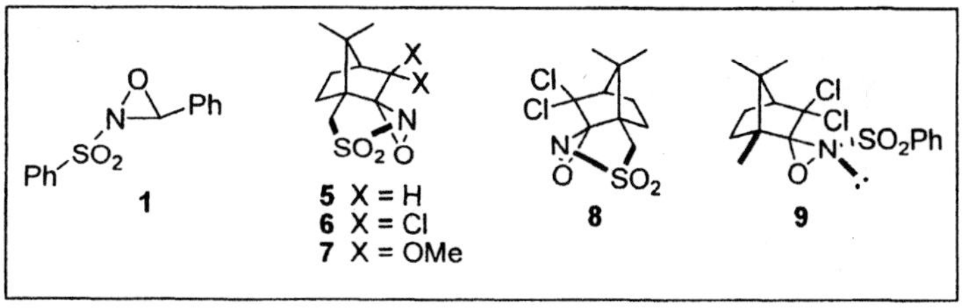

O N Ph SO_2 Ph **1**; X X N SO_2 O **5** X = H **6** X = Cl **7** X = OMe; Cl Cl N O SO_2 **8**; Cl Cl N SO_2Ph O **9**

1.2.3 Mechanism

Mechanistic interpretation for this reaction began with comparisons to alkene epoxidations using metal peroxides, dioxiranes, and oxaziridines. In 1982, Mimoun was the first to propose a general mechanism for the oxygen transfer from compounds containing a peroxide moiety (active site oxygen is part of a three-membered ring).[4] Mimoun's mechanistic interpretation can be applied to the oxidation of a sulfide using an oxaziridine reagent. In this case, coordination between the oxaziridine nitrogen and the substrate would lead to the formation of a four-membered ring as an intermediate.[5] This cyclic intermediate would decompose resulting in the formation of the sulfoxide and imine products.

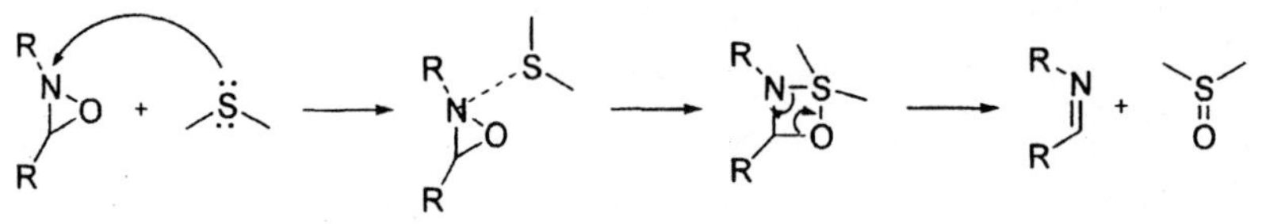

In the following years, studies conducted by Sharpless,[6,7] Bach,[8,9] Curci,[10] and others[11] relied on reaction kinetics to formulate support of a S_N2-type displacement by the nucleophilic substrate on the electrophilic oxygen atom of the three-membered ring. Similarly, the deoxygenation of oxaziridines, such as **1**, is kinetically consistent with the aforementioned S_N2 mechanism.

Davis extensively studied reactions involving **1** and found that all results exhibited properties that suggest the oxygen transfer by oxaziridines is S_N2 in nature.[12] When comparing the mechanism of the nucleophilic ring opening of oxaziridines to those of epoxides, aziridines, and thiiranes, it was found that attack at the nitrogen is favored when the substituent bound to the nitrogen is small and nonobtrusive. When this substituent is large as in **3**, nucleophilic attack at the oxygen is favored, implicating an S_N2 mechanism. Furthermore, when the rate of oxidation was monitored through the use of ^{1}H NMR, Davis found the reaction to be second-order. Hammett subsituent constants were used to determine the correlation of substituent electronic effects. Additionally, analyses of NMR spectra produced no tendencies that were characteristic of the formation of zwitterion intermediates, which were included in Mimoun's mechanism. NMR spectra suggest that both **A** and **B** are possible transition states for the oxygen transfer; however, transition state **B** is limited to cases when the nucleophile is anionic.

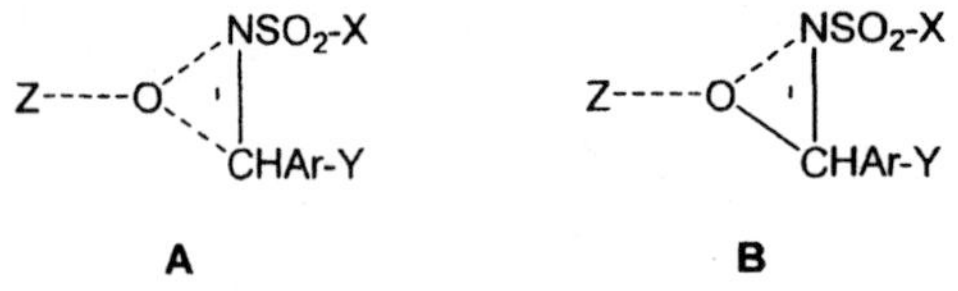

These detailed mechanistic studies have led to the following generalized mechanism of enolate oxation by oxaziridine reagents. In the hydroxylation of deoxybenzoin **10** to benzoin **15**, the enolate anion **11** is formed by kinetic deprotonation with LDA. The enolate anion then performs a nucleophilic attack on the oxygen of the oxaziridine **1** in a manner consistent with that proposed in transition state **B**. The hemiaminal **12** intermediate decomposes to form the alkoxide **14** and sulfonimine **13**. The alkoxide is then quenched to provide alcohol **15**.[2]

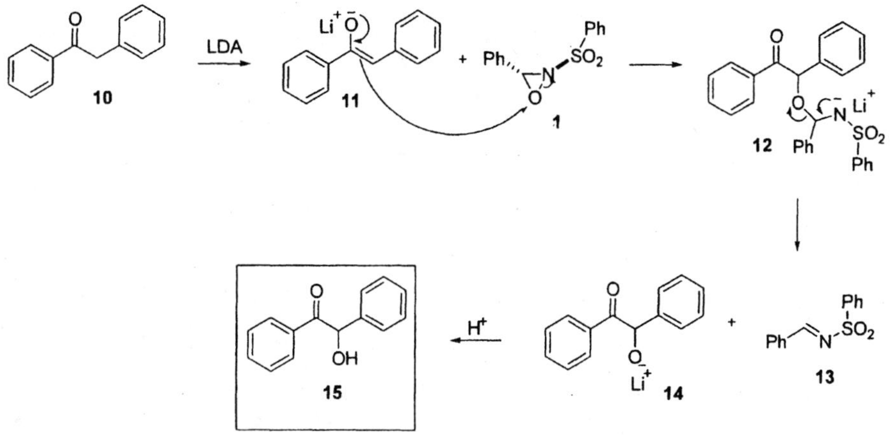

1.2.4 *Synthetic Utility*

Readers are directed to a thorough review on the chemistry of Davis oxaziridine reagents through 1992. While this work will occasionally be referenced, the primary focus of this section will be on the chemistry from 1992 to the present.

1.2.4.1 *Asymmetric oxidation via chiral auxiliaries*

Chiral auxiliaries have found abundant use in providing a template for efficient and highly diastereoselective enolate reactions. Concurrent with the development of chiral and nonracemic Davis oxaziridine reagents, the use of chiral auxiliaries to direct the stereoselectivity in these systems has been generalized.

The first example of the use of chiral auxiliaries for diastereoselective α-hydroxylation was published in 1985. Davis and co-workers[13] demonstrated the efficient and highly stereoselective formation of α-hydroxyamides using the (+)-(*S*)-2-pyrrolidinemethanol (R = H) chiral auxiliary. Thus, treatment of **16** with LDA was followed by enolate oxidation with (±)-**1** to provide **17** in high yield and high *de* (> 95%). An interesting reversal was noted when the sodium enolate of **16** was reacted under similar conditions. In this case, production of the (*R*)-product **18** was predominating with only slightly lower diastereoselectivity (93% *de*). In both cases, the chiral auxiliaries could be removed without racemization. The use of the related (+)-(*S*)-2-(methoxymethyl)pyrrolidine auxiliary (R = Me) provided inconsistent results with regard to solvent dependence, selectivity, and yield.[13]

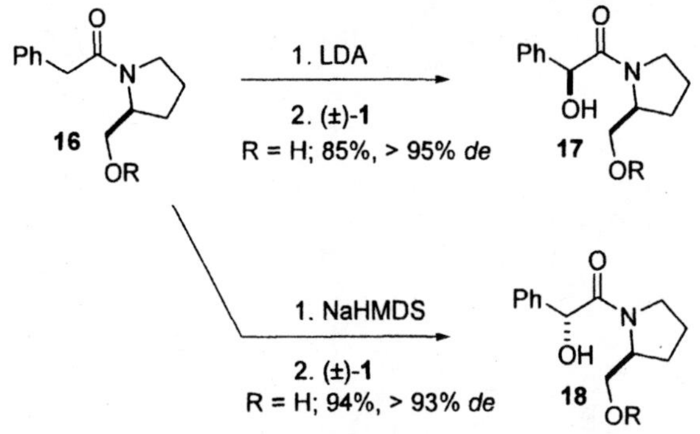

The synthesis of acyclic tertiary α-hydroxy acids poses an additional challenge, in that a specific enolate isomer must be generated prior to reaction with the oxaziridine reagent. As a means of overcoming this challenge, Davis and co-workers utilized a strategy of double stereodifferentiation, wherein the chiral enolate was reacted with an enantiopure oxaziridine. This strategy proved successful for the synthesis of **20** from **19** when either enantiomer of (camphorsulfonyl)oxaziridine **5** was employed as the oxidizing agent in the presence of HMPA.[14] Interestingly, the use of racemic **1** as the oxidizing agent resulted in only modest diastereoselectivity (55% *de*), implicating the necessity for chiral, nonracemic oxaziridines in this system.

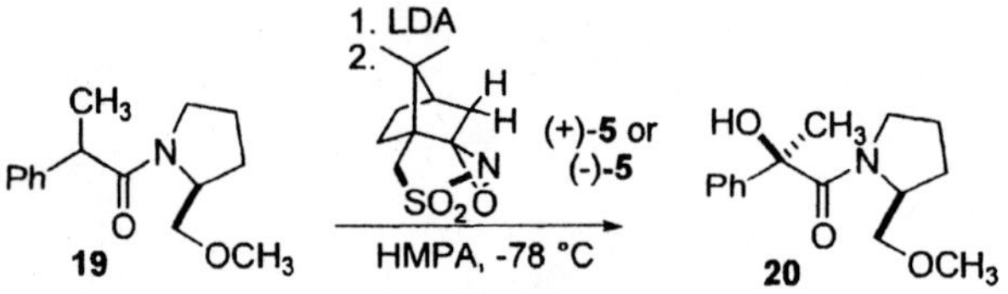

Expanding upon these results, Schultz *et al.* analyzed the oxidation of enolates produced via the Birch reduction of carboxylic acid derivatives.[15] It was found that when **21** was reduced and treated with (+)-**5**, **22** was obtained with only marginal yield and modest enantioselectivity. The enantioselectivity increased when **23** was deprotonated and then treated with (+)-**5**.

1. Li/NH_3, THF, *t*-BuOH
2.(+)-**5**
57%, 85% de

1. KH, THF, rt
2. (+)-**5**, -78 °C
47%, 94% de

21 **22** **23**

While more plentiful, alcohol-based chiral auxiliaries have been limited in their ability to direct the diastereoselective hydroxylation for the preparation of tertiary α-hydroxy acids. Among these, the best results in this series were obtained when oxidation of the enolate of chiral ester substrate **24** with (+)-**5** yielded (*S*)-**25**.[14b] The use of (–)-**5** as the hydroxylating agent, provided a reversal in stereoselectivity, providing (*R*)-**25**. Interestingly, when substoichiometric amounts (0.5 equiv) of (+)-**5** were used, stereoselectivity improves (94% *de*), a fact attributed to the matching of the enolate geometry to the oxidant. This speculation is credible, as evidenced by the fact that oxidation with 0.50 equivalent of (–)-**5** produces (*S*)-**25** in only 37% *de* in a stereochemically mismatched case.

LDA, THF
(+)-**5**, -78 °C
76%, 66% *de*

24 **25**

Among chiral enolates, those derived from oxazolidinone carboximides, as developed in the Evans laboratories,[16,17] have shown the most generality in directing the hydroxylation with the racemic Davis oxaziridine **1**. Pioneered in 1985, a variety of carboximide derivatives have been hydroxylated with consistently high levels of diastereoselectivity.[18] The high levels of selectivity are due, in part, to the facile and exclusive formation of the *Z*-enolate under these conditions. Importantly, these reactions are necessarily quenched by the organic soluble camphor sulfonic acid (CSA) to avoid formation of a product arising from intramolecular attack of the alkoxide oxygen at the oxazolidinone carbonyl under standard aqueous acidic workup methods.

1. NaHMDS, THF, -78 °C
2. **1**, THF, -78 °C
3. CSA, THF

68%, 92% *de*

1. NaHMDS, THF, -78 °C
2. **1**, THF, -78 °C
3. CSA, THF

86%, 98% *de*

1. NaHMDS, THF, -78 °C
2. **1**, THF, -78 °C
3. CSA, THF

83%, 90% *de*

The generality of this method has been demonstrated in a number of total synthesis efforts.[19,20] In their studies aimed at the elucidation of the structure of capensifuranone, the Williams group utilized the 4-phenyl auxiliary for introduction of the hydroxyl substituent in **27**.[21] While the stereocenter was eventually destroyed, this reaction is notable in that high stereoselectivity can be obtained with the 4-phenyl auxiliary, in addition to the auxiliaries used above.

26

1. NaHMDS, THF, -78 °C
2. **1**, THF, -78 °C
3. CSA, THF, -78 °C

93%, 10:1 *dr*

27

In two separate syntheses of the microtubule stabilizing antitumor agent epothilone B (**28**), the C_{15} stereocenter was prepared using diastereoselective hydroxylation mediated by the carboximide auxiliary. White and co-workers relied upon the 4-benzyl auxiliary to effect the diastereoselectivity in the preparation of **30**.[22a] Meanwhile, the generality of this reaction protocol is showcased by the high functional group tolerance demonstrated in the hydroxylation of **31** to give **32**.[22b]

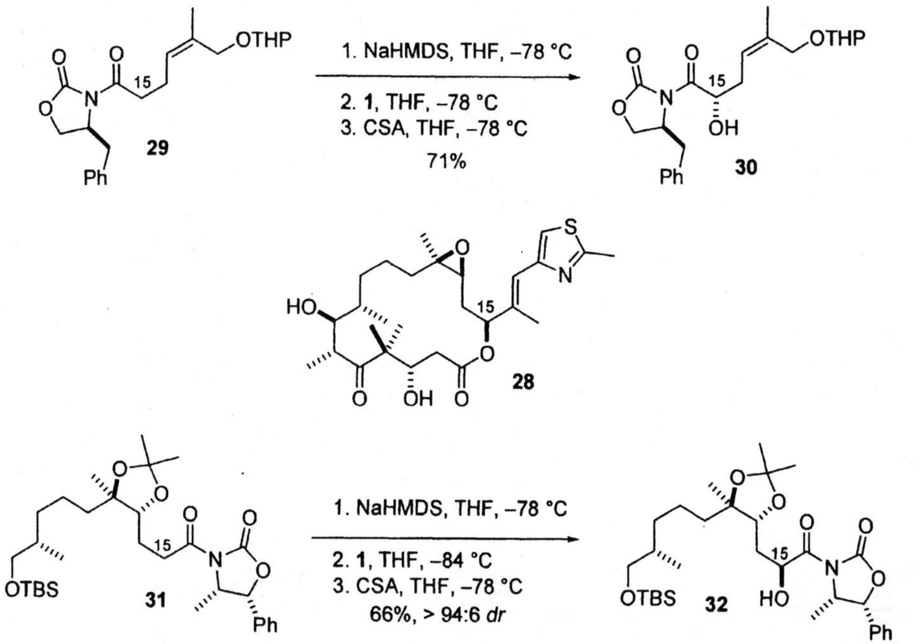

29 → **30**

1. NaHMDS, THF, –78 °C
2. **1**, THF, –78 °C
3. CSA, THF, –78 °C
71%

28

31 → **32**

1. NaHMDS, THF, –78 °C
2. **1**, THF, –84 °C
3. CSA, THF, –78 °C
66%, > 94:6 *dr*

1.2.4.2 Substrate directed diastereoselective hydroxylation

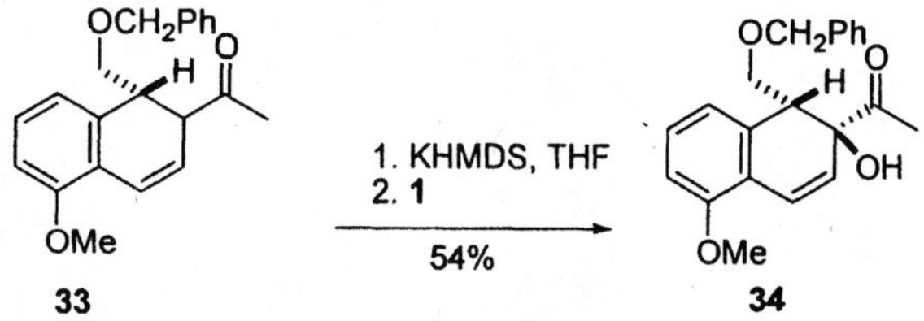

33 → **34**

1. KHMDS, THF
2. **1**
54%

Several examples exist wherein the chirality of the substrate serves to influence the degree of diastereoselectivity obtained on oxidation with racemic and nonracemic oxaziridines. These examples are different from those above, since an auxiliary whose sole purpose is to direct the stereoselectivity is not present. The majority of these examples rely on cyclic stereocontrol to direct the facial selectivity. For example, in Meyers synthesis of the AB-ring of aklavinone, oxidation of the thermodynamic enolate derived from **33** resulted in the production of tertiary α-hydroxy ketone **34** as a single diastereomer in modest yield.[23] The stereoselectivity is rationalized by invoking a transition state wherein pseudo axial addition to the least hindered top face occurs preferentially, with addition to the bottom face blocked by the neighboring benzyl ether.

Similarly, selectivity was observed in Weinreb's efforts toward the synthesis of the microbial immunosuppressive agent FR901483.[24] In this case, axial addition was favored by reaction of the lithium enolate of amide **35** with racemic **1** to produce **36**. An interesting reversal of stereoselectivity was observed when, on slight alteration of the synthetic sequence, the Boc-protected amide was subjected to similar conditions. For reasons not fully understood, equatorial alcohol **37** was produced in a 53% yield, the structure of which was confirmed by X-ray crystal analysis.

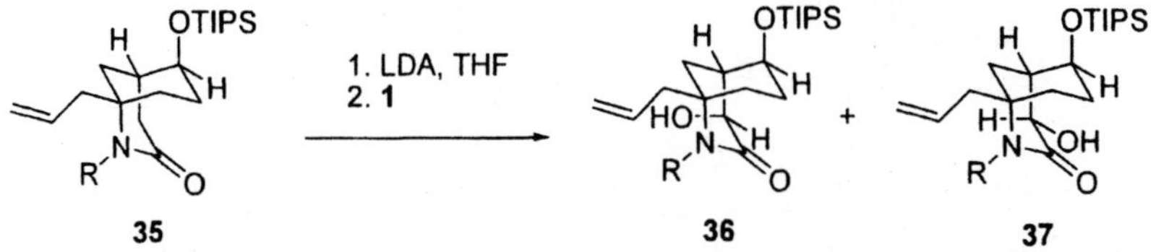

1. LDA, THF
2. **1**

35 → **36** R = Bn, 60% + **37** R = Boc, 53%

During the synthesis of baccatin III derivatives, Baldelli and co-workers used the diastereoselective 14β-hydroxylation of **38** to form **39** using **1** as the hydroxylating reagent. As part of this study, several solvents, bases, and oxaziridine reagents were tested for this reaction, of which a majority produced high yields.[25] Optimum results were obtained using KO^{t}Bu as the base in a THF/DMPU (83:17) solvent system, though this is likely system dependent. The stereochemistry obtained was rationalized by a folded conformation of the terpene skeleton which precluded attack of the bulky oxaziridine reagents from the α-face. This example illustrates the generality of this approach and demonstrates its potential for use with highly congested and heavily functionalized substrates.

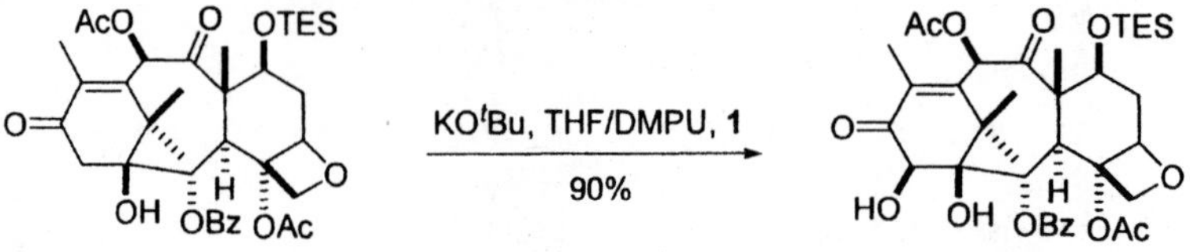

KO^{t}Bu, THF/DMPU, **1**
90%

38 → **39**

Using a similar strategy, Paquette and co-workers used this method for the α-oxygenation of **40** in the synthesis of precursors of the antitumor agent Taxol.[26] The α-hydroxy ketone **41** was produced by quenching the potassium enolate of **40** with **1**, yielding a 5:1 mixture of the *exo* and *endo* epimers of **41**.

1. $KN(SiMe_3)_2$, THF
2. **1**, -78°C
52%, 5:1 *dr*
PMPO **40** → PMPO **41**

While the majority of examples of substrate control deal with cyclic sterocontrol, there are a few examples where diasteroselectivity is induced in an acylic system.[27] A notable example of this was demonstrated during the synthesis of a fragment of tubulysin, by Wipf and co-workers.[28] Utilizing a Davis reagent in their synthesis of an α-hydroxy-γ-amino acid, the enolate of the γ-amino acid derivative **42** was reacted with **1** to form the α-hydroxy derivative **43** as a single diastereomer in good yield. The stereoselective reaction has precedence in literature and likely involves a highly chelated dianionic species.[29]

1. NaHMDS, THF
2. **1**, THF, 1 h
-78°C, 66%
42 → **43**

Further examples of acyclic stereocontrol in related amino acid systems make use of chiral and nonracemic oxaziridine reagents to induce high levels of stereocontrol. In 1992, Davis and coworkers synthesized the methyl ester of the Taxol C_{13} side chain using this method.[30] Following enolization, the dianion of **44** was reacted with the Davis reagent **5** to yield the α-hydroxy β-amino acid **45** in 49% yield. While the yield was marginal in this particular example, the 86:14 ratio of stereoisomers produced is impressive in this acyclic system.

1. LDA, LiCl, -42°C
2. **5**, -78°C
44 → **45**

A similar protocol has been generalized in the Davies laboratories, relying on a highly diastereoselective tandem conjugate addition/hydroxylation strategy for synthesis of α-hydroxy-β-amino acid derivatives.[31–37] For example, the enolate produced via diastereoselective conjuagate addition of lithium benzylamide **47** to cinnamate **46** was quenched with (+)-(camphorsulfonyl)oxaziridine (**5**) to give **48** in impressive yield and high diastereoselectivity.

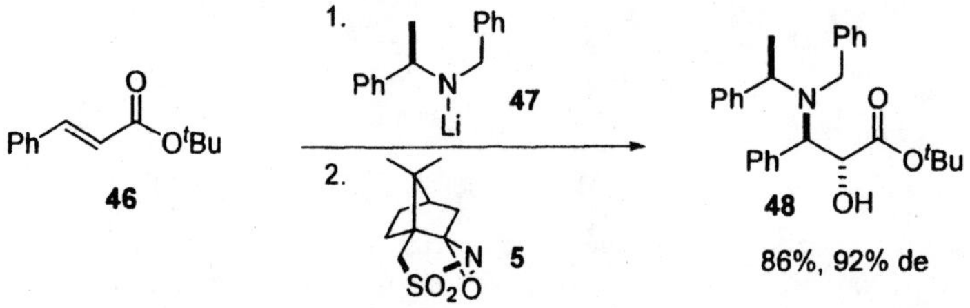

The above protocol was applied during efforts to assign the stereochemistry of the ACE-inhibitor microginin.[38] Following a similar procedure, conjugate addition of **47** to **49** was followed by oxidation with **5** to provide **50** in 63% yield. The (2*S*)-isomer was isolated as a by-product in 4% yield, giving the reaction an overall diastereoselectivity of around 15:1, similar to that obtained above.

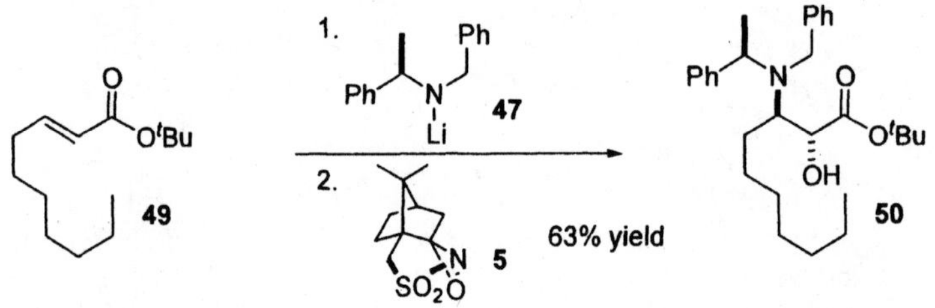

When the above transformation was conducted in a stepwise manner, β-amino acid derivative **51** was deprotonated with LDA and reacted with **5** to produce **50** in 65% yield and ~ 92% *de*.[38]

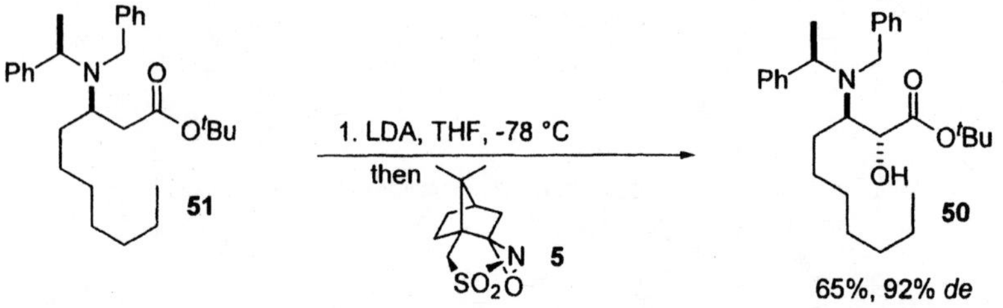

The syntheses of α-hydroxy β-amino acids has also been accomplished using a perhydropyrimidin-4-one template to direct the stereoselectivity of the hydroxylation.[39] For example, the enolate of perhydropyrimidin-4-one **52** was diastereoselectively hydroxylated to obtain **53**.[40] Examination of several oxidizing agents and bases revealed LiHMDS and the oxaziridine **5** as ideal choices for this transformation. Compound **53** can be hydrolyzed under acidic conditions to produce the enantiomerically pure α-hydroxy β-amino acid.

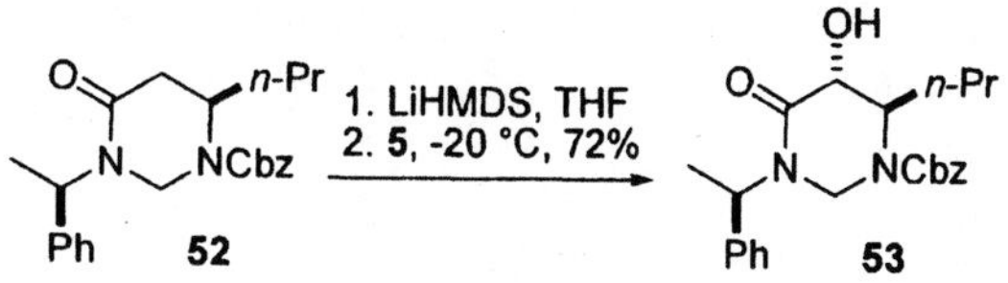

Substrate directed hydroxylation with Davis reagents has also found utility in the synthesis of architecturally complex natural products.[41–44] For example, efforts for the synthesis of (+)-spongistatin 1 by Smith and co-workers relied on hydroxylation of the enolate of **54** with Davis reagent **7**, which was followed by epimerization of the C_5 to produce **55** in high yield.[45]

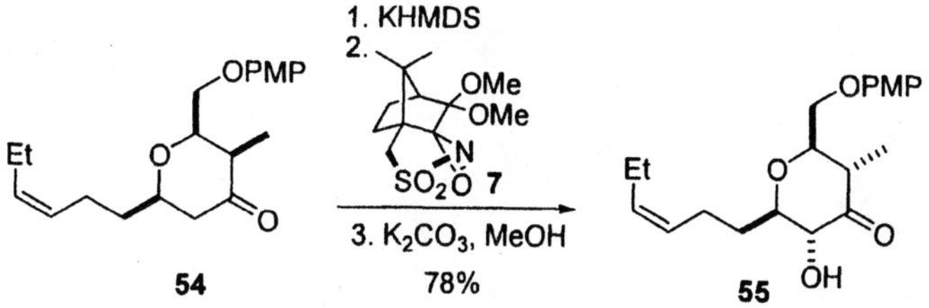

In the syntheses of (−)-lepadins A, B, and C, which exhibit strong cytotoxicity against human cancer cells, enolate oxidation relied on the chiral, nonracemic dichloro oxaziridine **6** to deliver the product **57** in a stereoselective manner.[46] Studies on this reaction demonstrated a higher stereoselectivity for the lithium enolate in comparison to a similarly generated sodium enolate. Their results also indicated that the reaction between **56** and **6** constituted a stereochemically matched case, whereas the use of racemic oxaziridine reagent **1** as well as use of the alternate enantiomer of **6** resulted in greatly reduced diastereoselectivity.

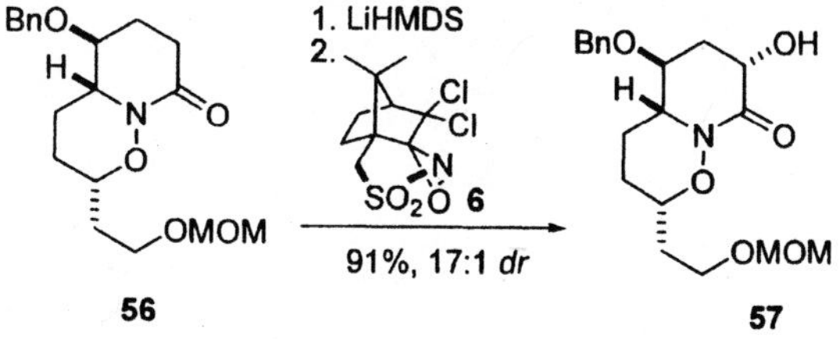

1.2.4.3 Enantioselective oxidation

Enantioselective oxidation using Davis chiral oxaziridine reagents is not as well developed as its diastereoselective counterpart. However, a number of simple enolates have been selectively hydroxylated, primarily in the Davis group.[47] After screening a number of bases and oxaziridines, optimum conditions generally utilize NaHMDS as the base with one of the (camphorsulfonyl)oxaziridine reagents (**5** or **6**). The impressive enantioselectivities for this protocol are demonstrated in the formation of **58**,[48] **59**,[49] and **60**.[50] Extension of this protocol to more complex ketones, or toward the synthesis of tertiary α-hydroxy ketones, is limited by the inability to selectively form a single enolate isomer. While hydroxylation of aromatic ketones has been well-documented,[51,52] the majority of more complicated enantioselective hydroxylations are done on cyclic ketones or lactones.[53]

NaHMDS, 5
58 84%, 95% *ee*

NaHMDS, 6
59 61%, 95% *ee*

NaHMDS, 5
60 71%, 89% *ee*

1. LiHMDS
2. 5
82%, 86% ee
61 → **62**

Several examples exist in the literature in which cyclic ketone enolates are enantioselectively hydroxylated by chiral, nonracemic Davis oxaziridine reagents. In contrast to their acyclic counterpart, the enolate geometry is fixed in cyclic systems. During the preparation of enantiomerically pure (–)-blebbistatin, the enolate of the quinolone **61** was reacted with the Davis reagent **5** to afford the optically enriched **62** with 82% yield and 86% *ee*.[54] The related reagent **6** was used in the synthesis of (+)-*o*-trimethylbrazilin, which was accomplished by converting the enolate of **63** to the enantiomerically pure **64** in 50% yield.[55] Similar procedures were used during the synthesis of the AB ring segment of daunomycin.[56]

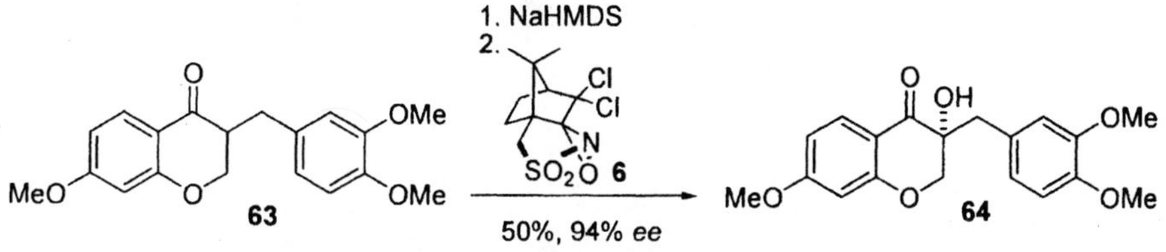

1. NaHMDS
2. 6
50%, 94% *ee*

A related procedure was utilized for the syntheses of Sch 42427 and ER-30346, azole antifungals, and (–)-tricycloillicinone; however, slight modifications in the experimental procedures were explored resulting in improved yield and *ee*. During the conversion of the enolate of ketone **65** to the chiral α-hydroxy ketone **66**, Gala and DiBenedetto observed that the metal salts used in the reaction greatly influenced both the yield and the enantioselectivity.[52] Furuya and Terashima were also able to increase both yield and *ee* by carrying out the oxidation of the lithium enolate of **67** using Davis reagent **6** in a mixture of THF and DMF, which increased both the yield and *ee* of **68** by over 30% each.[56]

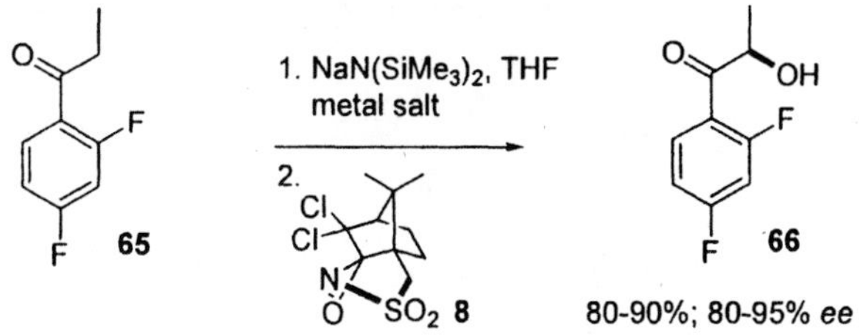

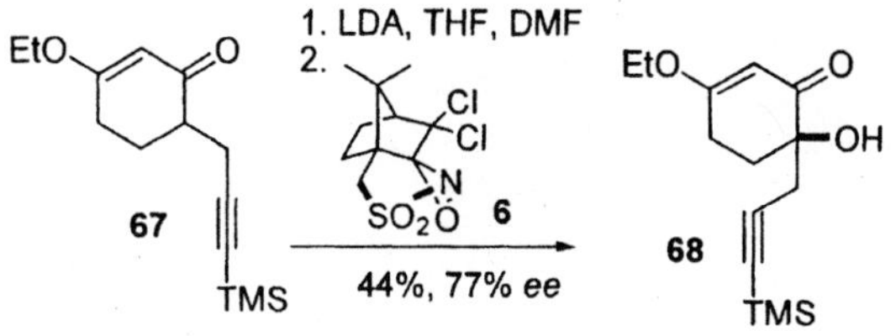

1. $NaN(SiMe_3)_2$, THF metal salt
2. 8
80-90%; 80-95% *ee*

1. LDA, THF, DMF
2. 6
44%, 77% *ee*

The utility of Davis chiral oxaziridine reagents has been more recently applied to the synthesis of optically active α-hydroxy phosphonates. Two groups have been largely responsible for developments in this area. Principally, Wiemer and co-workers have demonstrated the highly enantioselective hydroxylation of a series of benzyl phosphonates. As shown below for **69**, hydroxylation makes use of oxaziridine **6**, proceeding in moderate yield to give **70**, with high levels of enantioselectivity. The reaction works equally well with a variety of 4-substituted benzylphosphonates.[57] In an extension of this reaction to a more conformationally flexible molecule, the oxidation of dimethylfarnesylphosphate also proceeded with moderate enantioselectivity.[58]

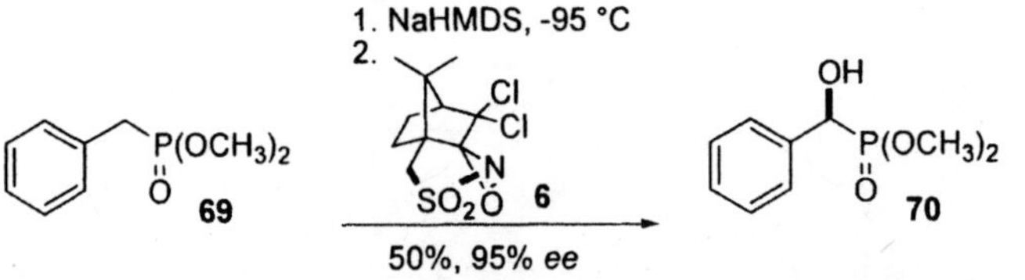

1. NaHMDS, -95 °C
2. 6
50%, 95% *ee*

Finally, Schmidt and coworkers have similarly applied the hydroxylation of phosphonates for synthesis of some sialyltransferase inhibitors. Thus, the stabilized enolate of **71** was selectively oxidized using dichloro(camphorsulphonyl)oxaziridine (**6**) to produce the α-hydroxy diallylphosphonate in low yield but as a single stereoisomer. The reaction was also found to be effective for compounds containing a variety of ring substituents.[59,60]

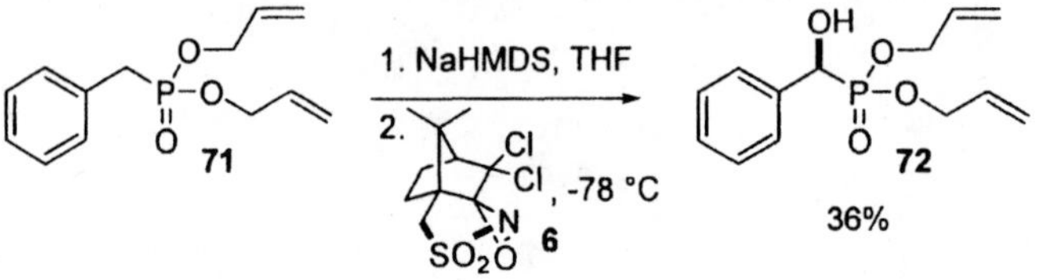

1. NaHMDS, THF
2. 6, -78 °C
36%

1.2.4.4 *Heterocyclic asymmetric oxidation*

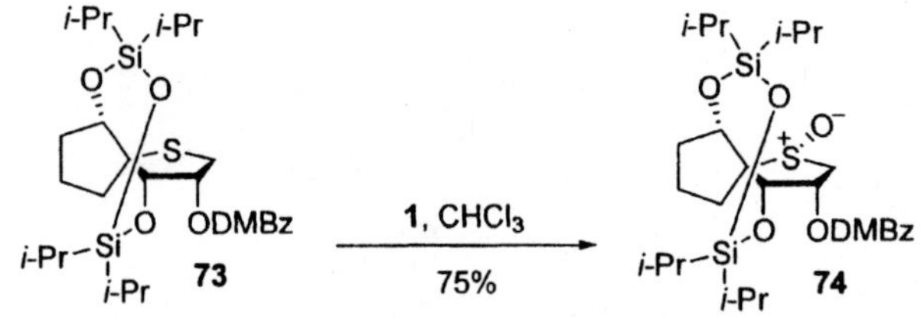

1, $CHCl_3$
75%

Davis chiral oxaziridine reagents have found ample synthetic utility in the asymmetric oxidation of sulfides to sulfoxides, providing excellent yields and high enantiomeric excess. Sulfoxides have seen growing importance in organic synthesis and the versatility of oxaziridine reagents enables them to be synthesized cleanly and efficiently.[61] The sulfide **73** was oxidized using the Davis reagent **1**, which could easily transfer an oxygen atom to the α-face of the tetrahydrothiopene ring to form the major product **74** in 75% yield.[62] Similarly, the asymmetric oxidation of **75** using the Davis reagent **6** in carbon tetrachloride produced the sulfoxide intermediate **76** in 95% yield and 75% *ee*.[63]

75 → (**6**, CCl_4) → **76**

Sulfenimines undergo asymmetric oxidations to form sulfinimines via a reaction with Davis reagents. The sulfenimine **77** was oxidized by **9** to yield the sulfinimine **78** in 82% yield and 97% *ee*. Yields and enantiomeric excess varied based on the oxaziridine reagent used.[64]

77 → (**9**, 82%, 97% *ee*) → **78**

1.2.5 Experimental

79 → (1. LDA; 2. **7**; THF, -78 °C) → **80**

(–)-(*R*)-2-Ethyl-5,8-dimethoxy-2-hydroxy-1,2,3,4-tetrahydronaphthalen-1-one (80)[65]
A solution of 58.5 mg (0.25 mmol) of tetralone **79** in 2 mL of THF was added dropwise to a stirred and cooled –78 °C solution of 0.3 mL (0.30 mmol) of a 1 M solution of LDA in 2 mL of THF. After the mixture was stirred at –78 °C for 30 min a solution of 116 mg (0.4 mmol) of (+)-**7** in 5 mL of THF was added dropwise. The reaction was monitored by TLC, warmed to 0 °C as required, and quenched by the addition of 3 mL of saturated aqueous NH_4Cl after 0.5–1 h. The reaction mixture was extracted with diethyl ether (3 × 25 mL), and the combined organic extracts were washed with H_2O (20 mL) and brine (20 mL) and dried

($MgSO_4$). The solvent was removed under reduced pressure to give a white solid (170 mg), which was purified by preparative TLC (eluant pentane/ether/CH_2Cl_2, 3:1:1, $R_f = 0.3$) to give 40.6 mg (66%) of **80**: mp 74–75 °C (lit. colorless oil); IR and NMR spectral data are consistent with reported values. Anal. Calcd for $C_{14}H_{18}O_1$: C, 67.20; H, 7.20. Found: C, 67.09; H, 7.22.

1.2.6 References

1. Davis, F. A.; Vishwakarma, L. C.; Billmers J. M.; Flinn, J. *J. Org. Chem.* **1984**, *49*, 3241–3.
2. [R] Davis, F. A.; Chen, B.-C. *Chem. Rev.* **1992**, *92*, 919–34.
3. Christoffers, J.; Baro, A.; Werner, T. *Adv. Synth. Catal.* **2004**, *346*, 143–51.
4. Mimoun, H. *Angew. Chem. Int. Ed. Engl.* **1982**, *21*, 734–50.
5. Mimoun, H.; Mignard, M.; Brechot, P.; Saussie, L. *J. Am. Chem. Soc.* **1986**, *108*, 3711–8.
6. [R] Sharpless, K. B.; Verhoeven, T. R. *Aldrichimica Acta* **1979**, *12*, 63.
7. [R] Behrens, C. H.; Sharpless, K. B. *Aldrichimica Acta* **1983**, *16*, 67.
8. Lang, T. J.; Wolber, G. J.; Bach, R. D. *J. Am. Chem. Soc.* **1981**, *103*, 3275–82.
9. Bach, R. D.; Wolber, G. J.; Coddens, B. A. *J. Am. Chem. Soc.* **1984**, *106*, 6098–9.
10. Curci, R.; Giannattasio, S.; Sciacovel, O.; Troisi, L. *Tetrahedron* **1984**, *40*, 2763–71.
11. Pitchen, P.; Dunach, E.; Deshmukh, M. N.; Kagan, H. B. *J. Am. Chem. Soc.* **1984**, *106*, 8188–93.
12. Davis, F. A.; Billmers, J. M.; Gosciniak, D. J.; Towson, J. C. *J. Org. Chem.* **1986**, *51*, 4240–45.
13. Davis, F. A.; Vishwakarma, L. C.; *Tetrahedron Lett.* **1985**, *26*, 3539–42.
14. (a) Davis, F. A.; Ultaowski, T. G.; Haque, M. S. *J. Org. Chem.* **1987**, *52*, 5288–90. (b) Davis, F. A.; Reddy, G. V.; Chen, B.-C.; Kumar, A.; Haque, M. S. *J. Org. Chem.* **1995**,*60*, 6148–53.
15. Schultz, A. G.; Harrington, R. E.; Holoboski, M. A. *J. Org. Chem.* **1992**, *57*, 2973–6.
16. Evans, D. A. *Aldrichimica Acta* **1982**, *15*, 23.
17. Evans, D. A.; Ennis, M. D.; Mathre, D. J. *J. Am. Chem. Soc.* **1982**, *104*, 1737–9.
18. Evans, D. A.; Morrissey, M. M.; Dorow, R. L. *J. Am. Chem. Soc.* **1985**, *107*, 4346–8.
19. Fürstner, A.; Radkowski, K.; Wirtz, C.; Goddard, R.; Lehmann, C. W.; Mynott, R. *J. Am. Chem. Soc.* **2002**, *124*, 7061–9.
20. Evans, D. A.; Gage, J. R. *J. Org. Chem.* **1992**, *57*, 1958–61.
21. Williams, D. R.; Nold, A. L.; Mullins, R. J. *J. Org. Chem.* **2004**, *69*, 5374–82.
22. (a) White, J. D.; Carter, R. G.; Sundermann, K. F. *J. Org. Chem.* **1999**, *64*, 684–5. (b) Mulzer, J.; Karig, G.; Pojarliev, P. *Tetrahedron Lett.* **2000**, *41*, 7635–38.
23. Meyers, A. I.; Higashiyama, K. *J. Org. Chem.* **1987**, *52*, 4592–7.
24. Kropf, J. E.; Meigh, I. C.; Bebbington, M. W. P.; Weinreb, S. M. *J. Org. Chem.* **2006**, *71*, 2046–55.
25. Baldelli, E.; Battaglia, A.; Bombardelli, E.; Carenzi, G.; Fontana, G.; Gambini, A.; Gelmi, M. L.; Guerini, A.; Pocar, D. *J. Org. Chem.* **2003**, *68*, 9773–9.
26. Elmore, S. W.; Paquette, L. A. *J. Org. Chem.* **1995**, *60*, 889–96.
27. Narasaka, K.; Ukaji, Y.; Watanabe, K. *Bull. Chem. Soc. Jpn.* **1987**, *60*, 1457–64.
28. Wipf, P.; Takada, T.; Rishel, M. J. *Organic Lett.* **2004**, *6*, 4057–60.
29. Hanessian, S.; Schaum, R. *Tetrahedron Lett.* **1997**, *38*, 163–6.
30. Davis, F. A.; Reddy, R. T.; Reddy, R. E. *J. Org. Chem.* **1992**, *57*, 6387–9.
31. Bunnage, M. E.; Davies, S. G.; Goodwin, C. J. *J. Chem. Soc., Perkin Trans. I* **1993**, 1375–6.
32. Bunnage, M. E.; Chernega, A. N.; Davies, S. G.; Goodwin, C. J. *J. Chem. Soc., Perkin Trans. I* **1994**, 2373–84.
33. Bunnage, M. E.; Davies, S. G.; Goodwin, C. J. *Synlett* **1993**, 731–2.
34. Bunnage, M. E.; Davies, S. G.; Goodwin, C. J. *Tetrahedron* **1994**, *50*, 3975–86.
35. Bunnage, M. E.; Davies, S. G.; Goodwin, C. J. *J. Chem. Soc., Perkin Trans. I* **1994**, 2385–91.
36. Davies, S. G.; Epstein, S. W.; Ichihara, O.; Smith, A. D. *Synlett* **2001**, *10*, 1599–601.
37. Davies, S. G.; Epstein, S. W.; Garner, A. C.; Ichihara, O.; Smith, A. D. *Tetrahedron: Asymmetry* **2002**, *13*, 1555–65.
38. Bunnage, M. E.; Burke, A. J.; Davies, S. G.; Goodwin, C. J. *Tetrahedron: Asymmetry* **1995**, *6*, 165–76.
39. Escalante, J.; Juaristi, E. *Tetrahedron Lett.* **1995**, *36*, 4397–400.
40. Cardillo, G.; Tolomelli, A.; Tomasini, C. *Tetrahedron* **1995**, *51*, 11831–40.
41. Dounay, A. B.; Forsyth, C. J. *Org. Lett.* **1999**, *1*, 451–3.

42. Roush, W. R.; Barda, D. A. *Tetrahedron Lett.* **1997**, *38*, 8785–8.
43. Hitotsuyanagi, Y.; Nishimura, K.; Ikuta, H.; Takeya, K.; Itokawa, H. *J. Org. Chem.* **1995**, *60*, 4549–58.
44. Yu,W.; Jin, Z. *J. Am. Chem. Soc.* **2001**, *123*, 3369–70.
45. Smith, A. B., III; Sfouggatakis, C.; Gotchev, D. B.; Shirakami, S.; Bauer, D.; Zhu, W.; Doughty, V. A. *Org. Lett.* **2004**, *6*, 3637–40.
46. Ozawa, T.; Aoyagi, S.; Kibayashi, C. *J. Org. Chem.* **2001**, *66*, 3338–47.
47. Davis, F. A.; Haque, M. S.; Przelawski, R. M. *J. Org. Chem.* **1989**, *54*, 2021–4.
48. Davis, F. A.; Haque, M. S. *J. Org. Chem.* **1986**, *51*, 4083–5.
49. Davis, F. A.; Weismiller, M. *J. Org. Chem.* **1990**, *55*, 3715–7.
50. Davis, F. A.; Sheppard, A. C.; Chen, B.-C.; Haque, M. S. *J. Am. Chem. Soc.* **1990**, *112*, 6679–90.
51. Matsunaga, N.; Kaku, T.; Ojida, A.; Tasaka, A. *Tetrahedron: Asymmetry* **2004**, *15*, 2021–8.
52. Gala, D.; DiBenedetto, D. J. *Tetrahedron: Asymmetry* **1997**, *8*, 3047–50.
53. Davis, F. A.; Haque, M. S.; Ultaowski, T. G.; Towson, J. C. *J. Org. Chem.* **1986**, *51*, 2402–4.
54. Lucas-Lopez, C.; Patterson, S.; Blum, T.; Straight, A. F.; Toth, J.; Slawin, A. M. Z.; Mitchison, T. J.; Sellers, J. R.; Westwood, N. J. *Eur. J. Org. Chem.* **2005**, *9*, 1736–40.
55. Davis, F. A.; Chen, B. C. *J. Org. Chem.* **1993**, *58*, 1751–3.
56. Furuya, S.; Terashima, S. *Tetrahedron Lett.* **2003**, *44*, 6875–8.
57. Pogatchnik, D. M.; Wiemer, D. F. *Tetrahedron Lett.* **1997**, *38*, 3495–8.
58. Cermak, D. M.; Du, Y.; Wiemer, D. F. *J. Org. Chem.* **1999**, *64*, 388–93.
59. Skropeta, D.; Schwörer, R.; Schmidt, R. R. *Bioorg. Med. Chem. Lett.* **2003**, *13*, 3351–4.
60. Skropeta, D.; Schmidt, R. R. *Tetrahedron: Asymmetry* **2003**, *14*, 265–73.
61. Davis, F. A.; Lal, S. G. *J. Org. Chem.* **1988**, *53*, 5004–7.
62. Paquette, L. A.; Dong, S. *J. Org. Chem.* **2005**, *70*, 5655–64.
63. Padmanabhan, S.; Lavin, R. C.; Durant, G. J. *Tetrahedron: Asymmetry* **2000**, *11*, 3455–7.
64. Davis, F. A.; Reddy, R. E.; Szewczyk, J. M.; Reddy, G. V.; Portonovo, P. S.; Zhang, H.; Fanelli, D.; Reddy, R. T.; Zhou, P.; Carroll, P. J. *J. Org. Chem.* **1997**, *62*, 2555–63.
65. Davis, F. A.; Kumar, A.; Chen, B.-C. *J. Org. Chem.* **1991**, *56*, 1143–5.

Richard J. Mullins and Michael T. Corbett

1.3 Midland Reduction

1.3.1 Description

The Midland reduction is the enantioselective reduction of a ketone (**1**) to an optically active alcohol (**2**) using the commercially available reagent alpine borane (**3**).[1]

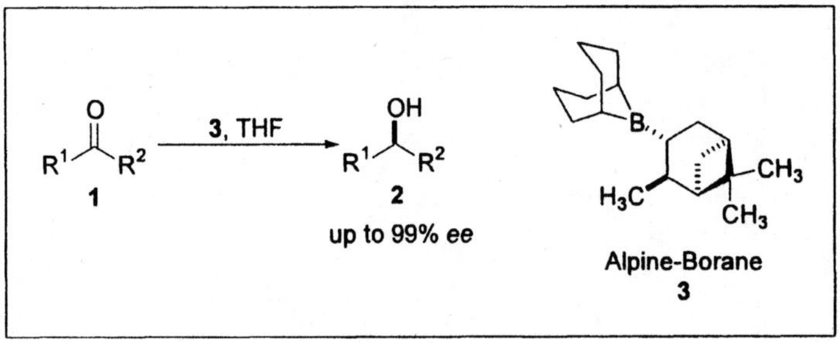

1.3.2 Historical Perspective

In 1977, Midland and co-workers found that *B*-alkyl-9-borabicycle[3.3.1]nonanes (**4**) were unique among trialkyl borane adducts (**5**) in that they rapidly reduced benzaldehyde (**6a**) at room temperature.[2]

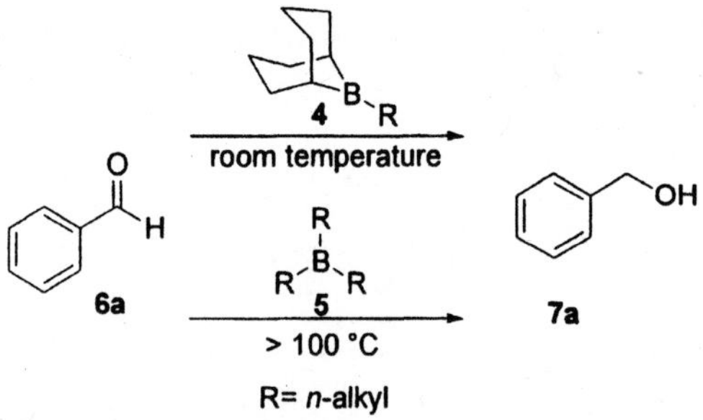

Extension of this methodology to an enantioselective variant soon followed. In 1979 Midland showed that by using the chiral reagent derived from hydroboration of α-pinene by 9-BBN (**3**), deuterium labeled benzaldehyde (**6b**) could be reduced to enantiomerically enriched alcohol **7b** in 98% *ee.*[3] Subsequent studies found that **3** was also useful for the enantioselective reduction of acetylinic ketones (**8**) to propargylic alcohols (**9**).[4]

6b → 3, THF, 98% ee → 7b

8 → 3, THF → 9

R^1, R^2 = alkyl, aryl; up to 99% ee

1.3.3 Mechanism and Stereochemical Rationalization

Initial debate over the mechanism of the Midland reduction centered around the idea that this reduction could reasonably proceed *via* either a one-step (Path A) or two-step (Path B) mechanism as shown below. However, mechanistic investigations by Midland showed that the rate of the reaction was dependent on the concentration of the aldehyde, thus lending support to the reaction proceeding *via* Path A[5]. The subsequent development of the enantioselective variant of this reaction using **3** essentially eliminated Path B as a possible mechanism because it is not consistent with the optically active alcohols produced in the reaction. Thus, Path A is widely accepted as the mechanism for this reaction.

Path A

Path B

The reduction of ketones with **3** affords chiral alcohols with a predictable stereochemical outcome. Transition state **10** is favored over **11** because the smaller group of the ketone (R_S) rather than the large group (R_L) is "axial" and eclipsing the methyl group of the pinene subunit.[6] While there is still some debate over the nuances of the transition state of this reduction, transition state **10** is a reasonable model that accurately predicts the stereochemical outcome of the reduction of ketones with **3**.

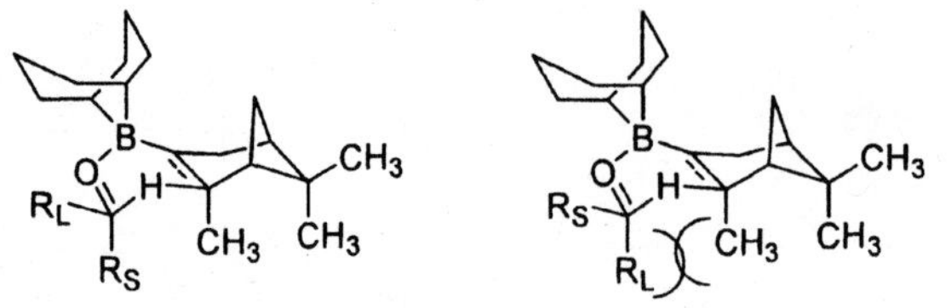

10
Favored Transition State

11
Disfavored Transition State

1.3.4 Variations and Improvements

While Midland's initial studies focused on the reduction of aryl aldehydes and alkynyl ketones, expansion of the scope of this methodology was problematic. Attempted reduction of a variety of aldehydes afforded the corresponding alcohols only after long reaction times and in low enantiomeric excess. The moderate increase in the steric bulk of these non-acetylenic ketones is believed to account for the marked decrease in reaction rate. Furthermore, the slow reduction of these ketones is problematic because it allows for the retrohydroboration of **3** into more reactive and achiral 9-BBN. The reduction of the ketone with 9-BBN is then much more facile and leads to optically inactive products (Path B, section 1.3.3). The initial solution to this problem was developed by H. C. Brown and Pai.[7] They found that by carrying out the reaction neat, instead of at 0.5 M in THF,the rate of the bimolecular reduction reaction increased enough so that the unimolecular retrohydroboration reaction was less competitive. Subsequently, Midland found that even better results could be obtained by increasing the pressure in the reaction vessel.[8,9] It was proposed that the reason for the increase in both rate and selectivity was due to the fact that the bimolecular reduction process (Path A, Section 1.3.3) should be favored at high pressure while the dissociative retrohydroboration reaction (Path B) should be suppressed. Indeed, Midland found that by increasing the pressure to 6000 atm the reduction of a variety of ketones, including acetophenone (**12**), was achieved with neat alpine borane (**3**) to the corresponding alcohol (**13**) at noticeably faster rates and with improved enantioselectivity.

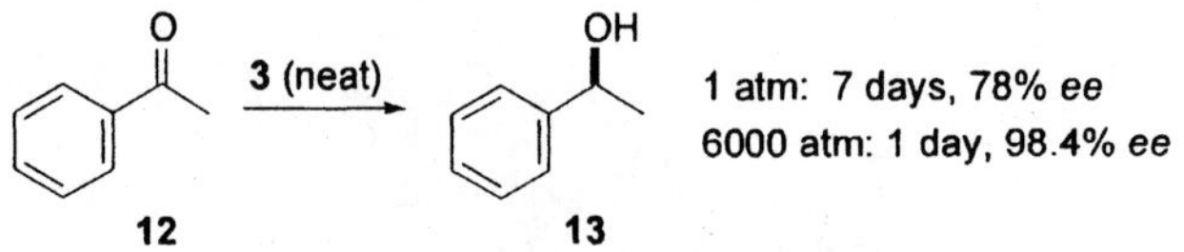

3 (neat)

1 atm: 7 days, 78% *ee*
6000 atm: 1 day, 98.4% *ee*

12

13
80% yield

The use of alpine-borane (**3**) has also been extended to include the reduction of acyl cyanides (**14**) to optically active β-amino alcohols (**15**),[10] as well as the reduction of α-keto esters (**16**) into the corresponding α-hydroxy esters (**17**).[11]

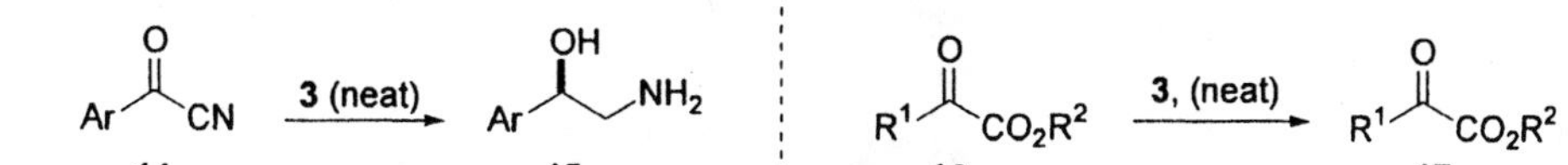

14 → **15**; **16** → **17**

Ar = C_6H_5, *p*-ClC_6H_4, *o*-ClC_6H_4, *m*-$CH_3C_6H_4$ — 74 – 86% yield, 85 – 90% *ee*

R^1 = CH_3, C_6H_5; R^2 = CH_3, C_2H_5, *i*-C_3H_7, *t*-C_4H_9 — 50 – 98% yield, 79 – 100% *ee*

1.3.5 Synthetic Utility

Alpine borane (the Midland reagent, **3**) has found broad use in the synthesis of complex natural products. As early as 1980, only one year after Midland's seminal publication, Johnson and co-workers used **3** for the reduction of ketone **18** to afford alcohol **19** in 75% yield and 97% *ee*.[12] This material was used to complete the synthesis of **20**, a cyclization precursor in Johnson's total synthesis of hydrocortisone acetate.

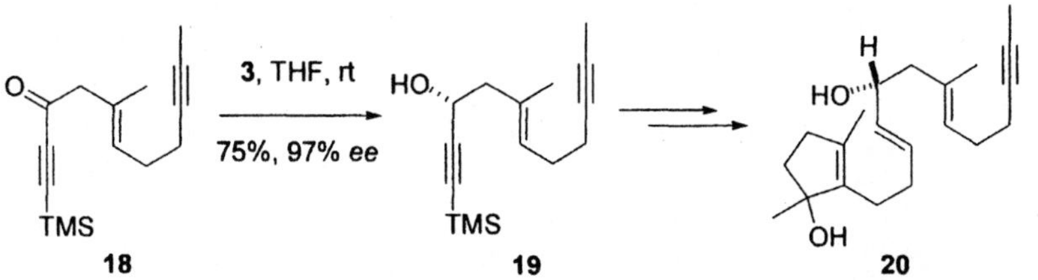

18 **19** **20**

The Midland reduction has also been used in the large-scale synthesis of chiral glycines. Deuterium labeled anisaldehyde was reduced with **3** to provide deuteriated arylmethyl alcohol **21** in 82% *ee*.[13] This alcohol was then converted in 4 steps to *N*-Boc-glycine (**22**).

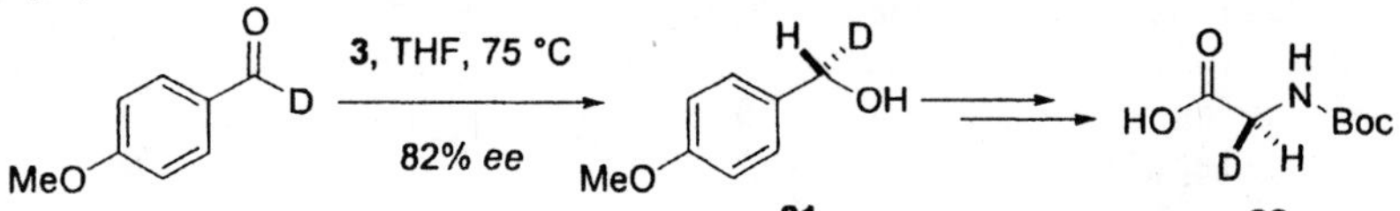

21 **22**

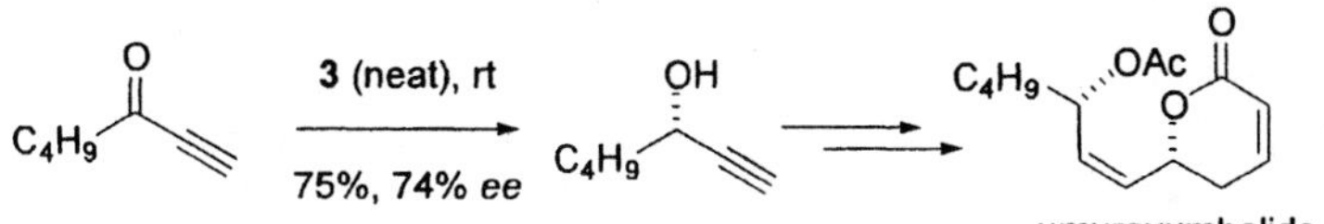

23

umuravumbolide
24

More recently, alpine-borane (**3**) was used as the first step in the total synthesis of umuravumbolide (**24**). Reduction of 1-heptyn-3-one with **3** produced **23** in 75% yield and 74% *ee*.[14]

The synthesis of the DE ring system of upenamide (**28**) utilized a Midland reduction to set the initial stereocenter in the synthesis. Conversion of **25** to **26** proceeded in 74% yield and 93% *ee*.[15] This material was carried on through a series of transformations to achieve a synthesis of the DE ring system (**27**).

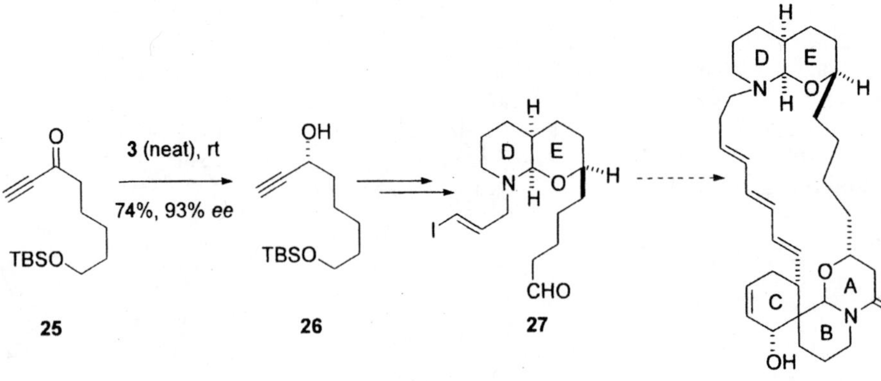

3 (neat), rt

74%, 93% *ee*

25 **26** **27**

upenamide
28

1.3.6 Experimental

Propargylic alcohol (26)[15]
Alkynone **25** (5.09 g, 20.0 mmol) was added to neat (*R*)-alpine-borane (7.23 g, 28 mmol) at 0 °C. The ice bath was removed after 30 min and the red solution was stirred at room temperature for 8 h at which time GC analysis showed complete consumption of starting material. Excess alpine-borane was quenched by the addition of propionaldehyde (1.08 mL, 15 mmol) with external cooling (0 °C) followed by stirring at room temperature for 1 h. The solution was concentrated and subject to high vacuum at 40 °C for 4 h with stirring to remove α-pinene. The residue was diluted with THF (10 mL) and solutions of NaOH (3 M, 7.5 mL) and H_2O_2 (30%, 7.5 mL) (caution: highly exothermic) was added dropwise at 0 °C. The resulting mixture was stirred for 4 h at 40 °C and extracted with ether (3 × 60 mL). The combined organic extracts were washed with brine, dried ($MgSO_4$), filtered, and concentrated. The residue was purified by flash chromatography (hexanes:ethyl acetate, 8:1) to give 3.78 g (74% yield, 93% *ee*) of 14 as a light-yellow oil.

1.3.7 References

1 [R] Midland, M. M. *Chem. Rev.* **1989**, *89*, 1553 (and references therein).
2. Midland, M. M.; Tramontano, A.; Zderic, S. A. *J. Orgmet. Chem.* **1977**, *134*, C17.
3. Midland, M. M.; Tramontano, A.; Zderic, S. A. *J. Orgmet. Chem.* **1978**, *156*, 203.
4. Midland, M. M.; Greer, S.; Tramontano, A.; Zderic, S. A. *J. Am. Chem. Soc.* **1979**, *101*, 2352.
5. Midland, M. M.; McDowell, D. C.; Hatch, R. L.; Tramontano, A. *J. Am. Chem. Soc.* **1980**, 867.
6. Midland, M. M.; Zderic, S. A. *J. Am. Chem. Soc.* **1982**, *104*, 525.
7. Brown, H. C.; Pai, G. G. *J. Org. Chem.* **1982**, *47*, 1606.
8. Midland, M. M.; McLoughlin, J. I. *J. Org. Chem.* **1984**, *49*, 1316.
9. Midland, M. M.; McLoughlin, J. I.; Gabriel, J. *J. Org. Chem.* **1989**, *54*, 159.
10. Midland, M. M.; Lee, P. E. *J. Am. Chem. Soc.* **1985**, *107*, 3237.
11. Brown, J. C.; Pai, G. G.; Jadhav, P. K. *J Am. Chem. Soc.* **1984**, *106*, 1531.
12. Johnson, W. S.; Frei, B.; Gopalan, A. S. *J. Org. Chem.* **1981**, *46*, 1512.
13. Ramalingam, K.; Nanjappan, P.; Kalvin, D. M.; Woodard, R. W. *Tetrahedron*, **1988**, *44*, 5597.
14. Ready, M. V. R.; Rearick J. P.; Hoch, N.; Ramachandran, P V. *Org Lett.* **2001**, *3*, 19.
15. Kiewel, K.; Luo, Z.; Sulikowski, G. A. *Org. Lett.* **2005**, *7*, 5163.

Julia M. Clay

1.4 Noyori Catalytic Asymmetric Hydrogenation

1.4.1 Description

The demand to produce enantiomerically pure pharmaceuticals, agrochemicals, flavors, and other fine chemicals from prochiral precursors has advanced the field of catalytic asymmetric hydrogenation.[1] In 2002 worldwide sales of single enantiomer pharmaceutical products approached $160 billion.[2]

Homogeneous catalytic asymmetric hydrogenation has become one of the most efficient methods for the synthesis of chiral alcohols, amines, α- and β-amino acids, and many other important chiral intermediates. Specifically, catalytic asymmetric hydrogenation methods developed by Professor Ryoji Noyori are highly selective and efficient processes for the preparation of a wide variety of chiral alcohols and chiral α-amino acids.[3] The transformation utilizes molecular hydrogen, BINAP (2,2'-bis(diphenylphosphino)-1,1'-binaphthyl) ligand and ruthenium(II) or rhodium(I) transition metal to reduce prochiral ketones **1** or olefins **2** to their corresponding alcohols **3** or alkanes **4**, respectively.[4]

functionalized ketone
1

H_2, (*R*)- or (*S*)-BINAP,
Metal [Ru(II) or Rh(I)]

3

functionalized olefin
2

4

This unique asymmetric transformation has become one of the most efficient, practical, and atom-economical methods for the construction of chiral compounds from simple prochiral starting material.[5] The transformation can offer either (*R*)- or (*S*)-stereoisomer, can have exquisite substrate-to-catalyst (S/C) ratio (100,000:1), and by selecting the appropriate substrate and catalyst, generality can often exceed the merits of a biotransformation. Furthermore, the reaction can proceed with high turnover number (TON), turnover frequency (TOF), and enantiomeric excess (*ee*). The operations of the reaction, isolation, separation, and purification are simple, easily performed, and well-suited for mass production.[6]

1.4.2 Historical Perspective

In 1968, W. S. Knowles[7] and L. Horner[8] reported independently the first homogeneous catalyzed asymmetric hydrogenation of olefins with chiral monodentate tertiary phosphine Rh-complexes, albeit in low enantiomeric excess (3–15% *ee*).[9] A major breakthrough came in 1971, when H. B. Kagan developed a Rh(I)-complex derived from a C_2 chiral diphosphine ligand of tartaric acid.[10] The Kagan Rh(I)-complex asymmetrically hydrogenated dehydro amino acids leading to phenylalanine in 72% *ee*. Subsequently, the pioneering work of the Knowles group at Monsanto established a method for the industrial synthesis of *L*-DOPA, a drug for treating Parkinson's disease, using DIPAMP-Rh(I) catalyzed asymmetric hydrogenation as a key step.[11] The successful and practical synthesis of *L*-DOPA constitutes a fundamental progression in asymmetric hydrogenation technology, and for his discovery Knowles shared the 2001 Nobel Prize in Chemistry with R. Noyori and K. B. Sharpless.[12a]

In 1980, Professor Noyori and the late H. Takaya consequently designed and synthesized a bidentate phosphine ligand BINAP (2,2′-bis(diphenylphosphino)-1,1′-binaphthyl) **5**, that contained an atropisomeric 1,1′-binaphthyl structure as a chiral element for use in transition metal catalyzed asymmetric hydrogenation reactions (Figure 1).

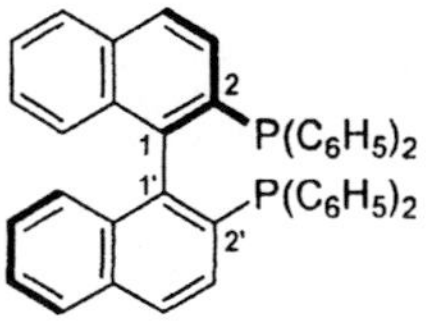

5 *(R)*-BINAP

Figure 1 (*R*)- 2,2′-Bis(diphenylphosphino)-1,1′-binaphthyl **5**

Rhodium(I) complexes of BINAP enantiomers are remarkably effective in various kinds of asymmetric catalytic reactions,[4c] including the enantioselective hydrogenation of α-(acylamino)acrylic acids or esters[4c] to provide amino acid derivatives, and the enantioselective isomerization of allylic amines to enamines.[13] Furthermore, Noyori's discovery of the BINAP-Ru(II) complex was a major advance in stereoselective organic synthesis. The scope and application of the BINAP transition metal catalyst system is far reaching. These chiral Ru complexes serve as catalyst precursors for the highly enantioselective hydrogenation of a range of functionalized ketones and olefins.[4b] For his contribution to asymmetric hydrogenation, Noyori shared the 2001 Nobel Prize in Chemistry with W. S. Knowles and K. B. Sharpless.[12b]

1.4.3 Mechanism

The BINAP ligand **5** (Figure 1) has numerous unique features. The diphosphine is characterized by full aromatic substitution, which exerts steric influence, provides polarizability, and enhances the Lewis acidity of the metal complex. The BINAP molecule is conformationally flexible and can accommodate a wide variety of transition metals by rotation about the binaphthyl C(1)-C(1′) pivot and C(2 or 2′)-P bonds without a serious increase in torsional strain. The framework of the chiral ligand determines enantioselectivity but can also alter the reactivity of the metal complex. In addition, the BINAP binaphthyl groups are axially dissymmetric possessing C_2 symmetry,[14] resulting in the production of an excellent asymmetric environment.[1f]

CO_2H, NHCOPh — H_2, **7** (*R*)-BINAP-Rh → CO_2H, NHCOPh

6 **8** (*S*) 97% yield, > 99% *ee*

Ph_2, P, H, OMe, Rh, OMe, H, P, Ph_2, +, ClO_4^-

7

In 1980, Noyori and co-workers reported cationic BINAP-Rh(I) catalyzing the hydrogenation of α-(acylamino)acrylic acids and esters to give amino acid derivatives in > 90% *ee*.[4c] For instance, catalyst **7** converts (*Z*)-α-(benzamido)cinnamic acid **6** to (*S*)-*N*-benzoylphenylalanine **8** in > 99% *ee* and 97% yield [substrate/catalyst molar ratio (S/C) = 100, 4 atm, room temp., 48 h, EtOH]. This reaction gives excellent enantioselectivity, but proved to be less than ideal because of the *unsaturated dihydride mechanism,* which has been thoroughly investigated by Halpern, Brown, and co-workers.[15] The catalytic asymmetric hydrogenation reaction mechanism of enamide **9** using C_2-chiral diphosphine Rh complex **7** to yield chiral α-amino acids **13** is shown in Figure 2. The Rh complex **7** forms a mixture of two diastereomeric substrate complexes with **9**, which leads to the *S* or *R* hydrogenation product, depending on the *Re*/*Si* face selection at the α-carbon C(2) in **9**. Hence, the enantioselectivity is determined by the relative equilibrium ratio and reactivity of diastereomer **10**. Under the reaction conditions, the major and more stable diastereomer is consumed much more slowly than the less stable minor isomer because of lower reactivity towards H_2. The thermodynamically favored *Si*-**10** diastereomeric Rh complex, leading to the

R-enantiomer of **13**, is weakly reactive and thus before hydrogenation can occur, it is converted to the highly reactive diastereomer *Re*-**10**. Diastereomer *Re*-**10** ultimately gives the *S*-enantiomer **13** via decoordination and recoordination of substrate **9**. Consequently, the observed enantioselectivity is a result of the delicate balance of the stability and reactivity of diastereomeric **10**. Because of this inherent mechanistic problem, the optimum conditions leading to high enantioselectivity are obtainable only by careful reagent and reaction condition choices. The reaction must be conducted under low substrate concentration and low hydrogen pressure. The BINAP-Rh catalyzed reaction occurs very slowly, because the reactive substrate Rh-complex **10** is present in very low concentration under these hydrogenation conditions. Unfortunately, the scope of the olefinic substrate is narrow. Thus, the BINAP-Rh catalyzed hydrogenation suffers from a mechanistic limitation and therefore remains far from ideal. Respectable enantioselection is obtainable only with Rh complexes of the ligand DuPhos, certain monodentate phosphates, and phosphoamidites.[16,17]

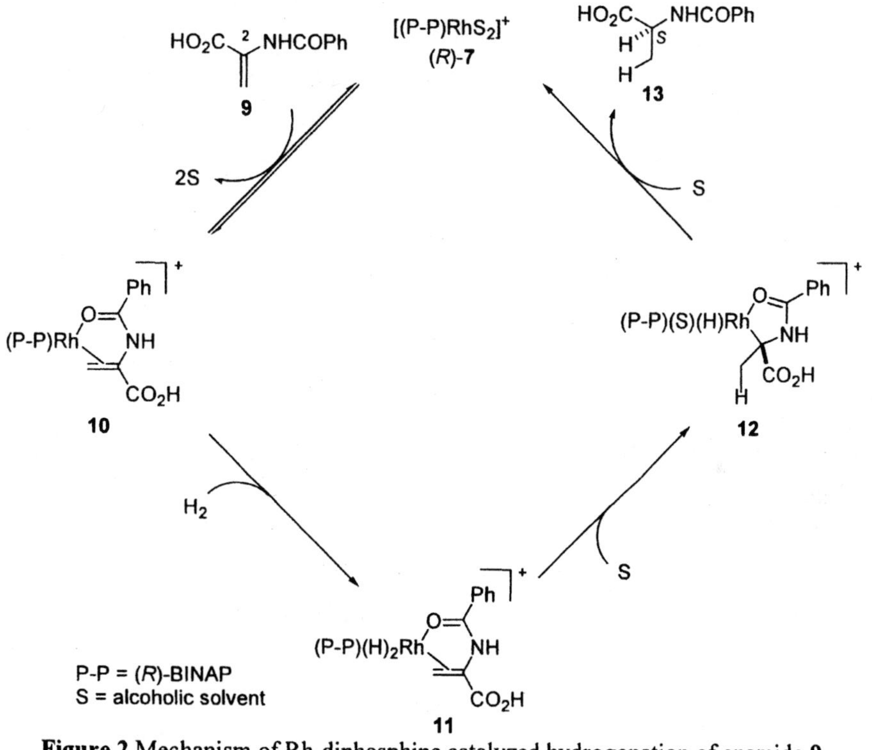

P-P = (*R*)-BINAP
S = alcoholic solvent

Figure 2 Mechanism of Rh-diphosphine catalyzed hydrogenation of enamide **9**

A breakthrough in catalytic asymmetric hydrogenation came when the Rh(I) metal was replaced by Ru(II). Hydrogenation of methyl (*Z*)-α-(acetamido)-cinnamate **14**, in the presence of $Ru(OCOCH_3)_2$[(*R*)-BINAP] **15**,[4b,18] affords methyl (*R*)-α-(acetamido)cinnamate **16** in 92% *ee* and > 99% yield [substrate/catalyst molar ratio (S/C) = 200, 1 atm, 30 °C, 24 h, MeOH].[19]

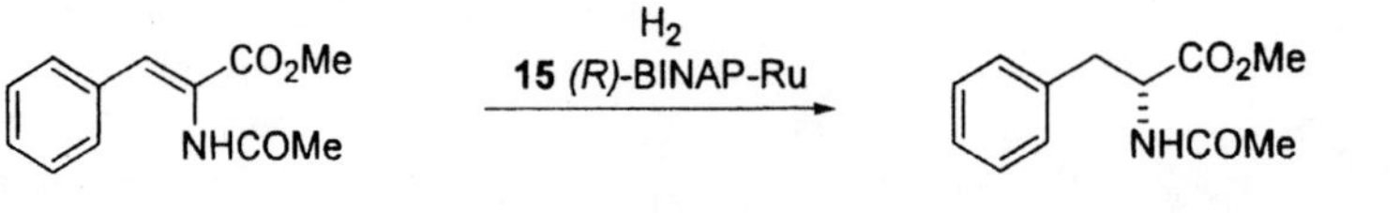

The asymmetric induction is opposite of that obtained with the (*R*)-BINAP-Rh **7** catalyst.[19,20] This contrasting behavior was found to be caused by an *unsaturated monohydride mechanism*, which is facilitated by the Ru monohydride catalyst **18** formed by the heterolytic cleavage of H_2 by precatalyst **17** (Figure 3).[19] Importantly, the metal hydride species **18** in this reaction is generated before substrate coordination, unlike the Rh chemistry involving the *unsaturated dihydride mechanism* (Figure 2).[15] The stereochemistry of the product is determined by the cleavage of the Ru–C bond in **21** by H_2. The enantiomeric ratio corresponds well to the relative stability of the diastereomeric substrate RuH complexes *Si*- and *Re*-**20**. The major *Si*-**20** diastereomer is converted to the *R* hydrogenation product by migratory insertion followed by hydrogenolysis. The two hydrogen atoms incorporated in product **22** are from two different H_2 molecules, as confirmed by isotope labeling experiments.[3] This result stands in contrast to the standard Rh-catalyzed reaction, which uses only one H_2 molecule per product.[3]

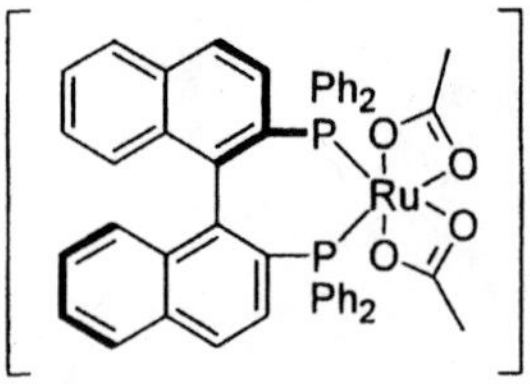

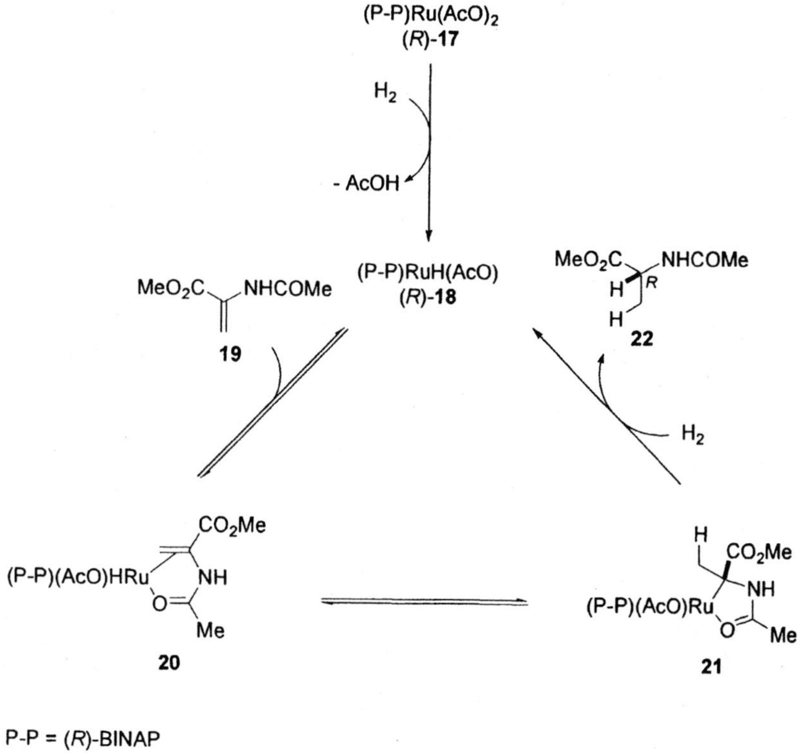

(P-P)Ru(AcO)$_2$
(*R*)-**17**

H_2

- AcOH

(P-P)RuH(AcO)
(*R*)-**18**

MeO$_2$C NHCOMe
19

MeO$_2$C NHCOMe
H *R*
H
22

H_2

(P-P)(AcO)HRu CO$_2$Me NH O Me
20

H CO$_2$Me NH (P-P)(AcO)Ru O Me
21

P-P = (*R*)-BINAP

Figure 3 Mechanism of Ru-complex catalyzed hydrogenation of enamide **19**

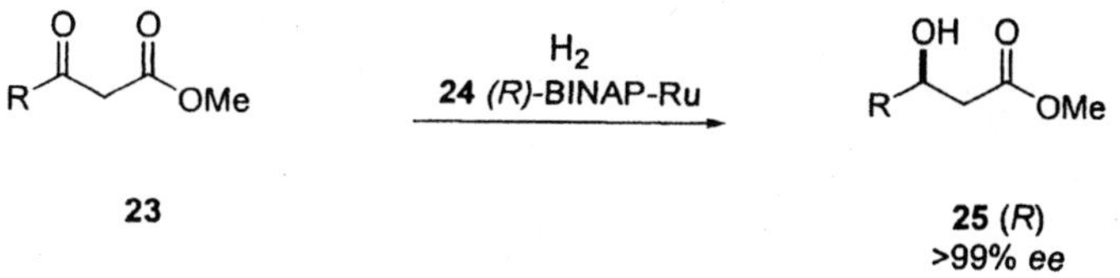

R–CO–CH$_2$–CO–OMe (**23**) → H_2, **24** (*R*)-BINAP-Ru → R–CH(OH)–CH$_2$–CO–OMe (**25** (*R*), >99% *ee*)

β-Keto esters are effectively reduced by halogen containing chiral precatalysts, including RuX_2(BINAP) **24** (X = Cl, Br, I),[21,22] [RuX(BINAP)(arene)]Y (X = halogen, Y = halogen or BF_4),[23] $[NH_2(C_2H_5)_2][(RuCl(BINAP))_2(\mu\text{-}Cl)_3]$,[24] and other *in situ* formed halogen containing BINAP-Ru complexes.[25] Moreover, various β-keto esters are hydrogenated in alcoholic solvent with an S/C of up to 10,000 to give chiral β-hydroxy esters in high *ee*. Interestingly, the $Ru(OCOCH_3)_2$(BINAP) **15** catalyst, although excellent for the asymmetric

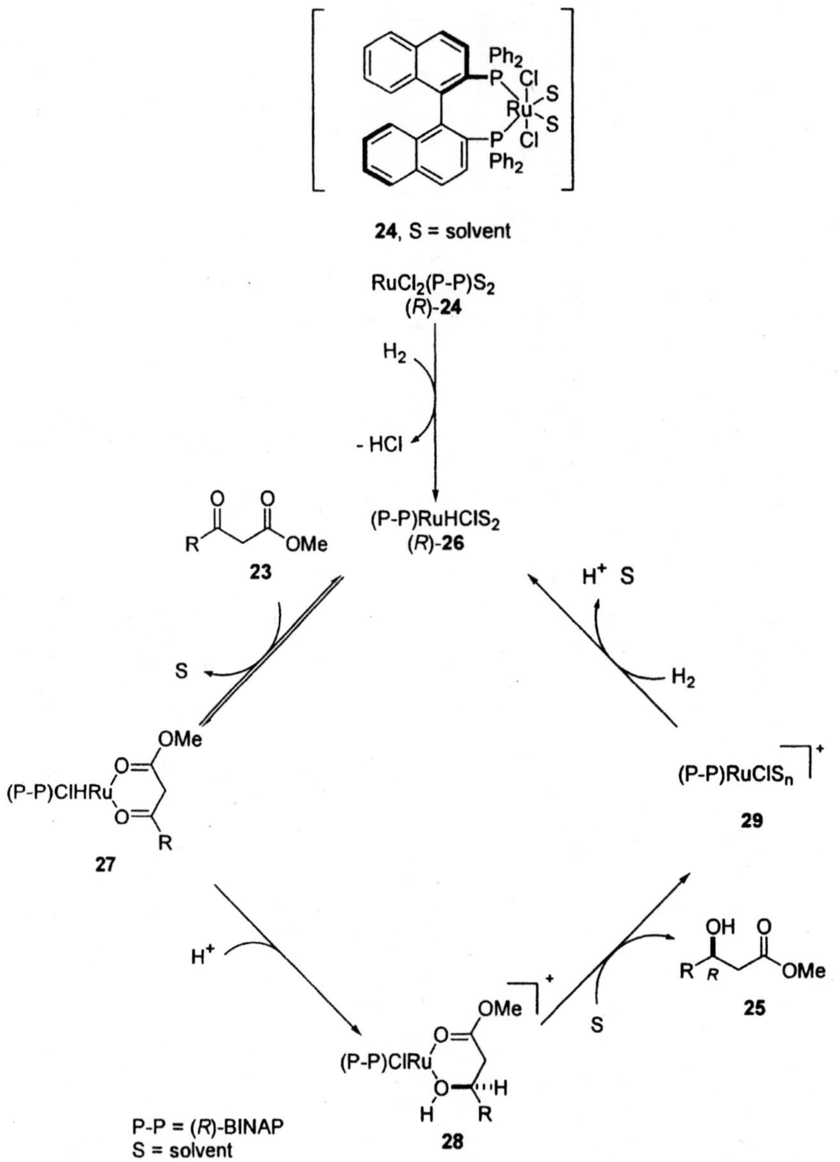

24, S = solvent

$RuCl_2(P\text{-}P)S_2$
(*R*)-**24**

H_2

- HCl

$(P\text{-}P)RuHClS_2$
(*R*)-**26**

23

OMe

H^+ S

S

H_2

$(P\text{-}P)ClHRu$

27

$(P\text{-}P)RuClS_n$ $^+$

29

H^+

$(P\text{-}P)ClRu$

28

25

S

P-P = (*R*)-BINAP
S = solvent

Figure 4 Mechanism of Ru-complex catalyzed hydrogenation of β-keto ester **23**

hydrogenation of functionalized olefins, is ineffective in reactions of structurally similar ketones such as β-keto esters.

A mechanistic model for the reduction of β-keto ester **23** is represented in Figure 4, wherein the actual catalyst is RuHCl **26** which is formed by the reaction between **24** and H_2.[26] Initially, **26** interacts reversibly with β-keto ester **23** to form the σ-type chelate complex **27**, in which metal-to-carbonyl hydride transfer is geometrically difficult. However, after protonation at the oxygen, the electrophilicity of the carbon is increased and the geometry is converted from σ to π, which facilitates the hydride migration. The hydroxy ester ligand in the resulting product **28** is liberated by solvent. The cationic Ru complex **29** cleaves H_2 and regenerates **26**. Enantioselectivity of > 99:1 is achieved in the hydride transfer step **27** to **28**. A key step is carbonyl protonation of **27**, caused by HCl generated in the induction step **24** to **26**.[27] The list of potential substrates includes various ketones possessing a directive functional group such as a dialkylamino, hydroxyl, alkoxyl, siloxyl, keto, halogeno, alkoxycarbonyl, alkylthiocarbonyl, dialkylamino-carbonyl, phosphoryl, and sulfoxyl group, among other possibilities.[1c,4a,28]

1.4.4 *Variations and Improvements*

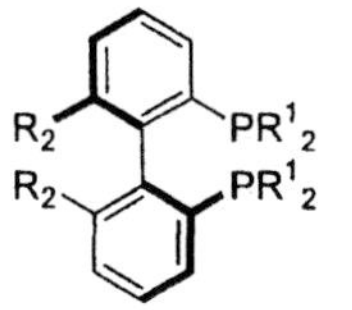

30 *(S)*-BIPHEMP: R^1 = Ph; R_2 = CH_3
31 *(S)*-BIPHEP: R^1 = Ph; R_2 = OCH_3

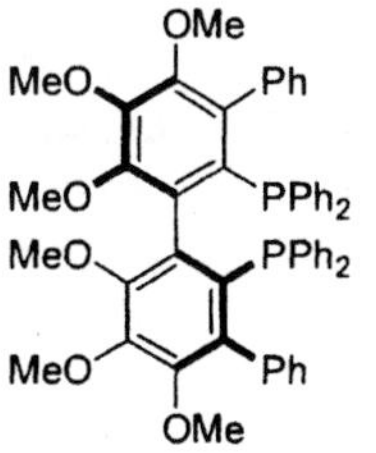

32 *(S)-o*-Ph-HexaMeO-BIPHEP

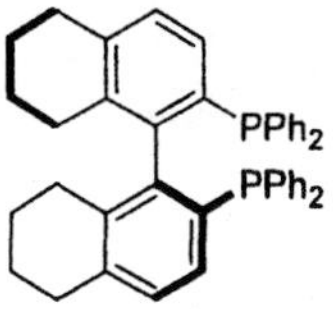

33 *(S)*-H_8-BINAP

Ever since Professor Noyori demonstrated that BINAP transition metal complexes were highly effective for asymmetric hydrogenation, a vast number of chiral ligands and catalysts have been developed by many researchers in academia and industry.[1d] In particular, many researchers have devoted their efforts to designing and developing new efficient and selective chiral phosphorus ligands. A major feature in the design of the new chiral phosphorus ligands is the ability to tune the steric and electronic properties of ligands within a given scaffold. Modification of the electronic and steric properties of BIPHEMP **30** and MeO-BIPHEP **31** led to the development of new and efficient atropisomeric ligands for ruthenium-catalyzed asymmetric hydrogenation.[1a] In addition, Zhang *et al.* have recently disclosed an *ortho*-substituted BIPHEP ligand, *o*-Ph-HexaMeO-BIPHEP **32**, for the rhodium-catalyzed asymmetric hydrogenation of cyclic enamides.[29] Takaya has found that a modified BINAP ligand (H_8-BINAP **33**) provides better enantioselectivity than BINAP in Ru-catalyzed hydrogenation of unsaturated carboxylic acids.[30]

The chiral biaryl bisphosphine ligand SEGPHOS **34**, developed by Takasago, possesses a smaller dihedral angle than BINAP. The ligand has provided greater enantioselectivity over BINAP in Ru-catalyzed hydrogenation of a wide variety of carbonyl compounds.[31]

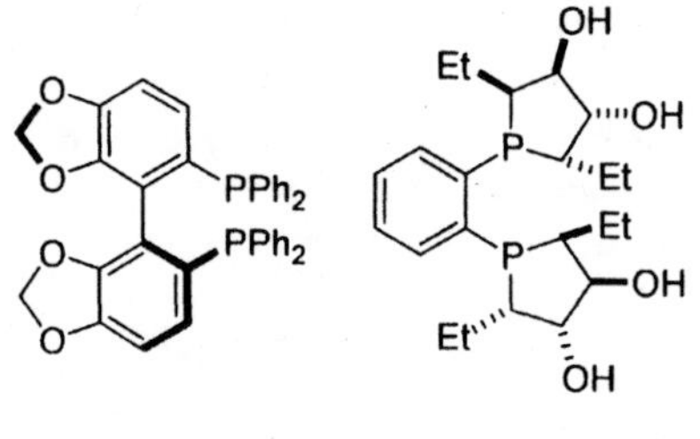

34 *(S)*-SEGPHOS **35** KetalPhos

A number of chiral bisphosphane ligands have emerged based on the modification of DuPhos and BPE ligands,[1a] which have proved successful for the rhodium-catalyzed asymmetric hydrogenation of functionalized olefins and ketones. The ligand with four hydroxyl groups, KetalPhos **35**,[32] enables the hydrogenation to be carried out in aqueous solution with high enantioselectivity.

Zhang has developed a series of 1,4-diphosphane ligands with a conformationally rigid 1,4-dioxane backbone, as exemplified by T-Phos **36** and SK-Phos **37**. These ligands have proved highly efficient and selective (> 99% *ee*) for the asymmetric hydrogenation of arylenamides and MOM-protected β-hydroxyl enamides.[33]

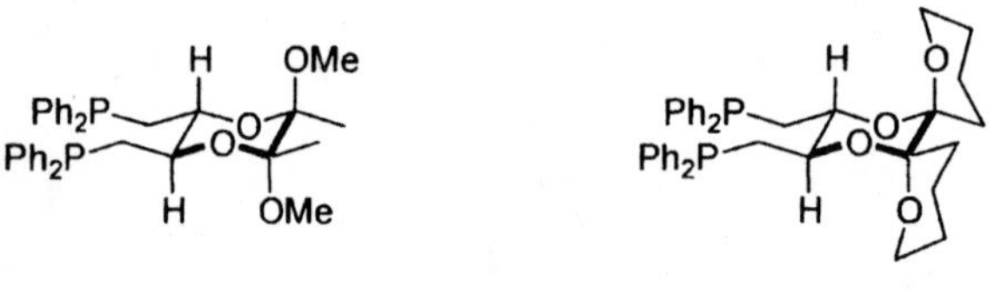

36 *(R,R,R,R)*-T-Phos **37** *(R,R,R,R)*-SK-Phos

Many chiral ferrocene-based bisphosphane ligands with great structural variations have been developed recently. Togni and Spindler introduced non-C_2-symmetric ferrocene-based Josiphos type ligands.[34] Josiphos **38** has been found to be effective for Rh-catalyzed hydrogenation of α-acetamidocinnamate, dimethyl itaconate, and β-keto esters. A class of non-C_2-symmetrical ferrocene-based 1,5-diphosphane ligands (TaniaPhos **39**) has also been developed by Knochel.[35] These ligands have been effectively used in Rh- and Ru-catalyzed asymmetric hydrogenation of functionalized α-(acylamino)acrylic acids and esters, β-keto esters, and imines. The TaniaPhos type ligands (**39**), which have a MeO group at the stereogenic carbon, have shown excellent applications in hydrogenation of several ketone and olefin substrates.[36]

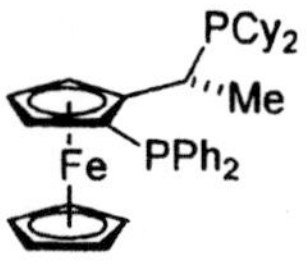

PCy_2 Me Fe PPh_2

38 (*R*)-(*S*)-Josiphos

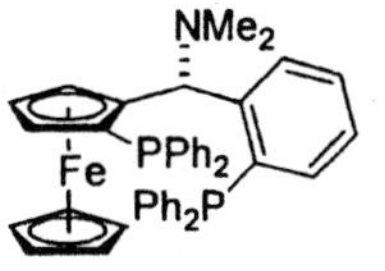

NMe_2 Fe PPh_2 Ph_2P

39 TaniaPhos

Although the first P-chiral bisphosphane (DIPAMP) was developed by Knowles over 30 years ago, the discovery of new efficient P-chiral bisphosphanes has been slow partly because of the difficulties in ligand synthesis. Pye and Rossen have developed a planar chiral bisphosphine ligand, [2.2]-PHANEPHOS **40**, based on a paracyclophane backbone.[37] The ligand has shown excellent enantioselectivity in Rh- and Ru-catalyzed hydrogenations.

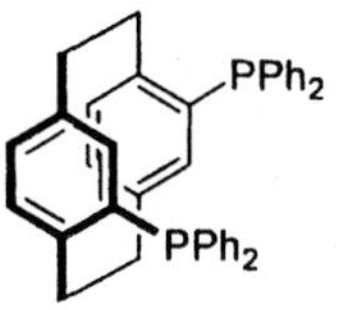

PPh_2 PPh_2

40 (*S*)-[2,2]PHANEPHOS

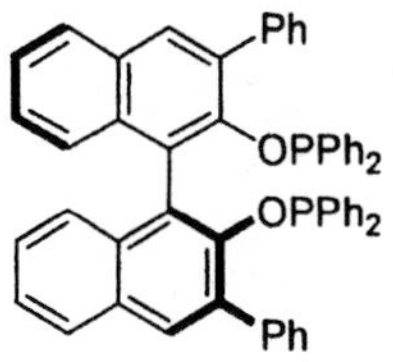

Ph $OPPh_2$ $OPPh_2$ Ph

41 (*S*)-Ph-*o*-BINAPO

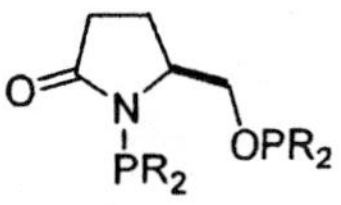

O N PR_2 OPR_2

42 (*S*)-Cy,Cy-oxoProNOP: R = Cy
43 (*S*)-Cp,Cp-oxoProNOP: R = Cp

The discovery of highly efficient bisphosphinites, bisphosphonites, and bisphosphites for asymmetric hydrogenation has been relatively slow compared to chiral bisphosphane ligands, due to their greater conformational flexibility and instability. Zhang has recently reported on a series of *o*-BINAPO ligands with substituents at the 3 and 3′ positions of the binaphthyl group. The ligand Ph-*o*-BINAPO **41** is an efficient ligand for hydrogenation of α-dehydroamino acid derivatives.[38] The *o*-BINAPO ligands have also been applied in Ru-catalyzed hydrogenation of β-aryl-β-(acylamino)acrylates and up to 99% *ee*'s have been obtained.[39]

Several efficient amidophosphine- and aminophosphine-phosphinite ligands have been reported by Agbossou and Carpentier.[40] Amidophosphine-phosphinite ligands (*S*)-Cy,Cy-oxoProNOP **42** and (*S*)-Cp,Cp-oxoProNOP **43** are efficient ligands for Rh-catalyzed hydrogenation of dihydro-4,4-dimethyl-2,3-furandione, and up to 98% *ee*'s have been obtained.

1.4.5 Synthetic Utility

Professor Ryoji Noyori developed and utilized the BINAP molecule as a chiral ligand to effect stereoselectivity in transition metal catalyzed asymmetric hydrogenation. The catalytic asymmetric hydrogenation reaction has been applied to a number of diverse functionalized ketones and olefins. The following illustrations represent the synthetic utility of the asymmetric hydrogenation reaction using the BINAP ligand, as well as new chiral phosphorus ligands.

1.4.5.1 Hydrogenation of dehydroamino acid derivatives

A number of natural and unnatural amino acids are now available in >90% *ee* due to catalytic asymmetric hydrogenation of (*Z*)- or (*E*)-α-(acylamino)acrylic acids or esters, and this methodology is gaining practical significance in industry and academics.[1f] In general, high enantioselectivity is achieved for *Z*-isomeric substrates of α-dehydroamino acids, where as hydrogenation of the *E*-isomeric substrates usually proceeds with slow rates and poor enantioselectivity.[11b,41] Interestingly, the DuPhos-Rh system provides excellent enantioselectivity for both *Z*- and *E*-isomeric substrate **44** and gives a single hydrogenation product **45** regardless of starting *Z*- and *E*-isomeric substrate.[41b]

CO_2Me NHAc or CO_2Me NHAc **44** —— H_2, (*R,R*)-Pr-DuPhos-Rh ——→ CO_2Me NHAc **45** >99% *ee*

Hydrogenation of β,β-disubstituted α-dehydroamino acids **46** remains a challenging problem. The less bulky Me-DuPhos- or Me-BPE-type ligands provide excellent enantioselectivity to give a variety of β,β-disubstituted products **47**.[42]

CO_2Me NHAc **46** —— H_2, (*R,R*)-Me-BPE-Rh ——→ CO_2Me NHAc **47** 98.2% *ee*

The asymmetric hydrogenation of β,β-disubstituted α-dehydroamino acid **48**, in which the β-substituents are nonequivalent, provides the opportunity to selectively construct two contiguous stereogenic centers as seen in **49**. The Me-DuPhos and Me-BPE ligands facilitate the rhodium catalyzed hydrogenation of the *E*- and *Z*-isomers with excellent enantioselectivity.[42]

H_2, (*R,R*)-Me-BPE-Rh

48 → **49**, 90.6% *ee*

Furthermore, the asymmetric hydrogenation of the *E*- or *Z*-isomer of β-(acetylamino)-β-methyl-α-dehydroamino acids (**50** and **51**, respectively) with the Me-DuPhos-Rh catalyst provides either diastereomer of *N,N*-protected 2,3-diaminobutanoic acid derivatives (**52** and **53**) with excellent enantioselectivity.[43]

H_2, (*R,R*)-Me-DuPhos-Rh

50 (*Z*) → **52**, 96% *ee*

H_2, (*R,R*)-Me-DuPhos-Rh

51 (*E*) → **53**, > 98% *ee*

1.4.5.2 Hydrogenation of enamides

The Ru-BINAP system has shown excellent enantioselectivity in hydrogenation of (*Z*)-*N*-acyl-1-alkylidenetetrahydroisoquinolines **54**. Thus, a series of chiral isoquinoline products **55** can be efficiently synthesized.[44]

54 → 55 (H_2, (*R*)-BINAP-Ru)

55 99.5% *ee*

Rhodium catalyzed hydrogenation of enamides has attracted much attention recently. With the development of more efficient chiral phosphorus ligands, extremely high *ee*'s can be obtained. Hydrogenation of tetra-substituted enamides, such as β,β-dimethyl-α-phenyl enamide derivative **56**, has been reported with *t*-Bu-BisP* and *t*-Bu-MiniPhos providing amide **57** with excellent *ee*.[45]

56 → **57** (H_2, (*R,R*)-*t*-Bu-BisP*-Rh)

57 99% *ee*

1.4.5.3 Hydrogenation of (β-acylamino) acrylates

Asymmetric hydrogenation of (β-acylamino) acrylates provides β-amino acids, an important constituent in many chiral drugs.[46] Many Rh and Ru complexes with chiral phosphorus ligands such as BINAP,[47] DuPhos,[48] BICP,[49] BDPMI,[50] and MalPHOS[51] are effective for the hydrogenation of (*E*)-β-alkyl (β-acylamino)acrylates. However, only a few chiral ligands, such as BDPMI[39,50] or TangPhos,[52] can hydrogenate (*Z*)-β-alkyl (β-acylamino)acrylate **58** in over 93% *ee*.

58 → **59** (H_2, catalyst)

catalyst: (*S*)-Xylyl-*o*-BINAPO-Ru, **59** (99% *ee*)[39]
catalyst: (*S,S,R,R*)-TangPhos-Rh, **59** (93.8% *ee*)[52b]

1.4.5.4 *Hydrogenation of enol esters*

In contrast to many examples of highly enantioselective hydrogenations of enamides, only a few successful demonstrations exist for the asymmetric hydrogenation of enol esters. One possible reason is the acyl group of an enol ester has weaker coordinating ability to the metal catalyst than the corresponding enamide substrate. Both Rh and Ru complexes associated with chiral phosphorus ligands such as DuPhos[53] and C_2-TunaPhos,[54] respectively, are effective for the asymmetric hydrogenation of α-(acyloxy)acrylate **60**.

R' OAc → (H_2, catalyst) → R' * OAc

60 **61**

catalyst: (*R*,*R*)-Et-DuPhos-Rh, **61** (>99% *ee*, *R*-confign., R' = CO_2Et)
catalyst: (*S*)-C_2-TunaPhos-Ru, **61** (97.7% *ee*, *S*-confign., R' = 1-Np)

1.4.5.5 *Hydrogenation of α-β-unsaturated carbonyls*

Significant progress has been achieved in the asymmetric hydrogenation of α,β-unsaturated carboxylic acids with chiral Ru catalysts. In the case of hydrogenation of **62**, high hydrogenation pressure and low temperature are required to achieve good enantioselectivity of (*S*)-2-(4-fluorophenyl)-3-methylbutanoic acid **63**, a key intermediate in the synthesis of the calcium antagonist Mibefradil.[55]

F, CO_2H → (H_2, $Ru(OAc)_2$, [(*R*)-BIPHEMP]) → F, CO_2H

62 **63**
94% *ee*

Many chiral phosphorus ligands with Ru complexes have achieved excellent enantioselectivity in the hydrogenation of α,β-unsaturated esters, amides, lactones, and ketones. The Ru-BINAP system is efficient for hydrogenation of 2-methylene-γ-butyrolactone **64** and 2-methylene-cyclopentanone **66**.[56,57] With a dicationic (*S*)-di-*t*-Bu-MeOBIPHEP-Ru complex under high hydrogen pressure, 3-ethoxy pyrrolidinone **68** is hydrogenated to give (*R*)-4-ethoxy-γ-lactam **69** in 98% *ee*.[58]

H_2
(*S*)-BINAP-Ru

64 → **65**
95% *ee*

H_2
(*R*)-BINAP-Ru

66 → **67**
98% *ee*

H_2
(*S*)-di-*t*-Bu-
MeO-BIPHEP-Ru

68 → **69**
98% *ee*

1.4.5.6 Hydrogenation of unsaturated alcohols

Asymmetric hydrogenation of allylic and homoallylic unsaturated alcohols was not very efficient until the discovery of the BINAP-Ru catalyst. With Ru(BINAP)$(OAc)_2$ as catalyst, geraniol **70** and nerol **72** are successfully hydrogenated to give (*S*)- or (*R*)-citronellol (**71** and **73**, respectively) in high overall yield with good enantioselectivity of 98 and 99% *ee*.[59]

H_2
(*S*)-BINAP-Ru

70 → **71**
99% *ee*

H_2
(*S*)-BINAP-Ru

72 → **73**
98% *ee*

1.4.5.7 Hydrogenation of ketones

Hydrogenation of α-keto esters and amides has been studied with Rh and Ru catalysts. Several neutral Rh catalysts with chiral ligands such as MCCPM[60] and Cy,Cy-oxoProNOP[61] have demonstrated excellent reactivity and enantioselectivity in the hydrogenation of α-keto esters and amides. A cationic (BoPhoz)-Rh complex efficiently hydrogenates the cyclic α-keto ester, dihydro-4,4-dimethyl-2,3-furandione **74**, with a high turnover number to afford α-hydroxy ketone **75**.[62]

H_2
(BoPhoz)-Rh

74

75
97.2% *ee*

Asymmetric hydrogenation of β-keto esters has been very successful using chiral Ru catalysts and a detailed review on this subject is available.[1e] The BINAP-Ru catalyst gives high enantioselectivity on a variety of β-keto esters.[22a] Furthermore, a Josiphos-Rh complex is found to be effective for hydrogenation of ethyl 3-oxobutanoate **76** to afford β-hydroxy ketone **77** with good enantioselectivity.[34]

H_2
(*R*)-(*S*)-Josiphos-Rh

76

77
97% *ee*

The asymmetric hydrogenation of unfunctionalized ketones is a more challenging task, due to the lack of secondary coordination to the metal.[26,63] Enantioselective hydrogenation of simple aromatic and aliphatic ketones, **78** and **80**, respectively, has been achieved with a XylBINAP-Ru complex in the presence of a chiral diamine such as daipen.[64]

H_2
trans-$RuCl_2$[(*S*)-Xyl-BINAP][(*S*)-daipen]

78

79
99% *ee*

cyclo-hexyl(C=O)Me → H_2, trans-$RuCl_2$[(S)-Xyl-BINAP][(S)-daipen] → cyclo-hexyl(CH-OH)Me

80 81
85% ee

1.4.5.8 *Hydrogenation of imines*

Currently, only a few efficient chiral catalytic systems are available for hydrogenation of imines. The recent development of *trans*-[$RuCl_2$(bisphosphine)(1,2-diamine)] complexes has provided promise in this area.[65] High enantioselectivity has been reported in hydrogenation of acetophenone *N*-arylimine derivatives **82** using a *trans*-[$RuCl_2$(Et-DuPhos)][dach] system and up to 94% *ee* has been obtained under basic conditions.[65] A *trans*-[$RuCl_2$(MeO-BIPHEP)][anden] complex has also shown promise in the asymmetric hydrogenation of cyclic imines **84** with moderate enantioselectivity (88% *ee*).[65]

Ph(C=N-Ph)Me → H_2, trans-$RuCl_2$[(R,R)-Et-DuPhos][(R,R)-dach] → Ph(CH-NHPh)Me

82 83
94% ee

84 → H_2, trans-[$RuCl_2$((S)-MeO-BIPHEP)((S,S)-anden] → 85

84 85
88% ee of product sign[65]

1.4.5.9 *Hydrogenation via dynamic kinetic resolution*

Many stereoselective reactions suffer from the disadvantage of producing the desired chiral product in 50% maximal yield. The lability of 2-substituted 3-oxo carboxylic esters to undergo facial epimerization, coupled with the high chiral recognition of the BINAP-Ru(II) complex, provides the possibility of stereoselective hydrogenation utilizing dynamic kinetic resolution.[66] If the racemization of enantiomers **86** and **87** is rapid with respect to hydrogenation, then the hydrogenation would form one isomer selectively among the four possible stereoisomeric hydroxyl esters (**88**).

H_2
(*R*)-BINAP-Ru

86 → *syn*-**88**

H_2
(*R*)-BINAP-Ru

87 → *anti*-**88**

The stereoselective transformation constitutes an ideal asymmetric catalysis, which is capable of converting racemic starting material to a single chiral product possessing stereo-defined vicinal stereogenic centers in 100% yield. In dichloromethane containing a (*R*)-BINAP-Ru complex,[23] racemic ketone **89** is hydrogenated with high *anti* diastereoselectivity, to give a 99:1 mixture of the *trans*-hydroxyl ester *anti*-**90** (92% *ee*) quantitatively.[66]

H_2
$[RuCl(C_6H_6)((R)\text{-BINAP})]\text{-Cl}$
(*anti*/*syn* ratio 99/1)

89 → *anti*-**90** 92% *ee* + *syn*-**90** 93% *ee*

1.4.6 Experimental

H_2
(*S*-BINAP)$RuCl_2$
110 atm, 45 °C, MeOH

91 → **92** 76% yield, 96% *ee*

5-Benzyloxy-3-hydroxy-pentanoic acid *tert*-butyl ester (92)[67]

A sample of $[((S)\text{-BINAP})RuCl_2]_2 \cdot NEt_3$ was prepared from $RuCl_2 \cdot COD$ (0.020 g, 0.07 mmol) and (*S*)-BINAP as previously described.[68] A solution of **91** (10.06 g, 36 mmol) in methanol (20 mL) was degassed with N_2 and then added to a Schlenk vessel containing the catalyst. Stirring the mixture for 30 minutes gave a homogeneous orange solution. The mixture was acidified with 2 N HCl (0.24 mL), transferred by canula to a 125-mL pressure reaction vessel (Parr No. 4651), and heated to 45 °C. The vessel was pressurized to 110 atm with H_2, and the temperature was maintained for 24 hours. The mixture was concentrated and purified by SiO_2 chromatography (20% ethyl acetate/hexane) to give alcohol **92** (7.63 g, 76% yield, 96% *ee*) as a slightly yellow oil.

1.4.7 References

1. (a) [R] Chi, Y.; Tang, W.; Zhang, X. Rhodium-Catalyzed Asymmetric Hydrogenation. In *Modern Rhodium-Catalyzed Organic Reactions*, (Evans, P. A. Ed.), WILEY-VCH Verlag GmbH & Co. KgaA, Weinheim, 2005: pp 1–31. (b) [R] Zanotti-Gerosa, A.; Hems, W.; Groarke, M.; Hancock, F. *Platinum Metals Rev.* **2005**, *49*, 158–165. (c) [R] Ohkuma, T.; Noyori, R. In *Comprehensive Asymmetric Catalysis*, (Jacobsen, E. N.; Pfaltz, A.; Yamamoto, H. Eds.), Springer, Berlin, 1999, Vol. I, and 2004, Supplement I. (d) [R] Tang, W.; Zhang, X. *Chem. Rev.* **2003**, *103*, 3029–3069. (e) [R] Ager, D. J.; Laneman, S. A. *Tetrahedron: Asymmetry* **1997**, *8*, 3327–3355. (f) [R] Noyori, R.; Takaya, H. *Acc. Chem. Res.* **1990**, *23*, 345–350.
2. Rouhi, A. M. *Chem. Eng. News* **2003**, *81*, 45–55.
3. Noyori, R.; Kitamura, M.; Ohkuma, T. *PNAS* **2004**, *101*, 5356–5362.
4. (a) [R] Noyori, R.; Takaya, H. *Acc. Chem. Res.* **1990**, *23*, 345–350. (b) Ohta, T.; Takaya, H.; Noyori, R. *Inorg. Chem.* **1988**, *27*, 566–569. (c) Miyashita, A.; Yasuda, A.; Takaya, H.; Toriumi, K.; Ito, T.; Souchi, T.; Noyori, R. *J. Am. Chem. Soc.* **1980**, *102*, 7932–7934.
5. (a) [R] Sumi, K.; Kumobayashi, H. *Topics Organomet. Chem.* **2004**, *6*, 63–95. (b) [R] Takaya, H.; Ohta, T.; Noyori, R. Rhodium-Catalyzed Asymmetric Hydrogenation. In *Catalytic Asymmetric Synthesis*, (Ojima, I. Ed.), VCH: New York, 1993: pp 1–31.
6. Anderson, N. G. *Practical Process Research and Development*; Academic Press: London, 2000.
7. Knowles, W. S.; Sabacky, M. J. *Chem. Commun.* **1968**, 1445–1446.
8. Horner, L.; Siegel, H.; Büthe, H. *Angew. Chem. Int. Ed. Engl.* **1968**, *7*, 942.
9. (a) Korpiun, O.; Lewis, R. A.; Chickos, J.; Mislow, K. *J. Am. Chem. Soc.* **1968**, *90*, 4842. (b) Horner, L. *Pure Appl. Chem.* **1964**, *9*, 225–244.
10. Dang, T. P.; Kagan, H. B. *J. Chem. Soc. Chem. Commun.* **1971**, 481.
11. (a) Knowles, W. S.; Sabacky, M. J.; Vineyard, B. D. *J. Chem. Soc. Chem. Commun.* **1972**, 10–11. (b) Vineyard, B. D.; Knowles, W. S.; Sabacky, M. J.; Bachman, G. L. Weinkauff, D. J. *J. Am. Chem. Soc.* **1977**, *99*, 5946–5952.
12. (a) Knowles, W. S. *Angew. Chem. Int. Ed. Engl.* **2002**, *41*, 1998–2007. (b) Noyori, R. *Angew. Chem. Int. Ed. Engl.* **2002**, *41*, 2008–2022.
13. (a) Tani, K.; Yamagata, T.; Otsuka, S.; Akutagawa, S. Kumobayashi, H.; Taketomi, T.; Takaya, H.; Miyashita, A.; Noyori, R. *J. Chem. Soc. Chem. Commun.* **1982**, 600. (b) Tani, K.; Yamagata, T.; Akutagawa, S.; Kumobayashi, H.; Taketomi, T.; Takaya, H.; Miyashita, A.; Noyori, R.; Otsuka, S. *J. Am. Chem. Soc.* **1984**, *106*, 5208–5217.
14. [R] Whitesell, J. K. *Chem. Rev.* **1989**, *89*, 1581–1590.
15. (a) Landis, C. R.; Halpern, J. *J. Am. Chem. Soc.* **1987**, *109*, 1746–1754. (b) Brown, J. M.; Chaloner, P. A. *J. Am. Chem. Soc.* **1980**, *102*, 3040–3048. (c) Chan, A. S.; Halpern, J. *J. Am. Chem. Soc.* **1980**, *102*, 838–840.
16. (a) Burk, M. J. *Acc. Chem. Res.* **2000**, *33*, 363–372. (b) van den Berg, M.; Minnaard, A. J.; Schudde, E. P.; van Esch, J.; de Vries, A. H. M.; de Vries, J. G.; Feringa, B. L. *J. Am. Chem. Soc.* **2000**, *122*, 11539–11540. (c) Claver, C.; Fernandez, E.; Gillon, A.; Heslop, K.; Hyett, D. J.; Martorell, A.; Orpen, A. G.; Pringle, P. G. *Chem. Commun.* **2000**, 961–962. (d) Reetz, M. T.; Mehler, G. *Angew. Chem. Int. Ed.* **2000**, *39*, 3889–3890.
17. Gridnev, I. D.; Higashi, N.; Asakura, K.; Imamoto, T. *J. Am. Chem. Soc.* **2000**, *122*, 7183–7194.
18. Kitamura, M.; Tokunaga, M.; Noyori, R. *J. Org. Chem.* **1992**, *57*, 4053–4054.
19. Kitamura, M.; Tsukamoto, M.; Bessho, Y.; Yoshimura, M.; Kobs, U.; Widhalm, M.; Noyori, R. *J. Am. Chem. Soc.* **2002**, *124*, 6649–6667.

20. Ikariya, T.; Ishii, Y.; Kawano, H.; Arai, T.; Saburi, M.; Yoshikawa, S.; Akutagawa, S. *J. Chem. Soc. Chem. Commun.* **1985**, 922–924.
21. Wiles, J. A.; Bergens, S. H.; Young, Y. G. *J. Am. Chem. Soc.* **1997**, *119*, 2940–2941.
22. (a) Noyori, R.; Ohkuma, T.; Kitamura, M.; Takaya, H.; Sayo, N.; Kumobayashi, H.; Akutagawa, S. *J. Am. Chem. Soc.* **1987**, *109*, 5856–5858. (b) Kitamura, M.; Ohkuma, T.; Inoue, S.; Sayo, N.; Kumobayashi, H.; Akutagawa, S.; Ohta, T.; Takaya, H.; Noyori, R. *J. Am. Chem. Soc.* **1988**, *110*, 629–631.
23. Mashima, K.; Kusano, K.; Ohta, T.; Noyori, R.; Takaya, H. *J. Chem. Soc. Chem. Commun.* **1989**, 1208–1210.
24. (a) King, S. A.; DiMichele, L. In *Catalysis of Organic Reactions*, (Scaros, M. G.; Prunier, M. L., Eds.), Dekker: New York, 1995: pp 157–166. (b) Ohta, T.; Tonomura, Y.; Nozaki, K.; Takaya, H.; Mashima, K. *Organometallics* **1996**, *15*, 1521–1523.
25. (a) Genet, J. P. In *Reduction in Organic Synthesis*, Abdel-Magit, A. F., Ed.; Am. Chem. Soc.: Washington, DC, 1996: pp 31–51. (b) Kumobayashi, H.; Miura, T.; Sayo, N.; Saito, T.; Zhang, X. *Synlett* **2001**, 1055–1064.
26. Noyori, R.; Ohkuma, T. *Angew. Chem. Int. Ed.* **2001**, *40*, 40–73.
27. (a) King, S. A.; Thompson, A. S.; King, A. O.; Verhoeven, T. R. *J. Org. Chem.* **1992**, *57*, 6689–6691. (b) Kitamura, M.; Yoshimura, M.; Kanda, N.; Noyori, R. *Tetrahedron* **1999**, *55*, 8769–8785.
28. (a) Ohkuma, T.; Noyori, R. In *Transition Metals for Organic Synthesis*, (Beller, M.; Bolm, C., Eds.), Wiley: Weinheim, Germany, 1998: Vol. 2, pp. 25–69. (b) Ohkuma, T.; Kitamura, M.; Noyori, R. In *Catalytic Asymmetric Synthesis*, (Ojima, I., Ed.), Wiley: New York, 2000: 2nd Ed., pp. 1–110. (c) Noyori, R. *Science* **1990**, *248*, 1194–1199.
29. Tang, W.; Chi, Y.; Zhang, X. *Org. Lett.* **2002**, *4*, 1695–1698.
30. (a) Zhang, X.; Mashima, K.; Koyano, K.; Sayo, N.; Kumobayashi, H.; Akutagawa, S.; Takaya, H. *Tetrahedron Lett.* **1991**, *32*, 7283–7286. (b) Zhang, X.; Mashima, K.; Koyano, K.; Sayo, N.; Kumobayashi, H.; Akutagawa, S.; Takaya, H. *J. Chem. Soc., Perkin Trans. 1* **1994**, 2309–2322.
31. Wan, X.; Sun, Y.; Luo, Y.; Li, D.; Zhang, Z. *J. Org. Chem.* **2005**, *70*, 1070–1072.
32. Dai, Q.; Wang, C.-J.; Zhang, X. *Tetrahedron* **2006**, *62*, 868–871.
33. Li, W.; Waldkirch, J. P.; Zhang, X. *J. Org. Chem.* **2002**, *67*, 7618–7623.
34. Togni, A.; Breutel, C.; Schnyder, A.; Spindler, F.; Landert, H.; Tijani, A. *J. Am. Chem. Soc.* **1994**, *116*, 4062–4066.
35. (a) Ireland, T.; Grossheimann, G.; Wieser-Jeunesse, C.; Knochel, P. *Angew. Chem., Int. Ed. Engl.* **1999**, *38*, 3212–3215. (b) Ireland, T.; Tappe, K.; Grossheimann, G.; Knochel, P. *Chem. Eur. J.* **2002**, *8*, 843–852.
36. Lotz, M.; Polborn, K.; Knochel, P. *Angew. Chem. Int. Ed. Engl.* **2002**, *41*, 4708–4711.
37. (a) Pye, P. J.; Rossen, K.; Reamer, R. A.; Tsou, N. N.; Volante, R. P.; Reider, P. J. *J. Am. Chem. Soc.* **1997**, *119*, 6207–6208. (b) Pye, P. J.; Rossen, K.; Reamer, R. A.; Volante, R. P.; Reider, P. J. *Tetrahedron Lett.* **1998**, *39*, 4441–4444. (c) Burk, M. J.; Hems, W.; Herzberg, D.; Malan, C.; Zanotti-Gerosa, A. *Org. Lett.* **2000**, *2*, 4173–4176.
38. Zhou, Y.-G.; Zhang, X. *Chem. Commun.* **2002**, 1124–1125.
39. Zhou, Y.-G.; Tang, W.; Wang, W.-B.; Li, W.; Zhang, X. *J. Am. Chem. Soc.* **2002**, *124*, 4952–4953.
40. Agbossou, F.; Carpentier, J.-F.; Hapiot, F.; Suisse, I.; Mortreux, A. *Coord. Chem. Rev.* **1998**, *178–180*, 1615–1645.
41. (a) Scott, J. W.; Keith, D. D.; Nix, G., Jr.; Parrish, D. R.; Remington, S.; Roth, G. P.; Townsend, J. M.; Valentine, D., Jr.; Yang, R. *J. Org. Chem.* **1981**, *46*, 5086–5093. (b) Burk, M. J.; Feaster, J. E.; Nugent, W. A.; Harlow, R. L. **1993**, *115*, 10125–10138.
42. Burk, M. J.; Bedingfield, K. M.; Kiesman, W. F.; Allen, J. G. *Tetrahedron Lett.* **1999**, *40*, 3093–3096.
43. Robinson, A. J.; Stanislawski, P.; Mulholland, D.; He, L.; Li, H.-Y. *J. Org. Chem.* **2001**, *66*, 4148–4152.
44. (a) Noyori, R.; Ohta, M.; Hsiao, Y.; Kitamura, M.; Ohta, T.; Takaya, H. *J. Am. Chem. Soc.* **1986**, *108*, 7117–7119. (b) Kitamura, M.; Hsiao, Y.; Noyori, R.; Takaya, H. *Tetrahedron Lett.* **1987**, *28*, 4829–4832. (c) Kitamura, M.; Hsiao, Y.; Ohta, M.; Tsukamoto, M.; Ohta, T.; Takaya, H.; Noyori, R. *J. Org. Chem.* **1994**, *59*, 297–310.
45. Gridnev, I. D.; Yasutake, M.; Higashi, N.; Imamoto, T. *J. Am. Chem. Soc.* **2001**, *123*, 5268–5276.
46. (a) Tang, T.; Ellman, J. A. *J. Org. Chem.* **1999**, *64*, 12–13. (b) Hoekstra, W. J.; Maryanoff, B. E.; Damiano, B. P.; Andrade-Gordon, P.; Cohen, J. H.; Costanzo, M. J.; Haertlein, B. J.; Hecker, L. R.; Hulshizer, B. L.; Kauffman, J. A.; Keane, P.; McComsey, D. F.; Mitchell, J. A.; Scott, L.; Shah, R. D.; Yabut, S. C. *J. Med. Chem.* **1999**, *42*, 5254–5265.
47. Lubell, W. D.; Kitamura, M.; Noyori, R. *Tetrahedron: Asymmetry* **1991**, *2*, 543–554.
48. Heller, D.; Holz, J.; Drexler, H.-J.; Lang, J.; Drauz, K.; Krimmer, H.-P.; Börner, A. *J. Org. Chem.* **2001**, *66*, 6816–6817.

49. Zhu, G.; Chen, Z.; Zhang, X. *J. Org. Chem.* **1999**, *64*, 2127–2129.
50. Lee, S.-g.; Zhang, Y. J. *Org. Lett.* **2002**, *4*, 2429–2431.
51. Holz, J.; Monsees, A.; Jiao, H.; You, J.; Komarov, I. V.; Fischer, C.; Drauz, K.; Börner, A. *J. Org. Chem.* **2003**, *68*, 1701–1707.
52. (a) Tang, W.; Zhang, X. *Org. Lett.* **2002**, *4*, 4159–4161. (b) Tang, W.; Zhang, X. *Angew. Chem., Int. Ed. Engl.* **2002**, *41*, 1612–1614.
53. Burk, M. J. *J. Am. Chem. Soc.* **1991**, *113*, 8518–8519.
54. Wu, S.; Wang, W.; Tang, W.; Lin, M.; Zhang, X.; *Org. Lett.* **2002**, *4*, 4495–4497.
55. Crameri, T.; Foricher, J.; Scalone, M.; Schmid, R. *Tetrahedron: Asymmetry* **1997**, *8*, 3617–3623.
56. Ohta, T.; Miyake, T.; Seido, N.; Kumobayashi, H.; Takaya, H. *J. Org. Chem.* **1995**, *60*, 357–363.
57. Ohta, T.; Miyake, T.; Seido, N.; Kumobayashi, H.; Akutagawa, S.; Takaya, H. *Tetrahedron Lett.* **1992**, *33*, 635–638.
58. Schmid, R.; Broger, E. A.; Cereghetti, M.; Crameri, Y.; Foricher, J.; Lalonde, M.; Müller, R. K.; Scalone, M.; Schoettel, G.; Zutter, U. *Pure Appl. Chem.* **1996**, *68*, 131–138.
59. Takaya, H.; Ohta, T.; Sayo, N.; Kumobayashi, H.; Akutagawa, S.; Inoue, S.-i.; Kasahara, I.; Noyori, R.; *J. Am. Chem. Soc.* **1987**, *109*, 1596.
60. Inoguchi, K.; Sakuraba, S.; Achiwa, K. *Synlett* **1992**, 169–178.
61. Carpentier, J.-F.; Mortreux, A. *Tetrahedron: Asymmetry* **1997**, *8*, 1083–1099.
62. Boaz, N. W.; Debenham, S. D.; Mackenzie, E. B.; Large, S. E. *Org. Lett.* **2002**, *4*, 2421–2424.
63. Fehring, V.; Selke, R. *Angew. Chem., Int. Ed. Engl.* **1998**, *37*, 1827–1830.
64. Ohkuma, T.; Koizumi, M.; Doucet, H.; Pham, T.; Kozawa, M.; Murata, K.; Katayama, E.; Yokozawa, T.; Ikariya, T.; Noyori, R. *J. Am. Chem. Soc.* **1998**, *120*, 13529–13530.
65. (a) Cobley, C. J.; Henschke, J. P.; *Adv. Synth. Catal.* **2003**, *345*, 195–201. (b) Abdur-Rashid, K.; Lough, A. J.; Morris, R. H.; *Organometallics* **2000**, *19*, 2655–2657.
66. Noyori, R.; Ikeda, T.; Ohkuma, T.; Widhalm, M.; Kitamura, M.; Takaya, H.; Akutagawa, S.; Sayo, N.; Saito, T.; Taketomi, T.; Kumobayashi, H. *J. Am. Chem. Soc.* **1989**, *111*, 9134–9135.
67. Rychnovsky, S. C.; Hoye, R.C. *J. Am. Chem. Soc.* **1994**, *116*, 1753–1765.
68. Ikariya, T.; Ishii, Y.; Kawano, H.; Aria, T.; Saburi, M.; Yoshikawa, S.; Akutagawa, S.; *J. Chem. Soc., Chem. Commun.* **1985**, 922–924.

Manjinder S. Lall

1.5 Sharpless Asymmetric Hydroxylation Reactions

1.5.1 Description

The Sharpless asymmetric hydroxylation can take one of two forms, the initially developed asymmetric dihydroxylation (AD)[1] or the more recent variation, asymmetric aminohydroxylation (AA).[2] In the case of AD, the product is a 1,2-diol, whereas in the AA reaction, a 1,2-amino alcohol is the desired product. These reactions involve the asymmetric transformation of an alkene to a vicinally functionalized alcohol mediated by osmium tetraoxide in the presence of chiral ligands (e.g., $(DHQD)_2$–PHAL or $(DHQ)_2$–PHAL). A mixture of these reagents (ligand, osmium, base, and oxidant) is commercially available and is sold under the name of AD-mix β or AD-mix α (*vide infra*).

AD

$K_2OsO_2(OH)_4$
K_2CO_3, $K_3Fe(CN)_6$
$(DHQD)_2$–PHAL
or
$(DHQ)_2$–PHAL

R_1, R_2, HO, OH

AA

$K_2OsO_2(OH)_4$
$(DHQ)_2$–PHAL
XNClNa or AcNBrLi

R_1, R_2, NHX, OH or R_1, R_2, HO, NHX

1.5.2 Historical Perspective

Makowka[3] first reported the use of osmium tetroxide for the dihydroxylation of alkenes. Later, Hofmann[4] showed this process could be made catalytic in osmium when conducted in the presence of chlorate. While both Milas[5] and Criegee[6] simultaneously reported that peroxides could be used to recycle the osmium, it was Criegee who reported the observation that amines, such as pyridine, dramatically accelerated the rate of the *cis*-hydroxylation reaction. The Criegee conditions became the standard method for osmium tetraoxide-catalyzed oxidation of alkenes until VanRheenen and co-workers[7] published improved conditions that employed tertiary amine oxides as oxidants. The first example that this dihydroxylation process could be carried out in an asymmetric fashion was reported in 1980, when Hentges and Sharpless[8] reported the chirality transfer from optically active cinchona alkaloids acting as chiral amine ligands. Refinements to this process ultimately led to the reaction conditions currently used and to reagents that are commercially available.[9] The AA was an offshoot of this work, when it was determined that imido analogs of osmium tetraoxide could react with alkenes to produce amino alcohols by a *cis*-addition reaction.[10] In October of 2001, the Royal Swedish Academy of Sciences awarded the Nobel Prize in Chemistry for the development of catalytic asymmetric syntheses. K. Barry Sharpless was awarded half of the prize "for his work on chirally catalyzed oxidation reactions."[11]

1.5.3 Mechanism

In general, the mechanism for the AD reaction is depicted in Figure 1. Coordination of a ligand to osmium tetraoxide **1** generates complex **2**. This species then reacts with alkene **3** producing osmium glycolate **4** that can then decompose to the desired 1,2-diols **5** and the reduced osmium species **6**. Catalytically active **2** can be regenerated from **6** by an external oxidant, such as ferricyanide.

Figure 1

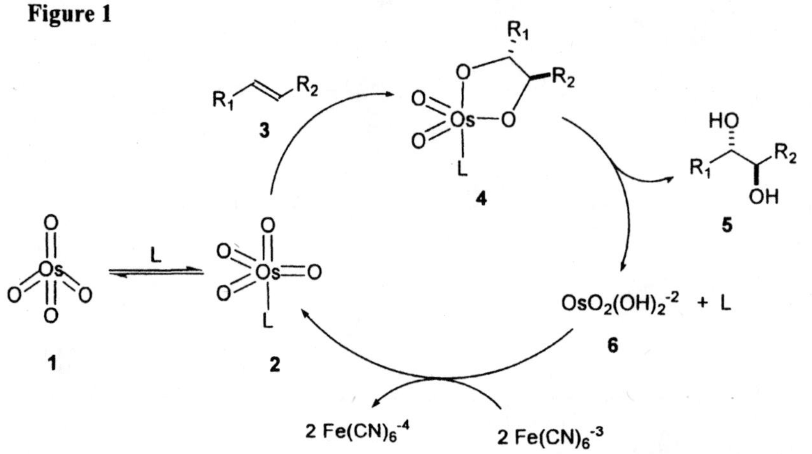

The specific details of how the transformation of **1** to **2** to **4** occurs still remain to be fully characterized. However, numerous studies have been reported in an attempt to provide the mechanistic details. The first rationalization, proposed by Boseken[12] and Criegee,[6,13] was based on the similarity between osmylation and permanganate oxidations of alkenes. A concerted [3 + 2] reaction of the osmium tetraoxide and the alkene could produce the observed product. It was proposed that the ligand acceleration of this reaction was related to the rehybridization from tetrahedral to trigonal bipyramidal at the osmium center upon ligand coordination (**1** to **2**). The decrease in the O–Os–O bond angle would reduce the strain in transition-state **7** on the way to formation of the five-membered ring intermediate **8**.

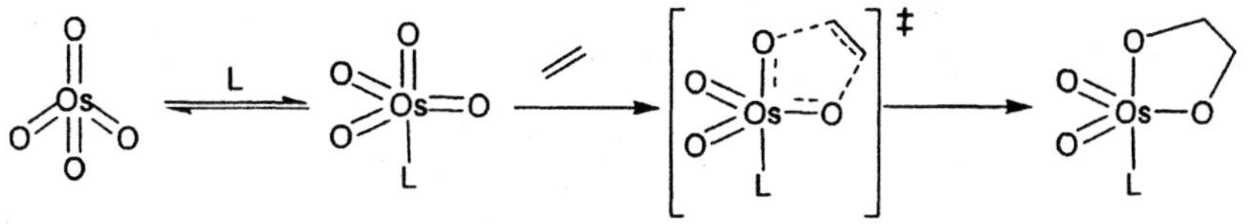

1 2 7 8

This mechanism fell out of favor with the observation of a nonlinear Eyring relationship between % *ee* and temperature.[14] As a result, this rationalization was replaced by a mechanism that required a stepwise pathway through an osmaoxetane **10**.[15] Ligand acceleration in the [2 + 2] mechanism could occur if rearrangement of osmaoxetane **10**, *via* intermediate **11**, to osmaglycolate **8** is facilitated by coordination of a ligand. The electrophilic nature of osmium in this species is also consistent with initial attack of the alkene at the metal center to form **9** and is inconsistent with an alkene π-bond attack of two partial negatively charged oxygen atoms in the [3 + 2] mechanism.

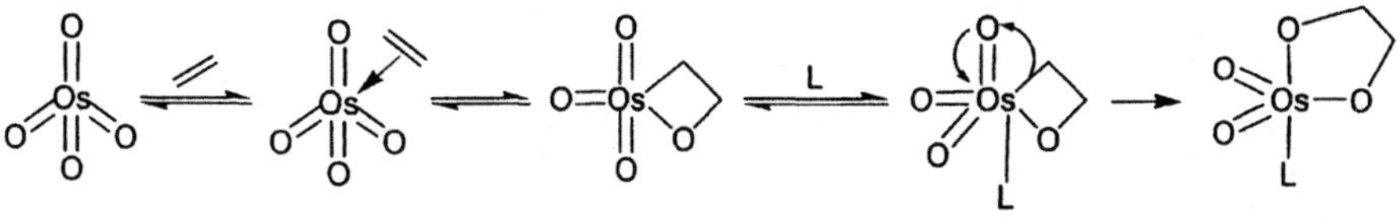

1 9 10 11 8

By direct analogy, the mechanism for the AA can be adapted from the work associated with the AD reaction.[2] The point of departure results from the change in the reagents required for the AA reaction. The catalytically active species is likely the imidotrioxoosmium (VIII) **12** (Figure 2). This complex is formed *in situ* from osmium tetraoxide and the stoichiometric nitrogen source (e.g., chloramines). Addition of alkene **3** produces complex **13**, the result of the asymmetric aminohydroxylation. The desired product **15** is released from the metallic center along with regeneration of the catalytic species by addition of another equivalent of the nitrogen source to convert **13** to **14** followed by hydrolysis of **14** to generate **15**.

Figure 2

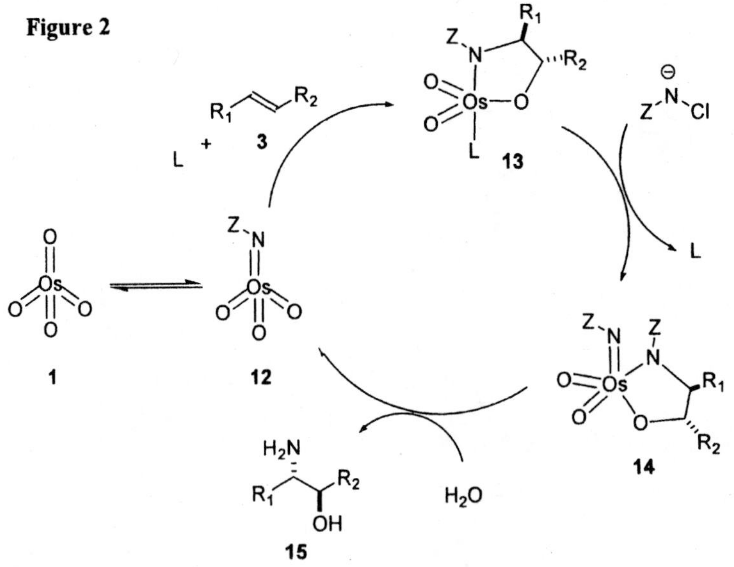

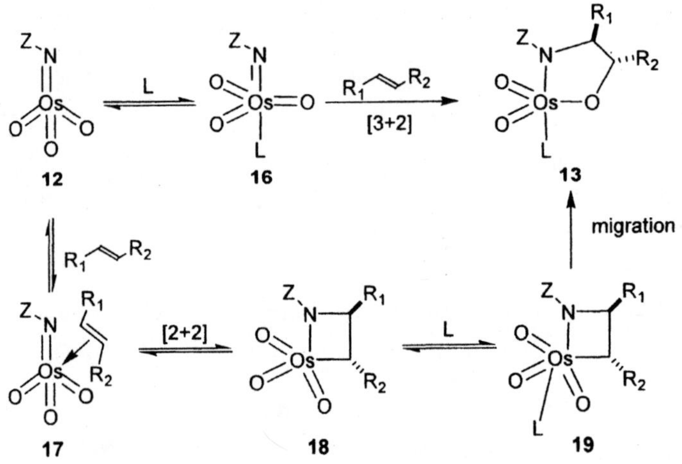

For the generation of **13**, all the uncertainty that was found in the AD reaction can also be translated to this variation of the reaction.[16] The dichotomy between a [2 + 2] and a [3 + 2] mechanistic process remains for the AA reaction. The formation of **13** can be rationalized using a [3 + 2] mechanism in which **16** undergoes the cycloaddition after coordination of a ligand to **12**. Alternatively, the [2 + 2] pathway could proceed by π-bond complex **17** followed by the cycloaddition reaction to afford **18**. Ligand complexation then precedes bond migration to convert **19** to **13**.

Computational chemistry approaches aimed at resolving this mechanistic dichotomy have only made the situation less clear.[17] Furthermore, one cannot study this reaction using the standard principle of microscopic reversibility as these reactions are irreversible. Consequently, the chemistry of rhenium complexes has been applied to those of osmium to provide some insight as to what possibilities are accessible for this metal.[18]

1.5.4 Variations and Improvements or Modifications

To improve the asymmetric induction in these reactions, numerous ligands were evaluated (over 500 have been tested in the Sharpless labs).[19] Within the cinchona alkaloid family, over 75 derivatives were screened. The best ligands have been found to be analogs of dihydroquinine **20** and dihydroquinidine **23**. The result of these studies are the DHQ **21** and DHQD **22** ligands, respectively.

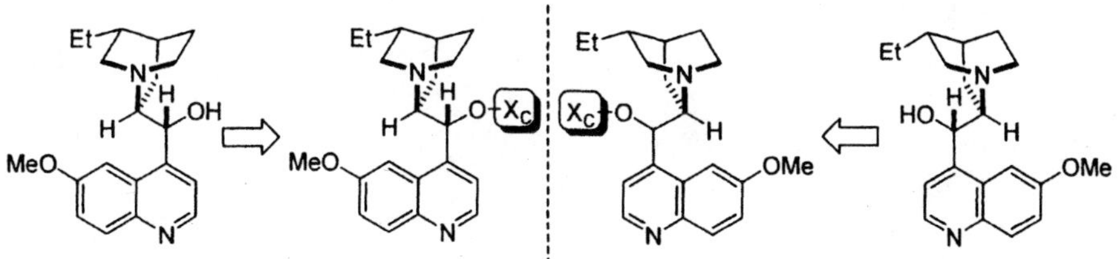

dihydroquinine **20**   DHQ **21**   DHQD **22**   dihydroquinidine **23**

Figure 3

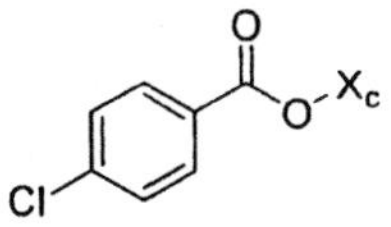

Chlorobenzoate (CLB) 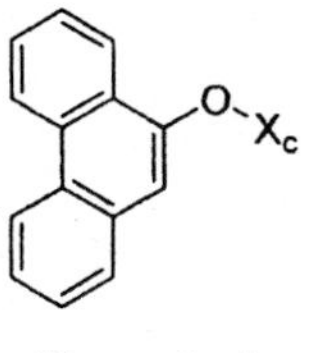   Phenanthryl ether (PHN) 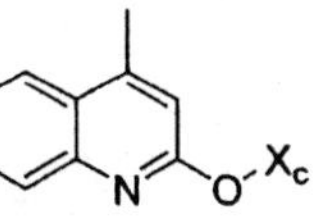   4-methyl-2-quinolyl ether (MEQ)

The nature of the chiral auxiliary has given rise to two generations of ligands. The first generation ligands consist of a single chiral auxiliary bonded to an aryl scaffold (Figure

3). The second generation ligands (Figure 4) consist of the aryl scaffold bonded to two chiral auxiliaries resulting in a C_2-symmetric system. These modifications were based on the desire to improve the chiral induction and broaden the scope of alkene substitution patterns that could be accommodated by these catalysts.

Figure 4

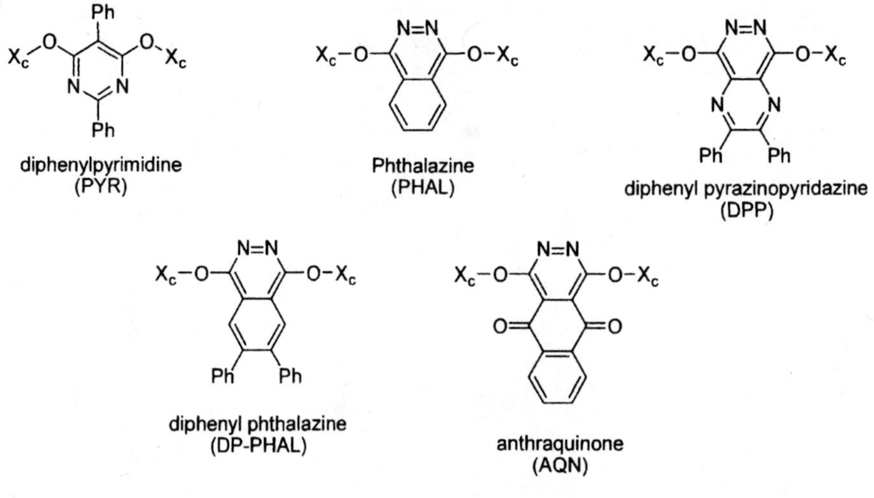

diphenylpyrimidine (PYR)

Phthalazine (PHAL)

diphenyl pyrazinopyridazine (DPP)

diphenyl phthalazine (DP-PHAL)

anthraquinone (AQN)

The most common and commercially available variation of this reaction makes use of the PHAL-based ligands, **24** (AD-mix β) and **25** (AD-mix α). Both reagents are a mix of an osmium source ($K_2OsO_2(OH)_4$), an oxidant ($K_3Fe(CN)_6$),[20] and a base (K_2CO_3) in combination with the chiral ligand. For AD-mix β this ligand is $(DHQD)_2$–PHAL **24** and for AD-mix α this ligand is $(DHQ)_2$–PHAL **25**.

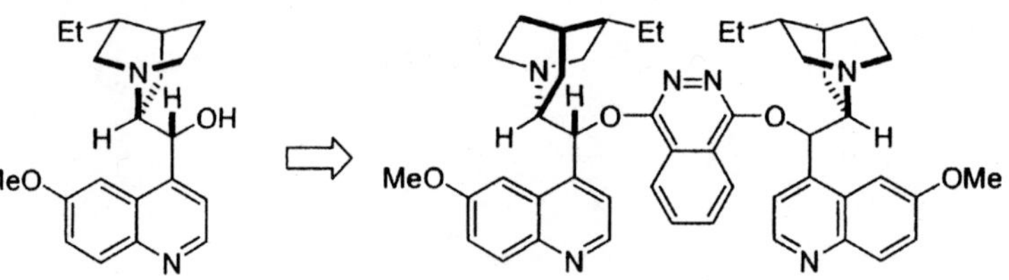

dihydroquinine

20

$(DHQD)_2$–PHAL

24

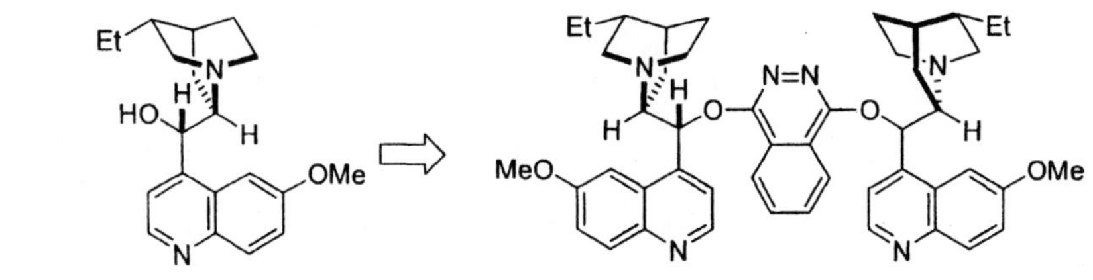

dihydroquinidine

23

$(DHQ)_2$-PHAL

25

The lack of resolution by which mechanism the osmylation proceeds has resulted in two models to rationalize the face selectivity of the AD reaction. The commonality of these two predictive models resides in the basic principle that a chiral binding pocket is formed from the ligand's aromatic groups. However, the shape and location of this pocket in the complex is not identical.

The Corey group at Harvard, based on a [3+2] mechanistic pathway, has proposed a U-shaped binding pocket constructed from two parallel methoxyquinoline moieties contained in the second generation Sharpless ligands (*vide supra*).[21] Thus, chiral ligand **26** coordinates an osmium center through one of the bicyclo[2.2.2]octane moieties. This places the oxidant in close proximity to the bound substrate as depicted in **27**. The chiral volume generated by complex **26** permits the alkene substrate to come in contact in only one possible orientation **28**, thereby inducing chirality in the product.

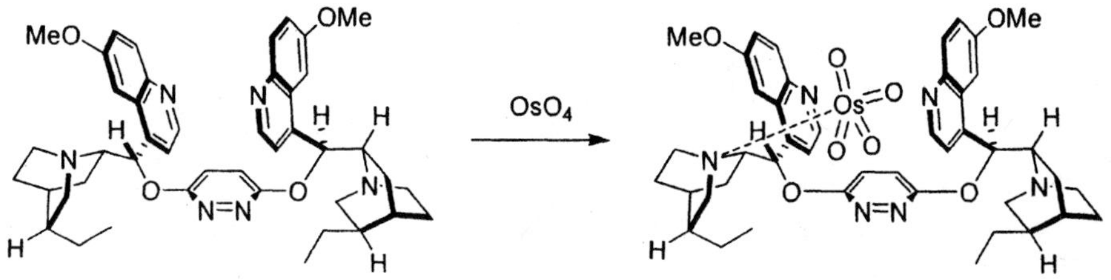

OsO_4

26

27

28

The Sharpless group has proposed a more L-shaped conformation for the methoxyquinoline moieties of the ligand.[22] Coordination of the osmium tetraoxide and binding of the alkene then generates complex **29**. As before, the chiral environment in which the alkene is held results in the asymmetry imparted during the course of this transformation.

29

Table 1.

alkene class	R_1CH=CH$_2$	R_1R_2C=CH$_2$	cis-R_1CH=CHR_2	trans-R_1CH=CHR_2	R_1R_2C=CHR_3	R_1R_2C=CR_3R_4
preferred ligand	R_1 = aromatic DPP, PHAL	R_1, R_2 = aromatic DPP, PHAL	Acyclic IND	R_1, R_2 = aromatic DPP, PHAL	PHAL, DPP, AQN	PYR, PHAL
	R_1 = aliphatic AQN	R_1, R_2 = aliphatic AQN	Cyclic PYR, DPP, AQN	R_1, R_2 = aliphatic AQN		
	R_1 = branched PYR					

The utility of the ligands for inducing asymmetry for different alkene substitution patterns has been explored. Since the binding pocket will vary in size and preference for which substituents it will best interact, a systematic analysis of this relationship has been conducted. There are six possible substitution patterns that can be found for an alkene and the best combination of substitution pattern and ligand for maximizing asymmetric induction is shown in Table 1. While there are specific ligands that are best for certain alkene substitution patterns, the PHAL ligand is the one that appears to have the best selectivity, in general.

Conducting the AD reaction at pH 12 was found to improve the reaction rates for internal alkenes.[23] Additionally, the need for hydrolysis adjuvants, such as methanesulfonamide, could be omitted.

Additional modifications include the immobilization of the catalyst on an insoluble surface. Silica-anchored ligands have been reported based on the DPP **33** and PYR **34** cores.[24] Their use in AD reactions were comparable to the untethered versions of the chiral ligands. Alkenes substituted with alkyl or aryl groups and with internal and terminal double bonds gave diol products with yields ranging 51–93% and optical purities of 61–99% *ee*. The differentiation between **33** or **34** was the ability to readily recover and recycle the chiral ligands.

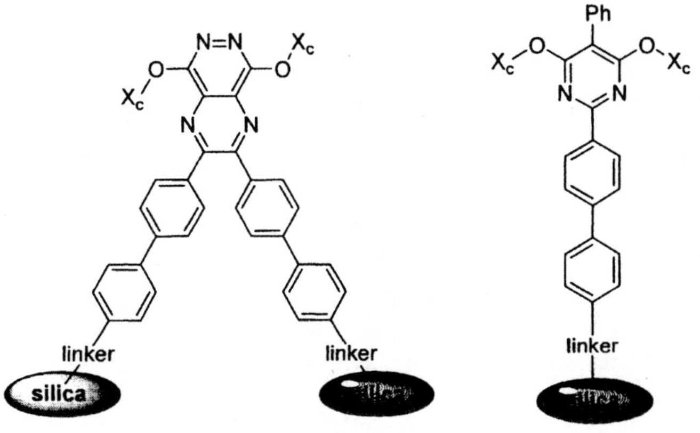

N=N X$_c$ O O X$_c$ N N linker linker silica Ph X$_c$ O O X$_c$ N N linker

33 **34**

For the AA reaction, the smaller the substituent on the nitrogen source, the more efficient the reaction became. Thus, there are three variations of this reaction based on the nitrogen source, sulfonamide **30**, carbamate **31**, or amide **32**.

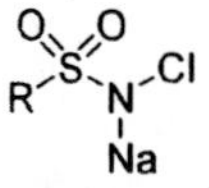

30 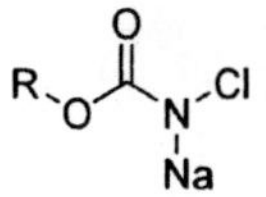 31 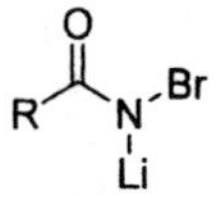 32

Polymer-based derivatives of PHAL **35** have been investigated in the AA reaction.[25] The utility of this ligand was evaluated relative to the free form using the reaction of cinnamate **36** to amino alcohol **37**. This conversion, using the chiral ligand **28**, afforded **37** in 96% yield and 96% *ee*, while **35** effected this transformation in 91% yield and 87% *ee*.

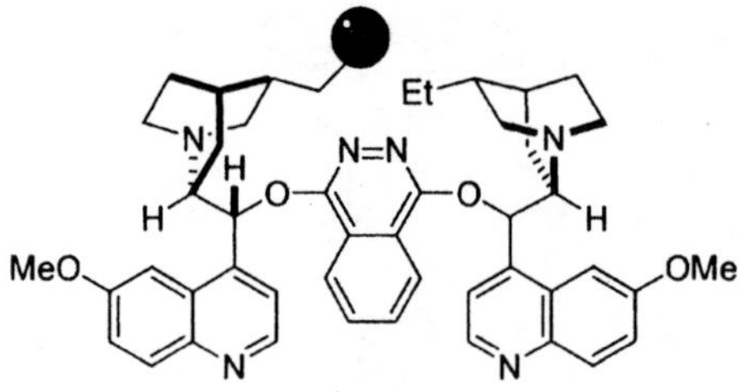

35

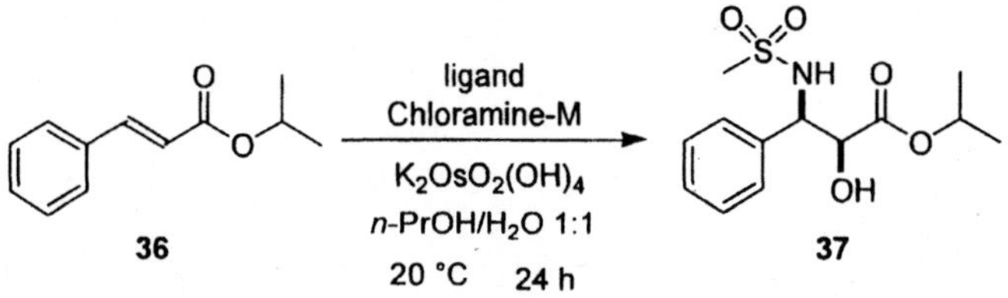

36

ligand
Chloramine-M
$K_2OsO_2(OH)_4$
n-PrOH/H_2O 1:1
20 °C 24 h

37

Despite the uncertainty in the exact mechanism for the AD reaction, a useful mnemonic to predict the direction of asymmetric dihydroxylation has been established.[19] If one places the carbon–carbon double bond of the alkene along the east–west axis of a compass, the substituents on the alkene point to the off-directional positions into the four quadrants shown in Figure 5. The SE quadrant is sterically most demanding, so there is sufficient room for only a hydrogen substituent. The diagonal position, NW, is slightly more accessible and can accept small substituents. The NW quadrant is more accommodating and will fit medium-sized groups. The largest groups are typically placed into the SW quadrant, but the nature of the substituent that can be accommodated is ligand dependent. For example, aromatic groups are favored at this position when employing PHAL ligands, whereas PYR ligands show a preference for aliphatic groups at this position (*vide supra*).

Figure 5

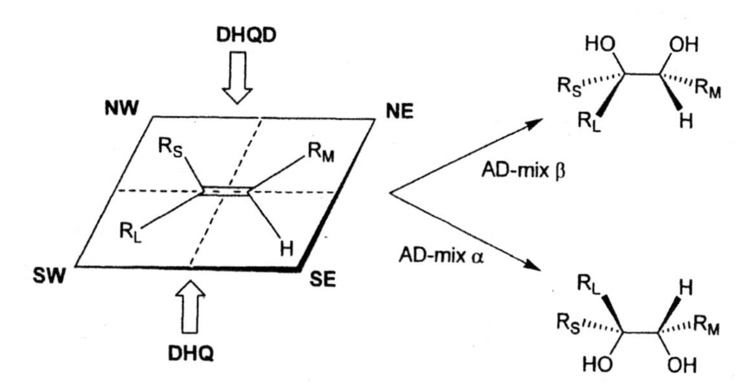

Just as with the AD reaction, a mnemonic for the AA reaction has also been put forward (Figure 6). The same model predicts the identical sense of enantiofacial selectivity indicating the chiral ligands are behaving in a similar manner.

Figure 6

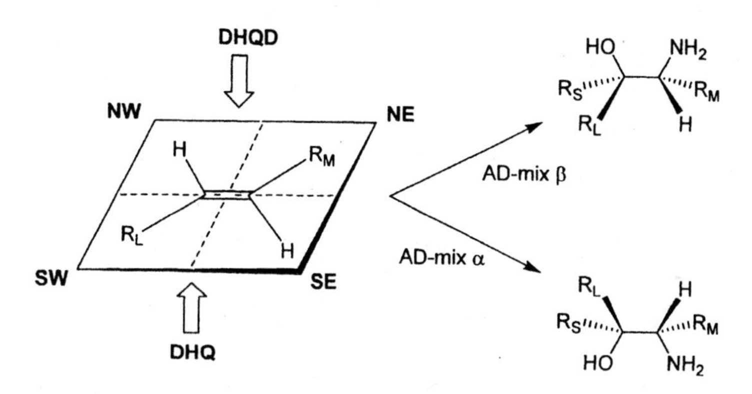

1.5.5 Synthetic Utility

Azasugars, in general, have shown promise as anti-cancer and anti-viral agents. The preparation of pyrrolidine azasugar **40** took advantage of the AD reaction.[26] Diene **38** could be mono-dihydroxylated using the AD-mix α system to produce **39** in excellent optical purity. This compound was then taken on in three steps to **40** in an overall yield of 60%.

AD-mix α
$NaHCO_3$
$MeSO_2NH_2$
H_2O/t-BuOH 1:1
0 °C, 16 h
70% 97% *ee*

38 39 40

The AD reaction was central in the preparation of (+)-*cis*-sylvaticin **41**,[27] a natural product found to have potent anti-tumor activity. The ability of this compound to inhibit ATP production by blockade of the mitochondrial complex I was thought to be the origin of this biological outcome. The AD reaction, in this example, exploited the preference of this reaction for the oxidation of 1,2-*trans*-alkenes over monosubstituted alkenes. The *E,E*-isomer of tetradecatetraene **42** could be chemoselectively dihydroxylated at both internal alkenes, while the terminal alkenes remained untouched. Thus, **43** was generated in excellent chemical yield.

41

AD-mix α
37%
98% *ee*
> 90% *de*

42 43

The AA reaction has been used in the preparation of the paclitaxel **44** side-chain **47**.[28] The structure activity relationship (SAR) associated with the anti-cancer activity of the scaffold embedded in this substituent is wellknown. Starting with the commercially available methyl cinnamate **45**, on a one-third mole scale, the advanced intermediate **46** could be prepared in one step with essentially no workup. The ready access to **46** enabled the preparation of **47** which could be used in the semi-synthesis of **44**.

44

45 → **46** ⇒ **47**

2.5% $(DHQ)_2$-PHAL
2.0% $K_2OsO_2(OH)_4$
3.5 eq TsNClNa–$3H_2O$
t-BuOH/H_2O 1:1
69% 82% *ee*

A tethered aminohydroxylation (TA) reaction extends the utility of carbamate versions of the nitrogen source.[29] Combining the nitrogen source with allylic or homoallylic alcohols enabled formation of products not previously accessible. Thus, exposure of **48** to the osmium reagent, in the absence of a chlorinating agent and base, led to cyclized amino alcohol **49**. A similar result was observed for homoallylic alcohol derivative **50**.

48 → **49**

$K_2OsO_2(OH)_4$
PrOH, H_2O, RT
74%

50 → **51**

$K_2OsO_2(OH)_4$
PrOH, H_2O, RT
63%

If the reaction was conducted with 1 equivalent of osmium and in the presence of TMEDA, then the intermediate osmium azaglycolate **52** could be isolated. The structure of **52** was confirmed by single crystal X-ray analysis. This observation demonstrated that the reaction does indeed occur by *syn*-addition across the alkene and provides strong evidence that the Os(VI) species is oxidized by a *N*-sulfonyl derivative.

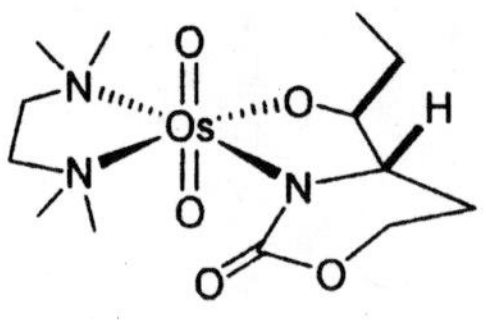

52

1.5.6 Experimental

The two examples below provide representative experimental protocols for the AD[1i] and AA reactions, respectively. There are several confirmed methods that have been reported in *Organic Syntheses*.[30]

(2*R*)-4-(*N*,*N*-Diethylcarbamoyloxy)-5,5-difluoropent-4-en-1,2-diol (54)[31]

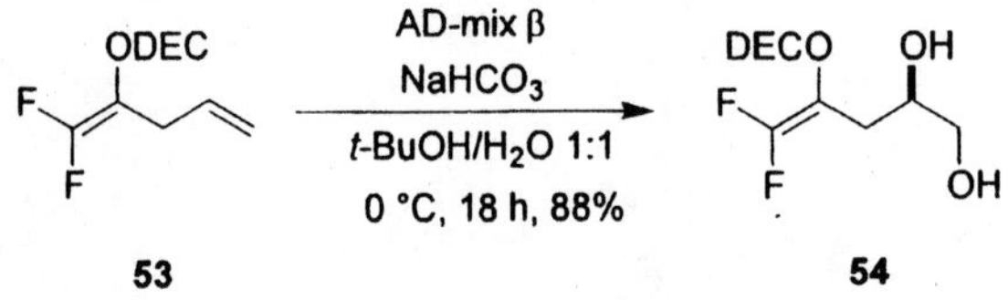

AD-mix β
$NaHCO_3$
t-BuOH/H_2O 1:1
0 °C, 18 h, 88%

53 **54**

A mixture of AD-mix β (11.7 g, 1.41g/mol) and $NaHCO_3$ (1.8 g, 21.4 mmol) in *t*-BuOH/H_2O (82 mL, 1:1 v/v) at room temperature was stirred vigorously until a solution was obtained. The reaction was cooled to 0 °C and diene **53** (1.82 g, 8.32 mmol) was added. Once the reaction was complete, as judged by TLC, the yellow mixture was quenched with NaS_2O_3 (12 g) and stirred an additional 30 min to produce a grey solution. This was diluted with DCM (5 mL) and the phases were separated. The aqueous layer was back extracted with DCM (3 × 20 mL) and the combined organic phases were dried and concentrated *in vacuo*. The crude product was purified by silica gel chromatography to afford **54** (1.85 g, 88%) as a colorless oil.

(2*S*,3*R*)-*tert*-Butyl-2-hydroxy-3-(*N*-benzyloxycarbonyl)-aminobutanoate 56.[32]

Bn–O–C(=O)–NH_2
$K_2OsO_2(OH)_4$
$(DHQD)_2$–PHAL
t-BuOCl, NaOH
n-PrOH/H_2O
5h, 67%, 99% *ee*

55 → **56**

To a solution of benzyl carbamate (6.6 g, 43.6 mmol) in *n*-propanol (56 mL) was added a solution of NaOH (1.71 g, 42.9 mmol) in 105 mL of water followed by freshly prepared *tert*-butyl hypochlorite (4.7 mL, 43.9 mmol). To this mixture was then added a solution of $(DHQD)_2$–PHAL (0.54 g, 0.56 mmol) in 49 mL of *n*-propanol. Finally, **55** (2.0 g, 14.1 mmol) and $K_2OsO_2(OH)_4$ (0.20 g, 0.56 mmol) were added. Once the reaction was completed, partially judged by the change in colour from the initially light green coloured reaction to a light yellow colour, the mixture was diluted with 30 mL of ethyl acetate. The aqueous phase was extracted with ethyl acetate (4 × 60 mL) and the combined organic phases were washed with brine, dried (Na_2SO_4), and concentrated. The crude product was purified by flash chromatography to afford **56** (2.92 g, 67%) as a white solid.

1.5.7 References

1. [R] (a) Schroder, M. *Chem. Rev.* **1980**, *80*, 187–213. [R] (b) Lohray, B. B. *Tetrahedron: Asymm.* **1992**, *3*, 1317–1349. [R] (c) Johnson, R. A.; Sharpless, K. B. *Catal. Asymmetric Synth.* **1993**, *227*, 227–272. [R] (d) Kolb, H. C.; VanNieuwenhze, M. S.; Sharpless, K. B. *Chem. Rev.* **1994**, *94*, 2483–2547. [R] e) Berrisford, D. J.; Bolm, C.; Sharpless, K. B. *Angew. Chem. Int. Ed. Engl.* **1995**, *34*, 1059–1070. [R] (f) Beller, M.; Sharpless, K. B. *Appl. Homogen. Catal. Organomet. Compounds* **1996**, *2*, 1009–1024. [R] (g) Becker, H.; Sharpless, K. B. *Asymm. Oxid. Reac.* **2001**, 81–104. [R] (h) Kolb, H. C.; Sharpless, K. B. in *Transition Metals for Organic Synthesis*, eds Beller, M.; Bolm, C. 2nd Ed., **2004**, Wiley-VCH, pp 275-298. [R] (i) Noe, M. C.; Letavic, M. A.; Snow S. I. *Org. React.* **2005**, *66*, 109–625.
2. [R] (a) Reiser, O. *Angew. Chem. Int. Ed. Engl.* **1996**, *35*, 1308–1309. [R] (b) O'Brien, P. *Angew. Chem. Int. Ed.* **1999**, *38*, 326–329. [R] (c) Schlingloff, G.; Sharpless, K. B. *Asymm. Oxid. Reac.* **2001**, 104-114. [R] (d) Bodkin, J. A.; McLeod, M. D. *J. Chem. Soc., Perkin Trans. 1* **2002**, 2733–2746. [R] (e) Nilov, D.; Reiser, O. *Adv. Synth. Catal.* **2002**, *344*, 1169–1173. [R] (f) Nilov, D.; Reiser, O. *Org. Synth. Highlights V* **2003**, *118*, 118–124. [R] (g) Kolb, H. C.; Sharpless, K. B. in *Transition Metals for Organic Synthesis*, eds Beller, M.; Bolm, C. 2nd Ed., **2004**, Wiley-VCH, pp 309-326.
3. Makowka, O. *Chem. Ber.* **1908**, *41*, 943–944.
4. (a) Hofmann, K. A. *Chem. Ber.* **1913**, *45*, 3329–3336. (b) Hofmann, K. A.; Ehrhart, O.; Schnieder, O. *Chem. Ber.* **1913**, *46*, 1657–1668.
5. (a) Milas, N. A.; Sussman, S. *J. Am. Chem. Soc.* **1936**, *58*, 1302–1304. (b) Milas, N. A.; Sussman, S. *J. Am. Chem. Soc.* **1937**, *59*, 2345–2347.
6. (a) Criegee, R. *Ann.* **1936**, *522*, 75–96. (b) Criegee, R.; Marchand, B.; Wannowius, H. *Ann.* **1942**, *550*, 99–133.
7. VanRheenen, V.; Kelly, R. C.; Cha, D. Y. *Tetrahedron Lett.* **1976**, 1973–1976.
8. Hentges, S. G.; Sharpless, K. B. *J. Am. Chem. Soc.* **1980**, *102*, 4263–4265.
9. Aldrich

10. (a) Sharpless, K. B.; Patrick, D. W.; Truesdale, L. K.; Biller, S. A. *J. Am. Chem. Soc.* **1975**, *97*, 2305–2307. (b) Herranz, E.; Biller, S. A.; Sharpless, K. B. *J. Am. Chem. Soc.* **1978**, *100*, 3596–3598. (c) Herranz, E.; Sharpless, K. B. *J. Org. Chem.* **1978**, *43*, 2544–2548. (d) Patrick, D. W.; Truesdale, L. K.; Biller, S. A.; Sharpless, K. B. *J. Org. Chem.* **1978**, *43*, 2628–2638.
11. Sharpless, K. B. *Angew. Chem. Int. Ed.* **2002**, *41*, 2024–2032.
12. Bosecken, J.; de Graff, C. *Rec. Trav. Chim.* **1922**, *41*, 199–207.
13. (a) Criegee, R. *Angew. Chem.* **1937**, *50*, 153–155. (b) Criegee, R. *Angew. Chem.* **1938**, *51*, 519–520.
14. Gobel, T.; Sharpless, K. B. *Angew. Chem. Int. Ed. Engl.* **1993**, *32*, 1329–1331.
15. (a) Sharpless, K. B.; Teranishi, A. Y.; Backvall, J.-E. *J. Am. Chem. Soc.* **1977**, *99*, 3120–3128. (b) Nelson, D. W.; Gypser, A.; Ho, P. T.; Kolb, H. C.; Kondo, T.; Kwong, H.-L. McGrath, D. V.; Rubin, A. E.; Norrby, P.-O.; Gable, K. P.; Sharpless, K. B. *J. Am. Chem. Soc.* **1997**, *119*, 1840–1858. [R] (c) Jorgensen, K. A.; Schiott, B. *Chem. Rev.* **1990**, *90*, 1483–1506.
16. (a) Sharpless, K. B.; Patrick, D. W.; Truesdale, L. K.; Biller, S. A. *J. Am. Chem. Soc.* **1975**, *97*, 2305–2307. (b) Sharpless, K. B.; Chong, A. O.; Oshima, K. *J. Org. Chem.* **1976**, *41*, 177–179. (c) Sharpless, K. B.; Teranishi, A. Y.; Backvall, J.-E. *J. Am. Chem. Soc.* **1977**, *99*, 3120–3128. (d) Chong, A. O.; Oshima, K.; Sharpless, K. B. *J. Am. Chem. Soc.* **1977**, *99*, 3420–3426. (e) Herranz, E.; Sharpless, K. B. *J. Org. Chem.* **1978**, *43*, 2544–2548. (f) Herranz, E.; Biller, S. A.; Sharpless, K. B. *J. Am. Chem. Soc.* **1978**, *100*, 3596–3598. (g) Patrick, D. W.; Truesdale, L. K.; Biller, S. A.; Sharpless, K. B. *J. Org. Chem.* **1978**, *43*, 2628–2638. (h) Herranz, E.; Sharpless, K. B. *J. Org. Chem.* **1980**, *45*, 2710–2713. (i) Hentges, S. G.; Sharpless, K. B. *J. Org. Chem.* **1980**, *45*, 2257–2259. (j) Li, G.; Chang, H.-T.; Sharpless, K. B. *Angew. Chem. Intl. Ed. Engl.* **1996**, *35*, 451–454.
17. (a) Veldkamp, A.; Frenking, G. *J. Am. Chem. Soc.* **1994**, *116*, 4937–4946. (b) Norrby, P.-O.; Kolb, H. C.; Sharpless, K. B. *J. Am. Chem. Soc.* **1994**, *116*, 8470–8478. (c) Norrby, P.-O.; Becker, H.; Sharpless, K. B. *J. Am. Chem. Soc.* **1996**, *118*, 35–42. (d) Dapprich, S.; Ujaque, G.; Maseras, F.; Lledos, A.; Musaev, D. G.; Morokuma, K. *J. Am. Chem. Soc.* **1996**, *118*, 11660–11661. (e) DelMonte, A. J.; Haller, J.; Houk, K. N.; Sharpless, K. B.; Singleton, D. A.; Strassner, T.; Thomas, A. A. *J. Am. Chem. Soc.* **1997**, *119*, 9907–9908.
18. (a) Gable, K. P.; Juliette, J. J. J. *J. Am. Chem. Soc.* **1995**, *117*, 955–962. (b) Gable, K. P.; Juliette, J. J. J. *J. Am. Chem. Soc.* **1996**, *118*, 2625–2633.
19. (a) Jacobsen, E. N.; Marko, I.; Mungall, W. S.; Schroder, G.; Sharpless, K. B. *J. Am. Chem. Soc.* **1988**, *119*, 1968–1970. (b) Sharpless, K. B.; Amberg, W.; Beller, M.; Chen, H.; Hartung, J.; Kawanami, Y.; Lubben, D.; Manoury, E.; Ogino, Y.; Shibata, T.; Ukita, T. *J. Org. Chem.* **1991**, *56*, 4585–4588. (c) Sharpless, K. B.; Amberg, W.; Bennani, Y. L.; Crispino, G. A.; Hartung, Y.; Jeong, K.-S.; Kwong, H.-L.; Morikawa, K.; Wang, Z.-M.; Xu, D.; Zhang, X.-L. *J. Org. Chem.* **1992**, *57*, 2768–2771. d) Crispino, G. A.; Jeong, K.-S.; Kolb, H. C.; Wang, Z.-M.; Xu, D.; Sharpless, K. B. *J. Org. Chem.* **1993**, *58*, 3785–3786. (e) Kolb, H. C.; Andersson, P. G.; Bennani, Y. L.; Crispino, G. A.; Jeong, K.-S.; Kwong, H.-L.; Sharpless, K. B. *J. Am. Chem. Soc.* **1993**, *115*, 12226–12227. (f) Becker, H.; King, S. B.; Taniguchi, M.; Vanhessche, K. P. M.; Sharpless, K. B. *J. Org. Chem.* **1995**, *60*, 3940–3941. (g) Iwahima, M.; Kinsho, T.; Smith, A. B., III *Tetrahedron Lett.* **1995**, *36*, 2199–2202. (h) Becker, H.; Sharpless, K. B. *Angew. Chem. Int. Ed. Engl.* **1996**, *35*, 448–451. (i) Vanhessche, K. P. M.; Sharpless, K. B. *J. Org. Chem.* **1996**, *61*, 7978–7979.
20. Minato, M.; Yamamoto, K.; Tsuji, J. *J. Org. Chem.* **1990**, *55*, 766–768.
21. (a) Corey, E. J.; Noe, M. C.; Sarshar, S. *Tetrahedron Lett.* **1994**, *35*, 2861–2864. (b) Corey, E. J.; Noe, M. C.; Grogan, M. J.; *Tetrahedron Let.* **1994**, *35*, 6427–6430. (c) Corey, E. J.; Guzman-Perez, A.; Noe, M. C. *J. Am. Chem. Soc.* **1994**, *116*, 12109–12110. (d) Corey, E. J.; Guzman-Perez, A.; Noe, M. C. *J. Am. Chem. Soc.* **1995**, *117*, 10805-10816. (e) Corey, E. J.; Noe, M. C. *J. Am. Chem. Soc.* **1996**, *118*, 319–329. (f) Corey, E. J.; Noe, M. C.; Ting, A. Y. *Tetrahedron Lett.* **1996**, *37*, 1735–1738. (g) Corey, E. J.; Noe, M. C.; Grogan, M. J. *Tetrahedron Lett.* **1996**, *37*, 4899–4902.
22. (a) Kolb, H. C.; Andersson, P. G.; Bennani, Y. L.; Crispino, G. A.; Jeong, K.-S.; Kwon, H.-L.; Sharpless, K. B. *J. Am. Chem. Soc.* **1993**, *115*, 12226–12227. (b) Becker, H.; Ho, P. T.; Kolb, H. C.; Loren, S.; Norrby, P.-O.; Sharpless, K. B. *Tetrahedron Lett.* **1994**, *35*, 7315–7318. (c) Kolb, H. C.; Andersson, P. G.; Sharpless, K. B. *J. Am. Chem. Soc.* **1994**, *116*, 1278–1291. (d) Norrby, P.-O.; Kolb, H. C.; Sharpless, K. B. *J. Am. Chem. Soc.* **1994**, *116*, 8470–8478. (e) Norrby, P.-O.; Becker, H.; Sharpless, K. B. *J. Am. Chem. Soc.* **1996**, *118*, 35–42.
23. Mehltretter, G. M. Dobler, C.; Sundermeier, U.; Beller, M. *Tetrahedron Lett.* **2000**, *41*, 8083–8087.
24. Bolm, C.; Maischak, A.; Gerlach, A. *Chem. Commun.* **1997**, 2353–2354.

25. Mandoli, A.; Pini, D.; Agostini, A.; Savadori, P. *Tetrahedron: Asymm.* **2000**, *11*, 4039–4042.
26. Lindstrom, U. M.; Ding, R.; Hidestal, O. *Chem. Commun.* **2005**, 1773–1774.
27. Donohoe, T. J.; Harris, R. M.; Burrows, J.; Parker, J. *J. Am. Chem. Soc.* **2006**, *128*, 13704–13705.
28. Li, G.; Sharpless, K. B. *Acta Chem. Scand.* **1996**, *50*, 649–651.
29. Donohoe, T. J.; Chughtai, M. J.; Klauber, D. J.; Griffin, D.; Campbell, A. D. *J. Am. Chem. Soc.* **2006**, *128*, 2514–2515.
30. (a) McKee, B. H.; Gilheany, D. G.; Sharpless, K. B. *Org. Synth.* **1992**, *70*, 47–53. (b) Oi, R.; Sharpless, K. B. *Org. Synth.* **1996**, *73*, 1–12. (c) Gonzalez, J.; Aurigemma, C.; Truesdale, L. *Org. Synth.* **2002**, *79*, 93–102.
31. Cox, L. R.; DeBoos, G. A.; Fullbrook, J. J.; Percy, J. M.; Spencer, N. *Tetrahedron: Asymm.* **2005**, *16*, 347–359.
32. Kandula, S. R. V.; Kumar, P. *Tetrahedron* **2006**, *62*, 9942–9948.

Paul Galatsis

CHAPTER 2 Reduction 85

2.1 Eschweiler–Clark Reductive Alkylation of Amines

2.1.1 Description

The Eschweiler–Clarke reaction is the reductive methylation of amines **1**, both primary and secondary, using formaldehyde (**2**) and formic acid (**3**).[1,2] This represents a specific application of the Leuckart–Wallach reaction.

R–NH₂ (**1**) + HCHO (**2**) + HCOOH (**3**) → R–NMe₂ (**4**)

In contrast to other methods for the methylation of primary amines, for example, the use of methyl iodide, this method leads to good yields of the *N,N*-dimethyl tertiary amines, without the formation of quaternary salt products.[3] In comparison to reductive amination techniques, the reagents for the Eschweiler–Clarke reaction are cheap, and the reaction is easily scaled up.[4]

2.1.2 Historical Perspective

In 1905, Eschweiler reported that the reaction of formaldehyde with primary and secondary amines in the presence of formic acid produced, on the application of high temperature, the methylated products.[1] It was not until 1933, however, that Clarke *et al.* elucidated the mechanism of the reaction.[2]

Eschweiler illustrated in his report several examples, including a benzyl amine and diamines such as piperazine and ethylenediamine, and noted the complete methylation of the amines, without quaternisation.[1]

2.1.3 Mechanism

The mechanism of the Eschweiler–Clarke reaction proceeds *via* the formation of an imine, followed by reduction by formic acid. That the methylation is attributable to the formaldehyde and the reduction to the formic acid has been confirmed using ^{14}C-labeled isomers of each in a series of studies.[5] Thus, the amine reacts with formaldehyde to produce an imine, and this is then reduced with the loss of carbon dioxide by formic acid. In the case of primary amines this process is then repeated to produce a tertiary *N,N*-dimethyl amine.

In some cases the formation of carbonyl derivatives of the amines can reduce the useful yield.[6] These products arise from the hydrolysis of the isomerised Schiff base **14**. Interestingly, it has been shown that the final product is not derived from these carbonyl by-products. Cope *et al.* showed that, in the case of methylation of optically active amines in which the asymmetric carbon is alpha to the amine, complete retention of configuration was seen even when the appropriate carbonyl product was formed.[7] Thus the isomerized Schiff base **14** must be hydrolyzed faster that it isomerizes to Schiff base **8**.

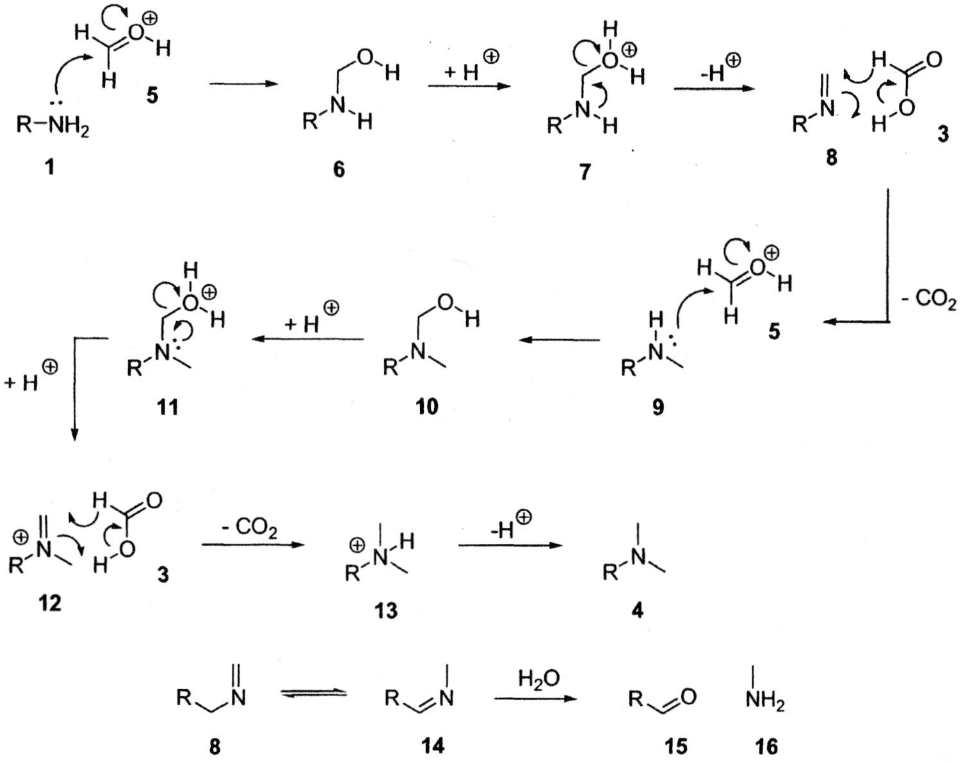

Other side reactions are possible, usually leading to formamide derivatives.[3] In addition, Alder *et al.* have reported that in the Eschweiler–Clarke reaction of polyamines, methylation fragments of the original polyamine are frequently isolated.[4]

2.1.4 Variations

2.1.4.1 N-Alkylation of amides

Lemaire *et al.* report an extension of the Eschweiler–Clarke reaction to allow the *N*-alkylation of amides.[8] Although many other methods for this reaction do exist,[9-11] they tend to be inefficient on secondary and cyclic amides[11] or compounds with low N-H acidity.[9] In addition, this modification of the Eschweiler–Clarke reaction allows *N*-alkylation with a secondary alkylating agent, a reaction that would normally be impossible due to the competing elimination process.[9] In the reaction, reductive alkylation is effected with an aldehyde or ketone (see Leuckart–Wallach reaction) under hydrogen with Pd/C and sodium sulphate in a non-protic solvent.[8]

R, NH2 O 17 or n N H O 18 n = 1,3,... R1 R2 O 19 O O Pd/C/Na2SO4 H2, 40bar 100 °C R H N R1 O R2 20 or n N O R2 R1 21 n = 1,3,... 81 - 95%

2.1.4.2 *Solvent- and formaldehyde-free conditions*

Kosma *et al.* have developed a modification of the Eschweiler–Clarke reaction that avoids the use of toxic formaldehyde, and concentrated formic acid.[12] In place of these reagents they propose the use of paraformaldehyde and oxalic acid, in solvent-free conditions; in addition, this process requires no purification steps.

O N H 22 HO O O OH ·2H2O 23 H O n H 24 100 °C O N 25 100%

2.1.4.3 *Reductive cleavage of cyclic aminol ethers to n,n-dialkylamino-derivatives*

Ph O N 26 O H OH reflux, 7 h Ph OH N 27 96%

Page and Heaney *et al.* have reported the reductive ring opening of five- and six-membered aminol ethers derived from formaldehyde as a general procedure to give *N*-methyl-derivatives in high yields.[13]

2.1.4.4 *Exceptions and Unusual Results*

α-Amino amides

Chen and Sung have recently reported a special case where the Eschweiler–Clarke methylation of α-amino amides leads to cyclocondensation products.[14]

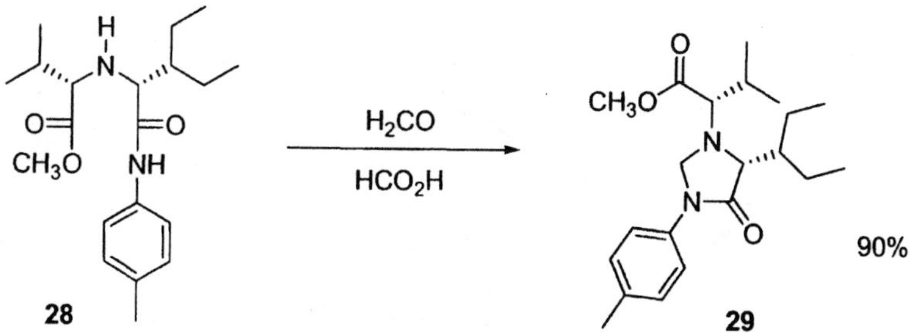

CH_3O NH **28** $\xrightarrow[HCO_2H]{H_2CO}$ CH_3O **29** 90%

Unexpected aromatic methylation of anilines

Bobowski has noted that in the case of electron-rich anilines, aromatic methylation can take place during the Eschweiler–Clarke reaction.[15] It appears that the donation of the aniline lone pair is mechanistically involved in the production of a quinoid-like intermediate that is reduced *in situ*.

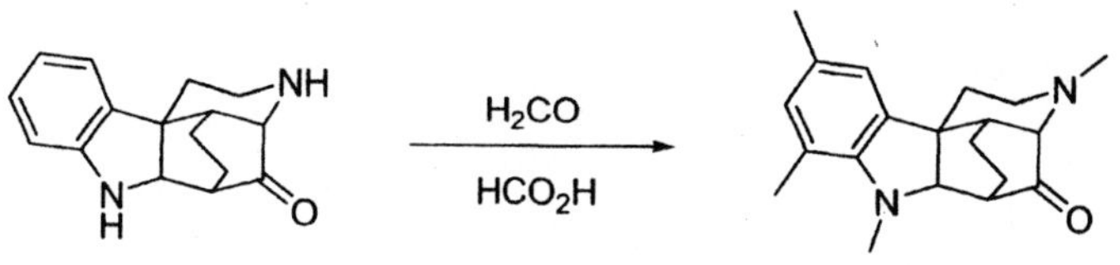

30 $\xrightarrow[HCO_2H]{H_2CO}$ **31**

2.1.5 Synthetic Utility

H_2CO, HCO_2H: 32 → 33

H_2CO, HCO_2H: 34 → 35

H_2CO, HCO_2H: 36 → 37

H_2CO, HCO_2H: 38 → 39

The Eschweiler–Clarke reaction represents a synthetically useful method for the methylation of amines that avoids quaternisation and is easy to scale up. The reaction conserves stereocentres present elsewhere, and has been shown to be possible with less toxic reagents, in solvent-free conditions.

Recently, the application of microwave conditions to reactions that normally proceed thermally, but with long reaction times,[16] has led to a much-hastened procedure for the Eschweiler–Clarke reaction.[17,18] Harding *et al.*[18] and Torchy *et al.*[17] have implemented a microwave assisted Eschweiler–Clarke reaction with a 1:1:1 ratio of amine : formalin:formic acid with severely reduced reaction times of 1–3 minutes, in yields of 70–100 %.

In their studies toward the synthesis of antioxidants, Wijtmans *et al.* have carried out several methylations of 2-aminopyridines, both cyclic and acyclic, in quantitative yields using the Eschweiler–Clarke procedure.[19]

H_2CO, HCO_2H, *i*PrOH: **40** → **41**

H_2CO, HCO_2H, *i*PrOH: **42** → **43**

It is also interesting to note that the Eschweiler–Clarke reaction can be used to form a "methylene bridge." This has been noted many times and in particular can be viewed as a special case of the Pictet–Spengler reaction.[20]

H_2CO, HCO_2H: **44** → **45**

This reaction has been used by Kametani *et al.* in the formation of the berberine bridge in their synthesis of codamine.[21]

H_2CO, HCO_2H, 100 °C, 5h: **46** → **47** (88%)

The "bridging partner" need not be aromatic, as has been reported by Cope *et al.*[22]

48 → 49 (H_2CO, HCO_2H)

2.1.6 *Experimental*

H_2CO, HCO_2H

5-Bromo-1,4,6-trimethyl-2,3-dihydro-1H-pyrrolo[2,3-b]pyridine[19]

A solution of 5-bromo-4,6-dimethyl-2,3-dihydro-1H-pyrrolo[2,3-b]pyridine (1.1 g, 4.56 mmol) in a mixture of 10 mL 37% aqueous HCHO and 10 mL of 90% HCO_3H was refluxed for 18 h. On cooling, the mixture was reduced *in vacuo* and basified with 15 M aqueous KOH. The suspension was cooled in an ice bath, and the precipitate filtered, washed with water, and dried. Chromatography (3:1 hexanes: EtOAc) gave the product as white needles.

2.1.7 *References*

1. Eschweiler, W. *Chem. Ber.* **1905**, *38*, 880
2. Clarke, H. T.; Gillespie, H. B.; Weishaus, S. Z. *J. Am. Chem. Soc.* **1933**, *55*, 4571.
3. Pine, S. H.; Sanchez, B. L. *J. Org. Chem.* **1971**, *36*, 829.
4. Alder, R.; Coleclough, D.; Mowlam, R. W. *Tetrahedron Lett.* **1991**, *52*, 7755.
5. Tarpey, W.; Hauptmann, H.; Tolbert, B. M.; Rapoport, H. *ibid* **1950**, *72*, 5126.
6. Parham, W. E.; Hunter, W. T.; Hanson, R.; Lahr, T. *J. Am. Chem. Soc.*, **1952**, *74*, 5646.
7. Cope, A. C.; Ciganek, E.; Fleckenstein, L. J.; Meisinger, M. A. P. *J. Am. Chem. Soc.* **1960**, *82*, 4651.
8. Fache, F.; Jacquot, L.; Lemaire, M. *Tetrahedron Lett.* **1994**, *20*, 3313.
9. Gajdo, T.; Zwierak, A. *Synthesis* **1981**, 1005
10. Yamawaki, J.; Ando, T.; Hanafusa, T. *Chem. Lett.* 1981, 1143
11. Watanabe, Y.; Ohta, T.; Tsuji, Y. *Bull. Chem. Soc. Jpn.* **1983**, *56*, 2647.
12. Rosenau, T.; Potthast, A.; Rohrling, J.; Hofinger, A.; Sixta, H.; Kosma, P. *Synth. Comm.* **2002**, *3*, 457.
13. Page, P. C. B.; Heaney, H.; Rassias, G. A.; Reignier, S.; Sampler, E. P.; Talib, S. *Synlett* **2000**, *1*, 104.
14. Chen, F.-L.; Sung, K. *J. Heterocyclic Chem.* **2004**, *41*, 697.
15. Bobowski, G. *J. Org. Chem.* **1985**, *50*, 929.
16. Pine, S. H.; Sandoz, B. L. *J. Org. Chem.* **1971**, *36*, 829.
17. Torchy, S.; Barby, D.; *J. Chem. Res. (S)* **2001**, 292.
18. Harding, J. R.; Jones, J. R.; Lu, S.-Y.; Wood, R. *Tetrahedron Lett.* **2002**, *43*, 9487.
19. Wijtmans, M.; Pratt, D. A.; Brinkhorst, J.; Serwa, R.; Valgimgli, L.; Pedulli, G. F.; Porter, N. A. *J. Org. Chem.* **2004**, *69*, 9215.
20. Castillon, J. A. *J. Am. Chem. Soc.* **1952**, *74*, 558.
21. Kametani, T.; Terui, T.; Agui, H.; Fukumoto, K. *J. Org. Chem.* **1970**, *35*, 3808.
22. Cope, A. C; Burrow, W. D. *J. Org. Chem.* **1965**, 2163.

Alice R. E. Brewer

2.2 Gribble Reduction of Diaryl Ketones

2.2.1 Description

The Gribble reduction of diaryl ketones refers to the sodium borohydride–trifluoroacetic acid ($NaBH_4$/TFA) reduction of diaryl ketones (**1**) to diarylmethanes (**2**).[1] The reaction is very general and is applicable to the reduction of diarylmethanols (**3**) and triarylmethanols (**4**) to the respective diarylmethanes (**2**) and triarylmethanes (**5**).[2]

$$\underset{\mathbf{1}}{Ar-\overset{O}{\overset{\|}{C}}-Ar} \xrightarrow[CF_3CO_2H]{NaBH_4} \underset{\mathbf{2}}{Ar-CH_2-Ar} \xleftarrow[CF_3CO_2H]{NaBH_4} \underset{\mathbf{3}}{Ar-\overset{OH}{\overset{|}{C}}H-Ar}$$

$$\underset{\mathbf{4}}{Ar_3COH} \xrightarrow[CF_3CO_2H]{NaBH_4} \underset{\mathbf{5}}{Ar_3CH}$$

Ar = aryl, hetaryl

The use of $NaBH_4$ in carboxylic acid media has been extensively reviewed.[3–10]

Since hydrogen gas is evolved in these reactions, it is important to use pellets of $NaBH_4$ to minimize the surface area in contact with the TFA, and to employ inert gas conditions.[11]

2.2.2 Historical Perspective

Following the discovery of the ability of $NaBH_4/RCO_2H$ to effect the reduction and *N*-alkylation of indoles,[12] Gribble and co-workers found that diaryl ketones are smoothly reduced to diarylmethanes using $NaBH_4$/TFA.[1] A year earlier it was found that diarylmethanols and triarylmethanols are similarly reduced to the respective hydrocarbons.[2]

The reduction is slow with 4-nitrobenzophenone giving a mixture of 4-nitrodiphenylmethane (43%) and 4-nitrodiphenylmethanol (57%). Further treatment of the latter affords 4-nitrodiphenylmethane in 84% yield.[1] The reduction is also poor with sterically crowded benzophenones, as **6**–**8** are essentially recovered unchanged with $NaBH_4$/TFA. Likewise, 4,4′-bis(dimethylamino)benzophenone (Michler's ketone) and decafluorobenzophenone fail to react under these conditions.

$$\text{R-C}_6\text{H}_4\text{-CO-C}_6\text{H}_5 \xrightarrow[\text{TFA, 15-20°C}]{\text{NaBH}_4} \text{R-C}_6\text{H}_4\text{-CH}_2\text{-C}_6\text{H}_5$$ (ref. 1)

R	% yield	R	% yield
H	92	CN	90
Me	89	NMe_2	82
OMe	88	NHPh	93
OH	90	CO_2H	73
F	82	CO_2Me	93
Br	94	NO_2	43 (+ 57% alcohol)

6 7 8 (ref. 1)

2.2.3 *Mechanism*

The mechanism of this reduction is presumed to involve reduction of the carbonyl group to the hydroxyl group, followed by TFA-promoted solvolysis to a doubly benzylic carbocation **9**, and subsequent borohydride (or acyloxyborohydride) quenching of the carbocation to give product hydrocarbon. The highly colored carbocations are visible as transient species.[2]

$$\underset{\mathbf{1}}{\text{Ar-C(=O)-Ar}} \xrightarrow[\text{TFA}]{\text{NaBH}_4} \left[\underset{\mathbf{3}}{\text{Ar-CH(OH)-Ar}} \xrightarrow{\text{TFA}} \underset{\mathbf{9}}{\text{Ar-}\overset{+}{\text{C}}\text{H-Ar}} \right] \xrightarrow{\text{BH}_4^-} \underset{\mathbf{2}}{\text{Ar-CH}_2\text{-Ar}}$$

Consistent with this mechanism is the observed facile reduction of diarylmethanols and triarylmethanols with $NaBH_4$/TFA.[2] Other examples of products from the corresponding alcohols are **10–15**. The corresponding cations leading to **10–15** are spectacularly colored in TFA (e.g., scarlet-red, red-orange, purple, dark blue, and blue-green).

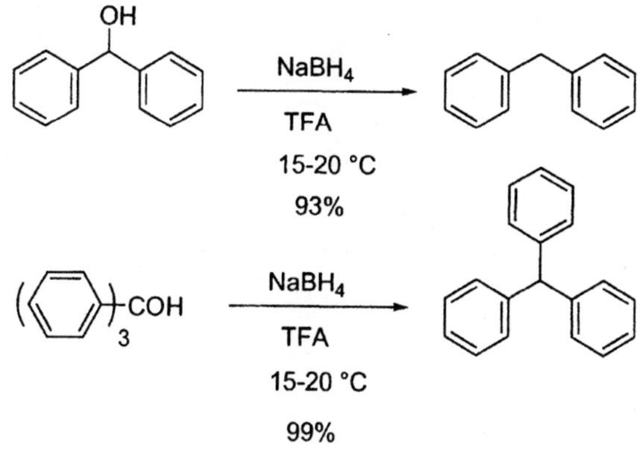

OH

NaBH$_4$

TFA

15-20 °C

93%

(ref. 2)

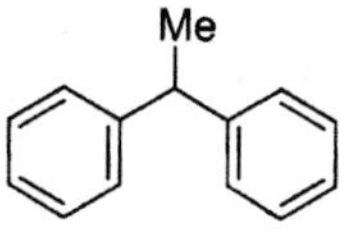

$(C_6H_5)_3COH$

NaBH$_4$

TFA

15-20 °C

99%

Me | Ph | Me

Me Me

10, 97% 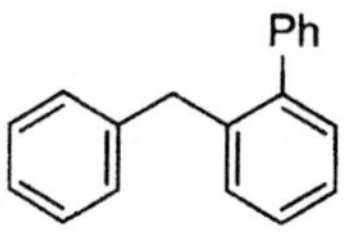 | **11**, 93% 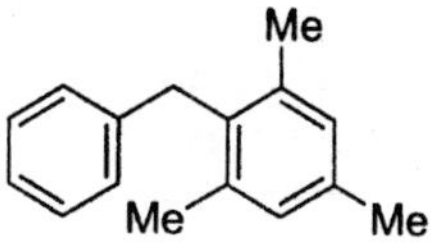 | **12**, 90%

Me | Me

Me Me

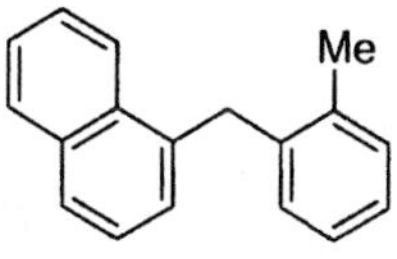

13, 97% 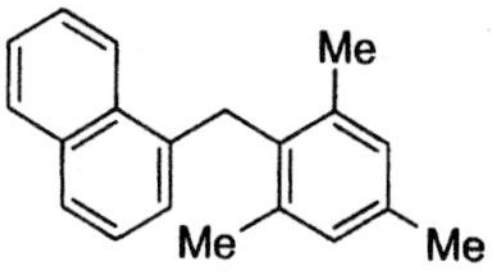 | **14**, 89% 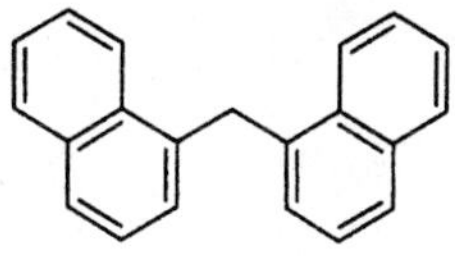 | **15**, 94%

The reaction takes a different course with monobenzylic alcohols (e.g., Friedel–Crafts dimerization or no reaction) and is very slow or fails in acetic acid. Some examples of aberrant behavior are shown in Scheme 1.

Scheme 1.

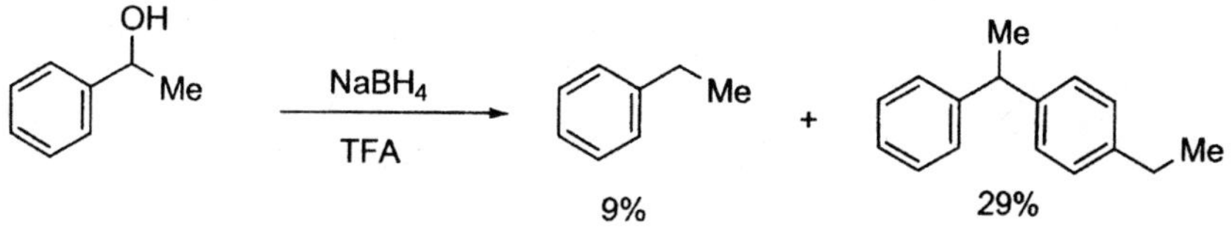

OH

Me

NaBH$_4$

TFA

Me

+

Me

Me

9%

29%

HO Me Me — $NaBH_4$ / TFA → Me Me (38-45%) + Me Me Me (24-32%) (ref. 2)

OH Ph — $NaBH_4$ / TFA → Ph (70%) + Ph Ph Ph (30%)

2.2.4 Variations and Improvements

The two main operational variations are (1) to add a solution of the substrate in dichloromethane to a mixture of $NaBH_4$ pellets in TFA acid at 15–20 °C under nitrogen or argon, and (2) to add the $NaBH_4$ pellets to a solution of the substrate in TFA at 15–20 °C under nitrogen or argon.[1,2] Powdered $NaBH_4$ can be used but the greater surface area of this form of $NaBH_4$ presents a greater danger of fire from the evolved hydrogen.[2,11] The author has never experienced a fire using $NaBH_4$ pellets in conducting hundreds of such reactions in his laboratory.

Whereas monobenzylic ketones and alcohols are not usually cleanly reduced with $NaBH_4$/TFA, suitably activated substrates do undergo reduction to the corresponding hydrocarbons. An early example was provided by Snieckus in the reduction of tetralone **16** to **17**, following TFA–promoted C-3 epimerization.[13] This deoxygenation also occurs readily with 3-acyl-1-(phenylsulfonyl)indoles **18**.[14] Interestingly, the corresponding 3-formylindole (**18**, $R_1 = R_2 = R_3 =$ H) affords a complex mixture of uncharacterized products with $NaBH_4$/TFA, as does 2-acetyl-1-(phenylsulfonyl)indole.[14a] However, several formylazulenes are reduced to the corresponding methyl derivatives with $NaBH_4$/TFA (e.g., **20** to **21**).[15]

O O O CONMe2 OMe OMe **16** — 1. TFA, 50 °C; 2. $NaBH_4$, 0 °C to rt; 76% → O O CONMe2 OMe OMe **17** (ref. 13)

R_1 = Me, Et, Ph, $CH_2CH_2CO_2H$, $CH_2CH_2CO_2Et$
R_2 = H, Cl, Br, I
R_3 = H, Br

18 → 19: $NaBH_4$, TFA, 0 °C to rt, 65-99% (ref. 14)

20 → 21: $NaBH_4$, TFA, CH_2Cl_2, 0 °C to 25 °C, 90% (ref. 15)

Likewise, activated benzylic alcohols can undergo deoxygenation depending on the level of activation as demonstrated by Nutaitis (22→23; 24→25).[16] In such cases the variation was used that involves the addition of TFA to a mixture of the substrate alcohol and $NaBH_4$ in THF, a procedure that generates the more reactive $NaBH_3(OCOCF_3)$ species.[6]

22 → 23: $NaBH_4$, TFA, THF (ref. 16)

R	% 23	R	% 23
4-NMe$_2$	78	4-NO$_2$	0
4-OH	45	2-OMe	0
4-SMe	41	4-Cl	0
4-OMe	19	4-Me	0
3,4-diOMe	20	4-Ph	0
2,5-diOMe	0		

24 → 25 ($NaBH_4$, TFA THF) (ref. 16)

Entry	R_1	R_2	R_3	R_4	% **25**
1	Bu	H	H	OMe	91
2	Bu	H	OMe	H	0
3	Bu	H	OMe	OMe	83
4	Bu	Et	H	OMe	82
5	Bu	H	H	NMe_2	90
6	Me	Me	H	Me	68
7	Me	H	H	H	0

A comparison between entries 1 and 2 dramatically illustrates the expected effect of the methoxy group on stabilizing the intermediate carbocation. In some cases (e.g., entry 6) olefin formation is observed.[16]

Despite the enormous success of the combination of $NaBH_4$/TFA to reduce aryl ketones and alcohols to hydrocarbons (*vide infra,* Section 2.2.5), there are notable failures as shown below (**26–33**), most of which involve monobenzylic substrates with little or no special activation. For these substrates either complex mixtures obtain or no reaction occurs. The profound effect that the presence of a basic nitrogen atom can have on reactions involving $NaBH_4$ in carboxylic acid media was observed some years ago[12,24] and may pertain to most of these cases (i.e., **27**, **28**, **30–32**).

26[17] ($NHCOCF_3$, COPh) **27** (X = CH, N; R = H, Cl)[18] **28** (X = CH, N; R = H, Cl)[18] **29**[19] ($(CH_2)_3Br$)

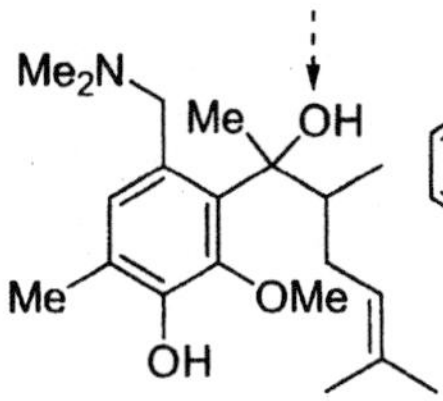

30[20] 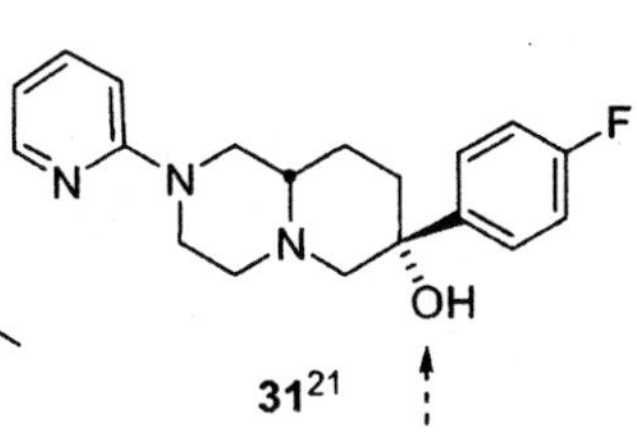 **31**[21] 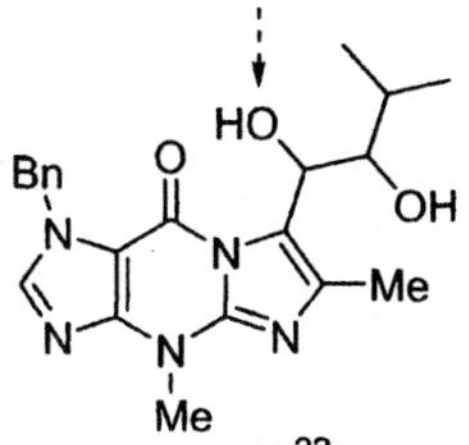 **32**[22] 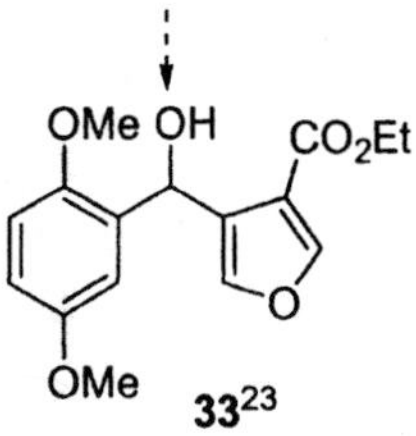 **33**[23]

Other reaction substrate variations that will not be covered in this chapter include the $NaBH_4$/TFA reductive cleavage of 1,4-epoxy-1,4-dihydronaphthalenes[25] and benzylic ozonides,[26] the tandem solvolysis–reduction of triarylmethyl chlorides,[27] the cleavage of ferrocenyl ketone hydrazines,[28] and the reduction of benzophenones to diarylmethanols (!).[29]

2.2.5 *Synthetic Utility*

In addition to the $NaBH_4$/TFA reduction of simple benzophenones presented earlier,[1] there are numerous examples of more complex diaryl ketones undergoing this transformation. The following compounds result from $NaBH_4$/TFA reduction of the respective diaryl ketones. The range of functional group tolerance is noteworthy, the yields are very good to excellent, and this reduction has proved to be very useful in synthesis.

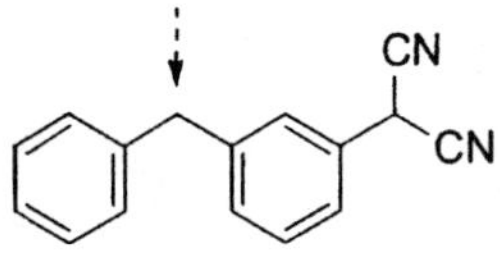

34 (51%)[30] 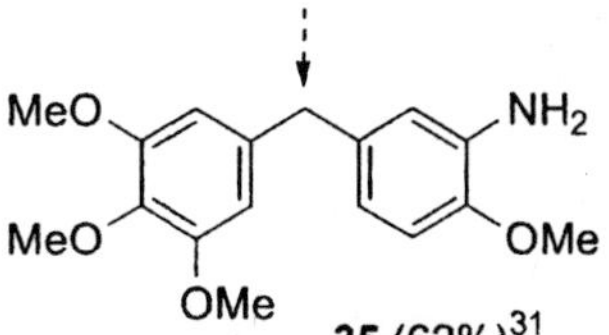 **35** (62%)[31] 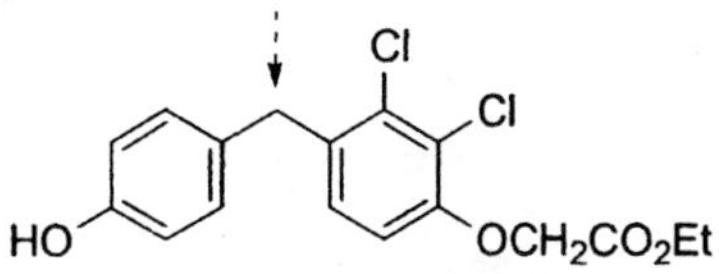 **36** (69%)[32]

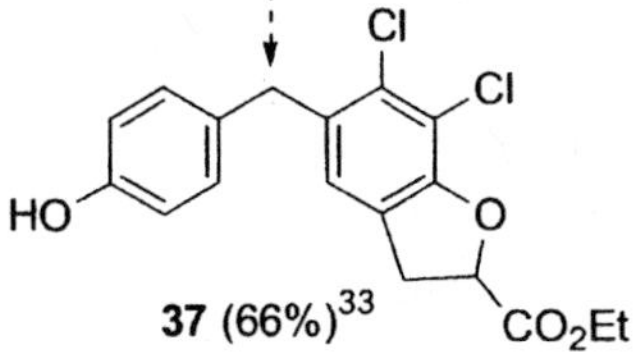

37 (66%)[33] 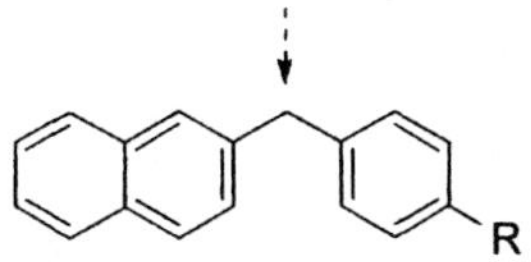 **38** (R = H, Cl, CN, OMe; 63-95%)[34] 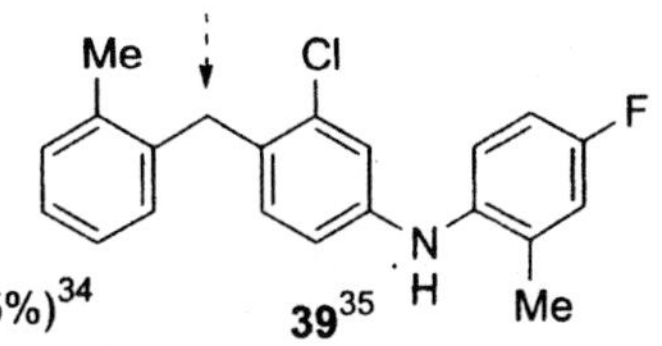 **39**[35]

The reaction leading to amine **39**[35] also affords some of the expected *N*-trifluoroethyl analog,[12,43] and some reduction of the isoquinoline ring is observed with **40**.[36] The reduction of three benzophenone-type carbonyl groups in a tricalix[5]arene proceeded in 95% yield.[44]

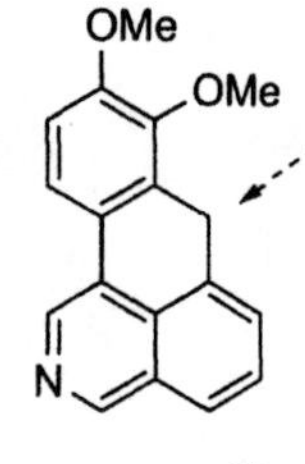

OMe
OMe
N
40 (75%)[36]

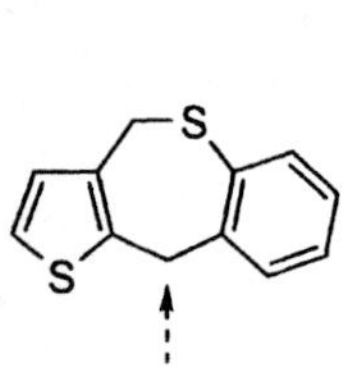

S
S
41 (89%)[37]

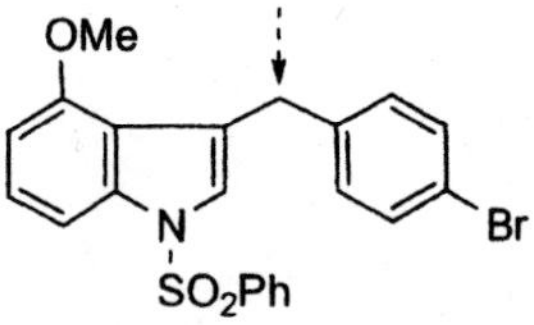

OMe
N
SO2Ph
Br
42[38]

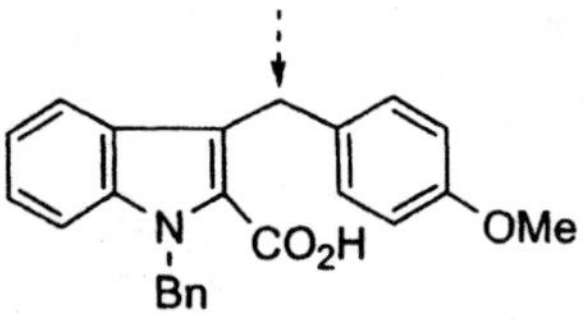

N
Bn
CO2H
OMe
43 (90%)[39]

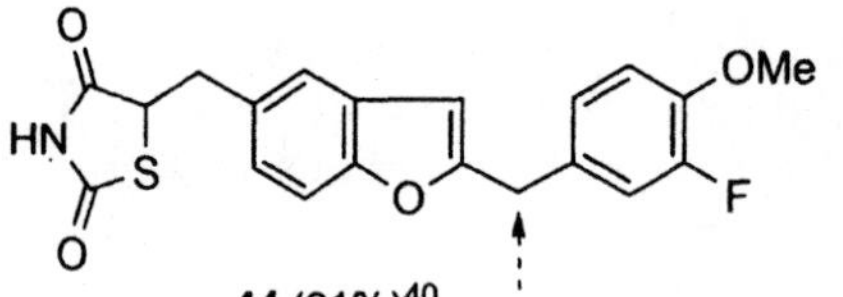

O
HN
S
O
O
OMe
F
44 (91%)[40]

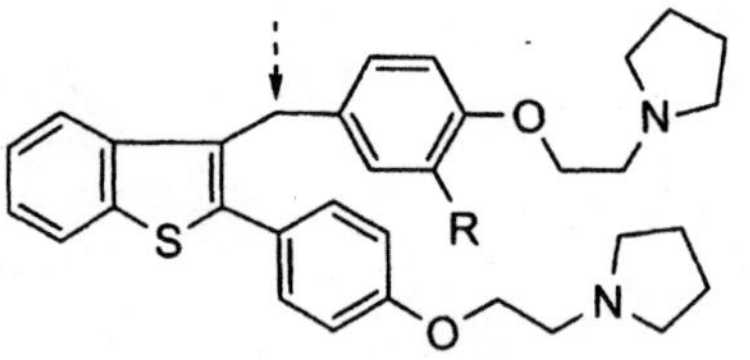

S
O
N
R
O
N
45 (R = OH, NO2; 10-24%)[41]

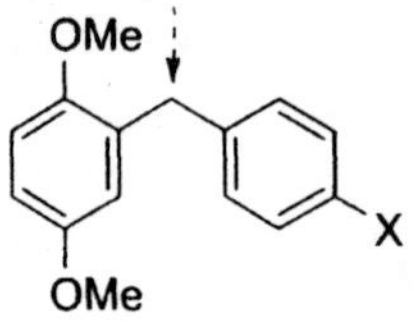

OMe
OMe
X
46 (X = Cl, Br; 61-63%)[42]

Fewer examples exist of successful monobenzylic ketone reduction but several are known. Indeed, the application of $NaBH_4$/TFA deoxygenation of keto porphyrins, and of the corresponding alcohols, has been very successful (e.g., **47→48**).[45] Similarly, pyrrole **49** is cleanly reduced to **50**,[46] and both carbonyl groups in **51** are deoxygenated leading to **52** with the more reactive $NaBH_3OCOCF_3$ to reduce also the amide functionality.[47] Tricyclic thiepins and oxepins **53** are cleanly reduced to **54**, as are the corresponding alcohols.[48]

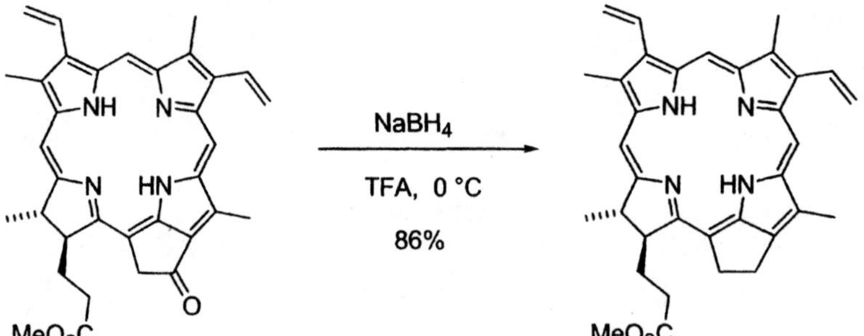

NH N
N HN
O
MeO2C
47

$NaBH_4$
TFA, 0 °C
86%

NH N
N HN
MeO2C
48

(ref. 45)

49 → **50**: $NaBH_4$, TFA, 0 °C to rt, 65-99% (ref. 46)

51 → **52**: $NaBH_3OCOCF_3$, THF, reflux, 60% (ref. 47)

53 (X = S, O) → **54**: $NaBH_4$, TFA, –10 °C, 73-74% (ref. 48)

Acyl and diacyl ferrocenes are reduced to the corresponding alkyl ferrocenes using $NaBH_4$/TFA,[49] although the contrary result has also been reported.[50] An interesting deoxygenation is the conversion of hydroxyphenanthrene **55** to **56**, presumably involving TFA-promoted tautomerization to a ketone intermediate.[51] A similar low-yield deoxygenation of 1- and 2-naphthol was observed by us nearly 30 years ago.[52]

55 → **56**: $NaBH_4$, TFA, 0 °C, 20% (ref. 51)

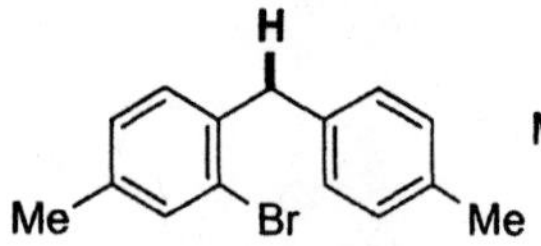

57 (94%)[53] 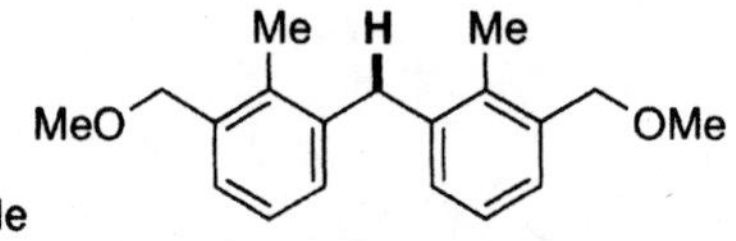 **58** (94%)[54]  **59**[55]

R_1	R_2	R_3	R_4	% **59**
H	OMe	OMe	Me	93
OMe	OMe	H	H	95
H	OMe	OMe	Me	96
H	H	H	Me	88

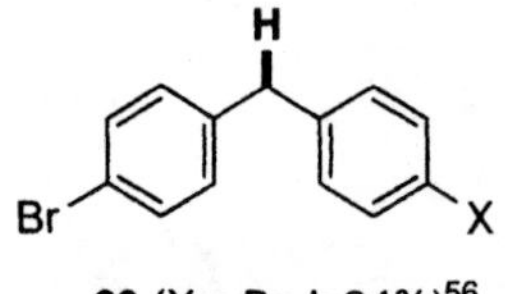

60 (X = Br, I; 94%)[56] 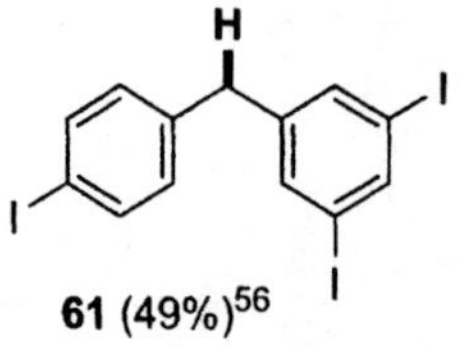 **61** (49%)[56]

Numerous applications of the $NaBH_4$/TFA deoxygenation of diarylmethanols and related benzylic alcohols have been reported, especially those involving diarylmethanols. In addition to the examples shown earlier,[2] diarylmethanes **57**–**64** have been prepared from the corresponding alcohols with $NaBH_4$/TFA. The site of reduction is shown as a bold **H**.

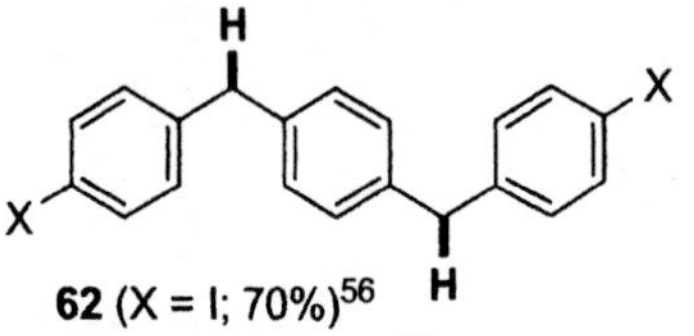

62 (X = I; 70%)[56]
63 (X = Br; 100%)[57]

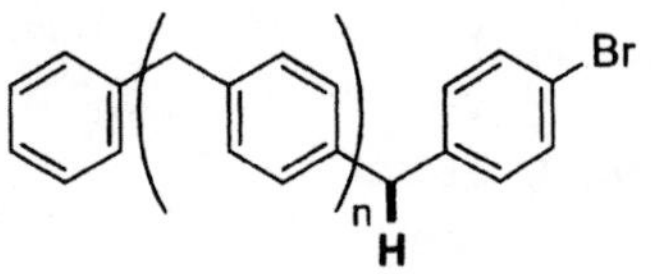

64 (n = 1-4; 89-95%)[57]

As we have seen, many functional groups are tolerant of $NaBH_4$/TFA (e.g., **65**–**68**) and naphthyl groups also serve as suitable substrates (e.g., **69**–**71**).

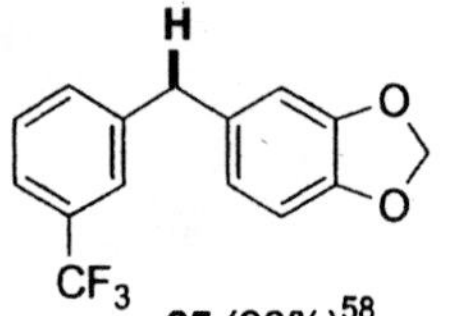

65 (90%)[58] 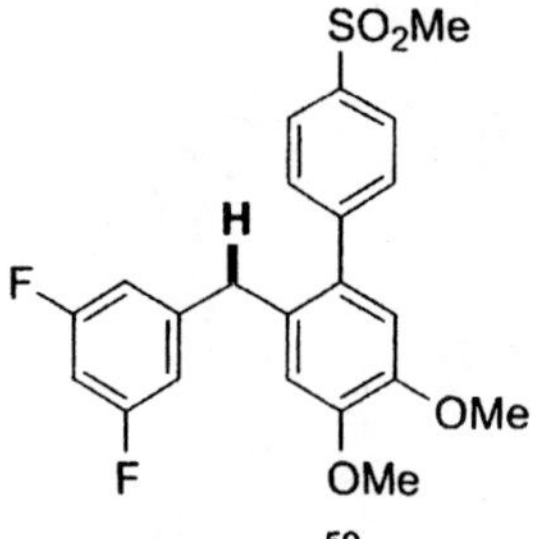 **66** (93%)[59] 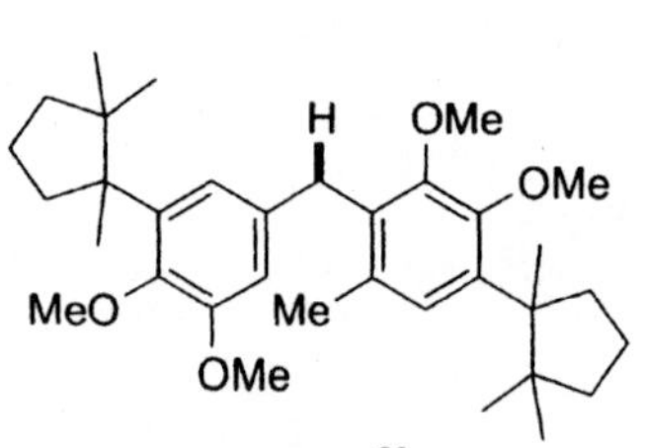 **67** (83%)[60]

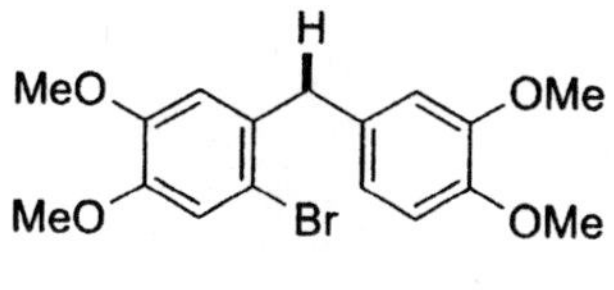

68 (97%)[61]

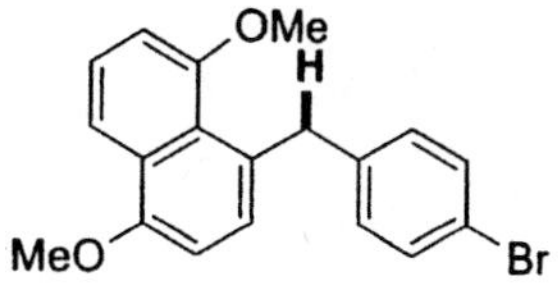

69 (93%)[62]

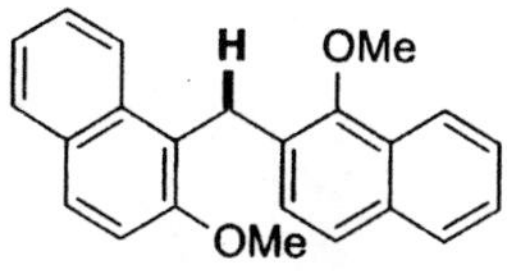

70 (71%)[63]

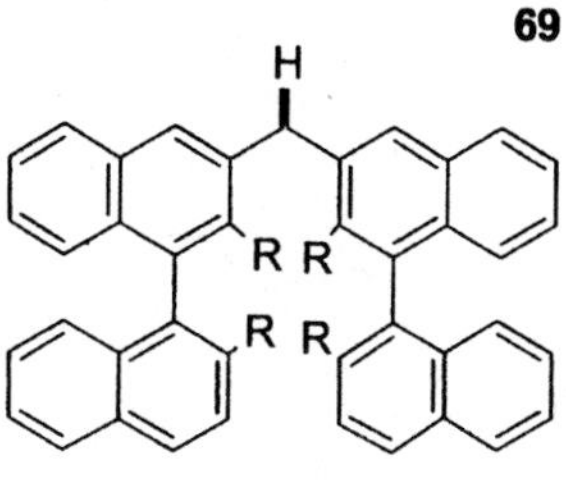

71 (R = OH; 50%)[64]

In addition to intermolecular carbocation captures that were shown earlier (Scheme 1),[2] one must also be wary of intramolecular ambushes by, for example, adjacent phenyl groups (e.g., **72**→**73**; **74**→**75**).

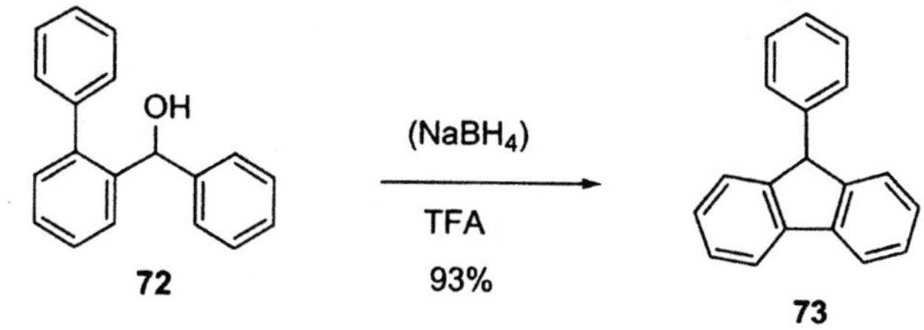

72 —($NaBH_4$), TFA, 93%→ **73** (ref. 2)

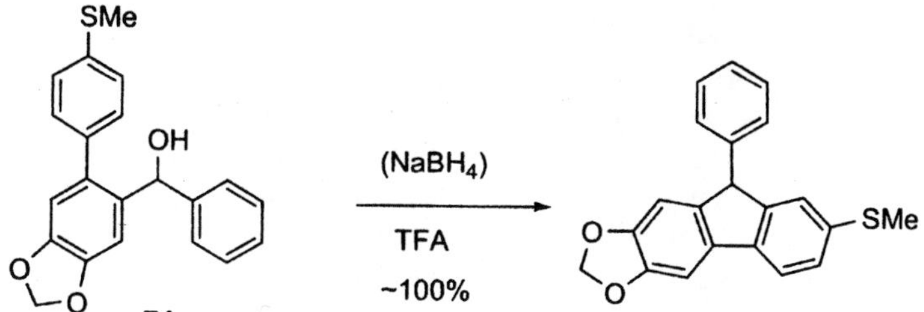

74 —($NaBH_4$), TFA, ~100%→ **75** (ref. 59)

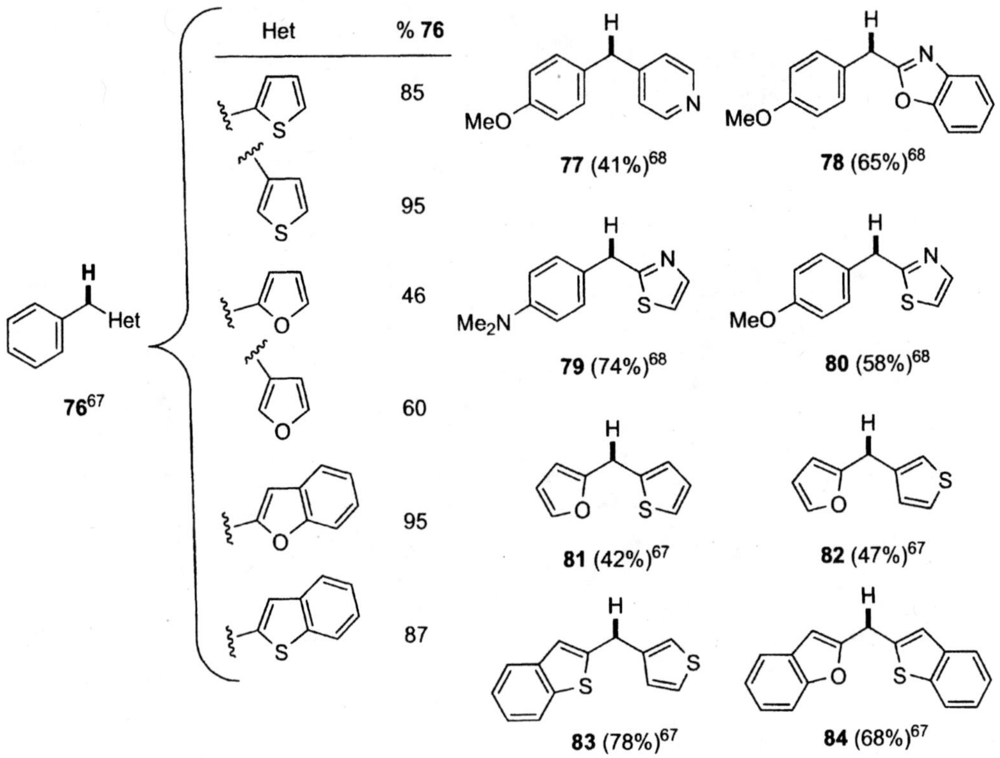

Het	% 76
	85
	95
	46
	60
	95
	87

76[67]

77 (41%)[68] **78** (65%)[68]

79 (74%)[68] **80** (58%)[68]

81 (42%)[67] **82** (47%)[67]

83 (78%)[67] **84** (68%)[67]

The $NaBH_4$/TFA deoxygenation reaction has been extended to more complex systems such as cavitands wherein four doubly benzylic hydroxyls are reduced (80% yield),[65] to cyclophanes,[66] and to numerous substrates having one or two heterocyclic rings. Nutaitis and co-workers have explored these heterocyclic carbinols in great depth.[67,68] Selected examples are shown (**76–84**), and it is observed that π-excessive heterocycles tolerate the $NaBH_4$/TFA conditions and are reduced nicely, but π-deficient heterocycles with a basic nitrogen atom are difficult substrates to reduce. For example, product **76** is not obtained where the heterocycle is 3-pyridyl, 2-oxazolyl, 2-benzothiazolyl, 2- and 4-(1-methylimidazolyl), and 2-(1-methylbenzimidazolyl) and starting material is recovered unchanged.[67] With a more activating phenyl group reduction can be achieved with these π-deficient heterocycles (e.g., **77–80**).[68] Nutaitis has extended this methodology to dihetarylmethanols resulting in reduction products **81–84**.[67]

Other studies of $NaBH_4$/TFA reduction of heterocyclic methanols have furnished **85-88** as well as a key intermediate en route to quinone-linked porphyrin.[73] Despite these successes a few failures have been reported involving heterocyclic methanols (imidazole[74] and thiophenes[75]), in addition to the inherent limitations discovered by Nutaitis.[67,68]

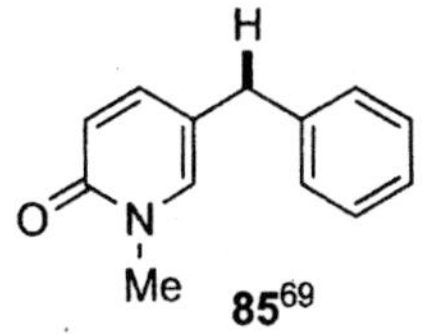

85[69] 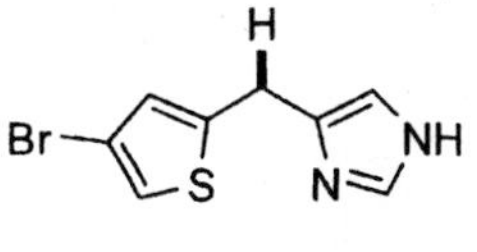 86 (30%)[70] 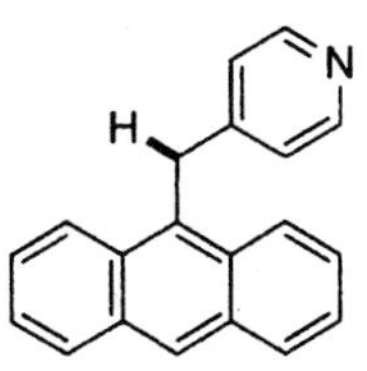 88 (~100%)[72]

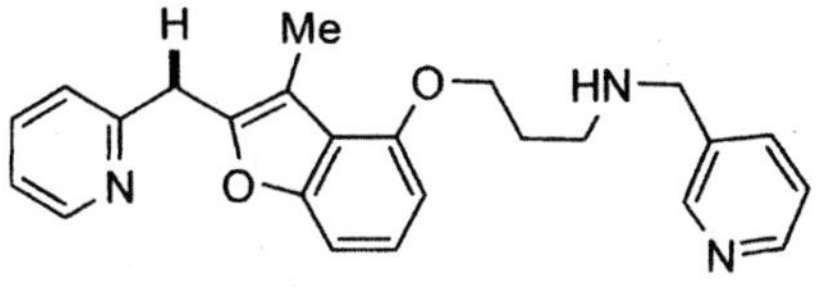

87 (41%)[71]

The discovery that triphenylmethanol is reduced essentially quantitatively to triphenylmethane with $NaBH_4$/TFA[2] has led to several applications of this facile reduction, including the synthesis of deuterated and C-13 labeled triphenylmethanes.[76] Other product examples are **89–93**. Attempts to secure **91** on a large scale (25 g) led to alkene formation,[79] which is often a side reaction of appropriate substrates.[2] For example, a conjugated alkene accompanies **92**.[80] Alcohol **94** failed to react with $NaBH_4$/TFA, perhaps a consequence of the proximate nitrogen and electron-deficient benzene rings prohibiting carbocation formation.[82]

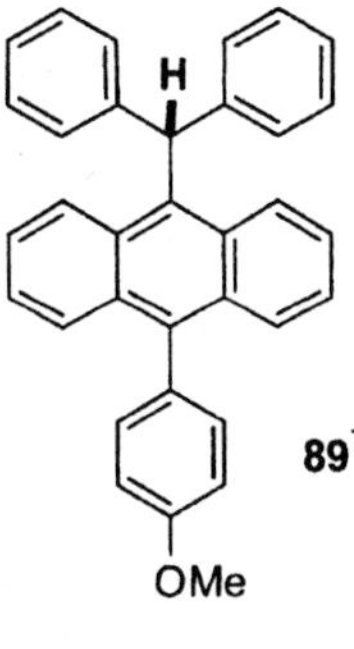

89[77] 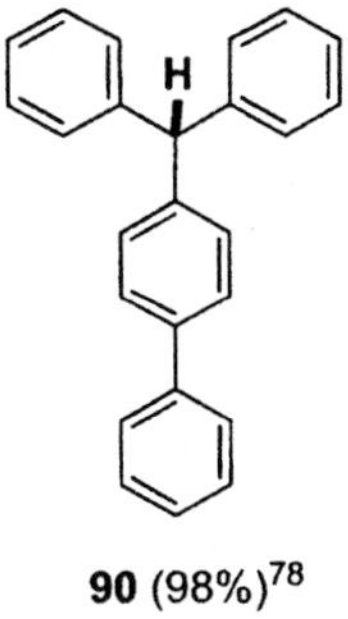 90 (98%)[78] 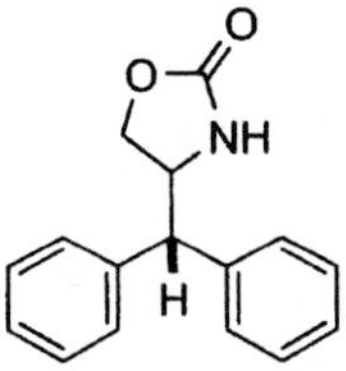 91 (72%)[79] 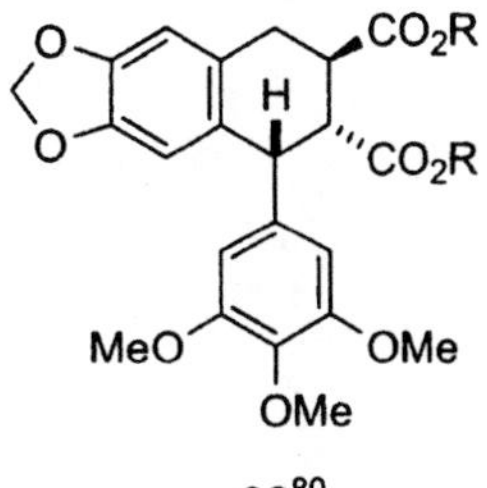 92[80]

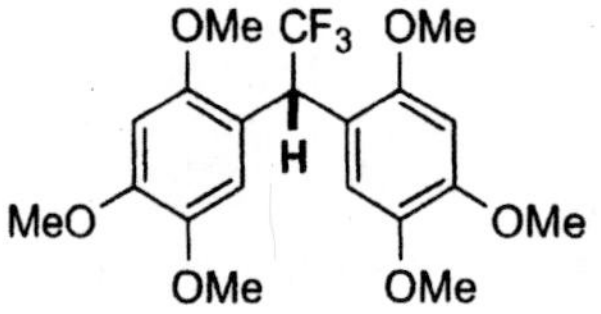

93 (78%)[81]

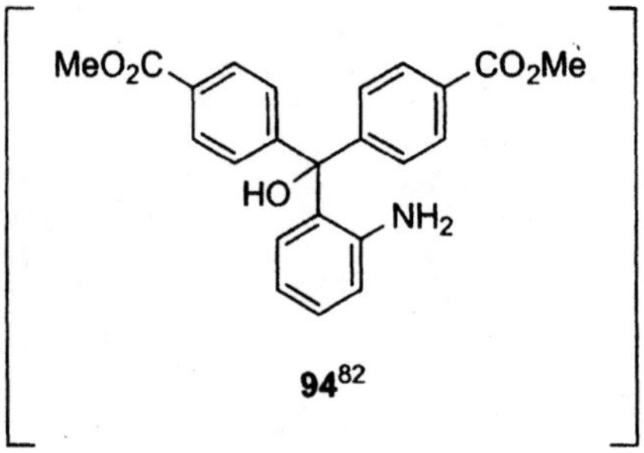

94[82]

Several heterocycle-containing triarylmethanols have been deoxygenated with $NaBH_4$/TFA and some of the resulting products are shown (**95**–**100**). Other combinations of heterocyclic rings have been studied in this reduction.[83,84] The preparation of **98** was performed on a 100 kg scale,[85] but the tris-heterocycle **99** failed to afford the expected hydrocarbon.[86]

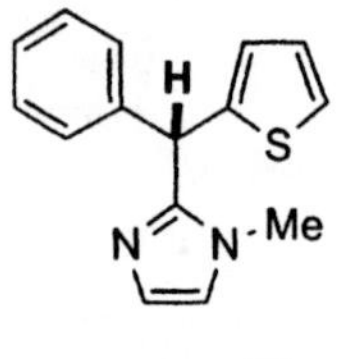

95 (73%)[83]

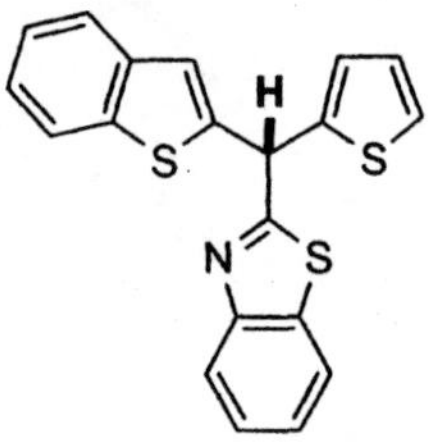

96 (39%)[83]

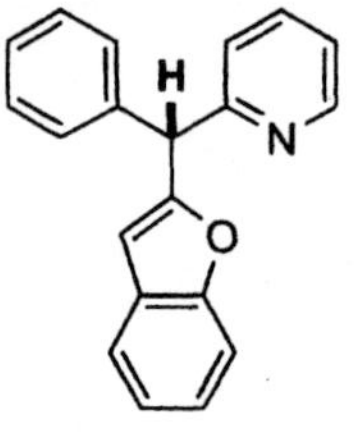

97 (70%)[84]

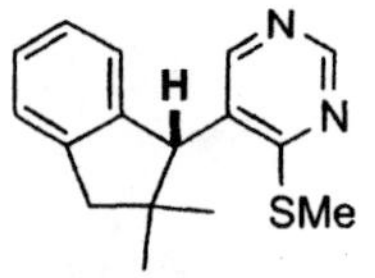

98 (94%)[85]

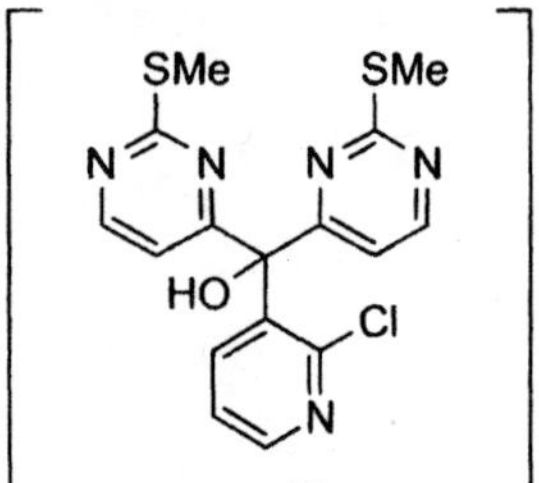

99[86]

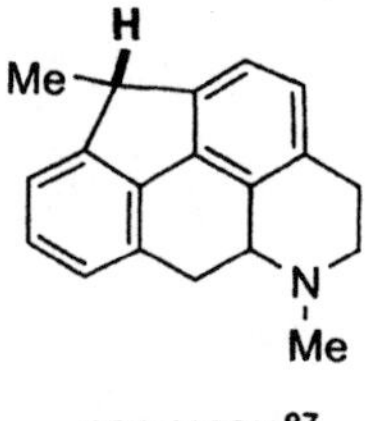

100 (46%)[87]

As introduced in Section 2.2.4, monobenzylic alcohols are inherently poor substrates for reduction using $NaBH_4$/TFA. However, notable exceptions have been discovered and this tactic has a synthetic niche as illustrated with reduction products **101–105**. Diastereo-selective introduction of deuterium was reported in the studies involving tetrahydroisoquinoline **102**.[89] Note that alkene formation from the carbocation intermediates leading to **104** and **105** would be prohibitive (Bredt's rule and ring strain). The $NaBH_4$/TFA reduction of ferrocenyl alcohols has also been described,[92] as has the reduction of monobenzylic alcohols that are also further activated by an adjacent amide nitrogen (hydroxylactams).[93]

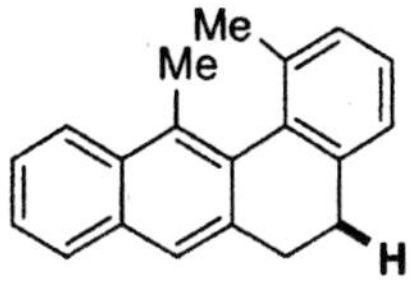

101 [88] 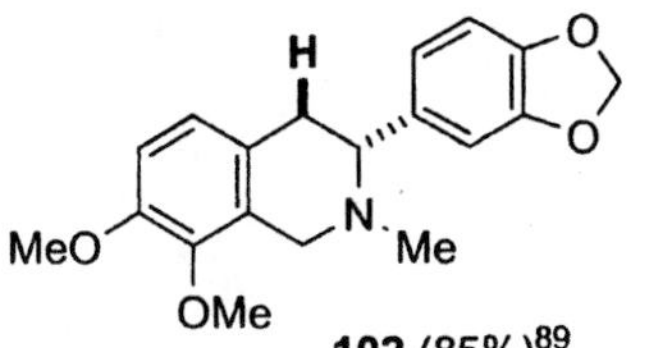 **102** (85%)[89] 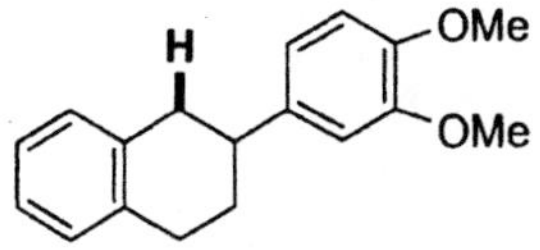 **103** (80%)[89]

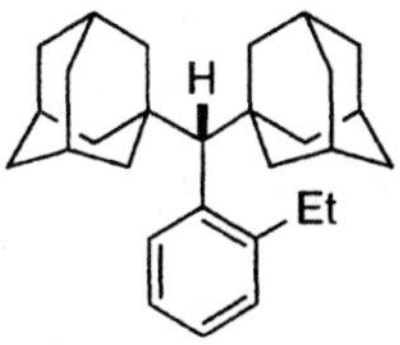

104 (98%)[90] 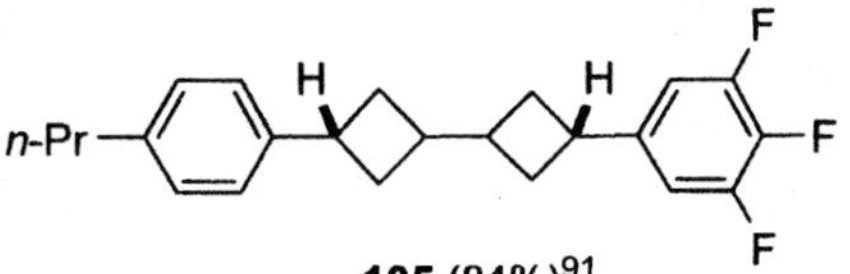 **105** (84%)[91]

A few examples are known wherein a single heterocyclic ring provides the necessary carbocation stabilization for ionization and reduction to occur. In addition to the deoxygenation of hydroxy porphyrins,[94] two indole examples are known wherein a benzylic hydroxyl group is removed (**106**, **107**).

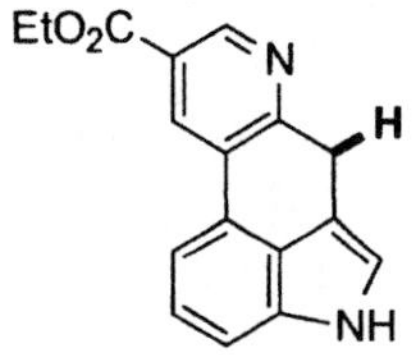

106 (41%)[95] 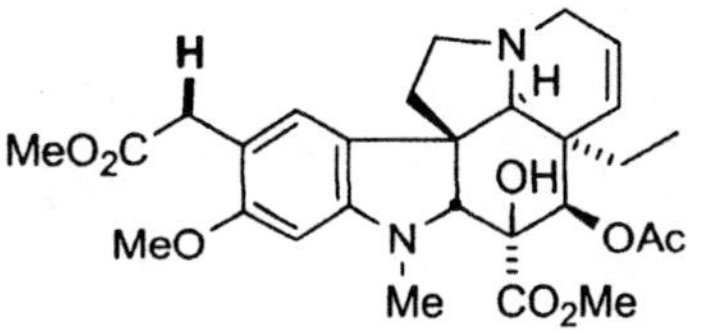 **107** (72%)[96]

2.2.6 *Experimental*

Due to the liberation of hydrogen gas from **all** reactions involving $NaBH_4$ and carboxylic acids, proper precautions must be observed. The use of $NaBH_4$ pellets and inert gas conditions is strongly recommended. The addition of powdered $NaBH_4$ to TFA poses the greatest danger of ignition.

Diphenylmethane[2]

To magnetically stirred TFA (50 mL) at 15–20 °C under nitrogen was added $NaBH_4$ pellets (1.76 g, 46 mmol) over 30 min. To this mixture (pellets not completely dissolved) at 15–20 °C was added diphenylmethanol (2.00 g, 10.9 mmol) in portions over 15 min. The mixture was stirred under nitrogen at 20 °C for 12 h (to complete dissolution of the $NaBH_4$). The mixture was diluted with water, basified with NaOH (pellets), and extracted with ether. The ether layer was washed (H_2O then brine), dried (Na_2SO_4), and concentrated in vacuo to give 1.77 g of a light yellow oil. Distillation gave pure diphenylmethane as a colorless oil (1.70 g, 93%), bp 63–64 °C/0.4 Torr.

4-(Diphenylmethyl)-1,1′-biphenyl (90)[78]

A mixture of α,α-diphenyl-1,1′-biphenyl-4-methanol (1.34 g, 4 mmol) and $NaBH_4$ powder (1.52 g, 40 mmol) was added in portions over 5 min to TFA (50 mL) at 0 °C under a stream of argon with vigorous stirring. Additional TFA (6 mL) was added, and 40 min later the solution was evaporated. To the residue was added H_2O (100 mL) and the solution was extracted with hexane (300 mL). The organic layer was washed with saturated aqueous $NaHCO_3$, dried (Na_2SO_4), and evaporated to give the title compound (1.25 g, 98%) as a pale yellow solid, mp 112–113 °C (hexane); ^1H NMR ($CDCl_3$) d 7.58 (d, 2H, 8.4 Hz), 7.51 (d, 2H, 8.4 Hz), 7.43 (t, 2H, 7.7 Hz), 7.3–7.4 (m, 5H), 7.1–7.3 (m, 7H), 5.59 (s, 1H). Anal. Calcd for $C_{25}H_{20}$: C, 93.71; H, 6.29. Found: C, 93.76; H, 6.21.

1-Bromo-5-methyl-2-(4-methylbenzyl)benzene (57)[53]

Crushed $NaBH_4$ pellets (2.10 g, 55.4 mmol) were added to TFA (49 mL) cooled in an ice bath under nitrogen. A solution of (2-bromo-4-methylphenyl)-(4-methylphenyl)methanol (2.69 g, 9.23 mmol) in CH_2Cl_2 (25 mL) was then added dropwise. After the addition, the mixture was stirred at rt for 14 h until all $NaBH_4$ dissolved. The solution was diluted with H_2O and made basic with NaOH pellets. The product was extracted into CH_2Cl_2, the solution was washed, dried, and concentrated. The residue was chromatographed to give the title compound as a colorless liquid (2.40 g, 94%); ^1H NMR ($CDCl_3$) d 7.21 (s, 1H), 7.07–6.99 (s, 6H), 4.02 (s, 2H), 2.30 (s, 3H), 2.28 (s, 3H). Anal. Calcd for $C_{15}H_{15}{}^{79}Br$: 274.0357. Found: 274.0369.

3-Ethyl-2-chloro-1-(phenylsulfonyl)indole[14a]

To magnetically stirred TFA (25 mL) at 0 °C under N_2 was added $NaBH_4$ pellets (18 mmol, 3 pellets) over 30 min. To this mixture was added dropwise over 30 min a solution of 3-acetyl-2-chloro-1-(phenylsulfonyl)indole (0.50 g, 1.50 mmol) in CH_2Cl_2 (25 mL). The mixture was stirred overnight at 25 °C, diluted with H_2O (75 mL), and made basic by the addition of NaOH pellets at 0 °C. The layers were separated and the aqueous layer was extracted with additional CH_2Cl_2. The organic layer was washed (H_2O, brine), dried (Na_2SO_4), and concentrated in vacuo to give after flash chromatography (hexane:CH_2Cl_2, 60:40) the title compound (0.35 g, 73%), mp 82–83 °C; ^1H NMR ($CDCl_3$) d 8.4–7.2 (m, 7H), 2.7 (q, 2H, 7

Hz), 1.1 (t, 3H, 7 Hz). Anal. Calcd for $C_{16}H_{14}NO_2SCl$: C, 60.09; H, 4.41; N, 4.38. Found: C, 60.15; H, 4.43; N, 4.33.

2.2.7 *References*

1. Gribble, G. W.; Kelly, W. J.; Emery, S. E. *Synthesis* **1978**, 763.
2. Gribble, G. W.; Leese, R. M.; Evans, B. E. *Synthesis* **1977**, 172.
3. [R] Gribble, G. W. *East. Org. Chem. Bull.* **1979**, *51* (1), 1.
4. [R] Gribble, G. W.; Nutaitis, C. F. *Org. Prep. Proc. Int.* **1985**, *17*, 317.
5. [R] Nutaitis, C. F. *J. Chem. Ed.* **1989**, *66*, 673.
6. [R] Gribble, G. W. *Encycloped. Reagents Org. Syn.* **1995**, *7*, 4649, 4655, 4658.
7. [R] Gribble, G. W. *Chemtech* **1996**, *26* (12), 26.
8. [R] Gribble, G. W. *ACS Symposium Series* **1996**, *641*, 167.
9. [R] Gribble, G. W. *Chem. Soc. Rev.* **1998**, *27*, 395.
10. [R] Gribble, G. W. *Org. Proc. Res. Dev.* **2006**, *10*, 1062.
11. Liddle, J. *Chem. Brit.* **2000**, 19.
12. Gribble, G. W.; Lord, P. D.; Skotnicki, J.; Dietz, S. E.; Eaton, J. T.; Johnson, J. L. *J. Am. Chem. Soc.* **1974**, *96*, 7812.
13. Mpango, G. B.; Snieckus, V. *Tetrahedron Lett.* **1980**, *21*, 4827.
14. (a) Ketcha, D. M.; Lieurance, B. A.; Homan, D. F. J.; Gribble, G. W. *J. Org. Chem.* **1989**, *54*, 4350. (b) Ketcha, D. M.; Gribble, G. W. *J. Org. Chem.* **1985**, *50*, 5451.
15. (a) Song, T.; Hansen, H.-J. *Helv. Chim. Acta* **1999**, *82*, 309. (b) Song, J.; Hansen, H.-J. *Helv. Chim. Acta* **1999**, *82*, 1690. (c) Singh, G.; Linden, A.; Abou-Hadeed, K.; Hansen, H.-J. *Helv. Chim. Acta* **2002**, *85*, 27.
16. Nutaitis, C. F.; Bernardo, J. E. *Synth. Commun.* **1990**, *20*, 487.
17. Varle, D. L. *Tetrahedron Lett.* **1990**, *31*, 7583.
18. Sharifian, A.; Parang, K.; Zorrieh-Amirian, H.; Nazarinia, M.; Shafiee, A. *J. Heterocycl. Chem.* **1994**, *31*, 1421.
19. Bodwell, G. J.; Li, J.; Miller, D. O. *Tetrahedron* **1999**, *55*, 12939.
20. Saá, J. M.; Llobera, A. *Tetrahedron Lett.* **1987**, *28*, 5045.
21. Saleh, M. A.; Compernolle, F.; Toppet, S.; Hoornaert, G. J. *Tetrahedron* **1994**, *50*, 1811.
22. Itaya, T.; Watanabe, N.; Iida, T.; Kanai, T.; Mizutani, A. *Tetrahedron* **1995**, *51*, 6419.
23. Piggott, M. J.; Wege, D. *Tetrahedron* **2006**, *62*, 3550.
24. Gribble, G. W.; Johnson, J. L.; Saulnier, M. G. *Heterocycles* **1981**, *16*, 2109.
25. Gribble, G. W.; Kelly, W. J.; Sibi, M. P. *Synthesis* **1982**, 143.
26. Fujisaka, T.; Nojima, M.; Kusabayashi, S. *J. Org. Chem.* **1985**, *50*, 275.
27. Ogata, F.; Takagi, M.; Nojima, M.; Kusabayashi, S. *J. Am. Chem. Soc.* **1981**, *103*, 1145.
28. (a) Enders, D.; Peters, R.; Lochtman, R.; Raabe, G. *Angew. Chem. Int. Ed.* **1999**, *38*, 2421. (b) Enders, D.; Peters, R.; Lochtman, R.; Raabe, G.; Runsink, J.; Bats, J. W. *Eur. J. Org. Chem.* **2000**, 3399.
29. Poirot, M.; De Medina, P.; Delarue, F.; Perie, J.-J.; Klaebe, A.; Faye, J.-C. *Bioorg. Med. Chem.* **2000**, *8*, 2007.
30. Wityak, J.; Hobbs, F. W.; Gardner, D. S.; Santella, J. B. III, Petraitis, J. J.; Sun, J.-H.; Favata, M. F.; Daulerio, A. J.; Horiuchi, K. Y.; Copeland, R. A.; Scherle, P. A.; Jaffe, B. D.; Trzaskos, J. M.; Magolda, R. L.; Trainor, G. L.; Duncia, J. V. *Bioorg. Med. Chem. Lett.* **2004**, *14*, 1483.
31. Liou, J.-P.; Chang, J.-Y.; Chang, C.-W.; Chang, C.-Y.; Mahindroo, N.; Kuo, F.-M.; Hsieh, H.-P. *J. Med. Chem.* **2004**, *47*, 2897.
32. Plattner, J. J.; Martin, Y. C.; Smital, J. R.; Lee, C.-M.; Fung, A. K. L.; Horrom, B. W.; Crowley, S. R.; Pernet, A. G.; Bunnell, P. R.; Kim, K. H. *J. Med. Chem.* **1985**, *28*, 79.
33. Lee, C.-M.; Parks, J. A.; Bunnell, P. R.; Plattner, J. J.; Field, M. J.; Giebisch, G. H. *J. Med. Chem.* **1985**, *28*, 589.
34. Utley, J. H. P.; Rozenberg, G. G. *Tetrahedron* **2002**, *58*, 5251.
35. Ottosen, E. R.; Sørensen, M. D.; Björkling, F.; Skak-Nielsen, T.; Fjording, M. S.; Aaes, H.; Binderup, L. *J. Med. Chem.* **2003**, *46*, 5651.
36. Sattelkau, T.; Qandil, A. M.; Nichols, D. E. *Synthesis* **2001**, 262. For the reduction of isoquinoline, see Gribble, G. W.; Heald, P. W. *Synthesis* **1975**, 650.
37. Daich, A.; Decroix, B. *J. Heterocycl. Chem.* **1992**, *29*, 1789.
38. Zhang, X.; Urbanski, M.; Patel, M.; Cox, G. G.; Zeck, R. E.; Bian, H.; Conway, B. R.; Beavers, M. P.; Rybczynski, P. J.; Demarest, K. T. *Bioorg. Med. Chem. Lett.* **2006**, *16*, 1696.
39. Miki, Y.; Tsuzaki, Y.; Matsukida, H. *Heterocycles* **2002**, *57*, 1645.

40. Huff, B. E.; Leffelman, C. L.; LeTourneau, M. E.; Sullivan, K. A.; Ward, J. A.; Stille, J. R. *Heterocycles* **1997**, *45*, 1363.
41. Sall, D. J.; Briggs, S. L.; Chirgadze, N. Y.; Clawson, D. K.; Gifford-Moore, D. S.; Klimkowski, V. J.; McCowan, J. R.; Smith, G. F.; Wikel, J. H. *Bioorg. Med. Chem. Lett.* **1998**, *8*, 2527.
42. Waterlot, C.; Hasiak, B.; Couturier, D.; Rigo, B. *Tetrahedron* **2001**, *57*, 4889.
43. Gribble, G. W.; Nutaitis, C. F.; Leese, R. M. *Heterocycles* **1984**, *22*, 379.
44. Wang, J.; Gutsche, C. D. *J. Org. Chem.* **2002**, *67*, 4423.
45. (a) Smith, N. W.; Smith, K. M. *Energy Fuels* **1990**, *4*, 675. (b) Abraham, R. J.; Rowan, A. E.; Smith, N. W.; Smith, K. M. *J. Chem. Soc., Perkin Trans. 2* **1993**, 1047. (c) Jeandon, C.; Ocampo, R.; Callot, H. J. *Tetrahedron Lett.* **1993**, *34*, 1791. (d) Jeandon, C.; Ocampo, R.; Callot, H. J. *Tetrahedron* **1997**, *53*, 16107. (e) Mettath, S.; Shibata, M.; Alderfer, J. L.; Senge, M. O.; Smith, K. M.; Rein, R.; Dougherty, T. J.; Pandey, R. K. *J. Org. Chem.* **1998**, *63*, 1646.
46. Alazard, J.-P.; Millet-Paillusson, C.; Boyé, O.; Guénard, D.; Chiaroni, A.; Riche, C.; Thal, C. *Bioorg. Med. Chem. Lett.* **1991**, *1*, 725.
47. Gribble, G. W.; Pelcman, B. *J. Org. Chem.* **1992**, *57*, 3636.
48. Kurokawa, M.; Uno, H.; Itogawa, A.; Sato, F.; Naruto, S.; Matsumoto, J. *J. Heterocycl. Chem.* **1991**, *28*, 1891.
49. Bhattacharyya, S. *J. Chem. Soc., Perkin Trans. 1* **1996**, 1381.
50. Bhatt, J.; Fung, B. M.; Nicholas, K. M. *Liq. Crystals* **1992**, *12*, 263.
51. Banerjee, A. K.; Acevedo, J. C.; González, R.; Rojas, A. *Tetrahedron* **1991**, *47*, 2081.
52. Gribble, G. W., unpublished results.
53. Lai, Y.-H.; Peck, T.-G. *Aust. J. Chem.* **1992**, *45*, 2067.
54. Mitchell, R. H.; Lai, Y.-H. *J. Org. Chem.* **1984**, *49*, 2534.
55. Almeida, W. P.; Costa, P. R. R. *Synth. Commun.* **1996**, *26*, 4507.
56. Tour, J. M.; Rawlett, A. M.; Kozaki, M.; Yao, Y.; Jagessar, R. C.; Dirk, S. M.; Price, D. W.; Reed, M. A.; Zhou, C.-W.; Chen, J.; Wang, W.; Campbell, I. *Chem. Eur. J.* **2001**, *7*, 5118.
57. Gribble, G. W.; Nutaitis, C. F. *Tetrahedron Lett.* **1985**, *26*, 6023.
58. Lages, A. S.; Silva, K. C. M.; Miranda, A. L. P.; Fraga, C. A. M.; Barreiro, E. J. *Bioorg. Med. Chem. Lett.* **1998**, *8*, 183.
59. Khanapure, S. P.; Garvey, D. S.; Young, D. V.; Ezawa, M.; Earl, R. A.; Gaston, R. D.; Fang, X.; Murty, M.; Martino, A.; Shumway, M.; Trocha, M.; Marek, P.; Tam, S. W.; Janero, D. R.; Letts, L. G. *J. Med. Chem.* **2003**, *46*, 5484.
60. Bringmann, G.; Pabst, T.; Henschel, P.; Michel, M. *Tetrahedron* **2001**, *57*, 1269.
61. Maddaford, S. P.; Charlton, J. L. *J. Org. Chem.* **1993**, *58*, 4132.
62. Corrie, J. E. T.; Papageorgiou, G. *J. Chem. Soc., Perkin Trans. 1*, **1996**, 1583.
63. Kasturi, T. R.; Sattigeri, J. A.; Pragnacharyulu, P. V. P.; Cameroon, T. S.; Pradeep, B. *Tetrahedron* **1995**, *51*, 3051.
64. Vogl, E. M.; Matsunaga, S.; Kanai, M.; Iida, T.; Shibasaki, M. *Tetrahedron Lett.* **1998**, *39*, 7917.
65. Anduini, A.; Giorgi, G.; Pochini, A.; Secchi, A.; Ugozzoli, F. *J. Org. Chem.* **2001**, *66*, 8302.
66. Mitchell, R. H.; Lai, Y.-H. *J. Org. Chem.* **1984**, *49*, 2541.
67. Nutaitis, C. F.; Patragnoni, R.; Goodkin, G.; Neighbour, B.; Obaza-Nutaitis, J. *Org. Prep. Proc. Int.* **1991**, *23*, 403.
68. Nutaitis, C. F.; Swartz, B. D. *Org. Prep. Proc. Int.* **2005**, *37*, 507.
69. Hartmann, R. W.; Reichert, M. *Arch. Pharm. Pharm. Med. Chem.* **2000**, *333*, 145.
70. Boyd, R. E.; Rasmussen, C. R.; Press, J. B.; Raffa, R. B.; Codd, E. E.; Connelly, C. D.; Li, Q. S.; Martinez, R. P.; Lewis, M. A.; Almond, H. R.; Reitz, A. B. *J. Med. Chem.* **2001**, *44*, 863.
71. Kawasaki, K. *et al., Bioorg. Med. Chem. Lett.* **2003**, *13*, 87.
72. MacQueen, D. B.; Eyler, J. R.; Schanze, K. S. *J. Am. Chem. Soc.* **1992**, *114*, 1897.
73. Osuka, A.; Zhang, R. P.; Maruyama, K.; Yamazaki, I.; Nishimura, Y. *Bull. Chem. Soc. Jpn.* **1992**, *65*, 2807.
74. Wu, X.; Mahalingam, A. K.; Alterman, M. *Tetrahedron Lett.* **2005**, *46*, 1501.
75. (a) Parakka, J. P.; Cava, M. P. *Tetrahedron* **1995**, *51*, 2229. (b) Wex, B.; Kaafarani, B. R.; Neckers, D. C. *J. Org. Chem.* **2004**, *69*, 2197.
76. Currie, G. J.; Bowie, J. H.; Massy-Westropp, R. A.; Adams, G. W. *J. Chem. Soc., Perkin Trans. 2* **1988**, 403.
77. Takagi, M.; Nojima, M.; Kusabayashi, S. *J. Chem. Soc., Perkin Trans. 1* **1979**, 2941.
78. Saito, S.; Ohwada, T.; Shudo, K. *J. Org. Chem.* **1996**, *61*, 8089.
79. Sibi, M. P.; Deshpande, P. K.; La Loggia, A. J.; Christensen, J. W. *Tetrahedron Lett.* **1995**, *36*, 8961.

80. Bogucki, D. E.; Charlton, J. L. *J. Org. Chem.* **1995**, *60*, 588. For another example, see Manning, L. E.; Peters, K. S. *J. Am. Chem. Soc.* **1985**, *107*, 6452.
81. Broidy, J. M.; Marshall, G. L. *Synthesis* **1982**, 939.
82. Carde, R. N.; Hayes, P. C.; Jones, G.; Cliff, C. J. *J. Chem. Soc., Perkin Trans. 1* **1981**, 1132.
83. Nutaitis, C. F.; Obaza-Nutaitis, J. *Org. Prep. Proc. Int.* **1997**, *29*, 315.
84. Nutaitis, C. F.; Greshock, T. J.; Houghton, S. R.; Moran, L. N.; Walter, M. A. *Org. Prep. Proc. Int.* **2002**, *34*, 332.
85. Dietsche, T. J.; Gorman, D. B.; Orvik, J. A.; Roth, G. A.; Shiang, W. R. *Org. Proc. Res. Dev.* **2000**, *4*, 275.
86. Anderson, R. J.; Hill, J. B.; Morris, J. C. *J. Org. Chem.* **2005**, *70*, 6204.
87. Linnanen, T.; Brisander, M.; Unelius, L.; Sundholm, G.; Hacksell, U.; Johansson, A. M. *J. Med. Chem.* **2000**, *43*, 1339.
88. Levy, L. A.; Kumar, S. *Tetrahedron Lett.* **1983**, *24*, 1221.
89. Vicario, J. L.; Badía, D.; Carrillo, L.; Anakabe, E. *Tetrahedron Asym.* **2003**, *14*, 347.
90. Lomas, J. S.; Vaissermann, J. *J. Chem. Soc., Perkin Trans. 2* **1997**, 2589.
91. (a) Dehmlow, E. V.; Kinnius, J.; Buchholz, M.; Hannemann, D. *J. Prakt. Chem.* **2000**, *342*, 409. (b) Kinnius, J.; Dehmlow, E. V. *J. Prakt. Chem.* **2000**, *342*, 421.
92. (a) Vicard, D.; Gruselle, M.; Jaouen, G.; Nefedova, M. N.; Mamedyarova, I. A.; Sokolov, V. I.; Vaissermann, J. *J. Organomet. Chem.* **1994**, *484*, 1. (b) Malezieux, B.; Gruselle, M.; Troitskaya, L. L.; Sokolov, V. I.; Vaissermann, J. *Organometallics* **1994**, *13*, 2979. (c) Gruselle, M.; Malezieux, B.; Sokolov, V. I.; Troitskaya, L. L. *Inorg. Chim. Acta* **1994**, *222*, 51.
93. (a) Osante, I.; Collado, M. I.; Lete, E.; Sotomayor, N. *Synlett* **2000**, 101. (b) Osante, I.; Collado, M. I.; Lete, E.; Sotomayor, N. *Eur. J. Org. Chem.* **2001**, 1267.
94. (a) Bauder, C.; Ocampo, R.; Callot, H. J. *Synlett* **1990**, 335. (b) Bauder, C.; Ocampo, R.; Callot, H. J. *Tetrahedron* **1992**, *48*, 5135.
95. Hendrickson, J. B.; Wang, J. *Org. Lett.* **2004**, *6*, 3.
96. Fahy, J.; du Boullay, V. T.; Bigg, D. C. H. *Bioorg. Med. Chem. Lett.* **2002**, *12*, 505.

Gordon W. Gribble

2.3 Luche Reduction

2.3.1 Description

The Luche reduction involves a selective 1,2-reduction of α,β-unsaturated aldehydes and ketones using sodium borohydride in combination with cerium(III) chloride, resulting in the formation of allylic alcohols.[1]

$$\mathbf{1} \xrightarrow[\text{MeOH}]{NaBH_4,\ CeCl_3 \cdot 7H_2O} \mathbf{2}$$

2.3.2 Historical Perspective

As of the late 1970s, a simple, reliable, and efficient method for the selective conversion of α,β-unsaturated ketones to allylic alcohols was notably absent from the literature.[2] Prior to Luche's discovery, previous attempts for selective reduction suffered from limitations such as contamination by the product of 1,4-addition, slow reactivity, expensive reagents, and often the requirement of inert reaction conditions. During the same time period, the utility of lanthanide reagents in synthetic organic chemistry was just being discovered. Previously, lanthanides had been used as shift reagents in NMR spectroscopy, and as catalysts in petrochemical reactions, epoxide rearrangements, optical resolution, and reactions of secondary amines with acetonitrile. This emergence prompted Luche to examine the influence of lanthanides on the sodium borohydride reduction of α,β-unsaturated ketones.

Of all the reagents tested during this time ($CeCl_3$, $SmCl_3$ and $EuCl_3$), Luche's method, using $CeCl_3$ in combination with $NaBH_4$, proved to be the most effective way to exclusively achieve the 1,2-addition of hydride to an α,β-unsaturated ketone. Several advantages of this reaction were noted in the preliminary communication of this reaction. Primarily, the conditions of the reaction do not affect carboxylic acids, esters, amides, acyl halides, nitriles, or nitro groups. Also, the reaction can be done at room temperature and takes little more than five minutes. Steric hindrance has no effect on the rate of the reduction and therefore allows the preferential 1,2-addition on most molecules. Additionally, the presence of water has little effect on the selectivity of the reaction, thus allowing the use of the lanthanides in their commercially available hydrate form. Most notable among the results reported by Luche is the reaction of 2-cyclopentenone (**1**). In the absence of $CeCl_3$, the reaction proceeds to give exclusively the completely saturated pentanone. However, in the presence of $CeCl_3$, the selectivity is reversed, as demonstrated by the production of a 97:3 ratio of 2-cyclopentenol

(**2**) to cyclopentanone. Recently, $LiBH_4$, a more reactive reducing agent, has been used in combination with $CeCl_3$, resulting in better stereoselectivity in an isolated case.[3]

2.3.3 Mechanism

The mechanism of the Luche reduction is complicated by a lack of understanding of the exact nature of the agent involved in the reduction process. However, some elegant studies by Luche have resulted in the following mechanistic interpretation, which seems consistent with the regio- and stereoselectivity seen in the reaction.[4] The major role of the Ce^{3+} cation appears to be as a catalyst for the degradation of BH_4^- by the methanolic solvent to give a reducing agent with increased hardness (hard and soft acids and bases [HSAB] theory) relative to borohydride itself. This increased hardness of the species **3** results in 1,2-addition since the 2-position of the enone system is understood to be harder. As well, the high propensity for axial hydride addition under these conditions can be justified with this explanation, since harder reagents tend to favor axial attack on cyclohexanones.[5]

$$^-BH_{4-n}(OR)_n \quad + \quad ROH \xrightarrow{Ce^{3+}} {^-BH_{4-p}(OR)_p} \; (\mathbf{3})$$

$$n = 0, 1, 2; \; p = n + 1$$

A common interpretation of this reaction invokes the use of the cerium(III) chloride as a Lewis acid that bonds to the carbonyl system, to make 1,2-addition favored. However, lanthanoid ions are known to preferentially bind to alcohols rather than carbonyl groups.[6] Additionally, if the cerium did bind with the carbonyl, one would expect that as dilution with methanol was increased, the observed regioselectivity, presumably due to this complexation, should be reduced. However, experimental results do not support this hypothesis. Since cerium complexation with methanol would result in increased acidity of the methanolic proton, the following scheme appears to be the more likely course of the reaction.

RO⁻----B(OR)(OR)(OR(H))–H----C=O----H–O(R)–Ce³⁺ ⟶ allylic alcohol (OH)

2.3.4 Synthetic Utility

2.3.4.1 Regioselectivity

During the course of developing a synthetic method for the central eight-membered ring of Taxol, the preparation of alcohol **5** was required.[7] Thus, the Luche reduction of ketone **4** proceeded smoothly to give **5** with undefined stereochemistry. This reaction best demonstrates the enhanced regioselectivity obtained in the Luche process. In addition to having no substitution at the β-carbon of the enone system, this ketone suffers from significant steric hindrance by the geminal dimethyl substituents at C_5. While 1,4-addition would normally be favored for this molecule, the use of $CeCl_3$ and $NaBH_4$ overrides this preference, giving complete 1,2-selectivity. The heightened selectivity for 1,2-addition is completely general and is one of the defining features of the Luche reduction.[8,9]

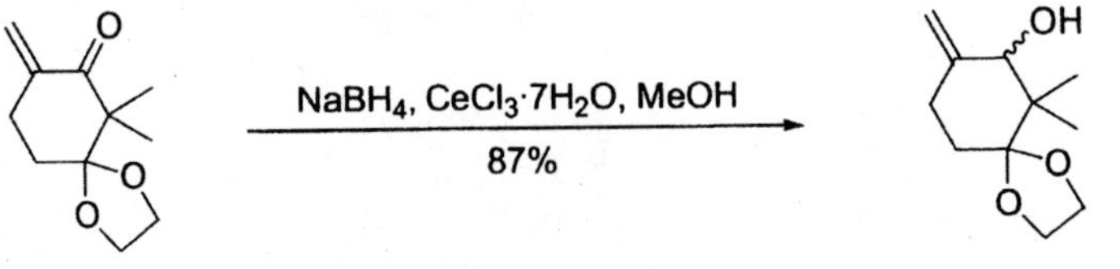

$NaBH_4$, $CeCl_3 \cdot 7H_2O$, MeOH
87%

4 5

The pronounced regioselectivity is further deomonstrated by selective 1,2-reduction of **6**, a precursor to *cis*-2,7-dihydroxycycloheptanone.[10] Once again, reduction of the sterically hindered ketone occurs preferentially over 1,4-hydride addition. In addition to the exclusive regioselectivity, this reaction was completely stereoselective in the production of **7**. The stereochemical result is rationalized by considering conformation **A**. For steric reasons, the hydride reagent is relegated to attack at the *exo*-surface of the π-system.

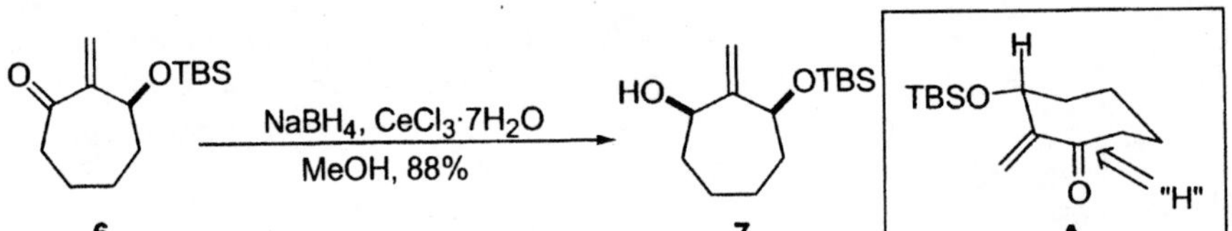

$NaBH_4$, $CeCl_3 \cdot 7H_2O$
MeOH, 88%

6 7 A

2.3.4.2 *Cyclic stereocontrol*

The stereoselectivity seen in the reduction of the seven-membered ketone above has proved to be general. This particular result, rationalized by the propensity for pseudoaxial hydride addition under Luche conditions, proved useful in the convergent total synthesis of Gymnocin-A, a polycyclic ether toxin isolated from the red tide dinoflagellate—*Karenia mikimotoi*.[11] The Luche reduction was used for the production of allylic alcohol **9** from ketone **8** in 84% yield. Luche conditions were similarly applied to the synthesis of the related polycyclic ether toxin gambierol.[12]

8 → 9: $NaBH_4$, $CeCl_3 \cdot 7H_2O$, MeOH, 84%

The majority of examples found in the literature regarding the utilization of the Luche reduction deals with reduction of six-membered enones.[13–15] This is due to the reliable and predictable stereochemical outcome of these reactions. In the total synthesis of phytuberin, a phytoalexin of the *Solanum* genus, compound **10** was produced *via* conjugate addition to (–)-α-santonin.[16] Unfortunately, all attempts to reduce the C_3 carbonyl group stereoselectively to provide the α-hydroxyl group (as in **11**) failed, possibly due to the highly sterically hindered β-face of the molecule. Instead, the β-alcohol was obtained with complete stereoselectivity *via* axial addition under Luche conditions. Subsequently, stereochemical inversion was accomplished using a Mitsunobu reaction to give the desired α-alcohol in 82% overall yield.

10 → 11: 1. $NaBH_4$, $CeCl_3$, MeOH, 91%; 2. Ph_3P, HCO_2H, DEAD, THF, 96%; 3. K_2CO_3, MeOH, 0 °C, 94%

A key step in the preparation of a new gibberellin, GA_{99}, involved the reduction under Luche conditions of a $\Delta^{1(10)}$-2-one derivative to furnish the critical 2β-stereochemistry.[17] While the 2-oxo group in **12** was resistant to reduction, presumably due to *in situ* ketol formation, the methoxymethyl ester derivative was reduced satisfactorily to the desired 2β-epimer **13** using the combination of $CeCl_3$ and $NaBH_4$. This combination efficiently and selectively delivered the hydride to the more hindered α-face of the A-ring. This example further demonstrates the propensity for hydride delivery *via* axial addition under Luche conditions.

12 → 13: 1. MOMCl, DMAP, *i*-Pr_2NEt, CH_2Cl_2, 90%; 2. $NaBH_4$, $CeCl_3$, MeOH, –78 °C to rt, 75%

Axial hydride delivery in the Luche reduction was employed as a key step in the total synthesis of (–)-mintlactone (**16**).[18] Utilizing Luche conditions, (+)-pulegone (**14**) was reduced stereoselectively to give *cis*-pulegol (**15**) in 98% yield. Ozonolysis was followed by acylation and an intramolecular Wittig–Horner reaction to complete the short, efficient synthesis of the target molecule.

14 → ($NaBH_4$, $CeCl_3{\cdot}7H_2O$, MeOH, 98%) → 15 → (O_3, CH_2Cl_2, 95%) →

→ (1. $(EtO)_2P(O)CH(Me)CO_2H$, DCC, CH_2Cl_2, 85%; 2. NaH, THF, reflux, 90%) → 16

Luche conditions have also been utilized in the synthesis of an electrophilic galactose equivalent, specifically for preparation of an *O*-linked glycopeptide isostere in the laboratory of Randall L. Halcomb.[19] To accomplish this goal, per-acetylated cyclohexone **17** was reduced using $NaBH_4$ and $CeCl_3$ to produce psuedoequatorial cyclohexenol **18.** The reaction produced a 30:1 ratio of diastereomers favoring the product of axial hydride addition.

17 → ($NaBH_4$, $CeCl_3$, 81%) → 18

Stereoselective Luche reduction of dihydropyrans has been widely exploited because of its synthetic practicality.[20–23] In all cases, an overwhelming preference for axial attack is observed. An elegant example of this comes from the work of Hodgson and co-workers in their efforts toward the synthesis of the C_{58}–C_{71} fragment of palytoxin.[24] Thus, Luche reductions of the *meso*-dipyranone **19** were highly stereoselective and gave the diol **20** as a 95:5 ratio of diastereomisomers. The selectivity of this overall transformation is remarkable and reflects the high level of stereo- and regioselectivity for each individual reduction.

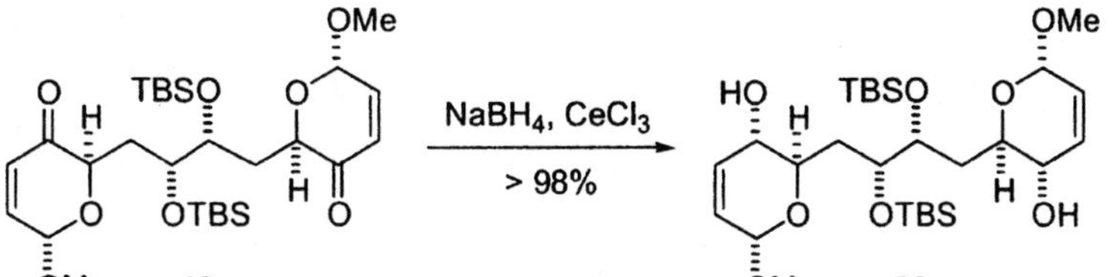

19 20

Similar results were observed in the synthetic efforts toward the C_{16}–C_{27} segment of bryostatin 1.[25] The synthesis of the fragment utilized a Luche reduction as the final step. This particular reaction using $NaBH_4$ and $CeCl_3$ reduced from the least hindered side of the C_{20} ketone in **21** producing a 90% yield of the single stereoisomer **22**. The resulting C_{17}–C_{27} bryostatins produced in these studies have shown potential anti-cancer properties.

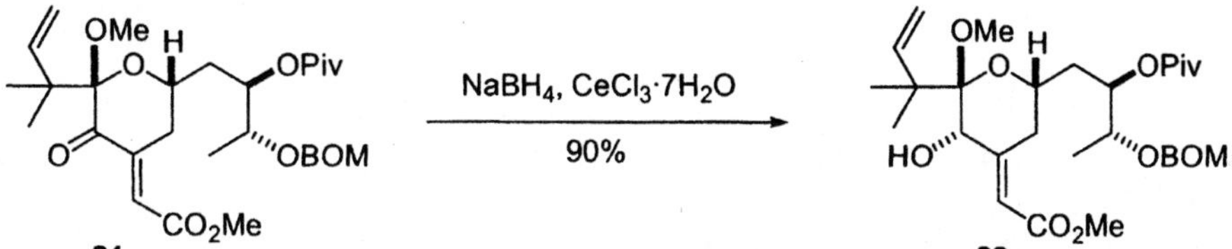

21 22

The opposite stereochemical outcome is observed when reduction is performed on the fully oxidized **23**.[26] In their synthesis of uncommon sugars from furanaldehyde, Wang and co-workers discovered that the stereochemical outcome of the Luche reduction can be manipulated by changing the temperature of the reaction. As shown below, Luche conditions successfully reduced the fully unsaturated ketone producing intermediate **24** in 75% yield. At 0 °C, a separable mixture of the two lactones **24** and **25** was obtained in a 2:1 ratio. However, execution at −78 °C solely yielded lactone **24**.

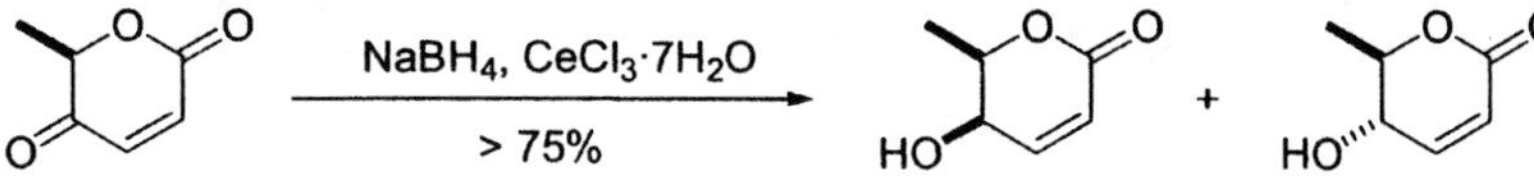

23 24 25

at 0 °C, **24**:**25** = 2:1
at −78 °C, only **24** isolated

An interesting, and general, reversal of stereoselectivity occurs in the reduction of ketones contained in 2,6-disubstituted aminal ring systems, such as **26**. The reduction of **26** efficiently produced the single diasteromer **27**, an intermediate in the syntheses of (±)-azimic acid and (±)-deoxocassine.[27] Speculation on the origin of the stereoselectivity suggests the presence of a sterically hindered axial approach of the hydride reagent due to the bulky, pseudoaxially oriented 2,6-substituents. Thus, in what has proved to be quite general for this system, equatorial addition seems to be the favored course of reaction.[28–31] This generality is further demonstrated in the similar reduction of **28** to give **29**, an intermediate in the syntheses of *D*- and *L*-deoxymannojirimycin.[32]

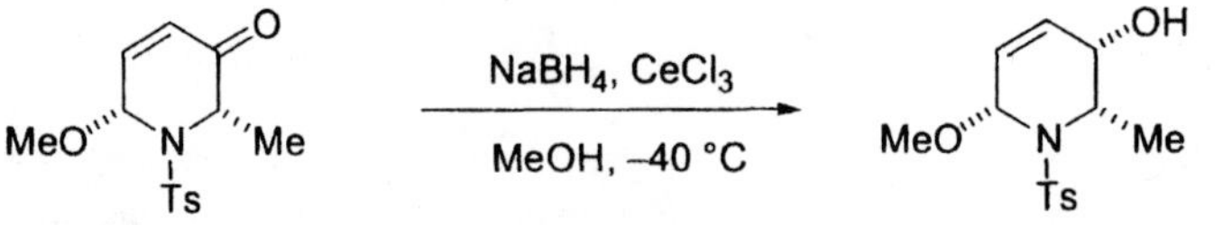

$NaBH_4$, $CeCl_3$; MeOH, –40 °C: **26** → **27**

$NaBH_4$, $CeCl_3$; $MeOH/CH_2Cl_2$, –78 °C; 86%: **28** → **29**

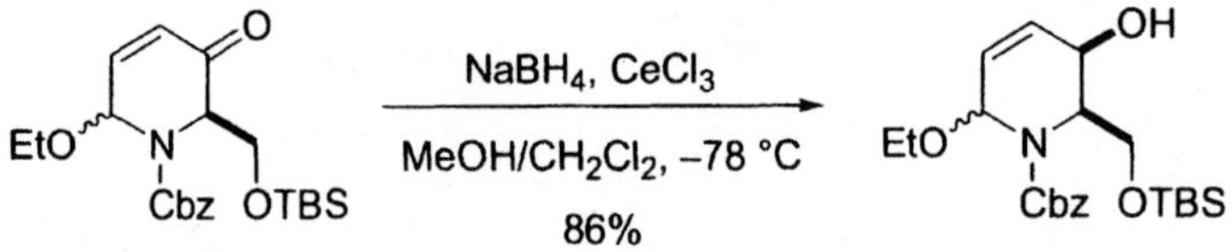

While not as prevalent as reduction of six-membered cyclic ketones, stereoselective reduction in five-membered rings is also possible.[33–35] One example of this reaction was realized during efforts to synthesize C_2-deuterated ribonucleosides.[36] As a solution to problems they encountered using more traditional hydride reducting agents ($NaBH_4$), the Luche reduction was found to be successful ($NaBD_4$ and $CeCl_3$ in THF). The workup required quenching the reaction with acetic acid which reproducibly yielded 90% of the desired alcohol.

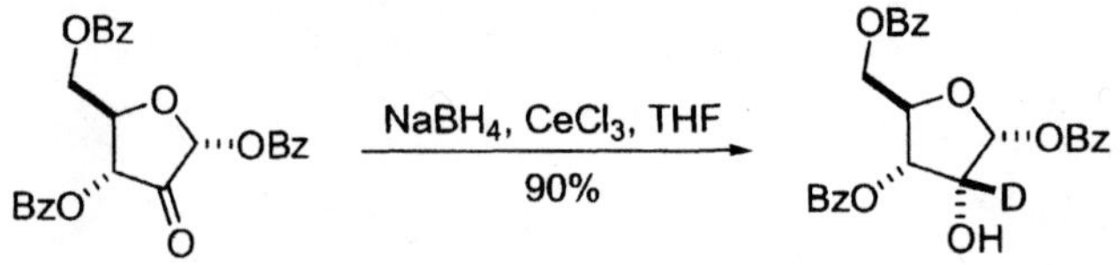

$NaBH_4$, $CeCl_3$, THF; 90%

Cerium(III) chloride as well as some other lanthanide salts invert the stereoselectivity of the reduction of bicyclo[3.1.0]hexane-2,4-diones and bicyclo[3.1.0]hexane-2-ones compared to sodium borohydride in methanol alone. This reversal leads to the more favored attack of the hydride from the most hindered, concave face of these molecules to give *exo*-4-hydrobicyclo[3.1.0]hexan-2-ones and *exo*-bicyclo[3.1.0]hexan-1-ols, respectively, in excellent yields.[37] If Luche conditions are used on **30** at low temperatures, complete stereoselectivity is induced in the production of **31**.

30 → 31: $NaBH_4$, $CeCl_3$, MeOH, 88%

Luche conditions have also found synthetic utility in the preparation of steroid derivatives such as the shark repellent pavoninin-5 and aragusterol H, an antiproliferative agent.[38] During the synthesis of pavoninin-5, the intermediate enone **32** is reduced to the enantiopure allylic alcohol **33** in 98% yield. β-Delivery of the hydride is rationalized by blocking of the α-face by the bulky chlosterol side chain. Additionally, the flattened D-ring reduces the steric hindrance of the C-18 methyl group.

32 → 33: $NaBH_4$, $CeCl_3$, 98%

2.3.4.3 Acyclic stereocontrol

Aside from the well-documented ability of the Luche reduction to provide stereocontrol in cyclic systems, acyclic stereocontrol is also viable through this process.[39–41] A notable example of this was demonstrated in the synthesis of (+)-cannabisativine, a unique natural product found in the common marijuana plant.[42] This synthesis necessitated a stereoselective Luche reduction to produce the diol **36** as a single diastereomer. The reaction proceeded in 96% yield and with 95% *de*. The pronounced diastereoselectivity can be attributed to Cram's rule, in which the hydride ion is delivered from the least sterically hindered side of the intermediate **34**. Reduction *via* the chelated intermediate **35** would also account for the observed stereochemical outcome.

→ 36: $NaBH_4$, $CeCl_3 \cdot H_2O$, MeOH, 96%

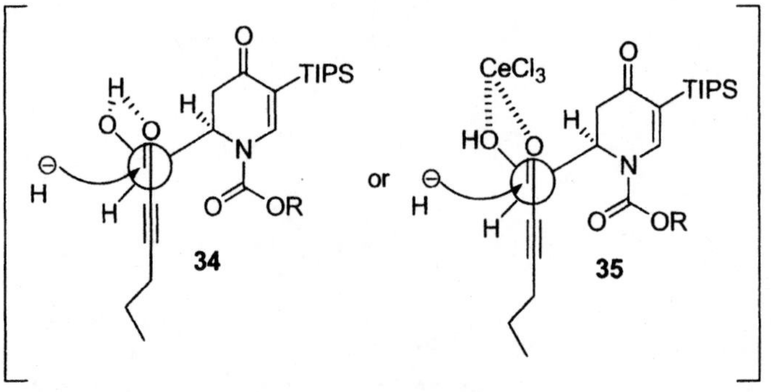

Reduction of α,β-epoxy ketones under the Luche conditions with $NaBH_4/CeCl_3$ in MeOH has been shown to provide anti- (or erythro) α,β-epoxy alcohols stereoselectively and in high yield.[43] As a means of comparison, the stereoselectivity using only $NaBH_4$ and MeOH produced a 29:71 ratio of diastereomers **37** and **38**. In contrast, standard Luche conditions gave a much improved ratio of 7:93 (**37**:**38**). The stereoselectivity is likely achieved through a chelated intermediate, since similar selectivity was obtained using the related reducing agent, $Zn(BH_4)_2$.

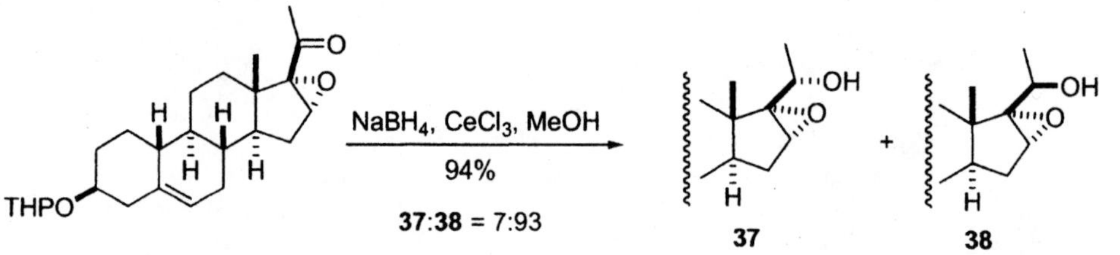

2.3.5 *Experimental*

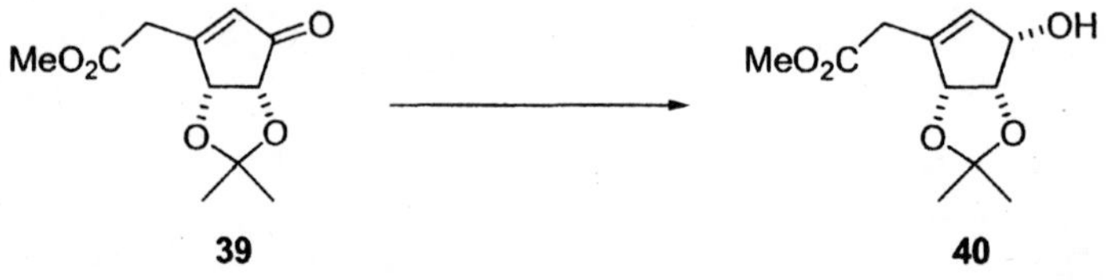

Methyl-((3a*S*,4*S*,6a*R*)-4,6a-dihydro-4-hydroxy-2,2-dimethyl-3a*H*-cyclopenta[*d*][1,3]-dioxol-6-yl)acetate (40)[33]

To a solution of **39** (0.63 g, 2.8 mmol) and $CeCl_3 \cdot 7H_2O$ (0.89 g, 2.4 mmol) in MeOH (15 mL) at 0 °C was added portion-wise $NaBH_4$ (0.15 g, 3.9 mmol). The mixture was stirred at the same temperature for 1.5 h before the reaction was quenched with H_2O (10 mL). Methylene chloride (30 mL) was added to the mixture, and the organic phase was separated. The aqueous phase was extracted with CH_2Cl_2 (2 × 15 mL). The combined organic phases were washed with brine and dried (anhydrous Na_2SO_4). Evaporation of the solvent afforded **40** as a clean product (0.64 g, 100%) as determined by NMR with no further purification necessary: ^{1}H NMR (250 MHz, $CDCl_3$) δ 5.69 (s, 1H), 5.01 (d, J = 5.4 Hz, 1H), 4.77 (t, J = 5.4 Hz, 1H), 4.58 (m, 1H), 3.70 (s, 3H), 3.23 (s, 2H), 2.80 (br, 1H), 1.41 (s, 6H); ^{13}C NMR (62.9 MHz, $CDCl_3$) δ 171.0, 138.3, 133.3, 112.6, 84.3, 78.0, 73.5, 52.1, 33.4, 27.7, 26.8. Anal. ($C_{11}H_{16}O_5$) C, H.

2.3.6 References

1. Li, Jie J. *Name Reactions: A Collection of Detailed Reaction Mechanisms*, Springer-Verlag, Heidelberg, germany, 2002, p215.
2. Luche, J.-L. *J. Am. Chem. Soc.* **1978**, *100*, 2226.
3. Fuwa, H.; Kainuma, H.; Satake, M.; Sasaki, M. *Biorg. Med. Chem. Lett.* **2003**, *13*, 2519.
4. Gemal, A. L.; Luche, J.-L. *J. Am. Chem. Soc.* **1981**, *103*, 5454.
5. Huet, J.; Maroni-Barnaud, Y.; Ahn, N. T.; Seyden-Penne, J. *Tetrahedron Lett.* **1976**, *17*, 159.
6. [R] Cockerill, A. F.; Davies, G. L. O.; Harden, R. C.; Rackham, D. M. *Chem. Rev.* **1973**, *7*, 553.
7. Iwamoto, M.; Miyano, M.; Utsugi, M.; Kawada, H.; Nakada, M. *Terahedron Lett.* **2004**, *45*, 8653.
8. Barluenga, J.; Fañanás, F. J.; Sanz, R.; García, F.; García, N. *Tetrahedron Lett.* **1999**, *40*, 4735.
9. Liu, C.; Burnell, D. J. *Tetrahedron Lett.* **1997**, *37*, 6573.
10. Paquette, L. A.; Hartung, R. E.; Hofferberth, J. E.; Vilotijevic, I.; Yang, J. *J. Org. Chem.* **2004**, *69*, 2454.
11. Tsukano, C.; Ebine, M.; Sasaki, A. *J. Am. Chem. Soc.* **2005**, *127*, 4326.
12. Fuwa, H.; Okamura, Y.; Natsugari, H. *Tetrahedron.* **2004**, *60*, 5341.
13. Gaul, C.; Njardarson, J. T.; Shan, D.; Dorn, D. C.; Wu, K. D.; Tong, W. P.; Huang, X. Y.; Moore, M. A.; Danishefsky, S. J. *J. Am. Chem. Soc.* **2004**, *126*, 11326.
14. Li, M.; Scott, J.; O'Doherty, G. A. *Tetrahedron Lett.* **2004**, *45*, 1005.
15. Moreno-Dorado, F. J.; Guerra, F. M.; Aladro, F. J.; Bustamante, J. M.; Jorge, Z. D.; Massanet, G. M. *Tetrahedron* **1999**, *55*, 6997.
16. Prangé, T.; Rodríguez, M. S.; Suárez, E. *J. Org. Chem.* **2003**, *68*, 4422.
17. Mander, L. N.; Owen, D. J. *Tetrahedron Lett.* **1996**, *37*, 723.
18. Pandey, R. K.; Upadhyay, R. K.; Shinde, S. S.; Kumar, P. *Synthetic Commun.* **2004**, *34*, 2323.
19. Whalen, L. J.; Halcomb, R. L. *Org. Lett.* **2004**, *6*, 3221.
20. Hodgson, R.; Majid, T.; Nelson, A. *J. Chem. Soc., Perkin Trans. I*, **2002**, 1444.
21. Harding, M.; Hodgson, R.; Nelson, A. *J. Chem. Soc., Perkin Trans. I*, **2002**, 2403.
22. Harding, M.; Hodgson, R.; Majid, T.; McDowall, K. J.; Nelson, A. *Org. Biomol. Chem.* **2003**, *1*, 338.
23. Hodgson, R.; Majid, T.; Nelson, A. *J. Chem. Soc., Perkin Trans. I*, **2002**, 1631.
24. Hodgson, R.; Nelson, A. *Org. Biomol. Chem.* **2004**, *2*, 373.
25. Keck, G. E.; Yu, T.; McLaws, M. D. *J. Org. Chem.* **2005**, *70*, 2543.
26. Zhu, L.; Talukdar, A.; Zhang, G.; Kedenburg, J. P.; Wang, P. G. *Synlett*, **2005**, *10*, 1547.
27. Cassidy, M. P.; Padwa, A. *Org. Lett.* **2004**, *6*, 4029.
28. Koulocheri, S. D.; Magiatis, P.; Skaltsounis, A.-L.; Haroutounian, S. A. *Tetrahedron*, **2000**, *56*, 6135.
29. Yang, C.-F.; Xu, Y.-M.; Liao, L.-X.; Zhou, W.-S. *Tetrahedron Lett.* **1998**, *39*, 9227.
30. Park, K. H.; Yoon, Y. J.; Lee, S. G. *J. Chem. Soc., Perkin Trans. I* **1994**, 2621.
31. Wu, X.-D.; Khim, S.-K.; Zhang, X.; Cederstrom, E. M.; Mariano, P. S. *J. Org. Chem.* **1998**, *63*, 841.
32. Haukaas, M. H.; O'Doherty, G. A. *Org. Lett.* **2001**, *3*, 401.
33. Yang, M.; Schneller, S. W.; Korba, B. *J. Med. Chem.* **2005**, *48*, 5043.
34. Hirose, T.; Sunazuka, T.; Yamamoto, D.; Kojima, N.; Shirahata, T.; Harigaya, Y.; Kuwajima, I.; Ōmura, S. *Tetrahedron* **2005**, *61*, 6015.
35. Moon, H. R.; Lee, H. J.; Kim, K. R.; Lee, K. M.; Lee, S. K.; Kim, H. O.; Chun, M. W.; Jeong, L. S. *Bioorg. Med. Chem. Lett.* **2004**, *14*, 5641.
36. Cook, G. P.; Greenberg M. M. *J. Org. Chem.* **1994**, *59*, 4704.
37. Krief, A.; Surleraux, D. *Synlett* **1991**, 273.
38. Williams, J. R.; Gong, H.; Hoff, N.; Olubodun, O. I.; Carroll, P.J. *Org. Lett.* **2004**, *6*, 269.

39. Hutton, G.; Jolliff, T.; Mitchell, H.; Warren, S. *Tetrahedron Lett.* **1995**, *36*, 7905.
40. Bruyère, H.; Ballereau, S.; Selkti, M.; Royer, J. *Tetrahedron*, **2003**, *59*, 5879.
41. Lysenko, I. L.; Bekish, A. V.; Kulinkovich, O. G. *Russian J. Org. Chem.* **2002**, *38*, 875.
42. Kuethe, J. T.; Comins, D. L. *J. Org. Chem.* **2004**, *69*, 5219.
43. Li, K.; Hamann, L. G.; Koreeda, M. *Tetrahedron Lett.* **1992**, *33*, 6569.

Richard J. Mullins, John J. Gregg, and George A. Hamilton

2.4 Meerwein–Ponndorf–Verley Reduction

2.4.1 Description

The Meerwein–Ponndorf–Verley (MPV) reaction[1–3] is the reduction of a carbonyl (**1**) to an alcohol (**2**) using an aluminum alkoxide catalyst, generally Al(O*i*-Pr)$_3$, and an alcoholic solvent such as *i*-PrOH. The alcoholic solvent serves as the hydride source for the reduction and in the course of the reaction is oxidized to the corresponding carbonyl.

O OH Al(O*i*-Pr)$_3$ OH O
R^1 R^2 *i*-PrOH R^1 R^2
1 **2**

2.4.2 Historical Perspective

The Meerwein–Ponndorf–Verley reduction is so named because of the simultaneous and independent contributions from the labs of Meerwein, Ponndorf and Verley. The first report to appear in the literature was from Meerwein and Schmidt in 1925 who showed that an aldehyde could be reduced to a primary alcohol by Al(OEt)$_3$ in an ethanolic medium.[4] Independently, Verley demonstrated that butyraldehyde could be reduced by geraniol and Al(OEt)$_3$.[5] The following year, Ponndorf extended this reaction to include the reduction of ketones by using an easily oxidized secondary alcohol, such as *i*-PrOH, as the hydride source and Al(O*i*-Pr)$_3$ as the metal catalyst.[6]

2.4.3 Mechanism

In Meerwein's initial publication he recognized that the reduction of aldehydes (**3**) to alcohols (**4**) with Al(OEt)$_3$ and EtOH was a reversible process.[4] As the reaction proceeded EtOH (**5**) was oxidized to acetaldehyde (**6**) and the reaction slowed; however, when acetaldehyde was allowed to evaporate over time, full conversion of the starting aldehyde **3** to alcohol **4** was observed. This implied that EtOH and acetaldehyde were in equilibrium with each other and that the aluminum alkoxide promoted their interconversion. The reverse of the MPV reduction, in this case the oxidation of EtOH to acetaldehyde, is known as the Oppenauer oxidation.[3]

O OH Al(OEt)$_3$ OH O
R H EtOH R H H
3 **5** **4** **6**

OH Al(OEt)$_3$ O
H
5 **6**

The mechanism of the MPV reductions has been extensively studied[7–12] and while pathways involving radical intermediates[11] and aluminum hydrides have been proposed, the most widely accepted mechanism proceeds *via* a direct hydrogen transfer between the alcohol and the carbonyl functionality of **7** (a ketone in this case). Coordination of the aluminum alkoxide with ketone **7** is thought to form complex **8**, subsequent hydride transfer *via* a concerted six-centered transition state, and decomplexation affords the desired alcohol product along with acetone. Computational modeling[12] and experiments measuring k_H/k_D[8,12] both suggest that hydride transfer is the rate determining step of the reaction.

Al(O*i*-Pr)$_3$

rate-determining step

7 **8** **9**

2.4.4 *Variations and Improvements*

The original conditions for the MPV reduction (*i*-PrOH, Al(O*i*-Pr)$_3$) had the disadvantage of requiring long reaction times. Often superstoichiometric amounts of both the alcohol and aluminum alkoxide were required to drive the reaction to completion. However, in 1977 Rathke discovered that the use of a small amount of protic acid, such as HCl or trifluoroacetic acid, significantly increased the rate of the Oppenauer oxidation.[13] Akamanchi and Noorani subsequently showed that this same concept could be applied to the MPV reduction. In one example, *p*-nitrobenzaldehyde (**10**) was reduced to the corresponding alcohol (**11**) using *i*-PrOH and Al(O*i*-Pr)$_3$ in 15 minutes when 2% of trifluoroacetic acid was used. In the absence of trifluoroacetic acid the reaction proceeded to only 36% conversion in 15 minutes.[14]

Al(O*i*-Pr)$_3$ (1 eq)
benzene, rt
15 min.

10 **11**

without TFA 36% conversion
with TFA (0.03 eq) 100% conversion

A variety of other metal alkoxides have been utilized in the MPV reduction. The most notable of these are the trivalent lanthanide alkoxides. Kagan and co-workers were the first to publish the use of these reagents for MPV reductions. They showed that a variety of ketones and aldehydes, such as 2-octanone (**12**), could be reduced using *i*-PrOH and catalytic SmI_2(O*t*-Bu) in good yield in 24 hours.[15]

$SmI_2(Ot\text{-}Bu)$ (0.1 eq)
i-PrOH (4 eq)
THF, 65 °C
86%
12 → 13

There has also been significant effort toward the development of an enantioselective MPV reduction.[16,17] The first reported example of an enantioselective MPV reduction was the reduction of (**14**) by *rac*-$Al(Oi\text{-}Bu)_3$ in the presence of an enantiomerically pure alcohol (**15**) to afford alcohol **16** in 22 % *ee*.[18]

14 + **15**
rac-$Al(Oi\text{-}Bu)_3$ (0.5 eq)
22% *ee*
→ **16**

Since this initial finding there have been significant improvements in the degree of asymmetric induction observed in these reactions, most notably due to the use of chiral Lewis acids. In one example Evans and co-workers developed a chiral samarium catalyst (**17**) that was capable of reducing aryl ketones (**18**) in the presence of *i*-PrOH with high levels of enantioselectivity.[19]

18 (Ar, Me)
17 (0.05 eq),
i-PrOH (25 eq)
THF
→ **19** (Ar, *, Me)
36–96% yield
92–97% *ee*

17: Ph, Bn, N, Ph, O—Sm—O, I

In 2002 the Nguyen lab demonstrated that Al(O*i*-Pr)((*R*)-BINOL) was an effective catalyst for the asymmetric reduction of aryl ketones (**20**).[20] This methodology was recently extended to include the enantioselective reduction of imines (**22**) to the corresponding amine (**23**).[21]

20 (R)
(*R*)-BINOL (0.1 eq)
$AlMe_3$ (0.1 eq)
i-PrOH (4 eq)
toluene
→ **21** (OH, *, R)
43–99% yield
30–80% *ee*

(R)-BINOL (1.2 eq)
$AlMe_3$ (1.2 eq)
i-PrOH (4 eq)
toluene

22 R^1, R^2 = alkyl, aryl → **23** 79–95% yield, 94–96% *ee*

2.4.5 Synthetic Utility

In Woodward's synthesis of reserpine (**24**),[22–24] quinone (**25**) is reduced to the corresponding diol which undergoes rapid lactonization to form the desired product (**26**).[23,24]

25 → $Al(Oi\text{-}Pr)_3$, i-PrOH → **26** → → **24**

28 → SmI_2 (0.1 eq), i-PrOH (10 eq), THF, 98% → **29** → → **27**

The synthesis of cytovaricin (**27**) by Evans utilized the $SmI_2(Ot\text{-}Bu)$ reagent developed by Kagan to accomplish the reduction of ketone (**28**) to alcohol (**29**) in 98% yield.[25]

2.4.6 Experimental

$Al(Oi\text{-}Pr)_3$, *i*-PrOH

25 → **26**

MPV Reduction of dione 25[24]

The dione adduct **25** (440 mg) and aluminum isopropoxide (1.83 g) were dissolved in dry isopropanol (8 mL). The mixture was boiled gently, and acetone and isopropanol were distilled slowly from the reaction mixture through a short column. From time to time isopropanol was added to maintain a constant volume. After one hour no more acetone could be detected in the distillate. The reaction mixture was then concentrated under reduced pressure, treated with ice-cold 2 N hydrochloric acid, and extracted with methylene chloride/ether (1:3). The organic phase was washed with sodium bicarbonate and saturated sodium chloride solution, and concentrated. The residue was crystallized twice from acetone/ether, to give 230 mg of colorless lactone **26**.

SmI_2 (0.1 eq), *i*-PrOH (10 eq), THF, 98%

28 → **29**

Samarium catalyzed MPV reduction of ketone 28[25]

To an argon-blanketed flask containing 6.32 g (1 1.2 mmol) of a solution of ketone **28** in 38 mL of THF at 23 °C was added 8.60 mL (6.75 g, 112 mmol) of freshly distilled and argon-degassed isopropyl alcohol, followed by 16.8 mL (assumed 0.10 M in THF, 1.68 mmol) of samarium diiodide. After 3 h, the dark-blue solution was diluted with 300 mL of diethyl ether and was extracted with 200 mL of saturated aqueous $NaHCO_3$. The aqueous layer was extracted with 2 × 100 mL of diethyl ether, and the combined organic layers were sequentially washed with 150 mL of saturated aqueous sodium sulfite and 150 mL of brine, dried (Na_2SO_4,), filtered, concentrated, and purified by flash chromatography 6.20 g (98%) of the desired equatorial alcohol **29**.

2.4.7 References

1. [R] Wilds, A. L. *Org. React.* **1944**, *2*, 178 (and references therein).
2. [R] Kellogg, R. M. in *Comp. Org. Syn.*; (Trost, B. M., Fleming, I., Eds.), Pergamon Press: Oxford, 1991; Vol. 8, p 88 (and references therein).
3. [R] de Graauw, C. F.; van Bekkum P. H.; Huskens, J. *Synthesis* **1994**, 1007.
4. Meerwein, H. Schmidt, R. *Liebigs Ann. Chem.* **1926**, *444*, 221.
5. Verley, A. *Bull. Soc. Chim. Fr.* **1925**, *37*, 537.
6. Ponndorf, W. *Angew. Chem.* **1926**, *39*, 138.
7. Woodward, R. B.; Wendler, N. L.; Brutschy, F. J. *J. Am. Chem. Soc.* **1945**, *67*, 1425.
8. Williams, E. D.; Krieger, K. A.; Day, A. R. *J. Am. Chem. Soc.* **1953**, *75*, 2404.
9. Moulton, W. N.; Van Atta, R. E.; Ruch, R. R. *J. Org. Chem.* **1961**, *26*, 290.
10. Shiner, V. J.; Whittaker, D. *J. Am. Chem. Soc.* **1963**, *85*, 2337.
11. Ashby, E. C. *Acc. Chem. Res.* **1988**, *21*, 414.
12. Cohen, R.; Graves, C. R.; Nguyen, S. T.; Martin, J. M. L.; Ratner, M. A. *J. Am. Chem. Soc.* **2004**, *126*, 14796.
13. Kow, R.; Nygren, R.; Rathke, M. W. *J. Org. Chem.* **1977**, *42*, 826.
14. Akamanchi, K. G.; Noorani, V. R. *Tetrahedron Lett.* **1995**, *36*, 5085.
15. Namy, J. L.; Souppe, J.; Collin, J.; Kagan, H. B. *J. Org. Chem.* **1984**, *49*, 2045.
16. [R] Nishide, K.; Node, M. *Chirality*, **2002**, *14*, 759 (and references therein).
17. [R] Graves, C. R.; Campbell, E. J.; Nguyen, S. T. *Tetrahedron Assym.* **2005**, *16*, 3460 (and references therein)
18. Doering, W. E.; Young, R. W. *J. Am. Chem. Soc.* **1950**, *72*, 631.
19. Evans, D. A.; Nelson, S. G.; Gagne, M. R.; Muci, A. R. *J. Am. Chem. Soc.* **1993**, *115*, 9800.
20. Campbell, E. J.; Zhou, H.; Nguyen, S. T. *Angew. Chem. Int. Ed.* **2002**, *41*, 1020.
21. Graves, C. R.; Scheidt, K. A.; Nguyen, S. T. *Org. Lett.* **2006**, *8*, 1229.
22. Woodward, R. B.; Bader, F. E.; Bickel, H.; Frey, A. J.; Kierstead, R. W. *J. Am. Chem. Soc.* **1956**, *78*, 2023.
23. Woodward, R. B.; Bader, F. E.; Bickel, H.; Frey, A. J.; Kierstead, R. W. *J. Am. Chem. Soc.* **1956**, *78*, 2657.
24. Woodward, R. B.; Bader, F. E.; Bickel, H.; Frey, A. J.; Kierstead, R. W. *Tetrahedron,* **1958**, *2*, 1.
25. Evans, D. A.; Kaldor, S. W.; Jones, T. K.; Clardy, J.; Stout, T. J. *J. Am. Chem. Soc.* **1990**, *112*, 7001.

Julia M. Clay

2.5 Staudinger Reaction

2.5.1 Description

Azides react with tertiary phosphines under mild conditions to form iminophosphoranes (**1**), which are versatile synthetic intermediates that have found a number of applications.[1–10] Upon treatment with water, an iminophosphorane will hydrolyze to form a primary amine and phosphine oxide in what is referred to as the Staudinger reaction or Staudinger reduction.

$$R{-}N_3 \xrightarrow{PPh_3} \underset{\mathbf{1}}{R{-}N{=}PR'_3} \xrightarrow{H_2O} R{-}NH_2 + R'_3P{=}O$$

In the literature, the term "Staudinger reaction" is actually associated with any process that involves the conversion of an azide to the iminophosphorane as its initial step. Hydrolysis is the typical fate of this intermediate, but iminophosphoranes (sometimes called phosphazenes) can also undergo a host of synthetically useful tandem reactions, most notably aza-Wittig-type reactions with carbonyl containing compounds.[1–4,6–10]

2.5.2 Historical Perspective

The reaction is named after its inventor, Hermann Staudinger, who first described the reaction in 1919.[5] The first major review of the Staudinger reaction appeared in 1981, and even though this review appeared over 60 years after Staudinger and Meyer's seminal 1919 *Helvetica Chimica Acta* article, its lack of reference to applications in synthetic organic chemistry illustrated that the power of this reaction took quite some time to be realized. The decades following this 1981 review would see an increased interest in the transformation and numerous new applications of the classical Staudinger reaction would be published in the following decades.[9] In fact, just one decade later, a second review by the same authors appeared, this time describing some of these emerging synthetic utilities that would continue to garner interest and expand the scope of the reaction.[10]

2.5.3 Mechanism

The first step in the Staudinger reaction is the nucleophilic attack of the phosphane's phosphorus atom onto the terminal nitrogen (N-3) of the azide to yield a phosphazide intermediate (**2**). Select phosphazides have been isolated and characterized but typically these intermediates are unstable and readily eliminate dinitrogen gas to generate iminophosphoranes **1** in a fashion suggested below. While the phosphazides that have been isolated have been found by X-ray analysis to exist in the *s-trans* configuration, it's the *s-cis* configuration that would allow for the nitrogen and phosphorus bond to form due to proximity. A number of computational studies have been aimed at studying the reaction pathway of the Staudinger reaction and the geometry of the resulting phosphazide intermediate which seem to agree upon a pathway such as that depicted here.[11–13] Nucleophilic attack of the phosphazide's N-1 onto the phosphorus results in a highly strained four-membered ring that immediately extrudes nitrogen and the iminophosphorane. It is important to note that azides and the subsequent Staudinger reaction intermediates have many resonance structures, but only those resonance structures that serve to better explain reactivity are drawn.

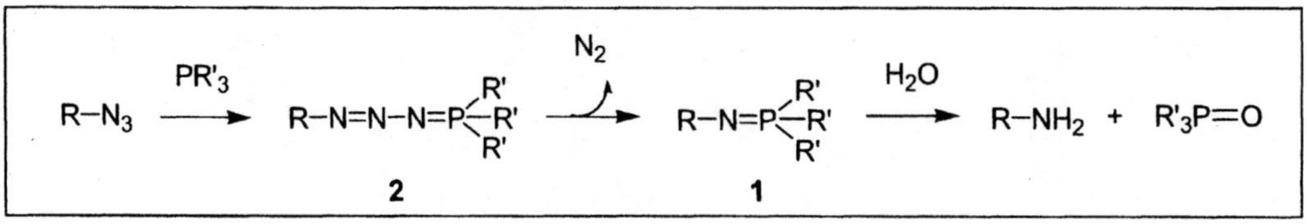

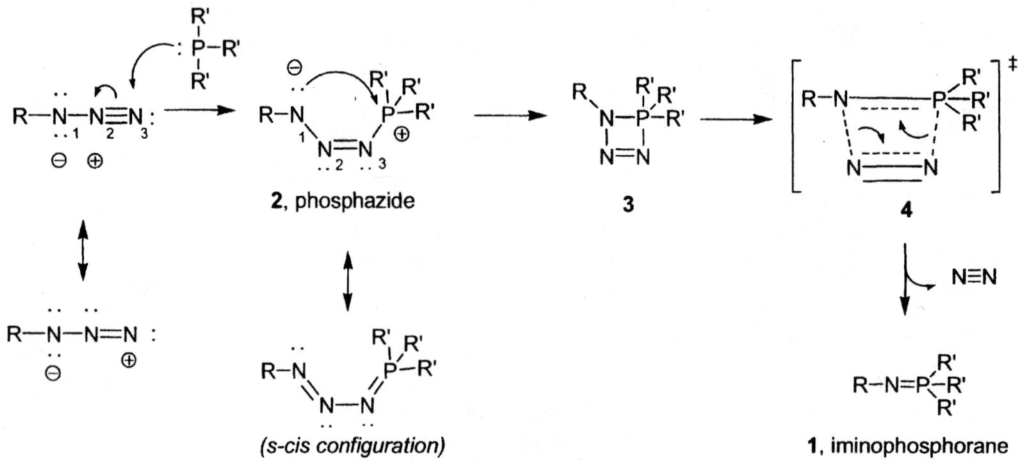

R–N$_3$ →(PR'$_3$) R–N=N–N=PR'$_3$ (**2**) →(−N$_2$) R–N=PR'$_3$ (**1**) →(H$_2$O) R–NH$_2$ + R'$_3$P=O

2, phosphazide

(s-cis configuration)

3

4

N≡N

1, iminophosphorane

Upon exposure to water, nucleophilic attack of water onto the iminophosphorane phosphorus atom leads to a cascade of proton transfers that ultimately yield the primary amine and phosphine oxide.

Hydrolysis:

$R{-}\ddot{N}{=}PR'_3$ (1) + $H{-}\ddot{O}{-}H$ → $R{-}N(H){-}PR'_3(OH)$ (5) → $R{-}NH_2$ + $O{=}PR'_3$

If the Staudinger reaction of an alkyl azide containing a pendant carbonyl group is heated under anhydrous conditions, the iminophosphorane can nucleophilically attack the carbonyl leading to cyclic imines (*via* the aza-Wittig mechanism) and liberate phosphine oxide as shown below. The reaction also works intermolecularly.

6 —[2 + 2]→ 7 → 8 + $O{=}PR'_3$

2.5.4 *Variations and Improvements*

Although the classical Staudinger reaction is technically the simple conversion of an azide to an iminophosphorane, this name reaction is most often associated with the subsequent hydrolysis to afford primary amines (also called the Staudinger reduction). The French team of Vaultier, Knouzi, and Carrie published several papers on the exploration of the Staudinger reduction in the 1980s that seemed to kick-start interest in this area.[14,15] They discovered that water can be added together with the azide and phosphine in a one-pot process (as opposed to a two-step process shown below) without a drop in yield. This improvement made this a transformation that is technically easier to perform in the laboratory.[15] The team applied the protocol to a series of functionalized primary alkyl, secondary alkyl, tertiary alkyl, and aryl azides and all were reduced in very good to excellent yield (see amines **10–15**). The azides were treated with 1 equivalent of triphenylphosphine and a slight excess of water in THF. The mild and chemoselective nature of this reaction became apparent in that nitro groups, olefins, esters, ketals, and so on were all unaffected during this study. This study also revealed the order of reactivity for azides being primary > secondary > tertiary. Primary and secondary azides can convert at room temperature but secondary azides require longer reaction times and/or heat. Tertiary azides, being the most hindered, required both longer reaction time and heat. The authors conducted competition experiments to directly compare the reactivities of various azides using 1:1 mixtures of two amines in the presence of 0.5 equivalent of triphenylphosphine at room temperature for 12 hours. In primary versus tertiary and secondary versus tertiary experiments, the primary and secondary azides were reduced while the tertiary azide was unaffected. In a primary versus secondary experiment, after 12 hours there was only 20% primary azide remaining while 80% of the secondary azide remained.

When an azidoketone was treated with triphenylphosphine the authors noted that the iminophosphorane had reacted intramolecularly with the carbonyl to form a cyclic imine, a transformation which would later become an important variation of the reaction.

two-step protocol: formation of iminophosphorane followed by hydrolysis

R–N₃ + PPh₃ (1:1) —THF→ [R–N=PPh₃ **9**] —H₂O→ R–NH₂

one-step protocol: water added to reaction mixture at onset

R–N₃ —PPh₃ (1 eq), THF/H₂O→ R–NH₂

10, 90% **11**, 88% **12**, 95% **13**, 79%

14, 87% **15**, 92%

One of the reasons that the Staudinger reduction is such a powerful synthetic tool is the fact that azides can be introduced into a molecule fairly easily.[1,2] In fact, some of the improvements to the Staudinger reduction over the years have included many one-pot protocols converting alkyl bromides, alkyl chlorides, alcohols, and acetates to azides followed by the *in situ* reduction to the corresponding amines using the Staudinger reaction.

Numerous examples of the use of the Staudinger reduction in the synthesis of complex biologically active compounds appear in the literature (see Section 2.5.5) and synthetically useful variations have been developed.

2.5.4.1 Variations to facilitate phosphine oxide removal

There have been many attempts to develop alternative phosphines as well as various solid-phase methods in order to avoid the issue of the removal of the triphenylphosphine oxide by-product.[16–18] The advantage of a solid-supported phosphine reagent is that a simple filtration would remove the oxide and thus simplify the workup. One example of this approach is the non-cross-linked polystyrene supported triphenylphosphine reagent (**16**) whose synthesis and reactivity was recently described and found to be superior to a traditional PEG-supported triphenylphosphine resin and better than triphenylphosphine itself.[16] This reagent allowed for high loading (1 mmol/g) and was shown to react rapidly in tandem Staudinger/aza-Wittig reactions with aldehydes and ketones. The authors speculate that the faster reaction rate observed with **16** in comparison to triphenylphosphine could be due to the more electron-rich phosphine (due to the *p*-alkoxy).

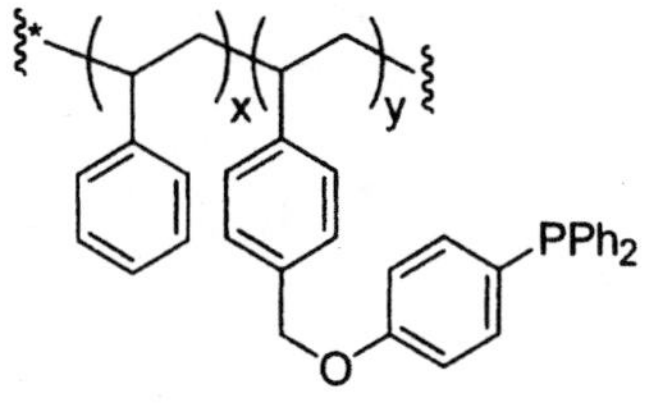

16

Another approach to simplify the Staudinger reaction and make it amenable to solution-phase parallel synthesis comes from the Merck Research Laboratories. In this work, a fluorous-tethered triphenylphosphine reagent (**18**) was found to effectively facilitate the reduction of a variety of azides and the spent fluorous reagent was easily removed using Fluoro*Flash*™ SPE column chromatography.[17] The azide group of 4-azidobenzoic acid **17** was converted to the corresponding aniline in 93% isolated yield and very good yields were also obtained for a series of complex functionalized and hindered azides. Amines **20** thru **25** were synthesized using this reagent and serve to highlight the high functional group compatibility this reaction has.

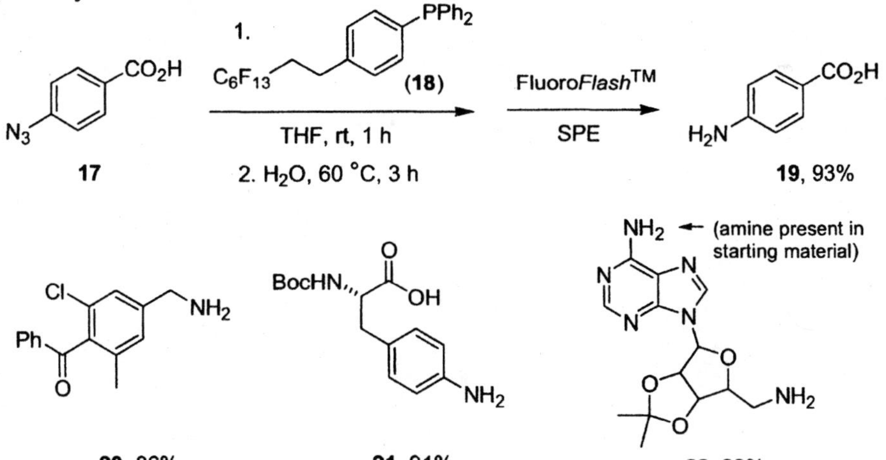

1. C_6F_{13} (**18**) PPh$_2$; THF, rt, 1 h; 2. H_2O, 60 °C, 3 h; Fluoro*Flash*™ SPE

17 → **19**, 93%

20, 86% **21**, 91% **22**, 88% (amine present in starting material)

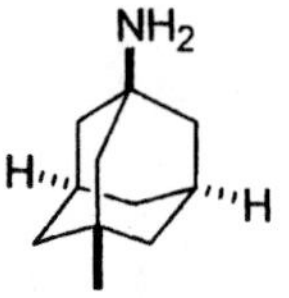

23, 92% 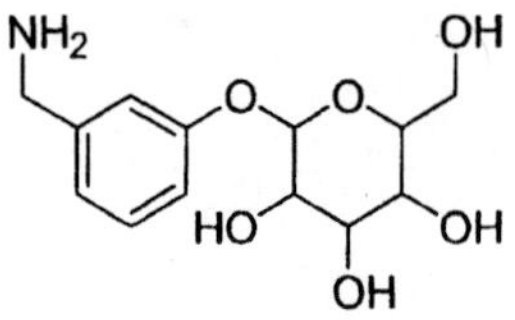 **24**, 80% 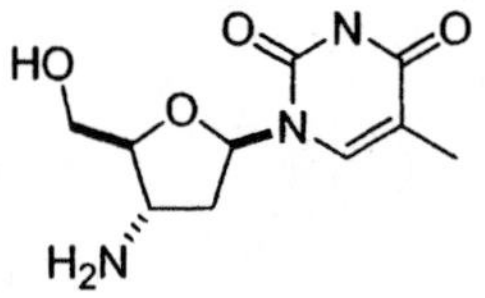 **25**, 82%

2.5.4.2 One-pot transformations to protected amines

Carbamates are one of the most widely used protecting groups for amines and the development of one-pot conversions of azides to carbamates such as Boc, Cbz, and allyl further expanded the scope of utility of the Staudinger reduction. Treatment of the Staudinger reaction iminophosphorane intermediate **26** with 1.05 equivalents of 2-*t*-(butoxycarbonyloximino)-2-phenylacetonitrile (Boc-on) led to Boc-protected amines **28** thru **32** in 87–100% overall yield.[19] This process utilizes trimethylphosphine, which is more reactive than triphenylphosphine and provides an easier work-up because trimethylphosphine oxide is water soluble. It should be noted that the use of di-*t*-butyldicarbonate (Boc_2O) has been investigated but leads to lower yields in comparison.[20]

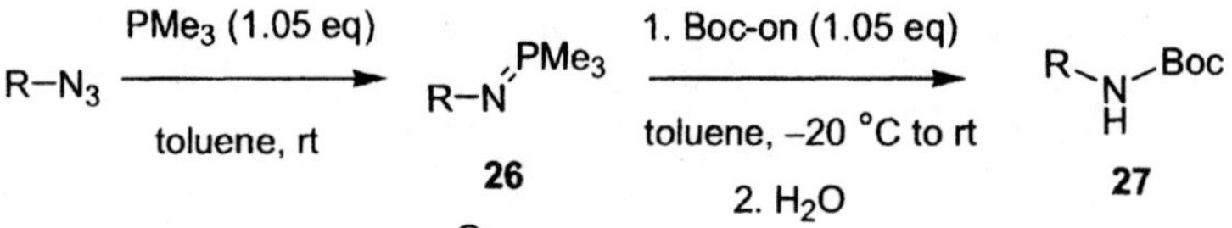

$R{-}N_3$ → (PMe_3 (1.05 eq), toluene, rt) → **26** → (1. Boc-on (1.05 eq), toluene, –20 °C to rt; 2. H_2O) → **27**

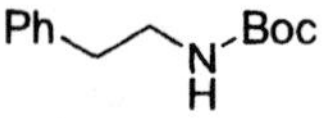

28, 100% 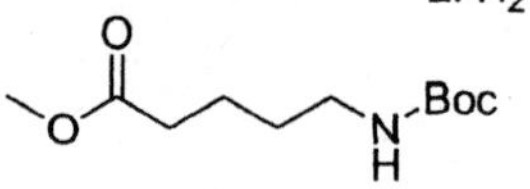 **29**, 96% 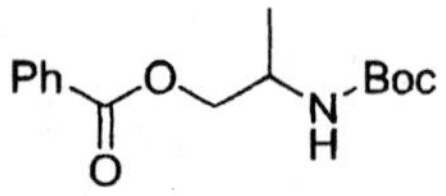 **30**, 90%

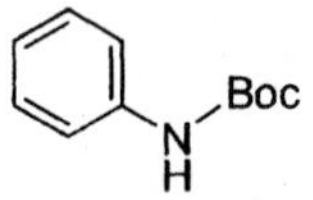

31, 95% 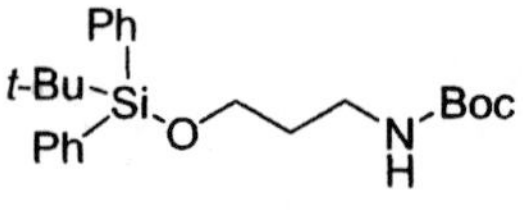 **32**, 92%

The same authors also published a simple one-pot protocol for the azide to carbamate transformation using a variety of chloroformates.[21] This preparation is excellent for azide to carbamate transformations (such as Cbz, Troc, and Alloc) that would clearly not be feasible via the one-pot catalytic hydrogenation/protection preparations due to functional group incompatibility. Trimethylphosphine is the phosphine of choice once again. As the rapid and room temperature conversion of **33** to intermediate **34** illustrates, the use of trimethylphosphine allows for excellent yields under mild conditions.

$ClCO_2R = ClCO_2Bn, ClCO_2Me, ClCO_2Et, ClCO_2CH_2CCl_3$, and $ClCO_2CH_2CH{=}CH_2$

2.5.4.3 One-pot transformations to monomethylamines

Two synthetically useful one-pot protocols for the preparation of monomethylamines from azides were reported by Suzuki in 2001 and illustrated below.[22] Monomethylamines are synthetically challenging motifs to prepare cleanly but in this discovery, paraformaldehyde is added to the Staudinger reaction iminophosphorane intermediate **26** followed by sodium borohydride reduction which results in a conversion of azides to the corresponding monomethylamines. This transformation actually takes place through an intermolecular tandem Staudinger/aza-Wittig reaction and the resulting imine is reduced by the $NaBH_4$. In another method, the addition of iodomethane followed by hydrolysis afforded monomethylamines such as **39–41** in high yield.

Method A:

Method B:

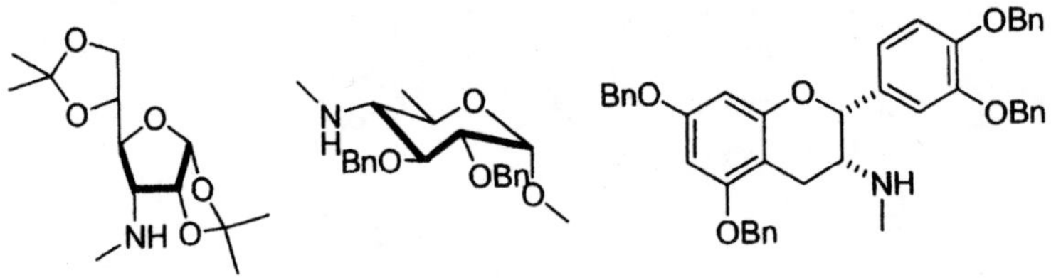

	39	40	41
Yield *(Method A)*:	68%	80%	86%
Yield *(Method B)*:	81%	76%	81%

2.5.4.4 One-pot tandem Staudinger/aza-Wittig reactions with carbonyls

Under strictly anhydrous conditions, the iminophosphorane intermediate that is formed as a result of the Staudinger reaction can react with aldehydes and ketones in an intermolecular fashion (as in the synthesis of imine **36** described above) or intramolecularly with a variety of carbonyl containing functional groups to afford a host of products. Nitrogen containing ring systems such as cyclic imines (**44**) represent just one of the many products one can prepare and the reaction is particularly well suited for the facile synthesis of five, six, and seven-membered rings. In addition to aldehydes and ketones, carboxylic acids, esters, thio-esters, and amides can also react in an intramolecular fashion to trap an iminophosphorane to afford a variety of heterocycles. Examples from the current literature are described in Section 2.5.5.

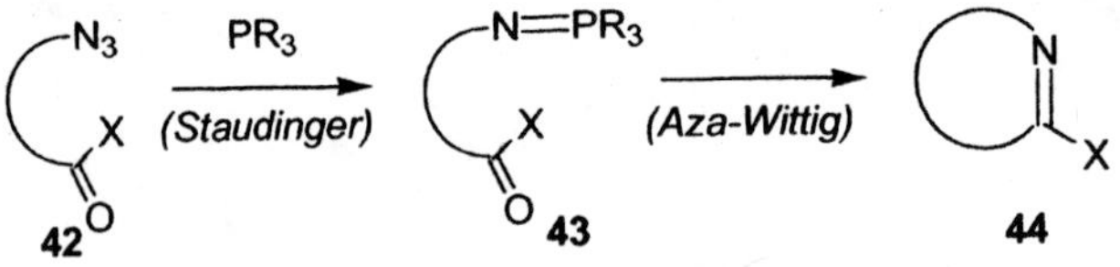

42 → PR_3 *(Staudinger)* → **43** → *(Aza-Wittig)* → **44** 5, 6, 7-membered rings

2.5.5 Synthetic Utility

There are several factors that determine the fate of the Staudinger reaction's iminophosphorane intermediate which, if appropriately set, can result in a huge variety of synthetic applications. One factor is reaction conditions: if the Staudinger reaction is run in the presence of water, hydrolysis to the amine will rapidly become the fate of the iminophosphorane intermediate. If the reaction is run under anhydrous conditions, the reactive intermediate can be involved in tandem reactions (typically with warming of the reaction mixture). The presence, position, and nature of substituents on the azide component play a role as well. Appropriately positioned carbonyl groups and hydroxyl groups on the azide, for example, can react with the iminophosphorane resulting in ring closure. Lastly, the iminophosphorane can react intermolecularly with a number of functionalized compounds and many three-component reactions have been utilized to afford a variety of heterocycles. As mentioned earlier, the variety of heterocycles that can be assembled using the Staudinger reaction is enormous and has been reviewed elsewhere.[1–4,6–10] Another utility comes from the fact that azides can react with carboxylic acids or esters to form amides (or lactams intramolecularly) in the presence of phosphines. The field of chemical biology has recently embraced this application of the Staudinger reaction as a means by which to prepare bioconjugates and termed the process the "Staudinger ligation."[23] The Staudinger ligation has been reviewed elsewhere and will not be discussed here. This review will provide examples of the most widely used applications found in the current literature and is by no means comprehensive.

2.5.5.1 Preparation of primary amines

PMe_3, H_2O: **45-I**, R = N_3 → **45-II**, R = NH_2

PMe_3, H_2O: **45-III**, R = N_3 → **45-IV**, R = NH_2

45

The literature is filled with examples of the use of the Staudinger reduction in the synthesis of biologically relevant molecules. Its mild reaction conditions and wide functional group compatibility make it an excellent method by which to introduce a primary amine into a complex structure. Nicolaou, for example, has used the reaction twice in his group's effort toward putting the pieces in place for a method to synthesize the highly complex and synthetically challenging naturally occurring antiobiotic thiostrepton.[24] The two N–H groups that are highlighted in compound **45** represent amides whose amine precursors were introduced via the Staudinger reduction. The amines were introduced during two separate steps in the multistep convergent synthesis of this analog of the natural product.

The Chang research group has tuned the regioselectivity of the Staudinger reduction toward the synthesis of several aminoglycoside antibiotics including the kanamycin and neomycin classes.[25] The synthesis of pyrankacin (**49**), a new class of aminoglycoside antibiotics, is another powerful example of the use of the Staudinger reaction in the synthesis of challenging biologically active molecules.[26] The first use of the reaction is the regioselective low temperature, one-pot reduction/Boc protection of **46** to **47**. Taking into account that electron-deficient azides have greater reactivity in the Staudinger reaction as compared to electron-rich azides and the steric environment surrounding the various azido groups, the authors speculate that the presence of the 4-chlorobenzoyl protecting group set up the proper environment to allow for the observed regioselectivity. Tri-azide **48** was then elaborated in several steps to afford pyrankacin. The remaining steps involved reducing the remaining azides to their corresponding amines upon addition of trimethylphosphine in aqueous THF, the simultaneous Cbz deprotection and alkene reduction, and lastly ion exchange chromatography.

PMe_3 (1.1 eq)
Boc-on (2.4 eq)
toluene, –78 °C to rt
84%

46 → **47** → steps →

1. PMe_3, H_2O THF, 50 °C
2. H_2, $Pd(OH)_2$
3. Dowex 1-X8 (Cl^-)

48

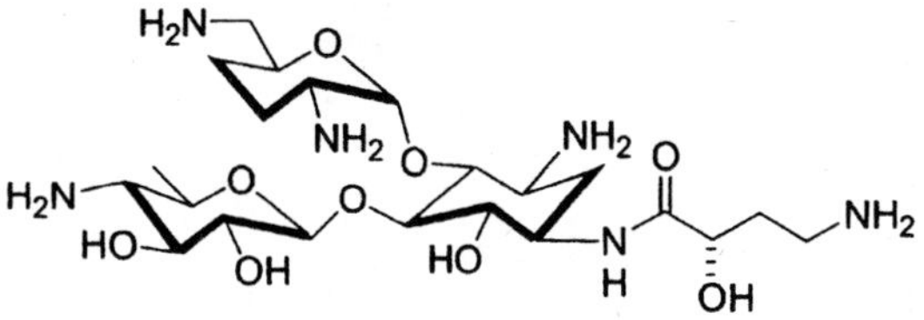

49, pyrankacin (5 HCl)

Benzodiazepines represent a common structural motif in drug discovery. A two-step method for the introduction of an amino group into the 3 position of 1,4-benzodiazepin-2-ones described by Merck involved selective positioning of a 3-azido group via deprotonation followed by trisyl azide quench.[27] This afforded 3-azidobenzodiazepine **51** that was then subjected to the Staudinger reduction to allow for the generation of multi-gram quantities of amine **52** in excellent yield. The reduction was carried out using 3 equivalents of triphenylphosphine in THF/H_2O at room temperature. Amines **53–56** are representative examples from his work and the yields given are for the reduction step.

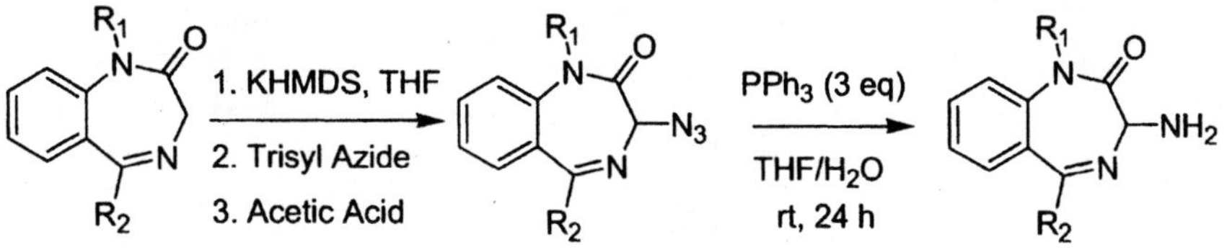

1. KHMDS, THF
2. Trisyl Azide
3. Acetic Acid

PPh_3 (3 eq)
THF/H_2O
rt, 24 h

50 **51** **52**

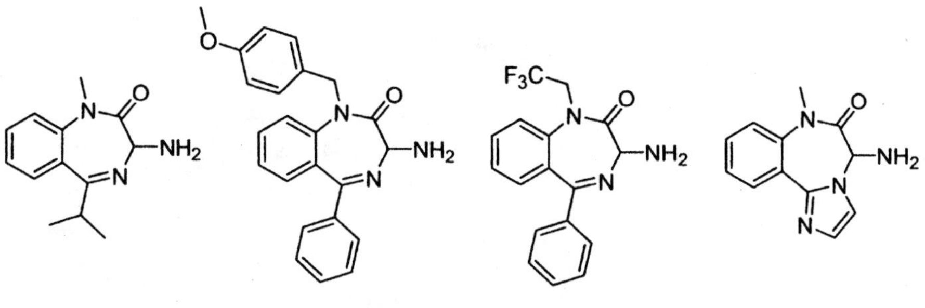

53, 97% **54**, 88% **55**, 91% **56**, 95%

In another example of the use of the Staudinger reduction toward the synthesis of medicinally active compounds, Danishefsky's group treated advanced intermediate **57** with trimethylphosphine and water in THF to afford epothilone analog **58** in 79% yield. This work was part of an effort to synthesize new analogs of the potent cytotoxic macrolide in pursuit of new anticancer therapeutics.[28]

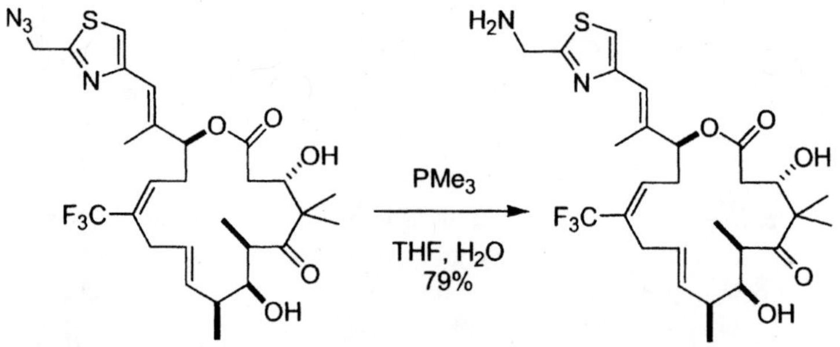

PMe_3, THF, H_2O, 79%

57 → 58

The Staudinger reduction has found its way into the synthesis of many biologically relevant molecules in both academia and industry. The following example, published in 2004 from scientists at Abbott Laboratories, comes from their work toward synthesizing new farnesyltransferase inhibitors. Quinolone **62** is one example from this publication.[29] In the synthesis of this particular inhibitor, bromide **59** was reacted with sodium azide to afford the corresponding azide in 70% yield. Addition of an excess of triphenylphosphine to compound **60** in refluxing THF/H_2O delivered the desired amine **61** in 83% yield. Reductive amination then afforded the final target compound **62**.

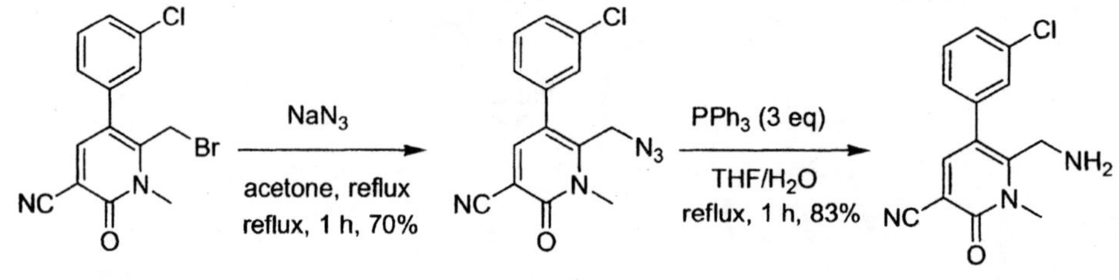

NaN_3, acetone, reflux, reflux, 1 h, 70%; PPh_3 (3 eq), THF/H_2O, reflux, 1 h, 83%

59 → 60 → 61

Scientists at Pfizer recently reported a synthesis of a series 3′-aminoadenosine-5′-uronamides that function as human-selective adenosine A_3 receptor agonists.[30] Compound **64** is an example of a new highly potent and water-soluble analog.[31] The 3′-amino group is introduced in this series *via* a Staudinger reduction using triphenylphosphine in the last step of the synthesis. No yield was reported however.

1. AcOH, MeOH

2. $NaBH_3CN$

62

1. PPh_3 (1.5 eq)

THF, 0 °C to rt

2. H_2O

63

64

In another 2006 example, the Staudinger reaction was utilized in the synthesis of a series of novel oxazolidinones with antibacterial activity.[32] The publication reports that analogs in this new series have superior activity compared to linezolid (marketed as Zyvox™). The Staudinger reaction is used to convert azides **65a** and **65b** to amines **66a** and **66b** and, once in-hand, these intermediates are further elaborated. The nitrile was converted to its corresponding *N*-hydroxyacetamidine and the primary amino group converted to a series of carbamates, amides, thioamides, and thiocarbamates that were then screened against Gram-positive and Gram-negative organisms to measure their antibacterial activity. Two examples (**67** and **68**) of final targets are illustrated below.

PPh_3, H_2O

THF, 45 °C

65a: R^1 = F, R^2 = H

65b: R^1 = R^2 = H

66a: R^1 = F, R^2 = H (78%)

66b: R^1 = R^2 = H (81%)

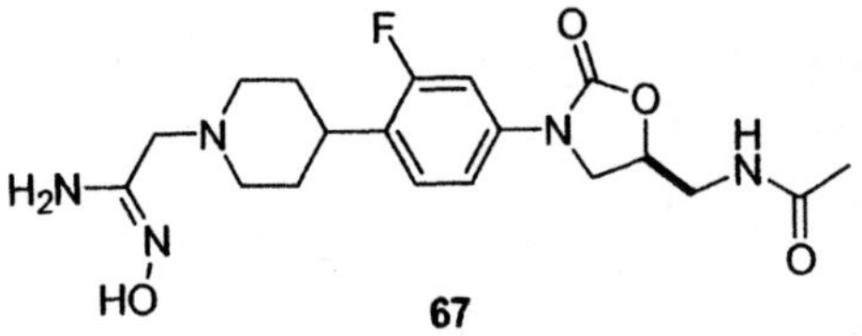

67

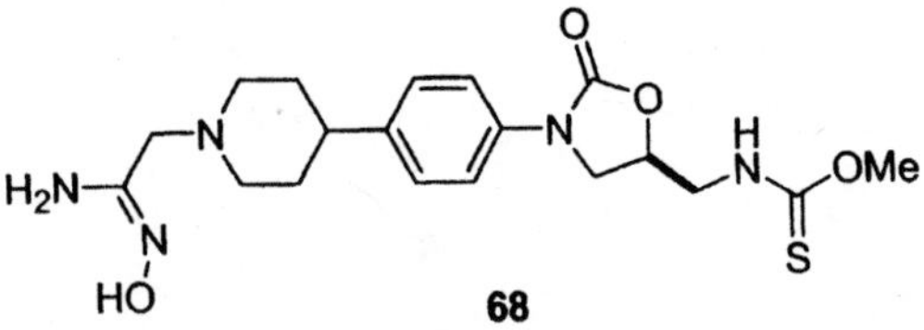

68

2.5.5.2 Tandem intramolecular Staudinger/aza-Wittig to form imines

As mentioned earlier, the Staudinger iminophosphorane intermediate, acting as an aza-ylide, can undergo both intermolecular aza-Wittig transformations with carbonyl containing compounds as well as intramolecular aza-Wittig reactions that can afford a variety of ring systems. There are an enormous number of variations of this application of the Staudinger reaction and an equally enormous number of examples found in the literature. In order to be brief, examples herein will be limited to the imine formation reaction, which is quite broadly found in the literature. The application of the Staudinger reaction to the area of heterocycle synthesis has been reviewed by two very prolific scientists in this arena, Shoji Eguchi and Pedro Molina, and these references are highly recommended.[6–8]

In the first total synthesis of quinine **73**, Gilbert Stork and co-workers constructed the piperidine ring system by refluxing azidoketone **69** in the presence of 1 equivalent of triphenylphosphine to form cyclic imine **71** in 81% yield via iminophosphorane intermediate **70**.[33] The imine was then reduced stereospecifically *via* an axial hydride delivery to afford piperidine **72** using sodium borohydride. The desired synthesis of alkaloid **73** was completed a few steps later.

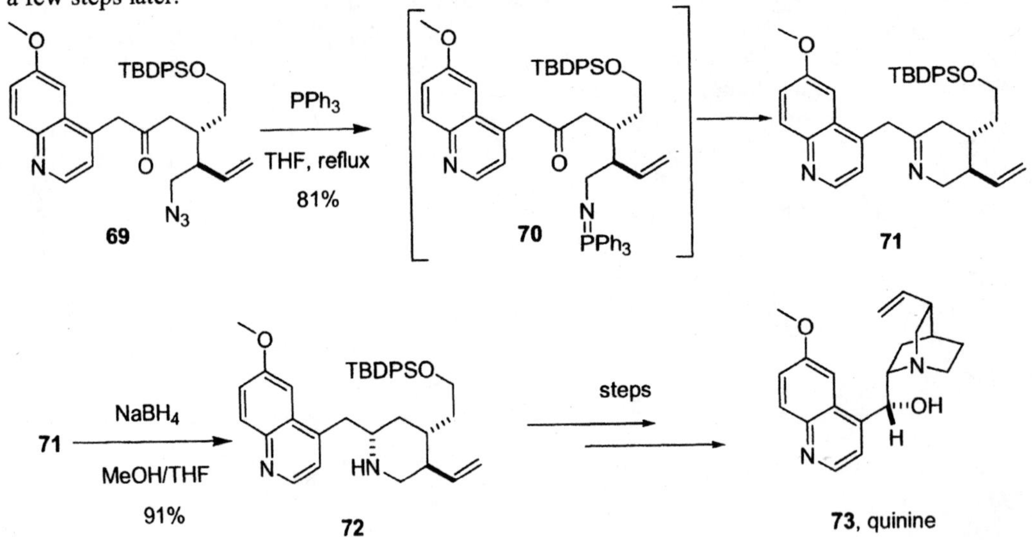

TBDPSO N_3 **69** — PPh_3, THF, reflux, 81% → [**70** N=PPh_3] → **71**

71 — $NaBH_4$, MeOH/THF, 91% → **72** — steps → **73**, quinine

An example of the use of the tandem Staudinger/aza-Wittig reaction to form a seven-membered ring comes from the total synthesis of the alkaloid (–)-stemospironine (**76**).[34] In this example, azidoaldehyde **74** was treated with triphenylphosphine to form a seven-membered cyclic imine which was then followed by an *in situ* sodium borohydride reduction to afford **75**. Treatment of the resulting amine with iodine initiated the formation of the pyrrolidino butyrolactone system of the final target.

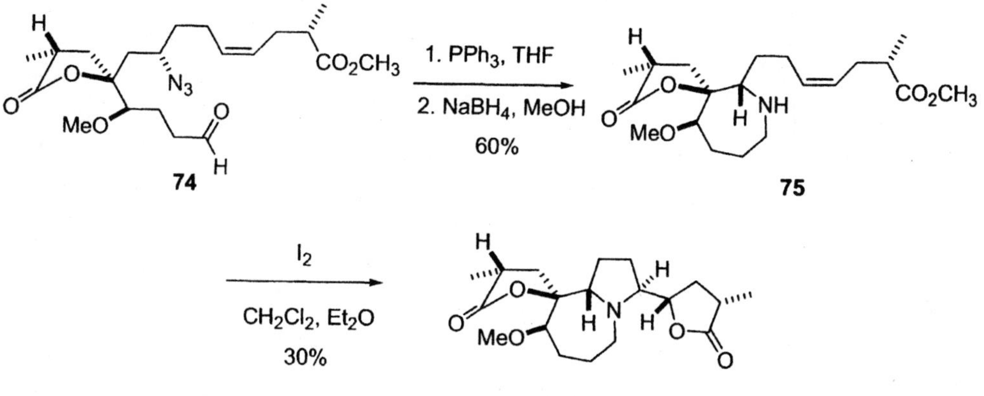

1. PPh_3, THF
2. $NaBH_4$, MeOH
60%

74 → **75**

I_2
CH_2Cl_2, Et_2O
30%

76, (–)-stemospironine

Jiang and co-workers have utilized the tandem intramolecular Staudinger/aza-Witig reaction as the key step toward their synthesis of marine natural product hamacanthin B (**79**) and the antipode of hamacanthin A (**82**).[35,36] In these examples, tributylphosphine was utilized to generate intermediate iminophosphoranes that immediately, upon heating, cyclized with the appropriately positioned carbonyl to afford the central ring systems **78** and **81**.

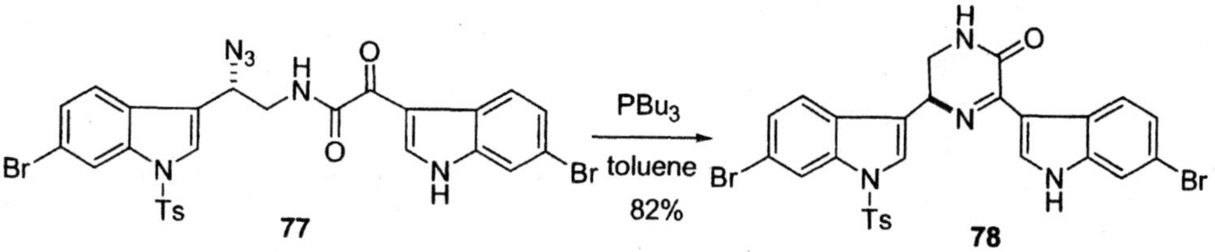

PBu_3
toluene
82%

77 → **78**

L-Selectride
75%

79, hamacanthin B

PBu3
toluene
97%

80

81

NaOH
88%

82, antipode of hamacanthin A

An interesting example of the construction of a five-membered ring *via* the intramolecular Staudinger/aza-Wittig reaction can be found in the work of the Forsyth group on the synthesis of the thiazoline ring contained within the aparatoxin family of potent cytotoxic marine natural products.[37] Their work involved the study of the mild Staudinger/cyclization reaction on model systems such as **83**, shown below, that afforded the desired ring system in very good yields.[38,39] Once their methodology for the synthesis of 2,4-disubstituted thiazolines was well developed and understood, construction of a complex viscinal azido thioester **85** followed by addition of triphenylphosphine and heating to 50 °C in anhydrous THF led to the total synthesis of apratoxin A (**87**).

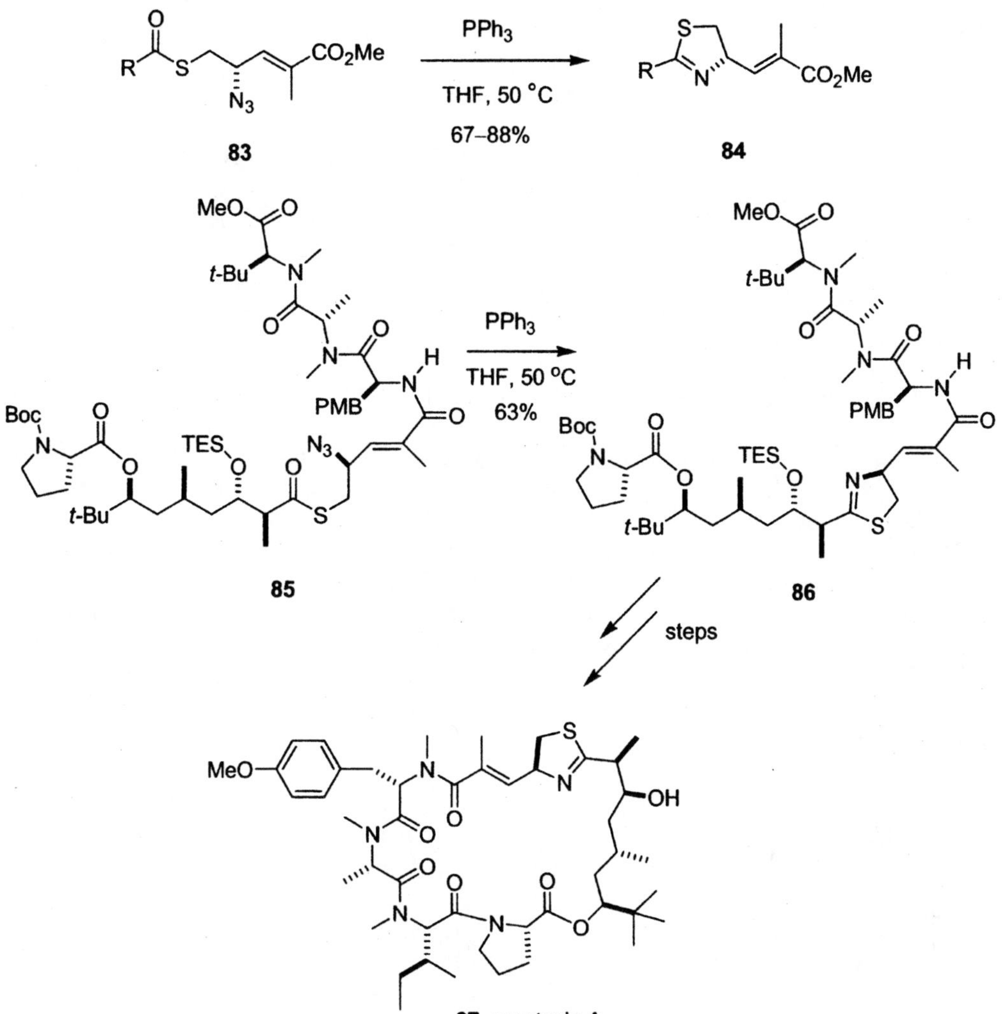

Iminophosphoranes react with carboxylic acids and esters to form amides. The intramolecular version of the process is more facile than the intermolecular version and is an excellent way to prepare large ring systems. Being less reactive than ketones and aldehydes, however, carboxylic acid derivatives require heat and longer reaction times in comparison. For example, the Staudinger/aza-Wittig reaction has been used quite often as the key step toward the construction of seven-membered ring systems of benzodiazepines such as **91**.[40] In this example, tributylphosphine was used to convert azido-ester **88** to iminophosphorane **89**. Compound **89** was then heated in a sealed tube at 140 °C to afford iminoether **90** which was then converted to the desired bis-lactam **91** upon heating in water.

toluene, 140 °C
sealed tube, 24 h

H_2O, reflux
30 min, 91%

90

91

PBu_3
CH_2Cl_2
rt

88, X = N_3

89, X = N=PBu_3

The first total synthesis of the pharmacologically active natural products (–)-benzomalvin A and benzomalvin B also involved intramolecular Staudinger/aza-Wittig reactions of azides with carboxylic acid derivatives.[41] The synthesis of (–)-benzomalvin A (**96**) is an excellent example of the power of this transformation in that both the seven-membered benzodiazepin and six-membered quinazolinone ring systems were constructed in this manner. The transformation of **92** to **94** took place without isolation of the iminophosphorane intermediate in excellent yield using tributylphosphine to generate the iminoether followed by immediate hydrolysis. The second key ring system was constructed by treating azido-bis-amide **95** with 1.1 equivalents of triphenylphosphine to afford natural product **96** in 98% yield.

92

PBu_3 (1.1 eq)
toluene, rt, 2.5 h
then
reflux, 5 h

93

THF/H_2O/TFA
rt, 7 h

(87% from **92**)

94

1. KHMDS, THF
2. 2-azidobenzoyl-chloride, 82%

95

PPh_3 (1.1 eq)
toluene, rt
overnight
then
reflux, 8 h
98%

96, (–)-benzomalvin A

In this last example, the 13-membered ring of the macrocyclic spermine alkaloid (–)-ephedradine A (orantine, **100**) was formed as a result of the Staudinger/aza-Wittig sequence.[42] Treatment of azide **97** bearing an activated ester with triphenylphosphine resulted in the successful formation of the 13-membered iminoether **98** under refluxing toluene conditions. Hydrolysis then afforded the desired lactam **99** in 73% yield. Removal of protecting groups finally revealed the natural product.

PPh_3 (1.1 eq), toluene, reflux; CH_3CN/H_2O, reflux, 73% from **97**

97 (Pfp - pentafluorophenyl) **98** **99**

99 steps → **100**, (–)-ephedarine A (orantine)

2.5.6 *Experimental*

The Staudinger reaction is operationally simple and is run under mild conditions. Solvents of choice are usually toluene or THF. The formation of the iminophosphorane, **1**, is typically rapid, quantitative, and can be followed by thin layer chromatography. Scanning the literature, it is clear that a number of different phosphines can be employed, but triphenylphosphine, tributylphosphine, and trimethylphosphine are the most common. Reaction times, reaction temperature, and stoichiometry vary and seem to be depended upon the steric and electronic environment of the particular azide.

$$R{-}N_3 \xrightarrow[\text{THF/H}_2\text{O}]{\text{PPh}_3\text{ (1 eq)}} R{-}NH_2 \quad \mathbf{10{-}15}$$

General procedure for the synthesis of examples 10–15[15]

To a 1 M solution of the azide in THF is added 1 equivalent of triphenylphosphine, one boiling stone to regulate the evolution of nitrogen gas, and 1.5 equivalents of water. The mixture was stirred for 12 hours, the THF was concentrated *in vacuo,* and the amine isolated by several different methods (a–c) depending upon the particular product.

Method a: The crude reaction mixture was dissolved in anhydrous benzene and dry HCl gas bubbled through. The precipitated HCl salt was collected via centrifugation and then recrystallized.

Method b: The crude residue was dissolved in a 1:1 mixture of diethyl ether and petroleum ether. The precipitated triphenylphosphine oxide was filtered and carefully rinsed. The residue was then purified by bulb-to-bulb distillation.

Method c: The crude residue was dissolved in benzene and the amine extracted with 0.3 M aqueous HCl. The aqueous layer was washed with benzene, neutralized with 1 M aqueous NaOH, saturated with sodium chloride, and extracted into diethyl ether. The ether layer was then dried over sodium sulfate and the resulting amine purified by distillation or recrystallization.

101 $\xrightarrow[\text{THF/H}_2\text{O, rt, 24 h}]{\text{PPh}_3\text{ (3 eq)}}$ **53**

Preparation of amine 53[27]

Treatment of a solution of azide **101** (100 mg, 0.339 mmol) in THF (1 mL) and water (70 μL) with triphenylphosphine (306 mg, 1.17 mmol) for 24 hours at room temperature followed by extractive work-up gave amine **53** (86 mg, 97%).

$$R{-}N_3 \xrightarrow[\text{toluene, rt}]{\text{PMe}_3\text{ (1.05 eq)}} R{-}N{=}PMe_3\ (\mathbf{26}) \xrightarrow[\text{toluene, }-20\ ^\circ\text{C to rt; 2. H}_2\text{O}]{\text{1. Boc-on (1.05 eq)}} R{-}NH{-}Boc\ (\mathbf{28\text{-}32})$$

General procedure for the synthesis of examples 28–32[19]

To a solution of the azide (1 mmol) in toluene (3 mL), under argon at room temperature, was added trimethylphosphine (1.05 mL of a 1 M solution in toluene). After 1 hour the flask was cooled to –20 °C and a solution of Boc-on (259 mg, 1.05 mmol) in toluene (1 mL) was added via canula. After stirring for 5 hours at room temperature, 50–100 mL of dichloromethane or diethyl ether was added and the solution extracted with water (3 ×) and brine. The organic phase was dried, concentrated, and filtered through a small pad of alumina (99:1 dichloromethane/methanol) to afford the desired carbamates **28** through **32**.

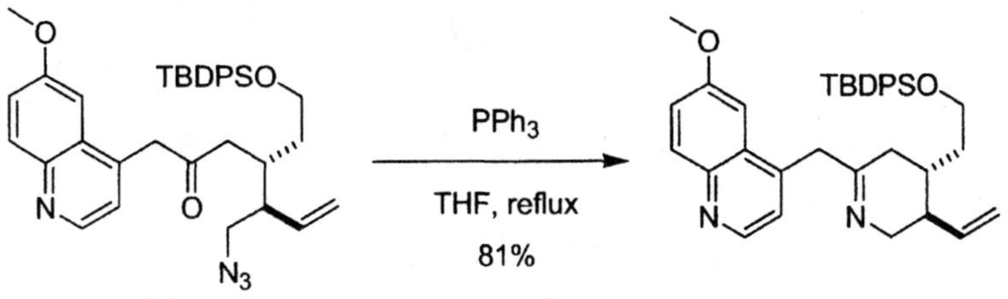

Preparation of 71[33]

To a solution of azidoketone **69** (130 mg, 0.22 mmol) in THF (20 mL) was added triphenylphosphine (62 mg, 0.24 mmol). The reaction mixture was refluxed for 3 hours, and the THF removed under vacuum. The residue was purified by flash chromatography (50% ether/hexanes/1% triethylamine) to give **71** (89 mg, 81%).

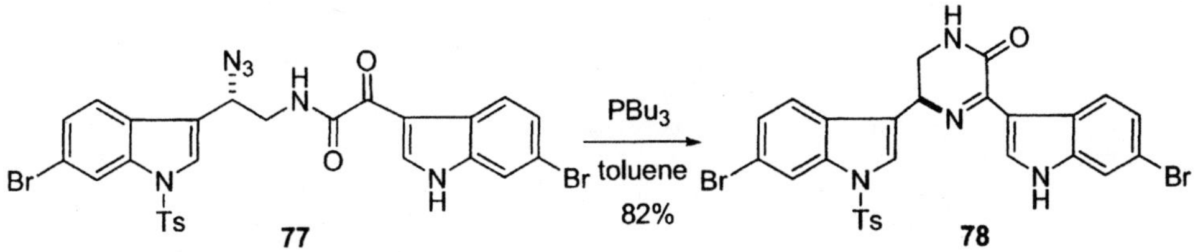

Preparation of 78[36]

To a solution of azide **77** (330 mg, 0.48 mmol) in dry toluene (25 mL) was added tributyphosphine (202 mL, 0.80 mmol). The mixture was stirred at room temperature for 2 hours and then warmed to reflux for 20 hours under an argon atmosphere. After the removal of toluene, the residue was subjected to flash chromatography (silica, hexane/ethyl acetate 1:1 to 1:2) to give compound **78** (270 mg, 82%) as a yellow solid.

95

PPh_3 (1.1 eq)

toluene, rt
overnight
then
reflux, 8 h
98%

96, (–)-benzomalvin A

Preparation of (–)-benzomalvin A (96)[41]

To a solution of **95** (159 mg, 0.374 mmol) in toluene (10 mL) was added triphenylphosphine (108 mg, 0.41 mmol, 1.1 eq) in toluene (20 mL) at room temperature. The reaction was stirred at ambient temperature overnight then refluxed for 8 hours. The mixture was concentrated under reduced pressure to afford a solid residue, which was then purified by silica gel chromatography (1:1 ethyl acetate/hexane) to give **96** (140 mg, 0.367 mmol, 98%).

2.5.7. References

1. [R] Scriven, E. F. V.; Turnbull, K. *Chem. Rev.* **1988**, *88*, 297.
2. [R] Brase, S.; Gil, C.; Knepper, K; Zimmermann, V. *Angew. Chem. Int. Ed.* **2005**, *44*, 5188.
3. [R] Barluenga, J.; Palacios, F. *Org. Prep. Proc. Int.* **1991**, *23*, 1.
4. [R] Valentine, D. H.; Hillhouse, J. H. *Synthesis* **2003**, 317.
5. Staudinger, H.; Meyer, J. *Helv. Chim. Acta.* **1919**, *2*, 635.
6. [R] Eguchi, S.; Matsushita, Y.; Yamashita, K. *Org. Prep. Proc. Int.* **1992**, *24*, 209.
7. [R] Molina, P; Vilaplana, M. J. *Synthesis* **1994**, 1197.
8. [R] Fresneda, P. M.; Molina, P. *Synlett* **2004**, 1.
9. [R] Gololobov, Y. G.; Zhmurova, I. N.; Kasukhin, L. F. *Tetrahedron* **1981**, 37, 437.
10. [R] Gololobov, Y. G.; Kasukhin, L. F. *Tetrahedron* **1992**, 48, 1353.
11. Alajarin, M.; Coneas, C.; Rzepa, H. S. *J. Chem. Soc. Perkin Trans. 2* **1999**, 1811.
12. Tian, W. Q.; Wang, Y. A. *J. Org. Chem.* **2004**, *69*, 4299.
13. Tian, W. Q.; Wang, Y. A. *J. Chem. Theory Comput.* **2005**, *1*, 353.
14. Vaultier, M; Knouzi, N.; Carrie, R. *Tetrahedron Lett.* **1983**, *24*, 763.
15. Knouzi, N.; Vaultier, M; Carrie, R. *Tetrahedron Lett.* **1985**, *26*, 815.
16. Charette, A. B.; Boezio, A. A.; Janes, M. K. *Org. Lett.* **2000**, *2*, 3777.
17. Lindsley, C. W.; Zhao, Z.; Newton, R. C.; Leister, W. H.; Strauss, K. A. *Tetrahedron Lett.* **2002**, *43*, 4467.
18. Bosanac, T.; Wilcox, C. S. *Org. Lett.* **2004**, *6*, 2321.
19. Ariza, X.; Urpi, F.; Viladomat, C.; Vilarrasa, J. *Tetrahedron Lett.* **1998**, *39*, 9101.
20. Alfonso, C. A. M. *Tetrahedron Lett.* **1995**, *36*, 8857.
21. Ariza, X.; Urpi, F.; Viladomat, C.; Vilarrasa, J. *Tetrahedron Lett.* **1999**, *40*, 7515.
22. Kato, H.; Ohmori, K.; Suzuki, K. *Synlett* **2001**, 1003.
23. Kohn, M.; Breinbauer, R. *Angew. Chem. Int. Ed.* **2004**, *43*, 3106.
24. Nicolaou, K. C.; Nevalainen, M.; Zak, M.; Bulat, S.; Bella, M.; Safina, B. S. *Angew. Chem. Int. Ed.* **2003**, *23*, 3418.
25. Li, J.; Chen, H-N.; Chang, H.; Wang, J.; Chang, C.-W., T. *Org. Lett.* **2005**, *7*, 3061.
26. Rai, R.; Chen, H-N.; Czyryca, P. G.; Li, J.; Chang, C-.W. *Org. Lett.* **2006**, *8*, 887.
27. Butcher, J. W.; Liverton, N. J.; Selnick, H. G.; Elloit, J. M.; Smith, G. R. *Tetrahedron Lett.* **1996**, *37*, 6685.
28. Rivikin, A. Yoshimura, F.; Gabarda, A. E.; Cho, Y. S.; Chou, T-C.; Dong, H.; Danishefsky, S. J. *J. Am. Chem. Soc.* **2004**, *126*, 10913.
29. Li, Q.; Claiborne, A.; Li, T.; Hasvold, L.; Stoll, V. S.; Muchmore, S.; Jakob, C. G.; Gu, W.; Cohen, J.; Hutchins, C.; Frost, D.; Rosenberg, S. H.; Sham, H. L. *Bioorg. Med. Chem. Lett.* **2004**, *14*, 5367.

30. DeNinno, M. P.; Masamune, H.; Chenard, L. K.; DiRico, K. J.; Eller, C.; Etienne, J. B.; Tickner, J. E.; Kennedy, S. P.; Knight, D. R.; Kong, J.; Oleynek, J. J.; Tracey, W. R.; Hill, R. J. *J. Med. Chem.* **2003**, *46*, 353.
31. DeNinno, M. P.; Masamune, H.; Chenard, L. K.; DiRico, K. J.; Eller, C.; Etienne, J. B.; Tickner, J. E.; Kennedy, S. P.; Knight, D. R.; Kong, J.; Oleynek, J. J.; Tracey, W. R.; Hill, R. J. *Bioorg. Med. Chem. Lett.* **2006**, *16*, 2525.
32. Takhi, M.; Murugan, C.; Munikumar, M.; Bhaskarreddy, K. M.; Singh, G.; Sreenivas, K.; Sitaramkumar, M.; Selvakumar, N.; Das, J.; Trehan, S.; Iqbal, J. *Bioorg. Med. Chem. Lett.* **2006**, *16*, 2391.
33. Stork, G.; Niu, D.; Fujimoto, A.; Koft, E. R.; Balkovec, J. M.; Tata, J. R.; Dake, G. R. *J. Am. Chem. Soc.* **2001**, *123*, 3239.
34. Williams, D. R.; Fromhold, M. G.; Earley, J. D. *Org. Lett.* **2001**, *3*, 2721.
35. Jiang, B.; Yang, C-G.; Wang, J. *J. Org. Chem.* **2001**, *66*, 4865.
36. Jiang, B.; Yang, C-G.; Wang, J. *J. Org. Chem.* **2002**, *67*, 1369.
37. Chen, J.; Forsyth, C. J. *Org. Lett.* **2003**, *5*, 1281.
38. Chen, J.; Forsyth, C. J. *J. Am. Chem. Soc.* **2003**, *125*, 8735.
39. Hartung, R.; Paquette, L. *Chemtracts-Organic Chemistry* **2004**, *17*, 72.
40. Molina, P.; Diaz, I.; Tarraga, A. *Tetrahedron* **1995**, *51*, 5617.
41. Sugimori, T.; Okawa, T.; Eguchi, S.; Kakehi, A.; Yashima, E.; Okamoto, Y. *Tetrahedron* **1998**, *54*, 7997.
42. Kurosawa, W.; Kan, T.; Fukuyama, T. *J. Am. Chem. Soc.* **2003**, *125*, 8112.

Donna M. Iula

2.6 Wharton Reaction

2.6.1 *Description*

The Wharton reaction is the transformation of α,β-epoxy ketones **1** by hydrazine to allylic alcohols **2**.[1–4] It is also known as the Wharton transposition, Wharton rearrangement, and Wharton reduction.

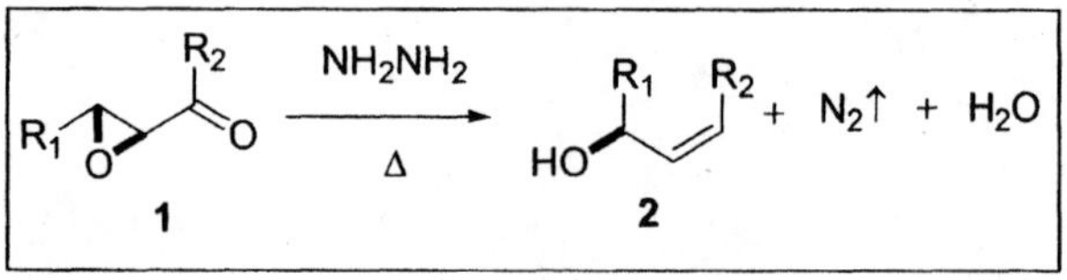

2.6.2 *Historical Perspective*

In September 1961, P. S. Wharton of the University of Wiscosin at Madison and his student, D. H. Bohlen, published a *Communication to the Editor* in the *Journal of Organic Chemistry* titled *Hydrazine Reaction of α,β-Epoxy Ketones to Allylic Alcohols*.[5] In that paper, they reported transformation of 4β,5-epoxy-3-coprostanone (**3**) to allylic alcohol **4** by treating **3** with two to three equivalents of hydrazine hydrate and 0.2 equivalent of acetic acid in ethanol. A couple months later, Wharton himself published another *JOC Communication* titled *Stereospecific Synthesis of 6-Methyl-*trans*-5-cyclodecenone,* delineating the utility of the aforementioned reaction.[6] After that, the organic chemistry community began to take notice because methods for this kind of transformation were not numerous. By now, not only has the Wharton reaction found widespread utility in organic synthesis, it is also immortalized in the pantheon of named reactions in organic chemistry.

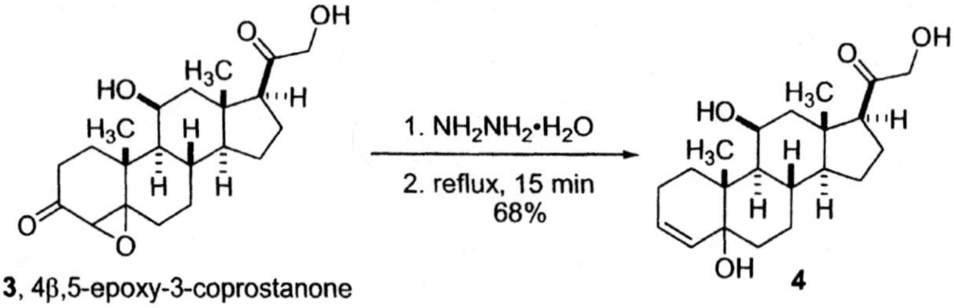

2.6.3 *Mechanism*

The mechanism of the Wharton reaction is analogous to that of the Wolff–Kishner reduction. Addition of hydrazine to α,β-epoxy ketone **1** gives rise to hydroxyl-hydrazine **6**, which dehydrates to afford the intermediate, hydrazone **7**. Tautomerization of hydrazone **7** then leads to diazene **8**, a common intermediate as the Wolff–Kishner reduction. But different from the Wolff–Kishner reduction, the diazene **8** pushes open the adjacent epoxide with concomitant release of nitrogen gas, giving rise to allylic alcohols **2**. According to the mechanism, allylic alcohols **2** retains the configuration of the α,β-epoxy ketone **1**, a fact confirmed by numerous experimental data.

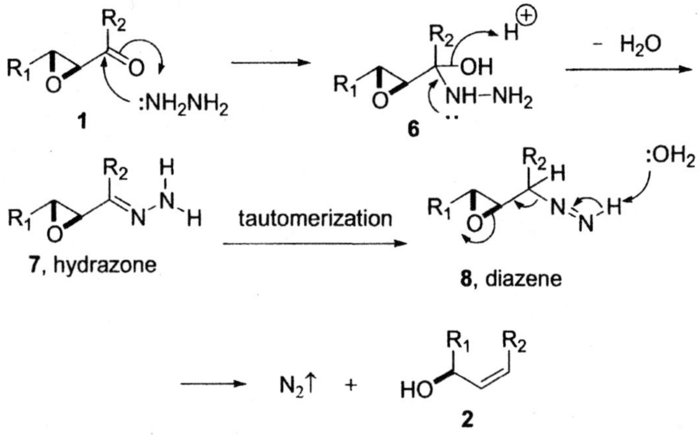

1 :NH2NH2 → 6 (R2, H⊕, OH, NH–NH2) — H2O →

7, hydrazone — tautomerization → 8, diazene (:OH2)

→ N2↑ + HO (R1, R2) 2

2.6.4 Variations and Improvements

No sooner than Wharton's landmark publications did Dodson of G. D. Searle discover a side reaction of the Wharton reaction.[7] Treatment of 16α,17-epoxypregnenolone (**9**) with hydrazine hydrate resulted in not only the expected allylic alcohol **10** (ca. 25% yield), but also a small amount of pyrazole **11** (ca. 10% yield) as well. It was speculated that the mechanism involved an intramolecular S_N2 displacement of the epoxide by the diazene (*the anchimeric process*) followed by dehydration.[8]

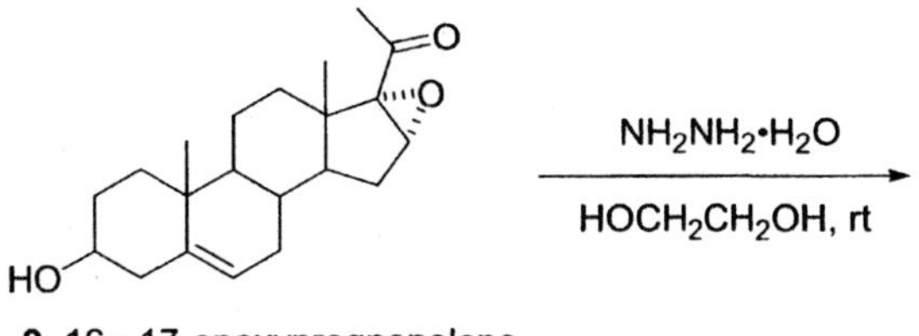

$NH_2NH_2{\cdot}H_2O$ / $HOCH_2CH_2OH$, rt

9, 16α,17-epoxypregnenolone

Another deviation from the "normal" course of the Wharton reaction was recorded by Ohloff and Uhde of Firmenich in 1970.[9] When olefinic epoxyketone **12** was treated with hydrazine hydrate, not only was the expected allylic alcohol **13** obtained, a bicyclic product **14** was also isolated. The authors proposed that both products went through a common intermediate, vinyl anion **17**.

However, Gilbert Stork further scrutinized the reaction in 1977 and rejected the vinyl carbanion intermediacy because such a species (**17**) could not survive in the methanolic medium and would not be expected to add to an unactivated tri-substituted oelfin.[10] He proposed that the cyclization went directly through a concerted collapse of **16** to give or involve a free radical intermediate **18** formed spontaneously from the unstable diazene **16**.

An apparent failure of the Wharton reaction of a sterically hindered epoxyketone **19** was recorded.[11] Instead of the expected allylic alcohol, its isomer, allylic alcohol **20**, and allylic ether **21** were isolated. The Wharton allylic alcohol could be otherwise synthesized using the Barton modification that involving an epoxy-zanthate and then a radical rearrangement of the resulting epoxymethylene radical.[12]

19 → 20, R = H, 43%; 21, R = Et, 40% ($NH_2NH_2{\bullet}H_2O$, EtOH, rt, overnight)

In 1989, Dupuy and Luche rendered a great service to the synthetic community by publishing their careful, systematic investigation and improvement of the Wharton reaction.[2] For stable epoxyhydrazones such as **12**, they found that treatment of **12** with 2 equivalents of hydrazine hydrate and a drop of acetic acid in dichloromethane gave the corresponding epoxyhydrazone **15** in nearly quantitative yields. Although treatment of **15** with *n*-BuLi and LDA at –100 to –78 °C suffered very low yields, potassium diisopropylamine (KDA) and potassium *t*-butoxide afforded the desired allylic alcohol **13** in 68% and 76% yields, respectively. As far as unstable epoxyhydrazones were concerned, the classic Wharton conditions seemed inadequate in providing satisfactory results. A possible solution discovered by Dupuy and Luche was replacing epoxyhydrazone with a stable derivative, such as epoxysemicarbazole **23**, in which case allylic alcohol **24** was obtained in 47% yield, whereas only 35% yield was obtained when using epoxyhydrazone. Last but not least, Dupuy and Luche discovered that free hydrazine consistently gave better yields than hydrazine hydrate. They prepared hydrazine *in situ* by treating the hydrazine salts with triethylamine in acetonitrile.

22 → 23 → 24

Unfortunately, for whatever reason, Dupuy and Luche's improvements have not caught on. Most publications surveyed by this reviewer still used the "classic" Wharton conditions.

2.6.5 *Synthetic Utility*

Although yields for the Wharton reaction varied considerably, it has found widespread utility in steroids, terpenoids, alkaloids, prostagladins, and many other substrates.

The Wharton reaction owes its genesis to steroid chemistry. So it is fitting that it enjoys extensive applications in steroid chemistry. When epoxyketone **25** was refluxed in a mixture of hydrazine and KOH, isomeric allylic alcohols **26** and **27** were obtained in 41% and 26% yields, respectively.[13]

NH_2NH_2, KOH

THF, reflux, 67%

25

26, 41%

27, 26%

Synthetic utility of the Wharton reaction in alkaloids can be exemplified by the oxygen transposition of epoxyketone **28** to allylic alcohol **29**.[14] Many similar cases have been reported.[15,16]

$NH_2NH_2 \cdot H_2O$, MeOH

HOAc, rt, 52%

28

29

In terpenoid chemistry, the Wharton reaction of epoxyketone **30** provided an expedient entry to allylic alcohol **31**, which was an intermediate toward scyphostatin, a natural product.[17]

TESO, 30, $NH_2NH_2 \cdot H_2O$, MeOH, HOAc, rt, 59%, TESO, OH, 31

Advantage was taken of the Wharton reaction to carry out an oxygen transposition of **32** to provide **33**, which was only two steps away from 11-deoxy-prostaglandin $F_{1\alpha}$ after desilylation and hydrolysis.[18]

TBDMSO, CO_2CH_3, 32, $NH_2NH_2 \cdot HCl$, Et_3N, (dimethylamino)ethanol, 32%

TBDMSO, CO_2CH_3, OH, 33

2.6.6 *Experimental*

TESO, 30, $NH_2NH_2 \cdot H_2O$, MeOH, HOAc, rt, 59%, TESO, OH, 31

6-Hydroxy-10-triethylsilyloxy-1-oxaspiro[4,5]dec-7-en-2-one (31)[17]

To a solution of **30** (571.8 mg, 1.83 mmol) in methanol (90 mL) at room temperature was added drop-wise $NH_2NH_2 \cdot H_2O$ (0.18 mL, 3.7 mmol) and AcOH (0.21 mL, 3.7 mmol). The reaction mixture was stirred at room temperature for 10 min. After diluting with saturated $NaHCO_3$ and saturated NH_4Cl, the reaction was extracted with EtOAc. The organic layer was washed, dried over $MgSO_4$, and evaporated. The resulting residue was purified by column chromatography (silica gel, hexane-EtOAc 3:1) to give **31** (319.5 mg, 59%) as a colorless oil.

2.6.7 References

1. [R] Morris, D. G. *Chem. Soc. Rev.* **1982**, *11*, 379–434.
2. [R] Dupuy, C.; Luche, J. L. *Tetrahedron* **1989**, *45*, 3437–3444.
3. [R] Chamberlin, A. R.; Sall, D. *Comp. Org. Syn.* **1991**, *8*, 927–929.
4. [R] Li, J. J. *Name Reactions, A Collection of Detailed Reaction Mechanisms*, Springer-Verlag. *Third Edition*, **2006**, 616–617.
5. Wharton, P. S.; Bohlen, D. H. *J. Org. Chem.* **1961**, *26*, 3615.
6. Wharton, P. S. *J. Org. Chem.* **1961**, *26*, 4781.
7. Benn, Walter R.; Dodson, R. M. *J. Org. Chem.* **1964**, *29*, 1142.
8. Coffen, David L.; Korzan, D. G. *J. Org. Chem.* **1971**, *36*, 390.
9. Ohloff, G.; Uhde, G. *Helv. Chim. Acta* **1970**, *53*, 531.
10. Stork, G.; Williard, P. G. *J. Am. Chem. Soc.* **1977**, *99*, 7067.
11. Thomas, A. F.; Di Giorgio, R.; Guntern, O. *Helv. Chim. Acta* **1989**, *72*, 767.
12. Barton, D. H. R.; Motherwell, R. S. H.; Motherwell, W. B. *J. Chem. Soc., Perkin Trans. 1* **1981**, 2323.
13. Di Filippo, M.; Fezza, F.; Izzo, I.; De Riccardis, F.; Sodano, G. *Eur. J. Org. Chem.* **2000**, 3247.
14. Kim, G.; Chu-Moyer, M. Y.; Danishefsky, S. J. *J. Am. Chem. Soc.* **1990**, *112*, 2003.
15. Majewski, M.; Lazny, R. *Synlett* **1996**, 785.
16. Majewski, M.; Lazny, R.; Ulaczyk, A. *Can. J. Chem.* **1997**, *75*, 754.
17. Takagi, R.; Tojo, K.; Iwata, M.; Ohkata, K. *Org. Biomol. Chem.* **2005**, *3*, 2031.
18. Yamada, K.-i.; Arai, T.; Sasai, H.; Shibasaki, M. *J. Org. Chem.* **1998**, *63*, 3666.

Jie Jack Li

CHAPTER 3 Oxidation 159

3.1 Baeyer–Villiger Oxidation

3.1.1 Description

The Baeyer–Villiger oxidation is an oxygen insertion reaction whereby an aldehyde, cyclic or acyclic ketone **1** is treated with a suitable oxidant, e.g., an organic peracid **3**, to form the corresponding formate, lactone or ester **2**. The oxidation takes place in two discrete steps: (1) the attack of the peroxide on the carbonyl to provide the Criegee intermediate, followed by (2) rearrangement to the product ester.

1
or suitable oxidant
Criegee Intermediate
2

R_1 = alkyl, aryl, H
R_2 = alkyl or aryl
R_m = alkyl or aryl

A variety of oxidants including inorganic or organic peracids, oxygen or hydroperoxides with metal catalyst or promoter, and isolated or partially purified enzymes with co-factors or whole cells have been successfully employed for this transformation. Key to the successful use of this reaction is having high predictability of the migrating group. It is empirically accepted that the migratory aptitude (the relative ease of a group to migrate) is in most cases based on the group's ability to stabilize a positive charge although it is very important to not underestimate conformational, steric, and electronic factors which may influence the migration.

3.1.2 Historical Perspective

menthone, **4**
$KHSO_5$
82%
5
dioxirane, **6**
dimer, **7**

Baeyer and Villiger reported the oxidative ring cleavage of cyclic ketones (terpenes) in 1899.[1] A subsequent article[2] appeared describing in more detail the reaction of Caro's acid ($KHSO_5$; sold today as Oxone® a mixture of $KHSO_5$, $KHSO_4$, and K_2SO_4) with ketones, in which Baeyer and Villiger also proposed the structure of Caro's acid. In addition to discovering a new reaction, with the lack of analytical equipment in comparison to what is used today, it is striking that the product composition and structure could be determined. The mechanism proposed at that time postulated a dioxirane intermediate **6** and dimerization to form **7**. The dimerized intermediate was dismissed years later when two isolated dimers (derived from benzophenone and cyclohexanone) were shown not to convert to esters.[3] Over 50 years after the discovery of the reaction and much debate, the mechanism of the Baeyer–Villiger oxidation was elucidated by an enriched ^{18}O tracer experiment which distinguished three proposed mechanisms in a single experiment. The mechanisms were proposed by Wittig, Baeyer–Villiger, and Criegee.[4] The experiment and apparent differentiation of the three mechanisms is shown below. Wittig's proposed mechanism would lead to the formation of ester **13** and subsequently give rise to ^{18}O enriched phenol **15** (after reduction of **13**). The Criegee mechanism would give rise to ^{18}O enrichment in the carbonyl oxygen of ester **16**, thus leading to ^{18}O enriched benzyl alcohol **17** after reduction of the esters **16**. The original mechanism put forth by Baeyer and Villiger described going though a dioxirane intermediate that would give rise to a 1:1 distribution of ^{18}O enrichment in both the phenol and benzyl alcohol formed after reduction. The Criegee mechanism was supported by an enriched ^{18}O content in isolated benzyl alcohol **17** and no ^{18}O enrichment of the phenol **15**.

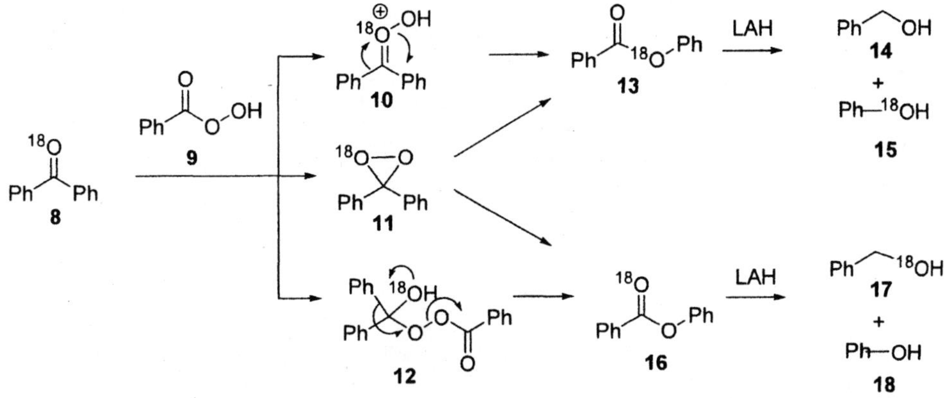

During this time, a variety of organic peracids were developed and used for this reaction which led to better yields. In addition, the reaction was determined to proceed with retention of configuration of the migrating group. A superb historical review has been written by Renz and Meunier on development and use of the Baeyer–Villiger oxidation.[3]

Predicting the migration ability of groups has also improved during the first 50 years and continues to be discussed today. The development of alternative reagents which are more selective and mild continues. The development and application of catalytic systems whether enzymatic or metal mediated, which will predictably provide asymmetric oxidation products, have been successfully applied but remains in its early stages.

3.1.3 Mechanism

3.1.3.1 General Mechanistic Considerations

The mechanism of the Baeyer–Villiger reaction is accepted to occur in two steps. Step 1 results from peracid **20** addition to a ketone or aldehyde **19** to form a tetrahedral intermediate **21** (Criegee intermediate). Step 2 is the subsequent rearrangement of the tetrahedral intermediate, which involves the collapse of electrons on the hydroxyl of **21** to form a carbonyl with concomitant migration of a migrating group and the ejection of the acid unit (formerly the peroxy unit) thus forming a lactone or ester **22**. The rearrangement requires the movement of six electrons.

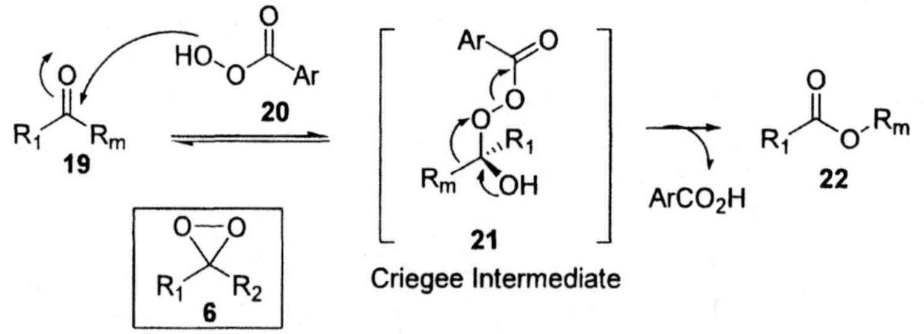

The establishment of the Criegee intermediate and mechanism was mentioned in the historical section. About 90 years after the discovery of the Baeyer–Villiger reaction, Curci and co-workers showed through ^{18}O tracer experiments, analogous to those employed by von Doering and Dorfman, that bis(TMS)peroxomonosulfate (bis-TMS Caro's acid) promoted the formation of dioxirane intermediates like **6** and led to Baeyer–Villiger products **22**.[5] It has been noted that dioxiranes are generated in reactions of ketones with Caro's acid at nearly neutral pH. Also, when the leaving group ($ArCO_2^-$) contained powerful EWG, the formation of intermediate **6** was noted. Curci noted that the dioxirane mechanism prevailed, although it was not exclusive in all substrates tried.

While Curci suggested a radical mechanism, all the data did not support such a mechanism.

Several co-workers have performed experiments showing that the Baeyer–Villiger reaction occurred with retention of configuration at the migrating center: first with racemic material comparing *cis* and *trans* 1-acetyl-2-methylcyclohexane then subsequently with a steroid backbone, (*S*)-3-phenyl-2-butanone, and an *exo*-norbornyl methyl ketone.[3,6] In each instance, 94–100% retention of configuration was proved to occur. As a rule, the rate determining step for the Baeyer–Villiger oxidation has been thought to be the rearrangement step; there are exceptions in which the addition to the ketone was shown to be rate determining.[6] Recently computations have supported that the rate determining step varied due to the variation in the keto-substituents, thus making the addition to the ketone rate determining.[7] A recent change in total synthesis was required when TFPAA could not attack the ketone due to steric reasons.[8]

3.1.3.2 Mechanism: Preference of Migrating Group

The Baeyer–Villiger reaction can be catalyzed by acid or base. Electron-withdrawing groups (EWGs) attached to Ar of peracid **20**, can facilitate the reaction (better leaving group), for example, *m*-chloroperbenzoic acid or trifluoromethyl peracetic acid. As mentioned in the previous section, one needs to consider both steps when trying to promote this reaction. Sheldon has summarized the preference for migration in Baeyer–Villiger reactions in the following order: tertiary > cylcohexyl > secondary > benzyl > phenyl > primary > methyl.[9] For diaryl unsymmetrical ketones the more electron-releasing group typically migrates. An ortho effect has been noted for diaryl ketones; if the aryl group contains an ortho substituent, migration is hindered relative to an aryl group that does not have ortho substituents. Oftentimes, dialkyl ketones require more powerful oxidizing agents than peracetic or perbenzoic acid. An exception to this rule is the Baeyer–Villiger oxidation of cyclobutanones which will undergo reaction using hydroperoxides. Other groups that one would expect to stabilize a cation have also been utilized to direct migration in a Baeyer–Villiger reaction. Having a β-silyl group was suggested to be more directing than a secondary alkyl group (as shown in the conversion of **23** into **24**),[10] while β-stannyl groups were shown to promote fragmentation of a substrate and proved more powerful in directing than the *t*-butyl group (conversion of **25** into **26**).[11]

O
SiMe$_3$
23
m-CPBA, Na_2HPO_4, CH_2Cl_2 99%
O O
SiMe$_3$
24

O
SnBu$_3$
25
1) *m*-CPBA, CH_2Cl_2, radical scav.
2) BnBr, DBU, PhH 93%
O
OBn
26

Contrarywise, electron-withdrawing substituents attached to potential migrating groups, specifically halogens, have been shown to inhibit migration when in competition with many different groups. In a series of non-, di-, and trifluorinated methyl alkyl ketones **27**, the halogenated side proved to be less prone to migration than methyl.[12] As one would suspect, due to the EWGs stabilizing the Criegee intermediate, the migration should be slower with these substrates. Only migration of the hexyl group was observed affording **28** under acidic conditions. Kitazume[12] described that the LUMO (lowest unoccupied molecular orbital) energy level at the reactive site of the most stable conformer of each of the Criegee intermediates predicted the migrating group. In contrast, under slightly basic conditions (buffered), Uneyama and co-workers reported[13] the increase in reaction rate depending on the number of α-fluorine atoms. Another experimental and computational article has recently appeared in which Grein and Crudden concluded that the introduction of a fluorine or chlorine substituent decreased the experimental and theoretical rate of reaction for *m*-CPBA promoted oxidations.[14]

m-CPBA, $BF_3 \cdot OEt_2$, $CHCl_3$, reflux: **27** → **28** + **29**

R	Conversion %	Yield % 28	Yield % 29
CH_3	100	99	0
CHF_2	90	87	0
CF_3	74	64	0
C_2F_5	6	6	0

Recently, the CF_3 group attached to the α-carbon was shown to retard migration, and a primary alkyl group migrated in preference to a secondary alkyl group containing the CF_3 group.[15] Contrary to this observation, substituents attached to carbon next to the ketone such as ether, acetate, *N*-alkyl, *N*-tosyl, and *N*-pthalimido are reported to facilitate migration.[6] Various benzoyl diethylphosphonates **30** were treated with benzoyl peroxide to give 70–85% yields of the benzoyl diethyl phosphates **31** in which the phosphoryl group migrates.[16]

$PhCO_3H$, EtOAc, PhH, 70–85%: **30** → **31**

R = H-, *p*-F-, *m*-Cl-, *p*-MeO-

The study to predict the migrating group in the Baeyer–Villiger reaction has been scrutinized by several authors. As previously stated, conformational, steric, and electronic factors have been noted as dictating which group migrates. These effects are called the primary stereoelectronic effect and the secondary stereoelectronic effect pictorially shown in **32**.[6,9] The primary stereoelectronic effect is the observation that for the migration to occur the peroxide bond is required to be antiperiplanar to the migrating group. Alternately, the primary stereoelectronic effect can be viewed as the alignment of the migrating group orbital with the σ* of the proximal oxygen of the peroxide, e.g. (**33**).

primary effect; secondary effect
32

primary effect: σ–σ* interaction
33

The secondary electronic effect is the observation that for the rearrangement to occur, an oxygen lone pair should be antiperiplanar to the migrating group. Both of these effects impact the regioselectivity of the reaction and illustrate that a stereoelectronically preferred orientation of atoms is required to promote the rearrangement in the Baeyer–Villiger reaction. This orientation may be hindered by a surrounding steric environment.

Examples of the primary stereoelectronic effect have been demonstrated in the literature. Chandrasekhar and Roy showed that rearrangement of 2-oxo-cyclohexyl-peroxyacetic acid **35**, derived from acid **34**, proceeded via intermediates **36** and **37** to **38** in 62% overall yield.[17] Migration of bond **a** was the only bond which migrated being antiperiplanar to the peroxide; no product of migration of bond **b** was observed. Computational studies recently confirmed the preference for antiperiplanar migration versus gauche migration.[18]

H_2O_2, CH_3CO_2H, 70 °C, 3 h

34 **35** **36** **37**

62%

38

More recently, Crudden and co-workers have published experiments which support the primary stereoelectronic effect and go one step further arguing that dipole–dipole interactions of the α- and α′-F biased the conformation of the peroxyester bond.[18] The "migration occurred primarily from the conformer in which the O–O bond of the peroxyester is oriented to minimize the overall dipole" (neutralization of dipoles). The carbon containing an equatorial F in Criegee intermediate **40** proved to migrate with 90% selectivity in comparison the carbon containing an axial F. Analysis of the configuration required for the rearrangement shows disalignment of dipoles in configuration **43**, in comparison to **41** which predicts the major product **44**.

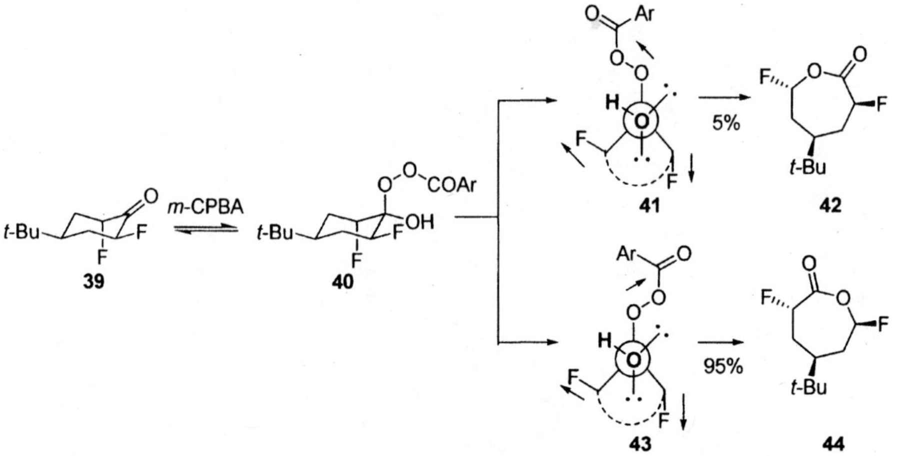

Noyori and co-workers have reported evidence for the secondary stereoelectronic effect.[20] Oxidation of ketone **45** with trifluoromethyl peracetic acid (TFMPAA) provided most reactive conformers **46** and **48**. In both of the preferred arrangements, a lone pair of electrons can be antiperiplanar to the migrating group, but due to non-bonding steric interactions, the α′-conformer is more populated than the α-conformer; and therefore, **47** predominates in the ratios shown in the table. The proximity of the R substituent to the hydroxyl of the Criegee intermediate became obvious after viewing **50**. Note that when R = *t*-butyl, no reaction was reported. Sterics may impact the attack on the ketone but most certainly impact the arrangement of atoms necessary to promote rearrangement.

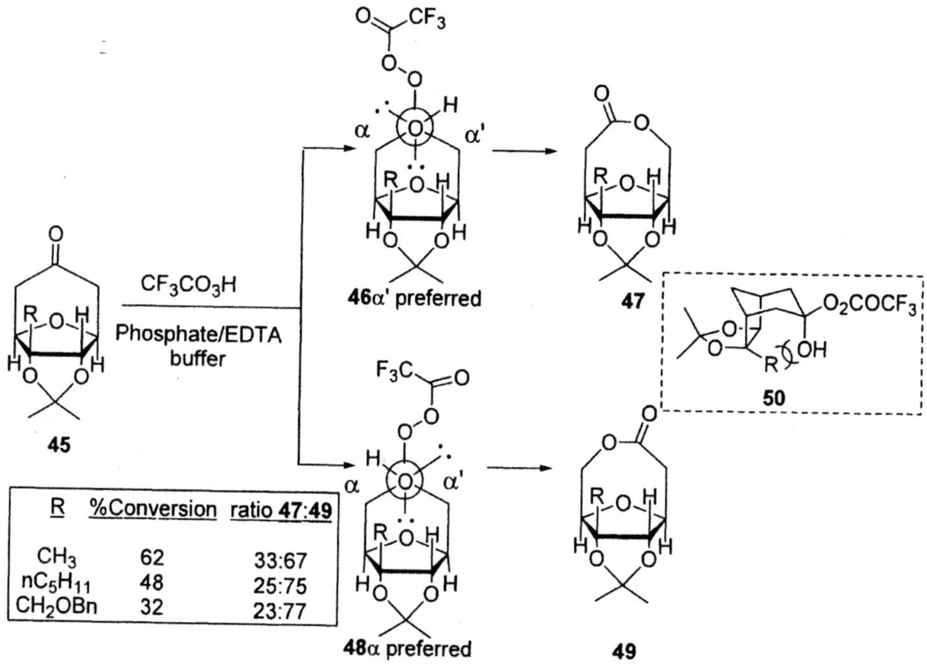

R	%Conversion	ratio 47:49
CH_3	62	33:67
nC_5H_{11}	48	25:75
CH_2OBn	32	23:77

While the primary and secondary stereoelectronic effects have been invoked to rationalize products,[21] there have been reports in series of polycyclic ketones in which regiochemistry of migration was not explained by these effects.[22]

3.1.4 Variations and Improvements

For over 100 years the Baeyer–Villiger reaction has been developed. The reaction was improved as chemists explored new and various oxidizing agents and conditions. Interestingly, while Baeyer and Villiger reported the formation of perbenzoic acid in their second aticle providing further support for the reaction that bears their names, they did not use this oxidant to promote the Baeyer–Villiger oxidation. As expected, many new reagents have been developed to improve selectivity. For example, oftentimes competing reactions may occur within the molecule which may contain an alkene, sulfur, pyridine, or amine which would be prone to epoxidation, *S*-oxidation, or *N*-oxidation, respectively. There have been two comprehensive reviews[6,23] on the Baeyer–Villiger oxidation of which one contained 420 pages of examples. In addition, reviews as recent as 2004 have appeared[9] as well as book chapters. The latest *Organic Reactions* publication included a list of oxidants, their preparation, and relative oxidizing power of a variety of peracids. Trifluoroperacetic acid (TFPAA), *p*-nitroperbenzoic acid (PNPBA), metachloroperbenzoic acid (*m*-CPBA), permaleic acid (MPMA), mono-perphthalic acid (MPTA) or its magnesium salt (MMPP), potassium peroxymonosulfate (Caro's acid, Oxone®), perfomic acid, peracetic acid (PAA), perbenzoic acid (PBA), silylated peracids, phenyl-peroxyselenic acids, sodium perborate, sodium percarbonate,[24] *N*-alkoxy-carbonyltriazole/H_2O_2,[25] and urea hydrogen peroxide complex (UHP)[26] have been commonly utilized to promote the Baeyer–Villiger reaction. Some of these reagents are commercially available while others must be prepared. To date, *m*-CPBA has been one of the most utilized oxidants to promote the Baeyer–Villiger reaction due to its reactivity, commercial availability, and relative ease of use. Protic acids, Lewis acids, or resin bound acids have been added to an organic or inorganic peracid to promote the reaction. Additives such as these have been suggested to promote the attack of peroxide on the carbonyl as well as to facilitate the rearrangement step.[6,9] Base with hydrogen peroxide or alkyl hydroperoxides have also been successfully employed to promote the reaction. Phase transfer catalysis has been employed in basic peroxide reactions, Oxone® reactions, and in using phenyl selenic acids. For some reagents like MMPP, a phase transfer catalyst was not required for a two-phase system.[27] Water has also been reported to accelerate the reaction.[28]

3.1.4.1 Metal Catalysis

A variety of metals have been used to promote the Baeyer–Villiger oxidation. Using catalytic metal complexes has allowed the use of inexpensive H_2O_2 or O_2. Yamada and co-workers reported the use of a Ni(II) complex with O_2 to promote the Baeyer–Villiger oxidation on simple ketones. In this work was reported the use of (dipivaloylmethanato)-nickel(II) to be a good complex with isovaleraldehyde or benzaldehyde under an atmosphere of O_2 to provide good conversions of ketones to lactones or esters. These conditions proved to be amenable to promote the Baeyer–Villiger oxidation of a variety of ketones in 33–96% yield and is exemplified by the conversion of ketone **51** into lactone **52**.[29]

$(CH_2Cl)_2$, O_2, rt, 21 h
Me_2CHCH_2CHO (3 equiv), $Ni(dpm)_2$ (1 mol%)
51 → **52**
dpm

Subsequent to this disclosure, Bolm reported the use of $Ni(oxa)_2$ or $Cu(OAc)_2$ to provide good yields of the lactone **52**. Again, a similar set of keto-substrates were oxidized in 23–93% yield using these O_2 promoted Baeyer–Villiger oxidation conditions.

The metal/ligand combination was the difference. Of course the $Ni(oxa)_2$ scaffold was utilized as a model knowing that this could be transformed into a chiral catalyst.[30]

Cat.	Yield %
$Cu(OAc)_2$	93
$Ni(oxa)_2$	90

Murahashi reported a similar oxidation using Fe_2O_3 catalyst on a number of simple substrates in 56–98% yield and on a very important carbapenam intermediate **54**.[31] Keto-lactam oxidation of **53** provided **54** in 93% yield with retention of configuration.

The mechanism for these metal-catalyzed, O_2 promoted oxidations was proposed to be metal-catalyzed O_2 oxidation of the aldehyde **55** to form a peroxy radical species **56** which adds to ketone **60** to provide a radical variant of a Criegee intermediate, **57**. This intermediate would then extract a proton from the aldehyde **55** to provide the normal Criegee intermediate **58**. Alternatively, the generated peroxy radical intermediate **56** abstracts a proton from an aldehyde molecule providing peracid **59** which attacks the ketone **60** to provide the Criegee intermediate **58**.

As previously aluded to, Bolm has used Cu with a chiral oxazoline ***S,S*-cat** to promote the asymmetric formation of lactones from racemic ketones. This catalyst worked well only on 2-phenylcyclohexanones.[32]

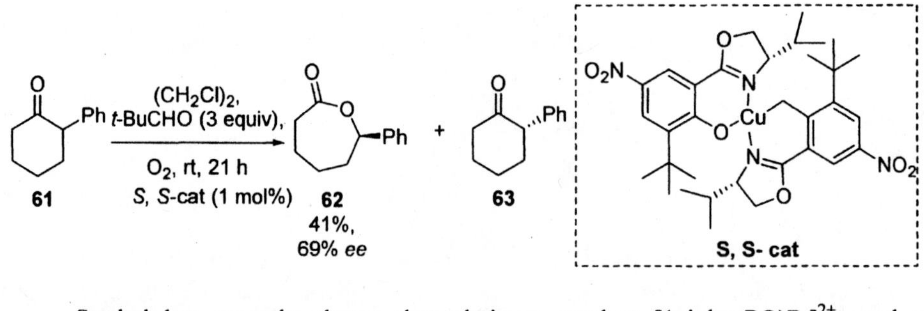

Strukul has recently shown that platinum catalyst $[(triphosPO)Pt]^{2+}$ used to promote the Baeyer–Villiger oxidation with H_2O_2 operated by Lewis acid catalysis and promoted the leaving of the OH group.[33] This dual catalysis property was considered unique to Pt catalysts. Chiral platinum systems have been applied to cyclic ketones to give moderate enantioselectivity of lactones.[32]

Katsuki recently reported the use of Zr-salen complexes to promote the oxidative desymmetrization of 2-Ph-cyclobutanone **64**. Katsuki utilized UHP and the C2-symmetric salen ligand, **Zr-salen,** which proved to be the best ligand screened. Reaction of ketone **64** with 5 mol% Zr-salen and UHP provided lactone **65** in good yield and 87% *ee*. These conditions also worked well with racemic bicyclooctanone **66** which provided a mixture of "normal" **67** and "abnormal" **68** Baeyer–Villiger products.[34] The remaining ketone was recovered in 92% *ee*.

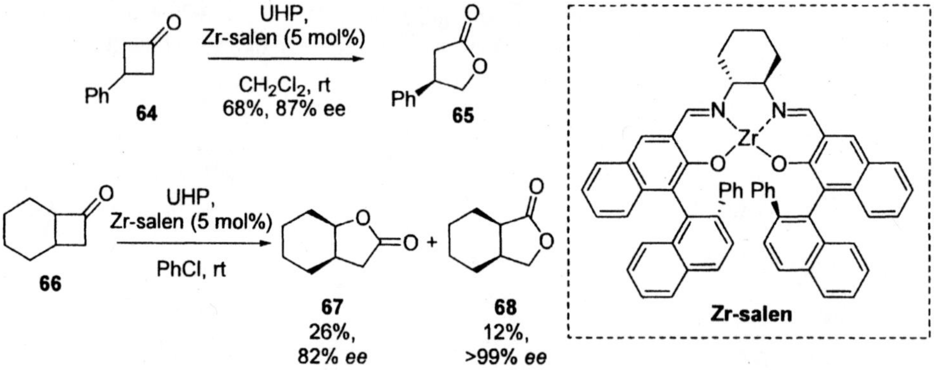

Molybdenum and rhenium complexes have also been reported to promote the Baeyer–Villiger oxidation using O_2. Although limited to cyclobutanones, Sharpless asymmetric epoxidation conditions have provided optically active lactones from hydroxymethylbutanone.[35]

Reaction of catalytic organoselenium with H_2O_2 has been used to promote Baeyer–Villiger oxidations via the formation of arylseleninic peracids. Ichikawa proposed that the transformation occurred by oxidation of the diselenide **70** to seleninic acid, which further oxidized to a seleninic peracid. The seleninic peracid then attacked the ketone (analogous to carbon peracids) forming a Criegee intermediate. Rearrangement of the intermediate then gave lactones or esters. Several simple ketones were reported with yields of 32–99% of the lactone as exemplified by the conversion of ketone **69** into lactone **71** in 85% yield using **70**.[36]

O
Ph
69
30% H_2O_2 (5 equiv)
CH_2Cl_2, rt, 24 h
85%
OTf
Se)$_2$
70, 5 mol%
O
O
Ph
71

Dubois recently reported a two cycle catalytic system whereby cyclohexanol was first oxidized to the ketone with an oxaziridine catalyst followed by subsequent reported oxidation catalyzed by the *in situ* generated seleninic peroxide to form ε-carprolactone in 70% yield.[37] UHP was used as peroxide source.

Binding of various metals to polymeric supports or clays has also been utilized to promote reported oxidations. Iron, tin, and tungsten have been reported on clays or on silicon supports. Oxone® on SiO_2 was shown to convert a variety of ketones into esters at room temperature (70–99% conversion; typically 98% yield based on conversion). For example, oxidation of **72** with $KHSO_5$–SiO_2 gave lactone **73** in nearly quantitative yield. The authors suggested that the reaction was acid catalyzed by $KHSO_4$ present at the active surface of the reagent.[38]

O
72
$KHSO_5$–SiO_2
CH_2Cl_2, rt
98%
O
O
73

Iron(III) on mesoporous silica (MCM-41) containing up to 1.8 wt% Fe using O_2 and benzaldehyde promoted Baeyer–Villiger oxidation of a variety of simple cyclic ketones to lactones in good conversion and yield.[39] Similarly, the reaction of a variety of ketones with Sn on MCM-41 and H_2O_2 gave lactones in good yield and fair conversion.[40] MCM-41 contains a larger diameter channel than Sn–beta, another support used to catalyze Baeyer–Villiger oxidations as well as other transformations. A comparison of Sn–beta/H_2O_2, Sn/MCM–41/H_2O_2, and *m*-CPBA has recently appeared. This work showcased yields using solid supports in comparison to *m*-CPBA. In particular, Sn/MCM–41/H_2O_2 proved superior in promoting the Baeyer–Villiger oxidation of vinyl aldehydes.[41] For example, safranal **74** gave 68% conversion and only one product, the desired Baeyer–Villiger oxidation product **75**; while *m*-CPBA gave 39% conversion of which only 7% of **75** was observed along with other oxidative by-products.

CHO
74
H_2O_2, Sn/MCM–41,
t-amylOH, 100 °C
15 min, 68%
OCHO
75

Lei recently reported using Sn–palygorskite (a natural clay mineral impregnated with Sn) and H_2O_2 to yield poor to good conversion and good selectivity of esters or lactones from cyclic or acyclic dialkyl ketone or alkyl aryl ketones.[42] In both of the Sn cases, it was proposed that Sn acts as a Lewis acid. This has recently been supported by computational work which suggested that it was less probable for the Sn species to activate the peroxide.[43] Tungsten oxide on MCM-48 with peracetic acid (WO_3/MCM–48/$MeCO_3H$) was reported to promote Baeyer–Villiger oxidation of several cyclic ketones in good conversion and yield.[44] The conversion of ketone **76** into lactone **77** is shown below for these recently-reported, Baeyer–Villiger reaction conditions.

76
cat. conditions
see table
77

Conditions	Yield % (conversion)
WO_3, MCM–48 $MeCO_3H$, 60 °C, 12 h	– (88%)
Fe/MCM–41 PhCHO (3 equiv), O_2, rt 15 h	75 (79%)
30% H_2O_2 (1.5 equiv), dioxane, 90 °C, 24 h Sn–palygorskite (cat.)	–(44%)

3.1.4.2 Enzyme Catalysis

While the Baeyer–Villiger oxidation has been known for quite some time, the application of enzymes for this transformation was not reported in the synthetic realm until 1988.[45] There are two types of flavin-mediated, Baeyer–Villiger monooxygenases (MOs) classified by their co-factors. *Type 1* enzymes are proteins using NADPH and are FAD dependent while *type 2* Baeyer–Villiger MOs use NADH and are FMN dependent. Oxygen typically is the oxidizing source. Enzymes impose their stereoselectivity by creating a preference for the formation of a Criegee intermediate which, due to stereoelectronic requirements for the Baeyer–Villiger oxidation (previously described), rearrange to a preferred stereoisomer. With purified or partially purified enzymes, co-factors are required, thus setting up a two enzyme system. Whole cell biotransformations circumvent the need for the addition of co-factors, tedious isolation of the enzyme, storage or stability of the isolated or partially isolated enzyme. CHMO (cyclohexane monooxygenase), CPMO (cyclopentane monooxygenase), and PAMO (phenylacetone monooxygenase) utilize FAD and have been used synthetically to perform a number of Baeyer–Villiger oxidations on several substrates. Additionally, enzymes were discovered from *Pseudomonas putida* that catabolize camphor (2,5-diketocamphane-1, 2-monooxygenase and 3,6-diketocamphane-1, 6-monooxygenase) and utilize FMN as prosthetic group.

The most well-studied Baeyer–Villiger MO is CHMO, which is derived from *Acinetobacter* NCIB 9871. This was cloned into yeast and used to map the Baeyer–Villiger oxidation of substituted cyclohexanones.[46] A variety of substituents at the 2-, 3-, or 4-position of cyclohexanone was shown to undergo selective Baeyer–Villiger oxidation to provide good to excellent % *ee* and good conversion and yield of both resolved ketone **ent-78** and rearranged lactone **79** in the bioengineered, whole cell.

R engineered baker's yeast
rac-78 **79** **ent-78**

R	79 Yield (ee)	ent-78 Yield (ee)
Me	25% (49%)	--
Et	39% (95%)	35% (≥98%)
n-Pr	27% (97%)	33% (92%)
i-Pr	21% (≥98%)	23% (96%)
allyl	30% (≥98%)	39% (≥98%)
n-Bu	30% (≥98%)	32% (98%)

Ottolina and co-workers[47] showed the oxidation of a variety of diones selectively using isolated enzymes. The enzyme system proved quite sensitive to changes in oxidation state and substitution in the substrate. In this instance, glucose dehydrogenase (GDH) was used as the second enzyme system to replenish NADPH. The Baeyer–

Villiger oxidation product **81** was isolated in good yield and excellent % *ee* and the recovered dione **ent-80** was enantiomerically enriched.

rac-80 → (O_2, H_3O^+, CHMO; NADPH/NADP$^+$; glucose 6-phosphate → gluconate 6-phosphate, GDH) → **ent-80**, 43% (80% *ee*) + **81**, 35% (99% *ee*)

The Baeyer–Villiger oxidation of a meso-diketone was recently reported by Gonzalo and co-workers using isolated PAMO.[48] Again, GDH was used as the reducing agent for NADP$^+$. For the reaction of dione **82**, conversion to **83** was 88% after 1.5 h. The authors suggested the utility of such chemistry would be to prepare ephedrine and pseudoephedrine.

82 → (O_2 → H_2O, PAMO; NADPH/NADP; Glucose-6-phos → Gluconate-6-phos, GDH; 1.5 h, 82% *ee*, 88% conversion) → **83** ⇢ Ephedrin, Pseudoephedrine

Alphand described the Baeyer–Villiger oxidation of azabicycloheptanone using different strains of CHMOs. A difference in selectivity was reported among the strains used. One strain gave preferentially the "normal" Baeyer–Villiger oxidation product **85**, while the TD63 strain preferred the "abnormal" Baeyer–Villiger adduct **86**. The authors were able to prepare the Geisman-Waiss lactone, which has been used in the preparation of pyrrolizidine natural products.[49]

whole cell biocat.

rac-84 → 85 + 86 + ent-84

Geisman-Waiss lactone

Microorganism	85 Yield (ee)	86 Yield (ee)	ent-84 Yield (ee)
A. calcoaceticus NCIMB 9871	36% (96% *ee*)	10% (93% *ee*)	41% (86% *ee*)
Acinetobacter TD63	11% (87% *ee*)	29% (98% *ee*)	45% (74% *ee*)

The above report brings up an interesting and useful point about the Baeyer–Villiger oxidation promoted by enzymes. With the stereoelectronic requirement of the Baeyer–Villiger oxidation, how could one see different Baeyer–Villiger products ("normal" and "abnormal" or enantiomeric substrates both reacting) derived from the same enzyme? Initially, some believed that it was due to the use of only partially purified enzymes or multiple pockets within the CHMO. The oxidation of enantiomerically enriched bicycloheptanone **87** was found to give the "normal" Baeyer–Villiger product **88**, while the treatment of the enantiomer **90** with the same enzyme provided instead the "abnormal" Baeyer–Villiger adduct **91**. It was established that both enantiomers **87** and **90** were processed at the same site of the enzyme. In comparing the two Criegee intermediates **89** and **92**, which must form from the products observed, it proved reasonable that both fit into the same active site only in different ways. The dotted migrating bonds of **89** and **92** give rise to the superimposition tricycle **93**. Treating tricycle **93** under Baeyer–Villiger conditions gave the predicted migration of the dotted bond and further understanding of Baeyer–Villiger oxidations in CHMOs.[35]

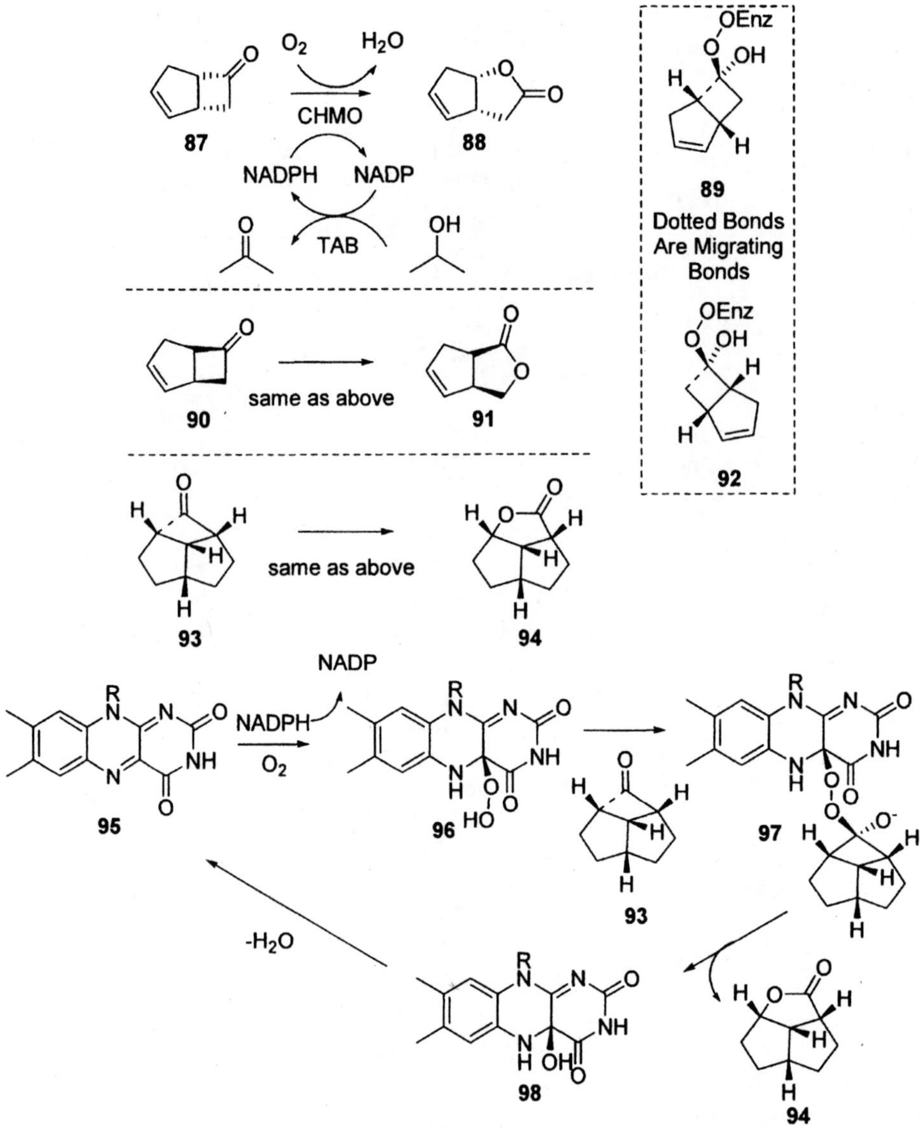

The proposed mechanism of flavin-type Baeyer–Villiger MOs is shown. First reduction of the flavin **95** and attack on O_2 provides **96**. Peroxo-flavin **96** attacks ketone **93** and forms the Criegee adduct **97**. Rearrangement provides lactone **94** along with hydroxylated flavin **98**. Dehydration returns to the beginning of the catalytic cycle.

Over the past few years, there have been reports of directed evolution of Baeyer–Villiger MOs and their application in synthetic chemistry. In the evolution and synthetic application of these mutated enzymes to previous and new substrates, Mihovilovic and co-workers have improved the selectivity of a variety of CHMOs and CPMOs toward the oxidation of bicyclic ketone **rac-87**. It was shown that **rac-87** provides 99% *ee* of each lactone **88** and **91** as a 1:1 mixture.[50] In addition, it was shown that high % *ee*'s and conversions could be obtained for a variety of substrates for the enzyme-promoted Baeyer–Villiger oxidation.[51]

Due to the importance and utility of Corey lactone **88**, some have strived to prove the utility of enzymes by showing that enzymatic transformations, like Baeyer–Villiger oxidations, can be scaled-up. To that end, Hilker, Alphand and Furstoss have worked diligently to design a system that would allow the preparation of large amounts of both of the lactones. They recently reported a 900 g run of **rac-88** to provide a 58% yield of Baeyer–Villiger oxidation products. This utilized novel resin substrate feeding and product removing technology and shows that large scale Baeyer–Villiger oxidations using enzymes can be done.[52]

rac-87 → Optipore L-493 resin; CHMO-whole cell; pH= 7, air, 37 °C; 58% → **88**, >98%ee + **91**, >98%ee

Application of dynamic kinetic resolution has also been reported in a whole-cell enzymatic Baeyer–Villiger process. While this was initially done using slightly basic conditions, Furstoss and Alphand have more recently reported the use of a weakly basic anion exchange resin to promote racemization of the slow oxidizing enantiomer **99** to the fast enantiomer **100**.[53] Baeyer–Villiger oxidation using recombinant *E. coli* to overexpress the CHMO from *A. calcoaceticus* provided excellent yield and % *ee* of lactone **101**.

O OBn 100 Lewatit MP62, EtOH, phosphate → Recomb *A. calcoaceticus* 84% (97% *ee*) O O OBn 101

O OBn 99

3.1.5 Synthetic Utility

Although some of the synthetic utility has been described above, there are many facets of the Baeyer–Villiger oxidation which have not been described at all—one of which is the late-stage revealing of a hydroxyl group in designing a synthesis. In the total synthesis of the marine spongiane diterpene, norrisolide **103**, Theodorakis and co-workers utilized the Baeyer–Villiger oxidation to install the C19 oxygen as the final step of the synthesis. They were able to achieve selectivity in the oxidation of this highly oxygenated, olefinic ring system providing a 60% yield of the natural product by the Baeyer–Villiger oxidation of the acetyl group of **102**.[54]

O O O O H 102 *m*-CPBA, $NaHCO_3$, → CH_2Cl_2, 0 °C, 4 h 60% AcO O O O H norrisolide, **103**

The oxidation of formyl groups accompanied by subsequent functionalization has showcased the utility of the Baeyer–Villiger oxidation. Franck reported the Baeyer–Villiger oxidation of naphthaldehydes as precursors to pharmacologically active quinone type compounds. This was exemplified by the conversion of aldehyde **104** into formate **105**, which after formate cleavage, was further oxidized to the quinone system **106**.[55]

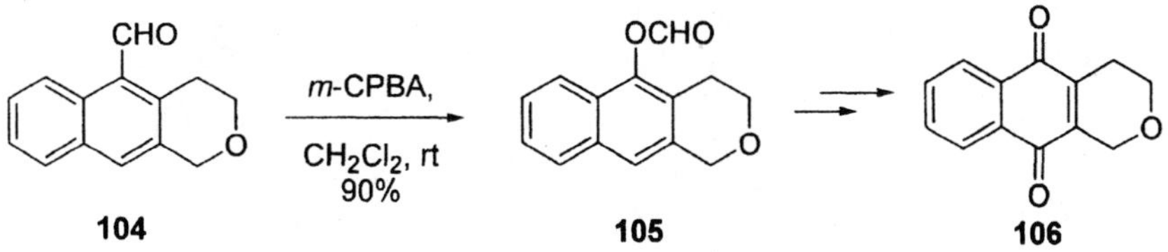

Chang and co-workers reported the Baeyer–Villiger oxidation of the formyl group of **107** to create a leaving group. Lewis acid promoted elimination led to aryl pyrrolines in good yield. Chang further illustrated the conversion of **109** into **110** which was further transformed into racemic baclofen **111**, a therapeutically useful GABA$_b$ agonist.[56]

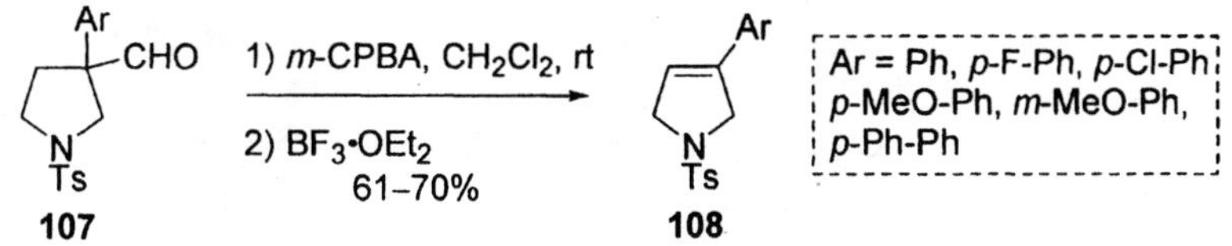

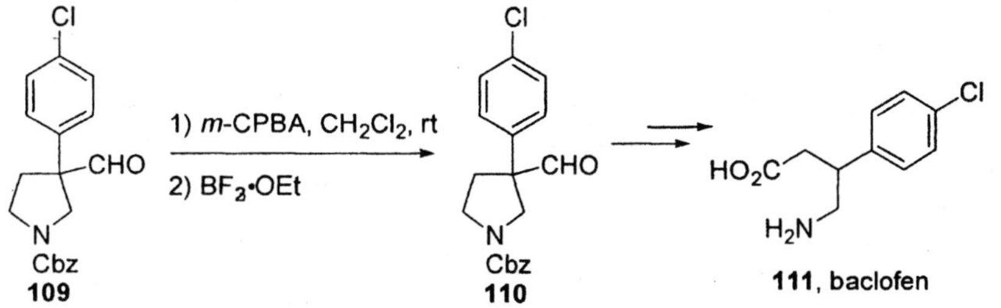

Use of the Baeyer–Villiger reaction to oxygenate a bridgehead has been done. Corey and Smith used this methodology in route to key intermediate **114** in the total synthesis of Gibberellic acid.[57] Baeyer–Villiger oxidation of **112** with dinitro-perbenzoic acid provided **113** in good yield. Corey noted that prior attempts to perform this transformation on a similar α-acetoxy bridgehead acetate had been reported to fail.

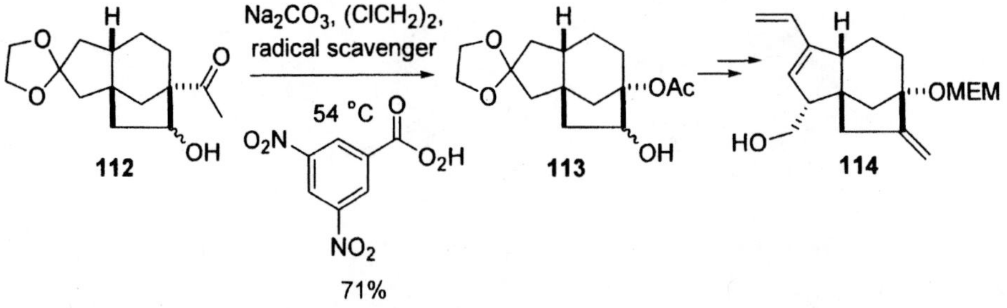

Knochel and co-workers[58] reported Baeyer–Villiger oxidation of α-trisubstituted aldehydes **116** in the preparation of tertiary chiral alcohols **117**. The fact that the reaction proceeds with retention of configuration of the migrating group permits the utility of this reaction. Several optically enriched, tertiary alcohols were prepared using this simple 3-step method from optically-enriched allylic tetrafluoroborates **115**.

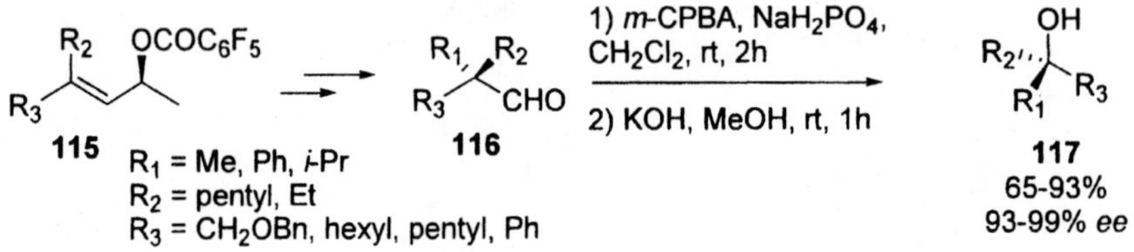

3.1.6 Experimental

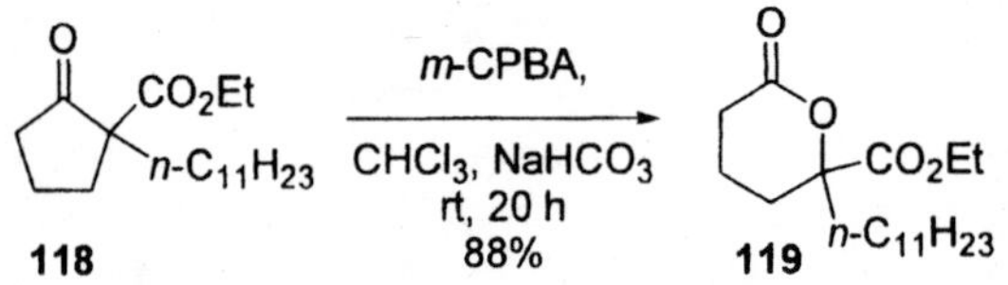

5-Ethoxycarbonyl-5-(*n*-undecyl)-δ-valerolactone (119)[59]

A solution of **118** (3.836 g, 12.36 mmol) in anhydrous $CHCl_3$ (100 mL) was treated with $NaHCO_3$ (1.922 g, 22.88 mmol) and *m*-CPBA (70%, 4.288 g, 17.39 mmol). The mixture was stirred at rt for 20 h, diluted with saturated aqueous $NaHCO_3$ solution (120 mL), vigorously stirred for 15 min, and extracted with CH_2Cl_2. The combined organic layers were washed in succession with water and brine, dried over Na_2SO_4, filtered, and evaporated at reduced pressure. The yellowish residue was chromatographed (SiO_2, petroleum ether–EtOAc, gradient, 20:1 to 10:1) to give **119** (3.53 g, 88% yield) as a colorless oil: IR (film) λ_{max} 1750 cm^{-1}; ^1H NMR ($CDCl_3$) δ 0.89 (t, J = 6.6 Hz, 3H),

1.19–1.34 (m, 19H), 1.61–2.21 (m, 8H), 2.43–2.64 (m, 2H), 4.23–4.31 (m, 2H); ^{13}C NMR ($CDCl_3$) δ 14.0, 14.1, 17.0, 22.6, 22.8, 28.6, 29.2, 29.2, 29.4, 29.4, 29.5, 30.4, 31.8, 38.6, 61.8, 86.0, 170.2, 171.9 (only 18 peaks shown in the spectrum); EI–MS: m/z (%) 326 (M^+), 253 (100), 225 (44), 97 (22), 71 (22), 57 (35), 55 (76), 43 (57), 41 (46). Anal. Calcd for $C_{19}H_{34}O_4$: C, 69.90; H, 10.50. Found: C, 70.04; H, 10.58.

3.1.7 References

1 Baeyer, A.; Villiger, V. *Ber Dtsch. Chem. Ges.* **1899**, *32*, 3625.
2 Baeyer, A.; Villiger, V. *Ber Dtsch. Chem. Ges.* **1900**, *33*, 858.
3 [R] Renz, M.; Meunier, B. *Eur. J. Org. Chem.* **1999**, 737.
4 Doering, W. von E.; Dorfman, E. *J. Am. Chem. Soc.* **1953**, *75*, 5595.
5 Camporeale, M.; Fiorani, T.; Troisi, L.; Adam, W.; Curci, R.; Edwards, J. O. *J. Org. Chem.* **1990**, *55*, 93.
6 [R] Krow, G. R. *Organic Reactions* **1993**, *43*, 251.
7 Reyes, L.; Castro, M.; Cruz, J.; Rubio, M. *J. Phys. Chem. A* **2005**, *109*, 3383.
8 Molander, G. A.; Czako, St. Jean, D. J. Jr. *J. Org. Chem.* **2006**, *71*, 1172.
9 [R] Ten Brink, G.-J.; Arends, I. W. C. E.; Sheldon, R. A. *Chem Rev.* **2004**, *104*, 4105.
10 Hudrlik, P. F.; Hudrlik, A. M.; Nagendrappa, G.; Yimenu, T.; Zellers, E. T.; Chin, E. *J. Am. Chem. Soc.* **1980**, *102*, 6894.
11 Bakale, R. P.; Scialdone, M. A.; Johnson, C. R. *J. Am. Chem. Soc.* **1990**, *112*, 6729.
12 Kitazume, T.; Kataoka, J. *J. Fluorine Chem.* **1996**, *80*, 157.
13 Kobayashi, S.; Tanaka, H.; Amii, H.; Uneyama, K. *Tetrahedron* **2003**, *59*, 1547.
14 Grein, F.; Chen, A C.; Edwards, D.; Crudden, C. M. *J. Org. Chem.* **2006**, *71*, 861.
15 Itoh, Y.; Yamanaka, M.; Mikami, K. *Org. Letters* **2003**, *5*, 4803.
16 Sprecher, M.; Nativ, E. *Tetrahedron Lett.* **1968**, *42*, 4405.
17 Chandrasekhar, S.; Deo Roy, C. *Tetrahedron Lett.* **1987**, *28*, 6371.
18 Crudden, C. M.; Chen, A. C.; Calhoun, L. A. *Angew. Chem. Int. Ed.* **2000**, *39*, 2852.
19 Snowden, M.; Bermudez, A.; Kelly, D. R.; Radkiewicz-Poutsma, J. L. *J. Org. Chem.* **2004**, *69*, 7148.
20 Noyori, R.; Sato, T.; Kobayashi, H. *Bull. Chem. Soc. Jpn.* **1983**, *56*, 2661.
21 [R] Krow, G. R. *Tetrahedron* **1981**, *37*, 2697.
22 Harmata, M.; Rashatasakhon, P. *Tetrahedron Lett.* **2002**, *43*, 3641.
23 [R] Hassal, C. H. *Organic Reactions* **1957**, *9*, 73.
24 Olah, G. A.; Wang, Q.; Trivedi, N. J.; Prakash, G. K. S. *Synthesis* **1991**, 739.
25 Tsunokawa, Y.; Iwasaki, S.; Okuda, S. *Chem Pharm Bull.* **1983**, *31*, 4578.
26 Cooper, M. S.; Heaney, H.; Newbold, A. J.; Sanderson, W. R. *Synthesis Lett.* **1990**, 533.
27 Ricci, M.; Battistel, E. *La Chimica e l'Industria* **1997**, 879.
28 Fringuelli, F.; Germani, R.; Pizzo, F.; Savelli, G. *Gazzetta Chimica Italiana* **1989**, *119*, 249.
29 Yamada, T.; Takahashi, K.; Kato, K.; Takai, T.; Inoki, S.; Mukaiyama, T. *Chem. Lett.* **1991**, 641.
30 Bolm, C.; Schlingloff, G.; Weickhardt, K. *Tetrahedron Lett.* **1993**, *34*, 3405.
31 Murahashi, S.-I.; Oda, Y.; Naota, T. *Tetrahedron Lett.* **1992**, *33*, 7557.
32 [R] (a) Bolm, C.; Palazzi, C.; Beckmann, O. In *Transition Metals for Organic Synthesis*; Beller, M.; Bolm, C. Ed.; Vol. 2 2nd Ed. Wiley-VCH, Weinheim, Germany; **2004**, p.267-274. [R] (b) Bolm, C.; Beckmann, O. In *Comprehensive Asymmetric Catalysis I-III* Springer-Verlag, Berlin, Germany; **1999**, *2*, p. 803–810
33 Brunetta, A.; Sgarbossa, P.; Strukul, G. *Catalysis Today* **2005**, *99*, 227.
34 Wantanabe, A.; Uchida, T.; Ito, K.; Katsuki, T. *Tetrahedron Lett.* **2002**, *43*, 4481.
35 [R] This is a two part review: (a) Kelly, D. R. *Chimica OGGI* **2000**, *18*, 33. (b) Kelly, D. R. *Chimica OGGI* **2000**, *18*, 52.
36 Ichikawa, H.; Usami, Y.; Arimoto, M. *Tetrahedron Lett.* **2005**, *46*, 8665.
37 Brodsky, B. H.; DuBois, J. *J. Am. Chem. Soc.* **2005**, *127*, 15391.
38 Gonzalez-Nunez, M. E.; Mello, R.; Olmos, A.; Asensio, G. *J. Org. Chem.* **2005**, *70*, 10879.

39 Kawabata, T.; Ohishi, Y.; Itsuki, S.; Fujisaki, N.; Shishido, T.; Takaki, K.; Zhang, Q.; Wang, Y.; Takehira, K. J. *Molecular Catalysis A: Chemical* **2005**, *236*, 99.
40 Corma, A.; Renz, M. *Collect. Czech. Chem. Commun.* **2005**, *70*, 1727.
41 Corma, A.; Iborra, S.; Mifsud, M.; Renz, M. *ARKIVOC* **2005**, *9*, 124.
42 Lei, Z.; Zhang, Q.; Luo, J.; He, X. *Tetrahedron Lett.* **2005**, *46*, 3505.
43 Boronat, M.; Corma, A.; Renz, M.; Sastre, G.; Viruela, P. M. *Chem. Eur. J.* **2005**, *11*, 6905.
44 Koo, D. H.; Kim, M.; Chang, S. *Org. Lett.* **2005**, *7*, 5015.
45 [R] Mihovilovic, M. D.; Rudroff, F.; Grotzl, B. *Current Organic Chemistry* **2004**, *8*, 1057.
46 Stewart, J. D.; Reed, K. W.; Martinez, C. A.; Zhu, J.; Chen, G.; Kayser, M. M. *J. Am. Chem. Soc.* **1998**, *120*, 3541.
47 Ottolina, G.; de Gonzalo, G.; Carrea, G.; Danieli, B. *Adv. Synth. Catal.* **2005**, *347*, 1035.
48 De Gonzalo, G.; Torres Pazmino, D. E.; Ottolina, G.; Fraaije, M. W.; Carrea, G. *Tetrahedron: Asymmetry* **2005**, *16*, 3077.
49 Luna, A.; Gutierrez, M.-C.; Furstoss, R.; Alphand, V. *Tetrahedron: Asymmetry* **2005**, *16*, 2521.
50 Mihovilovic, M. D.; Rudroff, F.; Grotzl, B.; Kapitan, P.; Snajdrova, R.; Rydz, J.; Mach, R. *Angew. Chem., Int. Ed.* **2005**, *44*, 3609.
51 Mihovilovic, M. D.; Rudroff, F.; Winninger, A.; Schneider, T.; Schulz, F.; Reetz, M. T. *Org. Lett.* **2006**, *8*, 1221.
52 Hilker, I.; Wohlgemuth, R.; Alphand, V.; Furstoss, R. *Biotechnology and Bioengineering* **2005**, *92*, 702.
53 Gutierrez, M.-C.; Furstoss, R.; Alphand, V. *Adv. Synth. Catal.* **2005**, *347*, 1051.
54 Brady, T. P.; Kim, S. H.; Wen, K.; Kim, C.; Theodorakis, E. A. *Chem. Eur. J.* **2005**, *11*, 7175.
55 Franck, R. W.; Gupta, R. B. *J. Org. Chem.* **1985**, *50*, 4632.
56 Chang, M.-Y.; Pai, C.-L.; Kung, Y.-H. *Tetrahedron Lett.* **2006**, *47*, 855.
57 Corey, E. J.; Gorzynski Smith, J. *J. Am. Chem. Soc.* **1979**, *101*, 1038.
58 Leuser, H.; Perrone, S.; Liron, F.; Kneisel, F. F.; Knochel, P. *Angew. Chem., Int. Ed.* **2005**, *44*, 4627.
59 Chen, Q.; Deng, H.; Zhao, J.; Lu, Y.; He, M.; Zhai, H. *Tetrahedron* **2005**, *61*, 8390.

Timothy T. Curran

3.2 Brown Hydroboration Reaction

3.2.1 Description

The Brown hydroboration reaction is the addition of B–H across a π-system (**1**) in an anti-Markovnikov fashion. Most commonly, this reaction utilizes BH_3•THF as the hydroborating reagent and is followed by an oxidation of the newly formed C–B bond to afford an alcohol product (**2**).[1]

R (1) → 1) BH_3•THF, 2) NaOH, H_2O_2 → R–OH (2)

3.2.2 Historical Perspective

In 1948 Hurd reported that diborane and ethylene, when heated together to 100 °C for four days in a sealed tube, generated triethylborane.[2] Ten years later H. C. Brown first demonstrated that hydroborations could proceed at room temperature if ethereal solvents were used.[3] He showed that when diborane was passed through a solution of diglyme it formed a BH_3•diglyme complex. This complex was capable of hydroborating hexene (**3**) to form trihexylborane (**4**) within minutes. Since that discovery in 1959, the hydroboration reaction using BH_3•THF has become a common transformation in modern synthetic organic chemistry.[4]

B_2H_6 → (diglyme) → BH_3•diglyme → (**3**, diglyme, rt, 10 minutes) → **4**

3.2.3 Mechanism

While there was initial debate between Pasto[5] and Brown[6] about the mechanism of the hydroboration reaction, Brown's proposed mechanism[6] is now generally accepted as the most likely pathway. Brown suggested that the reaction proceeds *via* an equilibrium between BH_3•THF (**5**) and free BH_3 (**6**). The free BH_3 then rapidly adds B–H across a π-system of the olefin (**7**) in an anti-Markovnikov fashion *via* an asynchronous, 4-centered transition state (**8**) to afford hydroboration product **9**.

O-BH_3 (5) ⇌ THF + BH_3 (6) → (R, 7) → [H–BH_2, R (8)]‡ → R–H_2B–O (9)

3.2.4 Variations and Improvements

Although BH_3•THF is the most commonly used hydroborating reagent, there are drawbacks to this reagent, which has led to the development of alternative hydroborating reagents. Variable regioselectivity is obtained in hydroborations when BH_3•THF is used. This is especially prevalent in reactions involving 1,2-disubstituted alkenes such as **10**. Over the last 45 years a variety of substituted boranes have been developed to improve the regioselectivity of hydroboration reactions.[7] By increasing the steric bulk around the borane, improved regioselectivity has been observed with sterically differentiated olefins, due to the preference of the large boron moiety to bond to the less sterically hindered end of the π-system in the transition state.

10 → 1) Borane, THF; 2) NaOH, H_2O_2 → 11 + 12

Borane	11 : 12
BH_3•THF	57 : 43
thexBH_2	66 : 34
Sia_2BH	97 : 3
9BBN-H	99 : 1

Alkynes were also found to be viable hydroboration substrates. Hydroboration of an alkyne, followed by an oxidative workup affords a carbonyl containing product. BH_3•THF and numerous other boranes have been used for the hydroboration of alkynes such as 1-phenyl-1-propyne (**13**).[8]

Ph—≡— (13) → 1) Borane, THF; 2) NaOH, H_2O_2 → 14 + 15

Borane	14 : 15
BH_3	74 : 26
9-BBN	65 : 35
ThexBH_2	43 : 57
Sia_2BH	19 : 81

Evans and co-workers developed rhodium catalyzed hydroboration reactions, which enable the use of catechol borane and other boronate esters as hydroborating agents to afford organoboronic esters as products.[9] These products have increased stability and can be used directly in palladium cross-coupling applications. However, the rhodium-catalyzed method is most effective for the hydroboration of monosubstitued olefins (i.e., **16**), as lower reactivity is observed with more substituted alkenes.

C_6H_{13} (16) → Catecholborane, $Rh(PPh_3)_3Cl$, 20 °C, 25 min → C_6H_{13} (17)

The first example of an enantioselective hydroboration used ipc_2BH as the hydroborating reagent and afforded high levels of enantioselectivity in the hydroboration of *cis* olefins, such as **18**.[10] However, ipc_2BH hydroborates more hindered olefins with only moderate enantioselectivity, but it was found that monoisopinocampheylborane ($ipcBH_2$) hydroborated trans-disubstituted (**20**)[11] and trisubstituted alkenes (**22**)[12] with relatively high asymmetric induction. While there have been numerous other terpene derived hydroborating reagents ipc_2BH and $ipcBH_2$ still serve as the benchmark for useful enantioselective hydroborating reagents in most cases.

1) $ipcBH_2$, THF, 0 °C
2) H_2O_2, NaOH
18 → **19** (OH), 98% *ee*

1) $ipcBH_2$, THF, 0 °C
2) H_2O_2, NaOH
20 → **21** (OH), 73% *ee*

1) $ipcBH_2$, THF, 0 °C
2) H_2O_2, NaOH
22 → **23** (OH), 66% *ee*

3.2.5 *Synthetic Utility*

The product of hydroboration reactions are organoboranes which represent an especially useful class of substrates due to their ability to undergo a wide variety of synthetic transformations, a small sampling of which are shown below. The most ubiquitous of these manipulations is the oxidation of a C–B bond to generate the corresponding alcohol (**24**→**25**) through the addition of base and hydrogen peroxide to an organoborane.[13] Aldehydes and esters can be accessed using $Li(CCl_2C(O)CH_3)$ (**24**→**26**) and N_2CHCO_2Et (**24**→**27**), respectively.[14-16] Furthermore, the C–B bond can also be directly transformed into a C–N bond (**24**→**28**) through the addition of BCl_3 and an alkyl azide in what formally amounts to a hydroamination of an alkene.[17] Conversion of the C–B bond into a C–halogen bond (**24**→**29**) can readily be accomplished *via* the use of base and an electrophilic halogen source. Lastly, organoboranes can act as Michael donors in Lewis acid-catalyzed Michael additions (**24**→**30**).

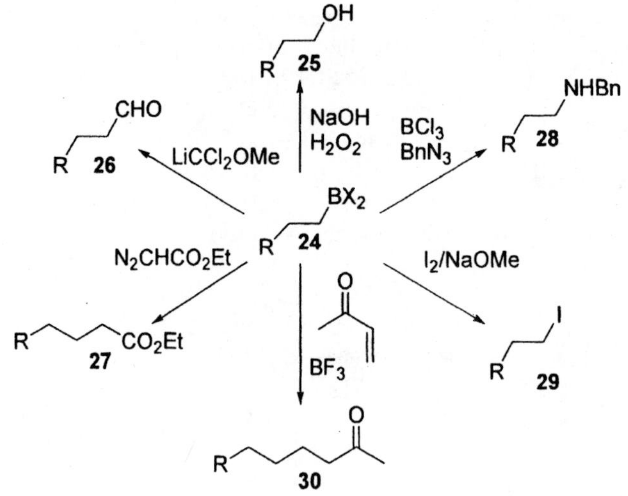

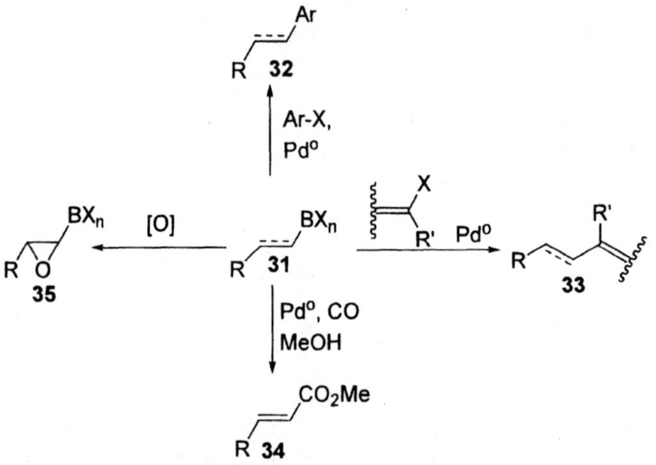

Recent advances in metal-catalyzed processes have further increased the utility of organoboranes. The coupling between aryl- or vinyl halides and alkenyl boranes (accessed from the hydroboration of an alkyne) using Suzuki cross-coupling reactions has found significant use in organic synthesis.[18] The more challenging problem of using alkyl organoboranes in cross-coupling reactions has received a great deal of attention and recently significant advances in this area have been made.[19–22] Palladium catalyzed carbonylations of organoboranes have also been demonstrated (**31**→**34**).[23] Lastly, because of the recent discovery of air and moisture stable organotrifluoroborate salts, the direct functionalization of alkenyl boranes is possible. Molander has shown that these potassium alkenyltrifluoroborate salts readily undergo epoxidation (**31**→**35**),[24] further increasing the utility of organoboranes in synthesis.

Recently Danishefsky and co-workers used a hydroboration reaction early in the synthesis of (–)-scabronine G (**38**).[25] Hydroboration of (**36**) using 9-BBN, followed by oxidative workup, afforded (**37**) in 92% yield as a single regioisomer.

9-BBN, THF; H_2O_2, NaOH, 0 °C; 92%

36 **37** **38** (–)-scabronine G

A hydroboration reaction was used to convert alkene **39** to a regioisomeric mixture of alcohols **40** and **41**. Product **41** was then carried on toward the total synthesis of Agelastatin A (**42**).[26]

BH_3•THF, THF; H_2O_2, NaOH

39 **40** **41** **42** agelastatin A

In 1979, Corey and co-workers utilized the hydroboration of **43** using disiamylborane to provide desired alcohol (**44**), after oxidative workup, in very high yield. This material was used to complete a total synthesis of gibberellic acid (**45**).[27]

disiamylborane (3 eq), THF, rt, 17 h; H_2O_2, NaOH; > 95%

43 **44**

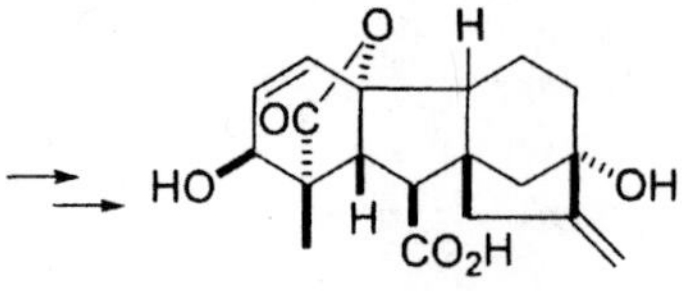

45

gibberellic acid

3.2.6 Experimental

Hydroboration of Tricycle 39[26]

A solution of the olefin **39** (46 mg, 0.98 mmol) in THF (1 mL) was cooled to 0 °C, and 1 M BH_3•THF complex in THF (0.98 mL, 0.98 mmol) was added. The reaction mixture was allowed to warm to room temperature, and after 4 h was cooled back to 0 °C and diluted with methanol (4 drops), 1.5 M NaOH (1 mL), and 30% aqueous H_2O_2 (0.5 mL). The reaction mixture was diluted with water and extracted with CH_2Cl_2 (3 × 20 mL). The combined organic fractions were dried over $MgSO_4$, filtered through a pad of silica gel (eluted with EtOAc), and evaporated to afford a mixture of alcohols **40** and **41** in 88% yield and not requiring any subsequent purification.

3.2.7 References

1. [R] Smith, K.; Pelter, A. *Hydroboration of C=C and Alkynes.* In *Comp. Org. Syn.* (Trost, B. M., Fleming, I., Eds.), Pergamon Press: Oxford, 1991; Vol. 8, p 703 (and references therein).
2. Hurd, D. T. *J. Am. Chem. Soc.* **1948,** *70,* 2053.
3. Brown, H. C.; Rao, B. S. S. *J. Am. Chem. Soc.* **1959,** *81,* 8428.
4. Brown, H. C.; Rao, B. S. S. *J. Org. Chem.* **1957,** *22,* 1136.
5. Pasto, D. J.; Lepeska, B.; Cheng, T. C. *J. Am. Chem. Soc.* **1972,** *94,* 6083.
6. Brown, H. C.; Chandrasekharan, J. *J. Am. Chem. Soc.* **1984,** *106,* 1863.
7. Brown, H. C.; Liotta, R.; Scouten, C. G. *J. Am. Chem. Soc.* **1976,** *98,* 5297 (and references therein).
8. Brown, H. C.; Scouten, C. G.; Liotta, R. *J. Am. Chem. Soc.* **1979,** *101,* 96.
9. Evans, D. A.; Fu, G. C.; Hoveyda, A. H. *J. Am. Chem. Soc.* **1992,** *114,* 6671.
10. Brown, H. C.; Desai, M. C.; Jadhav, P. K. *J. Org. Chem.* **1982,** *47,* 5065.
11. Brown, H. C.; Jadhav, P. K.; Mandal, A. K. *J. Org. Chem.* **1982,** *47,* 5074.
12. Brown, H. C.; Mahindroo, V. K.; Dhokte, U. P. *J. Org. Chem.* **1996,** *61,* 1906.
13. Johnson, J. R.; VanCampen, M. G. J. Am. Chem. Soc. 1938, 60, 121.
14. Li, N. S.; Yu, S.; Kabalka, G. W. *Organometallics* **1997,** *16,* 709.
15. Soundararajan, R.; Li, G. S.; Brown, H. C. *Tetrahedron Lett.* **1994,** *35,* 8957.
16. Brown, H. C.; Carlson, B. A. *J. Org. Chem.* **1973,** *38,* 2422.
17. Hupe, E.; Denisenko, D.; Knochel, P. *Tetrahedron* **2003,** *59,* 9187.
18. Miyaura, N.; Suzuki, A. *Chemical Reviews* **1995,** *95,* 2457.
19. Kirchhoff, J. H.; Netherton, M. R.; Hills, I. D.; Fu, G. C. *J. Am. Chem. Soc.* **2002,** *124,* 13662.
20. Netherton, M. R.; Dai, C. Y.; Neuschutz, K.; Fu, G. C. *J. Am. Chem. Soc.* **2001,** *123,* 10099.
21. Littke, A. F.; Dai, C. Y.; Fu, G. C. *J. Am. Chem. Soc.* **2000,** *122,* 4020.
22. Chemler, S. R.; Trauner, D.; Danishefsky, S. J. *Angew. Chem. Int. Ed.* **2001,** *40,* 4544.
23. Suzuki, A. *J. Organomet. Chem.* **1999,** *576,* 147.
24. Molander, G. A.; Ribagorda, M. *J. Am. Chem. Soc.* **2003,** *125,* 11148.
25. Walters, S. P.; Tian, Y.; Li, Y.-M.; Danishefsky, S. J. *J. Am. Chem. Soc.* **2005,** *127,* 13514.
26. Stien, D.; Anderson, G. T.; Chase, C. E.; Koh, Y.-H.; Weinreb, S. M. *J. Am. Chem. Soc.* **1999,** *121,* 9574.
27. Corey, E. J.; Smith, J. G. *J. Am. Chem. Soc.* **1979,** *79,* 1038.

Julia M. Clay

3.3 Burgess Dehydrating Reagent

3.3.1 Description

This is a mild and selective method of converting secondary and tertiary alcohols to their respective olefins by use of the stereospecific *cis*-dehydrating agent, methyl *N*-(triethylammoniumsulfonyl)carbamate (Burgess reagent, **1**).[1]

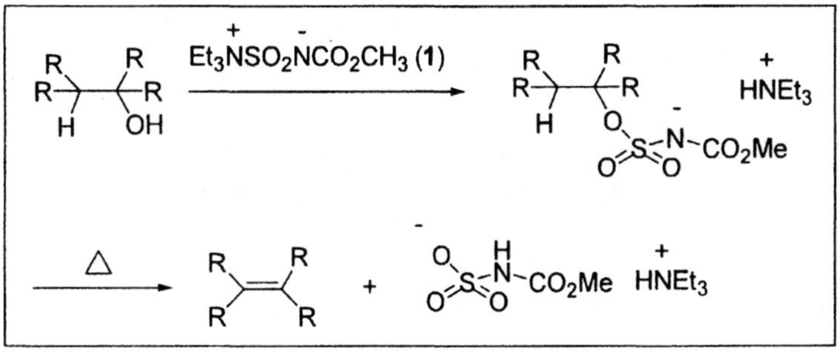

$$\mathrm{Et_3\overset{+}{N}SO_2\overset{-}{N}CO_2CH_3}\ (\mathbf{1})$$

The Burgess reagent is made by exposing chloromethanesulfonyl chloride to methanol at room temperature followed by the addition of triethylamine. Burgess reagent (**1**) is a white crystalline solid that is air and moisture sensitive.

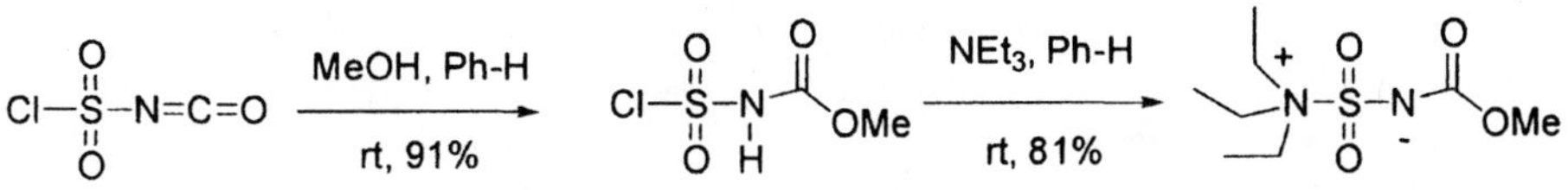

MeOH, Ph-H, rt, 91%; NEt$_3$, Ph-H, rt, 81%

1

The reagent is commercially available as well as a version that is supported on solid phase.

3.3.2 Historical Perspective

The reagent was first disclosed in 1968 by Edward Burgess.[1a] In 1970, Burgess reported that the reagent, when exposed to secondary and tertiary alcohols, gave the corresponding olefins (e.g. **2**–**4**). Furthermore, it was also observed that primary alcohols react with Burgess reagent (**1**) to yield urethanes **5**. Burgess reported that in many cases it was operationally convenient to exchange the triethylammonium counterion for a sodium cation of **2** in the more stable primary and secondary sulfamate ester salts. This allowed the dehydration reactions to occur at lower temperatures.[2]

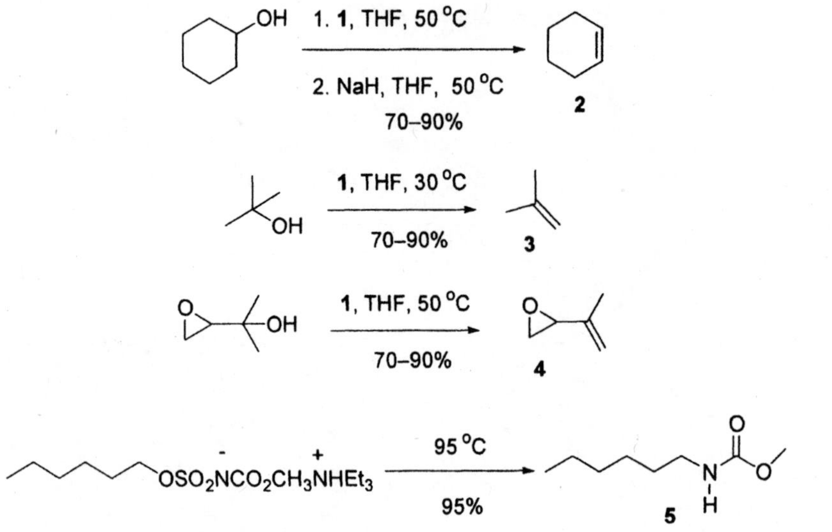

The use of Burgess reagent did not gain much attention until the late 1980s and early 2000s.

3.3.3 *Mechanism*

Burgess conducted a kinetic study of the dehydration of *erythro*- and *threo*-2-deuterio-1,2-diphenylethanols (**6**, **8**). The results were consistent with rate-limiting formation of an ion pair, followed by fast *syn*-β-proton transfer. Geometrical constraints require that the abstracted hydrogen is *syn* with respect to the leaving group.

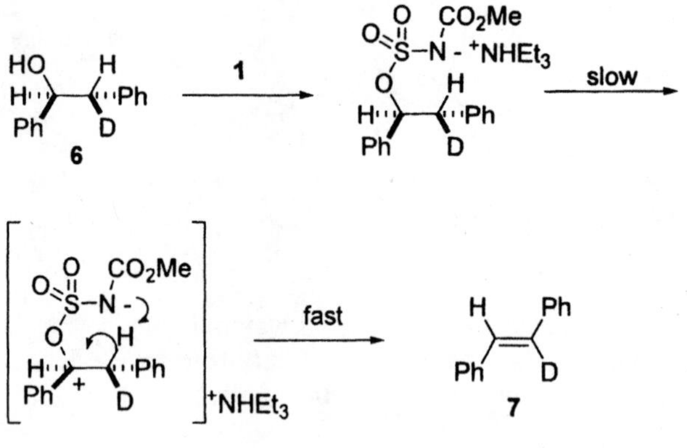

Therefore, *erythro* and *threo*-2-deuterio-1,2-diphenylethyl-*N*-carbomethoxy sulfamates, upon decomposition, give α-deuterio *trans*-stilbene (7) and protio *trans*-stilbene (**9**), respectively. This is consistent with *syn* elimination and an E1 mechanism. It is important to note that elimination can be accomplished in an *anti* fashion. However, it is still considered somewhat rare. The conversion of **10** to **11** is rationalized to be driven to *anti*-elimination by conformational effects.[3]

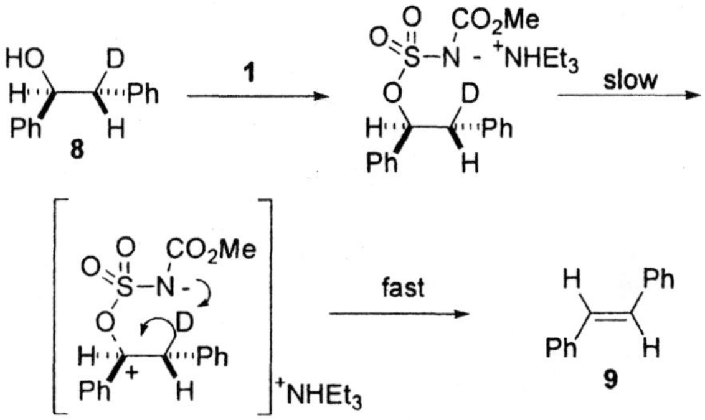

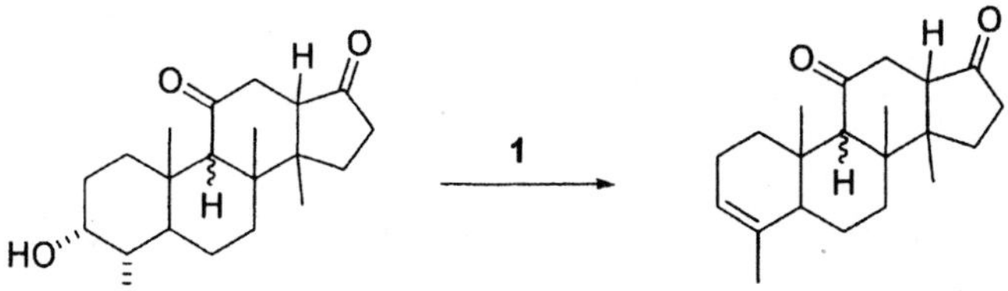

Double bond formation can sometimes be less predictable when the resulting carbocation is highly stabilized. For example, the substrate **12**, when exposed to Burgess reagent, produced **13** *via* a Wagner–Meerwein rearrangement.

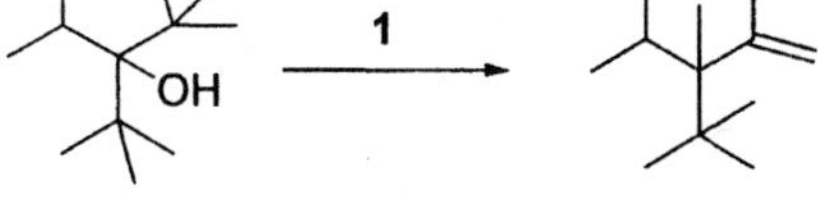

Allylic alcohols also can either undergo elimination (E1) or an S_N1 rearrangement. When **14** is exposed to **1** (sodium salt) and heated to 80 °C neat, followed by treatment with water, **15** is isolated in 96%. When **14** is conducted in triglyme at 75 °C, diene **16** is the major product (73%).

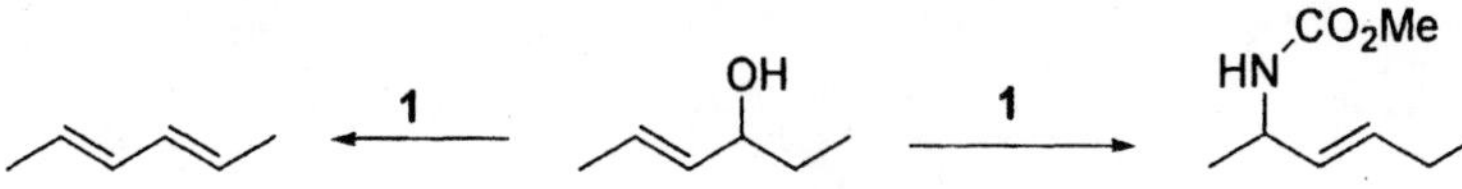

16 ← 1 — 14 — 1 → 15

In the majority of examples, the dehydration will occur according to Zaitsev's rule.

2.4 :1

As mentioned previously, primary alcohols, when subjected to **1**, form the corresponding carbamates, *via* an S_N2 mechanism, to form the corresponding olefins. For instance, **17** forms an *N*-carbomethoxysulfamate salt intermediate that rearranges to urethane **18** when heated. When primary sulfamate esters (**19**) are formed in a sterically hindered environment, the E1 mechanism prevails followed by rearrangement *via* Wagner–Meerwein, to form olefin **20**.[2]

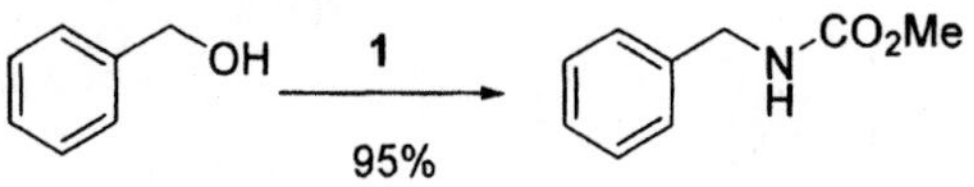

17 → (1, 95%) → **18**

19 OSO_2NCO_2MeNa → (110 °C, 84%) → **20**

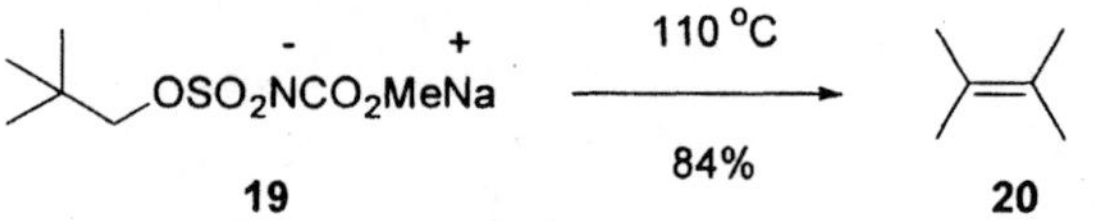

3.3.4 Variations and Improvements/Synthetic Utility

In 1992, Wipf reported the use of **1** to form 4,5-dihydrooxazoles (**22**) from β-hydroxy-α-amino acids (**21**).[4] This procedure was found to be general in scope. Furthermore, it provided compounds devoid of elimination, β-lactam, or aziridine side products. The configuration of the chiral center bearing the sulfonate leaving group was inverted. The reaction is characterized as mild, effective under neutral conditions, and useful for highly functionalized substrates that may contain easily epimerizable centers. This novel reaction was a key step for the total synthesis of the cyclic hexapeptide, westiellamide. During the total synthesis, Wipf demonstrated that by employing two intramolecular cyclizations, retention of configuration of the chiral center is obtained (**23**–**26**).[5]

21 → 22: 1, THF, 70 °C, 62%

23 → 24: 1, THF, 70 °C, 81%

24 → 25: 1. 0.3 N HCl/THF, 30 min; 2. K_2CO_3, pH 9.5, 2 h; 77%

25 → 26: 1, THF, 70 °C, 81%

Miller has used this method to invert the chiral center of a threonine residue while synthesizing the peptide fragment of pseudobactin (**27**–**28**).[6]

27 → 28: 1. **1**, THF, 70 °C; 2. 1 N HCl, THF; 3. K_2CO_3, pH 9.5; 70%

During the course of the total synthesis of (+)-curacin A (**30**), Wipf disclosed a nice conversion of an oxazoline (**29**) to a thiazoline (**30**) by double use of Burgess reagent.[7]

OMe

HO NH O H H

1

THF, 21 °C, 9 h

71%

OMe

O N H H

29

H_2S, MeOH

Et_3N, 35 °C, 20 h

66%

OMe

HO NH S H H

1

THF, 21 °C, 1 h

54%

OMe

S N H H

(+)-curacin A (**30**)

OH CO_2Me O NH NHBoc

31

1

THF, 70 °C

81%

CO_2Me O N NHBoc

22%

33

NiO_2

or

MnO_2

CO_2Me O N NHBoc

22–53%

34

CO_2Me O NHBoc

44%

32

Meyers has shown that during the total synthesis of (–)-bistatramide C, the valine-serine derived dipeptide **31**, when exposed to **1**, forms the expected oxazoline **33** analog along with the elimination compound **32**. **33** was further converted to the oxazole **34** with use of NiO_2 or MnO_2.[8]

This protocol of using **1** to develop oxazolines, oxazoles, and thiazolines has been used extensively for the creation of chiral bis-oxazoline ligands,[9] cyclic peptide synthesis, and natural products that contain these important heterocyclic moieties.[10]

Wipf has also developed a polyethyleneglycol-linked Burgess reagent (**35**). The major advantages for this polymer bound reagent are improved ease of handling and greater yields in the synthesis of labile oxazolines (**36**) and thiazolines (**37**).[11]

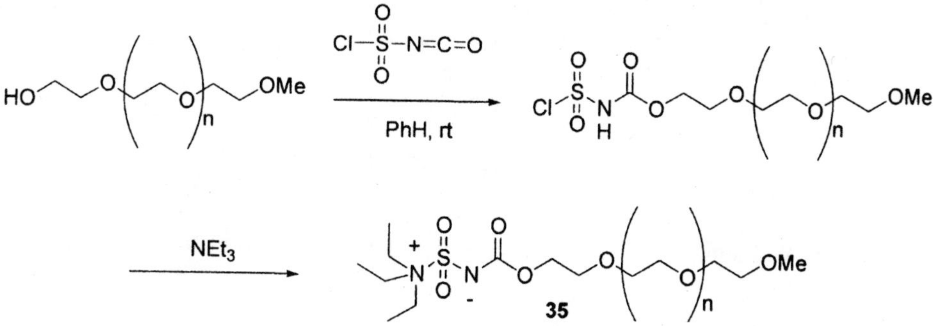

HO–PEG–OMe (n) → ($ClSO_2N=C=O$, PhH, rt) → $ClSO_2NHC(O)O$–PEG–OMe (n) → (NEt_3) → **35**

36 → (dioxane/THF (1:1), 85 °C, 3 h, X = O, S) → **37**

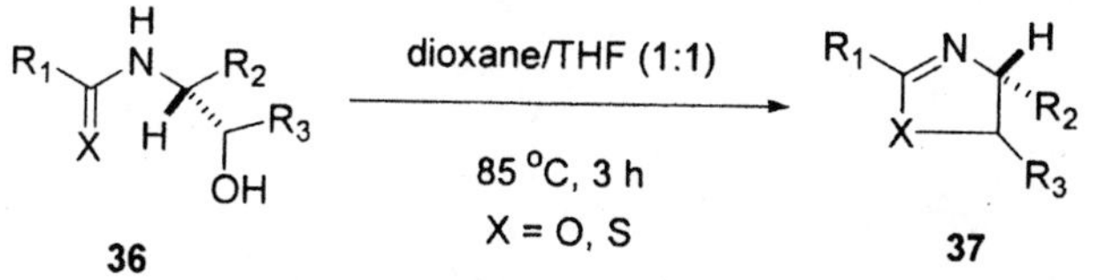

When ureas **38** were exposed to Burgess reagent, the production of carbodiimides (**39**) was achieved. This intermediate was mixed with amines to form 2,3,5-trisubstituted 4H-imidazolones (**40**).[12]

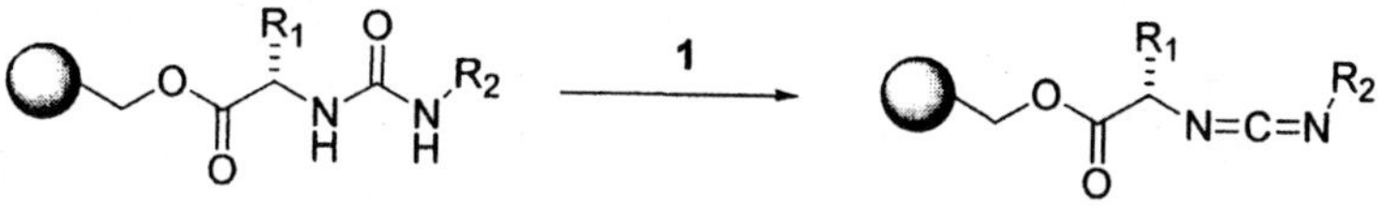

38 → (**1**) → **39**

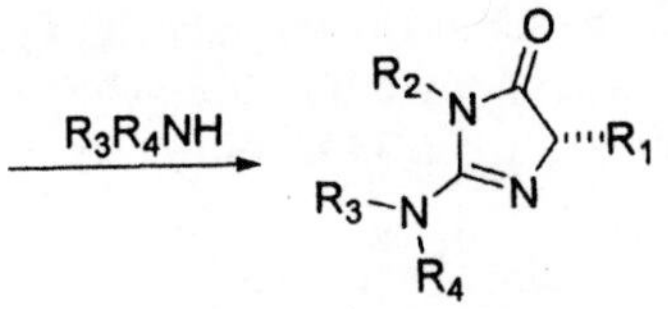

A new method of generating nitriles (**42**) from oximes (**41**) has been accomplished using Burgess reagent.[13]

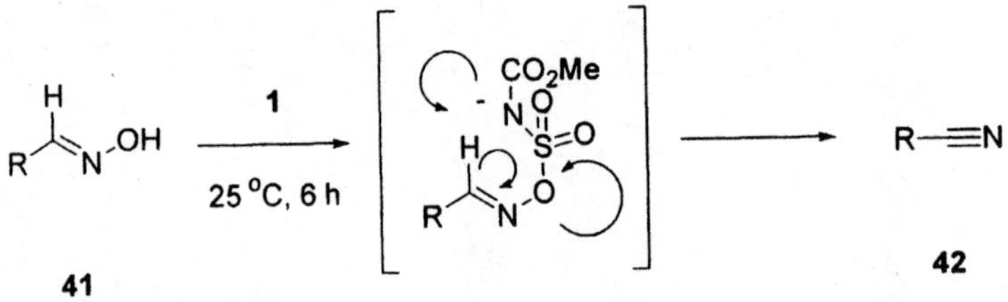

The dehydration is stereospecific (*syn*) and produces the requisite nitriles (**43–45**) in good yield.

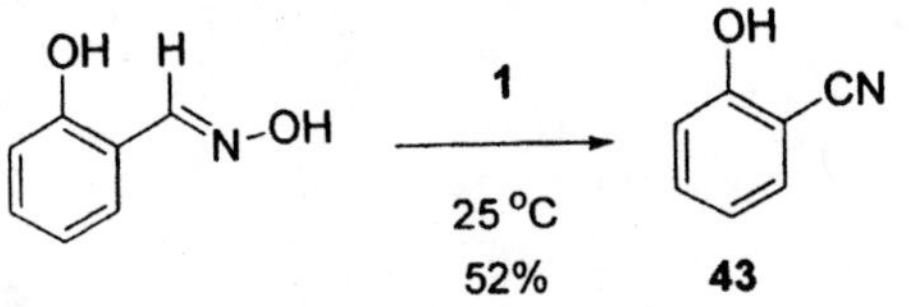

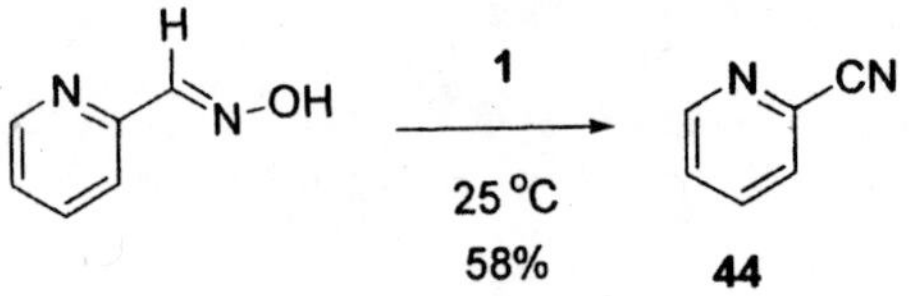

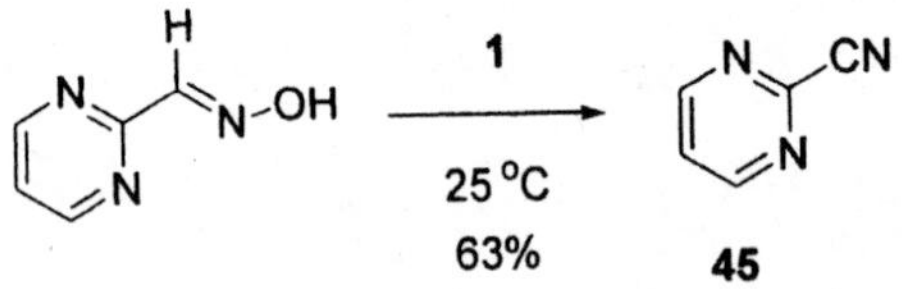

Burgess reagent is commonly known for dehydration reactions but has also recently been shown to act as an oxidizer. The conversion of benzoin to benzil in good yield under mild conditions has been achieved using Burgess reagent.[14] This method seems to be superior to previous methods of oxidizing this type of framework (**46–48**).

1 (3eq), 25 °C
DCM, 6 h
67%
46

1 (3eq), 25 °C
DCM, 6 h
63%
47

1 (3eq), 25 °C
DCM, 6 h
52%
48

It has been mentioned earlier that when Burgess reagent is treated with primary alcohols the carbamate is formed rather than the dehydration product. The development of a "benzyl Burgess reagent" (**49**) was shown to convert primary alcohols to the corresponding amines that are conveniently protected with the Cbz group.[15]

49

Below are a few examples (**50** and **51**), using this modified Burgess reagent.

77%
50

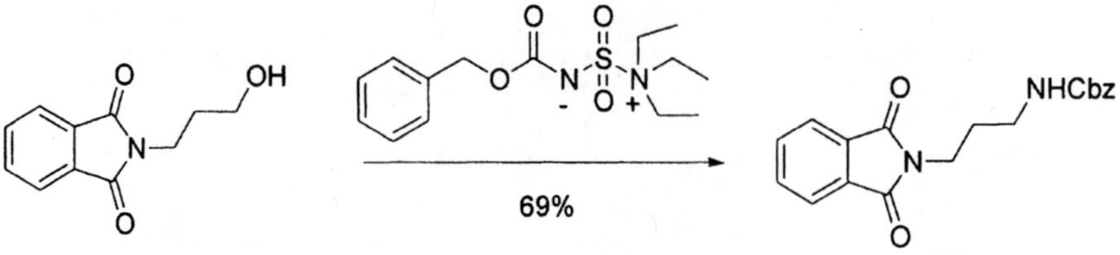

69%

51

Nicolaou has developed novel Burgess reagents (**49** and **52**) to form orthogonally protected sulfamidates in a regio- and stereoselective manner. The new Burgess reagents, when exposed to chiral diols (**53, 55**), afford cyclic sulfamidates (**54, 56**).[16]

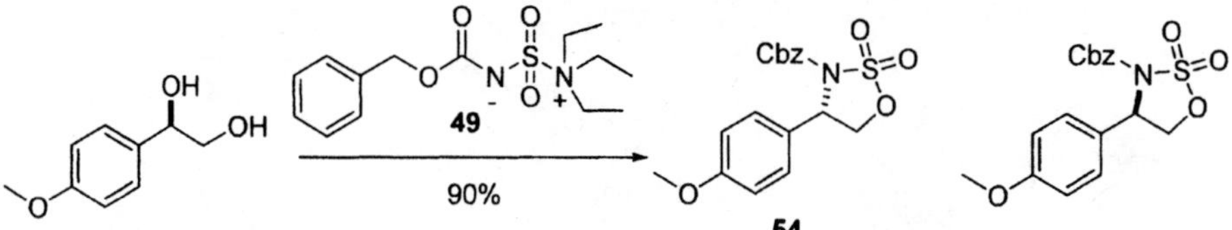

53 49 90% 54 98:2

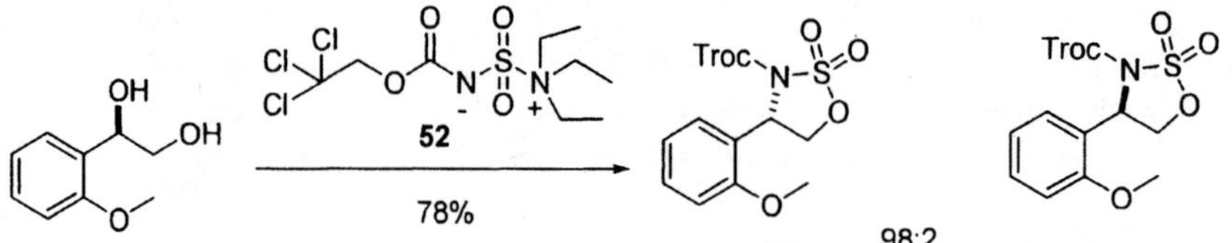

55 52 78% 56 98:2

Furthermore, the cyclic sulfamidates (**54**) have been shown to generate chiral β-amino alcohol **57** when subjected to aqueous hydrochloric acid in dioxane.

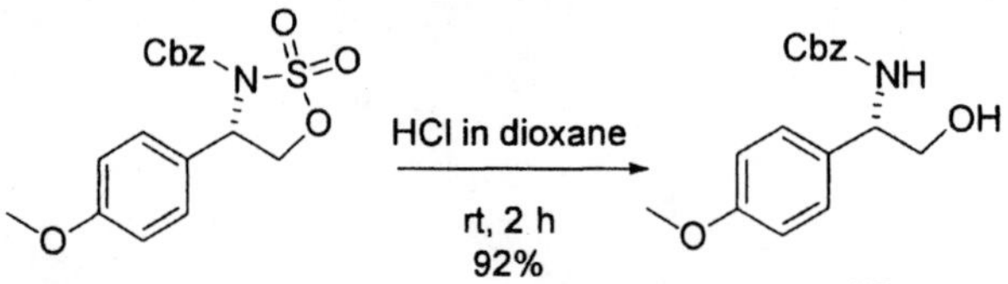

54 HCl in dioxane rt, 2 h 92% 57

Exposure of **1** to aliphatic expoxides (**58**) results in the corresponding sulfamidates (**59**). The sulfamidates can be further converted to *cis*-amino alcohols (**60**) with double inversion of configuration.[17]

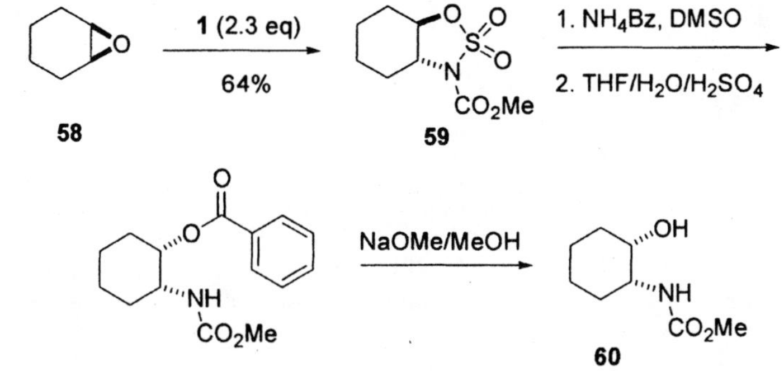

It was further shown that aromatic epoxides, when exposed to **1**, result in the formation of seven-membered ring systems (**61**–**62**).

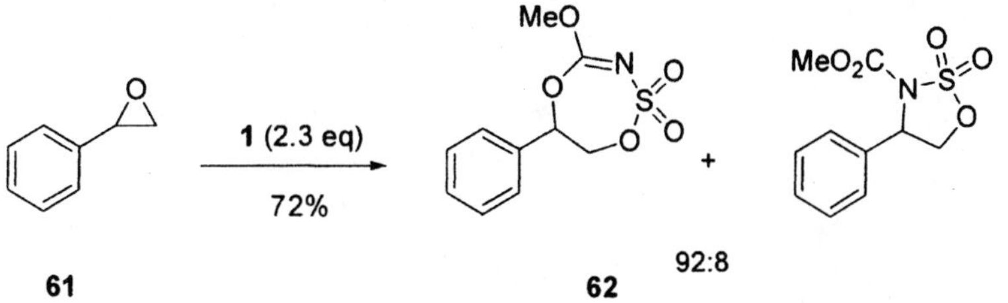

Hudlicky *et al.* created a chiral version of the Burgess reagent (**63**), that was exposed to meso-aliphatic epoxides, to afford a 1:1 mixture of *cis*-fused sulfamidate diastereoisomers (**64**, **65**, M* = menthol chiral auxillary) in modest yield.[18]

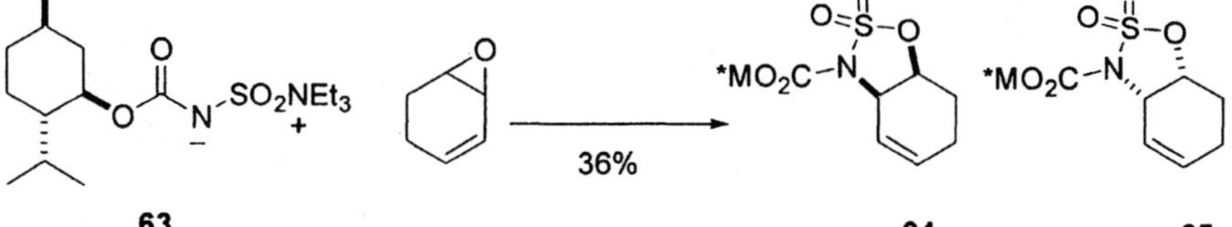

The *cis*-sulfamidates **64** and **65** were converted to *trans*-benzoates **66** and **67** *via* inversion at the oxygenated carbon with ammonium benzoate.

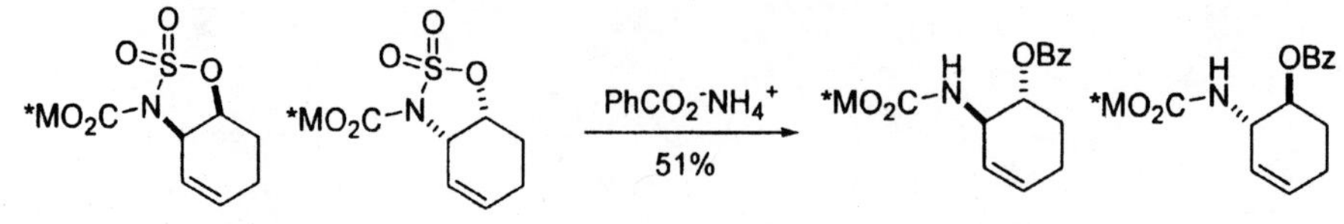

Using the concept of sulfamidate formation, Nicolaou used glucose derivatives to synthesize α-and β-glycosylamines (**68** and **69**) in a stereoselective manner using Burgess reagent.[19]

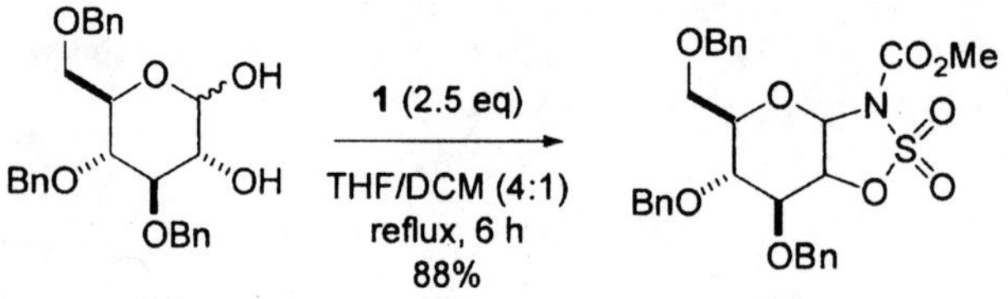

α/β (8:1)

Direct conversion of anomeric alcohols (**70**) to amines (**71**) was also accomplished using Burgess reagent to carbohydrate derivatives.

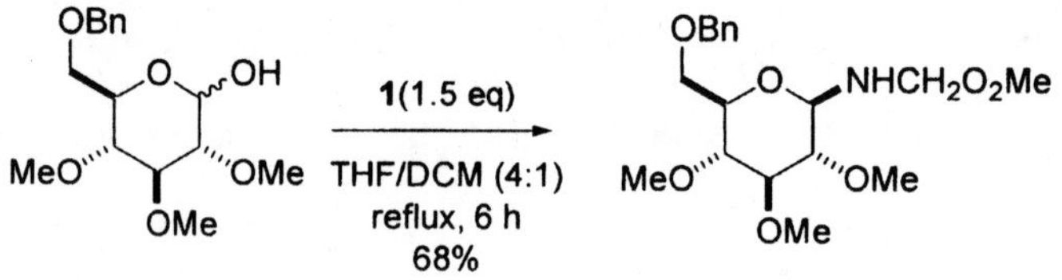

β/α > (20:1)

A convenient enantioselective synthesis of (*S*)-α-trifluoromethylisoserine *via* the cyclic sulfamidate was accomplished in high yield. Treatment of (*S*)-diol (**72**) with Burgess reagent in refluxing THF afforded the cyclic sulfamidate **73** in 53% yield. Hydrolysis of **73** with 6 N HCl gave (*S*)-α-CF_3 isoserine **74** as its hydrochloride salt.[20]

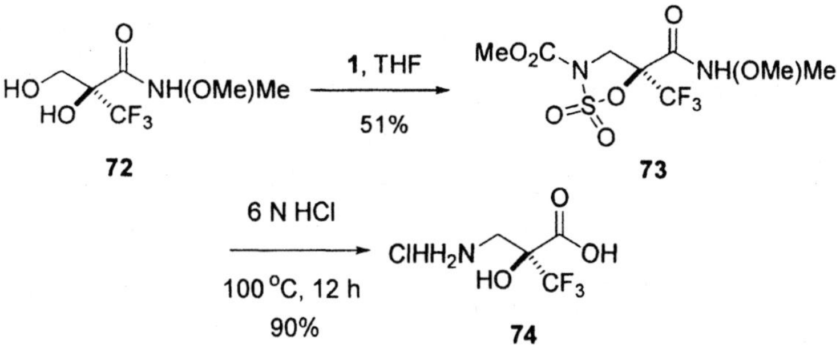

The formation of Burgess-type reagents (**75**) from chlorosulfonyl isocyanate (CSI) has recently been disclosed. The one-pot synthesis works well with primary, secondary, and tertiary alcohols and amines containing an active hydrogen.[21]

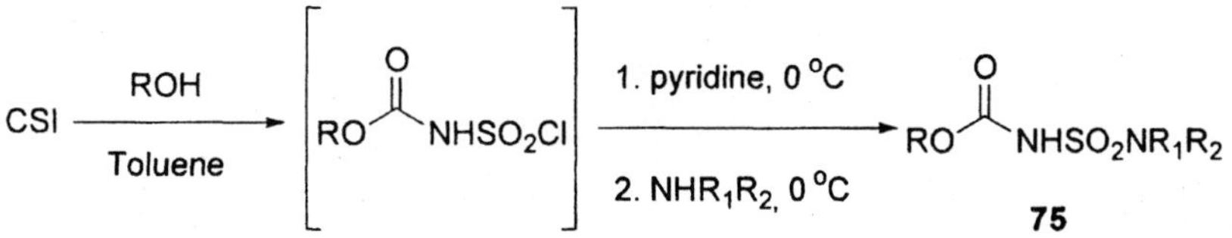

A one-pot conversion of Bayliss–Hillman adducts (**76**) into carbamates of unsaturated β-amino acids was accomplished using Burgess reagent.[22] When the sulfamate ester 77 is heated, the ester undergoes pyrolysis, extruding SO_3 to provide the carbamate **78**. When 77 is treated with NaH, followed by heating, an allylic rearrangement occurs to provide carbamate **79** in high yield. Examples with the phenyl ring substituted at the 4-OMe, 3-NO_2, 4-Cl, and 2-Cl position were shown to undergo either transformation in high yield. The 4-NO_2 phenyl substituent provided both products in ~ 20% yield.

When carboxylic acids (**80**) are exposed to **1** under microwave irradiation, mixed sulfocarboxy anhydrides (**81**) are formed.[23] Subsequent treatment of the anhydrides with amines at elevated temperatures (microwave) yield acyl ureas (**83**) and amides (**84**). The ratio of the two products appears to be temperature controlled. For instance, when the anhydride **81** is exposed to primary or secondary amines at 150 °C, the acyl ureas are formed *via* the urethane **82**. If **81** is exposed to primary or secondary amines at 80 °C, the corresponding amides are formed.

OH O OMe 1, THF OMe O NH O=S=O O O OMe 95 °C 85% OMe O NH O OMe

76 77 78

1. NaH, THF
90%
2. heat, H_2O

O OMe NH O OMe

79

R_1CO_2H **80** — **1** (1.4eq), DIEA, 80 °C, 5 min, microwave → [**81**] — R_2R_3NH (1.5eq), 150 °C, 8 min, microwave → [**82**] → **83**

R_1 O O=S=O N OMe (–) $^+$N

R_1 N H O O

R_1 N H NR_2R_3

R₁C(=O)OS(O₂)N⁻C(=O)OMe · ⁺NH(Et)(iPr)₂

81

R_2R_3NH (3 eq), 80 °C, 1 h, microwave

$R_1C(O)NR_2R_3$

84

3.3.5 Synthetic Utility via Dehydration of Secondary And Tertiary Alcohols

The Burgess reagent can dehydrate secondary and tertiary alcohols to the respective olefins under neutral conditions and low temperatures. This is in contrast to many other dehydrating type reactions/conditions that usually require acidic conditions that are accompanied with rearrangements and high temperatures. Thus the Burgess reagent has found great utility in the synthesis of molecules that contain multiple functionality.[24,1c,1d] For instance, Rigby used Burgess reagent for the formation of alkene **85** during the total synthesis of (+)-narciclasine.[25]

1, Benzene, reflux

85

Burgess reagent was found to be the only effective dehydroxylation method during a key intermediate for the total synthesis of fredericamycin A (**86–87**).[26] The dehydration protocol produced the alkene in quantitative yield. However, if compound **86** is exposed to Burgess reagent in refluxing benzene, no reaction occurs.

1, THF, reflux

86 **87**

Formation of the desired alkeneone (**88**) was accomplished with Burgess reagent during the synthesis of 7,11-dihydroxyguaianolides.[27]

1, Benzene
reflux
98%

88

Treatment of **89** with Burgess reagent in acetonitrile first provided the sulfamate **90** in 91% yield. Heating in toluene at 100 °C in the presence of NaH furnished minovine **91** (39%) and its isomer **92** in 53% yield.[28]

1, ACN
50 °C
1. NaH, THF
2. Toluene, 100 °C

89 **90**

39% 53%

91 **92**

Dehydration of seven-member rings can often lead to complications due to ring contraction *via* cationic rearrangement. After many dehydrating reagents were tried, Burgess reagent was found to convert **93** into **94** and **95** in a 1:2 ratio in 90% yield.[29]

1, Benzene
60 °C, 4 h
90%

1:2

93 **94** **95**

3.3.6 *Experimental*[30]

NC O H HO OBn **96** → **1**, Toluene, 50 °C, 60% → NC O H OBn **97**

A solution of **96** (150 mg, 0.55 mmol) in dry toluene (3 mL) was added under nitrogen to a suspension of **1** (133 mg, 0.59 mmol) in dry toluene (2 mL). The reaction mixture was stirred at 50 °C for two days and concentrated *in vacuo*. The residue was taken up in water (10 mL) and dichloromethane (25 mL) and the aqueous layer was separated and extracted with dichloromethane (3 × 25 mL). The combined organic layers were dried with $MgSO_4$ and concentrated *in vacuo*. The residue was purified by column chromatography (cyclohexane/ethyl acetate, 9:1) to give enol **97** as a white solid (84 mg, 60%). mp = 107 °C. $[\alpha]_D = -30.4$ (c = 0.35, $CHCl_3$); ^1H NMR (400 MHz, $CDCl_3$): δ = 1.34 (td, J = 13.8, 3.5, 1H), 1.72 (m, 1H), 1.84 (m, 1H), 2.02–2.13 (m, 2H), 2.23 (m, 1H), 2.51 (dt, J = 13.8, 3.5, 1H), 2.85 (d, J = 13.6, 1H), 4.49 (d, J = 10.3, 1H), 4.58 (d, J = 10.3, 1H), 4.99 (d, J = 8.2, 1H), 5.12 (t, J = 3.8, 1H), 7.27–7.46 (m, 5H).

3.3.7 *References*

1. [R] (a) Burgess, E. M.; Atkins, G. M. *J. Am. Chem. Soc.* **1968**, *90*, 4744. (b) Taibi, P.; Mobashery, S. in *Encyclepedia of Reagents for Organic Synthesis*, L. A. Paquette (Ed.), p. 3345, Wiley, New York, **1995**. (c) Lamberth, C., *J. Prakt. Chem.* **2000**, *342*, 518. (d) Mal, D.; Dey, S.; Khapli, S. *J. Indian Inst. Sci.* **2001**, *81*, 461.
2. Burgess, E. M.; Penton, H. R.; Taylor, E. A. *J. Org. Chem.* **1973**, *38*, 26.
3. O'Grodnick, J. S.; Ebersole, R. C.; Wittstruck, T.; Caspi, E. *J. Org. Chem.* **1974**, *39*, 2124.
4. (a) Wipf, P.; Miller, C. P. *Tetrahedron Lett.* **1992**, 33, 907. (b) Wipf, P.; Miller, C. P. *Tetrahedron Lett.* **1992**, 33, 6267.
5. Wipf, P.; Miller, C. P. *J. Am. Chem. Soc.* **1992**, *114*, 10975.
6. Miller, M. J.; Kolasa, T.; Okonya, J. F. *J. Org. Chem.* **1995**, *60*, 1932.
7. (a) Wipf, P.; Xu, W. *J. Org. Chem.* **1996**, *61*, 6556. (b) Muir, J. C.; Pattenden, G.; Ye, T. *J. Chem. Soc. Perkin Trans I* **2002**, 2243.
8. Meyers, A. L.; Aguilar, E. *Tetrahedron Lett.* **1994**, *35*, 2477.
9. Benaglia, M.; Benincori, T.; Mussini, P.; Pilati, T.; Rizzo, S.; Sannicolo, F. *J. Org. Chem.* **2005**, *70*, 7488.
10. (a) Ino, A.; Murabayashi, A. *Tetrahedron* **2001**, *57*, 1897. (b) Xia, Z.; Smith, C. D. *J. Org. Chem.* **2001**, *66*, 3459.
11. (a) Wipf, P.; Venkatraman, S. *Tetrahedron Lett.* **1996**, *37*, 4659. (b) Wipf, P.; Hayes, G. B. *Tetrahedron* **1998**, *54*, 6987.
12. Lange, U. E. W. *Tetrahedron Lett.* **2002**, *43*, 6857.
13. Jose, B.; Sulatha, M. S.; Pillai, M.; Prathapan, S. *Synth. Commun.* **2000**, 30, 1509.
14. Jose, B.; Unni, M. V. V.; Parathapan, S.; Vadakkan, J. J. *Syn. Commun.* **2002**, *32*, 2495.
15. Wood, M. R.; Kim, J. Y.; Books, K. M. *Tetrahedron Lett.* **2002**, *43*, 3887.
16. Nicolaou, K. C.; Huang, X.; Snyder, S. A.; Rao, P. B.; Bella, M.; Reddy, M. V. *Angew. Chem. Int. Ed.* **2002**, *41*, 834.
17. Rinner, U.; Adams, D. R.; Dos Santos, M. L.; Abboud, K. A.; Hudlicky, T. *Synlett* **2003**, 1247.

18. Leisch, H.; Saxon, R.; Sullivan, B.; Hudlicky, T. *Synlett* **2006**, 445.
19. (a) Nicolaou, K. C.; Snyder, S. A.; Nalbandian, A. Z.; Longbottom, D. A. *J. Am. Chem. Soc.* **2004**, *126*, 6234. (b) Nicolaou, K.C.; Snyder, S. A.; Longbottom, D. A.; Nalbandian, A. Z.; Huang, X. *Chem. Eur. J.* **2004**, *10*, 5581.
20. Avenoza, A.; Busto, J. H.; Jimeniz-Oses, G.; Peregrina, J. M. *J. Org. Chem.* **2005**, *70*, 5721.
21. Masui, Y.; Watanabe, H.; Masui, T. *Tetrahedron Lett.* **2004**, *45*, 1853.
22. Mamaghami, M.; Badrian, A. *Tetrahedron Lett.* **2004**, *45*, 1547.
23. Wodka, D.; Robbins, M.; Lan, P.; Martinez, R. L.; Athanasopoulos, J.; Makara, G. M. *Tetrahedron Lett.* **2006**, *47*, 1825.
24. (a) Nicolaou, K. C.; Zak, M.; Safina, B. S.; Lee, S. H.; Estrada, A. A. *Angew. Chem. Int. Ed.* **2004**, *43*, 5092. (b) Horiguchi, Y.; Nakamura, E.; Kuwajima, I. *J. Am. Chem. Soc.* **1989**, *111*, 6257. (c) Bartman, W.; Beck, G.; Granzer, E.; Jendralla, H.; Kerekjarto, B. V.; Wess, G. *Tetrahedron Lett.* **1986**, *27*, 4709.
25. Rigby, J. H.; Mateo, M. E. *J. Am. Chem. Soc.* **1997**, *119*, 12655.
26. Kita, Y.; Higuchi, K.; Yoshida, Y.; Iio, K.; Kitagaki, S.; Ueda, K.; Akai, S.; Fujioka, H. *J. Am. Chem. Soc.* **2001**, *123*, 3214.
27. Manzano, F. L.; Guerra, F. M.; Moren-Dorado, F. J.; Jorge, Z. D.; Massanet, G. M. *Org. Lett.* **2006**, *8*, 2879.
28. Yuan, Z. Q.; Ishikawa, H.; Boger, D. L. *Org. Lett.* **2005**, 7, 741.
29. Weyerstahl, P.; Marschall-Weyerstahl, H.; Penninger, J.; Walther, L. *Tetrahedron* **1987**, *43*, 5287.
30. Keller, L.; Dumas, F.; D'Angelo, J. *Eur. J. Chem.* **2003**, 2488.

Daniel D. Holsworth

3.4 Corey–Kim Oxidation

3.4.1 Description

The Corey–Kim oxidation uses *N*-chlorosuccinimide (NCS) and dimethylsulfide (DMS) followed by triethylamine (TEA) addition to convert primary and secondary alcohols to aldehydes and ketones, respectively.[1]

1. NCS, DMS
2. TEA

1 → 2

1. NCS, DMS
2. TEA

3 → 4

Aldehydes and ketones are prepared in this manner by premixing the NCS and DMS at low temperature (0 °C initially then cooling to –25 °C as in the original publication) before adding the alcohol substrate followed by TEA a short while later. The yields are generally very good to excellent and optimal solvents are toluene and CH_2Cl_2. Polar solvents should be avoided as methyl thiomethyl ethers can form in such an environment as a side-product and lower the yield of the reaction. There is a high degree of functional group compatibility with this mild reaction and overall reaction times typically vary from 30 minutes to 2 hours.

3.4.2 Historical Perspective and Improvements

toluene, 0 °C

5

1. 6
–25 °C, 2 h
2. TEA, 97%

7

In a 1972 communication, E. J. Corey and C. U. Kim described the first example of this mild oxidation method and expressed their hope that this discovery would find widespread use.[1] The oxidation of 4-*tert*-butylcyclohexanol (**6**) to corresponding ketone **7**, for example, was achieved in 97% yield by the *in situ* generated sulfonium chloride complex **5**. This complex is now commonly referred to as the "Corey–Kim Reagent." The complex is not stable at temperatures above 0 °C for extended periods of time and its rate of decay over time has been followed by continuous FT-IR.[2]

By this manner, several additional primary and secondary alcohols were oxidized to their corresponding aldehydes and ketones in high yield.[1]

1. NCS, DMS
toluene, –25 °C
2. TEA, 94%

8 → **9**

1. NCS, DMS
toluene, –25 °C
2. TEA, 91%

10 → **11**

Efforts to develop an odor-free variation of the Corey–Kim oxidation resulted in the development of several modifications that avoid the usage of odorous and volatile DMS. In 2002, a high molecular weight fluorous version was described and at about the same time a separate research group described their odorless version.[3,4] The fluorous version, in addition to using a less volatile sulfide, is environmentally friendly in that the fluorous reagent is recyclable. Isoborneol (**12**) was converted to ketone **14** in 88% yield and 72% of the polyfluorinated sulfide (**13**), referred to as "fluorous dimethyl sulphide," was recovered after continuous fluorous extraction. As in the original Corey–Kim communication, the NCS and sulfide were stirred at 0 °C before cooling to –25 °C, adding substrate, and ultimately TEA.

The other variation utilizes readily obtainable, inexpensive, and odorless dodecyl methyl sulfide (**16**) to afford oxidized products in excellent yields.[4] For example, after adding testosterone (**15**) to a stirring solution of NCS and **16** at –40 °C, the desired ketone **17** was obtained in 97% yield after TEA addition.

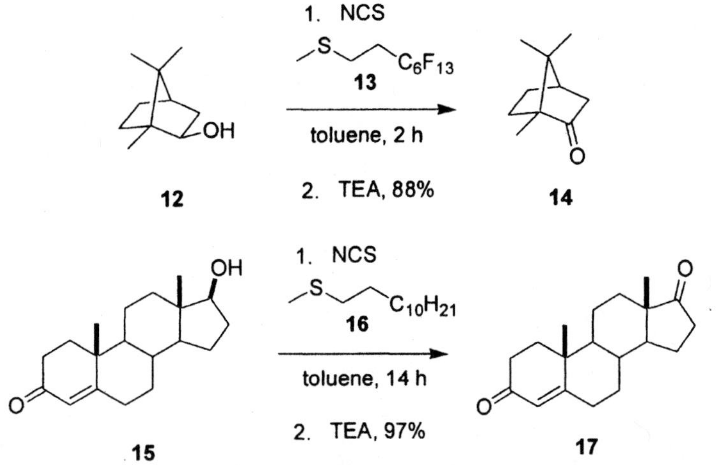

In an effort to follow modern trends toward green chemistry, a new odourless, water-soluble DMS equivalent (**19**) that is recyclable was introduced by the same group.[5] Up to 89% of this new morpholino sulfide can be recovered using simple acid/base extraction techniques. Yields using this new DMS alternative are excellent and vary from 85% to 99% depending upon substrate, stoichiometry, and solvent choice. For example, the transformations illustrated above (**12** to **14** and **15** to **17**) were repeated using sulfide **19** and yields of 94% and 91%, respectively, were reported. In addition, the highly successful oxidation of benzylic alcohol **18** and allylic alcohol **21** were carried out at low temperatures.

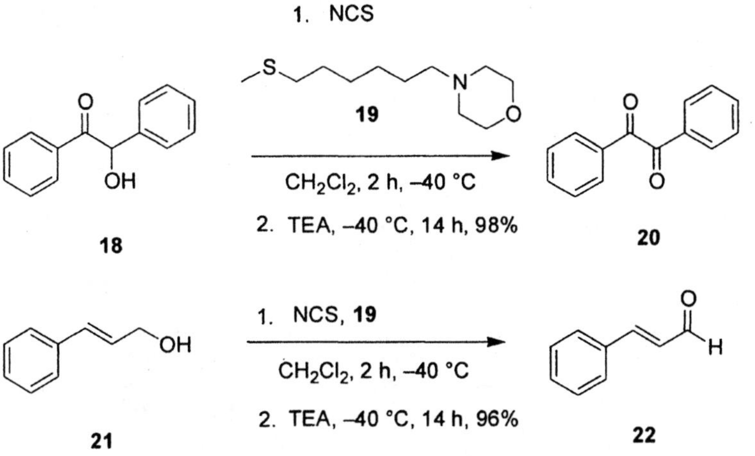

3.4.3 Mechanism

It is plausible that the first steps in the mechanism of the Corey–Kim oxidation involve the initial S_N2 reaction between the sulfide nucleophile and the NCS that bears the electrophilic chlorine to form an intermediate sulfonium chloride species and a succinimide anion that in turn attacks the newly formed electrophilic sulfonium to yield the Corey–Kim intermediate (**5**), or what is more commonly called "Corey–Kim reagent."[1] As shown below, the Corey–Kim intermediate is then attacked by the nucleophilic alcohol substrate to afford a succinimide anion and an oxonium species that is deprotonated by the succinimide anion to furnish alkoxysulfoxonium salt **23**. The role of triethyl amine is to then deprotonate **23**, setting off a cascade of events that ultimately liberate the desired product, DMS, and TEA•HCl. As illustrated below, a proton from one of the *S*-methyl groups is removed, providing ylide **24** that in turn intramolecularly deprotonates the proton alpha to the oxygen atom. The carbonyl group then forms, thus liberating the oxidized substrate and DMS.

In the case where a nonsymmetrical sulfide such as **19** is used, deuterium labeling studies using 2-deuteroisoborneol **25** indicate that it's the *S*-methyl group rather than the *S*-methylene that is deprotonated (second to last step of mechanism) and the fact that the deuterium label from the alcohol is transferred to this position supports the last step in the hypothesized sequence of events that is proposed. Spectral analyses confirm that the selectivity for *S*-methyl deprotonation is in fact 95:5.[4]

3.4.4 Synthetic Utility

3.4.4.1 Preparation of aldehydes and ketones

HO N Pf O — 1. NCS, DMS / 2. TEA, 84% → H O N Pf O — steps → HO O OH HN

26 27 28

Pf = 9-phenylfluoren-9-yl=

The mild reaction conditions of the Corey–Kim oxidation reaction make it an excellent choice when the oxidation of an alcohol that is contained within a complex synthetic intermediate is needed. Rapoport has shown this oxidation method to be useful to prepare aldehydes from primary alcohols in several multi-step syntheses. For example, treatment of alcohol **26** with NCS and DMS in toluene provided aldehyde **27** that was ultimately used to construct the (–)-enantiomer of the C-9-amino acid constituent (**28**) of immunosuppressant drug cyclosporine as well as the naturally occurring (+)-enantiomer after a slight modification.[6]

The same research group also utilized the Corey–Kim reaction to prepare aldehyde **30** from alcohol **29** in the total synthesis of polyhydroxylated indolizidine alkaloid castanospermine (**31**) and its C-6 epimer.[7]

1. NCS, DMS toluene, –25 °C
2. TEA, 92%

steps

29 **30** **31**

Pf = 9-phenylfluoren-9-yl

The [2 + 2]-cycloaddition reaction was used to explore synthetic routes toward key intermediates that could be applied in carbapenem antiobiotic synthesis and in the preparation of such intermediates, chemists at the Sagami Research Institute utilized the Corey–Kim reaction.[8] In this example, secondary alcohol **32** was converted to ketone **33** in 95% yield.

1. NCS, DMS toluene
2. TEA, 95%

32 **33**

3.4.4.2. Preparation of benzyl and allyl halides

While a variety of alcohols are excellent substrates for the traditional Corey–Kim oxidation, alcohols capable of forming stabilized carbocations such as allylic and benzylic alcohols can undergo a side-reaction to form alkyl chlorides. In fact, the Corey–Kim reagent and conditions, with slight modification, have been used to convert allylic and benzylic alcohols to their corresponding chlorides in very high yield as shown below.[9] Alkyl bromides can also be prepared if *N*-bromosuccinimide is used in place of *N*-chlorosuccinimide.

DMS, NCS
CH_2Cl_2
–25 °C, 4 h, 95%

34 **35**

OH → DMS, NCS / CH_2Cl_2 / −25 °C, 4 h, 95% → Cl

36 37

Referring back to the mechanism discussed earlier, if the unstable sulfoxonium intermediate **23** is allowed to decompose, that is, if triethyl amine is not added to trigger the carbonyl formation, then alkyl chlorides result.

3.4.4.3 Preparation of 1,3-dicarbonyls

The Corey–Kim conditions have also been applied to 3-hydroxycarbonyl compounds to afford 1,3-dicarbonyls and this variation is curious in that in some cases stable dimethylsulfonium dicarbonylmethylides are isolated which have to be further treated with zinc-acetic acid to afford the desired dicarbonyl product. In 1988, Yamauchi showed that the outcome of the addition of 3-hydroxycarbonyls to the Corey–Kim reagent varied depending upon the C-2 substitution pattern.[10] As illustrated by the examples below, if the C-2 position is unsubstituted such as in **38**, the dimethylsulfonium dicarbonylmethylide **39** was isolated whereas the desired oxidation product **42** was produced if there was at least one substituent (R_3 or R_4) present at C-2. Other cases of non-C-2-substituted 3-hydroxycarbonyls furnishing stable dimethylsulfonium methylides and their conversion to the desired diketones have been reported.[11]

Corey–Kim Reagent / CH_2Cl_2, −78 °C, 1 h / then Et_3N, −78 °C, 1 h / (0–98 %); Zn/AcOH / CH_2Cl_2 / 0 °C to rt, 12 h / (0–99 %)

38 39 40

Corey–Kim Reagent / CH_2Cl_2, −78 °C, 1 h / then Et_3N, −78 °C, 1 h / (52–99 %)

(R_3, R_4 = mono- or disubstituted)

41 42

Noting the greater reactivity of the newly formed dicarbonyl products for the Corey–Kim reagent present in excess, the authors propose the following mechanism in which the C-2 of the dicarbonyl enol ether tautomer attacks the electrophilic sulfonium atom of the Corey–Kim reagent giving adduct **43**, whose acidic proton is removed by TEA to generate the stable ylide **39**.[10]

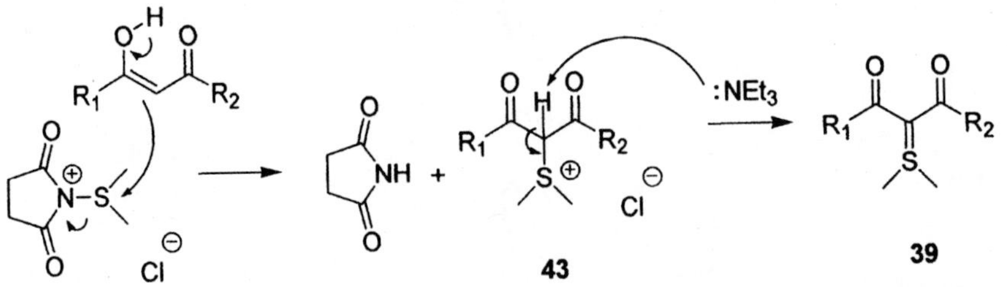

An example of the successful application of the Corey–Kim oxidation of a 3-hydroxyketone to yield a 1,3-diketone in natural product synthesis is illustrated by the total synthesis of (±)-jatrophone.[12] As shown below, the presence of the C-2 methyl group of **44** allowed for the clean conversion to desired diketone **45** in one step (i.e., no need for reductive desulferization). The Corey–Kim oxidation was then actually used a second time in the sequence to introduce another ketone and ultimately afforded the desired macrolide diterpene **46**.

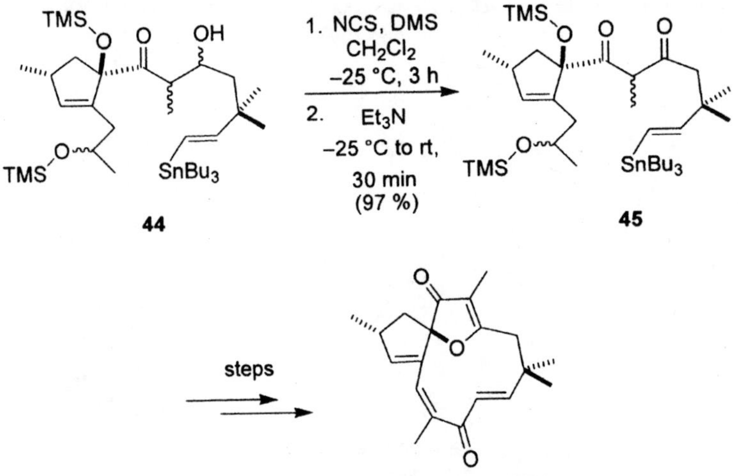

Jatrophone (**46**)

3.4.4.4 Preparation of α-hydroxyketones and α-diketones

Corey and Kim first described the use of NCS and DMS for the conversion of 1,2-diols into α-hydroxyketones in 1974, two years after their initial communication and gave the conversion of 1,2-diol **47** to α-hydroxyketone **48** in 86% yield as an example.[13] The authors comment that glycol cleavage, which could occur using the harsh oxidation conditions frequently employed at the time, was not observed because such an elimination process would involve a seven-membered cyclic transition state (B) whereas α-hydroxyketones form as a result of the proposed favored five-membered ring cyclic transition state (A) as shown below.

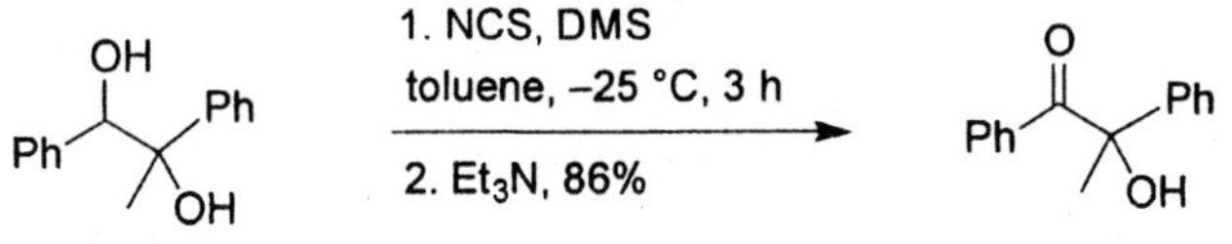

1. NCS, DMS toluene, –25 °C, 3 h
2. Et_3N, 86%

47 **48**

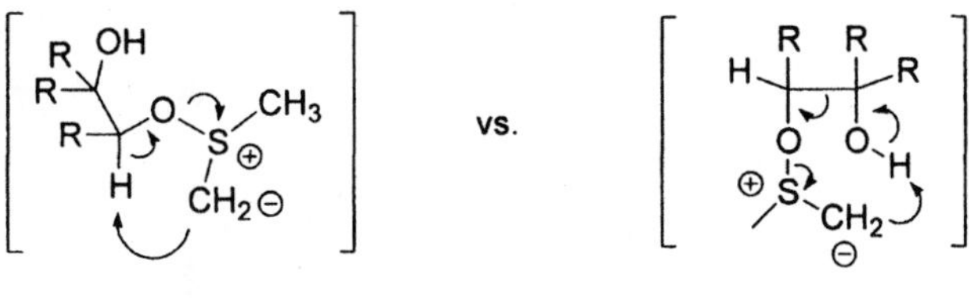

vs.

A **B**

The Corey–Kim protocol was successfully applied late in the multi-step total synthesis of ingenol (**51**) to convert diol **49** to α-ketol **50** where it was the less hindered hydroxyl group that was selectively oxidized.[14] The newly formed ketone was then used to construct the allylic alcohol moiety of ingenol after several steps.

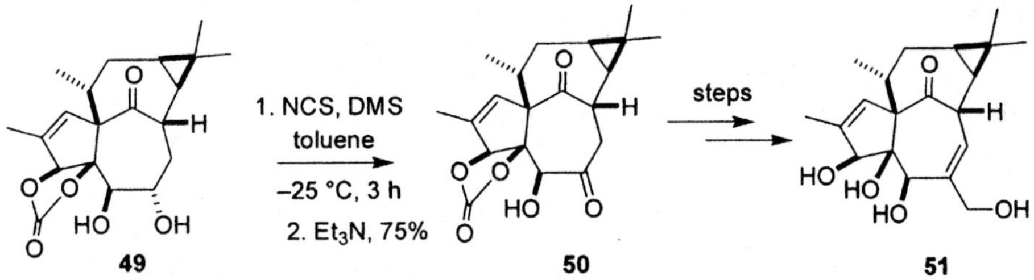

1. NCS, DMS toluene –25 °C, 3 h
2. Et_3N, 75%

steps

49 **50** **51**

An example of the construction of a 1,2-diketone from a 1,2-diol comes from the total synthesis of (±)-cephalotaxine (**54**) where the Corey–Kim conditions were used to convert diol **52** to diketone **53** in 89% yield while other oxidation methods were reported to have failed to affect this conversion.[15] The more accessible ketone was then converted to the enol ether and LAH reduction afforded the racemic natural product **54**.

1. NCS, DMS CH_2Cl_2, –42 °C
2. Et_3N, 89%

52 53 54

3.4.5 *Experimental*

1. NCS, DMS CH_2Cl_2
2. Et_3N, 92%

epi-44 *epi*-45

Synthesis of *epi*-45[12]

To a solution containing 0.25 g (0.187 mmol) of *N*-chlorosuccinimide in 35 mL of toluene at –25 °C was added 0.18 mL (2.45 mmol) of dimethylsulfide. The mixture was stirred at –25 °C for 30 minutes, followed by the addition of 0.35 g (0.46 mmol) of alcohol ***epi*-44** in 18 mL of toluene over 10 minutes. Stirring continued for 3 hours, followed by the addition of 0.28 mL (2.01 mmol) of triethylamine in 18 mL of toluene dropwise. The cold bath was removed, and the contents were warmed to room temperature and maintained for an additional 30 minutes. The mixture was diluted with water and extracted with ether. The organic layer was dried over sodium sulfate. Removal of solvent and column chromatography of the yellow oil on silica gel with 10% ethyl acetate/hexanes gave 0.32 g (92%) of product ***epi*-45** as a mixture of diastereomers.

1. $N-(CH_2)_6SCH_3$ (19)
NCS, CH_2Cl_2
2. TEA
–40 °C to rt
91%

55 56

General procedure for the Corey–Kim oxidation using recyclable 19[5]

To a solution of *N*-chlorosuccinimide (63.76 mg, 0.48 mmol) in anhydrous CH_2Cl_2 (2 mL) under nitrogen at –40 °C was added **19** (104 mg, 0.48 mmol) in CH_2Cl_2 (2 mL) dropwise. The reaction mixture was stirred at –40 °C for 30 min before the addition of alcohol **55** (0.32 mmol) in CH_2Cl_2 (2 mL). After the reaction had been stirred for 2 h at –40 °C, freshly distilled triethylamine (0.14 mL, 0.95 mmol) was added and the reaction mixture was stirred at the same temperature for a further period of 2.5 h. It was then allowed to warm to room temperature for 8 h with continued stirring before being poured into aq. 1 N HCl (60 mL) and extracted with ethyl acetate (3 × 30 mL). The organic component was washed again with aq. 1 N HCl (50 mL) brine and dried over sodium sulfate. The solvent was evaporated *in vacuo* to afford pure ketone **56** in 91% yield.

General procedure for the recovery of 19[5]

The aq. 1 N HCl solution collected after work-up was made alkaline (pH > 9) using aq. 5 M NaOH and extracted with diethyl ether (3 × 30 mL), dried (sodium sulfate) and concentrated followed be Kuegelruhr distillation (145 °C, 1.5 mmHg) to afford pure **19** (77%).

3.4.6. References

1. (a) Corey, E. J.; Kim, C. U. *J. Am. Chem. Soc.* **1972,** *94,* 7586. (b) Corey, E. J.; Kim, C. U.; Misco, P. F.; *Org. Syn.* **1978,** *58,* 122.
2. Chambournier, G.; Surjono, H.; Xiao, Z.; Naris, M. Poster Presentation, 225th ACS National Meeting, New Orleans, LA, **2003,** ORGN-403.
3. Crich, D.; Neelamkavil, S. *Tetrahedron* **2002,** *58,* 3865.
4. (a) Nishide, K.; Ohsugi, S.; Fudesaka, S.; Kodama, S.; Node, M. *Tet. Lett.* **2002,** *43,* 5177. (b) Ohsugi, S.; Nishide, K.; Oono, K.; Okuyama, K.; Fudesaka, S.; Kodama, S.; Node, M. *Tetrahedron* **2003,** *59,* 8398.
5. Nishide, K.; Patra, P. K.; Matoba, M.; Shanmugasundaram, K.; Node, M. *Green Chem.* **2004,** *6,* 142.
6. Lubell, W. D.; Jamison, T. F.; Rapoport, H. *J. Org. Chem.* **1990,** *55,* 3511.
7. Gerspacher, M.; Rapoport, H. *J. Org. Chem.* **1991,** *56,* 3700.
8. Ito, Y.; Kobayashi, Y.; Kawabata, T.; Takase, M.; Terashima, S. *Tetrahedron* **1989,** *45,* 5767.
9. Corey, E. J.; Kim, C. U.; Takeda, M. *Tetrahedron Lett.* **1972,** 4339.
10. Katayama, S.; Fukuda, K.; Watanabe, T.; Yamauchi, M. *Synthesis* **1988,** 178.
11. Pulkkinen, J. T.; Vepsalainen, J. J. *J. Org. Chem.* **1996,** *61,* 8604.
12. Gyorkos, A. C.; Stille, J. K.; Hegedus, L. S. *J. Am. Chem. Soc.* **1990,** *112,* 8465.
13. Corey, E. J.; Kim, C. U. *Tetrahedron Lett.* **1974,** 287.
14. Tanino, K.; Onuki, K.; Asano, K.; Miyashita, M.; Nakamura, T.; Takahashi, Y.; Kuwajima, I. *J. Am. Chem. Soc.* **2003,** *125,* 1498.
15. Kuehne, M. E.; Bornmann, W. G.; Parsons, W. H.; Spitzer, T. D.; Blount, J. F.; Zubieta, J. *J. Org. Chem.*, **1988,** *53,* 3439.

Donna M. Iula

3.5 Dess–Martin Periodinane Oxidation

3.5.1 Description

The Dess–Martin periodinane oxidation utilizes 1,1,1-triacetoxy-1,1-dihydro-1,2-benziodoxol-3(1H)-one (**2**) as an oxidant for the conversion of primary and secondary alcohols (**1**) to their corresponding aldehyde (**3**) and ketone products, respectively.

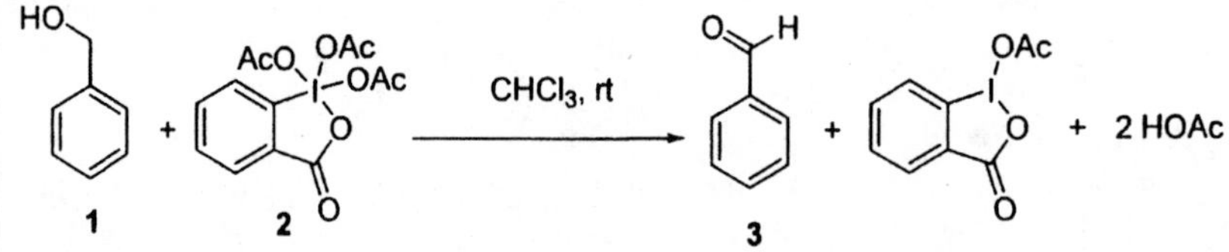

HO; AcO, OAc, OAc; + ; $CHCl_3$, rt; O, H; + ; OAc; + 2 HOAc; **1**; **2**; **3**

The oxidation procedure has found wide utility due to the mild conditions employed, high degree of functional group compatibility, ease of workup, inexpensive materials utilized, efficiency, and predictability.

3.5.2 Historical Perspective

The oxidation of primary and secondary alcohols to aldehydes and ketones, respectively, utilizing 1,1,1-triacetoxy-1,1-dihydro-1,2-benziodoxol-3(1H)-one (**2**) was first disclosed by D. B. Dess and J. C. Martin in 1983.[1] The synthesis of the iodo-oxidizing reagent **2** was readily prepared from 2-iodobenzoic acid (**4**). Formation of **5** with potassium bromate in sulfuric acid gave the cyclic tautomer of 2-iodoxybenzoic acid (IBX) (**5**) in 93% yield. Compound **5**, commonly referred to as IBX, is also an efficient oxidizing agent that has received much attention.[2,8,12] IBX is sparingly soluble in most organic solvents but is readily soluble in DMSO. IBX was subjected to acetic anhydride in acetic acid at 100 °C to form 1,1,1-triacetoxy-1,1-dihydro-1,2-benziodoxol-3(1H)-one (DMP, **2**) in 87% yield.[3]

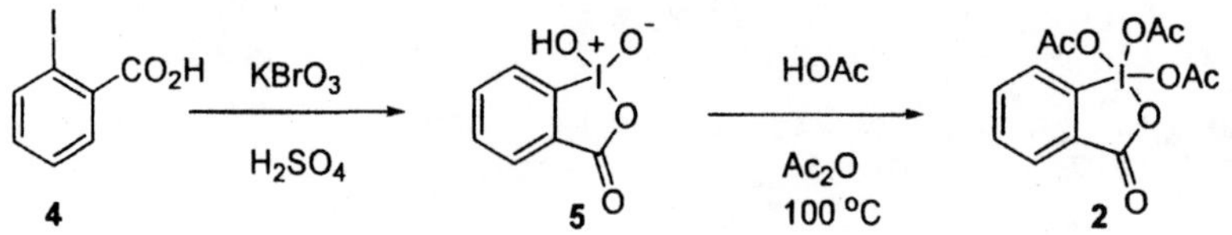

I, CO_2H; $KBrO_3$ / H_2SO_4; HO, +, O^-; HOAc / Ac_2O, 100 °C; AcO, OAc, OAc; **4**; **5**; **2**

DMP (**2**) is sparingly soluble in hexane or ether, but very soluble in chloroform, methylene chloride, and acetonitrile. DMP is commercially available and decomposes slowly. DMP is heat and shock sensitive and shows an exotherm when heated to temperatures of 130 °C or greater.

Using aliphatic and benzyl alcohols as examples, Dess and Martin demonstrated that oxidation occurred smoothly at room temperature, in methylene chloride, using a slight excess of **2** (1.05–1.1 eq). Although strong acid (i.e., TFA) was found to catalyze the reaction, it is usually not necessary. Below are a few examples of oxidations listed in the first disclosure.[1]

OH → O
2, $CHCl_3$, rt
TFA, 30 min
90%

OH → H, O
2, $CHCl_3$, rt
TFA, 60 min
93%

HO → O, H
2, $CHCl_3$, rt
60 min
91%

Oxidation of the above substrates were characterized by short reaction times and efficient conversion to expected products. Over-oxidation of substrates was not observed. Furthermore, the work up procedure was much easier and straightforward than other oxidation methods such as PCC, PDC, or Swern oxidation. The Dess–Martin reagent is also less toxic than the chromium-based oxidizing reagents.

3.5.3 Mechanism

The generally accepted mechanism is outlined below. The hydroxyl group of the substrate acts as the nucleophile to react with **2**, forming **6**, after ejection of one acetate group. The acetate molecule allieviates the positive charge formed on oxygen, providing one equivalent of acetic acid and producing intermediate **7**. Next, intermediate **7** collapses to form products *via* elimination of the alpha proton of the substrate and disassociation of an acetate ligand. The products produced from the collapse of **7** are the oxidized product, by-product **8**, and two equivalents of acetic acid. The rate of reaction has been shown to be dependent on rate of conversion of **7** to **8**. Studies from Dess and Martin[1,2] and Schreiber[4] have shown that the rate of dissociation of the acetate ligand of **7** is dependent upon the electron donating ability of the alkoxy substituent present in intermediate **7**. For example, Dess and Martin observed an increase in the rate of oxidation of ethanol when two equivalents of ethanol were added to **2** at room temperature to form **10**.

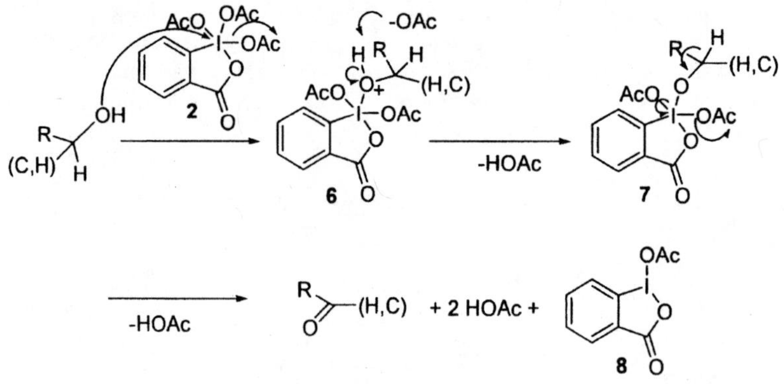

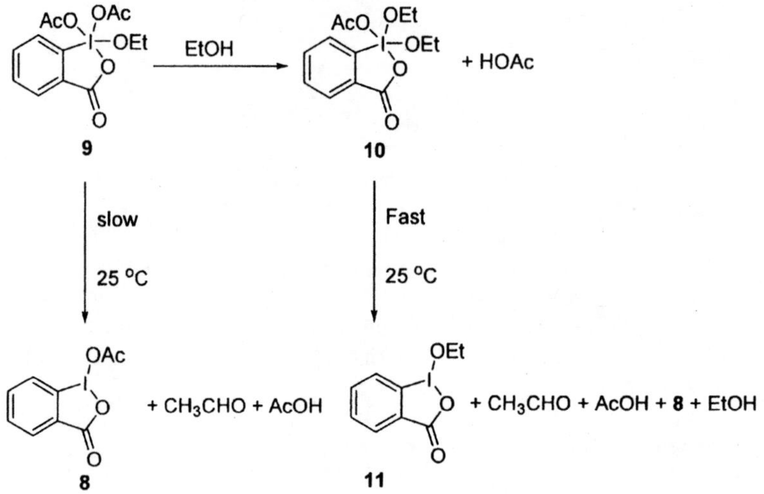

The dissociation rate of the acetate ligand of **10** was very fast (instantaneous) to form **11** and acetaldehyde, while the dissociation rate of the acetate ligand of **9** occurred in two or more hours at room temperature, forming **8** and oxidized product.

3.5.4 Variations and Improvements

In 1993, Ireland[5] reported an improved procedure for the preparation of the Dess–Martin reagent (**2**). At the time, the reagent was not commercially available and the original procedure outlined by Dess and Martin produced reagent that varied in activity from batch to batch. The key step of the synthesis was the conversion of **5** to **2**. By using a catalytic amount of *p*-toluenesulfonic acid, the acetylation of **5** occurred within 2 hours in 91% yield. This procedure was accomplished routinely on 100 g scale.

4 $\xrightarrow[H_2SO_4,\ 93\%]{KBrO_3}$ **5** $\xrightarrow[0.5\%\ TsOH,\ 80\ ^\circ C,\ 2\ h,\ 91\%]{Ac_2O}$ **2**

Since residual bromine associated with IBX can cause the material to be explosive at high temperature (> 150 °C), oxone in water has been used to oxidize 2-iodobenzoic acid (**4**) to make very pure IBX (**5**).[3]

In 1994, Schreiber[4] reported inconsistencies with batches of Dess–Martin reagent. In particular, an old lot of reagent was found to be an effective oxidizer, while newer batches were not as efficient. It was noted that when water was added to the Dess–Martin oxidation procedure, the rate of the reaction was accelerated. This observation led to the conclusion that the "old" Dess–Martin reagent was being partially hydrolyzed to compound **5**, *O*-iodoxybenzoic acid (IBX), which was a more effective oxidant.

2 $\xrightarrow{H_2O,\ CH_2Cl_2}$ **5**

IBX based oxidations are considered a variant of the Dess–Martin oxidation and many interesting chemistries have been conducted using this reagent. IBX is stable to moisture and oxidations can be performed in an open flask without use of inert atmosphere or dry solvents. The reagent dissolves readily in DMSO (up to 1.5 M) and is insoluble in sulfalone, *N,N'*-dimethylformamide, acetonitrile, chloroform, methylene chloride, acetone, and tetrahydrofuran. It is possible to use tetrahydrofuran as a co-solvent for compounds that do not readily dissolve in DMSO or when low temperatures are required.

Frigerio and Santagostino[6] have shown that **5**, in DMSO, oxidizes primary alchohols to aldehydes at room temperature without over oxidation. Secondary alcohols are cleanly oxidized to ketones. Sterically hindered alcohols are oxidized at room temperature in a few hours (**6** → **7**).

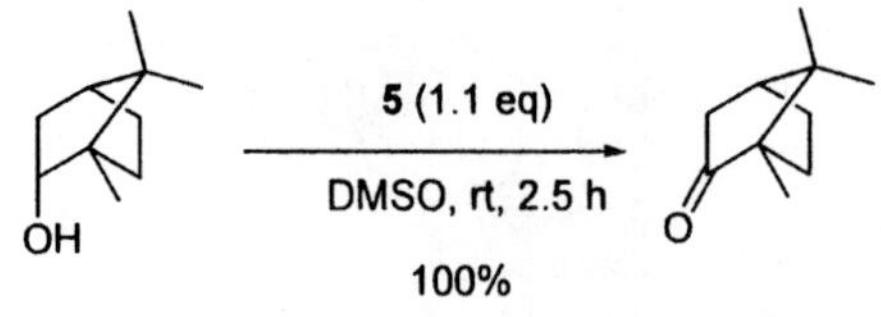

5 (1.1 eq)

DMSO, rt, 2.5 h

100%

6 → 7

IBX (5) oxidizes γ,δ-unsaturated alcohols to the corresponding carbonyls in good yield (**8** → **9**).

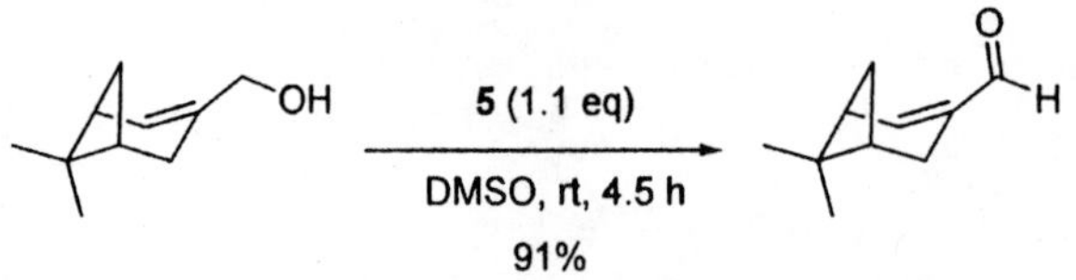

5 (1.1 eq)

DMSO, rt, 4.5 h

91%

8 → 9

Furthermore, the oxidation of chiral primary alcohols proceeds without epimerization (**10** → **11**).

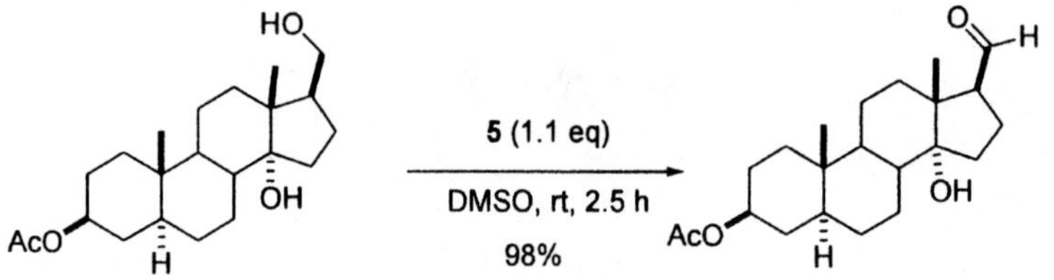

5 (1.1 eq)

DMSO, rt, 2.5 h

98%

10 → 11

It is important to note that tertiary alcohols do not interfere with IBX oxidations. Furthermore, it has been shown that pyridine and furan rings and double bonds, both conjugated and isolated are not affected with IBX.

Frigerio and Santagostino[6] also demonstrated that **5** smoothly oxidizes 1,2-diols to α-ketols or α-diketones without oxidative cleavage of the glycol bond (**12** → **13**; **14** → **15**). The IBX oxidations were conducted in DMSO solution at room temperature.

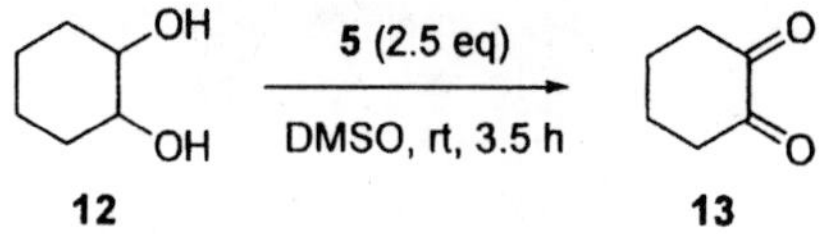

5 (2.5 eq)

DMSO, rt, 3.5 h

12 → 13

14 → 15: 5 (1.5 eq), DMSO, rt, 1.5 h

Use of **5** provides a new route for the conversion of 1,4-bisprimary diols or 1,4-primary-secondary diols to γ-lactols. The conversion of the primary-secondary diol (**16**) to the γ-lactol (**17**) implies that the oxidation of the primary hydroxyl group is considerably faster than the secondary hydroxyl moiety. This convenient and efficient process is superior to previous methods of forming the γ-lactol in one step.[7]

16 → 17: 5 (1.2 eq), DMSO, rt, 2 h

Nicolaou[8] has studied this oxidative agent extensively and has shown that IBX is useful in many oxidations other than simply converting alcohols to aldehydes and ketones. For example, IBX chemoselectively forms benzylic imines rapidly at room temperature in DMSO (**18** → **19**). It has also been demonstrated that primary amines can be oxidized to ketones (**20** → **21**) via the imine intermediate.

18 → 19: 5 (1.1 eq), DMSO, 25 °C, 10 min, 98%

20 → 21: 5 (1.1 eq), DMSO, 25 °C, 48 h, 79%

When the primary amine is "doubly benzylic" the yield increases with a shorter reaction time, indicative of facile benzylic oxidation/hydrolysis (**22** → **23**).

5 (1.1 eq), DMSO
25 °C, 5 h
99%

22 → 23

To gain insight into the mechanism of this reaction, *t*-butylamine was mixed with IBX (**5**) at room temperature for 48 hours in DMSO. Intermediate **24** was isolated from the reaction and not the "dehydrated" product.

24

This suggests that **24** represents the local energy minimum along the course of the reaction. Nicolaou has also shown that IBX conveniently cleaves dithioacetals and dithioketals to their corresponding carbonyl compounds (**25** → **26**; **27** → **28**) in high yield.

5 (2 eq), DMSO/H_2O (9:1)
25 °C, 0.3 h
99%

25 → 26

5 (2 eq), DMSO/H_2O (9:1)
25 °C, 2 h
84%

27 → 28

Nicolaou utilized IBX to effectively carry out oxidations adjacent to carbonyl functionalities (to form α,β-unsaturated carbonyl compounds) and at benzylic and related carbonyl centers to form conjugated aromatic carbonyl systems. For example, cycloheptanol (**29**) was converted to the synthetically challenging cycloheptadienone (**30**) in 76% yield.[9]

HO

5 (4 eq)

DMSO, 80 °C
76%

29 **30**

This protocol was employed to make tropinone (**31**) from cyclohepatanol in a "one-pot" reaction. The overall yield was 58%.

HO

5 (4 eq)

DMSO, 80 °C, 22h
then cool to 25°C,
add K_2CO_3 followed by
$MeNH_3^+Cl^-$, 3h
58%

29 **31**

Some other examples of this powerful oxidation protocol are shown below:

5 (4 eq)

DMSO, 70 °C, 2 h
89%

5 (3 eq)

DMSO, 80 °C, 48 h
69%

5 (4 eq)

DMSO, 70 °C, 12 h
84%

Benzylic methyl positions were shown to oxidize to the corresponding aldehydes using IBX, presumably through a single electron transfer (SET) mechanism. Over-oxidation to the corresponding carboxylic acid was not observed, even with electron-rich substrates. Some examples are shown below:

5 (3 eq)
DMSO, 85 °C, 12 h
85%
CHO

5 (3 eq)
Br
DMSO, 80 °C, 16 h
73%
Br
CHO

The rate of oxidation using IBX (**5**) has been determined to follow the trend: primary alcohols oxidize faster than secondary alcohols which oxidize much faster than the benzylic moieties.

Nicolaou further discovered that IBX can form heterocycles with certain *N*-aryl amides.[10] The mechanism is postulated to be SET based and the aromatic ring plays an key role in the initial SET process (compounds that do not have the *N*-aryl moiety do not undergo the cyclization process). THF in the solvent system is also postulated to quench the reaction by providing the source of hydrogen atom transfer. The utility of this protocol is outlined below:

CO_2Me
MeO
NH
O
96%
CO_2Me
MeO
NH
O

H
N
O
90%
H
O
N
H

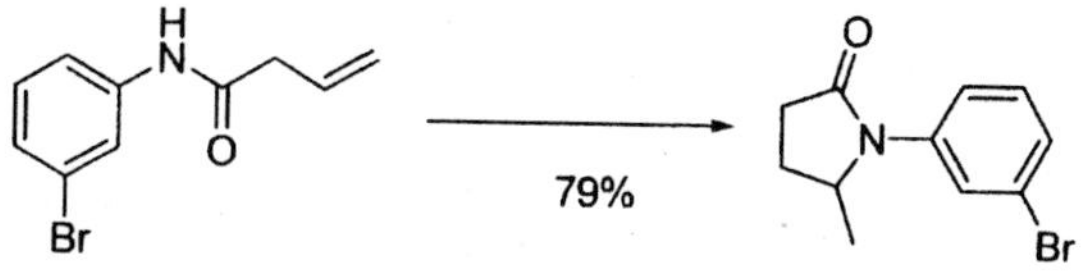

79%

Optimal conditions for the reactions above were to heat the anilides with IBX (2 eq) [in a mixed solvent system consisting of THF/DMSO (10:1)], in a pressure tube at 90 °C for 8–12 hours. The reaction mixtures were then cooled to room temperature and 2 equivalents of additional IBX was added, followed by reheating the mixture to 90 °C for 8 hours. The extra equivalents of IBX were used to ensure that reactions went to completion.

Applying the conditions listed above, Nicolaou demonstrated that carbamates (**32**, **33**) and ureas (**34**, **35**) could be cyclized to form oxazolidinones and cyclic ureas, respectively (see below).

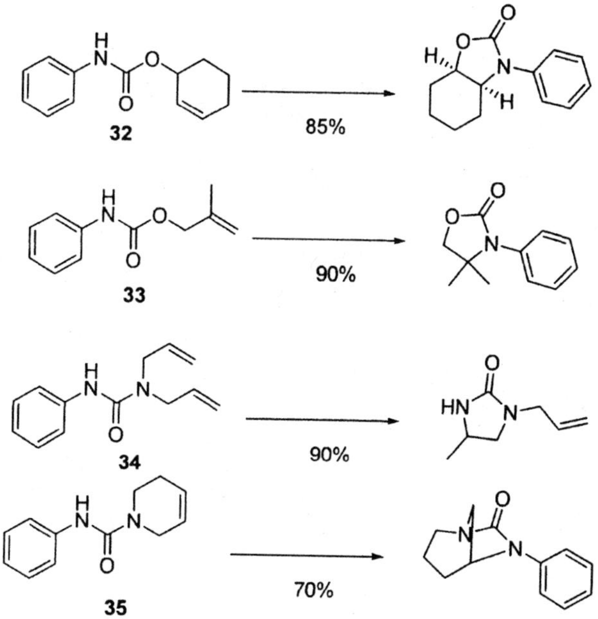

32 85%

33 90%

34 90%

35 70%

Ureas prepared from primary allylic amines slowly decomposed under the conditions listed without providing any cyclized products. Furthermore, the synthesis of amino sugars by the IBX-mediated cyclization protocol was also accomplished. This method was found to be stereoselective. A few examples are listed below:

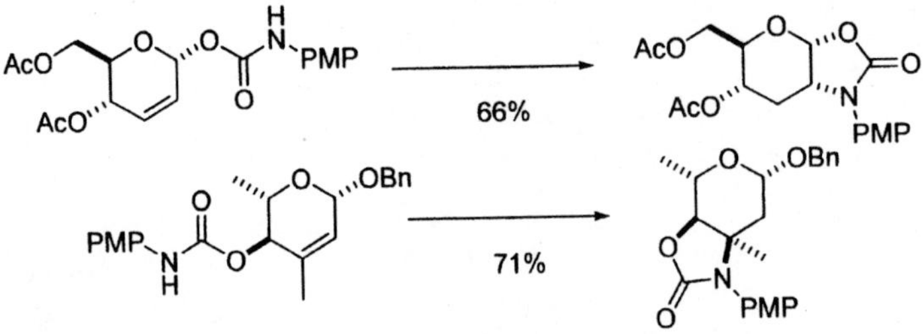

Another modification of the Dess–Martin periodinane oxidation is the use of ionic liquids such as hydrophilic [bmim]BF_4 and hydrophobic [bmim]BF_6 to increase the rate of reaction.[11] These solvents are more environmentally friendly than conventional organic solvents and avoid the use of high boiling polar organic solvents such as DMSO or DMF.

3.5.5 Synthetic Utility

Due to the mild reaction conditions, high chemoselectivity, and convenience of use, the Dess–Martin reagent is the method of choice for oxidation of complex, sensitive, and multifunctional alcohols.[12] For instance, it has been shown that hydroxyl groups can be oxidized in the presence of sulfur (**36** → **37**) with no racemization of the stereocenter.[13] Natalie *et al.*[14] have shown interesting selectivity with the DMP reagent. For example, oxidation of 1,4-dihydropyridine **38** to the corresponding pyridine analog (**39**) with DMP occurs in 55% yield. However, in the presence of a β-carbinol moiety (**40**), the alcohol is oxidized more rapidly than the dihydropyridine ring (**41**). It is also important to note that elimination of the hydroxyl group was not observed.

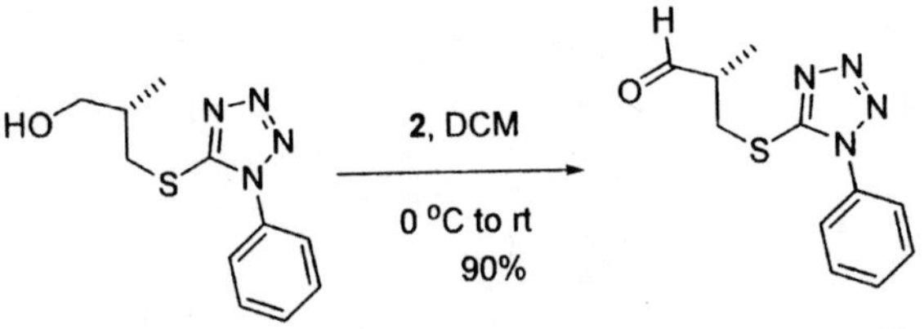

38 → 39: 2 (1.8 eq), DCM, rt, 2 h, 55%

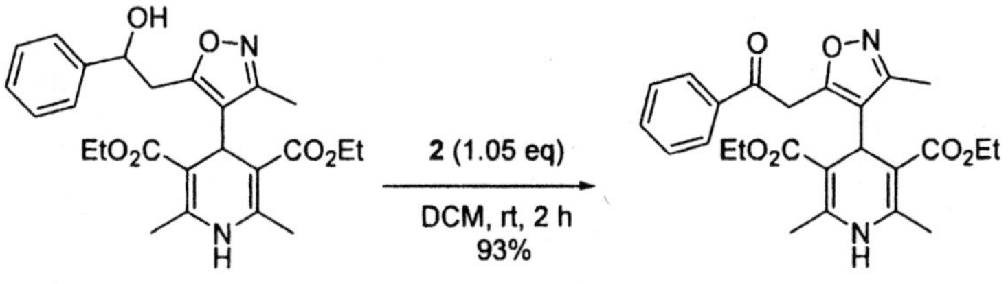

40 → 41: 2 (1.05 eq), DCM, rt, 2 h, 93%

When DMP was impregnated onto HNO_3/silica gel, "solventless" oxidations of 1,4-dihydropyridines under microwave irradiation (~ 5 min), produced the respective pyridines in excellent yield.[15]

The Dess–Martin periodinane oxidation is the method of choice for complex molecular systems. For example, during the total synthesis of the antitumor agent OSW-1, it was observed that oxidation of the alcohol (**42**) provided the aldehyde in 86% yield. Under Swern conditions, the aldehyde (**43**) was obtained in an inseparable mixture with another epimer.[16]

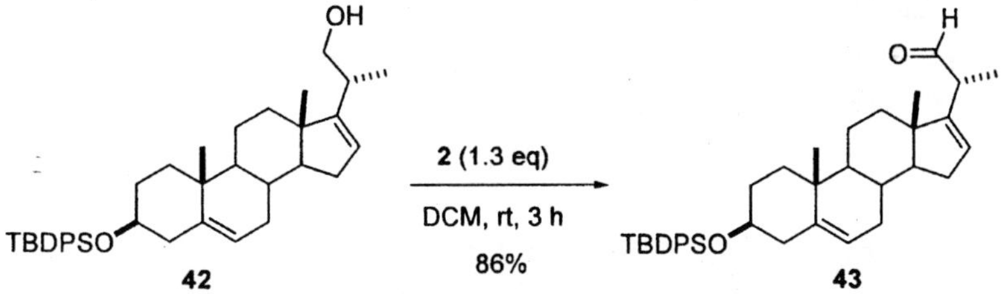

42 → 43: 2 (1.3 eq), DCM, rt, 3 h, 86%

The DMP conditions have been used successfully in β-lactam synthesis.[17] For instance, **44** was subjected to DMP (1.4 eq) to produce **45** in 72% yield. Previous attempts to convert the alcohol to the aldehyde with Swern conditions[18] and PCC[19] were unsuccessful. Only Jones oxidation conditions[20] provided the desired product, albeit in 15% yield.

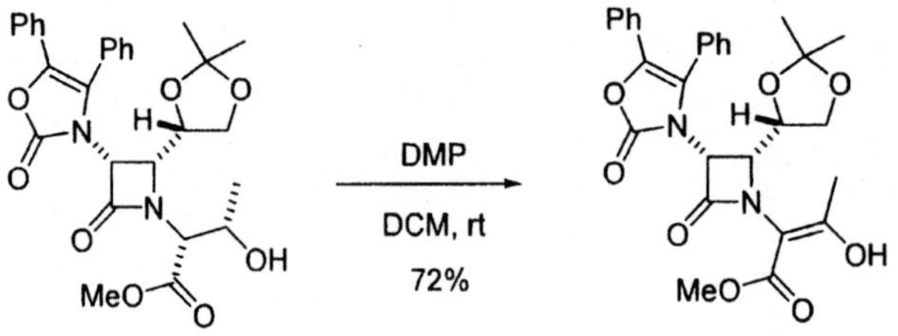

Another example where DMP reagent worked well while other oxidation methods proved unsuccessful was with the angucycline antibiotic framework.[21] Oxidation of the secondary alcohol (**46**) to the desired ketone (**47**) adduct proceeded smoothly using DMP. However, compound **47** was found to be labile under other oxidation conditions (Swern, PCC, PDC) causing the epoxide to open and subsequent aromatization of the "B" ring to occur.

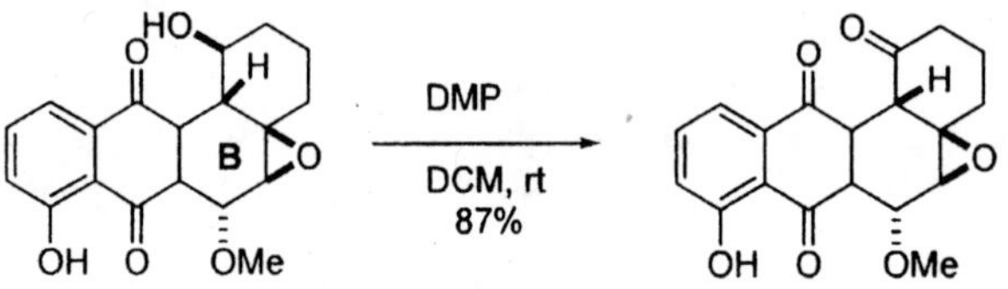

Similar to IBX, the synthetic utility of DMP goes beyond the standard selective oxidation of primary and secondary alcohols. A simple and mild method of decarboxylative bromination of α,β-unsaturated carboxylic acids has been developed using DMP in combination with tetraethylammonium bromide (TEAB) at room temperature.[22] It was observed that electron donating groups attached to the phenyl ring of the α,β-unsaturated carboxylic acids undergo fast decarboxylation/bromination in excellent yields (**48** → **49**).

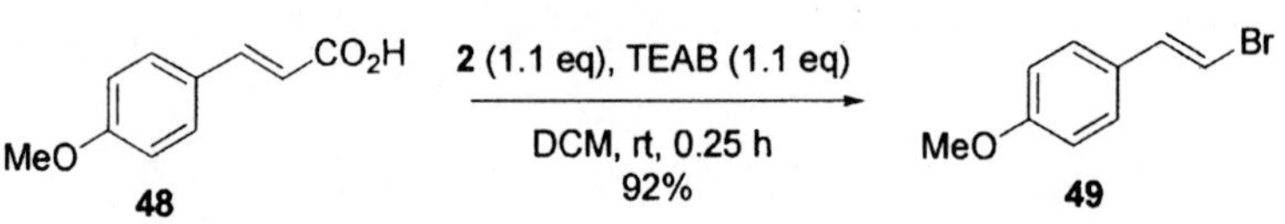

Electron withdrawing groups attached to the phenyl ring of α,β-unsaturated carboxylic acids undergo slow decarboxylation/bromination in moderate yields (**50** → **51**).

CO_2H **2** (1.1 eq), TEAB (1.1 eq) Br

DCM, rt, 14 h

65%

O_2N **50** O_2N **51**

Aliphatic groups next to α,β-unsaturated carboxylic acid moiety also undergo fast decarboxylation/bromination in good yields (**52** → **53**).

2 (1.1 eq), TEAB (1.1 eq)

CO_2H Br

DCM, rt, 0.5 h

82%

52 **53**

The formation of predominately "E" alkenes can be formed from a one-pot sequential deprotection–oxidation–Wittig reaction utilizing the DMP reagent. The deprotection of silyl ethers of aryl, allylic, propargylic, and unactivated alcohols (**54**) by acetic acid associated with the DMP reagent is followed by oxidation of the primary alcohols by DMP.[23] The resulting aldehydes are then converted directly to their corresponding α,β-unsaturated esters (**55**) in one-pot with stabilized phosphoranes. A general scheme is shown below:

2, CH_2Cl_2, rt

R O–$SiMe_3$ R R'

Ph_3P=$CHCO_2R'$

54 **55**

It was shown that activated silyl ethers proceed in high yield (**56** → **57**) whereas non-activated silyl ethers (**58** → **59**) proceed in moderate yields and require much longer reaction times.

O–$SiMe_3$ **2**, CH_2Cl_2, rt CO_2Et

Ph_3P=$CHCO_2Et$

20 h, 94%

56 **57** (E:Z = 20:1)

$OSiMe_3$ **2**, CH_2Cl_2, rt –CO_2Me

Ph_3P=$CHCO_2Me$

3 days, 70%

58 **59** (E:Z = 4.5:1)

Nicolaou[24] demonstrated that imides can be formed when primary and secondary amides are heated with DMP in a mixture of fluorobenzene and DMSO at 80–85 °C (**60** → **61**). It is envisioned that the amide oxygen atom of the substrate attacks the iodine atom of one equivalent of DMP (**2**), followed by expulsion of an acetate group and intramolecular rearrangement to give rise to imine intermediates that are further oxidized by an additional equivalent of DMP. An extensive range of functional groups are tolerated, including aromatic halides, olefins and acetates. It was also shown that carbamates are inert, as well as, benzylic positions to these oxidation conditions (**62** → **63**).

60 → **61**: 2 (2eq), Ph-F/DMSO, 85 °C, 86%

62 → **63**: 2 (2eq), Ph-F/DMSO, 85 °C, 98%

To showcase this method, the antibiotic fumaramidmycin (**64**), was synthesized in two steps from commercially available phenethylamine. Oxidation alpha to the nitrogen of the amide moiety was accomplished in high yield and high selectivity.

Monoethyl fumarate, EDC, DMAP, 73%

DMP (2 eq), Ph-F/CH_2Cl_2/H_2O (40:20:1), 85 °C, 14 h, 96% → **64**

Nicolaou also showed that DMP can react with β-carbonyl structural motifs to furnish vinylogous carbamates (**65**) and ureas (**66**). The reaction conditions have been shown to favor the *cis*-configured product, which is likely due to the stabilization provided by the intramolecular hydrogen bond present in the putative mechanistic intermediate. The hydrogen bond is anticipated to be reinforced by use of nonpolar solvents.

2 (5eq), Ph-F/DMSO, 85 °C

65, Z/E (25/1), 98%

BzHN — **2** (5eq), Ph-F/DMSO, 85 °C — BzHN

66, Z/E (25/1), 29%

Furthermore, upon exposure of DMP to primary amines at room temperature, the corresponding nitrile compounds were formed (**67**, **68**).[24]

MeO, NH_2 — **2** (2–3 eq), DCM, 25 °C, 95% — MeO, CN

67

H_2N — **2** (2–3 eq), DCM, 25 °C, 84% — CN

68

The production of polycyclic heterocycles via DMP oxidation of unsaturated anilides has been shown to be possible.[25] When substrate **69** was subjected to 4 equivalents of DMP and 2 equivalents of water at room temperature, the polyheterocycle **70** was produced. A small amount of the *p*-quinone by-product (**71**) was also observed.

2 (4eq)

H_2O (2 eq), DCM
23 °C, 15 h

69 → 70, 44% + 71, 7%

When the reaction is heated to reflux in benzene, the quinone by-product (**71**) is not observed, presumably due to the thermal instability of the quinone. A wide variety of functional groups on the aryl moiety are tolerated. These groups range from electron withdrawing nitro and trifluoromethyl groups, to halides and electron rich substituents. Electron rich groups such as the methoxy group did provide desired products, but in low yields. The reaction pathway involves a fleeting *o*-imidoquinone system that engages the proximate olefin in an inverse electron demand hetero-Diels–Alder reaction to furnish the polycycles.

As was stated previously, if the reaction is conducted a room temperature with electron donating substituents attached to the phenyl moiety, anilides are converted to p-quinones.

Nicolaou[26] took advantage of this "side reaction" to rapidly synthesize at metabolite (**BE-10988**) from a culture broth of a strain of *Actinomycetes*. This concise approach is another testament of the power and versatility of the DMP oxidation protocol.

2 (4 eq)

H_2O (2 eq), DCM
23 °C, 12 h

67%

NH_3

BE-10988

Urethanes (**72**) and ureas (**73**) can also be used to produce novel polycycles when exposed to DMP (**2**).

2 (2 eq)

H_2O (2 eq), benzene
reflux, 35 min

72

73

2 (2 eq)

H_2O (2 eq), benzene reflux, 30 min

56%

A rapid oxidative deoximination reaction using DMP (**2**) proceeds selectively in the presence of alcohols, *O*-methyl oximes, tosyl hydrazones, acid sensitive groups, and with electron rich and poor moieties in high yields. The reaction is conducted at room temperature, is rapid (within 20 min), with no over-oxidation observed due to the high chemoselectivity of DMP (**2**).[27]

2 (1.05 eq)

DCM, rt, 20 min

88-100%

DMP (**2**), like IBX (**5**), has shown to be a mild and effective method for the removal of thioacetals and thioketals.[28] For example, **74** was converted to the desired α,β-unsaturated ketone (**75**) in 91% yield after 12 hours. No detectable loss of the TBDPS ether or olefin isomerization was observed.

2 (2 eq)

CH_3CN/DCM/H_2O (8:1:1) rt, 2 h

91%

74

75

3.5.6 *Experimental*[2]

2, CH_2Cl_2, rt

2 h, 86%

76

77

A solution of cyclooctanol (**76**) (0.9 g, 7.0 mmol) in dichloromethane (10 mL) was added to a stirred solution of **2** (3.5 g, 8.3 mmol) in dichloromethane (30 mL). After 2 hours, the solution was poured into a separatory funnel containing 1 N aqueous sodium hydroxide (70 mL) and diethyl ether (150 mL). The layers were separated and the ether layer was extracted with water (2 × 50 mL). The ether layer was dried over magnesium sulfate and the ether removed under vacuum. Kugelohr distillation of the remaining oil gave cyclooctanone (**77**) (0.76 g, 6.02 mmol, 86%): mp = 34–37 °C.

3.5.7 References

1. Dess, D. B.; Martin, J. C. *J. Org. Chem.* **1983**, *48*, 4156.
2. Dess, D. B.; Martin, J. C. *J. Am. Chem. Soc.* **1991**, *113*, 7277.
3. Mullins, J. J.; Shao, P.; Boeckman, R. K. *Org. Syn.* **2004**, *Coll. Vol. 10*, 696.
4. Schreiber, S. L.; Meyer, S. D. *J. Org. Chem.* **1994**, *59*, 7549.
5. Ireland, R. E.; Liu, L. *J. Org. Chem.* **1993**, *58*, 2899.
6. Frigerio, M.; Santagostino, M. *Tetrahedron Lett.* **1994**, *35*, 8019.
7. Corey, E. J.; Palani, A. *Tetrahedron Lett.* **1995**, *36*, 3485.
8. Nicolaou, K. C.; Mathison, C. J. N.; Montagnon, T. *J. Am. Chem. Soc.* **2004**, *126*, 5192 (and references therein).
9. Nicolaou, K. C.; Montagnon, T.; Baran, P. S.; Zhong, Y.-L. *J. Am. Chem. Soc.* **2002**, *124*, 2245.
10. Nicolaou, K. C.; Baran, P. S.; Zhong, Y.-L.; Barluenga, S.; Hunt, K. W.; Kranich, R.; Vega, J. A. *J. Am. Chem. Soc.* **2002**, *124*, 2233.
11. Yadav, J. S.; Reddy, B. V. S.; Basak, A. K.; Narsaiah, A. V. *Tetrahedron* **2004**, *60*, 2131.
12. [R] Zhdankin, V.; Stang, P. J. *Chem. Rev.* **2002**, *102*, 2523 (and references therein).
13. Jasper, C.; Wittenberg, R.; Quitschalle, M.; Jakupovic, J.; Kirschning, A. *Org. Lett.* **2005**, *7*, 479.
14. Nelson, J. K.; Burkhart, D. J.; McKenzie, A.; Natalie, N. R. *Synlett* **2003**, 2213.
15. Heravi, M. M.; Dirkwand, F.; Oskooie, H. A.; Ghassemzadeh, M. *Heterocyclic Commun.* **2005**, *11*, 75.
16. Deng, S. D.; Yu, B.; Lou, Y.; Hui, Y. *J. Org. Chem.* **1999**, *64*, 202.
17. Niu, C.; Petterson, T.; Miller, M. J. *J. Org. Chem.* **1996**, *61*, 1014.
18. Manucuso, A. J.; Swern, D. *Synthesis* **1981**, 165.
19. Brown, H. C.; Garg, C. P.; Liu, K.-T. *J. Org. Chem.* **1971**, *36*, 387.
20. Bowers, A.; Halsall, T. G.; Jones, E. R. H.; Lemin, A. J. *J. Chem. Soc.* **1953**, 2548.
21. Larson, D. S.; O'Shea, M. D. *J. Org. Chem.* **1996**, *61*, 5681.
22. Telvekar, V. N.; Arote, N. D.; Herlekar, O. P. *Synlett* **2005**, 2495.
23. Deng, G.; Xu, B.; Liu, C. *Tetrahedron* **2005**, *61*, 5818.
24. Nicolaou, K. C.; Mathison, C. J. N. *Angew. Chem. Int. Ed.* **2005**, *44*, 5992.
25. Nicolaou, K. C.; Baran, P. S., Zhong, Y.-L., Saguita, K. *J. Am. Chem. Soc.* **2002**, *124*, 2212.
26. Nicolaou, K. C.; Baran, P. S.; Zhong, Y.-L.; Saguita, K. *J. Am. Chem. Soc.* **2002**, *124*, 2221.
27. Chaudhari, S. S.; Akamanchi, K. G. *Tetrahedron Lett.* **1998**, *39*, 3209.
28. Langille, N. F.; Dakin, L. A.; Panek, J. S. *Org. Lett.* **2003**, *5*, 575.

Daniel D. Holsworth

3.6 Tamao–Kumada–Fleming Oxidation

3.6.1 Description

The Tamao–Kumada and Fleming oxidation reactions involve the oxidation of specific silyl groups to provide a hydroxyl group with retention of configuration.[1]

KF, H_2O_2 / $KHCO_3$, DMF

OH

3.6.2 Historical Perspective

Research by Tamao and Kumada led to the 1983 discovery that tetra-coordinate organosilicon compounds could be easily cleaved by 30% hydrogen peroxide, with retention of configuration at the carbon center to give the corresponding alcohols. This reaction proved to be simplistic, highly regioselective, and compatible with a wide range of functional groups, while introducing the first successful procedure for the direct conversion of allylsilanes to allylic alcohols without allylic transposition.[2] The Tamao–Kumada oxidation also provided synthetic possibilities that were not available with trimethylsilyl compounds,[3] while benefiting from H_2O_2 being an inexpensive and mild reagent, providing for clean reaction conditions.[2]

The oxidative cleavage was compared to several oxidative methods, including those using MeOH/THF, *t*-BuOOH, *m*CPBA, and $NaHCO_3$, but was found to proceed best with the addition of DMF, KF, and $KHCO_3$.[2] Extensive studies performed by Tamao and Kumada indicated that the H_2O_2 oxidation proceeds smoothly only if the substrate carries at least one electron withdrawing group on the silicon. As shown below in a general case, treatment of **1** with the reagent mixture described above would result in the preparation of **2** with retention of configuration at the carbon stereocenter. The isolation of the desired alcohol is facile owing to the aqueous solubility of the alcohol by-products.

KF, H_2O_2 / $KHCO_3$, THF

1 → **2**

Shortly after the initial report by Tamao and Kumada, Fleming and co-workers reported a similar use of silanes as masked hydroxyl groups.[4] In contrast, the Fleming oxidation utilizes silanes containing a phenyl substituent rather than the electron withdrawing group required for oxidation under Tamao conditions. As a result, the Fleming protocol requires substitution of the phenyl substituent with a halogen, or other electron withdrawing group prior to the oxidation step. As shown below, treatment of **3** with HBF_4 results in the formation of a silyl fluoride, which is oxidized by *m*CPBA, in the presence of base to give alcohol **2**. Once more, the reaction proceeds with retention of configuration. Under the appropriate conditions, this two-step process can be achieved in a single pot.[5]

1. HBF_4
2. *m*CPBA, Et_3N

3 → **2**

An extensive review has been written which covers in detail the oxidation of the carbon–silicon bond under both Tamao and Fleming oxidation conditions.[6] One virtue of the use of these procedures as compared to the related oxidation of organoboron compounds is the fact that the silicon groups can often be carried through several synthetic steps while masking the hydroxyl functionality.

3.6.3 Mechanism

S, fast, S = solvent; H_2O_2, RDS; $-H_2O$; 2 ROH

Due to a lack of kinetic measurements under the appropriate conditions, the mechanism of the Tamao–Kumada oxidation is not well known. However, on the basis of their studies, Tamao and Kumada have suggested the following mechanism.[6] Most importantly, the silicon group is first activated for attack by basic peroxide. Following attack, an alkyl group transfers to the oxygen of peroxide, displacing a molecule of water and regenerating the pentacoordinate silicon geometry. Once all the alkyl groups have been oxidized, hydrolysis of the resulting silyl ether provides for the formation of the product alcohol.

The Fleming oxidation is a two-step process with the second step finding mechanistic similarities in the Tamao oxidation. Principally, treatment of the phenyl(dimethyl)silyl group with HBF_4 provides for an electrophilic addition of H^+ to the phenyl group to give **5**. Attack of the fluoride ion cleaves the carbon–silicon bond, restoring aromaticity in the formation of benzene.

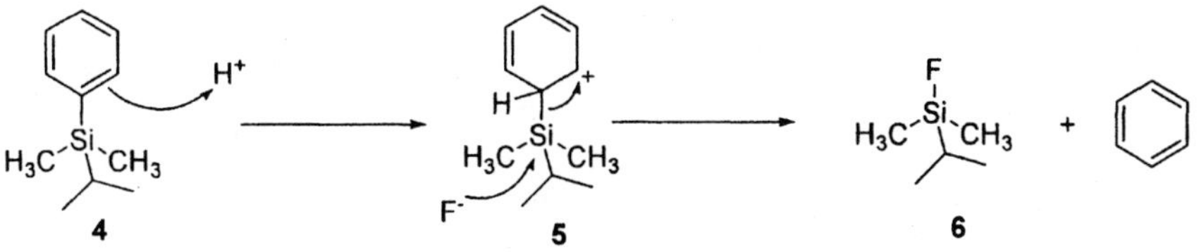

The newly formed Si—F bond in **6** enables the oxidative cleavage in the next step in a manner similar to what is observed in the Tamao oxidation. Thus, displacement of the fluoride with perbenzoate anion **7** results in **8**. Peroxide **8** can rearrange by a mechanism similar to the Baeyer–Villiger oxidation to yield **9**. This process can be repeated twice more resulting in **10**. Finally, hydrolysis of **10** produces the desired alcohol.

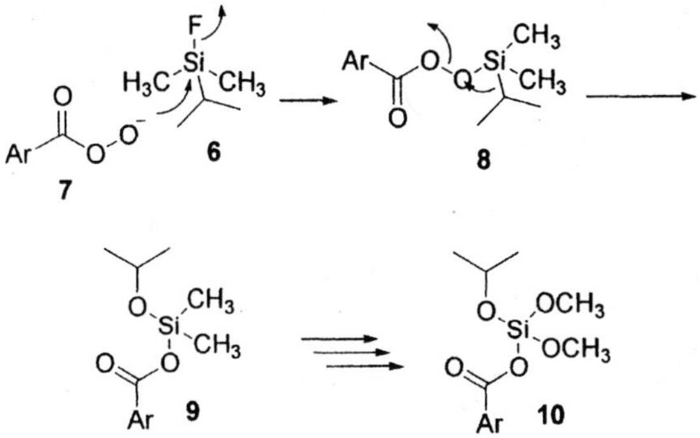

3.6.4 Variations and Improvements

Since the discoveries by Tamao and Fleming of their respective oxidative cleavage reactions, several other silicon reagents have been developed that act as hydroxyl group surrogates.[7] The figure above is not meant to be exhaustive, but provides representative examples of substrates which can undergo these oxidation reactions. The cleavage mechanism for oxidations using these substrates **11–23**[7–25] proceeds in a similar manner to that described above.

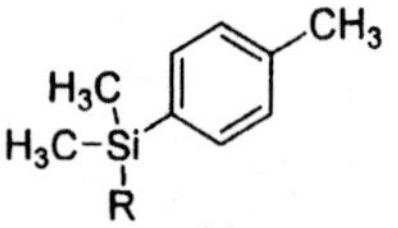

11[8,9] 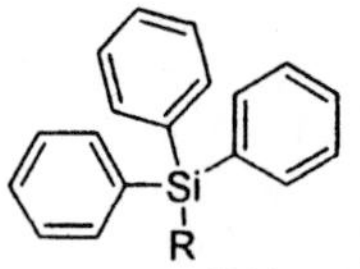 12[10,11] 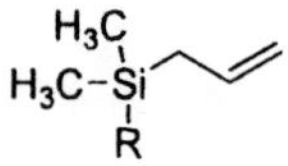 13[12,13] 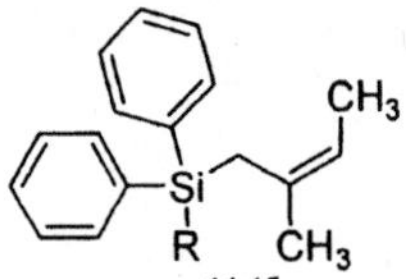 14[14,15]

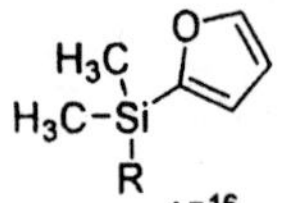

15[16] 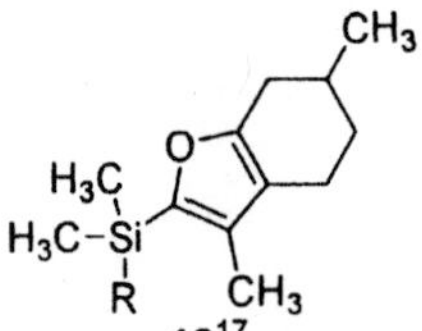 16[17] 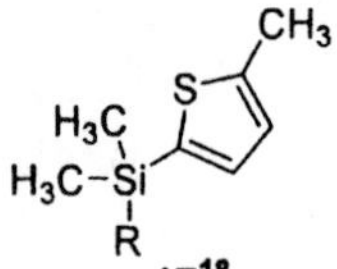 17[18] 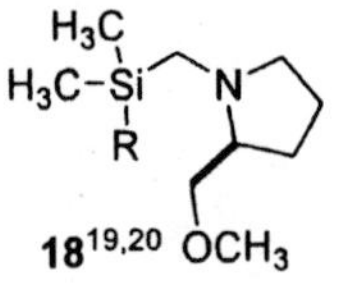 18[19,20]

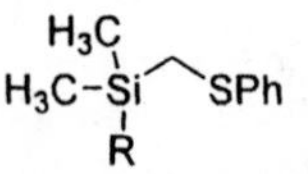

19[21] 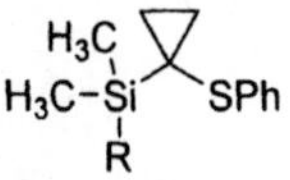 20[22] 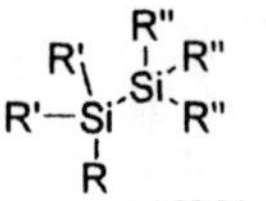 21[23,24] 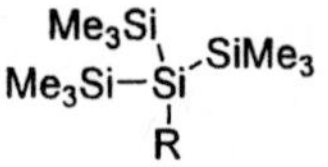 22[25] 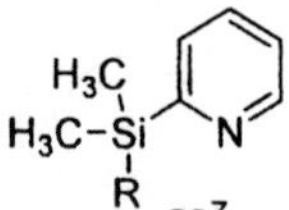 23[7]

3.6.5 Synthetic Utility

The Tamao–Kumada and Fleming oxidations have found wide synthetic utility in the oxidation of a variety of silyl groups.[6,26-29] However, the true virtue of these reactions lies in the manner in which the silyl group is introduced into the molecule to be oxidized. A number of creative strategies have been developed for this process, some of which will be highlighted below.

The palladium-catalyzed cyclization and hydrosilylation of functionalized dienes was recently introduced by the Widenhoefer group.[30] Unfortunately, general conditions were only achieved utilizing trialkylsilanes in the hydrosilylation procedure. As these are largely unsuitable for the Tamao or Fleming oxidation procedures, several different silanes were explored for use in this two-step process. Pentamethyldisiloxane (PMDS) was found to be amenable to both the palladium catalyzed cyclization/hydrosilylation and oxidation procedures.[31] Thus, high regio-, diasterio-, and enantioselectivity results when diene **24** is treated with PMDS using catalytic **26** and **27**. Following the oxidation, alcohol **25** is produced with high yield over the two-step process.

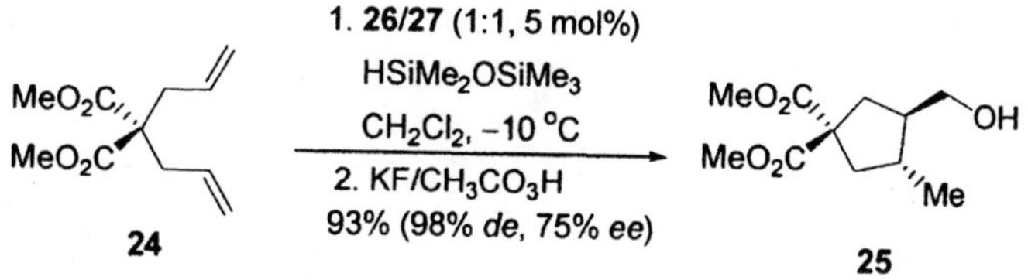

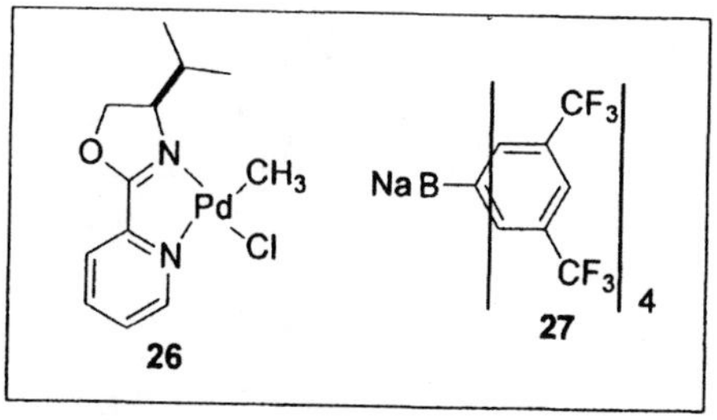

After considerable experimentation, a similar hydrosilylation protocol was used as a key step for the syntheses of jatrophatrione and citlalitrione by Paquette and co-workers.[32] Following the stereoselective reduction of a tricyclic ketone with lithium aluminum hydride to provide alochol **28**, silylation and platinum catalyzed hydrosilylation were effected to produce **29**. Finally, the carbon–silicon bond was successfully cleaved to generate diol **30** in an impressive 93% yield.

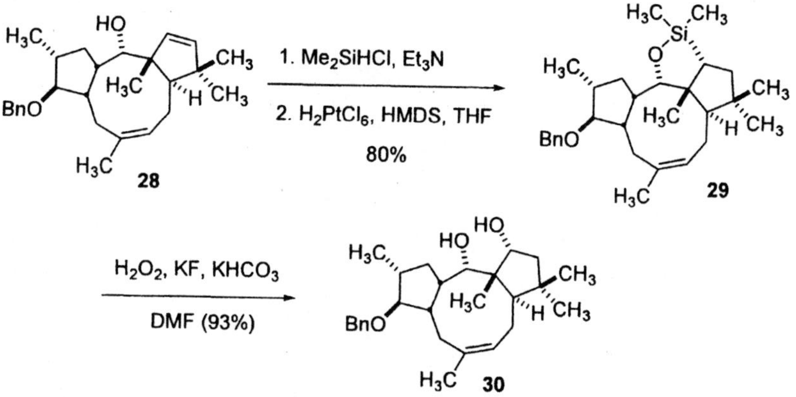

A highly general procedure relies on the use of a silicon tethered radical cyclization process to provide for introduction of a hydroxymethyl substituent using the Tamao conditions.[33-38] Pioneered by Nishiyama[39] the synthesis of regioisomeric diols, as in **33**, from readily available allyl silyl ethers, such as **31**, was achieved via radical cyclization and oxidization. The predominance of the 5-exo cyclization is further demonstrated by the formation of **36** by this same process.

31 → nBu$_3$SnH, AIBN; benzene, reflux

32 → 30% H_2O_2, KF; DMF, 60 °C; 85% → **33**

34 → nBu$_3$SnH, AIBN; benzene, reflux → **35** → 30% H_2O_2, KF; DMF, 60 °C; 83%; **36**:**37** 91:9

36 + **37**

Radical cyclization has also proved to be a useful tool for the Shishido group in their syntheses of triptoquinones B and C and triptocallol.[40] Triptoquinone B and C have been shown to exhibit inhibitory activity on interleukin-1 and are both candidates for the treatment of rheumatoid arthritis. Shishido has accomplished synthesis of all three of these compounds, implementing the radical cyclization of bromomethyldimethylsilyl ether **39**, derived from optically enriched alcohol **38**, to give **40**. Finally, the Tamao–Kumada oxidation step provides the natural product, triptocallol (**41**).

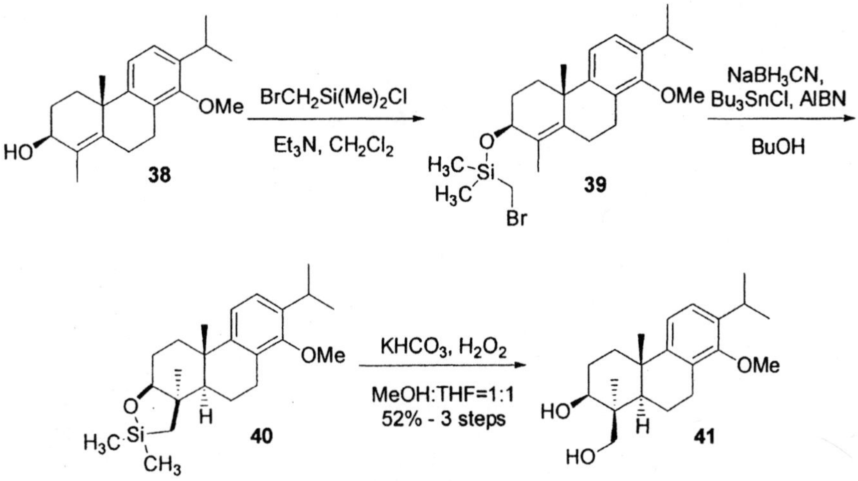

Finally, a radical cyclization/oxidation protocol was utilized as the final step in the Crimmins synthesis of the natural product (–)-talaromycin A.[41] Silylation of allylic alcohol **42** produced **43**, which cyclized upon treatment with Bu_3SnH and AIBN to provide **44**. Oxidation of the crude residue resulting from the cyclization reaction resulted in the formation of (–)-talaromycin A (**45**).

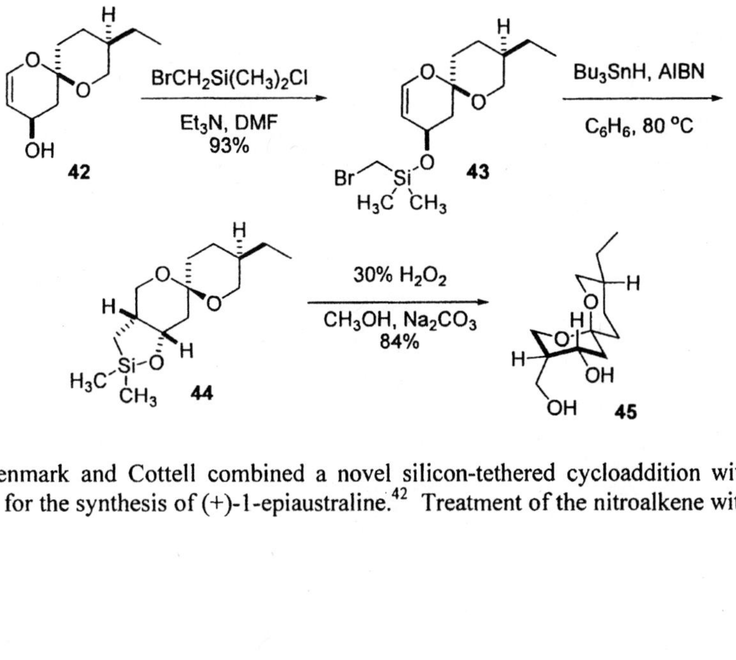

Denmark and Cottell combined a novel silicon-tethered cycloaddition with a later oxidation for the synthesis of (+)-1-epiaustraline.[42] Treatment of the nitroalkene with MAPh, an aluminum-based Lewis acid, resulted in a [4 + 2]/[3 + 2] tandem nitroalkene cycloaddition between **46** and **47** to give the cycloadduct **48**. Standard functional group transformations produced **49**, which was efficiently oxidized under Tamao conditions to generate **50**.

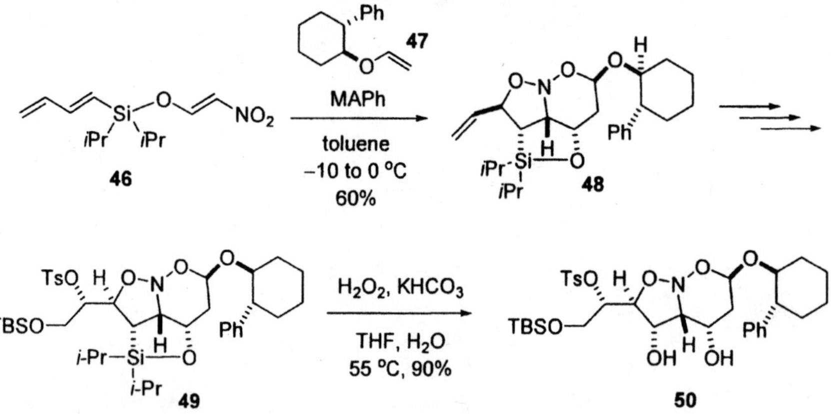

Several organometallic reagents have found utility as nucleophilic hydroxymethyl equivalents.[2,6] One such reagent, (2-pyridyldimethylsilyl)methyl lithium (**51**), has been developed by the Yoshida group.[43,44] This reagent, prepared by deprotonation of 2-trimethylsilyl pyridine, has found utility in alkylation and carbonyl addition reactions. As demonstrated in the reaction of **52** and **54**, under Tamao conditions, oxidation proceeds efficiently to provide alcohols **53** and **55**. It is noteworthy that the oxidation occurs with substrates prone to undergo a Peterson elimination reaction. Another noteworthy utility of this particular silane is found in its ability to serve as a phase tag.[45] Because of the basic properties of the pyridine nucleus, purification of 2-pyridylsilylated compounds is easily accomplished using acid–base extraction.

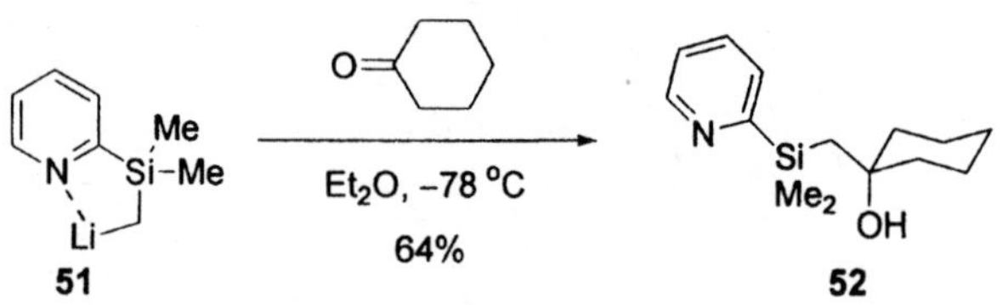

30% H_2O_2
$KHCO_3$, KF
MeOH/THF
50 °C
95%

HO OH

53

51 N Si–Me Me Li

Br Ph
Et_2O, –78 °C
84%

N Si Me Me Ph

54

30% H_2O_2
$KHCO_3$, KF
MeOH/THF
50 °C
98%

HO Ph

55

Weinreb and co-workers made use of a nucleophilic hyroxymethyl equivalent in the synthesis of the marine alkaloid lepadiformine.[46] A highly stereoselective addition to α-methoxybenzamide **56** was achieved using the silylmethyl cuprate derived from Grignard reagent **57**. Exposure of **58** to Fleming oxidation conditions provided **59** in high yield.

Si MgBr
57
BF_3-Et_2O
$CuBr_2$-Me_2S
Et_2O, -78 °C to rt
87%

MeO N Ph O
56

Si N H Ph O
58

1. BF_3-HOAc
CH_2Cl_2, rt
2. 35% H_2O_2
$NaHCO_3$
MeOH, THF
heat, 95%

HO N H Ph O
59

One advantage of a synthetic sequence requiring oxidation of the carbon–silicon bond is the large number of silyl groups which can be oxidized under either Tamao or Fleming conditions.[4] This aspect was particularly important in the Barrett synthesis of the natural product (–)-pramanicin, a compound that has been shown to be active against the acapsular form of *Cryptococcus neoformans,* a microorganism that is a causative agent of meningitis in AIDS patients.[47] During these synthesis efforts, several different silyl groups were employed in the Fleming oxidation process. Only when a silane was used which should be oxidized under Tamao conditions was successfully acheieved. Thus, treatment of **60** under standard conditions resulted in the preparation of **61**. Of note is the fact that the silyl group was introduced via a stereoselective conjugate addition, a concept which has found wide utility in the literature.[6,48,49]

m-CPBA, KHF_2
DMF, 0 to 25 °C
70%

60 → 61

3.6.6 *Experimental*

62 → 63

(2-Hydroxylmethyl-6-vinyl-1-azaspiro[4.5]dec-1-yl)-phenylmethanone (63)[46]

To a solution of allylsilane **62** (140 mg, 0.37 mmol) in CH_2Cl_2 (5 mL) was added $BF_3 \cdot 2AcOH$ (150 μL, 0.73 mmol) at room temperature. After 15 min, the mixture was diluted with saturated aqueous $NaHCO_3$ (5 mL) and extracted with CH_2Cl_2. The organic extracts were dried over $MgSO_4$ and concentrated under reduced pressure to give the fluorosilane. To the solution of the fluorosilane in a mixture of MeOH (5 mL) and THF (5 mL) was added $NaHCO_3$ (16 mg, 0.20 mmol) at room temperature. After 15 min, H_2O_2 (35 wt % solution in water, 500 μL) was added to the reaction mixture. After being refluxed overnight, the mixture was diluted with saturated aqueous NH_4Cl (10 mL) and extracted with CH_2Cl_2. The organic extracts were dried over $MgSO_4$ and concentrated under reduced pressure. The residue was purified by flash column chromatography (1:1 EtOAc/hexanes) to afford alcohol **63** (89 mg, 95%) as a white solid. Recrystallization from EtOAc/CH_2Cl_2 provided X-ray quality crystals: mp 169–170 °C; ^1H NMR (300 MHz, $CDCl_3$) δ 7.33–7.27 (m, 5 H), 5.77 (dt, J = 17.4, 8.9 Hz, 1 H), 3.61 (br t, J = 7.7 Hz, 1 H), 3.20–3.09 (m, 2 H), 2.89 (td, J = 13.0, 3.5 Hz, 1 H), 2.29 (br s, 1 H), 2.18 (br t, J = 7.2 Hz, 1 H), 1.84–1.67 (m, 6 H), 1.46–1.24 (m, 4 H); ^{13}C NMR (75 MHz, $CDCl_3$) δ 170.4, 139.7, 139.2, 128.7, 128.3, 126.1, 115.7, 69.6, 63.6, 62.3, 42.7, 37.4, 30.9, 29.8, 25.8, 24.9, 23.9; CIMS (APCI+) (relative intensity) 300.1 (MH^+, 100); HRMS (APCI+) calcd for $C_{19}H_{26}NO_2$ (MH^+) 300.1963, found 300.1958.

3.6.7 References

1. Li, Jie J. *Name Reactions: A Collection of Detailed Reaction Mechanisms*, Springer-Verlag, 2002.
2. Tamao, K.; Ishida, N.; Kumada, M. *J. Org. Chem.* **1983**, *48*, 2120.
3. Tamao, K.; Neyoshi, I. *Organometallics*, **1983**, *2*, 1694.
4. Fleming, I.; Henning, R.; Plaut, H. *Chem. Commun.* **1984**, 29.
5. Fleming, I.; Sanderson, P. E. J. *Tetrahedron Lett.* **1987**, *28*, 4229.
6. Jones, G. R.; Landais, Y. *Tetrahedron* **1996**, *52*, 7599.
7. Itami, K.; Mitsudo, K.; Yoshida, J. *J. Org. Chem.* **1999**, *64*, 8709.
8. Fleming, I.; Ghosh, S. K. *Chem. Commun.* **1992**, 1775.
9. Fleming, I.; Ghosh, S. K. *Chem. Commun.* **1992**, 1777.
10. Corey, E. J.; Chen. Z. *Tetrahedron Lett.* **1994**, *35*, 8731.
11. Knölker, H.; Wanzl, G. *Synlett*, **1995**, 378.
12. Tamao, K.; Ishida, N. *Tetrahedron Lett.* **1984**, *25*, 4249.
13. Magar, S. S.; Desai, R. C.; Fuchs, P. L. *J. Org. Chem.* **1992**, *57*, 5360.
14. Fleming, I.; Winter, S. B. D. *Tetrahedron Lett.* **1993**, *34*, 7287.
15. Fleming, I.; Winter, S. B. D. *Tetrahedron Lett.* **1995**, *36*, 1733.
16. Stork, G. *Pure Appl. Chem.* **1989**, *61*, 439.
17. Hunt, J. A.; Roush, W. R. *Tetrahedron Lett.* **1995**, *36*, 501.
18. Landais, Y.; Planchenault, D.; Weber, V. *Tetrahedron Lett.* **1995**, *36*, 2987.
19. Chan, T. H.; Pellon, P. *J. Am. Chem. Soc.* **1989**, *111*, 8737.
20. Chan, T. H.; Nwe, K. T. *J. Org. Chem.* **1992**, *57*, 6107.
21. Van Delft, F. L.; van der Marel, G. A.; van Boom, J. H. *Synlett* **1995**, 1069.
22. Angelaud, R.; Landais, Y.; Maignan, C. *Tetrahedron Lett.* **1995**, *36*, 3861.
23. Krohn, K.; Khanbabaee, K. *Angew. Chem. Int. Ed. Engl.* **1994**, *33*, 99.
24. Suginome, M.; Matsunaga, S.; Ito, Y. *Synlett*, **1995**, 941.
25. Kopping, B.; Chatgilialoglu, C.; Zehnder, M.; Giese, B. *J. Org. Chem.* **1992**, *57*, 3994.
26. Yoshida, J.; Itami, K.; Kamei, T. *J. Am. Chem. Soc.* **2001**, *123*, 8773.
27. Oshima, K.; Takaku, K.; Shinokubo, H. *Tetrahedron Lett.* **1996**, *37*, 6783.
28. Fuchs, P. L.; Kim, S.; Emeric, G. *J. Org. Chem.* **1992**, *57*, 7362.
29. Akiyama, T.; Asayama, K.; Fujiyoshi, S. *J. Chem. Soc., Perkin Trans. 2* **1998**, 3655.
30. Widenhoefer, R. A.; DeCarli, M. A. *J. Am. Chem. Soc.* **1998**, *120*, 3805.
31. Widenhoefer, R. A.; Pei, T. *Organic Lett.* **2000**, *2*, 1469.
32. Paquette, L. A.; Yang, J.; Long, Y. O. *J. Am. Chem. Soc.* **2003**, *125*, 1567.
33. Stork, G.; Kahn, M. *J. Am. Chem. Soc.* **1985**, *107*, 500.
34. Linker, T.; Maurer, M.; Rebien, F. *Tetrahedron Letters* **1996**, *37*, 8363.
35. Wicha, J.; Kurek-Tyrlik, A.; Zarecki, A. *J. Org. Chem.* **1990**, *55*, 3484.
36. Koreeda, M.; Hamann, L. *J. Am. Chem. Soc.* **1990**, *112*, 8175.
37. Shuto, S.; Yairo, Y.; Ichikawa, S.; Matsuda, A. *J. Org. Chem.* **2000**, *65*, 5547.
38. Wipf, P.; Graham, T. H. *J. Org. Chem.* **2003**, *68*, 8798.
39. Nishiyama, H.; Kitajima, T.; Matsumoto, M.; Itoh, K. *J. Org. Chem.* **1984**, *49*, 2298.
40. Shishido, K; Yamamura, I.; Fujiwara, Y.; Yamato, T.; Irie, O. *Tetrahedron Lett.* **1997**, *38*, 4121.
41. Crimmins, M.; O'Mahony, R. *J. Org. Chem.* **1989**, *54*, 1157.
42. Denmark, S.; Cottell, J. *J. Org. Chem.* **2001**, *66*, 4276.
43. Yoshida, J.; Itami, K.; Mitsuda, K. *Tetrahedron Lett.* **1999**, *40*, 5537.
44. Yoshida, J.; Itami, K.; Mitsudo, K. *J. Org. Chem.* **1999**, *64*, 8709.
45. Yoshida, J.; Itami, K.; Mitsudo, K.; Suga, S. *Tetrahedron Lett.* **1999**, *40*, 3403.
46. Weinreb, M.; Sun, P.; Sun, C. *J. Org. Chem.* **2002**, *67*, 4337.
47. Barrett, A. G. M.; Head, J.; Smith, M.; Stock, N.; White, A.; Williams, D. J. *J. Org. Chem.* **1999**, *64*, 6005.
48. Ager, D. J.; Fleming, I.; Patel, S. K. *J. Chem. Soc., Perkin Trans. 1* **1981**, 2520.
49. Fleming, I.; Higgins, D.; Lawrence, N. J.; Thomas, A. P. *J. Chem. Soc., Perkin Trans. 1* **1992**, 3331.

Richard J. Mullins, Sandy L. Jolley, and Amanda R. Knapp

3.7 Martin's Sulfurane Dehydrating Reagent

3.7.1 *Description*

Martin's sulfurane (**1**) enables quick and efficient elimination of secondary and tertiary alcohols **2** to yield alkenes **3** along with production of diphenyl sulfoxide (**4**) and alcohol **5**.[1,2] This highly reactive dehydrating reagent is effective at room temperature and below, and reactions are often complete in less than one hour. Primary alcohols rarely react to yield alkenes and instead are transformed into ethers. The title reagent is available from numerous commercial sources and can be prepared in large quantities according to several published procedures.[3]

F_3C CF_3 O S: O F_3C CF_3 **1** + OH R^4 R^3 H R^2 R^1 **2** ⟶ R^1 R^4 R^2 R^3 **3** + O S **4** + 2 F_3C CF_3 HO **5**

3.7.2 *Historical Perspective*

James C. Martin (known as J. C. to his friends), a professor at the University of Illinois at Urbana-Champaign, studied the structure and reactivity of hypervalent sulfur compounds during the 1960s and 1970s. One sulfur-containing molecule that proved especially interesting to Martin was perester **6**. In 1962 he described the decomposition of this compound into sulfoxide **8** and isobutylene (**9**) via sulfurane **7**.[4] Nine years later, Martin published the first report of the reagent that would eventually bear his name.[5] This remarkably stable, crystalline sulfurane (compound **1**) rapidly exchanged alkoxy ligands with added alcohols. Martin quickly exploited this reactivity to show that addition of alcohols containing β-hydrogens yielded sulfurane intermediates that dehydrated to provide alkene products in high yields (like his initial report of the conversion of **7**→**9**). Martin studied the reactivity of sulfurane **1** between 1971 and 1975 and discovered that, in addition to dehydrations, it promoted formation of cyclic ethers and sulfilimines and facilitated cleavage of secondary amides.

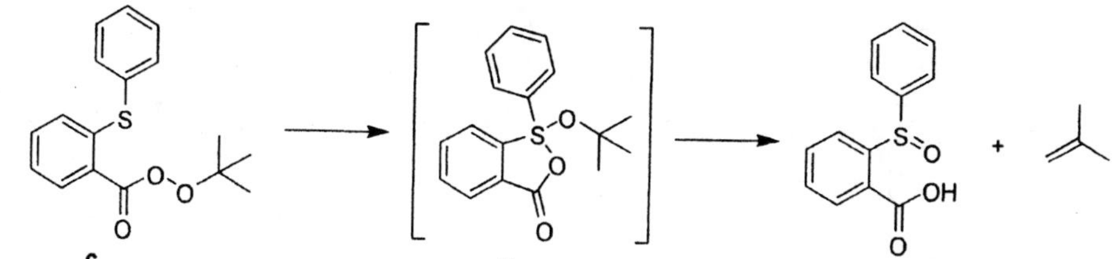

3.7.3 *Mechanism*[6]

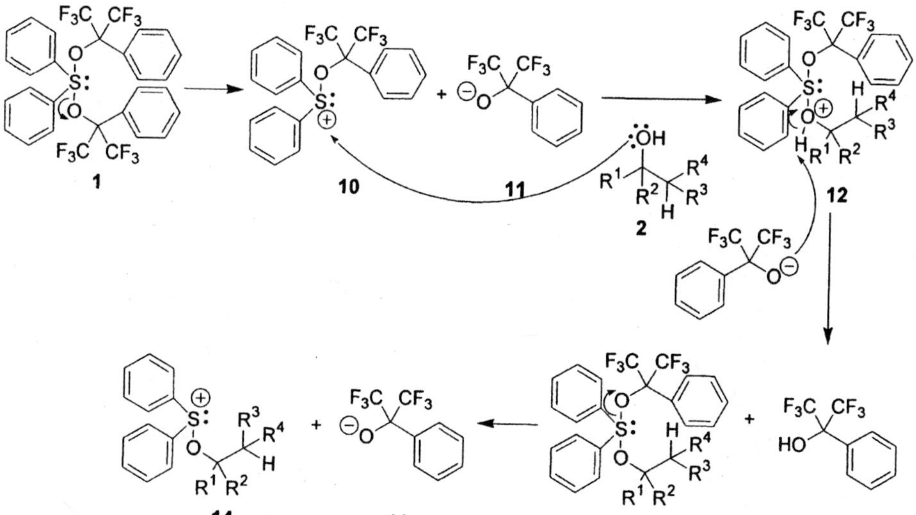

The mechanism commences with rapid exchange between the alkoxy ligands on sulfurane **1** and the reacting alcohol **2** to yield sulfurane **13** and alcohol **5**. Ionization of the other fluorinated alcohol yields alkoxide **11** and key alkoxysulfonium ion **14**. The structure of the alkoxy substituent in **14** (originally the starting alcohol) determines the ultimate course of the reaction. Reactions with tertiary alcohols follow option #1 below (R^1 and R^2 = C) and eliminate via an E1 pathway. Carbocation **15** is converted into alkene **3** upon reaction with alkoxide **11**. Secondary alcohol starting materials also yield the desired alkene products **3** but only after an E2 elimination (option #2). The requisite antiperiplanar geometry in this step enables stereospecific alkene generation from chiral secondary alcohols. Primary alcohols do not furnish alkenes and instead participate in S_N2 reactions that provide ethers like **16** (option #3).

Option #1
$R^1, R^2 = C$

14 4 15 11 3 5

Option #2
$R^1 = H, R^2 = C$

14 11 4 3 5

Option #3
$R^1, R^2 = H$

14 7 4 16

3.7.4 Variations and Improvements

Although widely used as a dehydration reagent, Martin's sulfurane is also known to facilitate amide cleavage reactions,[7] cyclic ether (including epoxide) formations,[8] and sulfilimine syntheses.[9] In the 1970s Martin demonstrated all of these transformations in his series of papers outlining the reactivity of the title sulfurane.

Secondary amides react selectively to yield esters and sulfilimines, which can then be converted into the corresponding amines. For example, amide **17** reacts with sulfurane **1** to furnish ester **19** and sulfilimine **18** which yields amine **21** upon reaction with either hydrochloric acid followed by base or hydrogen gas and catalytic palladium on carbon. Thus, this reaction can function both as an ester synthesis and an amine deprotection. Bulky secondary amides do not immediately undergo cleavage and instead yield imidates that, upon treatment with aqueous acid, furnish amines and esters. For example, combination of Martin's sulfurane (**1**) with amide **22** provides imidate **23** that can be easily converted into amine **25** and ester **24**. Primary amides provide *N*-acylsulfilimines, while tertiary amides are completely unreactive.[7]

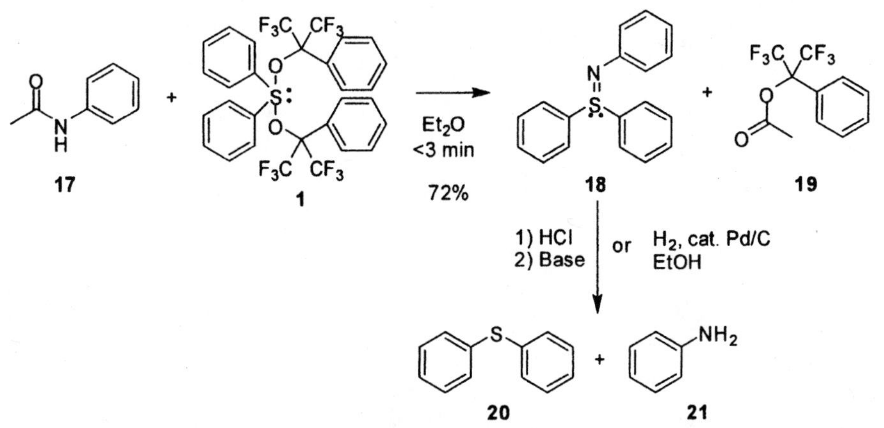

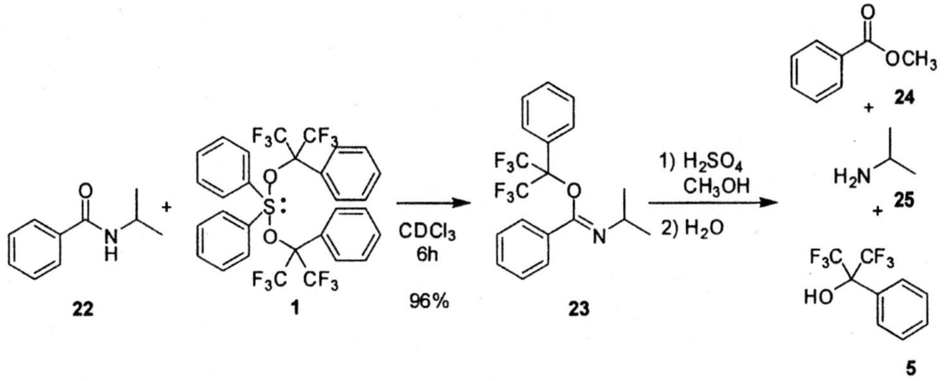

Sulfurane **1** reacts with a variety of diols to yield cyclic ethers; the most efficient transformations involve formation of epoxides and oxetanes. This process occurs via an intramolecular S_N2 mechanism and, thus, is only possible for 1,2-diols that can exist in an antiperiplanar orientation.[8] Eschenmoser successfully deployed this strategy to form epoxide **27** from diol **26**.[10]

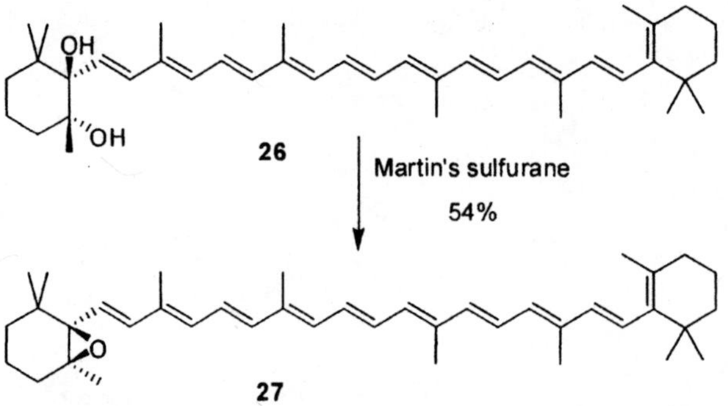

26

Martin's sulfurane

54%

27

Sulfilimines are produced upon reaction of Martin's sulfurane with primary amines.[9] This transformation was recently exploited for the synthesis of nucleoside sulfilimines.[11] Cytosine **28** was easily converted into the corresponding sulfilimine **29** in 88% yield after 1 h at room temperature.

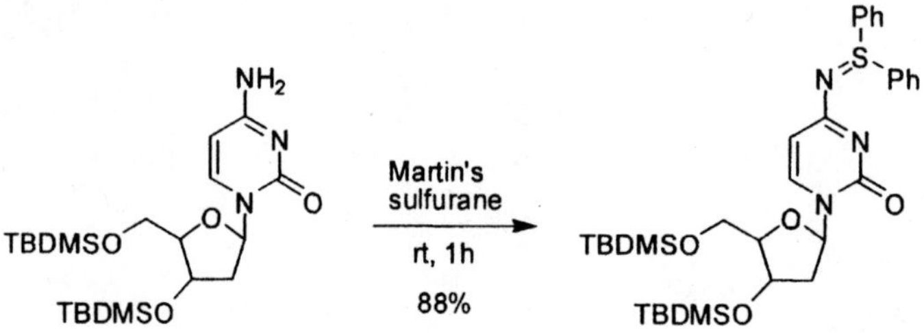

Martin's sulfurane

rt, 1h

88%

28 29

Shiori demonstrated that some β-hydroxy-α-amino acids furnish oxazolines upon exposure to Martin's sulfurane.[12] For example, amino acid **30** reacts in 94% yield to provide oxazoline **31**. This transformation proceeds via an intramolecular S_N2 reaction of the amide oxygen onto the carbon bearing the secondary alcohol.

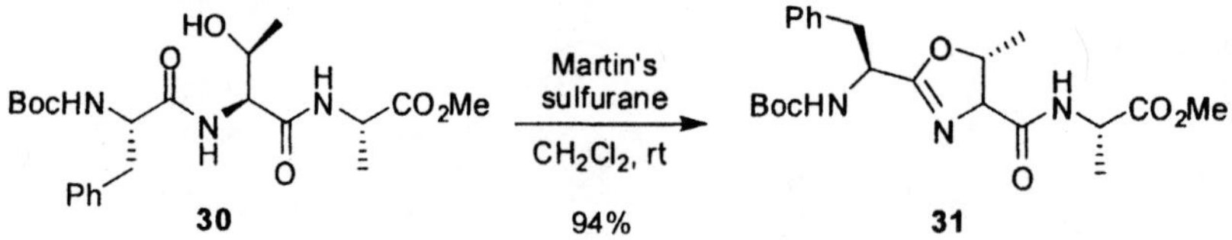

Martin's sulfurane

CH_2Cl_2, rt

94%

30 31

3.7.5 *Synthetic Utility*

Martin's sulfurane is frequently employed in organic synthesis and is an especially valuable reagent in natural product total syntheses. In this context, it has been used to produce acyclic alkenes and dienes along with a variety of cyclic alkenes.

3.7.5.1 *Acyclic alkenes*

Terminal alkenes, both mono- and disubstituted, have been prepared using the title reagent. In Wood's recently disclosed synthesis of welwitindolinone A isonitrile, he achieved a selective dehydration of the less hindered secondary alcohol in **32** to yield monosubstituted alkene **33**.[13] Mori obtained terminal disubstituted alkene **35** in 93% yield from phosphorylated tertiary alcohol **34** in the penultimate step in his total synthesis of acoradiene (**36**), the aggregation pheromone of the broad-horned flour beetle.[14]

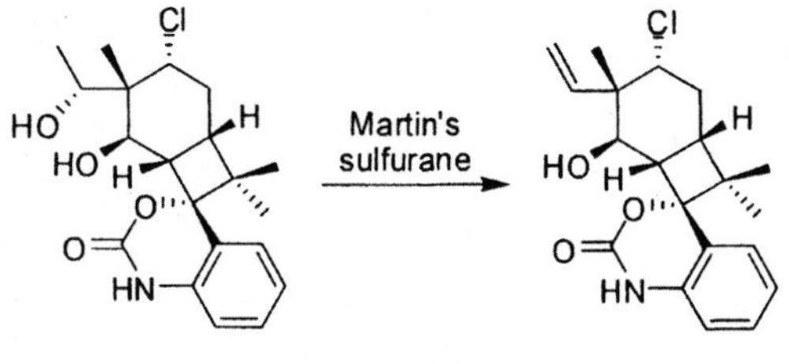

Martin's sulfurane

32 → 33

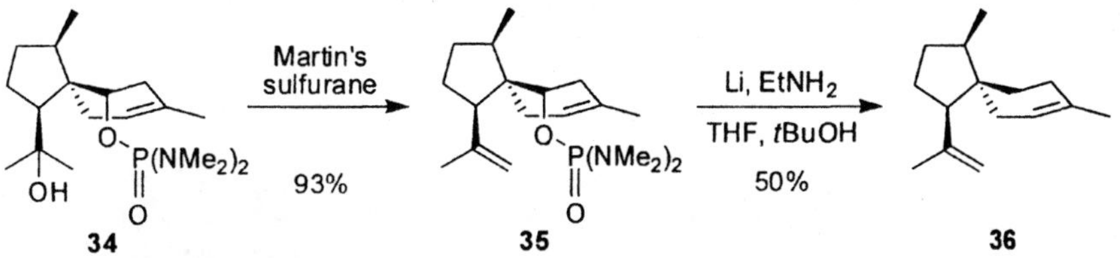

Martin's sulfurane, 93%; Li, $EtNH_2$, THF, *t*BuOH, 50%

34 → 35 → 36

Trans-disubstituted alkenes are readily available from secondary alcohols as demonstrated in several recent applications. A variety of norneolignans **38**, prepared and evaluated for antimalarial and antispasmodial activity, are attained after treatment of alcohols **37** with Martin's sulfurane and triethylamine in refluxing toluene for 1–48 h.[15] Regioselective syntheses of *trans* alkenes using **1** have been reported both by Movassaghi and Evans. Movassaghi used Martin's sulfurane in the late stages of his total synthesis of galbulimima alkaloid-13 to prepare unsaturated imine **40** from alcohol **39** in 81% yield.[16] Evans prepared alkene **42** in quantitative yield from alcohol **41** as part of an investigation into the synthesis of the phorboxazole nucleus.[17]

Martin's sulfurane
Et_3N, toluene
reflux, 1-48 h

37 → 38

Martin's sulfurane
PhH, rt
81%

39 → 40

Martin's sulfurane
CH_2Cl_2, 0 °C
100%
95:5 E:Z

41 → 42

Seconday and tertiary alcohols readily react with Martin's sulfurane to yield trisubstituted alkenes. These eliminations preferentially provide more highly substituted and conjugated olefins. In his recent synthesis of monopyrrolinone-based HIV-1 protease inhibitors, Smith used sulfurane **1** as part of a three-step carbonyl addition–elimination–hydrogenation sequence (**43**→**44**→**45**).[18] The sulfurane-mediated elimination step proceeded in 99% yield. Martin's sulfurane is the reagent of choice for introducing acyclic unsaturation at the Δ^{24} position from a C24 alcohol in a variety of sterols.[19] For example, **46** reacts quantitatively with Martin's sulfurane to furnish the desired alkene **47** in 97% yield and only 3% of the undesired Δ^{23}-isomer.

Martin's sulfurane induces stereospecific eliminations on amino acid residues in (*Z*)-methyldehydrobutyrine (**48**),[20] and (*Z*)-2-amino-2-butenoic acid (**50**),[21] respectively. Various other elimination conditions including the Burgess reagent, Ph_3P and DEAD, $SOCl_2$ and Et_3N, and DAST were investigated and Martin's sulfurane was superior. The requisite antiperiplanar geometry for the E2 elimination-mediated by the title reagent stereospecifically yields the desired *Z* stereoisomers. Shiori used the same strategy to dehydrate a depsipeptide fragment in 94% yield.[21]

Gais reported a detailed study of asymmetric eliminations using Martin's sulfurane to produce trisubstituted alkenes.[22] Using tertiary alcohols **52** containing three different chiral auxiliaries, Gais selectively generated both possible alkene stereoisomers (**53** and **54**). Ratios of the two products, which ranged from 99:1 to 14:86, depended on the structure of the chiral auxiliaries, reaction temperature, and solvent polarity. Like all tertiary alcohol eliminations with Martin's sulfurane, these reactions proceed via carbocation intermediates. Gais proposed that the selectivity arose from deprotonation of the most sterically accessible of the two prochiral protons α to the ester carbonyl.

HO O OAux* — Martin's sulfurane, $CHCl_3$, -60 °C, 82-91% → OAux* / Aux*O

52 **53** **54**

Martin's sulfurane has proved remarkably effective for the synthesis of acyclic alkenes; however several groups have reported failed elimination reactions using this reagent. Carreira attempted to form a trisubstituted alkene using Martin's sulfurane in his synthesis of spirotryprostatin B.[23] He ultimately achieved success by a revised synthetic approach based on a Julia–Kocienski olefination. In Martin's synthesis of ambruticin S, he reports successful model studies using Martin's sulfurane to yield vinyl sulfones, but in the real case none of the desired product was obtained.[24] Similar to Carreira's report, Martin's group was able to produce the desired alkene by employing a Julia olefination. Donohoe was unable to convert a tertiary alcohol into an exocyclic olefin using Martin's sulfurane, DAST, or the Burgess reagent and ultimately was forced to develop a completely different strategy to complete his synthesis of the 20S proteasome inhibitor *clasto*-lactacystin β-lactone.[25]

3.7.5.2 Acyclic dienes

Not surprisingly, Martin's sulfurane mediates the production of acyclic dienes from allylic and homoallylic alcohols. In Heathcock's synthesis of altohyrtin C, secondary homoallylic alcohol **55** reacted with sulfurane **1** to yield the desired (*E*)-alkene **56** in greater than 84% yield.[26] Interestingly, Tse generated diene **58** from tertiary allylic alcohol **57** in 95% yield with only 5% of the undesired trisubstituted diene isomer.[27]

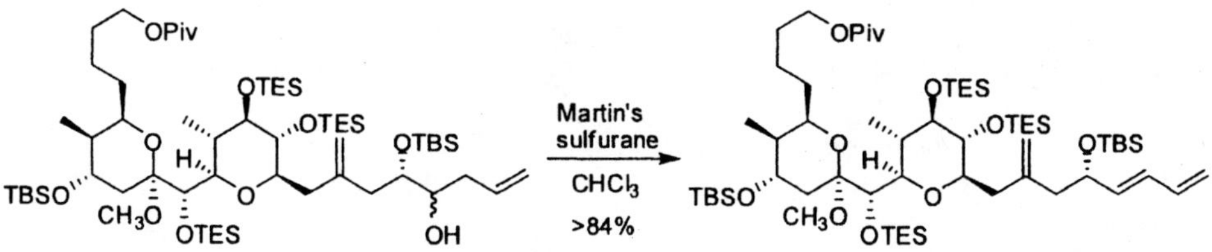

3.7.5.3 Macrocyclic alkenes and dienes

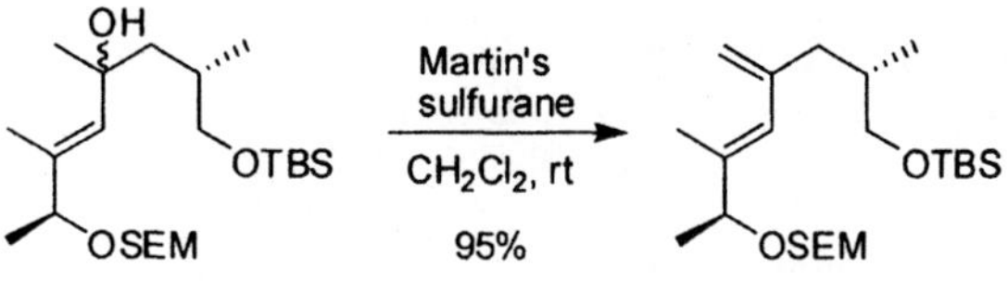

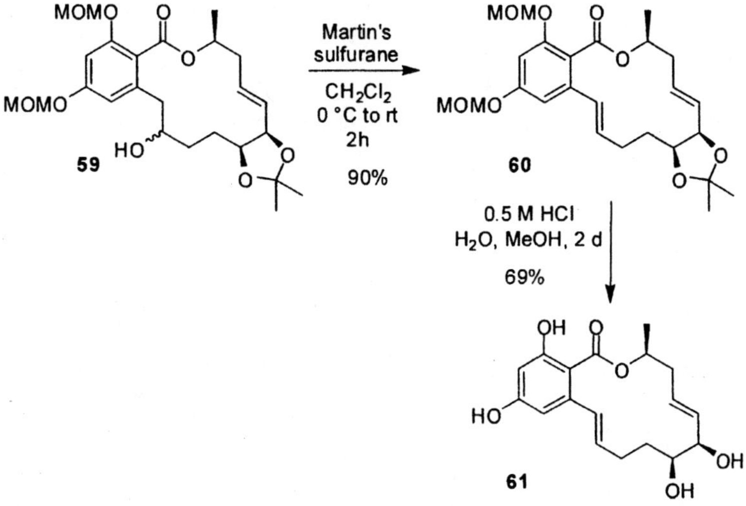

Martin's sulfurane is widely used for the production of cyclic alkenes. Surprisingly, it has seen limited use in the production of macrocyclic alkenes. In the penultimate step of Danishefsky's synthesis of aigialomycin (**61**), Martin's sulfurane reacts with homobenzylic alcohol **59** to provide conjugated alkene **60** in 90% yield.[28] The subsequent global deprotection step completed the total synthesis. Xu was unable to generate a macrocyclic diene using Martin's sulfurane and instead eliminated the starting tertiary alcohol using thionyl chloride and pyridine.[29]

3.7.5.4 Alkenes in eight- and nine-membered rings

Martin's sulfurane has been successfully used to generate alkenes in both the nine-membered ring of the neocarzinostatin chromophore and the eight-membered ring of taxusin. Myers completed his total synthesis of the neocarzinostatin chromophore (**63**) with a dehydration of alcohol **62** in 83% yield.[30] Myers also used Martin's sulfurane to introduce unsaturation into the five-membered ring in the neocarzinostating skeleton (see Section 3.7.5.6). Kobayashi and Hirama employed a sulfurane-mediated dehydration in their formal total synthesis of the neocarzinostatin chromophore.[31] They obtained late-stage synthetic intermediate **65** after treating tertiary alcohol **64** with Martin's sulfurane. Paquette generated a B-ring alkene (**67**) via a selective elimination of the less sterically hindered alcohol in **66** as part of his total synthesis of taxusin.[32] Surprisingly, Arseniyadis's group was unable to convert a secondary alcohol into an alkene in the eight -membered ring of the taxoid framework even with triethylamine in refluxing dichloromethane for 20 h.[33]

iPrO OMOM H3CO O OCH3 HN N Cl O O OTIPS OH O TESO O **62** —Martin's sulfurane, PhH, rt, 83%→ iPrO OMOM H3CO O OCH3 HN N Cl O O OTIPS O TESO O **63**

OTES O O TESO HO OCH3 O O O O **64** —Martin's sulfurane, CD_2Cl_2, 5 min, 100% (NMR)→ OTES O O TESO OCH3 O O O O **65**

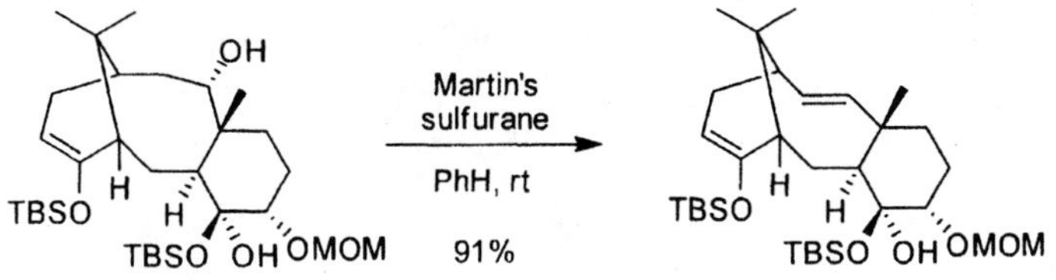

Martin's sulfurane
PhH, rt
91%

66 67

As part of his synthetic studies toward the synthesis of manzamine A, Winkler used Martin's sulfurane to form an alkene in an eight-membered ring.[34] Alcohol **68** was selectively converted into alkene **69** in 64% yield via elimination of the only hydrogen *trans* to the secondary alcohol.

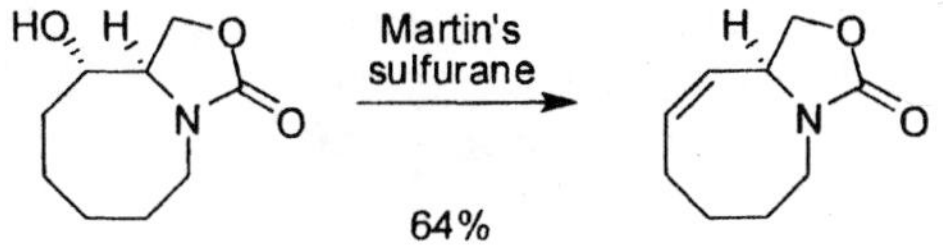

Martin's sulfurane
64%

68 69

3.7.5.5 Alkenes in six-membered rings

By far the most common cycloalkene generated using Martin's sulfurane is cyclohexene. Isolated six-membered ring alkenes and cyclohexenes in fused polycyclic systems are all generated with regularity using the title reagent.

Paquette reacted secondary alcohol **70** with Martin's sulfurane to furnish cyclohexene **71** in 78% yield.[35] In Nicolaou's total synthesis of everninomicin 13,384-1, he created several dihydropyrans from *anti* elimination of the corresponding secondary alcohols.[36] For example, alcohol **72** provides alkene **73** upon exposure to Martin's sulfurane and triethylamine in hot chloroform.

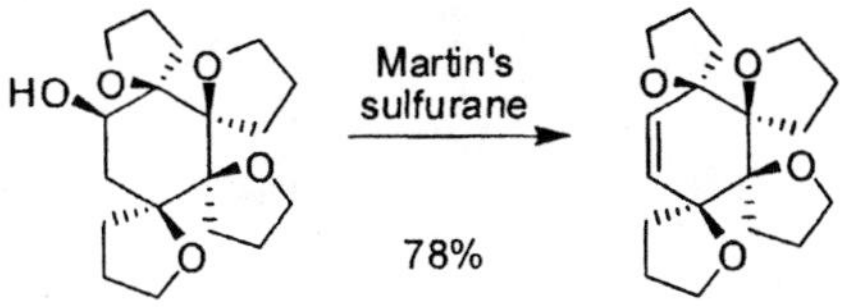

Martin's sulfurane
78%

70 71

Martin's sulfurane; Et_3N, $CHCl_3$, 50 °C, 2h; 87%

72 → 73

Several research groups have studied eliminations of tertiary alcohols on highly oxygenated cyclohexanes. As part of his synthesis of valienamine, Shing discovered that alcohol **74** provides only cyclohexene **75** in 90% yield with no evidence of the two other possible alkene products while exocyclic alkene **77** is the only product upon exposure of alcohol **76** to Martin's sulfurane.[37] Shing explains these results by proposing that the acetal in **76** blocks the ring protons for attack by the requisite base. While preparing difluorinated analogs of shikimic acid, Whitehead selectively generated cyclohexene **79** in 84% yield from tertiary alcohol **78**.[38] Usami reported a similar elimination of tertiary alcohol **80** to furnish cyclohexene **81** in 71% yield.[39]

74 → 75: Martin's sulfurane; PhH, reflux; 90%

76 → 77: Martin's sulfurane; PhH, reflux; 88%

78 → 79: Martin's sulfurane; CH_2Cl_2, rt; 84%

80 → 81: Martin's sulfurane; 71%

Martin's sulfurane has proved effective for introducing unsaturation into the steroid B-ring. Rao and Wang successfully generated alkene **83** from elimination of secondary alcohol **82**.[40] In another application for a polycyclic system, Heathcock applied this strategy in his total synthesis of isoschizogamine.[41] Treatment of secondary alcohol **84** provided late-stage synthetic intermediate **85** in 74% yield. Boger attempted an elimination of a tertiary alcohol in an alkaloid synthesis using Martin's sulfurane but was unable to obtain any of the desired alkene.[42]

TBDMSO ... OH **82** → Martin's sulfurane, Et_2O, rt, 80% → TBDMSO ... **83**

H_3CO, H_3CO, NO_2, N, O, OH **84** → Martin's sulfurane, CH_2Cl_2, rt, 74% → H_3CO, H_3CO, NO_2, N, O **85**

3.7.5.6 Alkenes in five-membered rings

Martin's sulfurane has been used to prepare a variety of cyclopentenes for use in the total syntheses of complex natural products. Myers effectively employed the title reagent to prepare synthetic intermediates for use in his total synthesis of the neocarzinostatin chromophore.[43] (For dehydrations to yield alkenes in the nine-membered ring of this compound, see Section 3.7.5.4.) One example is reaction of tertiary alcohol **86** to furnish alkene **87** in 93% yield. In Phillips's total synthesis of cylindramide A, he employed a conjugate addition–reduction–elimination sequence (**88**→**89**) to generate the bicyclo[3.3.0]octane structure found in the natural product.[44] Evans used Martin's sulfurane to convert secondary alcohol **90** into protected lepicidin A aglycon **91** in greater than 81% yield.[45] Subsequent installation of the two requisite sugar moieties completed the synthesis of lepicidin A.

Martin's sulfurane

CH_2Cl_2, rt

93%

TIPSO OTES HO O O O O

86 → **87**

1) Me_2CuLi, Et_2O -78 °C, 90%

2) $NaBH_4$, MeOH, 0 °C

3) Martin's sulfurane CH_2Cl_2, 0 °C

53% (2 steps)

OTIPS H

88 → **89**

Martin's sulfurane

CH_2Cl_2, 0 °C

>81%

OTIPS OH OTBS H

90 → **91**

3.7.6 *Experimental*

Martin's sulfurane

CH_2Cl_2, rt

84%

HO, CO_2CH_3 TBSO OCH_3 H_3CO

78 → **79**

Preparation of (2′*S*,3′*S*)-Methyl-3-*O*-*t*-butyldimethylsilyl-4-*O*,5-*O*-(2′,3′)-dimethoxy-butane-2′,3′-diyl)-shikimate (79)[38]

A solution of Martin's sulfurane (0.949 g, 1.41 mmol) in CH_2Cl_2 (5 mL) was slowly added under an atmosphere of nitrogen to a stirred solution of **78** (0.41 g, 0.94 mmol) in CH_2Cl_2 (10 mL) at rt. The resulting pale yellow solution was stirred for 24 h when the residual solvent was removed in vacuo to yield the crude product as a pale yellow oil. Purification by flash column chromatography (SiO_2; EtOAc-petroleum ether (40:60)) followed by recrystallization (CH_3OH-H_2O) furnished the title compound (**79**) as a colorless solid (0.33 g, 84%), mp 74–75 °C; (Found: C, 57.7; H, 8.7. $C_{20}H_{36}O_7Si$ requires C, 57.7; H, 8.7).

3.7.7 References

1. [R] Roden, B. A. Diphenylbis(1,1,1,3,3,3-hexafluoro-2-phenyl-2-propoxy)sulfurane. In *Encyclopedia of Reagents for Organic Synthesis*; Paquette, L. A., Ed.; Wiley: New York, 1995; 2201–2202.
2. Martin, J. C.; Arhart, R. J. *J. Am. Chem. Soc.* **1971**, *93*, 4327–4329.
3. (a) Martin, J. C.; Arhart, R. J.; Franz, J. A.; Perozzi, E. F.; Kaplan, L. J. *Org. Syn. Coll. Vol. 6*, 163–166. (b) Furukawa, N.; Ogawa, S.; Matsumura, K.; Shibutani, T.; Fujihara, H. *Chem. Lett.* **1990**, 979–982.
4. Bentrude, W. G.; Martin, J. C. *J. Am. Chem. Soc.* **1962**, *84*, 1561–1571.
5. Martin, J. C.; Arhart, R. J. *J. Am. Chem. Soc.* **1971**, *93*, 2341–2342.
6. Arhart, R. J.; Martin, J. C. *J. Am. Chem. Soc.* **1972**, *94*, 5003–5010.
7. (a) Franz, J. A.; Martin, J. C. *J. Am. Chem. Soc.* **1973**, *95*, 2017–2019. (b) Martin, J. C.; Franz, J. A. *J. Am. Chem. Soc.* **1975**, *97*, 6137–6144.
8. Martin, J. C.; Franz, J. A.; Arhart, R. J. *J. Am. Chem. Soc.* **1974**, *96*, 4604–4611.
9. Franz, J. A.; Martin, J. C. *J. Am. Chem. Soc.* **1975**, *97*, 583–591.
10. Eschenmoser, W.; Eugster, C. H. *Helv. Chim. Acta* **1978**, *61*, 822–831.
11. Slaitas, A.; Yeheskiely, E. *Phosphorus, Sulfur Silicon Relat. Elem.* **2004**, *179*, 153–171.
12. Yokokawa, F.; Shioiri, T. *Tetrahedron Lett.* **2002**, *43*, 8679–8682.
13. Reisman, S. E.; Ready, J. M.; Hasuoka, A.; Smith, C. J.; Wood, J. L. *J. Am. Chem. Soc.* **2006**, *128*, 1448–1449.
14. Kurosawa, S.; Bando, M.; Mori, K. *Eur. J. Org. Chem.* **2001**, 4395–4399.
15. Skytte, D. M.; Nielsen, S. F.; Chen, M.; Zhai, L.; Olsen, C. E.; Christensen, S. B. *J. Med. Chem.* **2006**, *49*, 436–440.
16. Movassaghi, M.; Hunt, D. K.; Tjandra, M. *J. Am. Chem. Soc.* **2006**, *128*, 8126–8127.
17. Evans, D. A.; Cee, V. J.; Smith, T. E.; Santiago, K. J. *Org. Lett.* **1999**, *1*, 87–90.
18. Smith, A. B.; Charnley, A. K.; Harada, H.; Beiger, J. J.; Cantin, L. D.; Kenesky, C. S.; Hirschmann, R.; Munshi, S.; Olsen, D. B.; Stahlhut, M. W.; Schleif, W. A.; Kuo, L. C. *Bioorg. Med. Chem. Lett.* **2006**, *16*, 859–863.
19. Dolle, R. E.; Schmidt, S. J.; Erhard, K. F.; Kruse, L. I. *J. Am. Chem. Soc.* **1989**, *111*, 278–284.
20. Samy, R.; Kim, H. Y.; Brady, M.; Toogood, P. L. *J. Org. Chem.* **1999**, *64*, 2711–2728.
21. Yokokawa, F.; Shioiri, T. *Tetrahedron Lett.* **2002**, *43*, 8673–8677.
22. Gais, H.-J.; Schmiedl, G.; Ossenkamp, R. K. L. *Liebigs Ann. Rec.* **1997**, 2419–2431.
23. Marti, C.; Carreira, E. M. *J. Am. Chem. Soc.* **2005**, *127*, 11505–11515.
24. Berberich, S. M.; Cherney, R. J.; Colucci, J.; Courillon, C.; Geraci, L. S.; Kirkland, T. A.; Marx, M. A.; Schneider, M. F.; Martin, S. F. *Tetrahedron* **2003**, *59*, 6819–6832.
25. Donohoe, T. J.; Sintim, H. O.; Sisangia, L.; Ace, K. W.; Guyo, P. M.; Cowley, A.; Harling, J. D. *Chem. Eur. J.* **2005**, *11*, 4227–4238.
26. Heathcock, C. H.; McLaughlin, M.; Medina, J.; Hubbs, J. L.; Wallace, G. A.; Scott, R.; Claffey, M. M.; Hayes, C. J.; Ott, G. R. *J. Am. Chem. Soc.* **2003**, *125*, 12844–12849.
27. Tse, B. *J. Am. Chem. Soc.* **1996**, *118*, 7094–7100.
28. (a) Geng, X. D.; Danishefsky, S. J. *Org. Lett.* **2004**, *6*, 413–416. (b) Yang, Z. Q.; Geng, X. D.; Solit, D.; Pratilas, C. A.; Rosen, N.; Danishefsky, S. J. *J. Am. Chem. Soc.* **2004**, *126*, 7881–7889.
29. Liu, P.; Xu, X. X. *Tetrahedron Lett.* **2004**, *45*, 5163–5166.
30. Myers, A. G.; Hogan, P. C.; Hurd, A. R.; Goldberg, S. D. *Angew. Chem. Int. Ed.* **2002**, *41*, 1062–1067.
31. Kobayashi, S.; Hori, M.; Wang, G. X.; Hirama, M. *J. Org. Chem.* **2006**, *71*, 636–644.

32. Paquette, L. A.; Zhao, M. *J. Am. Chem. Soc.* **1998**, *120*, 5203–5212.
33. Hamon, S.; Ferreira, M. D. R.; del Moral, J. Q.; Hernando, J. I. M.; Lena, J. I. C.; Birlirakis, N.; Toupet, L.; Arseniyadis, S. *Tetrahedron: Asymmetry* **2005**, *16*, 3241–3255.
34. Winkler, J. D.; Stelmach, J. E.; Axten, J. *Tetrahedron Lett.* **1996**, *37*, 4317–4318.
35. Paquette, L. A.; Ohmori, N.; Lowinger, T. B.; Rogers, R. D. *J. Org. Chem.* **2000**, *65*, 4303–4308.
36. (a) Nicolaou, K. C.; Mitchell, H. J.; Fylaktakidou, K. C.; Rodriguez, R. M.; Suzuki, H. *Chem. Eur. J.* **2000**, *6*, 3116–3148. (b) Nicolaou, K. C.; Fylaktakidou, K. C.; Mitchell, H. J.; van Delft, F. L.; Rodriguez, R. M.; Conley, S. R.; Jin, Z. D. *Chem. Eur. J.* **2000**, *6*, 3166–3185.
37. Kok, S. H. L.; Lee, C. C.; Shing, T. K. M. *J. Org. Chem.* **2001**, *66*, 7184–7190.
38. (a) Box, J. M.; Harwood, L. M.; Humphreys, J. L.; Morris, G. A.; Redon, P. M.; Whitehead, R. C. *Synlett* **2002**, 358–360. (b) Begum, L.; Box, J. M.; Drew, M. G. B.; Harwood, L. M.; Humphreys, J. L.; Lowes, D. J.; Morris, G. A.; Redon, P. M.; Walker, F. M.; Whitehead, R. C. *Tetrahedron* **2003**, *59*, 4827–4841.
39. Usami, Y.; Hatsuno, C.; Yamamoto, H.; Tanabe, M.; Numata, A. *Chem. Pharm. Bull.* **2004**, *52*, 1511.
40. Rao, P. N.; Wang, Z. *Steroids* **1997**, *62*, 487–490.
41. Hubbs, J. L.; Heathcock, C. H. *Org. Lett.* **1999**, *1*, 1315–1317.
42. Yuan, Z. Q.; Ishikawa, H.; Boger, D. L. *Org. Lett.* **2005**, *7*, 741–744.
43. Myers, A. G.; Glatthar, R.; Hammond, M.; Harrington, P. M.; Kuo, E. Y.; Liang, J.; Schaus, S. E.; Wu, Y.; Xiang, J. *J. Am. Chem. Soc.* **2002**, *124*, 5380–5401.
44. Hart, A. C.; Phillips, A. J. *J. Am. Chem. Soc.* **2006**, *128*, 1094–1095.
45. Evans, D. A.; Black, W. C. *J. Am. Chem. Soc.* **1993**, *115*, 4497–4513.

Kevin M. Shea

3.8 Oppenauer Oxidation

3.8.1 Description

The aluminum-catalyzed hydride shift from the α-carbon of an alcohol component **2** to the carbonyl carbon of a second component **1** is named the Meerwin–Ponndorf–Verley reduction or Oppenauer oxidation depending on the isolated product. If aldehydes or ketones **4** are the desired products, the reaction is viewed as the Oppenauer oxidation.[1–3]

$$R^1C(O)R^2 \;(\mathbf{1}) + R^3CH(OH)R^4 \;(\mathbf{2}) \xrightleftharpoons{Al(OR)_3} R^1CH(OH)R^2 \;(\mathbf{3}) + R^3C(O)R^4 \;(\mathbf{4})$$

The most extensive application of the Oppenauer oxidation has been in the oxidation of steroid molecules. The most common aluminum catalysts are aluminum *t*-butoxide, *i*-propoxide, and phenoxide. While only catalytic amounts of the aluminum alkoxide are theoretically required, in practice at least 0.25 mole of alkoxide per mole of alcohol is used. Acetone and methyl ethyl ketone have proved valuable hydride acceptors due to their accessibility and ease of separation from the product, whereas other ketones such as cyclohexanone and *p*-benzoquinone are useful alternatives, due to their increased oxidation potentials.[4] Although the reaction can be performed neat, an inert solvent to dilute the reaction mixture can reduce the extent of condensation, and, as such, benzene, toluene, and dioxane are commonly utilized. Oxidation of the substrate takes place at temperatures ranging from room temperature to reflux, with reaction times varying from fifteen minutes to twenty-four hours and yields ranging from 37% to 95%.

3.8.2 Historical Perspective

The reversible nature of the above reaction was demonstrated by Verley[5] in 1925 and shortly thereafter by Pondorff,[6] but it was not until 1937 that Oppenauer showed that unsaturated steroid alcohols **5** could be oxidized to the corresponding ketones **6** through action of aluminum *t*-butoxide in the presence of a large amount of acetone in excellent yields.[1]

$$\mathbf{5} \xrightarrow[\text{acetone, reflux}]{Al(Ot\text{-}Bu)_3} \mathbf{6}$$

The oxidation was accompanied by migration of the double bond from the β,γ to the α,β position. The resulting α,β-unsaturated compound has a characteristic absorption in the ultraviolet region of the spectrum and therefore the Oppenauer oxidation has been used as a test for localising double bonds in a variety of steroidal compounds.[7-10]

3.8.3 Mechanism

The generally accepted mechanism for the Oppenauer oxidation involves a cyclic transition state[11,12] as originally proposed by both Woodward[13] and Oppenauer.[14]

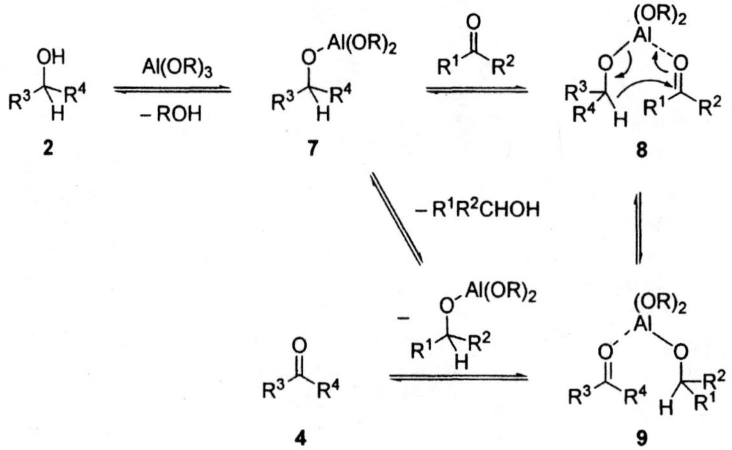

Exchange of the aluminum alcoholate with substrate **2** leads to alkoxide species **7** and subsequent coordination of the hydride acceptor (ketone) gives access to the cyclic transition state **8**. Intermolecular hydride transfer from the alkoxide to the acceptor gives rise to species **9**, from which the product, ketone **4**, is delivered, following decomplexation. The resulting aluminum species can reenter the catalytic cycle by further exchange of the alcoholate. In some cases, two moles of aluminum alkoxide are involved, one attacking the carbon and the other the oxygen, a conclusion that stems from a reaction order of 1.5 in alkoxide.[15] For simplicity the alkoxide has been depicted as a monomer; however, in reality, it exists as trimers and tetramers and it is these that react.[16,17]

3.8.4 Synthetic Utility

3.8.4.1 Saturated alcohols

Although it was originally implied that Oppenauer's method could only be applied to alcoholic groups activated by unsaturation, more recent studies have proved this to be incorrect. A variety of steroidal alcohols in which the double bond is three or more carbons removed from the hydroxyl group have been oxidized using benzene and acetone, including γ-chloestenol and α-ergostenol.[18,19] The steroid alcohol **10** bearing an acid-labile dienone group was converted to the corresponding ketone **11** in 55% yield.[20] Another example where the oxidation has been successfully applied to the saturated steroids is in the conversion of diene **12** to ketone **13**, a key intermediate in the synthesis of the cortical hormone 11-dehydrocorticosterone.[21]

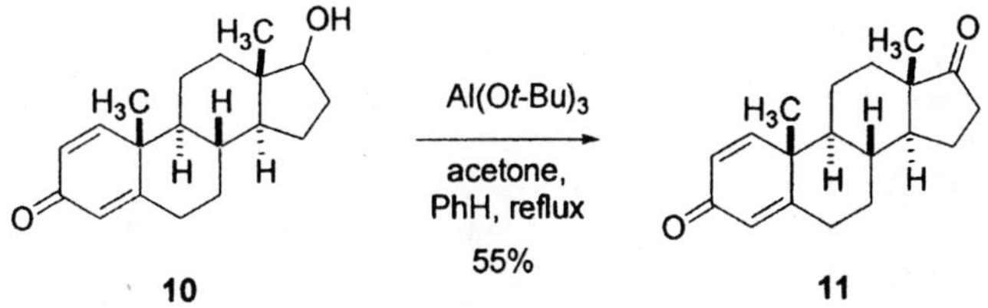

$Al(Ot\text{-}Bu)_3$, acetone, PhH, reflux, 55%: **10** → **11**

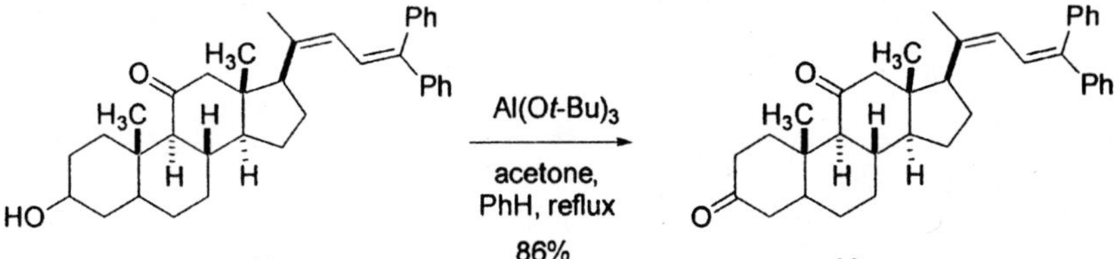

$Al(Ot\text{-}Bu)_3$, acetone, PhH, reflux, 86%: **12** → **13**

Among the non-steroidal alcohols applied to the Oppenauer oxidation are the *cis* and *trans* α-decalols (**14**), which give excellent yields of the corresponding α-decalones (**15**).[22] Oxidation of phenolic compounds bearing pendent aliphatic secondary alcohols can be readily performed without prior protection of the phenolic alcohol functionality.[23] Acid-sensitive acetal **16** is smoothly converted to the corresponding ketone **17**, by exploiting a modified experimental procedure.[3,24]

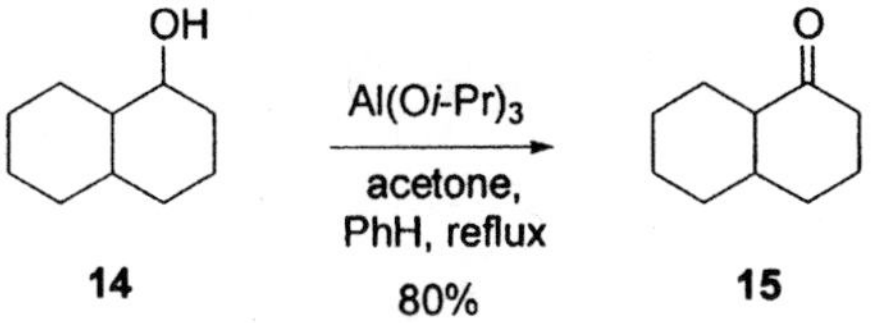

$Al(Oi\text{-}Pr)_3$, acetone, PhH, reflux, 80%: **14** → **15**

OH O O → O O O

$Al(Oi\text{-}Pr)_3$, benzophenone, 150 °C, Continuous distillation, 37%

16 17

3.8.4.2 Unsaturated alcohols

Following the pioneering studies of Oppenauer,[1] this procedure has been used to oxidize a plethora of unsaturated steroidal molecules and, in general, proves to be a superior method with respect to both yield and convenience to the previously utilized methods.[2,3] As such, the method has found use in the manufacture of a number of hormones, including testosterone (**6**, R = OH), progesterone (**6**, R = $COCH_3$), and desoxycorticosterone (**6**, R = $COCH_2OCOCH_3$).[25] As mentioned earlier, the oxidation of Δ^5-3-hydroxysteroids **5** is accompanied by migration of the double bond from the β,γ to the α,β position. In the oxidation of compounds bearing two conjugated double bonds (e.g., **18**) only the β,γ-double bond migrates (to give **19**).[1,26] In general, the Oppenauer oxidation is tolerant of a wide range of functionalities in the oxidation of unsaturated steroids, with substituents including allyl,[27] vinyl,[28] and ethynyl groups,[29] halides,[19] acetals,[30] mercaptals, ketals,[31] and epoxides.[32]

H_3C H HO → H_3C H O

$Al(Ot\text{-}Bu)_3$, acetone, PhH, reflux

18 19

A number of non-steroidal unsaturated alcohols have been oxidized using the Oppenauer procedure.[2,3] For example, bicyclic alcohol **20** was oxidized to the corresponding ketone **21**, with concomitant migration of the double bond.[33] The oxidation of the allylic alcohol derivatives such as **22** has been extremely useful in the preparation of a number of vitamin A analogs.[34–37] The reaction is also applicable to open-chain structures, for example octatrienol **23**, which gave access to the corresponding ketone in 80% yield.[38]

OH → O

$Al(Ot\text{-}Bu)_3$, acetone, PhH, reflux, 76%

20 21

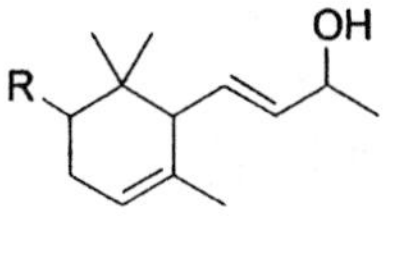

22

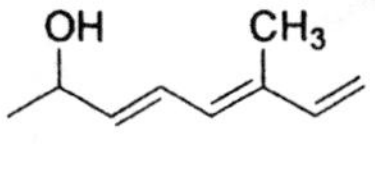

23

3.8.4.3 *Polyhydroxyl Compounds*

Simultaneous oxidation of two alcohol functionalities can be performed in both saturated and unsaturated compounds. For example, oxidation of diol **24** was achieved in 30% yield to give diketone **25**.[33] Steric factors can play a large effect and can result in other reaction processes, such as rearrangement.[3] Oxidation of saturated steroid **26** (R = $C_6H_{11}O_2$) furnishes diketone **27**, with inversion taking place at C-5,[39] whereas oxidation of unsaturated steroid **28** (R = C_8H_{17}) also yields diketone **27**, in which the double bond has been formally reduced.[40] The oxidation of steroids containing two *cis* alcohols at the C-3 and C-5 positions is hindered by the formation of an unreactive aluminum chelate.[41]

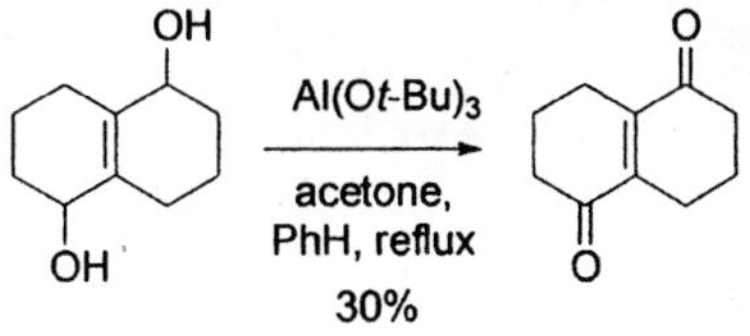

Al(O*t*-Bu)$_3$, acetone, PhH, reflux, 30%

24 → 25

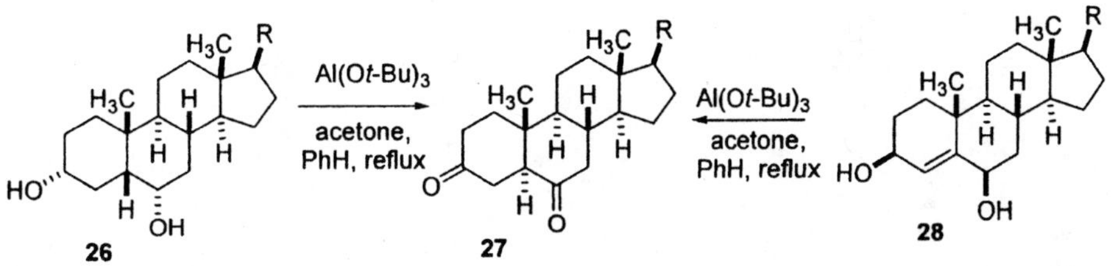

Al(O*t*-Bu)$_3$, acetone, PhH, reflux

26 → 27 ← 28

The Oppenauer oxidation has also proved particularly useful in the partial oxidation of polyhydroxyl compounds and the order of oxidation appears to almost be the reverse of chromic anhydride.[3] In many cases the selectivity is achieved by the unique steric environment of hydroxyl groups within the steroid molecules. Activation of one of the hydroxyl groups by unsaturation makes the differentiation even more facile.[42]

3.8.4.4 *Primary Alcohols*

An important advantage of the Oppenauer oxidation of primary alcohols to aldehydes, over several traditional methods, is that over-oxidation towards carboxylic acids is prohibited.

Initially, the application of the Oppenauer method to the oxidation of primary alcohols proved problematic, due to the condensation of the product aldehydes with the hydride acceptors.[3] The use of easily reducible oxidants like quinone and cyclohexanone, however, facilitated the conversion of primary alcohols to aldehydes.[43] For example, benzyl alcohol gave 50–60% of the corresponding aldehyde and furfuryl alcohol gave 20% of furfural. Aliphatic alcohols, however, gave very poor yields. Alternatively, by employing a modified procedure, which involved full conversion of the alcohol substrate to its aluminate, and slow distillation of the product in the presence of a high boiling hydride acceptor, a variety of primary alcohols were converted into the corresponding aldehydes.[44,45]

3.8.5 Variations

3.8.5.1 The "modified Oppenauer oxidation"

Although the traditional Oppenauer conditions utilized aluminum catalysts, alternative metal alkoxides, for example, chloromagnesium alkoxides, are competent in the transformation.[3] In 1945, Woodward devised a new system, which involved the use of potassium *t*-butoxide, and benzophenone for the oxidation of quinine (**29**) to quinone (**30**).[13] This was termed the "modified Oppenauer oxidation." The traditional aluminum catalytic system failed in this case due to the complexation of the Lewis–basic nitrogen to the aluminum centre. The synthetic flexibility of this procedure was extended by the use of more potent hydride acceptors.[46]

KO*t*-Bu
benzophenone
benzene

29 **30**

3.8.5.2 Catalytic Oppenauer oxidations

As stated earlier, theoretically, only catalytic amounts of the aluminum catalyst are required. Although aluminum has a high charge density and therefore good Lewis acidic character, it displays relatively poor ligand exchange abilities.[47] This limits its catalytic use. Although the addition of protic acids has been found to enable catalytic turnover, the aldol condensation sidereaction is prevalent.[48] The alkali metal alkoxides show high ligand exchange rates, but their low charge density or small coordination number causes a low activity.[47]

The hard-Lewis acidities, high ligand exchange rates and large coordination numbers of lanthanide complexes have enabled their application to both the Meerwin–Ponndorf–

Verley reduction and Oppenauer oxidation.[47] For example, 10 mol% of $Sm(Ot\text{-}Bu)I_2$ has been used for both oxidations and reductions in THF at 65 °C.[49] Another samarium(III) catalyst, formed *in situ*, has been used in the catalytic oxidation of octan-2-ol and several allylic alcohols.[50]

Following several reports on modified aluminum alkoxide catalysts,[51-53] a highly active aluminum catalyst **31** was reported by Maruoka and co-workers which allowed the conversion of carveol (**32**) to carvone (**33**) in excellent yield (94%).[54] The oxidation of a variety of allylic, benzylic and aliphatic primary and secondary alcohols, as well as terpenoids and steroids, was demonstrated.

$SO_2C_6F_{17}$ N AlMe O

OH **31**, 1 mol% toluene 21 °C, 0.5 h 1.2 eq. *t*-BuCHO 21 °C, 0.5 h 94% O

32 **33**

Other catalytic systems employed include ruthenium or iridium,[55,56] boron,[57] and heterogeneous catalysts.[47]

3.8.5.3 Cascade reactions

As mentioned previously, one common sidereaction of the Oppenauer oxidation is the aldol condensation of the product with the hydride acceptor. Simultaneous Oppenauer oxidation–aldol condensations have therefore been employed toward the synthesis of α,β-unsaturated carbonyl compounds.[3] For example, geraniol (**34**) in the presence of acetone and an aluminum alkoxide gave *ψ*-ionone (**35**) in good yield.[58]

CH_3 OH H_3C CH_3 $Al(Ot\text{-}Bu)_3$ acetone, PhH, reflux O CH_3 CH_3 H_3C CH_3

34 **35**

MgBr 36 + 37 → Et_2O, 25 °C, 5 min, 100% → 38 OMgBr → PhCHO, Et_2O, 25 °C, 5 h, 69% → 39

By utilizing the applicability of halomagnesium alkoxides in the Oppenauer oxidation, a one-pot, alkylation–oxidation has been reported using Grignard reagents.[59] For example, butylmagnesium bromide (**36**) and crotonaldehyde (**37**) gave access to oct-2-en-4-one (**39**), following addition of benzaldehyde as a hydride acceptor.

3.8.6 Experimental

14 → $Al(Oi\text{-}Pr)_3$, acetone, PhH, reflux, 80% → 15

cis-α-Decalone (15)[22]

To a solution of *cis*-α-decalol (**14**, 1.5 g, 9.9 mmol) in dry benzene (150 mL) and dry acetone (100 mL) was added aluminum *i*-propoxide (3.0 g, 14.7 mmol). The mixture, protected with a calcium chloride drying tube, was refluxed for 12 h. The reaction mixture was allowed to cool to ambient temperature and washed with 30% sulfuric acid (2 × 100 mL) and water until neutral. The organic extract was dried over anhydrous sodium sulfate and the solvent evaporated under reduced pressure. Fractional distillation of the product gave *cis*-α-decalone (**15**, 1.2 g, 80%) as a colorless oil (b.p. 116 °C (18 mmHg)); $n_D^{20°}$ 1.4939.

3.8.7 References

1. Oppenauer, R. V. *Rec. Trav. Chirn.* **1937**, *56*, 137.
2. [R] Bersin, T. The Meerwin–Ponndorf Reduction and Oppenauer Oxidation. In *Newer Methods of Preparative Organic Chemistry, English Ed.*; Interscience: New York, 1948; 125–158.
3. [R] Djerassi, D. The Oppenauer Oxidation. In *Organic Reactions* **1951**, *6*, 207.
4. Adkins, H.; Franklin, R. C. *J. Am. Chem. Soc.* **1941**, *63*, 2381.
5. Verley, A. *Bull. Soc. Chim. France* **1925**, *37*, 537.
6. Pondorff, W. *Angew. Chem.* **1926**, *39*, 138.

7. Jones, E. R. H.; Wilkinson, P. A.; Kerlogue, R. H. *J. Chem. Soc.* **1942**, 391.
8. Heilbron, I. M.; Jones, E. R. H.; Roberts, K. C.; Wilkinson, P. A. *J. Chem. Soc.* **1941**, 344.
9. Barton, D. H. R.; Jones, E. R. H. *J. Chem. Soc.* **1943**, 599.
10. Windaus, A. O. R.; Roosen–Runge, C. *Z. physiol. Chem.* **1939**, *260*, 184.
11. Shiner Jr, V. J.; Whittaker, D. *J. Am. Chem. Soc.* **1963**, *85*, 2337.
12. Warnhoff, E. W.; Reynolds-Warnhoff, P.; Wong, M. Y. H. *J. Am. Chem. Soc.* **1980**, *102*, 5956.
13. Woodward, R. B.; Wendler, N. L.; Brutschy, F. J. *J. Am. Chem. Soc.* **1945**, *67*, 1425.
14. Oppenauer, R. private communication.
15. Moulton, W. N.; Van Atta, R. E.; Ruch, R. R. *J. Org. Chem.* **1961**, *26*, 290.
16. Williams, E. D.; Krieger, K. A.; Day, A. R. *J. Am. Chem. Soc.* **1953**, *75*, 2404.
17. Shiner Jr, V. J.; Whittaker, D. *J. Am. Chem. Soc.* **1969**, *91*, 394.
18. Buser, W. *Helv. Chim. Acta.* **1947**, *30*, 1390.
19. Barton, D. H. R.; Cox, J. D. *J. Chem. Soc.* **1948**, 783.
20. Inhoffen, H. H.; Zühlsdorff, G.; Huang, Minlon. *Ber.* **1940**, *73*, 457.
21. Wettstein, A.; Meystre, Ch. *Helv. Chim. Acta.* **1947**, *30*, 1267.
22. English, J.; Cavaglieri, G. *J. Am. Chem. Soc.* **1943**, *65*, 1085.
23. Cornforth, J. W.; Robinson, R. *J. Chem. Soc.* **1949**, 1855.
24. Theimer, E. private communication.
25. *British Intelligence Subcommittee*, Final Report 996, H. M. Stationary Office, London, **1947**.
26. Windaus, A. O. R.; Kaufmann, H. P. *Ann.* **1939**, *542*, 220.
27. Butenandt, A.; Peters, D. *Ber.* **1938**, *71*, 2688.
28. Ruzicka, L.; Hofman, K.; Meldahl, H. F. *Helv. Chim. Acta.* **1938**, *21*, 597.
29. Ruzicka, L.; Hofman, K.; Meldahl, H. F. *Helv. Chim. Acta.* **1938**, *21*, 373.
30. Schindler, W.; Frey, H.; Reichstein, T. *Helv. Chim. Acta.* **1941**, *24*, 360.
31. Steiger, M.; Reichstein, T. *Helv. Chim. Acta.* **1938**, *21*, 177.
32. Julian, P. L.; Meyer, E. W.; Ryden, I. *J. Am. Chem. Soc.* **1950**, *72*, 369.
33. Campbell, W. P.; Harris, G. C. *J. Am. Chem. Soc.* **1941**, *63*, 2721.
34. Chanley, J. D.; Sobotka, H. *J. Am. Chem. Soc.* **1949**, *71*, 4140.
35. Heilbron, I.; Jones, E. R. H.; Richardson, R. W. *J. Chem. Soc.* **1949**, 292.
36. Heilbron, I.; Jones, E. R. H.; Lewis, R. W.; Richardson, R. W.; Weedon, B. C. L. *J. Chem. Soc.* **1949**, 742.
37. Heilbron, I.; Jones, E. R. H.; Lewis, R. W.; Weedon, B. C. L. *J. Chem. Soc.* **1949**, 2023.
38. Cheeseman, G. W. H.; Heilbron, I.; Jones, E. R. H.; Sondheimer, F.; Weedon, B. C. L. *J. Chem. Soc.* **1949**, 2031.
39. Gallagher, T. F.; Xenos, J. R. *J.Biol. Chem.* **1946**, *165*, 365.
40. Prelog, V.; Tagmann, E. *Helv. Chim. Acta.* **1944**, *27*, 1871.
41. Ehrenstein, M.; Johnson, A. R.; Olmstead, P. C.; Vivian, V. I.; Wagner, M. A. *J. Org. Chem.* **1950**, *15*, 264.
42. Jeanloz, R. W.; v. Euw, J. *Helv. Chim. Acta.* **1947**, *30*, 803.
43. Yamashita, M.; Matsumura, T. *J. Chem. Soc. Japan* **1943**, *64*, 506.
44. Schinz, H.; Lauchenauer, A.; Jeger, O.; Rüegg, R. *Helv. Chim. Acta.* **1948**, *31*, 2235.
45. Lauchenauer, A.; Schinz, H. *Helv. Chim. Acta.* **1949**, *32*, 1265.
46. Warnhoff, E. W.; Reynolds-Warnhoff, P. *J. Am. Chem. Soc.* **1963**, *28*, 1431.
47. [R] Graauw, C. F. d.; Peters, H. v. B.; Huskens, J. *Synthesis* **1994**, 1007.
48. Kow, R.; Nygren, R.; Rathke, M. W. *J. Org. Chem.* **1977**, *42*, 826.
49. Namy, J. L.; Souppe, J.; Collin, J.; Kagan, H. B. *J. Org. Chem.* **1984**, *49*, 2045.
50. Collin, J.; Namy, J. L.; Kagan, H. B. *Nouv. J. Chim.* **1986**, *10*, 229.
51. Akamanchi, K. G.; Chaudhari, B. A. *Tetrahedron Lett.* **1997**, *38*, 6925.
52. Ooi, T.; Miura, T.; Maruoka, K. *Angew, Chem. Int. Ed.* **1998**, *37*, 2347.
53. Ooi, T.; Miura, T.; Itagaki, Y.; Ichikawa, H.; Maruoka, K. *Synthesis*, **2002**, 279.
54. Ooi, T.; Miura, T.; Ichikawa, H.; Maruoka, K. *Org. Lett.* **2002**, *4*, 2669.
55. Almeida, M. L. S.; Kocovsky, P.; Backvall, J.–E. *J. Org. Chem.* **1996**, *61*, 6587.
56. Ajjou, A. N. *Tetrahedron Lett.* **2001**, *42*, 13.
57. Ishihara, K.; Kurihara, H. Yamamoto, H. *J. Org. Chem.* **1997**, *62*, 5664.
58. Batty, J. W.; Burawoy, A.; Harper, S. H.; Heilbron, I. M.; Jones, W. E. *J. Chem. Soc.* **1938**, 175.
59. Byrne, B.; Karras, M. *Tetrahedron Lett.* **1987**, *28*, 769.

Matthew J. Fuchter

3.9 Prilezhaev Reaction

3.9.1 Description

The *Prilezhaev reaction* is the formation of epoxides by the reaction of alkenes with peracids.

1 + 2 (R–C(=O)OOH) → 3 + 4 (R–C(=O)OH)

The *Prilezhaev reaction*[1–4] is a common reaction used in organic chemistry for the epoxidation of an alkene **1** by reaction with a peracid **2** to yield an epoxide **3**. Although the reaction is often used it is not commonly referred to as the *Prilezhaev reaction* in the literature.

3.9.2 Historical Perspective

Nikolai Aleksandrovich Prilezhaev (1872–1944) was a Russian chemist. A biographical reference was published in 1973 in *Zhurnal Obschei Khimii.*[5] The reaction was first published in 1910. The *Prilezhaev reaction* has wide utility and has been reviewed.[4,7]

3.9.3 Mechanism

The hydroxy oxygen of a peracid has a higher electrophilicity as compared to a carboxylic acid. A peracid **2** can react with an alkene **1** by transfer of that particular oxygen atom to yield an oxirane (an epoxide) **3** and a carboxylic acid **4**. The reaction is likely to proceed *via* a transition state as shown in **5** (butterfly mechanism),[5,6] where the electrophilic oxygen adds to the carbon—carbon doublebond and the proton simultaneously migrates to the carbonyl oxygen of the acid.[3,6] The rate of the reaction increases in the order: R = CH_3 < C_6H_5 < *m*-ClC_6H_4 < H < *p*-$NO_2C_6H_4$ < CO_2H < CF_3, which is related to the pK_a of acid (RCO_2H): 4.8; 4.2; 3.9; 3.8; 3.4; 2.9; 0. The lower the pK_a, the greater the reactivity is (i.e., the better the leaving group).

5 → 6 ← 7

The butterfly mechanism (usual representation) is illustrated in **5** and was described by Bartlett. The representation has been refined by Houk to a trans antiperiplanar arrangement of the O—O bond and reacting alkene, with n–p* stabilization by reacting lone pair in plane **7**.[6] The synchronicity of epoxide C—O bond formation and an overall transition state structure was postulated using *ab initio* calculations and experimental kinetic isotope effects.[7–9]

3.9.4 Variations and Improvements

m-Chloroperbenzoic acid is often used as the epoxidation reagent; it is commercially available, quite stable, and easy to handle. Various other peracids are unstable and have to be prepared immediately prior to use. The separation of the reaction products—that is, the oxidation product and the carboxylic acid—can usually be achieved by extraction with mild aqueous base.

The epoxidation reaction usually takes place under mild conditions and with good to very good yield. Functional groups that are sensitive to oxidation should not be present in the starting material; with carbonyl groups a *Baeyer–Villiger reaction* may take place.

In general, the stereochemistry of olefin is maintained and the reaction is diastereospecific. The reaction rate is insensitive to solvent polarity implying concerted mechanism without intermediacy of ionic intermediates. Generally the less hindered face of olefin is epoxidized. An example of the facial selectivity has been reported by Brown, and the less hindered face of **8** epoxidized to form **9** or **10** depending on the substitution of **8**.[8]

m-CPBA, CH_2Cl_2

8		**9**	**10**
R= H	20 min, 25 °C	99%	1%
R=CH_3	24 h, 25 °C	< 10%	90%

The chemoselectivity of the *Prilezhaev reaction* has been studied and the most electrophilic reagent or most nucleophilic C=C reacts fastest.

RO > R > EWG

The substitution and stereochemistry of the olefin has been studied and the most substituted double bond reacts fastest and the *cis* isomer reacts faster than the *trans* isomer.[11]

> > > ≥ >

Specific examples demonstrating the chemoselectivity are shown below. In the synthesis of **12** and **14**, the most substituted double bonds in substrate **11** and **13** are preferentially epoxidized.[12]

m-CPBA

–10 °C, 1 h

cis : *trans* 1:1

11 **12**

$C_6H_5CO_3H$

$CHCl_3$, 10 min

0 °C

13 **14**

In the total synthesis of reserpine, Woodward and co-workers selectively oxidized the double bond of **15** that is on the exposed convex face of the molecule. Attack from the concave side of **15** is hindered.[14]

HO_2C H O H OH

$C_6H_5CO_3H$

C_6H_6 - dioxane

25 °C, 24 h

HO_2C H O O H OH

15 **16**

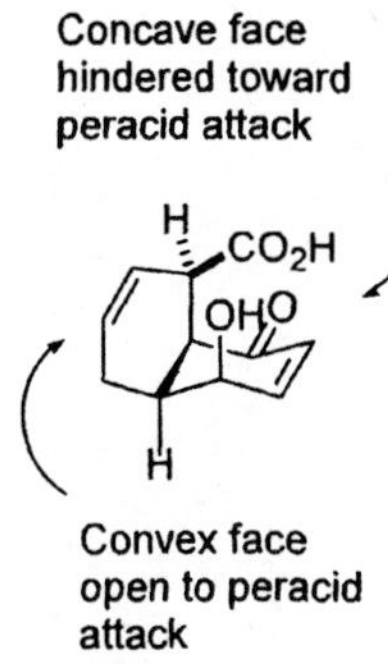
Concave face
hindered toward
peracid attack
H
CO_2H
OH O
H
Convex face
open to peracid
attack

17

Work has been done with salen catalysts to enable enantioselective epoxidations with *m*-CPBA. Highly enantioselective Kochi–Jacobsen–Katsuki epoxidation of unfunctionalized olefins with Mn(III)-(salen)-based chiral catalyst provides an efficient route to optically active epoxides. It has been noted that *m*-CPBA in the presence of *N*-methylmorpholine *N*-oxide at low temperature (e.g., –78 °C) suppresses bond rotation leading to high enantioselectivity even when the substrate is acyclic.[10,15]

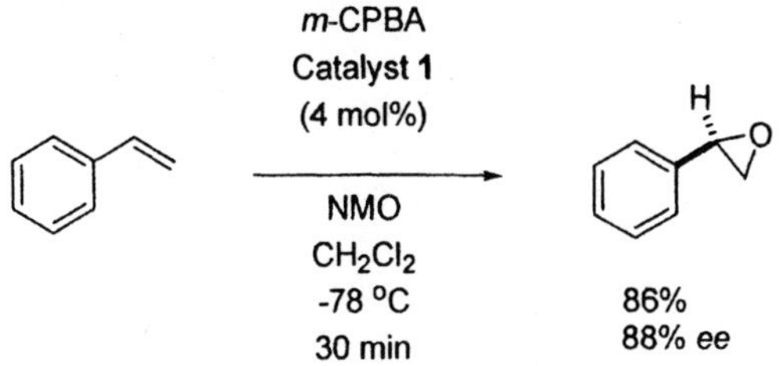
m-CPBA
Catalyst **1**
(4 mol%)
NMO
CH_2Cl_2
-78 °C
30 min
H
O
86%
88% *ee*

18 **19**

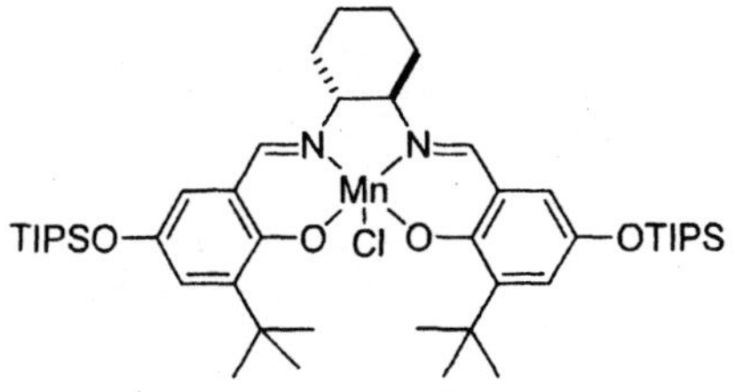
N
N
Mn
TIPSO
O
Cl
O
OTIPS

20

3.9.5 Synthetic Utility

In synthetic studies directed toward the natural product gymnodimine, Kishi and co-workers epoxidized compound **21** using *m*-CPBA to yield **22**.[17]

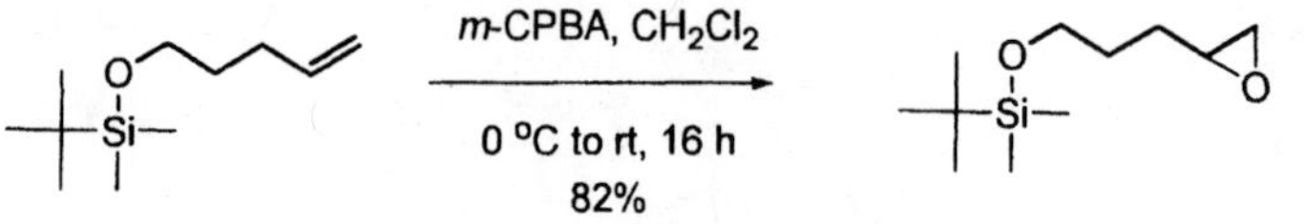

m-CPBA, CH_2Cl_2

0 °C to rt, 16 h

82%

21 **22**

In synthetic studies toward the natural poduct diazonamide A, Nicolaou and co-workers utilized several *Prilezhaev reactions*. Two examples are shown below; compounds **23** and **25** were epoxidized using *m*-CPBA to yield **24**, and **26**, respectively.[17]

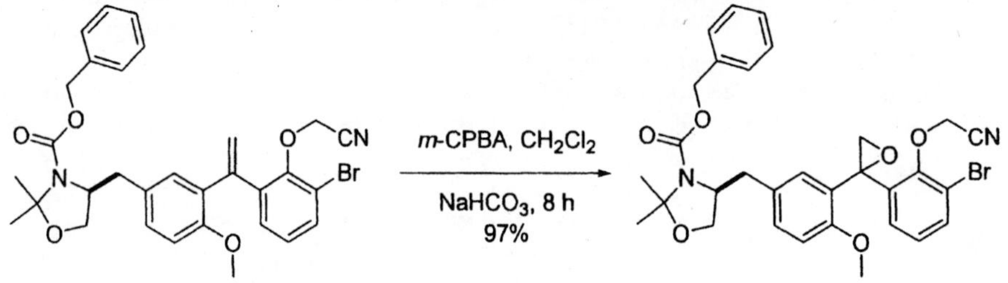

m-CPBA, CH_2Cl_2

$NaHCO_3$, 8 h

97%

23 **24**

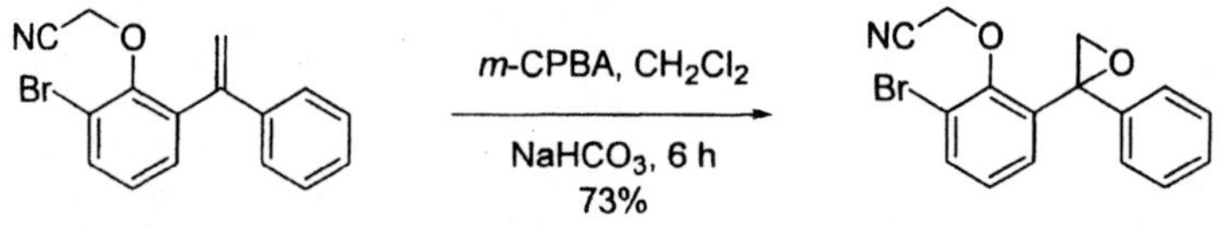

m-CPBA, CH_2Cl_2

$NaHCO_3$, 6 h

73%

25 **26**

In synthetic studies on isoprenoid aziridines, Coates and co-workers utilized several *Prilezhaev reactions* to synthesize epoxides which were reacted with diethylaluminum azide to form azidohydrins.[19] Two examples are shown below; compounds **27** and **29** were epoxidized using *m*-CPBA to yield **28** and **30**, respectively.

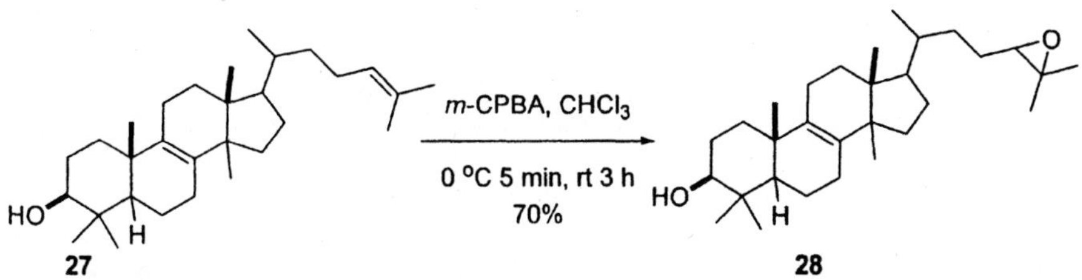

m-CPBA, $CHCl_3$; 0 °C 5 min, rt 3 h; 70% (27 → 28)

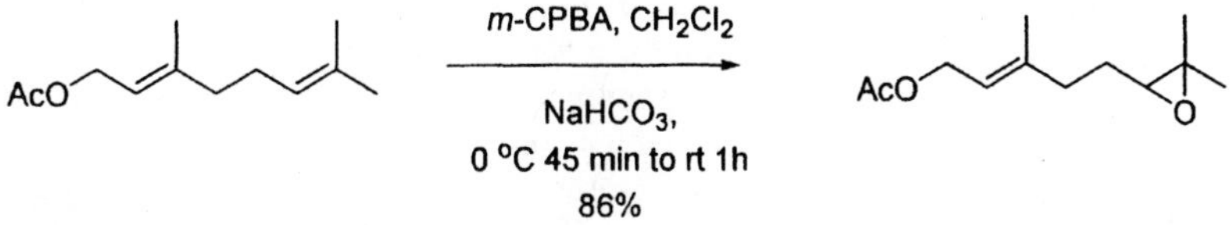

m-CPBA, CH_2Cl_2; $NaHCO_3$, 0 °C 45 min to rt 1h; 86% (29 → 30)

Martin and co-workers utilized the *Prilezhaev reactions* to synthesize epoxides **32** and **33**. Epoxidation with *m*-CPBA gave in 96% yield an inseparable mixture of epoxides **32**/**33** in a 2:3 ratio.[20]

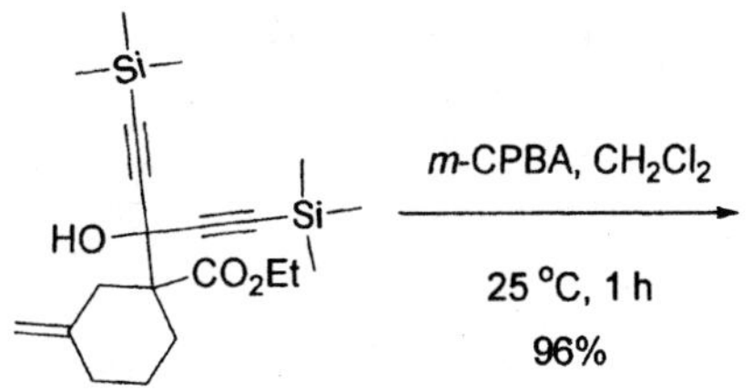

m-CPBA, CH_2Cl_2; 25 °C, 1 h; 96% (31 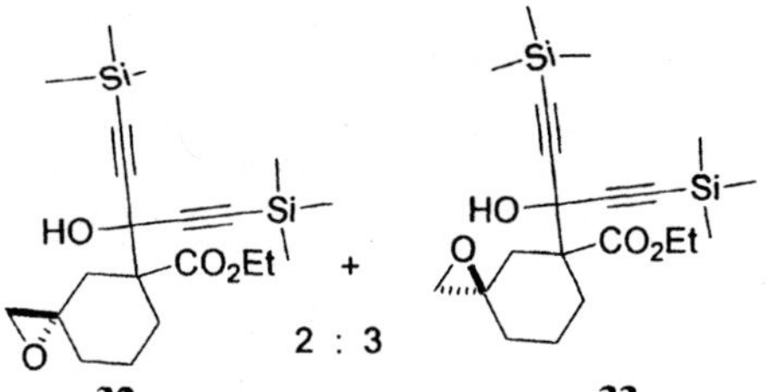 → 32 + 33, 2 : 3)

3.9.6 Experimental

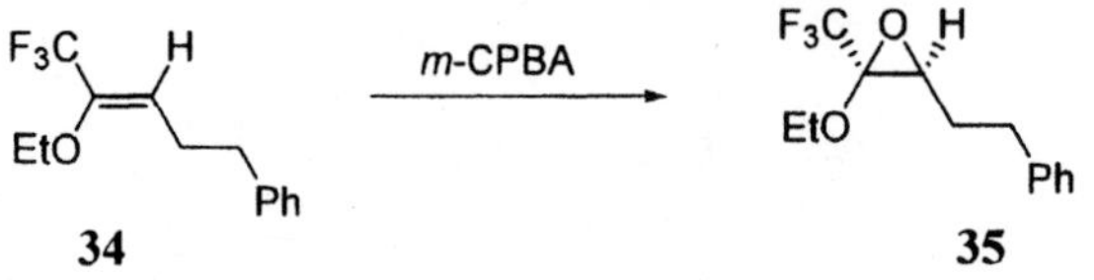

34 → (m-CPBA) → 35

1,1,1-Trifluoro-2-ethoxy-2,3-epoxy-5-phenylpentane (35)[9]

A 250-mL, round-bottomed flask, equipped for magnetic stirring and with a condenser fitted with a calcium chloride drying tube, is charged with 9.76 g (40 mmol) of 1,1,1-trifluoro-2-ethoxy-5-phenyl-2-pentene **34**, and a solution of 14.79 g (60 mmol) of 70% *meta*-chloroperoxybenzoic acid in 170 mL of dichloromethane. The resulting mixture is heated under reflux with stirring for 20 h, cooled, and concentrated to 50–60 mL under reduced pressure. The residual liquid is diluted with 200 mL of pentane and the supernatant liquid from the resulting suspension is passed through a short silica gel column. The residual solids are washed twice with 50 mL of a 10:1 mixture (v/v) of pentane-diethyl ether, and the wash solutions are passed through the same column. The column is then eluted with 100 mL of a 5:1 (v/v) mixture of pentane-diethyl ether. The combined eluants are concentrated under reduced pressure and the residual liquid is purified by bulb-to-bulb vacuum distillation (oven temperature 120–130 °C at 10 mm) to provide 9.36–9.88 g (90–95%) of pure 1,1,1-trifluoro-2-ethoxy-2,3-epoxy-5-phenylpentane (**35**) as a clear, colorless liquid, bp 90 °C (10 mm).[21]

3.9.7 References

1. Prilezhaev, N; *Ber. Dtsch. Chem. Ges.* **1910**, *42*, 4811–4815.
2. [R] Plesnicar, B. *The Chemistry of Peroxides,* Patai, S. Wiley, New York, **1983**, 521–584.
3. [R] Berti, C. *Top. Stereochem.* **1973**, *7*, 93–251.
4. [R] Plesnicar, B. *Oxidation in Organic Chemistry, Vol. C;* Trahanovsky, W. S. (ed.), Academic Press, New York, **1978**, 211–252.
5. Akhrem, A. A.; Prilezhaev, E. N. *Zh. Ob. Khim.* **1973**, *43*, 697–698.
6. Woods, K. W.; Beak, P. *J. Am. Chem. Soc.* **1991**, *113*, 6281–6283.
7. [R] Dryuk, V. G. *Russ. Chem. Rev.* **1985**, *54*, 986–1005.
8. Brown, H. C.; Kawakami, J. H.; Ikegami, S. *J. Am. Chem Soc.* **1970** *92*, 6914–6917.
9. [R] Adam, W.; Curci, R.; Edwards, J. O. *Acc. Chem. Res.* **1989**, *22*, 205–211.
10. Rebek, J., Jr.; Marshall, L.; McManis, J.; Wolak, R. *J. Org. Chem.* **1986, 51,** 1649–1653.
11. [R] Katsuki, T. *Curr. Org. Chem.* **2001**, *5*, 663–678.
12. Singleton, D. A.; Merrigan, S. R.; Liu, J.; Houk, K. N. *J. Am. Chem. Soc.* **1997**, 119, 3385–3386.
13. Huckel, W.; Schlee, H. *Chem. Ber.* **1955**, 88, 346–353.
14. Woodward, R. B.; Bader, F. E.; Bickel, H.; Frey, A. J.; Kierstead, R. W. *Tetrahedron* **1958**, 2, 1–57.
15. Knoell, W.; Tamm, C. *Helv. Chim. Acta* **1975**, 58, 1162–1171.
16. Palucki, M.; Pospisil, P. J.; Zhang, W.; Jacobsen, E.N. *J. Am. Chem. Soc.* **1994**, *116*, 9333–9334
17. Johannes, J. W.; Wenglowsky, S.; Kishi, Y. *Org. Lett.*, **2005** 7, 3997–4000.
18. Nicolau, K. C.; Snyder, S. A.; Huang, X.; Simonsen, K. B.; Koumbis, A. E.; Bigot, A. *J. Am. Chem. Soc* **2004**, *126*, 10162–10173.
19. Davis, C. E.; Bailey,J. L.; Lockner, J. W.; Coates, R. M. *J. Org. Chem*, **2003**, 68, 75–82.

20. Perez-Hernandez, N.; Febles, M.; Perez, C.; Perez, R.; Rodrıguez, M. L.; Foces-Foces, C.; Martın, J. D. *J. Org. Chem.* **2006**, 71, 79–85.
21. Begue, J. P.; Bonnet-Delpon, D.; Sdassi, H. *Tetrahedron Lett.* **1992**, *33*, 1879–1882.

Timothy J. Hagen

3.10 Rubottom Oxidation

3.10.1 *Description*

The Rubottom oxidation[1] is the peracid-mediated oxidation of trimethylsilyl enol ethers to afford α-silyloxy- or α-hydroxy aldehydes or ketones.[2,3] Use of an aqueous workup generally affords the hydroxy compounds, whereas nonaqueous workups provide the silyloxy derivatives. For example, the enolsilane **1** derived from cycloheptanone was converted to **2** in 77% yield by treatment with *m*-CPBA followed by workup with 10% aqueous sodium hydroxide. Omission of the aqueous workup afforded **3** in 85% isolated yield.[1a]

OSiMe3 **1** — *m*-CPBA / Workup (anhydrous or aqueous) → O, OR

2 R = H: 77%
3 R = $SiMe_3$: 85%

3.10.2 *Historical Perspective*

The first examples of enolsilane oxidations were described independently by Brook,[1b] Hassner,[1c] and Rubottom[1a] in late 1974–early 1975. Brook reported that oxidation of enolsilanes derived from cyclic and acyclic ketones with *m*-CPBA affords α-silyloxy ketones in good yields; subsequent hydrolysis of these products provided the corresponding alcohols. Rubottom noted that either α-silyloxy ketones or α-hydroxy ketones could be obtained depending on the nature of the workup (nonaqueous vs. aqueous). Hassner observed that enolsilanes derived from both aldehydes and ketones are suitable substrates for these transformations. Subsequent studies by Rubottom and others led to significant expansions of this methodology along with a more complete understanding of the mechanism of these reactions.[2,3]

3.10.3Me *chanism*

R, R, R, OSiMe3 — *m*-CPBA → R, R, R, O, OSiMe3, H+ **4** → R, R, OH, R, +, O—SiMe3 **5**

— $-H^+$ → R, R, OSiMe3, R, O **6** — H_3O^+ → R, R, OH, R, O **7**

The mechanism initially proposed for the Rubottom oxidation involved epoxidation of the enolsilane to afford intermediate silyloxyoxirane **4**. It was suggested that this intermediate undergoes acid-mediated cleavage to afford stabilized carbocation **5**, which is transformed to the α-silyloxy ketone **6** *via* 1,4-silicon migration. Hydrolysis of **6** by aqueous acid in a subsequent step generates the α-hydroxy ketone **7**.[1b,15] Attempts to provide support for this mechanism *via* isolation of intermediate silyloxyoxiranes derived from simple ketones proved difficult due to the lability of these compounds. However, Brook demonstrated that the related heterocyclic silyloxyoxirane **8** was isolable and was transformed to ketone **9** upon treatment with *p*-TsOH.[1b]

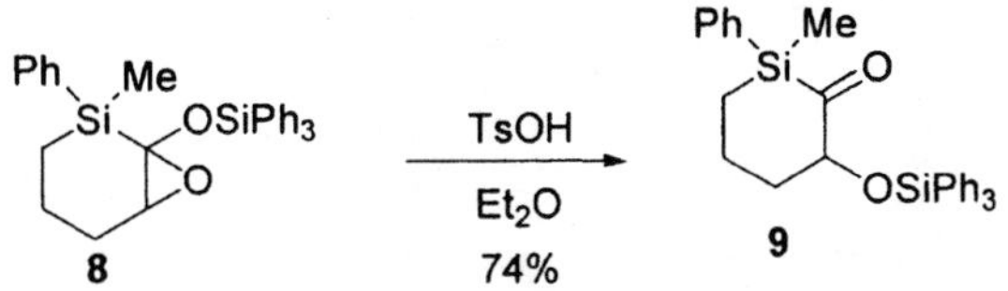

Further support for the mechanism described above was obtained in subsequent studies by several groups. Direct evidence for the initial epoxidation event in the Rubottom oxidation of an acyclic enolsilane was first obtained by Weinreb, who described the isolation of silyloxyoxirane **10** and demonstrated its conversion to α-silyloxy ketone **11** upon treatment with PPTS.[4] The isolation of a macrocyclic bis(silyloxyoxirane) has also been reported.[5]

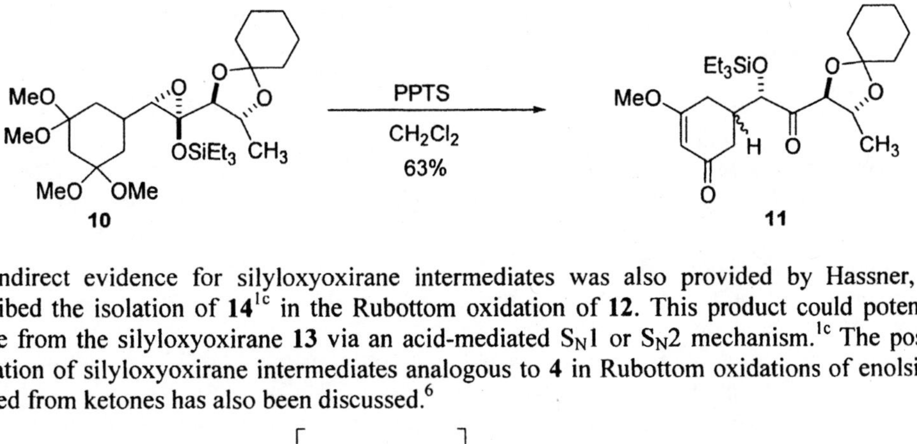

Indirect evidence for silyloxyoxirane intermediates was also provided by Hassner, who described the isolation of **14**[1c] in the Rubottom oxidation of **12**. This product could potentially derive from the silyloxyoxirane **13** via an acid-mediated S_N1 or S_N2 mechanism.[1c] The possible formation of silyloxyoxirane intermediates analogous to **4** in Rubottom oxidations of enolsilanes derived from ketones has also been discussed.[6]

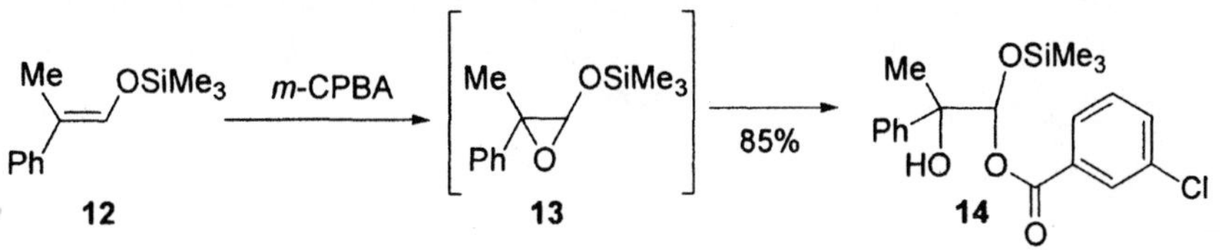

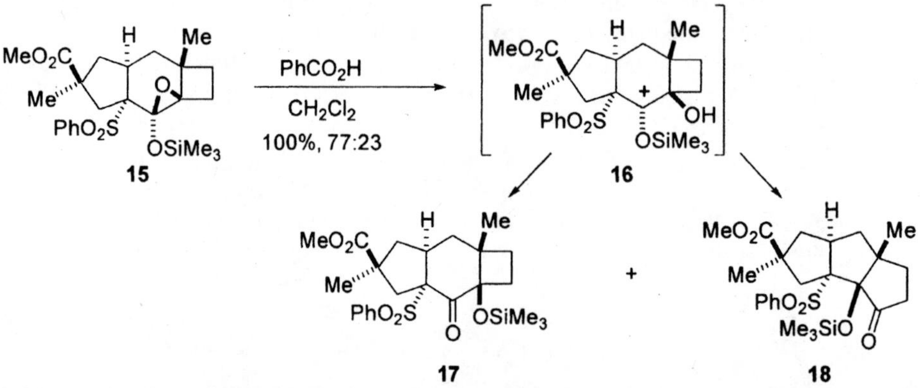

Evidence for the intermediacy of stabilized carbocations analogous to **5** was provided by Paquette, who examined the isolation and acid-mediated rearrangement of silyloxyoxirane **15**. Treatment of this species with benzoic acid led to the formation of two isomeric α-silyloxy ketones (**17** and **18**).[7] The formation of ketone **17** likely derives from a pinacol rearrangement of carbocation **16**.

3.10.4 Stereochemistry

The stereochemical outcome of the Rubottom oxidation reaction is generally believed to be controlled by the stereoselectivity of the epoxidation step, with the subsequent rearrangement to the α-hydroxy ketone occurring with retention of configuration at the α-stereocenter.[2,8] This issue was addressed further in an elegant study recently disclosed by Danishefsky.[9]

The Rubottom oxidation plays a key role in Danishefsky's synthesis of guanacastepine (see Section 3.10.6). During studies on the Rubottom oxidation of potential intermediate **19** a surprising stereochemical outcome was observed, as the hydroxyl group was installed on what appears to be the more hindered β-face of the molecule to generate product **21**. As shown below, one possible explanation for the stereochemical outcome of this transformation would involve epoxidation of the α-face followed by epimerization of the newly formed stereocenter via a deprotonation/reprotonation sequence (Path B).[10] However, when deuterated enolsilane **20** was subjected to the Rubottom oxidation conditions (DMDO followed by Ac_2O) product **22** was formed with complete retention of the deuterium label.[9] The result of this experiment suggests that the reaction proceeds as shown in Path A such that epoxidation occurs on the β-face and rearrangement proceeds with retention of configuration. In contrast, if the reaction were to occur via epoxidation of the α-face to generate **23** (Path B) with subsequent conversion to **22** through ring-opening (**23**–**24**), proton transfer (**24**–**25**), and deprotonation/reprotonation (**25**–**22**), generation of mixtures of **21** and **22** through partial (or complete) loss of deuterium would be expected.

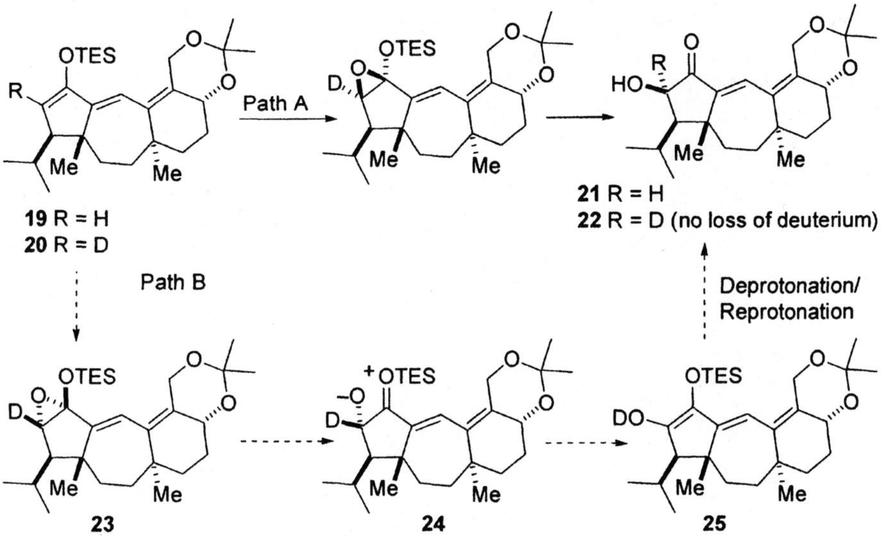

19 R = H
20 R = D
21 R = H
22 R = D (no loss of deuterium)
23
24
25

3.10.5 Variations and Improvements

Many variations of the Rubottom oxidation employ oxidants other than *m*-CPBA in order to execute the transformation under mild conditions or to allow for enantioselective synthesis. Use of dimethyl dioxirane (DMDO) for the oxidation of enolsilanes has become a popular alternative to traditional conditions for Rubottom oxidations. This mild oxidant has been used to facilitate the isolation of 2-silyloxyoxiranes, which are stable under the essentially neutral reaction conditions.[11] For example, treatment of **26** with DMDO at –40 °C afforded **27** in 99% yield.[11b] These compounds can subsequently be converted to 2-hydroxyketones, as described above, or can be used in other transformations.[12] Chiral dioxiranes generated *in situ* from chiral ketones and oxone have also been employed in enantioselective Rubottom oxidations developed independently by Shi[13a] and Adam.[13b] As shown above, enolsilane **28** was transformed to α-hydroxyketone **29** in 80% yield and 90% *ee*.[13a]

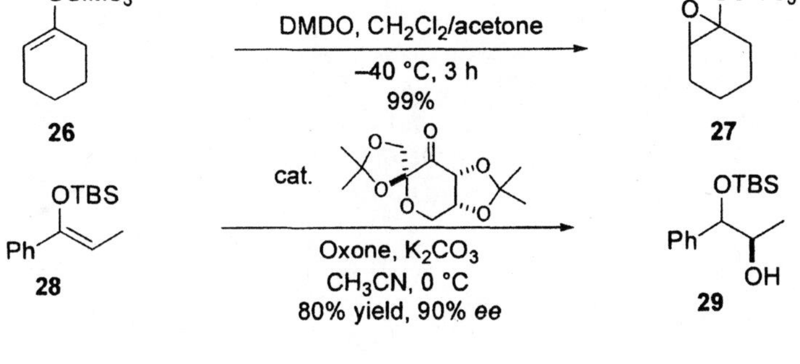

26 27 28 29

In addition to dioxiranes, a wide variety of other oxidants have also been employed in Rubottom oxidations including *t*-butyl hydroperoxide, oxone, and benzoyl peroxide.[2] However, use of these other oxidants is much less common than use of *m*-CPBA or DMDO. Chiral Mn(salen)complexes[14a] and the Sharpless asymmetric dihydroxylation system[14b] have also been used in asymmetric versions of the Rubottom oxidation.

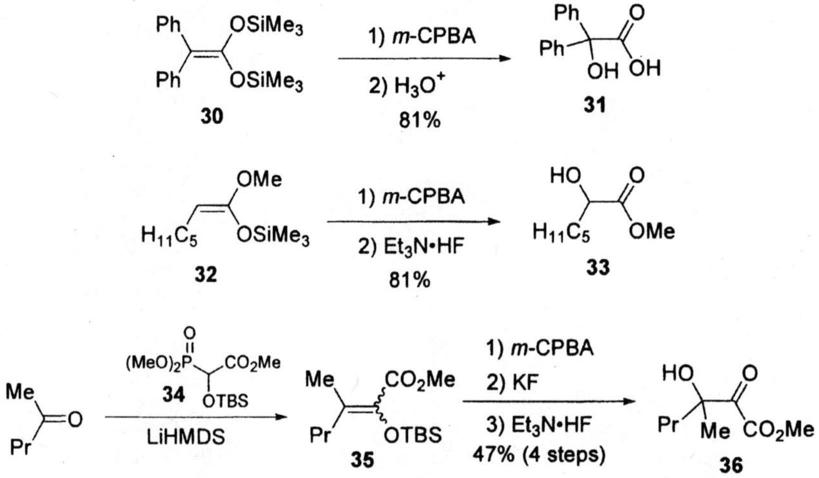

Rubottom oxidation reactions have been conducted on enolsilanes derived from a number of different carbonyl derivatives including carboxylic acids and esters.[15] For example, the Rubottom oxidation of bis(trimethylsilyl)ketene acetal **30** provided α-hydroxy carboxylic acid **31** in 81% yield. Use of alkyl trimethylsilyl ketene acetal substrates generates α-hydroxy esters, as seen in the conversion of **32** to **33**.[16] The synthesis of β-hydroxy-α-ketoesters (e.g., **36**) has been accomplished via Rubottom oxidation of enolsilanes such as **35** that are prepared via Horner–Wadsworth–Emmons reactions of aldehydes and ketones with 2-silyloxy phosphonoacetate reagent **34**.[17] The α-hydroxylation of enolsilanes derived from β-dicarbonyl compounds has also been described, although in some cases direct oxidation of the β-dicarbonyl compound is feasible without enolsilane formation.[18]

The oxidation of dienyl enolsilanes to the corresponding α-hydroxy ketones has also been developed and explored by Rubottom.[19] For example, diene **37** was cleanly converted to ketone **38** under standard conditions. As expected, high selectivity for oxidation of the more electron-rich double bond was observed. As illustrated by the conversion of **37** to **39**, acylation of the crude alcohols was achieved in high yields after removal of *m*-chlorobenzoic acid and the hexane solvent employed for the oxidation.

OTMS OH
38
1) m-CPBA
2) $Et_3N{\bullet}HF$
82%
OTMS
37
1) m-CPBA
2) $Et_3N{\bullet}HF$, Et_3N, Ac_2O
98%
OTMS OAc
39

A double hydroxylation of enolsilanes under modified Rubottom oxidation conditions has been developed by Nakamura and Kuwajima.[20] As shown below, treatment of enolsilane **40** with m-CPBA in the presence of excess $KHCO_3$ generates doubly oxidized product **41** in 72% yield. The mechanism of these transformations is believed to involve elimination/epoxide opening of the intermediate silyloxyoxirane **42** followed by a second oxidation of the resulting enolsilane **43**.

OTIPS
40
2.5 equiv m-CPBA
10 equiv $KHCO_3$
CH_2Cl_2, 0°C
72%
HO O OTIPS
41
m-CPBA
O OTIPS
42
$KHCO_3$
HO OTIPS
43
m-CPBA
HO OTIPS O

3.10.6 Synthetic Utility

The Rubottom oxidation has found widespread application in organic synthesis. A few recent examples of the use of this methodology for the construction of complex molecules are described below. As noted above, the stereoselectivity in these reactions is usually controlled by steric effects, which dictate the face-selectivity of the epoxidation step. The chemoselectivity is generally controlled by electronic effects, as the electrophilic oxidants react more rapidly with the electron-rich enol ether than with other double bonds in the substrate.

The regio- and stereoselective installation of a hydroxyl group adjacent to a ketone was accomplished through use of the Rubottom oxidation in Paquette's synthesis of (+)-epoxydictymene.[21] As shown below, the tricyclic ketone **44** was converted to hydroxy ketone **45** in good yield (72%) over three steps. A stereoselective and chemoselective Rubottom oxidation was also used by Taber in the synthesis of the antitumor agent (–)-fumagillin.[22] Treatment of enolsilane **46** with m-CPBA followed by addition of TBAF/NH_4Cl afforded the desired alcohol **48** in 55% yield. The buffered fluoride workup allowed for cleavage of the secondary triethylsilyl ether **47** generated in the oxidation without concomitant deprotection of the remote primary t-butyldimethylsilyl ether functionality.

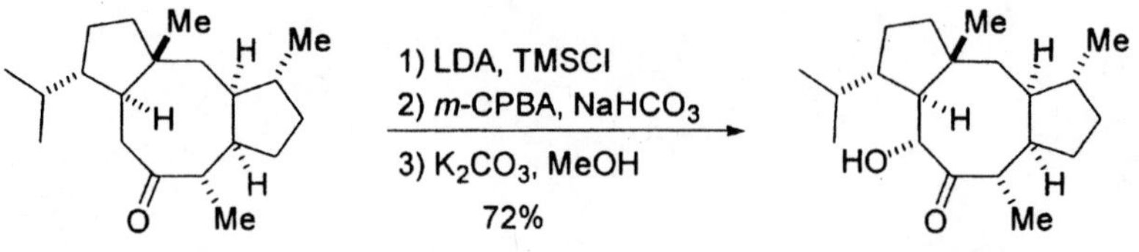

44 45

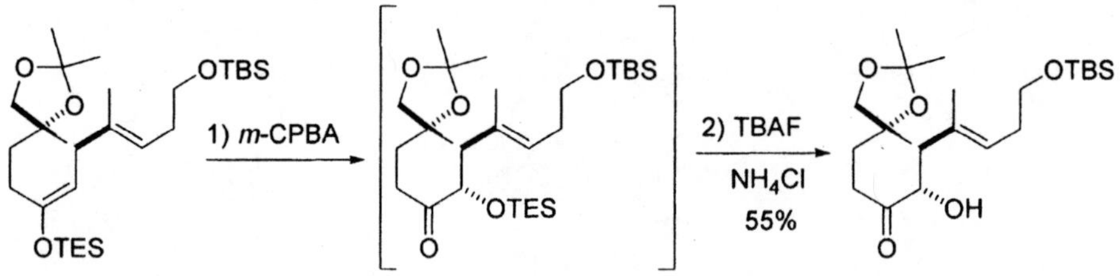

46 47 48

The Rubottom oxidation of enolsilane **49** was achieved in the presence of neighboring diene and allylic ether functionality to provide **51**, an intermediate in Crimmins's synthesis of (+)-milbemycin D.[23] The primary silyl ether product **50** was sufficiently labile that deprotection occurred upon slow chromatography on silica gel to yield **51**. Similarly, Danishefsky noted that electron-poor diene functionality was well tolerated in the stereoselective Rubottom oxidation of ketone **52**.[24] Enolsilane formation followed by DMDO expoxidation, rearrangement, and acylation afforded keto-alcohol **53** in 82–90% yield. This compound was subsequently converted to the natural product guanacastepene A.

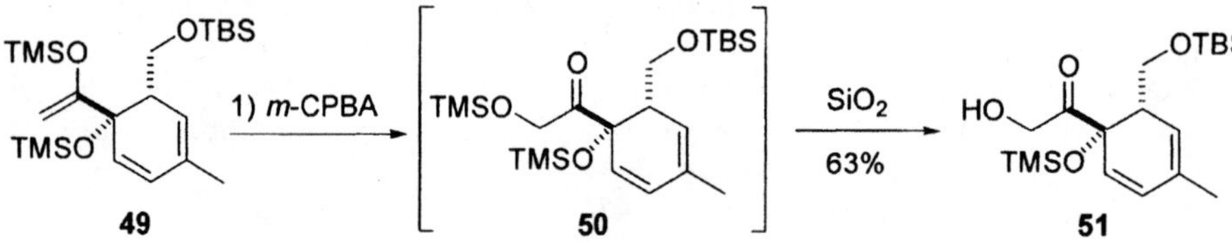

49 50 51

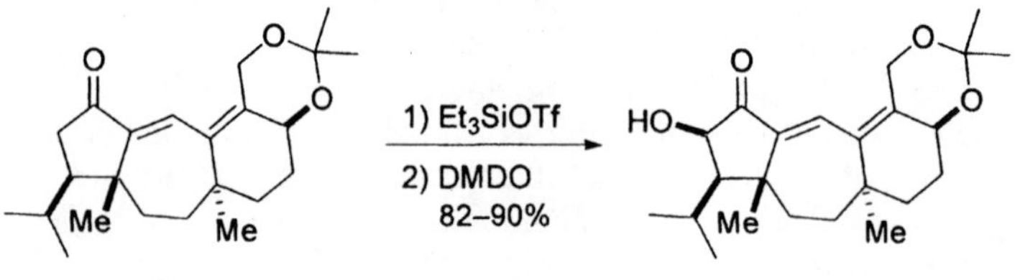

52 53

3.10.7 *Experimental*

OTMS 53 → 1) *m*-CPBA, 2) $Et_3N{\cdot}HF$, 66% → OTMS, OH 54

6-Hydroxy-3,5,5-timethyl-2-cyclohexen-1-one (54)[3]

An oven-dried flask was cooled under a stream of nitrogen and charged with dry hexanes (300 mL) and 4,6,6-trimethyl-2-trimethylsiloxycyclohexa-1,3-diene (**53**) (10.0 g, 47.5 mmol). The solution was cooled to –15 °C in an ice/CH_3OH bath and treated with a slurry of 85% *m*-CPBA (10.6 g, 52.3 mmol) in hexanes (50 mL). The resulting mixture was stirred for 15 min at –15 °C, then was warmed to 30 °C and stirred for an additional 2 h. The mixture was then filtered through a fritted funnel and concentrated *in vacuo*. The resulting oil was dissolved in anhydrous dichloromethane (150 mL) and treated with triethylammonium fluoride (11.5 g, 95.0 mmol). The reaction mixture was stirred at rt for 2 h, then was transferred to a separatory funnel and sequentially washed with saturated aqueous $NaHCO_3$ (2 × 100 mL), aqueous HCl (100 mL, 1.5 M), and saturated aqueous $NaHCO_3$ (2 × 50 mL). The organic phase was dried over anhydrous magnesium sulfate, filtered, and concentrated *in vacuo*. The crude product was purified by distillation (bp 73–75 °C, 1.3 Torr) followed by trituration with pentane. The resulting crystalline product was dried under a stream of nitrogen to afford 4.8 g (66%) of the title compound, mp 44.5–45 °C.

3.10.8 *References*

1 (a) Rubottom, G. M.; Vazquez, M. A.; Pelegrina, D. R. *Tetrahedron Lett.* **1974**, 4319–4322; (b) Brook, A. G.; Macrae, D. M. *J. Organomet. Chem.* **1974**,*7 7*, C19–C21. (c) Hassner, A.; Reuss, R. H.; Pinnick, H. W. *J. Org. Chem.* **1975**, *40*, 3427–3429.
2 [R] Chen, B. -C.; Zhou, P.; Davis, F. A.; Ciganek, E. *Org. React.* **2003**,*6 2*, 1–356.
3 [R] Rubottom, G. M.; Gruber, J. M.; Juve, H. D., Jr.; Charleson, D. A. *Org. Synth.* **1986**, *64*, 118–126.
4 Dodd, J. H.; Starrett, J. E., Jr.; Weinreb, S. M. *J. Am. Chem. Soc.* **1984**,*1 06*, 1811–1812.
5 Gleiter, R.; Staib, M.; Ackermann, U. *Liebigs Ann.* **1995**, 1655–1661.
6 Rubottom, G. M.; Gruber, J. M.; Boeckman, R. K., Jr.; Ramaiah, M.; Medwid, J. B. *Tetrahedron Lett.* **1978**, 4603–4606.
7 Paquette, L. A.; Lin, H. -S.; Gallucci, J. C. *Tetrahedron Lett.* **1987**,*2 8*, 1363–1366.
8 Jauch, J. *Tetrahedron* **1994**,*5 0*, 12903–12912.
9 Mandal, M.; Danishefsky, S. J. *Tetrahedron Lett.* **2004**,*4 5*, 3831–3833.
10 Magnus, P.; Ollivier, C. *Tetrahedron Lett.* **2002**, *43*, 9605–9609.
11 (a) Chenault, H. K.; Danishefsky, S. J. *J. Org. Chem.* **1989**, *54*, 4249–4250; (b) Adam, W.; Hadjiarapoglou, L.; Wang, X. *Tetrahedron Lett.* **1989**,*3 0*, 6497–6500.
12 Schaumann, E.; Tries, F. *Synthesis* **2002**, 191–194 and references cited therein.
13 (a) Zhu, Y. Tu, Y.; Yu, H.; Shi, Y. *Tetrahedron Lett.* **1998**, *39*, 7819–7822; (b) Adam, W.; Fell, R. T.; Saha-Moller, C. R.; Zhao, C. -G. *Tetrahedron: Asymmetry* **1998**,*9* , 397–401.
14 (a) Adam, W.; Fell, R. T.; Stegmann, V. R.; Saha-Moeller, C. R. *J. Am. Chem. Soc.* **1998**, *120*, 708–714. (b) Hashiyama, T.; Morikawa, K.; Sharpless, K. B. *J. Org. Chem.* **1992**, *57*, 5067–5068.
15 Rubottom, G. M.; Marrero, R. *J. Org. Chem.* **1975**, *40*, 3783–3784.

16 Rubottom, G. M.; Marrero, R. *Synth. Commun.* **1981**, *11*, 505–511.
17 Pujol, B.; Sabatier, R.; Driguez, P. -A.; Doutheau, A. *Tetrahedron Lett.* **1992**, *33*, 1447–1450.
18 [R] Christoffers, J.; Baro, A.; Werner, T. *Adv. Synth. Catal.* **2004**, *346*, 143–151.
19 Rubottom, G. M.; Gruber, J. M. *J. Org. Chem.* **1978**, *43*, 1599–1602.
20 Horiguchi, Y.; Nakamura, E.; Kuwajima, I. *Tetrahedron Lett.* **1989**, *30*, 3323–3326.
21 Paquette, L. A.; Sun, L. -Q.; Friedrich, D.; Savage, P. B. *Tetrahedron Lett.* **1997**, *38*, 195–198.
22 Taber, D. F.; Christos, T. E.; Rheingold, A. L.; Guzei, I. A. *J. Am. Chem. Soc.* **1999**, *121*, 5589–5590.
23 Crimmins, M. T.; Al-awar, R. S.; Vallin, I. M.; Hollis, W. G., Jr.; O'Mahoney, R.; Lever, J. G.; Bankaitis-Davis, D. M. *J. Am. Chem. Soc.* **1996**, *118*, 7513–7528.
24 Lin, S.; Dudley, G. B.; Tan, D. S.; Danishefsky, S. J. *Angew. Chem. Int. Ed.* **2002**, *41*, 2188–2191.

John P. Wolfe

3.11 Swern Oxidation

3.11.1 Description

The Swern oxidation is the oxidation of alcohols to the corresponding carbonyl compounds using oxalyl chloride and dimethylsulfoxide and a base, usually triethylamine, at low temperatures.

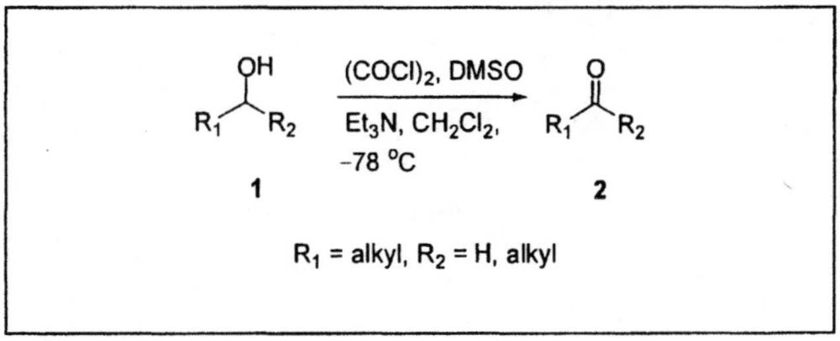

$(COCl)_2$, DMSO; Et_3N, CH_2Cl_2, -78 °C

1 → 2

R_1 = alkyl, R_2 = H, alkyl

3.11.2 Historical Perspective

Oxidations using dimethyl sulfoxide activated by various reagents began with discoveries by Kornblum and co-workers that primary tosylates and certain α-bromo ketones could be converted into aldehydes and glycoxals, respectively, by treatment with dimethyl sulfoxide as the oxidising agent.[1] This was followed by the discovery several years later by Pfitzner and Moffatt that alcohols could be oxidised to carbonyl compounds with dimethyl sulfoxide, dicyclohexylcarbodiimide (DCC) and phosphoric acid at room temperature.[2] Eventually the method developed by Swern and co-workers, involving activation of dimethyl sulfoxide with oxalyl chloride, came to be the most synthetically useful and widely applied of these mild oxidation procedures.[3–6]

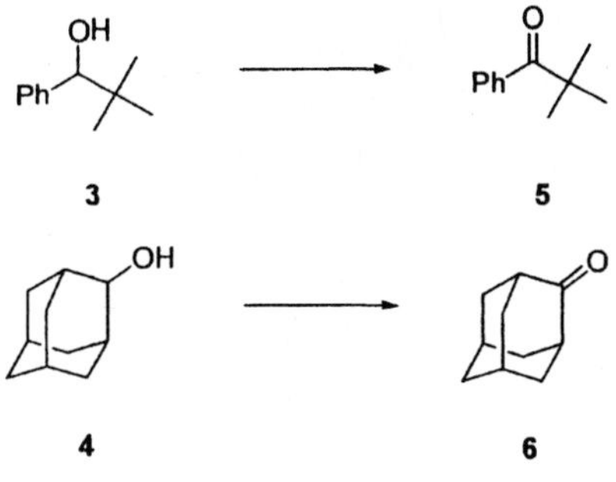

3 → 5

4 → 6

Initially, Swern *et al.* reported the oxidation of sterically hindered alcohols to carbonyls with a dimethyl sulfoxide-trifluoroacetic anhydride complex.[7] These reactants included primary alcohols such as 2,2-dimethyl-1-phenyl-propanol **3** and secondary alcohols, for example, 2-adamantanol **4**.

A few years later the Swern laboratory then developed an "activator" which they claimed to be the most successful in activating dimethyl sulfoxide toward oxidation, namely, oxalyl chloride. Since oxalyl chloride reacted violently and exothermically with dimethyl sulfoxide, successful activation required the use of low temperatures to form the initial intermediate.[6] Swern *et al.* reported the oxidation of long chain primary alcohols to aldehydes which was previously unsuccessful by first converting to the sulfonate ester (either mesylate or tosylate) and then employing the dimethyl sulfoxide–acetic anhydride procedure. They found that long-chain saturated, unsaturated, acetylenic and steroidal alcohols could all be oxidised with dimethyl sulfoxide–oxalyl chloride in high yields under mild conditions.

HO — 95% →

Cholesterol

3.11.3 Mechanism

The mechanism of the Swern oxidation has been studied in depth and the formation of an initial adduct **7** from the reaction between dimethyl sulfoxide and oxalyl chloride which then collapses to give a dimethylchlorosulfonium species **8** is clearly indicated by mechanistic studies.[3,8] Reaction of **8** with an alcohol then produces the alkyoxysulfonium ion **9** which upon treatment with an amine base gives the ylide **10.** Subsequent proton extraction gives the carbonyl product **2** with the release of dimethyl sulfide.

Formation of the dimethylchlorosulfonium species **8** and its reaction with alcohols and also the conversion of the ylide **10** to the product are both rapid at –78 °C. Additionally, formation of (methylthio)methyl ethers, a common side reaction in these procedures, is minimised. The reactive reagent **8** can also be formed directly by the reaction of dimethyl sulfide with chlorine[9a–d]; however, the high yields and convenient procedure using the Swern reagents makes it particularly advantageous.[5]

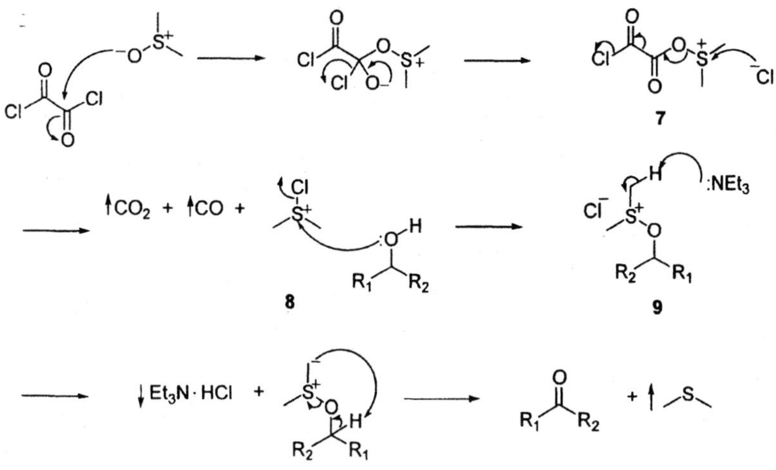

7

8

9

10

2

3.11.4 Variations

Modifications to the Swern procedure often include a change of base from the requisite triethylamine. In their synthesis of naphthoic acid, Eustache and co-workers attempted to use Swern conditions to oxidise the 1,5-diol **11** to the corresponding dicarbonyl **12.**[10] Although evidence of the formation of the keto aldehyde intermediate was verified through ^{1}H NMR, decomposition of the compound made further manipulations difficult. Eventually, a modified one-pot, two-step procedure was developed, involving oxidation followed by base-induced cyclisation to give the desired product.

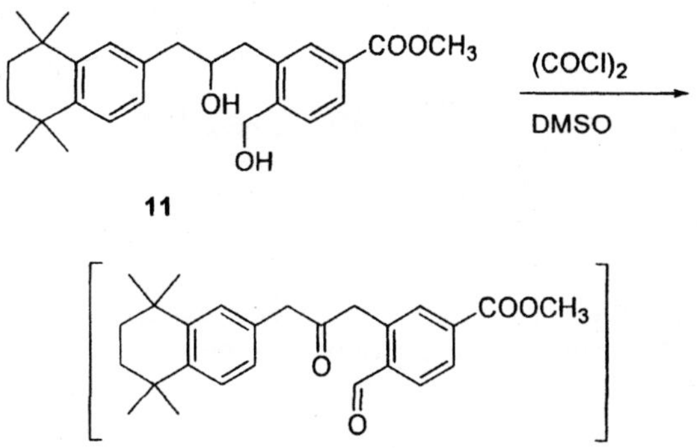

11

DBU

12

Similarly, diisopropylamine is sometimes substituted for Et_3N owing to its more hindered nature. Occasionally however, neither of these bases will provide the optimum yield for a reaction, as shown by Chrisman and Singaram.[11] A range of bases were compared in the oxidation of β-amino alcohols, such as **13**. The results are tabulated.

i. $(COCl)_2$ ii. DMSO CH_2Cl_2, -78 °C; tertiary amine -78 °C to rt 45 min to 24 h

13 → 14

Tertiary amine base	% recovered **13**	% yield **14**
N,*N*-Diisopropylethylamine	33	65
Triethylamine	24	68
N-Ethylpiperidine	5	93
DABCO	14	86
Pyridine	98	1

Thus the conditions for the Swern reaction are amenable to modifications to give optimum yields. The authors conclude that the optimum yields of the α-amino carbonyls were obtained when the steric demands of the tertiary amine were between Hunig's base and DABCO.

An unavoidable by-product of the Swern reaction is the volatile dimethylsulphide which, on account of its unpleasant smell, is a reagent regulated by offensive odour control laws. This makes large scale chemistry problematic, especially in industry. To overcome this, several methods exist to perform the Swern oxidation under odourless conditions. For example, Node *et al.* outline a protocol for the Swern oxidation which uses dodecyl methyl sulfoxide in place of methyl sulfoxide,[12] while Crich and co-workers have developed a fluorous Swern oxidation reaction that uses tridecafluorooctylmethyl sulfoxide **17**.[13a,b] This reagent can be recovered *via* a continuous fluorous extraction procedure and recycled by reoxidation with hydrogen peroxide. Additionally, the fluorous DMSO is crystalline, odourless and soluble in CH_2Cl_2 to –45 °C.

15 → ($NaBH_4$, Me_2S_2) → 16, fluorous DMS → (H_2O_2, MeOH) → 17, fluorous DMSO

Vederas *et al.* report one of the first examples of a recyclable, soluble polymer-supported sulphoxide which can be used as an alternative to DMSO without loss of efficiency.[14a,b]

18 ⇄ 19 ($NaIO_4$, 99%; $(COCl)_2$, Et_3N)

The optimum molecular weight of the poly(ethylene glycol) (PEG) polymer was determined to be 2000. At this weight the polymer was solid at room temperature and therefore can easily be precipitated from the reaction mixture. Alcohols such as benzoin and cholesterol were oxidised in comparable yields to those reported in the literature.

Although vigorously dry conditions are not necessary to instigate oxidation, alternate products have arisen through the use of wet dimethyl sulphoxide. For example, during Kende's synthesis of a taxane triene, Swern oxidation converted diol **20** to dialdehyde **21** in 85% yield.[15] When the dimethylsulphoxide used in the reaction was moist, however, up to 40% of a crystalline by-product was formed. X-ray analysis proved the crystalline product to be chloroaldehyde **22**. The preceding and subsequent stereochemical assignments could then be unambiguously confirmed.

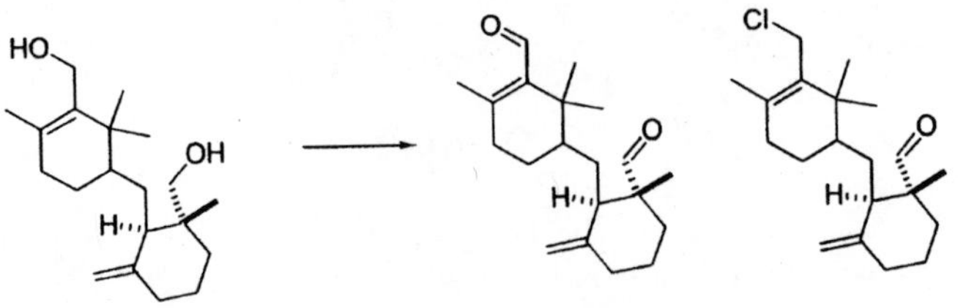

20 21 22

A more recent modification was reported by Giacomelli and De Luca wherein oxalyl chloride has been replaced by 2,4,6-trichloro[1,3,5]triazine (cyanuric chloride, TCT) as the activator.[16] The authors cite the dangerous toxicity and moisture sensitivity of oxalyl chloride as a motive for their work, the exclusion of which also means that the reaction can be carried out between –30 and 0 °C. Alcohols and *N*-protected α-amino alcohols were converted to the corresponding ketones and aldehydes.

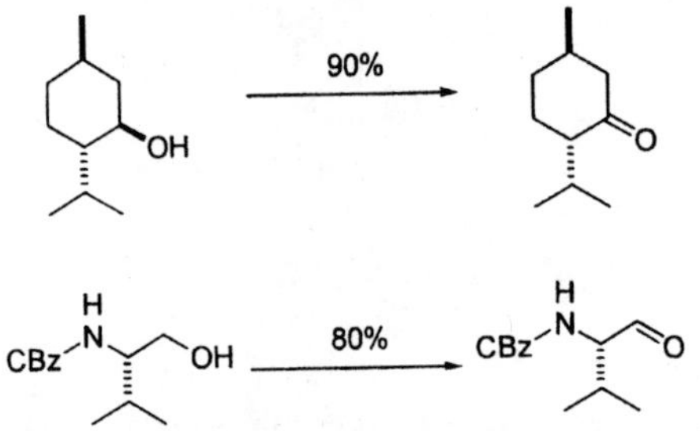

90%

80%

The mechanism of this modified procedure is thought to proceed in the same manner as that of the classic Swern oxidation with the formation of an oxosulphonium ylide. The structure of the oxidation reagent is depicted below.

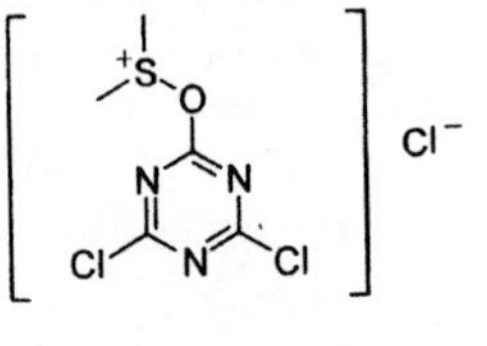

A mention should be made of the different activating agents that can be employed in oxidation reactions with DMSO. Some of these are name reactions in their own right; however, they are stated here as useful alternatives to the Swern reaction.

The Kornblum oxidation was the first reported oxidation reaction using DMSO. α-Bromo ketones such as **23** were converted to the corresponding glycoxals in good yield.[1] Less reactive halides, such as most benzylic halides and aliphatic halides, can be oxidised by conversion to the tosylate derivative with silver tosylate and then oxidising *in situ* with a sodium bicarbonate–DMSO mixture.[17,18]

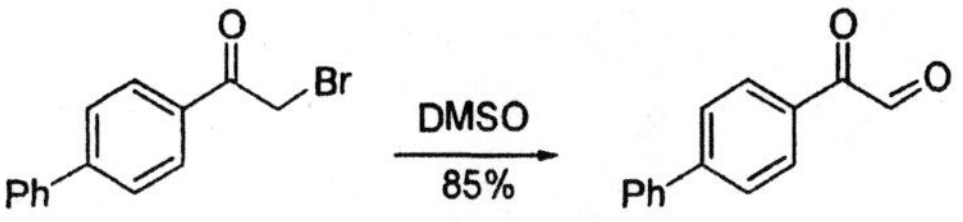

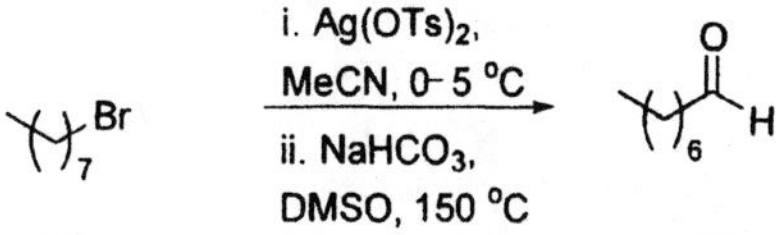

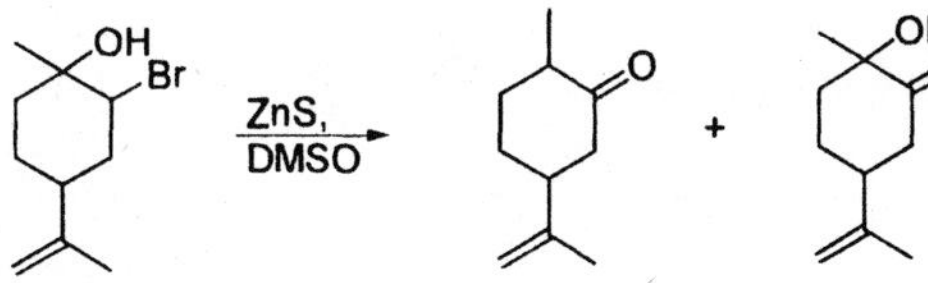

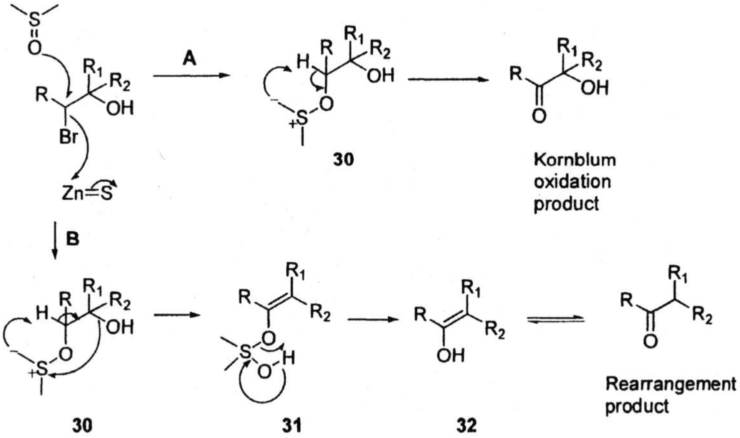

Variations of this method have been studied based on the halide to tosylate conversion which clearly indicates electrophilic catalysis, such as the silver salt–DMSO oxidation of halides.[19] A more recent example involves the direct conversion of *tert*-β-bromo alcohols to ketones using zinc sulfide and DMSO.[21] *tert*-β-Bromo alcohols cannot undergo oxidation to the corresponding carbonyl *via* a free-radical mechanism due to the absence of an α-proton.[20] A mixture of the Kornblum oxidation product and a product resulting from rearrangement of the sulfur ylide is obtained. The proposed mechanism involves the formation of the oxysulfonium ylide **30** which undergoes proton abstraction to give the expected oxidation product *via* path A. Path B shows the rearrangement of the ylide **30** to give an enol ether which tautomerises to give the resultant ketone product.

The Parikh–Doering protocol involves the use of a sulphur trioxide-pyridine complex and DMSO in the presence of triethylamine.[22,23] There are several advantages to this method: general applicability for the oxidation of primary and secondary alcohols, rapid reaction times at room temperature, and a particularly attractive feature is the selective oxidation of allylic alcohols to α,β-unsaturated carbonyl compounds. This can be achieved in the presence of secondary hydroxyl groups as in the case of 11β-hydroxy-4,17-pregnadien-3-one **33** which was oxidised to the corresponding α,β-unsaturated aldehyde **34** in 70% yield.[22] Additionally, production of the methyl thiomethyl ether derivative by-product which can arise from the use of acetic anhydride–DMSO oxidations is negligible and purification is straightforward, unlike that with the DCC–DMSO method.

33 → (SO_3-pyr; DMSO, Et_3N; 70%) → **34**

The Pfitzner–Moffatt oxidation utilises 1,3-dicyclohexylcarbodiimide as the DMSO activator in the presence of acid to afford oxidation of alcohols to carbonyls.[2,24,25] Initial work was carried out on steroids and thymidine residues although the procedure has since been found to be applicable to a large range of alcohols. In the following example, the Pfitzner–Moffatt method was used as an alternative to the Swern reaction which had resulted in the formation of an unwanted α-chloroketone (see later).[26] Other oxidation procedures were attempted (PDC, PCC, Dess–Martin periodinane); however; the Pfitzner–Moffatt gave the best, albeit modest, yield.

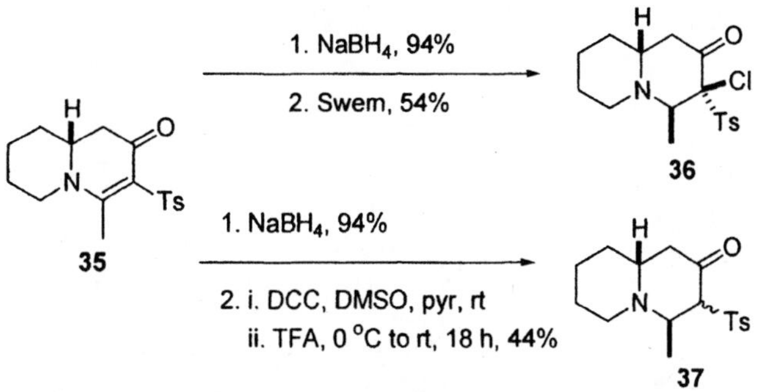

Alcohol oxidations using acid anhydrides, such as acetic anhydride and benzoic anhydride, and phosphorus pentoxide with DMSO, have also been found to proceed in mild conditions to give the corresponding carbonyl compounds in good yields.[27–29] The method is general and is especially useful for sterically hindered hydroxyl groups; however, the reactions often suffer from concomitant formation of methylthiomethyl ether by-products.

3.11.5 Synthetic Utility

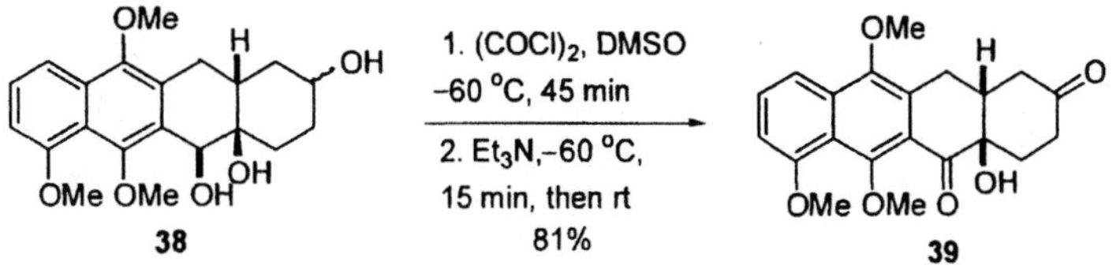

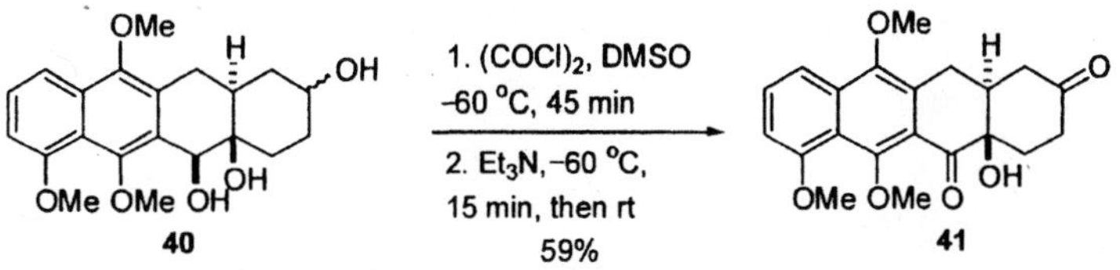

The Swern oxidation can be employed to oxidise secondary alcohols in the presence of tertiary alcohols. In the total synthesis of 11-deoxydaunomycinone, Ghera and Ben-David found that reagents developed specifically for this purpose, Cl_2–DMSO or NCS–Me_2S,[9b] failed to give the desired products.[30] Additionally, reagents which were reported to effect the monooxidation of primary–tertiary vicinal diols, SO_3–pyridine complex or the Fetizon reagent (Ag_2CO_3–celite)[31] gave only recovery of starting material. When the Swern conditions were employed, triols **38** and **40** were converted to the desired diketones in 81%

and 59% yield, respectively. The effectiveness of the reaction was found to depend significantly on the use of an appropriate excess of the oxidation reagents and defined reaction times.

Modifications of the Swern procedure can result in the chemoselective oxidation of alcohol functionalities in the presence of disulphide bonds, which themselves are highly susceptible to further oxidation.[32–34] Fang *et al.* report the first such examples of oxidizing secondary alcohols to the corresponding ketones without cleavage or oxidation of the disulphide bond.[35] Implementation of the standard Swern protocol, after experimenting with various other oxidants, gave the desired products; however, these reactions were low yielding and also resulted in side products through cleavage of the disulphide bond. Eventually, it was found that the yield was significantly improved if the reaction was quenched with water at low temperature instead of quenching at room temperature, as is standard in the Swern procedure. Consequently, the reaction was maintained at –78 °C for 2 h after addition of triethylamine, to ensure complete substrate consumption, and then a mixture of water and THF was added drop-wise while maintaining the reaction temperature at below –60 °C.

OH S S OH —96%→ O S S O

42 **43**

OTMS MeO_2C OTMS OTMS —75%→ OTMS MeO_2C O OTMS

44 **45**

OTES OTES —69%→ O OTES

46 **47**

Many examples exist wherein a silyl enol ether is deprotected and oxidised in a one-pot procedure using Swern conditions. Godfroid and co-workers report the selective oxidation of primary TMS or TES protected 1,2-diols, 1,3-triols, and polyhydroxy compounds to the corresponding silyloxy aldehydes even when using excess Swern reagents.[36]

Hindered silyl groups such as dimethyl-*t*-butylsilyl ethers have been shown to be completely stable to Swern conditions.[37]

TBDMSO 48 → 80% → TBDMSO 49

Employing the Swern conditions can also result in the formation of α,β-unsaturated compounds. For example, oxidation of the tryptamine derivative **50** resulted in the α,β-unsaturated aldehyde **52** instead of the expected product **51**.[38]

51 ← 50 → 52

Using one equivalent of the Swern reagents resulted in regeneration of the starting material while using excess reagent gave the unsaturated aldehyde. This presumably proceeded *via* the standard Swern oxidation of alcohol **50**, followed by removal of the acidic α-proton to generate the extended conjugated system as shown.

Another noteworthy side reaction arising from using the Swern conditions is the production of electrophilic chlorine which can give rise to α-chloroketones.[39,40] When a mixture of 2:1 β-hydroxy esters **53a,b** were subjected to standard Swern conditions, a single product was obtained in high yield. X-ray crystallography confirmed the structure to be chloroketone **54**.

$TMSCH_2CH_2O_2C$ 53a,b → Swern conditions, 80% → $TMSCH_2CH_2O_2C$ 54

The likely reactive species and source of electrophilic chlorine is thought to be **7** and/or its decomposition product **8**, which reacts with the enol of the ketone or β-ketoester, that is, the expected oxidised product. This side reaction can be avoided by employing one equivalent of the oxalyl chloride or by using either trifluoroacetic anhydride or acetic anhydride with DMSO.

7 8

In their synthesis of Baylis–Hillman adducts as potential alkylating agents, Lawrence *et al.* attempted the oxidation of diol **56** under standard Swern conditions, with unexpected results.[41] Instead of the required α-methylene-β-keto ester, the formation of an allylic chloride **57** was detected. The substitution is thought to take place *via* a conjugate-elimination process through enolate **59** involving the ester group, which is clearly implicated since the Swern oxidation of related alcohols such as **60** proceeds as expected.

DABCO™

55 56

DMSO, $(COCl)_2$, Et_3N

57

-HCl

58 59

60

The versatility and efficiency of the Swern conditions means that they can also be applied to various functional groups to effect different transformations other than just oxidations of alcohols. For example, primary amides have been converted to nitriles using the Swern reagents as dehydrating agents.[42] Although different activators of DMSO were tested, for example, the SO_3·pyridine complex and TFAA, oxalyl chloride was found to be the best based on the yield of the final product.

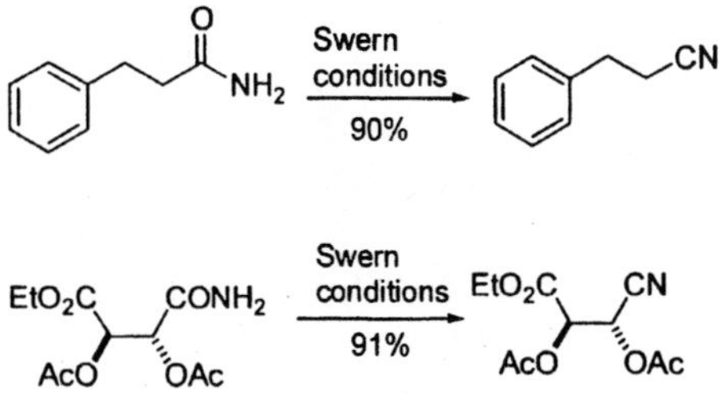

Swern conditions 90%

Swern conditions 91%

The Swern conditions were applied to a number of substrates with different functional and protecting groups, such as sugar derivatives, protected amino acid derivatives, tartaric acid derivatives, and optically active synthetic intermediates. Both acid-sensitive (epoxide, acetonide, silyl, NBoc, NCBz) and alkaline-sensitive groups (Ac, Bz, ester, silyl) were found to be completely unaffected. The proposed mechanism begins with tautomerisation of the amide to the hydroxy imine, followed by reaction with the dimethylchlorosulfonium species. The resulting oxysulfonium species **63** collapses when treated with base to regenerate DMSO and produce the nitrile.

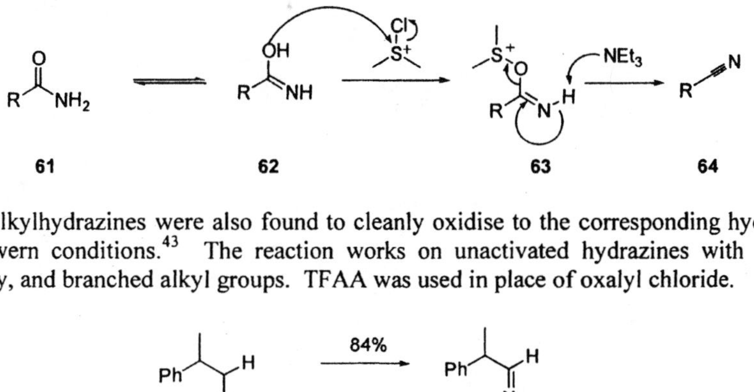

61 **62** **63** **64**

Alkylhydrazines were also found to cleanly oxidise to the corresponding hydrazones under Swern conditions.[43] The reaction works on unactivated hydrazines with primary, secondary, and branched alkyl groups. TFAA was used in place of oxalyl chloride.

84%

65 **66**

The Swern oxidation is also an excellent system for the oxidation of secondary amines into imines.[44–46] In particular, the method is compatible with small ring systems such as azirines.[47] Optically active 2*H*-azirine-2-carboxylic esters from the corresponding aziridines were synthesised by employing the Swern conditions and proceeded with complete retention of configuration at the C2 stereogenic centre.

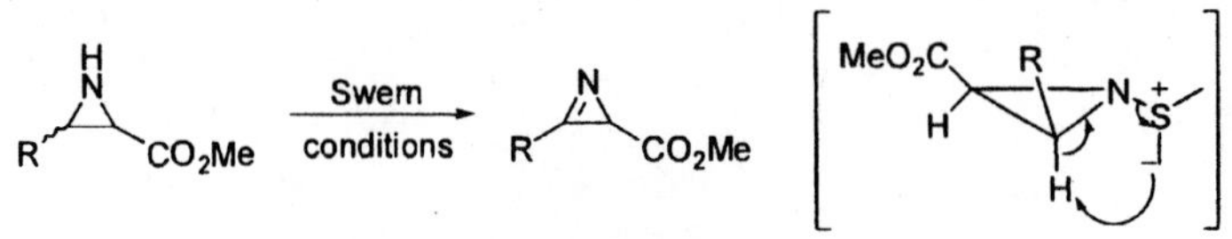

Interestingly, both *cis* and *trans* aziridine isomers gave the same azirine ester **68**, surprising as the C2 proton is more acidic than the C3 proton and would also result in conjugation with the ester functionality. The intermediate **69** is shown for the *cis* aziridine. R groups included alkyl chains and phenyl substituents and yields varied from 54% to 86% with higher yields obtained *via* the *trans* aziridines.

Mekonnen and co-workers present an intriguing route to 3-monosubstituted imidazo[1,2-b]pyridazine derivatives which are important motifs in several therapeutically interesting molecules.[48a–c] The key step involves a novel one-pot cyclisation/oxidation reaction under Swern oxidation conditions.[49]

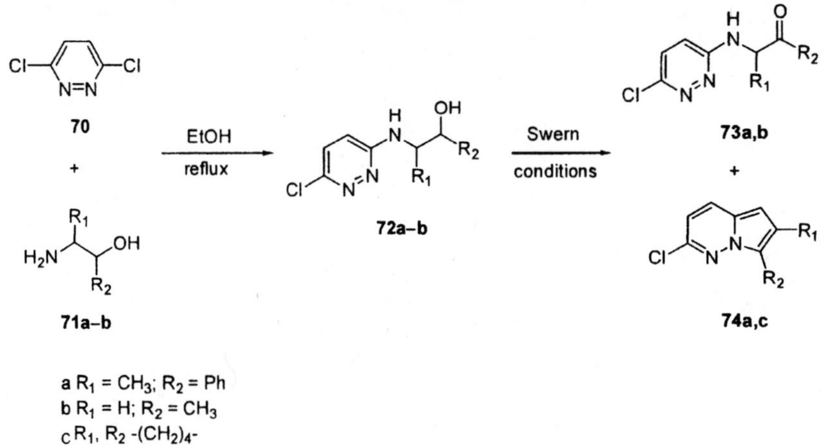

a $R_1 = CH_3$; $R_2 = Ph$
b $R_1 = H$; $R_2 = CH_3$
c R_1, R_2 -$(CH_2)_4$-

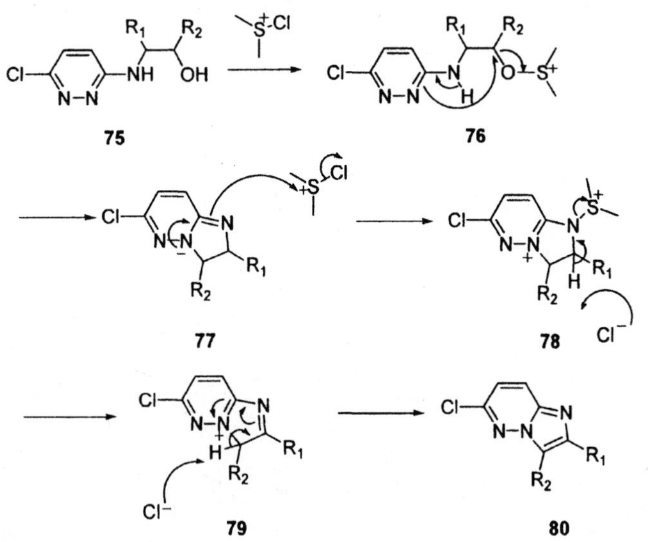

Condensation of dichloropyridazine **70** with various 2-hydroxyethylamines led to the monoadducts **72a–b**. Oxidation of the secondary hydroxyl functionality was then attempted using Swern oxidation conditions; this, however, unexpectedly resulted in imidazopyridazines **74a–b**. Since the endocyclic nitrogen of a heterocyclic amidine is reported to be a poor nucleophile toward carbonyl compounds, the formation of imidazopyridazine **74** could not be caused by the *in situ* cyclisation of the preformed ketones **73**. The authors also found that heating ketone **73b** under reflux in ethanol did not result in cyclisation product **74b**. Thus the mechanism is believed to occur *via* the intramolecular attack of a pyridazine nitrogen onto an activated tetrahedral sulfinium intermediate, with the Swern reagent acting as a Lewis acid.

The presence of an inductively electron-donating alkyl group resulted in formation of the ketone alone, while the geometrically more constrained cyclohexylidine **72c** gave exclusively the cyclisation product. This suggests that the oxidation of intermediates **72** is influenced by inductive effects as well as geometric considerations.

Kobayashi and co-workers report an interesting ring opening reaction which occurs when the Swern conditions are applied to pyrazine-fused bicycles.[50] *Exo-cis* diol **81** when treated with DMSO and oxalyl chloride gave alcohol **82** in 63%. The same product was obtained in the same yield using trifluoroacetic anhydride instead of oxalyl chloride.

Swern conditions 63%

81 → 82

R = CH=CH-CH=CH$_2$

A proposed mechanism, involving an intramolecular Cannizzaro reaction, is outlined below. The mechanism is not conclusive especially as the vicinal diols **87–90** have been reported to give the expected α-diketones under Swern conditions.[50–53]

83 → 84 → (-OH) 85 → (-CO$_2$) 86

87 88 89 90

3.11.6 Experimental

91 → (94%) 92

4-(2,2-Dibromovinyl)-5-[3-(4-methoxybenzyloxy)-propyl]-2,2-dimethyl-[1,3]-dioxolane (92)[54]

To a solution of oxalyl chloride (2.45 g, 1.69 mL, 19.30 mmol, in dry CH_2Cl_2 (100 mL) at –78 °C was added drop-wise dry DMSO (3.02 g, 2.74 mL, 38.65 mmol) in CH_2Cl_2 (20 mL). After 30 min, alcohol **91** (4.0 g, 12.89 mmol) in CH_2Cl_2 (20 mL) was added over 10 min giving a copious white precipitate. After stirring for 1 h at –78 °C the reaction mixture was brought to –60 °C and Et_3N (7.18 mL, 51.59 mmol) was added slowly and the reaction mixture was allowed to warm to room temperature over 30 min. The reaction mixture was poured into H_2O (150 mL) and the organic layer separated. The aqueous layer was extracted with CH_2Cl_2 (2 × 50 mL) and the combined organic layers were washed with H_2O (3 × 50 mL), brine (50 mL), dried (Na_2SO_4), and passed through a short pad of silica gel. The filtrate was concentrated to give the aldehyde **92** (3.9 g, 94%) as a pale yellow oil, which was used without further purification.

3.11.7 References

1. Kornblum, N.; Powers, J. W.; Anderson, G. J.; Jones, W. J.; Larson, H. O.; Levand, O.; Weaver, W. M. *J. Org. Chem.* **1957**, *79*, 6562.
2. Pfitzner, K. E.; Moffatt, J. G. *J. Am. Chem. Soc.* **1963**, 3027.
3. Omura, K.; Swern, D. *Tetrahedron* **1978**, *34*, 1651.
4. Mancuso, A. J.; Brownfain, D. S.; Swern, D. *J. Org. Chem.* **1979**, *44*, 4148.
5. Mars, M.; Tidwell, T. D. *J. Org. Chem.* **1984**, *49*, 788.
6. Mancuso, A. J.; Huang, S-L.; Swern, D. *J. Org. Chem.* **1978**, *43*, 2480.
7. Huang, S-L.; Omura, K.; Swern, D. *J. Org. Chem.* **1976**, *41*, 3329.
8. [R] Mancuso, A.; Swern, D. *Synthesis* **1981**, 165.
9. (a) Corey, E. J.; Kim, C. U. *J. Am. Chem. Soc.* **1972**, *94*, 7586; (b) Corey, E. J.; Kim, C. U. *Tetrahedron Lett.* **1974**, 287; (c) Corey, E. J.; Kim, C. U. *J. Org. Chem.* **1973**, *38*, 1233; (d) Gauvreau, J. R.; Poignant, S.; Martin, G. J. *Tetrahedron Lett.* **1980**, *21*, 1319.
10. Eustache, J.; Bernardon, J.-M.; Shroot, B. *Tetrahedron Lett.* **1987**, *28*, 4681.
11. Chrisman, W.; Singaram, B. *Tetrahedron Lett.* **1997**, *38*, 2053.
12. Ohsugi, S.; Nishide, K.; Oono, K.; Okuyama, K.; Fudesaka, M.; Kodama, S.; Node, M. *Tetrahedron* **2003**, *59*, 8393.
13. (a) Crich, D.; Neelamkavil, S. *J. Am. Chem. Soc.* **2001**, *123*, 7449; (b) Crich, D.; Neelamkavil, S. *Tetrahedron* **2002**, *58*, 3865.
14. (a) Liu, Y.; Vederas, J. C. *J. Org. Chem.* **1996**, *61*, 7856; (b) Harris, J. M.; Liu, Y.; Chai, S.; Andrews, M. D.; Vederas, J. C. *J. Org. Chem.* **1998**, *63*, 2407.
15. Kende, A. S.; Johnson, S.; Sanfilippo, P.; Hodges, J. C.; Jungheim, L. N. *J. Am. Chem. Soc.* **1986**, *108*, 3513.
16. De Luca, L.; Giacomelli, G. *J. Org. Chem.* **2001**, *66*, 7907.
17. Kornblum, N.; G. J.; Jones, Anderson, G. J. *J. Am. Chem. Soc.* **1957**, *81*, 4113.
18. Kornblum, N.; Frazier, H. W. *J. Am. Chem. Soc.* **1966**, *88*, 865.
19. Ganem, B.; Boeckman, R. K., Jr. *Tetrahedron Lett.* **1974**, 917.
20. Dolenc, D.; Harej, M. *J. Org. Chem.* **2002**, *67*, 312.
21. Bettadalah, B. K.; Gurudutt, K. N.; Srinivas, P. *J. Org. Chem.* **2003**, *68*, 2460.
22. Parikh, J. R.; Doering, W. von E. *J. Am. Chem. Soc.* **1967**, 5505;
23. Prestat, G.; Baylon, C.; Heck, M-P.; Grasa, G. A.; Nolan, S. P.; Mioskowski, C. *J. Org. Chem.* **2004**, *69*, 5770.
24. Pfitzner, K. E.; Moffatt, J. G. *J. Am. Chem. Soc.* **1965**, *87*, 5661.
25. Pfitzner, K. E.; Moffatt, J. G. *J. Am. Chem. Soc.* **1965**, *87*, 5670.
26. Back, T. G.; Hamilton, M. D.; Lim, V. J. J.; Parvez, M. *J. Org. Chem.* **2005**, *70*, 967.
27. Albright, J. D.; Goldman, L. *J. Am. Chem. Soc.* **1965**, *87*, 4214.
28. Onodera, K.; Hirano, S.; Kashimura, N.; *J. Am. Chem. Soc.* **1965**, *87*, 4651.
29. Albright, J. D.; Goldman, L. *J. Am. Chem. Soc.* **1967**, *89*, 2416.
30. Ghera, E.; Ben-David, Y. *J. Org. Chem.* **1988**, *53*, 2972.
31. Tanno, N.; Terashima, S. *Chem. Pharm. Bull.* **1983**, *21*, 811.
32. Allen, P., Jr.; Brook, J. W. *J. Org. Chem.* **1962**, *27*, 1019;
33. Folkins, P. L.; Harpp, D. N. *Tetrahedron Lett.* **1993**, *34*, 67.
34. Cogan, D. A.; Liu G.; Kim, K.; Backes, B. J.; Ellman, J. A. *J. Am. Chem. Soc.* **1998**, *120*, 8011.
35. Fang, X.; Bandarage, U. K.; Wang, T.; Schroeder, J. D.; Garvey, D. S. *J. Org. Chem.* **2001**, *66*, 4019.
36. A. Rodriguez, Nomen, M,; Spur, B. W.; Godfroid, J. J. *Tetrahedron Lett.* **1999**, *40*, 5161.
37. Afonso, C. M.; Barros, M. T. *J. Chem. Soc. Perkin Trans. 1* **1987**, 1221.

38. Bailey, P. D.; Cochrane, P. J.; Irvine, F.; Morgan, K. M.; Pearson, D. P. J.; Veal, K. T. *Tetrahedron Lett.* **1999**, *40*, 4593.
39. Smith, A. B., III.; Leenay, T. L.; Liu, H-J.; Nelson, L. A. K.; Ball, R. G. *Tetrahedron Lett.* **1988**, *29*, 49.
40. Beck, B. J.; Aldrich, C. C.; Fecik, R. A.; Reynolds, K. A., Sherman, D. H. *J. Am. Chem. Soc.* **2005**, *125*, 12551.
41. Lawrence, N. J.; Crump, J. P.; McGown, A. T.; Hadfield, J. A. *Tetrahedron Lett.* **2001**, *42*, 3939.
42. Nakajima, N.; Ubukata, M. *Tetrahedron Lett.* **1997**, *38*, 2099.
43. Dupont, J.; Bemish, R. J.; McCarthy, K. E.; Payne, E.; Pollard, E. B.; Ripin, D. H. B.; Vanderplas, B. C.; Watrous, R. M. *Tetrahedron Lett.* **2001**, *42*, 1453.
44. Keirs, D.; Overton, K. *J. Chem. Soc. Chem. Commun.* **1987**, 1660.
45. Jinbo, Y.; Kondo, H.; Taguchi, M.; Sakamato, F.; Tsukamoto, T. *J. Org. Chem.* **1994**, *59*, 6057.
46. Gaucher, A.; Olliveier, J.; Marguerite, J.; Paugam, R.; Salaün, J. *Can. J. Chem.* **1994**, *72*, 1312.
47. Gentilucci, L.; Grizjen, Y.; Thijs, L.; Zwanenburg, B. *Tetrahedron Lett.* **1995**, *36*, 4665.
48. (a) Sacchi, A.; Laneri, S.; Arena, F.; Abignente, E.; Gallitelli, M.; D'amico, M.; Filippelli, W.; Rossi, F. *Eur. J. Med. Chem.* **1999**, *34*, 1003; (b) Barlin, G. B.; Davies, L. P.; Harrison, P. W. J. *Aust. J. Chem.* **1997**, *50*, 61; (c) Kuwahara, M.; Kawano, Y.; Shimazu, H.; Ashida, Y.; Miyake, A. *Chem. Pharm. Bull.* **1996**, *44*, 122.
49. Raboisson, P.; Mekonnen, B.; Peet, N. P. *Tetrahedron Lett.* **2003**, *44*, 2919.
50. Kobayashi, T.; Kobayashi, S. *Molecules*, **2000**, *5*, 1062.
51. Russell, R. A.; Harrison, P. A.; Warrener, R. N. *Aust. J. Chem.* **1984**, *37*, 1035.
52. Wright, M. W.; Welker, M. E. *J. Org. Chem.* **1996**, *61*, 133.
53. Christl, M.; Kraft, A. *Angew. Chem. Int. Ed. Engl.* **1988**, *27*, 1369.
54. Kumar, P.; Naidu, S. V.; Gupta, P. *J. Org. Chem.* **2005**, *70*, 2843.

Nadia M. Ahmad

3.12 Wacker–Tsuji Oxidation

3.12.1 Description

The Wacker oxidation is one of the longest known palladium-catalyzed organic reactions. It is the industrial process where ethylene is oxidized to acetaldehyde with oxygen in the presence of a catalytic amount of palladium and copper salt as the redox co-catalyst.[1] There are plants that produce thousands of tons of acetaldehyde per year. Many reviews exist on this topic.[2–6]

O_2, $PdCl_2$ (cat)
$CuCl_2$ (cat), H_2O
O
H

Jiro Tsuji carried out many mechanistic and synthetic studies on the initial Wacker oxidation process.[7–11] It is now known as the Wacker–Tsuji oxidation for the oxidation of terminal olefin **1** to the corresponding methyl ketone **2** with oxygen in the presence of a catalytic amount of palladium and one equivalent of copper salt.[12–17] Nowadays, the Wacker–Tsuji oxidation is a standard methodology for transforming the terminal olefin to the corresponding methyl ketone.[17] The reaction is so widely used that Tsuji declared that a terminal olefin could be viewed as a masked methyl ketone.[11]

R **1**
O_2, $PdCl_2$ (cat)
1 eq. $CuCl_2$, DMF/H_2O
O
R **2**

3.12.2 Historical Perspective

In 1962, Juergen Smidt and co-workers at Wacker Chemie published an article titled "*The Oxidation of Olefins with Palladium Chloride Catalyst*,"[1] which heralded the birth of the Wacker oxidation. Although it was discovered as an industrial process, it is now widely used in organic synthesis in both laboratory and industrial scales. Since 1976, Jiro Tsuji's group at Tokyo Institute of Technology intensively investigated and expanded the reaction. Considering the audience of this book are mostly synthetic chemists rather than chemical engineers, this review will focus on the Wacker–Tsuji oxidation.

3.12.3 Mechanism

The mechanism of the Wacker–Tsuji oxidation has been intensively investigated by both academia and industry and it is now very well elucidated. In the 2002 book *Handbook of Orgamopalladium Chemistry for Organic Synthesis*, edited by Ei-ichi Negishi, Ptrick M.

Henry at Loyola University in Chicago wrote a comprehensive chapter on "*The Wacker Oxidation and Related Asymmetric Syntheses,*" in which he summarized important developments in deciphering the mechanism of the Wacker oxidation.[3] Accordingly, as shown in the catalytic cycle below, complexation of alkene **1** with the Pd(II) catalyst provides complex **3**. Nucleophilic attack of **3** by water gives rise to the key hydroxypalladation intermediate **4** as the *σ-bonded (β-hydroxyalkyl)-palladium(II)* species, which undergoes a hydride shift to deliver methyl ketone **2** and palladium hydride species L_nPdHCl. Subsequent reductive elimination of L_nPdHCl then generated the palladium(0) species $L_nPd(0)$ and HCl. Without reoxidation of Pd(0) to Pd(II), the reaction would have stopped there. But in the presence of Cu(II), L_nPdCl_2 is regenerated and $CuCl_2$ was concomitantly reduced to CuCl, which is reoxidized by molecular oxygen.

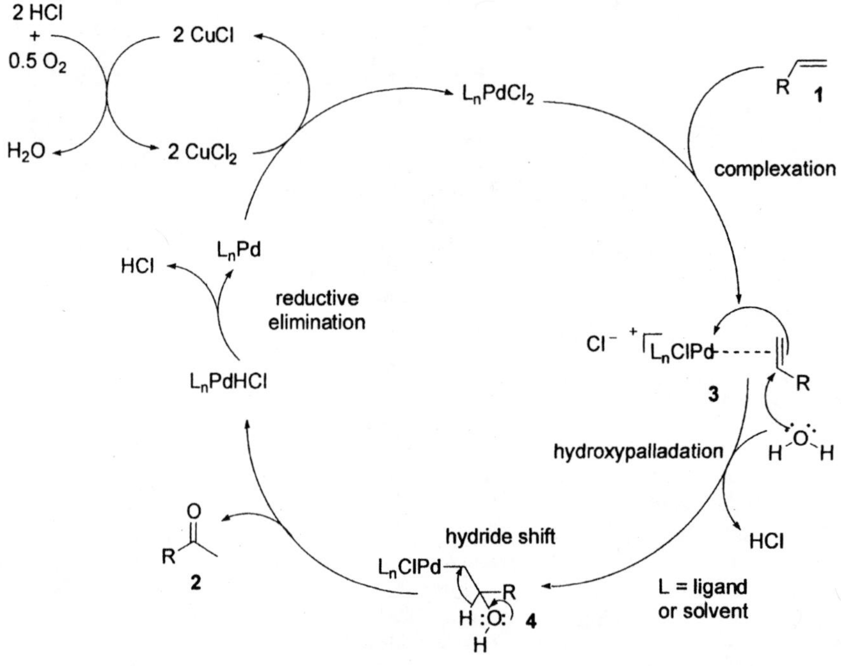

The aforementioned catalytic cycle is well accepted for its beauty and simplicity, but is far from definitive. As our understanding grows on this useful reaction, many alternatives have been proposed. For instance, a η^3-allylpalladium complex was suggested in 1981 as an intermediate of the Wacker oxidation.[18] In 1996, Murahashi's group reported the first isolation and characterization of a palladium–copper heterometallic complex bearing μ^4-oxo atom derived from molecular oxygen.[19] Thus heterometallic complexes **5** and **6** were isolated and characterized from the reaction of $PdCl_2(MeCN)_2$ and CuCl in HMPA. Not unexpectedly, treatment of 1-decene with **5** afforded 2-decanone in 99% yield, providing further credence to the theory. Similarly, a Cu-O-Pd-O-Cu cationic working catalyst was proposed for Cu(II)-Pd(II)-exchanged Y-zeolites for the heterogeneous Wacker oxidation.[20]

$$PdCl_2(MeCN)_2 + CuCl + HMPA \xrightarrow[ClCH_2CH_2Cl]{O_2}$$

$$\underset{\mathbf{5}}{Pd_6Cu_4Cl_{12}O_4(HMPA)_4} + \underset{\mathbf{6}}{[(HMPA)_2CuCl_2(PdCl_2)_2]_n}$$

3.12.4 Variations and Improvements

Numerous variations and improvements exist for the Wacker–Tsuji oxidation. Herein, only two major categories are discussed: anti-Markovnikov products and the Wacker-type oxidation.

3.12.4.1 Anti-Markovnikov products

Reversal of the regioselectivity has been observed for the Wacker–Tsuji oxidation of alkene systems. Mechanistically, the regiochemistry of the Wacker–Tsuji oxidation stems from the Markovnikov addition (which takes place for the vast majority of the cases) *versus* the anti-Markovnikov regioselectivity; although methyl-ketone **2** is often the major product and aldehyde **7** is the minor product. However, abnormal regiochemistry has been observed where aldehyde **7** is the major or even sole product in the Wacker–Tsuji oxidation.

R (**1**) —[O_2, $PdCl_2$ (cat); 1 eq. $CuCl_2$, DMF/H_2O]→ R–C(O)–CH$_3$ (**2**) + R–CH$_2$–CHO (**7**)

Feringa from The Netherlands was the first to report an abnormal Wacker oxidation in 1986.[21] Dec-1-ene was oxidized *in the presence of a tertiary alcohol* to a mixture of aldehyde **8** and methyl ketone **9**. Surprisingly, aldehyde **8** was isolated as the major product (**8/9** = 70/30). Even more strikingly, styrene was transformed exclusively to its corresponding phenylacetaldehyde under the same conditions. Feringa proposed that the aldehyde formation involved an oxygen transfer reaction *via* initial cycloaddition of the nitro-palladium complex, followed by β-hydrogen elimination.

air
$(MeCN)_2PdClNO_2$
$CuCl_2$, *t*-BuOH
30 °C, 3 h
CHO
8
9

Also in 1986, Bose's group in India observed a reversal of regiochemistry for the Wacker oxidation of 3-vinyl-4-substituted-2-azetidinones.[22] For instance, vinyl-lactam **11** was converted to the corresponding aldehyde **12** exclusively in 70% yield. The authors suggested that palladium coordinated with the carbonyl of the lactams as well as the olefin double bond, which influenced the regioselectivity of the hydration step. This was the first time the "heteroatom coordination" theory was put forward and it has been widely adopted since then.

$PdCl_2$, CuCl
O_2, 70%
11
12

$PdCl_2(MeCN)_2$
CuCl, HMPA
O_2, 52%
13
14
15
ClCu(HMPA)
16
17

Hosokawa *et al.* clearly spelled out the "heteroatom coordination" theory in explaining their region-selective formation of aldehydes.[20] Terminal olefin **13** was converted to a mixture of aldehyde **14** and methyl ketone **15**, with aldehyde **14** as the major product (**14**/**15** = 70/30). The authors explained that Pd(II) coordinated with both oxygen of the amide as well as the olefin as depicted in intermediate **16**, which blocked the normal Markovnikov hydration position and the addition of the "peroxide" took place at the terminal position as depicted in intermediate **17**. Interestingly, addition of water to the reaction would give the ketone **15** as the major product.

Pellissier and colleagues reported that terminal olefin **18** underwent a Wacker oxidation to give methylketone **19** as the major product in 85% yield and aldehyde **20** as the minor product in 7% yield.[23,24] However, when the configuration of the neighboring lactone was switched like in substrate **21**, the yield for the anti-Markovnikov addition product **23** was 35%. The authors proposed the assistance of the neighboring oxygen contributed to the regiochemistry.

$Pd(OAc)_2$, $HClO_4$, 10% benzoquinone, CH_3CN: **18** → **19**, 85% + **20**, 7%

$Pd(OAc)_2$, $HClO_4$, 10% benzoquinone, CH_3CN: **21** → **22**, 45% + **23**, 35%

In a unique case, 1,5-diene **24** was converted to the corresponding aldehyde **26** possibly *via* the intermediacy of **25**.[25] The authors believed that intermediate **25** was formed with the participation of the other double bond. In the same vein, diene **27** was converted to the corresponding aldehyde **28** in 73% yield.

O_2, $PdCl_2$, CuCl, DMF/H_2O, rt, 24 h, 60%: **24** → [**25**, HO, PdL_n] → **26** (OHC)

O_2, $PdCl_2$, CuCl
DMF/H_2O, rt, 24 h, 75%

27 28

Under standard catalytic conditions (10% $PdCl_2$, 1 eq. CuCl, O_2, in DMF/H_2O), the Wacker oxidation of 4-methoxystyrene proceeded to give a mixture of the two possible products. As expected, the Markovnikov product, methylketone **31** was predominate and only a small amount of aldehyde **30** was isolated (**31**:**30** = 8.4:1). However, Spencer and co-workers performed the reaction in the absence of the reoxidant CuCl and observed a reversal of the usual regioselectivity.[26] Thus, reaction of 4-methoxystyrene with 2 equivalents of $PdCl_2$ gave aldehyde **30** as the major product. The authors explained the regioselectivity by the involvement of a possible 4-palladium-styrene complex **29**.

2 eq. $PdCl_2$, O_2
DMF/H_2O (10:1)

29 30 31

30:**31** = 9.8:1

3.12.4.2 The Wacker-type oxidation

Wacker-Type Addition

Nuc + Pd(II)X_2 + R–CH=CH–R → (−X^-) Nuc–CHR–CHR–Pd(II)X

The Wacker-type addition is the *anti*-addition of (most commonly) a heteroatom and a Pd(II) species across a C–C double bond. The Wacker-type oxidations are Pd(II)-catalyzed transformations involving heteroatom nucleophiles and alkenes or alkynes as electrophiles.[27] In most of these reactions, the Pd(II) catalyst is converted to an inactive Pd(0) species in the final step of the process, and use of stoichiometric oxidants is required to effect catalytic turnover. For example, the synthesis of furan **33** from α-allyl-β-diketone **32** is achieved *via* treatment of the substrates with a catalytic amount of $Pd(OAc)_2$ in the presence of a stoichiometric amount of $CuCl_2$.[28] This transformation proceeds *via* Pd(II) activation of the alkene to afford **34**, which undergoes nucleophilic attack of the enol oxygen onto the alkene double bond to provide alkylpalladium complex **35**. β-Hydride elimination of **35** gives **36**, which undergoes isomerization and loss of Pd(H)(Cl) to yield the furan product **33**. The Pd(H)(Cl) complex undergoes reductive elimination of HCl to generate Pd(0), which is subsequently reoxidized to Pd(II) by $CuCl_2$.

Ph Ph **32** $Pd(OAc)_2$, $CuCl_2$ dioxane, 80 °C 72% Ph Ph O CH_3 **33**

32 $PdCl_2$ → **34** (OH, $PdCl_2$) → −HCl → **35** (ClPd) →

36 (H, Pd, Cl) tautomerization → **33** (CH_3) + Pd(H)(Cl) → Pd(0)

The synthesis of a variety of other heterocycles has been achieved using similar methodology. For example, benzopyran derivative **38** was readily prepared from 2-allyl phenol **37**;[29] note that the Pd(0) precatalyst is oxidized to Pd(II) by air before the reaction commences.

37 (OH) $Pd(dba)_2$ $KHCO_3$,air, DMSO/H_2O 80% **38**

Nitrogen heterocycles are also accessible from Wacker-type transformations, and the Hegedus indole synthesis is a quintessential example of the Wacker-type oxidation.[30] In 1976, Hegedus *et al.* described the synthesis of indoles using a Pd-assisted intramolecular amination of olefins, which tolerated a range of functionalities.[31,32] For example, addition of **39** to a suspension of stoichiometric $PdCl_2(CH_3CN)_2$ in THF produced a yellow precipitate, which upon treatment with Et_3N gave rise to indole **40** and deposited metallic palladium.

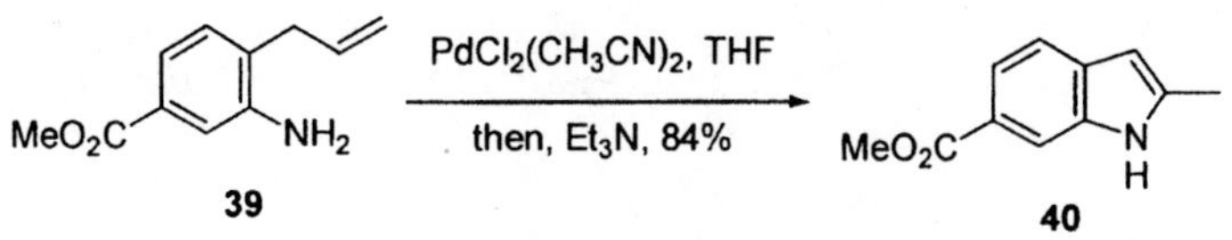

Hegedus proposed that the mechanism of this transformation proceeds through a Wacker-type reaction that is promoted by Pd(II). As shown below, coordination of olefin **39** to Pd(II) results in precipitate **41**, which upon treatment with Et_3N undergoes intramolecular *trans*-aminopalladation to afford intermediate **42**. As expected, the nitrogen atom attack occurs in a *5-exo-trig* fashion to afford **43**. β-Hydride elimination of **43** gives rise to exocyclic olefin **44**, which rearranges to indole **40**. The final step of this mechanism leads to the formation of catalytically inactive Pd(0). However, addition of oxidants such as benzoquinone allows for catalytic turnover.

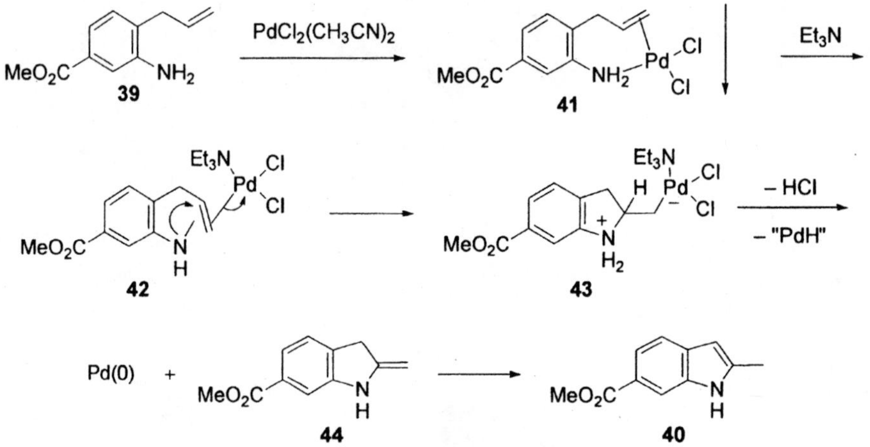

Another Wacker-type oxidation is the oxidative cyclization reaction, which is the intramolecular union of two arenes with formal loss of H_2 promoted by a Pd(II) species [typically $Pd(OAc)_2$].[33-36] In an early example of this transformation, treatment of diphenylamines **45** with $Pd(OAc)_2$ in acetic acid yielded carbazoles **46**.[37] The role of acetic acid in such oxidative cyclization processes is to protonate one of the acetate ligands, which affords a more electrophilic cationic Pd(II) species, thereby promoting the initial electrophilic palladation of the aromatic ring.

1–2 eq. $Pd(OAc)_2$

HOAc, reflux, 60–80%

45 → **46**

X = H, CH_3; R = CH_3, CH_3O, Cl, Br, NO_2, CO_2H

Overall, this transformation leads to the conversion of Pd(II) to Pd(0), which consumes one equivalent of expensive $Pd(OAc)_2$ in most cases. However, progress has been made toward the development of catalytic versions of this transformation, in which catalytic turnover is effected by employing a second oxidant that serves to convert Pd(0) back to Pd(II). For example, Knölker described the oxidative cyclization of **47** using *catalytic* $Pd(OAc)_2$ to afford indole derivative **48**.[38] The reoxidation of Pd(0) to Pd(II) was accomplished using excess cupric acetate in a manner analogous to the Wacker oxidation.[39]

0.1 $Pd(OAc)_2$, 2.5 $Cu(OAc)_2$

AcOH, reflux, 19 h, 71–73%

47 → **48**

Under the Wacker-type oxidation conditions, 4-trimethylsilyl-3-alkyn-1-ols **49** were converted to γ-butyrolactones **50**.[40] Similarly, palladium(II)-mediated oxidative cyclization of *N*-acyl aminoalkynes **51** provided a novel entry to γ-lactams **52**.[41]

Pd(II), $CuCl_2$, O_2

CH_3CN/1 N HCl (9:1)

34–74%

49 → **50**

R_1, R_2 = H, Ph, alkyl, cycloalkyl

Pd(II), $CuCl_2$, O_2

CH_3CN/1 N HCl (9:1)

34–74%

51 → **52**

R_1, R_2, R_3 = H, Ph, alkyl, cycloalkyl

R_4 = OCH_3, OBn, O*t*-Bu, CH_3

In their endeavor toward the total synthesis of zaragozic acids, Perlmuuter *et al.* in Australia took advantage of the intramolecular Wacker-type oxidation.[42,43] In 1999, they reported the synthesis of the dioxabicyclo[3.2.1]octane core utilizing a chiral auxiliary.[42] In 2004, they produced the dioxabicyclo[3.2.1]octane core **54** from the linear substrate diol **53** applying the classic Wacker oxidation conditions.[43]

5 mol% $PdCl_2$, $CuCl_2$

O_2, DME, 50 °C, 3 h, 89%

53 **54**

In 2003, Stoltz at CalTech described a palladium-catalyzed oxidative Wacker cyclization of *o*-allylphenols such as **55** in nonpolar organic solvents with molecular oxygen to afford dihydrobenzofurans such as **56**.[44] Interestingly, when (–)-sparteine was used in place of pyridine, dihydrobenzofuran **56** was produced asymmetrically. The *ee* reached 90% when $Ca(OH)_2$ was added as an additive. Stoltz considered it a "stepping stone to asymmetric aerobic cyclizations." In 2004, Muñiz carried out aerobic, intramolecular Wacker-type cyclization reactions similar to **55**→**56** using palladium-carbene catalysts.[45] Hiyashi *et al.* investigated the stereochemistry at the oxypalladation step in the Wacker-type oxidative cyclization of an *o*-allylphenol.[46] Like *o*-allylphenol, *o*-allylbenzoic acid **57** underwent the Wacker-type oxidative cyclization to afford lactone **58**.[47]

$Pd(TFA)_2$, MS 3 Å, O_2

pyr., Tol., 80 °C, 1 h, 87%

55 **56**

O_2, 5% $Pd(OAc)_2$

NaOAc, DMSO, 80%

57 **58**

3.12.5 Synthetic Utility

3.12.5.1 Methodology

Not surprisingly, the most prevalent utility of the Wacker–Tsuji oxidation is to convert terminal alkenes to the corresponding methyl ketones. Thus under the standard conditions, terminal alkene **59** was oxidized to methyl ketone **60** in 74% yield.[48] In addition, the

Wacker–Tsuji oxidation also works for carbohydrates as represented by transformation **61**→**62**.[49]

$PdCl_2$, CuCl, O_2
DMF/H_2O, rt, 24 h
74%

59 **60**

$PdCl_2$, O_2
DMF/H_2O (1:1)
rt, 24 h, 56%

61 **62**

So far we have only discussed the Wacker oxidation of terminal olefins; the regiochemistry for internal alkenes could be tricky because it often proceeds without regioselectivity. However, when allyl ether **63** bearing a neighboring oxygen function (β-methoxyl group) was subjected to the Wacker oxidation conditions, ketone **64** was isolated as the predominant product.[50] The regioselectivity can be easily accounted for using the "heteroatom coordination" theory. Moreover, the Wacker oxidation of homoallyl acetates showed similar regioselectivity. On the other hand, the regioselectivity of the Wacker oxidation of internal alkene **65** seemed to be governed by the availability of an allylic hydrogen.[51] In a study of competition between the formation of ketones **66** and **67**, increased electron on the aromatic ring density was shown to favor **66**. For instance, when X = 4-CF_3, the ratio was **67**/**66** = 1/19, but when X = 2,4,6-OCH_3, the ratio became **67**/**66** = 2.3/1. There must be a strong driving force for attack of the oxygen atom adjacent to the methyl group, and this was ascribed to an agnostic interaction between allylic hydrogen atoms and the palladium atom.

$PdCl_2$, $CuCl_2$
O_2, 50 °C, DMF
65%

63 **64**

air
$PdCl_2$, $CuCl_2$
DMF

65 **66** **67**

Palladium(II) complex supported on cyano-functionalized polyimide beads (PI-CN) was used in the Wacker oxidation of oct-1-ene to the corresponding methyl ketone **9**.[52] The PI-CN- supported Pd complex offered enhanced thermo-oxidative stability.

CN
Pd(II)
$CuCl_2$, H_2O, O_2
DMA, 62%
9

In 2002, Rawal developed a tandem reaction *via* multifunction palladium catalysis combining haloallylation and the Wacker–Tsuji oxidation.[53] For example, internal alkyne **68** first underwent a bromoallylation to give a 1,4-diene, which was subsequently oxidized to the methyl ketone **69** without addition of the palladium catalyst.

TBSO
68
1. 1 eq. allyl bromide
3 mol% $PdBr_2(PhCN)_2$
DME, 0 °C, to rt, 6 h
2. 1 eq. CuCl, O_2
DME/H_2O, rt, 24 h
66%
OTBS
Br
69

In Leighton's total synthesis of dolabelide D, the Wacker–Tsuji oxidation diene **70** was achieved chemoselectively to produce methyl ketone **71**.[54,55] Furthermore, addition of (–)-sparteine as a ligand prevented olefin isomerization and led to selective formation of methyl ketone **71** from the terminal olefin in good yield.[56]

OPMB
70
0.25 eq. $PdCl_2$
1.5 eq. $CuCl_2$
DMF/H_2O, rt, O_2
81%
PMBO
71

Spencer and gaunt developed a palladium-catalyzed amidation reaction.[57] When enone **72** was treated with carbamate Cbz-NH_2, in the presence of a catalytic amount of Pd(II), the Wacker addition intermediate underwent a protonolysis rather than the conventional β-hydride elimination process. As a consequence, amidation adduct **73** was obtained as the sole product.

72 + Cbz-NH$_2$ → (10 mol% PdCl$_2$(MeCN)$_2$; CH$_2$Cl$_2$, rt, 20 h, 96%) → 73

One of the exciting developments of the Wacker-type oxidation is the asymmetric synthesis of the reaction. For instance, using a new chiral bis(oxazoline) ligand L* = 3,3′-Disubstituted 2,2′-bis(oxazolyl)-1,1′-binaphthyls (boxax), a catalytic asymmetric Wacker-type cyclization converted allyl-phenol **55** to dihydrofuran **74** with 67% *ee*.[58]

55 → ([Pd(CH$_3$CN)$_4$](BF$_4$)$_2$, L*; 4 eq. *p*-benzoquinone; MeOH, 60 °C, 2 h, 90%) → 74

3.12.5.2 *Total synthesis of natural products*

Paquette was one of the first to apply the Wacker oxidation in the total synthesis. In his synthesis of 18-oxo-3-virgene, a constituent of the waxy surface resins of tobacco, a Wacker oxidation was deployed to convert a terminal alkene to the corresponding methyl ketone.[59] In an efficient total synthesis of (±)-laurene (**77**), alkene **75** was oxidized to keto-aldehyde **76** at ambient temperature.[60] In Smith's total synthesis of calyculin, a modified Wacker oxidation with substoichiometric cupric acetate transformed terminal alkene **78** to methyl ketone **79** without concurrent acetonide hydrolysis.[61,62]

75 → (10 mol% PdCl$_2$; 10 mol% CuCl$_2$; O$_2$, DME/H$_2$O (9:1); rt, 2 h, 82%) → 76 → → 77

78 → (O$_2$, 20 mol% Cu(OAc)$_2$; 10 mol% PdCl$_2$; DMA/H$_2$O (7:1); 84%) → 79

Classic Wacker–Tsuji oxidation was used in the total synthesis of copalol (**80**→**81**)[63] and a formal total synthesis of dysidiolide (**82**→**83**).[64]

PdCl$_2$, CuCl$_2$, O$_2$
DMF/H$_2$O, rt
overnight, 89%

80 → 81

PdCl$_2$, CuCl$_2$, O$_2$
DMF/H$_2$O,
65%

82 → 83

As mentioned in Section 3.12.4.1, the anti-Markovnikov product may be produced as a minor product, or a major product, or even the sole product. In the first total synthesis of 11-thia steroids, oxidation of terminal alkene **84** gave a mixture of ketone **85** and aldehyde **86** with the ketone as the major product (**86/85** = 8/1).[65] Similarly, in the total synthesis of macrosphelides H and G, macrolide **87** was oxidized with 5 equivalents of $PdCl_2$ to ketone **88** and aldehyde **89**, also with the ketone as the major product (**89/88** = 15/1).[66] Interestingly, the classic Wacker conditions (0.1 eq. $PdCl_2$, 1 eq. CuCl, DMF/H_2O = 10:1, O_2, rt, overnight) gave the ketone/aldehyde ratio **89/88** = 5/1 although no rationale emerged.

Pd(OAc)$_2$, HClO$_4$
benzoquinone
CH$_3$CN, H$_2$O, 12 h
96%

84 → 85 + 86

5 eq. $PdCl_2$, air

DMF/H_2O (7:1)
81%

87 → 88 + 89

In a synthesis toward cyclopenta[14,15]-19-norsteroids, a Wacker oxidation gave the methyl ketone and the aldehyde in a 1:1 ratio.[67] Therefore, 14β-acetonyl Δ^{15}-17-ketone **90** was oxidized to methyl ketone **91** and aldehyde **92** in a 1:1 ratio in 91% yield.

$PdCl_2$, CuCl, O_2

DMF/H_2O, 65 °C
91%

90 → 91 + 92

Another extreme of the regioselectivity of the Wacker oxidation was found in the oxidation of olefin **93**.[68] As shown below, in the synthesis of (±)-B(9α)-homo-C-nor-3-methoxy-12-oxa-17-vinylestra-1,3,5(10)-trienes, olefin **93** was converted to aldehyde **94** as the sole product. The regiochemical outcome could be readily explained by the "heteroatom coordination" theory.

$Pd(OAc)_2$, 0.3 $HClO_4$, benzoquinone, CH_3CN, H_2O, 74%

93 → 94

The Wacker–Tsuji oxidation has been applied in the synthesis of alkaloids as well. In an enantioselective total synthesis of (+)-monomorine I, alkene **95** was oxidized to ketone **96**, which was converted to (+)-monomorine I (**97**).[69] In addition, the synthesis of ABE tricyclic analogs of methyllycaconitine used a Wacker oxidation–aldol strategy to append the B ring to the AE fragment.[70]

$PdCl_2$, $CuCl_2$, O_2, DMF/H_2O, 80 °C, 91%

95 → 96 → 97

$PdCl_2$, CuCl, O_2, DMF/H_2O (8:1), rt, 24 h, 39%

98 → 99

Finally, in Mander's synthesis of the hexacyclic himandrin skeleton, amino-alkene **100** was oxidized to its corresponding methyl ketone, which was trapped by nucleophilic addition of the amine.[71] Dehydration of the adduct then delivered enamine **100**, which possessed all six rings of the himandrine skeleton.

$PdCl_2$, CuCl, O_2, n-Bu_4NCl, K_2CO_3, CH_3CN, Δ, 16 h, 85%

100 → 101

3.12.6 Experimental

Et-N bicyclic ketone **98** (CO_2Et) → **99**; $PdCl_2$, CuCl, O_2; DMF/H_2O (8:1); rt, 24 h, 39%

Ethyl (1*R,5*S**)-3-ethyl-9-oxobutyl)-3-azabicyclo[3.3.1]nonane-1-carboxylate** (99)[70]
A mixture of copper(I) chloride (445 mg, 4.50 mmol) and palladium(II) chloride (133 mg, 0.75 mmol) in DMF (10 mL) and water (2 mL) was stirred vigorously until the initially brownish solution became green (approximately 2 h). A solution of the terminal alkene **98** (1.10 g, 3.75 mmol) in DMF (5 mL) was then added and oxygen was bubbled through the mixture with stirring for 18 h. After this time, water (100 mL) was added and the mixture extracted with ethyl acetate (5 × 100 mL). The combined organic extracts were washed with brine (3 × 100 mL) then dried ($MgSO_4$) and the solvent removed *in vacuo* to leave the crude product. Purification by flash chromatography (4:1 hexane/EtOAc) afforded the title compound **99** (449 mg, 39%) as an orange oil.

3.12.7 References

1 Smidt, J.; Hafner, W.; Jira, R.; Sieber, R.; Sedlmeier, J. *Angew. Chem., Int. Ed. Engl.* **1962**, *1*, 80–88.
2 [R] Dedieu, A. *Catalysis by Metal Complexes* **1995**, *18* 167–95.
3 [R] Henry, P. M. in *Handbook of Orgamopalladium Chemistry for Organic Synthesis*; Negishi, E.-i. (Ed.), John Wiley & Sons: Hoboken, NJ, **2002**, pp. 2119–2139.
4 [R] Jira, R. Oxidations: Oxidation of olefins to carbonyl compounds (Wacker process). In *Applied Homogeneous Catalysis with Organometallic Compounds (2nd Edition);* Cornils, B.; Herrmann, W. A. (Eds.), Wiley-VCH: Weinheim, Germany, **2002**, *1*, pp386–405.
5 [R] Monflier, E.; Mortreux, A. Wacker-type oxidations. In *Aqueous-Phase Organometallic Catalysis* (2nd Edition) Cornils, B.; Herrmann, W. A. (Eds.), Wiley-VCH: Weinheim, Germany, **2004**, pp. 481–487.
6 [R] Monflier, E. Wacker-type Oxidations. in *Multiphase Homogeneous Catalysis* Cornils, B. (Ed), Wiley-VCH: Weinheim, Germany, **2005**, *1*, pp. 207–210.
7 Tsuji, J.; Shimizu, I.; Yamamoto, K. *Tetrahedron Lett.* **1976**, 2975.
8 Tsuji, J.; Yamamoto, K.; Shimizu, I. *J. Org. Chem.* **1980**, *45*, 5029.
9 Tsuji, J.; Nagashima, H. *Tetrahedron Lett.* **1982**, *23*, 2679.
10 Tsuji, J.; Nagashima, H.; Nemoto, H. *Org. Synth.* **1984**, *62*, 9–13.
11 [R] Tsuji, J. *Synthesis* **1984**, 369–384.
12 [R] Feringa, B. L. Wacker Oxidation. In *Transition Metals for Organic Synthesis* Beller, M.; Bolm, C. (eds), Wiley-VCH: Weinheim, Germany, **1998**, *2*, pp. 307–315.
13 [R] Takacs, J. M.; Jiang, X.-t. *Cur. Org. Chem.* **2003**, *7*, 369–396.
14 [R] Hintermann, L. Wacker-type oxidations. In *Transition Metals for Organic Synthesis (2nd Edition);* Beller, M.; Bolm, C. (Eds.) Wiley-VCH: Weinheim, Germany, **2004**, *2*, pp379–388.
15 [R] Tietze, L. F.; Ila, H.; Bell, H. P. *Chem. Rev.* **2004**, *104*, 3453–3516.
16 [R] Punniyamurthy, T.; Velusamy, S.; Iqbal, J. *Chem. Rev.* **2005**, *105*, 2329–2363.
17 [R] Tsuji, J. *Palladium Reagents and Catalysts*, John Wiley & Sons Ltd.: Chichester, **1996**.
18 Hamilton, R.; Mitchell, T. R. B.; Rooney, J. J. *J. Chem. Soc., Chem. Commun.* **1981**, 456.
19 Hosokawa, T.; Takano, M.; Murahashi, S.-I. *J. Am. Chem. Soc.* **1996**, *118*, 3990.
20 Hosokawa, T.; Ohta, T.; Kanayama, S.; Murahashi, S.-I. *J. Org. Chem.* **1987**, *52*, 1758.
21 Feringa, B. L. *J. Chem. Soc., Chem. Commun.* **1986**, 909.

22 Bose, A. K.; Krishnan, L.; Wagle, D. R.; Manhas, M. S. *Tetrahedron Lett.* **1986**, *27*, 5955.
23 Pellissier, H.; Michellys, P.-Y.; Santelli, M. *Tetrahedron* **1997**, *53*, 7577.
24 Pellissier, H.; Michellys, P.-Y.; Santelli, M. *Tetrahedron* **1997**, *53*, 10733.
25 Ho, T.-L.; Chang, M. H.; Chen, C. *Tetrahedron Lett.* **2003**, *44*, 6955.
26 Wright, J. A.; Gaunt, M. J.; Spencer, J. B. *Chem. Eur. J.* **2006**, *12*, 949.
27 [R] Balme, G.; Bouyssi, D.; Lomberget, T.; Monteiro, N. *Synthesis* **2003**, 2115–2134.
28 Han, X.; Widenhoefer, R. A. *J. Org. Chem.* **2004**, *69*, 1738.
29 Larock, R. C.; Wei, L.; Hightower, T. R. *Synlett* **1998**, 522.
30 [R] Hegedus, L. S. *Angew. Chem., Int. Ed. Engl.* **1988**, *27*, 1113–1126.
31 Hegedus, L. S.; Allen, G. F.; Waterman, E. L. *J. Am. Chem. Soc.* **1976**, *98*, 2674.
32 Hegedus, L. S.; Allen, G. F.; Bozell, J. J.; Waterman, E. L. *J. Am. Chem. Soc.* **1978**, *100*, 5800.
33 [R] Fujiwara, Y.; Jia, C. in *Handbook of Organopalladium Chemistry for Organic Synthesis*, Negishi, E.-i. (Ed.); John Wiley and Sons: Hoboken, 2002, vol. 2, pp 2859–2862.
34 [R] Åkermark, B.; Eberson, L.; Jonsson, E.; Pettersson, E. *J. Org. Chem.* **1975**, *40*, 1365–1367.
35 [R] Knölker, H.-J.; Fröhner, W. *J. Chem. Soc., Perkin Trans. 1* **1998**, 173–176.
36 [R] Knölker, H.-J.; O'Sullivan, N. *Tetrahedron* **1994**, *50*, 10893–10908.
37 Åkermark, B.; Oslob, J. D.; Heuschert, U. *Tetrahedron Lett.* **1995**, *36*, 1325.
38 Knölker, H.-J.; Reddy, K. R.; Wagner, A. *Tetrahedron Lett.* **1998**, *39*, 8267–8270.
39 Hagelin, H.; Oslob, J. D.; Åkermark, B. *Chem. Eur. J.* **1999**, *5*, 2413–2416.
40 Compain, P.; Gore, J.; Vatèle, J.-M. *Tetrahedron* **1996**, *52*, 10405.
41 Doan, H. D.; Gore, J.; Vatèle, J.-M. *Tetrahedron Lett.* **1999**, *40*, 6765.
42 Fallon, G. D.; Jones, E. D.; Perlmutter, P.; Selajarern, W. *Tetrahedron Lett.* **1999**, *40*, 7435.
43 Perlmutter, P.; Selajerern, W.; Vounatsos, F. *Org. Biomol. Chem.* **2004**, *2*, 2220.
44 Trend, R. M.; Ramtohul, Y. K.; Ferreira, E. M.; Stoltz, B. M. *Angew. Chem., Int. Ed.* **2003**, *42*, 2892.
45 Muñiz, K. *Adv. Synth. Catal.* **2004**, *346*, 1425.
46 Hiyashi, T.; Yamasaki, K.; Mimura, M.; Uozumi, Y. *J. Am. Chem. Soc.* **2004**, *126*, 3036.
47 Larock, R. C.; Hightower, T. R. *J. Org. Chem.* **1993**, *58*, 5298.
48 Brimble, M. A.; Elliott, R. J. R. *Tetrahedron* **2002**, *58*, 183.
49 Wood, A. J.; Holt, D. J.; Dominguez, M.-C.; Jenkins, P. R. *J. Org. Chem.* **1998**, *63*, 8522.
50 Tsuji, J.; Nagashima, H.; Hori, K. *Tetrahedron Lett.* **1982**, *23*, 2679.
51 Gaunt, M. J.; Yu, J.; Spencer, J. B. *Chem. Commun.* **2001**, 1844.
52 Ahn, J.-D.; Sherrington, D. C. *Macromolecules* **1996**, *29*, 4164.
53 Thadani, A. N.; Rawal, V. H. *Org. Lett.* **2002**, *4*, 4321.
54 Schmidt, D. R.; Park, P. K.; Leighton, J. L. *Org. Lett.* **2003**, *5*, 3535.
55 Park, P. K.; O'Malley, S. J.; Schmidt, D. R.; Leighton, J. L. *J. Am. Chem. Soc.* **2006**, *128*, 2796.
56 Cornell, C. N.; Sigman, M. S. *Org. Lett.* **2006**, *8*, 4117.
57 Gaunt, M. J.; Spencer, J. B. *Org. Lett.* **2001**, *3*, 25.
58 Uozumi, Y.; Kyota, H.; Kato, K.; Ogasawara, M.; Hayashi, T. *J. Org. Chem.* **1999**, *64*, 1620.
59 Paquette, L. A.; Wang, X. *J. Org. Chem.* **1994**, *59*, 2052.
60 Kulkarni, M. G.; Pendharkar, D. S. *J. Chem. Soc., Perkin Trans. 1* **1997**, 3127.
61 Smith, A. B., III; Cho, Y. S.; Friestad, G. K. *Tetrahedron Lett.* **1998**, *39*, 8765.
62 Smith, A. B.; Friestad, G. K.; Barbosa, J.; Bertounesque, E.; Hull, K. G.; Iwashima, M.; Qiu, Y.; Salvatore, B. A.; Spoors, P. G.; Duan, J. J.-W. *J. Am. Chem. Soc.* **1999**, *121*, 10468.
63 Toshima, H.; Oikawa, H.; Toyomasu, T.; Sassa, T. *Tetrahedron* **2000**, *56*, 8443. A very similar Acker oxidation was used in the total synthesis of (±)-nakamurol-A and its 13-epimer: Bonjoch, J.; Cuesta, J.; Diaz, S.; Gonzalez, A. *Tetrahedron Lett.* **2000**, *41*, 5669.
64 Paczkowski, R.; Maichle-Moessmer, C.; Maier, M. E. *Org. Lett.* **2000**, *2*, 3967.
65 Cachoux, F.; Ibrahim-Ouali, M.; Santelli, M. *Synlett* **2000**, 418.
66 Kobayashi, Y.; Wang, Y.-G. *Tetrahedron Lett.* **2002**, *43*, 4381.
67 Bull, J. R.; Mountford, P. G. *J. Chem. Soc., Perkin Trans. 1* **1999**, 1581.
68 Wilmouth, S.; Toupet, L.; Pellissier, H.; Santelli, M. *Tetrahedron* **1998**, *54*, 13805.
69 Ito, M.; Kibayashi, C. *Tetrahedron Lett.* **1990**, *31*, 5065.
70 Barker, D.; Brimble, M. A.; McLeod, M.; Savage, G. P.; Wong, D. J. *J. Chem. Soc., Perkin Trans. 1*, **2002**, 924.
71 O'Connor, P. D.; Mander, L. N.; McLachlan, M. M. *Org. Lett.* **2004**, *6*, 703.

Jie Jack Li

3.13 Woodward *cis*-Dihydroxylation

3.13.1 Description

The Woodward *cis*-dihydroxylation[1] involves the oxidation of an olefin to a *cis*-diol *via* a two-step reaction sequence. First, treatment of an olefin **1** with I_2, AgOAc and wet HOAc gives monoacetate **2**. Hydrolysis of **2** under basic conditions then affords *cis*-diol **3**.

I_2, AgOAc / HOAc, H_2O; KOH

1 2 3

3.13.2 Historical Perspective

In 1958, Woodward and Brutcher reported a study on the *cis*-dihydroxylation of steroid intermediate **4**.[1] They attempted to use OsO_4 as the oxidant, but obtained the undesired *cis*-diol diastereomer **5** with that reagent. Thus, they developed a two-step dihydroxylation sequence using I_2, AgOAc, and HOAc followed by KOH. Not only did this procedure work remarkably well, it afforded the desired diol diastereomer **8** in which dihydroxylation had occurred on the more sterically hindered β face of **4**. Thus, treatment of **4** with I_2, AgOAc, and HOAc afforded a mixture of acetates **6** and **7**. Basic hydrolysis of **6** and **7** with KOH in methanol then gave diol **8** in 71% yield after recrystallization.

4 → 5 (OsO_4)

4 → 6, 7 (I_2, AgOAc; HOAc, H_2O; 23 → 95 °C; 3.5 h)

6, R^1 = Ac, R^2 = H
7, R^1 = H, R^2 = Ac

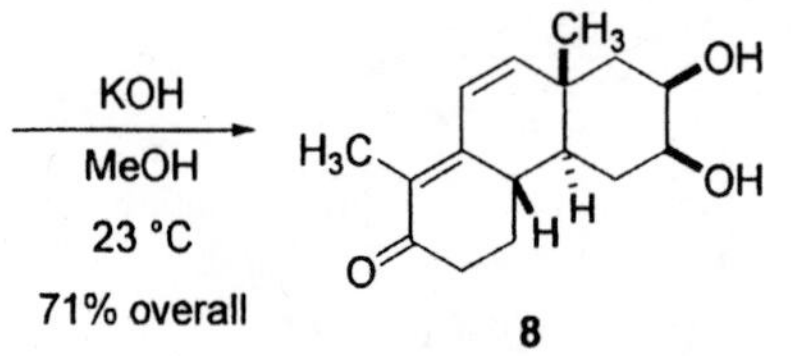

KOH, MeOH, 23 °C, 71% overall → 8

3.13.3 *Mechanism*

The proposed mechanism of the Woodward *cis*-dihydroxylation is based on the divergent stereochemical outcomes of the dihydroxylation of **4** with OsO_4 and I_2/HOAc.[1-4] It is proposed that the initial step in the dihydroxylation with I_2/HOAc is formation of iodonium ion **9** on the less hindered side of the steroidal skeleton, opposite from the angular methyl group. Nucleophilic opening of **9** with HOAc *via* axial attack gives iodoacetate **10**. When the reaction mixture is heated, cyclization of **10** occurs with displacement of the iodide to generate carbocation **11**. Capture of **11** with water then affords orthoester **12**, and collapse of **12** gives a mixture of acetates **6** and **7**. Finally, treatment of **6** and **7** with KOH in MeOH affects hydrolysis of the acetates affording diol **8**.

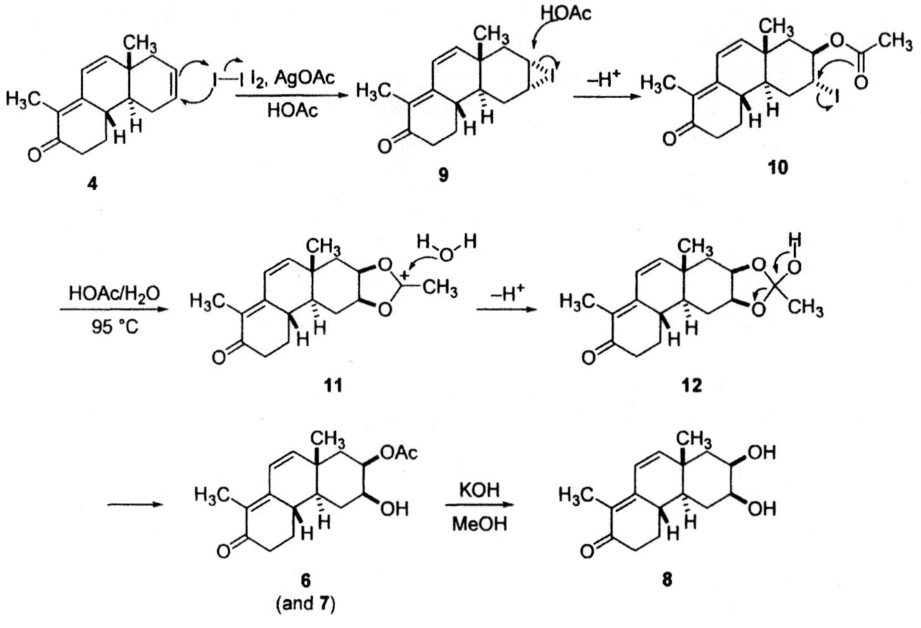

4 → (I_2, AgOAc, HOAc) → 9 → (HOAc, $-H^+$) → 10 → (HOAc/H_2O, 95 °C) → 11 → (H_2O, $-H^+$) → 12 → 6 (and 7) → (KOH, MeOH) → 8

3.13.4 Variations and Improvements

There have been several modifications to the Woodward *cis*-dihydroxylation. In one report, *N*-bromoacetamide was used in place of I_2, and in some cases this led to diol products with higher diastereoselectivity.[5] Efforts have also been focused on using different reagents in place of AgOAc. Reports in this area have included the use of diacetoxyiodine phosphonium salts,[6] polymer-supported diacetoxyiodine ammonium salts,[7,8] phosphinic acid,[9] and $Cd(OAc)_2$[10] as substitutes for AgOAc. It was recently reported that the combination of $NaIO_4$ (30 mol%), LiBr (20 mol%), and HOAc could be used to generate regioisomeric mixtures of acetates from olefins, which upon hydrolysis afforded diol products.[11]

13 → (I_2, TlOAc; HOAc, 80 °C, then H_2O, reflux) → 14 (OAc, OH) → (NaOH; H_2O, EtOH; reflux; 70–75% overall) → 15 (OH, OH)

13 → (I_2, TlOAc; HOAc, reflux) → 16 (OAc, OAc) → (NaOH; H_2O, EtOH; reflux; 65–70% overall) → 17 (OH, OH)

One of the most useful modifications of the Woodward dihydroxylation was published by Cambie and Rutledge.[12] This modification employs TlOAc in place of AgOAc and describes complementary procedures for the synthesis of *cis*-diols (Woodward reaction)[1] and *trans*-diols (Prévost reaction).[13] Thus, treatment of cyclohexene **13** with TlOAc and I_2 in HOAc at 80 °C, followed by the addition of water, affords acetate **14**. Hydrolysis of **14** then gives *cis*-diol **15** in 70–75% yield. On the other hand, exposure of cyclohexene **13** to TlOAc, I_2, and HOAc at reflux in the absence of water affords bis-acetate **16**, which upon hydrolysis gives *trans*-diol **17**.

The addition of water to the reaction mixture is required for the formation of *cis*-diol **15**. This observation can be rationalized by considering the proposed mechanism of the reaction. It is postulated that carbocation **18** is an intermediate along the reaction pathway (see intermediate **11** in Section 3.9.3). In the presence of water, **18** is presumably trapped to give orthoester **19**, which can then collapse to give *cis* mono-acetate **14**. However, in the absence of water attack of acetic acid on **18** leads to *trans* bis-acetate **16**. Thus, this modification serves as a nice compilation of methods for the synthesis of either *cis*- or *trans*-diols.

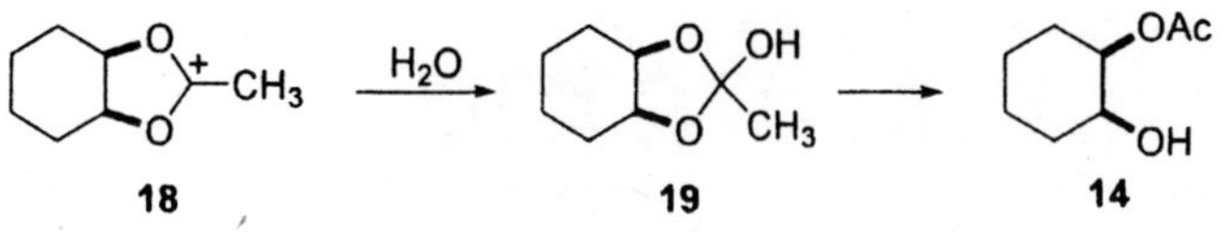

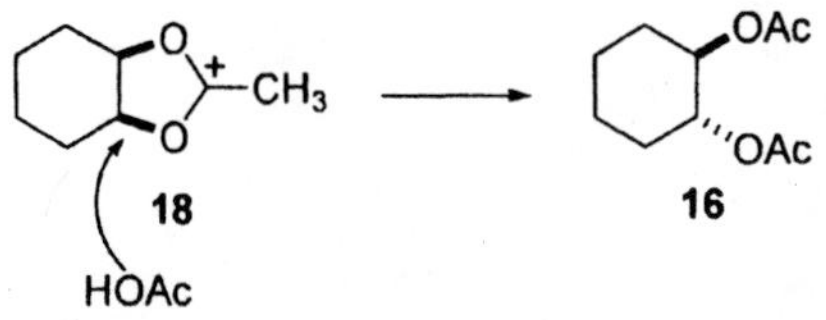

3.13.5 *Synthetic Utility*

Deslongchamps, in his synthesis of momilactone A (**24**), needed to convert olefin **20** to a diol *via* dihydroxylation on the more sterically hindered face of the molecule.[14] To accomplish this transformation, Deslongchamps used a variation of the Woodward *cis*-dihydroxylation. Thus, treatment of **20** with *N*-bromoacetamide (NBA) and AgOAc in HOAc generated bromide-acetate **21**, an intermediate in the modified Woodward *cis*-dihydroxylation. Oxidation of **21** and olefination of the resulting aldehyde then gave **22**. Exposure of **22** to HOAc and H_2O at 90 °C followed by K_2CO_3 in MeOH generated a diol intermediate, which cyclized under the reaction conditions to generate lactone **23**. The remaining hydroxy substituent in **23** was then used in a dehydration reaction to access momilactone A (**24**).

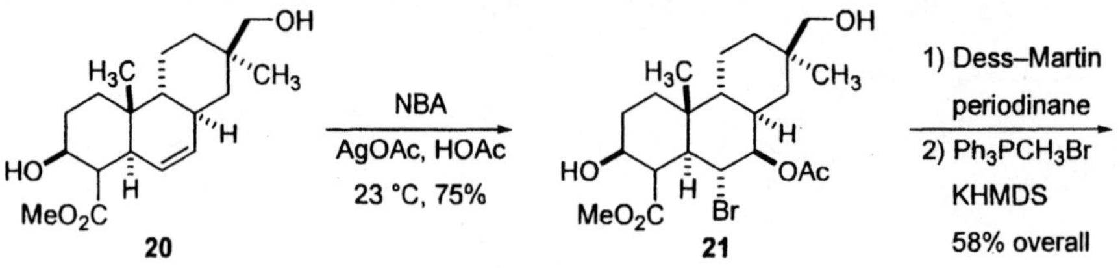

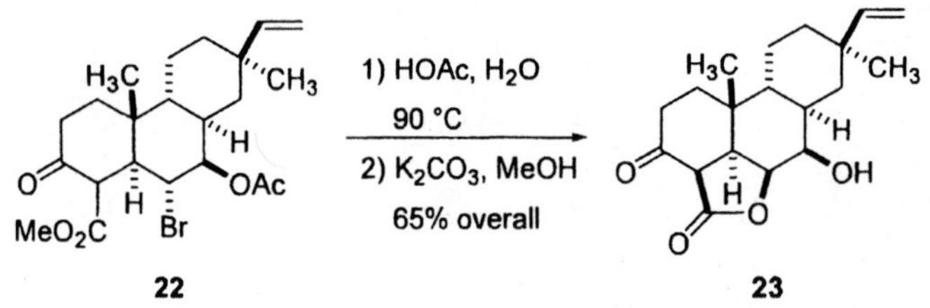

Burgess reagent
40–50%

momilactone A (**24**)

3.13.6 Experimental

I_2, AgOAc
HOAc, H_2O
23 → 95 °C
3.5 h

4

6, R^1 = Ac, R^2 = H
7, R^1 = H, R^2 = Ac

KOH
MeOH
23 °C
71% overall

8

***dl-anti-trans*-4,4a,4b,5,8,8a-Hexahydro-1,8a-dimethyl-6β-7β-dihydroxy-2(3*H*)-phenanthrone (8)**[1]

Ten grams (0.0438 mole) of *dl-anti-trans*-4,4a,4b,5,8,8a-hexahydro-1,8a-dimethyl-2(3*H*)phenanthrone **4** was dissolved in 200 mL of glacial acetic acid (analytical reagent, 99.5%) in a three-neck flask equipped with reflux condenser and thermometer. After addition of 16.43 g (0.0986 mole) of silver acetate, 11.69 g (0.0461 mole) of finely powdered iodine was added in small portions to the vigorously stirred reaction mixture over a 0.5 hour period at room temperature. When all of the iodine had been consumed (0.5 hour) 19.70 mL of aqueous glacial acetic acid (0.0438 mole of water, prepared by dilution of 2.0 mL of water up to 50 mL with glacial acetic acid) was added. The reaction mixture was then heated at 90–95 °C for three hours with vigorous stirring. At the end of this time it was cooled, treated with excess sodium chloride, and filtered free from insoluble salts. The precipitate was thoroughly washed with hot benzene. The filtrate was evaporated at the water-pump, taken up in methanol, filtered, neutralized with several milliliters of a methanolic potassium hydroxide solution, and then treated with 3.1 g of potassium hydroxide dissolved in methanol (to affect hydrolysis of acetates **6** and **7**). Hydrolysis was allowed to proceed overnight under nitrogen at room temperature. In the morning, the reaction product was neutralized carefully with dilute hydrochloric acid at ice-bath temperature. Methanol was removed at the water-pump and then under oil-pump vacuum to give a crystalline residue, which weighed 12.20 g. The crude glycol was dissolved in a large volume of ethyl acetate which was then concentrated until solid separated, at which time the solution was cooled gradually to ice-bath temperature attendant with the separation from solution of a large mass of fine needles. After filtration there remained 7.48 g (65%) of β-*cis*-glycol **8**, m.p. 184–185 °C. A second crop of β-*cis*-glycol, 0.71 g (6%) mp 181–183 °C, was obtained from the mother liquors making the overall yield 71%. After several recrystallizations from ethyl acetate the β-*cis*-glycol **8** melted at 184.6–185.2 °C (fine needles), g_{max}^{alc} 290 mμ (ε 26,986).

Note: For an experimental procedure using TlOAc, see ref. 12. For an experimental procedure using *N*-bromoacetamide, see ref. 14.

3.13.7 References

1. Woodward, R. B.; Brutcher, F. V. Jr. *J. Am. Chem. Soc.* **1958**, *80*, 209.
2. Ginsburg, D. *J. Am. Chem. Soc.* **1953**, *75*, 5746.
3. Mangoni, L.; Dovinola, V. *Tetrahedron Lett.* **1969**, *10*, 5235.
4. Whitesell, J. K.; Minton, M. A.; Flanagan, W. G. *Tetrahedron* **1981**, *37*, 4451.
5. Jasserand, D.; Girard, J. P.; Rossi, J. C.; Granger, R. *Tetrahedron Lett.* **1976**, *17*, 1581.
6. Kirschning, A.; Plumeier, C.; Rose, L. *Chem. Commun.* **1998**, 33.
7. Monenschein, H.; Sourkouni-Argirusi, G.; Schubothe, K. M.; O'Hare, T.; Kirschning, A. *Org. Lett.* **1999**, *1*, 2101.
8. Kirschning, A.; Jesberger, M.; Monenschein, H. *Tetrahedron Lett.* **1999**, *40*, 8999.
9. Muraki, T.; Yokoyama, M.; Togo, H. *J. Org. Chem.* **2000**, *65*, 4679.
10. Myint, Y. Y.; Pasha, M. A. *J. Chem. Res.* **2004**, 333.
11. Emmanuvel, L.; Shaikh, T. M. A.; Sudalai, A. *Org. Lett.* **2005**, *7*, 5071.
12. Cambie, R. C.; Rutledge, P. S. *Org. Synth.* CV 6, 348.
13. Prévost, C. *Compt. Rend.* **1933**, *196*, 1129.
14. Germain, J.; Deslongchamps, P. *J. Org. Chem.* **2002**, *67*, 5269.

Dustin J. Mergott

CHAPTER 4 OLEFINATION 333

4.1 Chugaev Elimination

4.1.1 Description

Alkyl xanthates **2** are prepared by treatment of an alcohol **1** with a base and carbon disulfide, followed by exposure of the resulting sodium salt with an alkylating agent.[1] Pyrolysis of the xanthate **2** to give an alkene **3** is called the Chugaev elimination.[2-4]

i) CS_2, NaOH; ii) R^3I — **1** → **2**; Heat, $- COS$, $-R^3SH$ — **2** → **3**

Although other xanthates have been used, methyl xanthates (R^3 = Me) are by far the most commonly employed in the Chugaev elimination. For the preparation of the xanthate, a variety of bases have been used including sodium hydride, sodium hydroxide, sodium amide, and sodium/potassium metal. In the case of pure chiral alcohol stereoisomers, epimerisation of the alcohol stereocentre, under the basic conditions, can occur (with the corresponding xanthates leading to different products). Purification of the xanthate, prior to pyrolysis, is often a problem and it is usual to pyrolyze the crude xanthate directly. Pyrolysis of the xanthate is often carried out by distillation. Depending on the pyrolysis temperature, pressure, and the boiling point of the olefin, the product will either distill with the other products (COS, thiol) or remain in the reaction flask.

4.1.2 Historical Perspective

4 → 200 °C → **5**

Chugaev discovered the formation of olefins from the pyrolysis of xanthates in 1899,[5] in connection with his studies on the optical properties of xanthates[6] and other compounds.[7] He subsequently employed the reaction in his structural investigations of terpenes and steroids, demonstrating its utility as an olefin-forming reaction and as a tool for structural determination. For example, he converted cholesterol into methyl cholesteryl xanthate (**4**) by action of carbon disulfide and methyl iodide and subsequently heated it to 200 °C to yield olefin **5**.[8]

The reaction is particularly useful for the conversion of sensitive alcohols to the corresponding olefins without the rearrangement of the carbon skeleton[9,10] and is analogous to the thermal decomposition of carboxylic esters of alcohols and related derivatives.[3]

4.1.3 Mechanism

The mechanism of the Chugaev elimination involves a cyclic, concerted transition state as originally proposed by Barton[11] and Cram.[12]

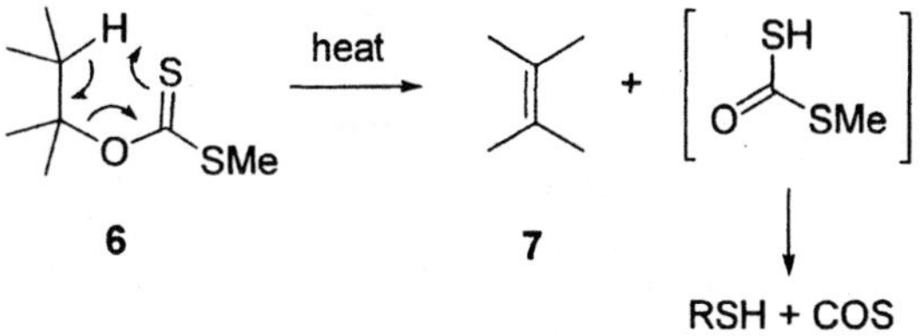

Upon heating, xanthate **6** eliminates *via* a pericyclic, *syn* elimination to yield olefin 7, along with a thiol and COS as by-products. The β-hydrogen is removed by the more nucleophilic and less hindered C=S sulfur atom. Confirmation of the sulfur atom involved in closing the ring was obtained by a study of ^{34}S and ^{13}C isotope effects.[13] The mechanism of the reaction is designated E_i in the Ingold terminology. It is important to note that there are some reports of an *anti* elimination pathway for the Chugaev elimination. However, it is likely in these examples that the starting material was impure, containing a mixture of stereoisomers and that the product distribution arose from *syn* elimination of each stereoisomer present.[2]

4.1.4 Synthetic Utility

4.1.4.1 Primary alcohols

Examples involving the pyrolysis of xanthates of primary alcohols are relatively few in number. The preparation of small molecular weight alkenes, for example, 1-pentene and isopropylethylene, using the Chugaev elimination has been demonstrated.[14] More recently, the heating of xanthate **8** furnished derivative **9**, containing an exocyclic double bond.[15] Compound **9** was subsequently converted to the insect-antifeedant dihydroclerodin.

dodecane
216 °C, 48 h
74%

8 9

4.1.4.2 *Secondary alcohols*

The Chugaev elimination of secondary alcohols may proceed in more than one direction to give structural isomers. In addition, the formed olefin may be *cis* or *trans* depending on the degree of substitution. At least three factors determine the direction of elimination: (a) the statistical, whereby the carbon atom carrying the greatest number of hydrogen atoms provides more chances for formation of the cyclic transition state; (b) the thermodynamic, whereby the more stable of the various possible olefins is preferred, and (c) the steric, which affects the energies of the possible transition states.[2]

The configuration of the olefin is determined, in part, by the stereochemistry of the xanthate. In acyclic compounds the β-hydrogen atom and the xanthate group must be co-planar, and this requirement determines the configuration of the double bond formed. For example, pyrolysis of the *S*-methyl xanthate of *erythro*-1,2-diphenyl-1-propanol (**10**) gave exclusively *trans* stilbene **11**, whereas *threo*-xanthate **12** gave exclusively *cis* stilbene **13**.[16] Interestingly, elimination took place at 130 °C for *erythro*-xanthate **10**, and at 145 °C and for *threo*- **12**, showing that the interference between the two phenyl groups increased the activation energy for the latter pyrolysis.

10 11

12 13

The utility of the Chugaev elimination for the formation of olefins without rearrangement of the carbon skeleton can be observed upon pyrolysis of the xanthate of alcohol **14**.[17] The desired vinyl cyclopropane (**15**) was isolated in reasonable yield (42%) along with a small amount of the rearranged xanthate. Alternatively, acid-mediated dehydration with sulphuric acid yielded a variety of rearranged products in low yield. Ester pyrolysis (of the acetate of **14**) also furnished a variety of compounds, with the major product being cyclopentene **16**.

In the pyrolysis of the xanthates of cyclic secondary alcohols the xanthate group and β-hydrogen atom must be *cis* to each other in order to achieve the co-planarity of the transition state. The reaction thus becomes a useful tool for determining the configuration of cyclic β-substituted alcohols and has been employed in the study of terpenoid compounds.[2] For example, pyrolysis of xanthate **17** smoothly gave 2-methylindene **18** in 80%.[18] In the Chugaev elimination of xanthate derived from substituted cyclohexanol **19**, the *trans* relationship of the β-hydrogen and the xanthate moiety favours the formation of isomer **20**, whereas for *cis* derivative **22**, formation of the more stable, conjugated product **21** is preferred.[19]

15 10%

i) Ac_2O
ii) Δ

16 60%

15 42%

i) CS_2, MeI
ii) Δ

CH_3
OH
14

CH_3
S
O
SMe

H_2SO_4

O

15 8%

17 → 18 (100 °C, 80%)

19 → 20, 96% + 21, 4% (Δ)

22 → 20, 12% + 21, 88% (Δ)

23 → 24 (220 °C, 100%)

24 → 25, 67% + 26, 33% (i) NaH, CS_2; MeI ii) 230 °C; 37% yield)

25 → 27 (Li, NH_3(l), 69%)

As mentioned previously, the Chugaev elimination has found particular use in the field of terpenoid chemistry, not only in structural studies, but also in synthesis. For example, a tandem Claisen rearrangement–ene reaction of geraniol derivative **23** gave access to alcohol **24**, which was dehydrated using the Chugaev method. This gave a mixture of olefins, which were separated and the desired compound **25** exposed to lithium metal in liquid ammonia, to furnish terpene **27**.[20]

Multiple Chugaev eliminations of several xanthates within the same molecule are possible. For example, triol **28** was smoothly converted to the corresponding methyl xanthates and subsequently heated to yield ellacene (**29**), a precursor to the polyquinenes.[21] The Chugaev elimination of xanthates derived from 1,2-diols, however, often yields other products such as cyclic thionocarbonates.[2]

OH OH OH

i) NaH, CS_2; MeI, 90%

ii) HMPA, 230 °C, 90%

28 **29**

4.1.4.3 Tertiary alcohols

The behaviour of tertiary alcohols in the Chugaev elimination mirrors that of primary and secondary systems; however, relatively few examples have been reported. The *S*-methyl xanthate of dimethylcyclopropylcarbinol **30** was pyrolyzed to give the olefin in 24% yield.[22]

S O SMe

xylene

130 °C

24%

30 **31**

The dehydration of cyclic systems is equally possible. As mentioned previously, the elimination is sensitive to steric effects. For example, in the case of cyclohexanol derivative **32a**, Chugaev elimination yielded 39% of **33a** and 10% **34a**. For **32b** and **32c**, 46% of **33b** and 6% of **34b**, and 36% of **33c** and 10% **34c** were received, respectively.[23] These results indicate the elimination proceeded as to avoid the formation of a double bond *exo* to the ring. Interestingly, the temperature required to affect the transformation was 200 °C for **32a** and **32b**, whereas it dropped to 100 °C for **32c**. This is due to the increasing bulk adjacent to the xanthate, decreasing the stability of the xanthate with elimination providing relief of the steric crowding.

32a, $R^1 = R^2 = H$
32b, $R^1 = CH_3$, $R^2 = H$
32c, $R^1 = R^2 = CH_3$

33a, $R^1 = R^2 = H$
33b, $R^1 = CH_3$, $R^2 = H$
33c, $R^1 = R^2 = CH_3$

34a, $R^1 = R^2 = H$
34b, $R^1 = CH_3$, $R^2 = H$
34c, $R^1 = R^2 = CH_3$

4.1.5 Variations

Due to the simplicity of the Chugaev elimination, few variations have been reported, although obviously it is possible to draw comparisons with other pyrolytic elimination reactions.[3] For all the example cases mentioned above, the *S*-methyl xanthates were used; however, other xanthates have been applied to the transformation. For example, the effect of the *S*-alkyl group on the stability of the xanthate for the (–)-menthyl xanthates has been reported.[24] An *S*-isopropyl group stabilizes the xanthate relative to an *S*-methyl group, whereas an *S*-benzyl group decreased stability. An *S*-*p*-nitrobenzyl group further decreased stability, suggesting that electron-withdrawing groups on the thiol sulfur atom decrease the activation energy for the Chugaev elimination. Rate studies on the decomposition showed that an increase in the electronegativity of the *S*-alkyl group decreased the stability of the xanthate in the following order: $C_2H_5 > CH_3 > C_6H_5CH_2 >$ *p*-$CH_3OC_6H_4CH_2 >$ *p*-$ClC_6H_4CH_2 > (C_6H_5)_2CH >$ *p*-$O_2NC_6H_4CH_2 > (C_6H_5)_3C$.[25] Replacement of the *S*-methyl portion by an amide group resulted in an increase in the stability of the xanthate.[26]

In terms of modified operational procedures for the transformation, in their route to (–)-kainic acid, Ogasawara and co-workers performed a tandem Chugaev elimination–intramolecular ene reaction and found sodium hydrogen carbonate to be a beneficial additive.[27] Derivative **35** was heated in refluxing diphenyl ether and in the presence of sodium hydrogen carbonate. This furnished tricyclic system **36**, which was converted to (–)-kainic acid. They later extended this procedure for the synthesis of the acromelic acids.[28]

$NaHCO_3$, Ph_2O
280 °C, 25 min
72%

35

36

4.1.6 *Experimental*

i) NaH, CS_2, THF; MeI, 18 h, 51%.
ii) dodecane, 216 °C, 48 h 74%

37 → **9**

(4a*S*,6*R*,7*S*,7a*R*,11a*R*,2′*S*,3a′*R*,6a′*S*)-7-(Hexahydrofuro[2,3-*b*]furan-2′-yl)-3,3,6,7-tetramethyl-11-methyleneoctahydronaphtho[1,8a-*d*][1,3]dioxine (9)[15]

To a stirred solution of the crude alcohol **37** (100 mg, 0.25 mmol) in THF (20 mL) was added sodium hydride (100 mg, 60% in mineral oil) at 0 °C. The reaction mixture was stirred for 2 h, followed by addition of CS_2 (1 mL), and the reaction mixture was stirred for an additional 1.5 h. After this period, MeI (0.5 mL) was added and the reaction mixture was allowed to come to room temperature and stirred overnight. Then ether (20 mL) was added, followed by ice-water (10 mL). The aqueous phase was extracted three times with ether. The combined organic layers were washed with brine, dried, and evaporated. The residue was purified by flash chromatography (10% EtOAc/hexanes) to give **8** (61 mg, 0.13 mmol, 51%) as a colorless oil: $n_D^{20^\circ}$ 9.7 (*c* 1.25 CH_2Cl_2).

A solution of degassed and freshly distilled dodecane (5 mL) and **8** (61 mg, 0.13 mmol) was heated for 48 h at reflux temperature (216 °C). Then the solvent was evaporated until 1 mL of the volume remained, followed by flash chromatography (10% EtOAc/hexanes) to give **9** (35 mg, 93 μmol, 74%) as a colorless oil: $n_D^{20^\circ}$ 13.2 (*c* 0.43 CH_2Cl_2).

4.1.7 *References*

1. Lee, A. W. M.; Chan, W. H.; Wong, H. C.; Wong, M. S. *Synth. Commun.* **1989**, *19*, 547.
2. [R] Nace, H. R. *Org. React.* **1962**, *12*, 57–100.
3. [R] Depuy, C. H.; King, R. W. *Chem. Rev.* **1960**, *60*, 444–457.
4. Smith, M. B.; March, J. In *March's Advanced Organic Chemistry: Reactions, Mechanisms and Structure*, 5th ed.; John Wiley & Sons: Hoboken, NJ, 2001; 1330.
5. Chugaev, L. A. *Ber.* **1899**, *32*, 3332.
6. Chugaev, L. A. *Ber.* **1898**, *31*, 1775.
7. Lowry, T. M. *J. Chem. Soc.* **1923**, *123*, 956.
8. Chugaev, L. A. *Ber.* **1909**, *42*, 4631.
9. Fomin, V.; Sochanski, N. *Ber.* **1913**, *46*, 244.
10. Stevens, P. G. *J. Am. Chem. Soc.* **1932**, *54*, 3732.
11. Barton, D. H. R. *J. Chem. Soc.* **1949**, 2174.
12. Cram, D. J. *J. Am. Chem. Soc.* **1949**, *71*, 3883.

13. Bader, R. F. W.; Bourns, A. N. *Can. J. Chem.* **1961**, *39*, 348.
14. Whitmore, F. C.; Simpson, C. T. *J. Am. Chem. Soc.* **1933**, *55*, 3809.
15. Meulemans, T. M.; Stork, G. A.; Macaev, F. Z.; Jansen, B. J. M.; Groot, A. D. *J. Org. Chem.* **1999**, *64*, 9178.
16. Cram, D. J.; Elhafez, F. A. A. *J. Am. Chem. Soc.* **1952**, *74*, 5828.
17. Overberger, C. G.; Borchert, A. E. *J. Am. Chem. Soc.* **1960**, *82*, 4896.
18. Alexander, E. R.; Mudrak, A. *J. Am. Chem. Soc.* **1950**, *72*, 1810.
19. Bordwell, F. G.; Landis, P. S. *J. Am. Chem. Soc.* **1958**, *80*, 6379.
20. Ho, T.-L.; Liu, S.-H. *J. Chem. Soc. Perkin Trans. 1.* **1984**, 615.
21. Fu, X.; Cook, J. M. *Tetrahedron Lett.* **1990**, *31*, 3409.
22. Van Volkenburgh, R.; Greenlee, K. W.; Derfer, J. M.; Boord, C. E. *J. Am. Chem. Soc.* **1949**, *71*, 172.
23. Benkeser, R. A.; Hazdra, J. J. *J. Am. Chem. Soc.* **1959**, *81*, 228.
24. McAlpine, I. M. *J. Chem. Soc.* **1932**, 906.
25. O'Conner, G. L.; Nace, H. R. *J. Am. Chem. Soc.* **1953**, *75*, 2118.
26. Chugaev, L. A. *Ber.* **1902**, *35*, 2473.
27. Nakagawa, H.; Sugahara, T.; Ogasawara, K. *Org. Lett.* **2000**, *2*, 3181.
28. Nakagawa, H.; Sugahara, T.; Ogasawara, K. *Tetrahedron Lett.* **2001**, *42*, 4523.

Matthew J. Fuchter

4.2 Cope Elimination Reaction

4.2.1 Description

Pyrolysis of the oxide of tertiary amine **1** yields olefin **2** and hydroxylamine **3**. This olefination method is known as the Cope elimination reaction.[1–3]

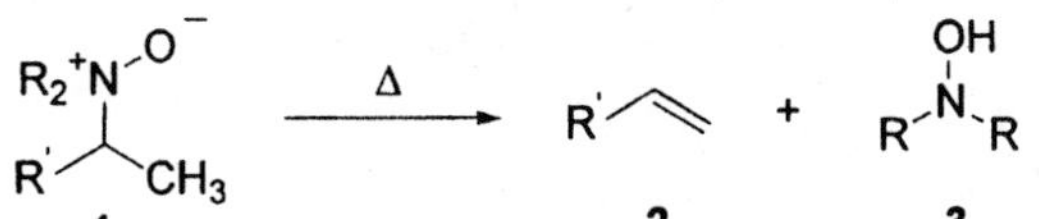

The amine oxide starting material may be prepared from the corresponding tertiary amine and 35% aqueous hydrogen peroxide in water or methanol. If a stronger oxidizing agent is required, 40% peroxyacetic acid or monoperoxyphthalic acid may be used. The excess peroxide is quenched, prior to concentration by addition of platinum black. Alternatively, common peroxy acids such as *m*-CPBA may be used for the oxidation, with the excess oxidant destroyed by addition of sodium bicarbonate. The solution of amine oxide is then concentrated under reduced pressure to give a syrup, which is then pyrolyzed by heating. Although the reaction resembles other pyrolytic eliminations, such as the Chugaev elimination and ester pyrolysis, often lower temperatures (< 150 °C) are required.

4.2.2 Historical Perspective

Although a few examples of the pyrolytic decomposition of amine oxides can be found in the early literature,[4–6] the synthetic and mechanistic importance of the reaction was first recognized by Cope and co-workers.[1] They showed that amine oxide **4** smoothly decomposed at 85–115 °C to styrene (**5**) and dimethyl hydroxylamine (**6**).

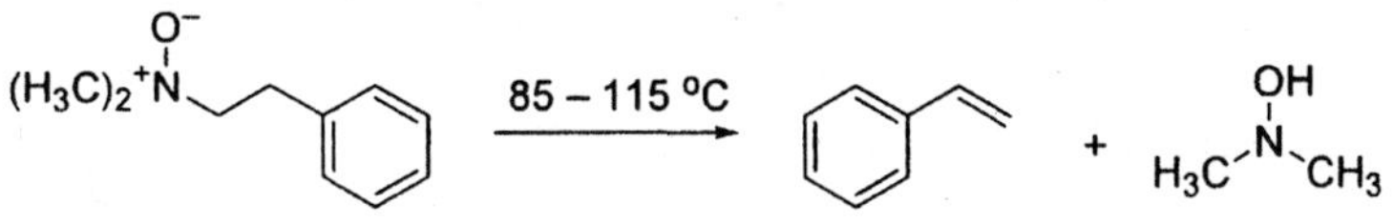

Since these pioneering studies, the reaction has proved a versatile method for introducing unsaturation under mild conditions. It can also be used as a method for accessing *N,N*-disubstituted hydroxylamine derivatives.

4.2.3 Mechanism

The mechanism of the Cope elimination involves a cyclic, concerted transition state.

Upon heating, amine oxide **7** eliminates *via* a pericyclic, *syn* elimination to yield olefin **8**, along with hydroxylamine **3** as by-products. The mechanism of the reaction is designated E_i in the Ingold terminology. The ring size of the cyclic five-membered transition state differs from the six-membered transition state of the Chugaev elimination or ester pyrolysis; however, the reactions all produce similar products. Interestingly, the Cope elimination displays enhanced stereochemical and orientational specificity compared to other pyrolysis methods. This is probably due to the relatively mild conditions of the reaction, derived from a high measure of resonance stabilization (and hence double bond character) in the transition state.[7]

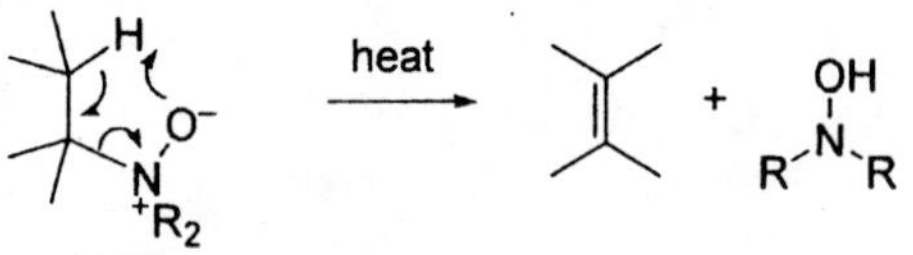

heat

7 **8** **3**

4.2.4 Synthetic Utility

4.2.4.1 Acyclic amines

Evidence for the *syn* elimination mechanism was obtained by the decomposition of *threo* and *erythro* derivatives of 2-amino-3-phenylbutane.[8] The *threo* isomer **9** reacts predominantly to give the *cis* conjugated alkene **10** in a selectivity of 400 to 1. For the *erythro* form **11**, the *trans* isomer **12** is favoured by a ratio of at least 20 to 1. The *threo* isomer **9** also reacts more readily due to the lower steric pressure in the transition state. In general, acyclic amines could undergo elimination to form either a *cis* or *trans* olefin, but the more stable *trans* form is obtained.[2]

Δ

9 **10**

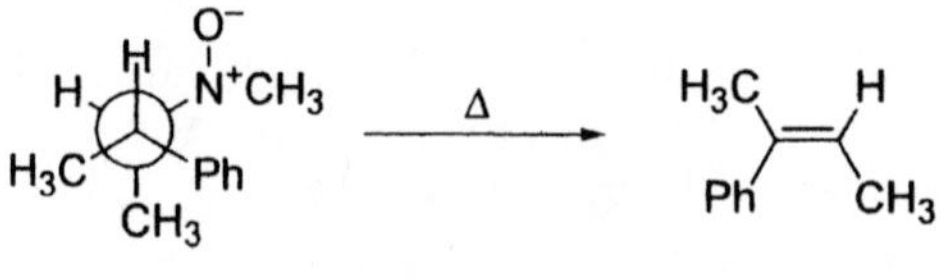

Δ

11 **12**

For simple alkyl-substituted amine oxides, the direction of elimination seems to be governed almost entirely by the number of hydrogen atoms at the various β positions. For example, the decomposition of amine oxide **13** yields 92% of 2-phenyl-2-butene (**14**) and 8% of 3-phenyl-1-butene (**15**).[8]

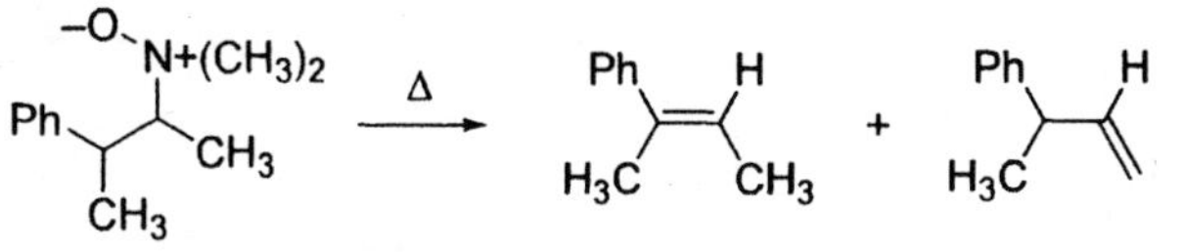

13 **14** **15**

For more complex examples, Cerero and co-workers used the Cope elimination to prepare enone **17**, a key intermediate in Paquette's approach to the taxanes, from amine oxide **16**.[9]

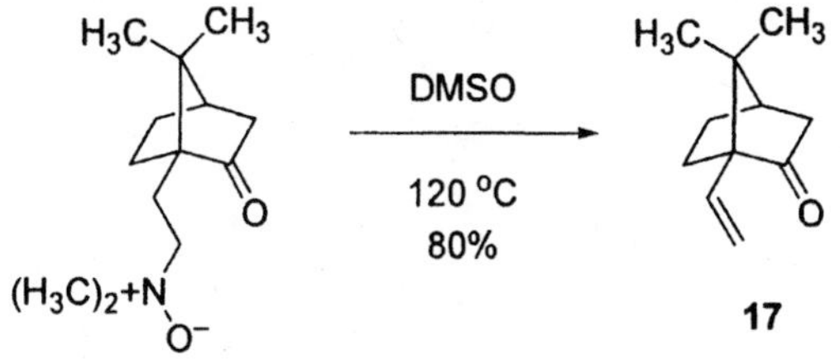

16 **17**

4.2.4.2 *Cyclic amines*

Alicyclic amine pyrolysis has been shown to follow the pattern of *cis* elimination. For example, in the case of menthyl and neomenthyl compounds **18** and **19**, respectively, the corresponding olefins **20** and **21** were obtained.[10,11] Neomenthyl compound **19** has only the *cis* hydrogen atom at the 2-position available and therefore only olefin **20** is formed, whereas menthyl compound **18** has *cis* hydrogen atoms at the 2- or 4-positions and both olefins **20** and **21** are isolated.

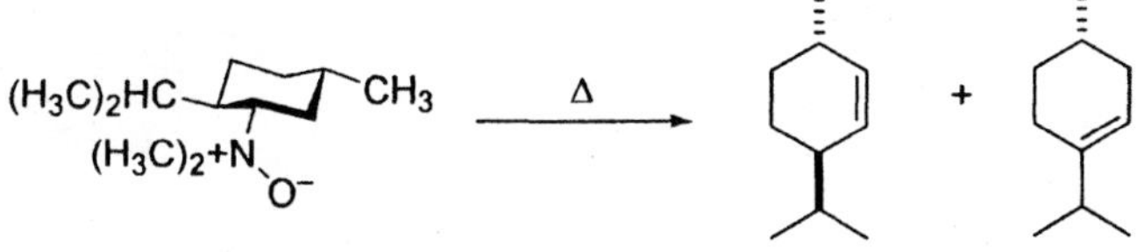

18 **20** **21**

64 : 36

$^{-}$O–N+(CH$_3$)$_2$ (H$_3$C)$_2$HC– –CH$_3$ **19** —Δ→ **20** + **21**

100 : 0

When an exocyclic branch in which the double bond may be formed is present, the product stability parallels the direction of elimination, except in the cyclohexyl compounds.[12] For example, cyclopentyl and cycloheptyl derivatives **22** and **24** give predominantly the endocyclic olefins **26** and **30**, whereas cyclohexyl derivative **23** gives **27** almost exclusively. For the cyclohexyl derivative, elimination to form an endocyclic product (**28**) would require elimination through a planar five-membered transition state, introducing unfavoured eclipsing interactions between groups at the 1, 2, 3, and 6 positions.

CH$_3$ N+(CH$_3$)$_2$ O$^-$ **22** —Δ→ **25** + **26** –CH$_3$

2.5 : 97.5

CH$_3$ N+(CH$_3$)$_2$ O$^-$ **23** —Δ→ **27** + **28** CH$_3$

97.2 : 2.8

CH$_3$ N+(CH$_3$)$_2$ O$^-$ **24** —Δ→ **29** + **30** –CH$_3$

15.2 : 84.8

OBn BnO OTBDPS BnO N+(CH$_3$)$_2$ –O **31** —145 °C, 63%→ OBn BnO OTBDPS BnO **32**

More recently, Vasella and co-workers used the Cope elimination to prepare derivative **32**, an important intermediate toward the synthesis of conformationally biased mimics of mannopyranosylamines.[13]

4.2.4.3 Heterocyclic amines

Pyrolysis of *N*-methylpiperidine oxide (**33**) results in no reaction; however, on increasing the ring size, seven- and eight-membered cyclic amines undergo ring opening readily. For example, derivatives **34** and **35** form the ring-opened products **36** and **37** in 53% and 79%, respectively.[14]

HO
N(CH3)
O⁻
N⁺–CH3
Δ
n
n

33, n = 1
34, n = 2
35, n = 3

36, n = 2
37, n = 3

N-Methyl-α-pipecoline oxide (**38**), which contains a six-membered ring, reacts to give a mixture of the elimination product **39** and saturated bicyclic system **40**.[14] Only the *trans* isomer forms the products; the *cis* isomer does not undergo elimination.

–O CH3
N⁺ CH3
Δ
OH
(H3C)N
+
O
N
(CH3)

38 **39** **40**

N-Methyl and *N*-ethyl tetrahydroquinoline oxide are reported to yield tetrahydroquinoline plus formaldehyde and acetaldehyde, respectively.[15]

4.2.5 Variations

4.2.5.1 Preparation of unsymmetrical hydroxylamines

As an alternative to an olefination method, the Cope elimination can be used as a preparative method for unsymmetrical hydroxylamines. Addition of unsymmetrical secondary amine **41** to α,β-unsaturated compounds gave access to compound **42**, which underwent oxidation and Cope elimination to yield the unsymmetrical hydroxylamine product **44**.[16]

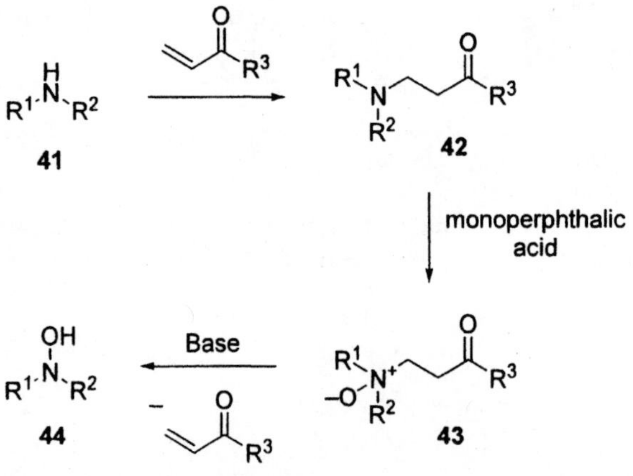

4.2.5.2 *Base-Catalyzed Elimination*

Another aspect to Rogers's work on the synthesis of unsymmetrical hydroxylamines (Section 4.2.5.1) is that the elimination step is base catalyzed (occurring at room temperature).[16] Salts of amine oxides derived from β-aminopropionic esters **45** or nitriles **46** undergo the reaction, which involves a retro-Michael addition, facilitated by the formal positive charge on nitrogen.

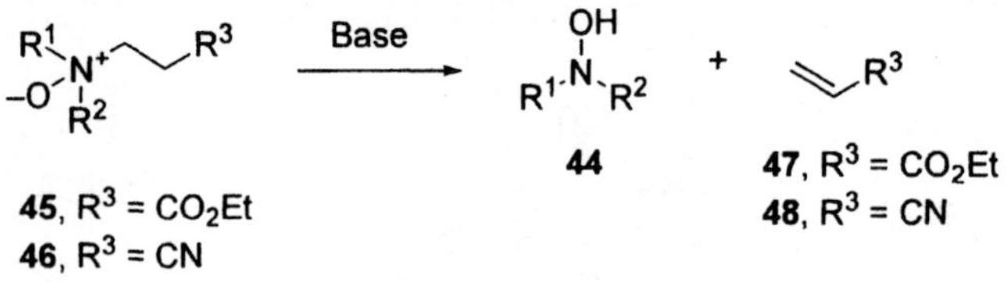

45, $R^3 = CO_2Et$
46, $R^3 = CN$
44
47, $R^3 = CO_2Et$
48, $R^3 = CN$

4.2.5.3 *The reverse cope elimination reaction*

As the name suggests, the reverse Cope elimination reaction involves formation of a tertiary amine oxide from a hydroxylamine and olefin.[17,18] For example, reaction of *N*-methylhydroxylamine (**49**) with 2,2-diphenyl-4-pentenal (**50**) in ethanol at room temperature gave amine oxide **51** in 51% yield, together with the expected nitrone **52**. It has been suggested the mechanism of the reverse Cope reaction could be a radical chain reaction[19,20]; however, more recent studies have confirmed the mechanism is analogous to the concerted Cope elimination.[17]

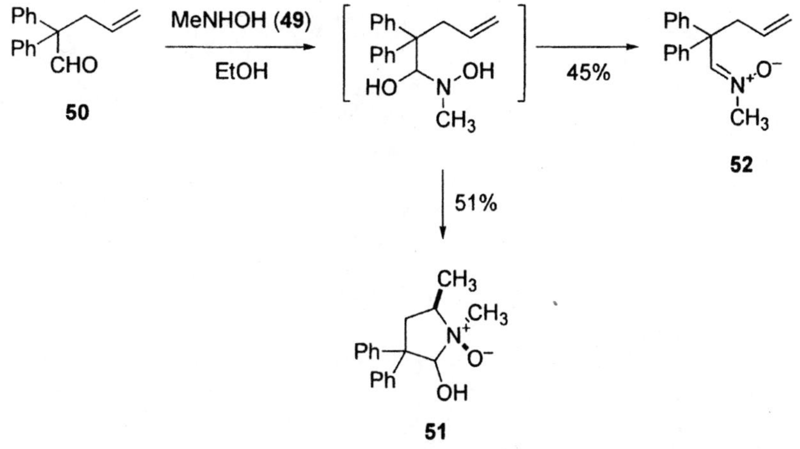

More elaborate examples have been reported; for example, the synthesis of bicyclic systems from two consecutive reverse Cope reactions have been demonstrated.[18] Thus, upon treatment with sodium cyanoborohydride (pH 4), 1,8-nonadien-5-one (**53**) underwent two consecutive reverse Cope reactions to yield the two epimeric *N*-oxides **54a** and **54b**. Reduction of these compounds with hexachlorodisilane gave indolizines **55a** and **55b** in 41% yield from **53**.

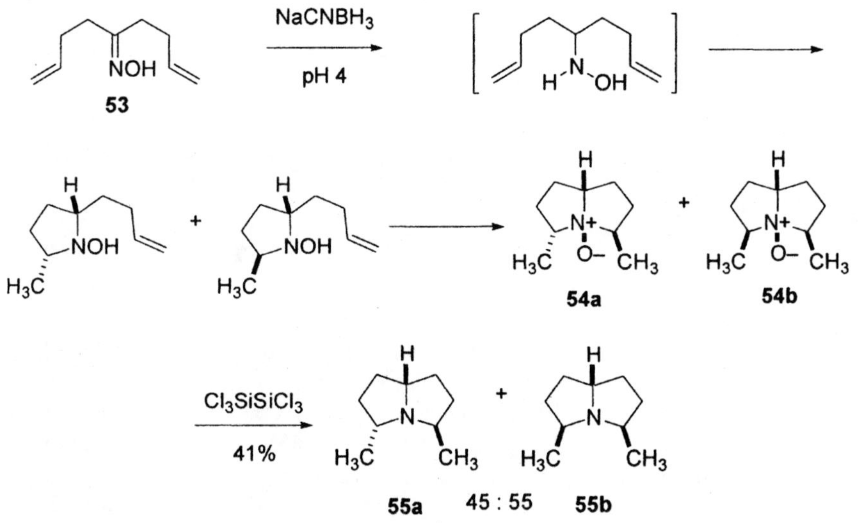

4.2.5.4 *Cope elimination for the cleavage of solid-phase linkers*

A general method for the parallel solid-phase synthesis of hydroxypiperazine derivatives **60** based on an oxidation–Cope elimination strategy of the polymer-bound phenethylamine linker with *m*-CPBA has been reported.[21] The phenethylamine linker proved useful due to its high stability in the oxidation step, allowing the solid-supported amine oxides **59** to be separated from the impurities, prior to cleavage of the products from the polymer backbone. This therefore gave access to the hydroxypiperazines **60** in high purity following elimination.

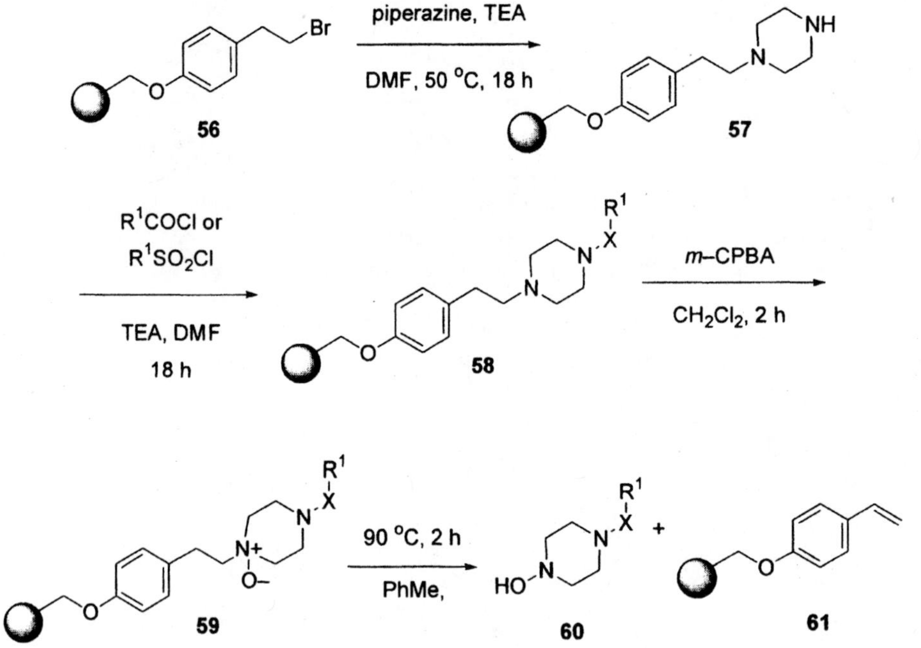

piperazine, TEA; DMF, 50 °C, 18 h

56 → 57

R^1COCl or R^1SO_2Cl; TEA, DMF, 18 h → 58

m–CPBA; CH_2Cl_2, 2 h

59 → 90 °C, 2 h; PhMe, → 60 + 61

4.2.5.5 *The sila-Cope elimination reaction*

Based on a silicon analo of the Cope elimination, Langlois and Boc devised a sila-Cope elimination for use in the synthesis of natural products.[22] They later used this methodology in the formal synthesis of Manzamine C.[23] Oxidation of intermediate **62** gave a diastereomeric mixture of *N*-oxides **63a** and **63b**, which were thermolyzed directly. For each diastereomer, *syn* elimination occurred; for **63a** the desired sila-Cope rearrangement occurred to give **64**, whereas for **63b** a normal Cope elimination occurred to give **65**. Reductive cleavage of the N–O bond gave access to amine **66**, which was further converted to cyclic amine **67**, a key intermediate to Manzamine C.

Bn H SiMe$_2$*t*-Bu
N
(CH$_2$)$_4$OSiMe$_2$*t*-Bu
H

62

m-CPBA
CH$_2$Cl$_2$, 0 °C, 2 h

Bn O– H SiMe$_2$*t*-Bu
N+
(CH$_2$)$_4$OSiMe$_2$*t*-Bu
H
+
–O Bn H SiMe$_2$*t*-Bu
N+
(CH$_2$)$_4$OSiMe$_2$*t*-Bu
H

63a **63b**

MeCN,
80 °C, 2 h

OSiMe$_2$*t*-Bu
Bn N
(CH$_2$)$_4$OSiMe$_2$*t*-Bu
+
OH (CH$_2$)$_4$OSiMe$_2$*t*-Bu
Bn N
SiMe$_2$*t*-Bu

64 (65%) **65** (35%)

Na-napthalene,
THF, 2 h

Bn
NH
(CH$_2$)$_4$OSiMe$_2$*t*-Bu

66

Ts
N

67

4.2.6 *Experimental*

1*L*-(1,4/5)-3,4,5-Tri-*O*-benzyl-1-([(*tert*-butyl)diphenylsilyloxy]methyl)cyclopent-2-ene-3,4,5-triol (32)[13]

A solution of **68** (780 mg, 1.11 mmol) in CH_2Cl_2 (3 mL) was treated at 0 °C with *m*-CPBA (206 mg, 1.19 mmol) and stirred at ambient temperature for 15 min. Evaporation, chromatography (basic Al_2O_3, activity I; EtOAc then CH_2Cl_2 : MeOH, 6 : 1), and drying *in vacuo* gave **31** (662 mg, 83%), which was immediately used for the next reaction; R_f (EtOAc : acetone, 1 : 1, Al_2O_3) = 0.5. Neat **31** (660 mg, 0.92 mmol) was heated at 145 °C for 15 min. Chromatography (hexane : EtOAc, 15 : 1 to 6 : 1) gave **32** (383 mg, 63%); R_f (EtOAc : cyclohexane, 8 : 1) 0.54; : $n_D^{20°}$ −7.8 (*c* 0.485, $CHCl_3$).

4.2.7 *References*

1. Cope, A. C.; Foster, T. T.; Towle, P. H. *J. Am. Chem. Soc.* **1949**, *71*, 3929.
2. [R] Cope, A. C.; Trumbull, E. R. *Org. React.* **1960**, *11*, 361.
3. [R] Depuy, C. H.; King, R. W. *Chem. Rev.* **1960**, *60*, 444.
4. Wernick, W.; Wolfenstein, R. *Ber.* **1898**, *31*, 1553.
5. Mamlock, L.; Wolfenstein, R. *Ber.* **1900**, *33*, 159.
6. Dodonov, Y. Y. *Zhur. Obshchei, Kim.* **1944**, *14*, 960.
7. [R] Cram, D. J. In *M. S. Newman, Steric Effects in Organic Chemistry*; John Wiley & Sons: Hoboken, NJ, 1956; 310–314.
8. Cram, D. J.; McCarthy, J. E. *J. Am. Chem. Soc.* **1954**, *76*, 5740.
9. Martinez, A. G.; Vilar, E. T.; Fraile, A. G.; Cerero, D. d. l. M. *Tetrahedron: Asymm.* **2002**, *13*, 17.
10. Cope, A. C.; Bumgardner, C. L. *J. Am. Chem. Soc.* **1957**, *79*, 960.
11. Cope, A. C.; Acton, E. M. *J. Am. Chem. Soc.* **1958**, *80*, 355.
12. Cope, A. C.; Bumgardner, C. L.; Schweizer, E. E. *J. Am. Chem. Soc.* **1957**, *79*, 4729.
13. Vasella, A.; Remen, L. *Helv. Chim. Acta.* **2002**, *85*, 1118.
14. Cope, A. C.; LeBel, N. A. *J. Am. Chem. Soc.* **1960**, *82*, 4656.
15. Dodonov, Y. Y. *J. Gen. Chem. U.S.S.R.* **1944**, *14*, 960.
16. Rogers, M. A. T. *J. Chem. Soc.* **1957**, *79*, 960.
17. Ciganek, E.; Read, J. M. *J. Org. Chem.* **1995**, *60*, 5795.

18. Ciganek, E. *J. Org. Chem.* **1995**, *60*, 5803.
19. House, H. O.; Manning, D. T.; Melillo, D. G.; Lee, L. F.; Haynes, O. R. *J. Org. Chem.* **1976**, *41*, 855.
20. House, H. O.; Lee, L. F. *J. Org. Chem.* **1976**, *41*, 863.
21. Seo, J.-s.; Kim, H.-w.; Yoon, C. M.; Ha, D. C.; Gong, Y.-D. *Tetrahedron* **2005**, *61*, 9305.
22. Bac, N. V.; Langlois, Y. *J. Am. Chem. Soc.* **1982**, *104*, 7666.
23. Vidal, T.; Magnier, E.; Langlois, Y. *Tetrahedron* **1998**, *54*, 5959.

Matthew J. Fuchter

4.3 Corey–Winter Olefin Synthesis

4.3.1 Description

The Corey–Winter olefin synthesis is a two-step transformation of a diol **1** to an olefin **3**.[1–3] A cyclic thionocarbonate **2** is prepared from the diol **1**, and subsequent heating of **2** with phosphite affords olefin **3**.

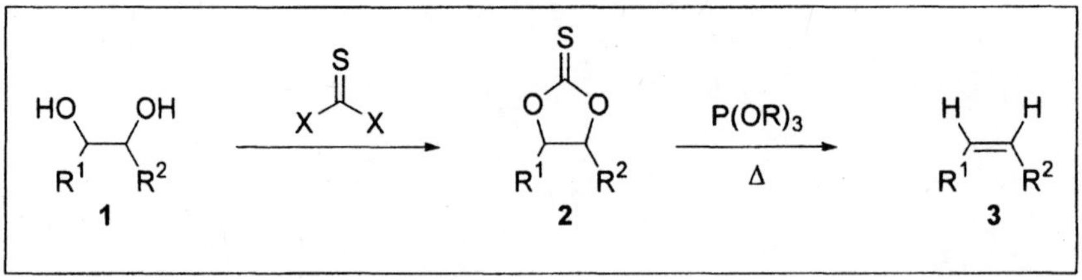

4.3.2 Historical Perspective

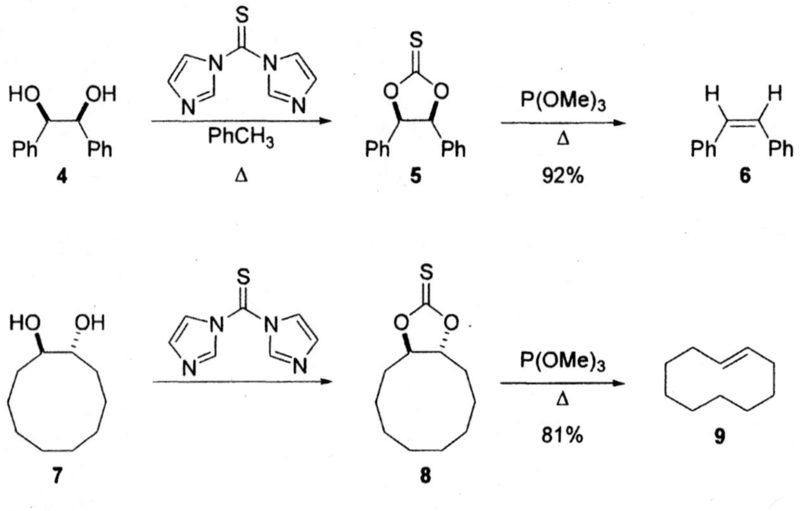

In 1963, Corey and Winter disclosed their discovery of a new method for the stereospecific synthesis of olefins from 1,2-diols.[1] Treatment of *meso*-hydrobenzoin **4** with carbonyldiimidazole in toluene at reflux gave the corresponding cyclic thionocarbonate **5** with quantitative conversion. Exposure of **5** to trimethylphosphite then afforded *cis*-stilbene **6** in 92% yield. *trans*-Stilbene could be obtained by starting with the corresponding *trans*-diol. This two-step reaction sequence could also be used for the synthesis of cyclic olefins. Treatment of *trans*-cyclodecanediol 7 with thionocarbonyldiimidazole gave the desired cyclic thionocarbonate **8**, and exposure of **8** to trimethylphosphite generated *trans*-cyclodecene **9** in 81% yield.

4.3.3 Mechanism

Formation of cyclic thionocarbonates from diols seems fairly straightforward mechanistically. However, the mechanism of decomposition of the thionocarbonates to generate olefins is more complicated.[1–8] Based on the original mechanistic hypothesis by Corey and Winter,[1] the conversion of thionocarbonate to olefin is believed to proceed *via* a carbene intermediate.[3–6] Nucleophilic attack on thionocarbonate **10** by trimethylphosphite gives zwitterion **11**, which loses trimethylthionophosphite to give carbene **12**. Cycloreversion[8] of **12** then affords olefin **13** in 85% yield along with carbon dioxide. This mechanistic hypothesis is supported by subsequent pyrolysis studies of norbornadienone ethylene ketals.[4]

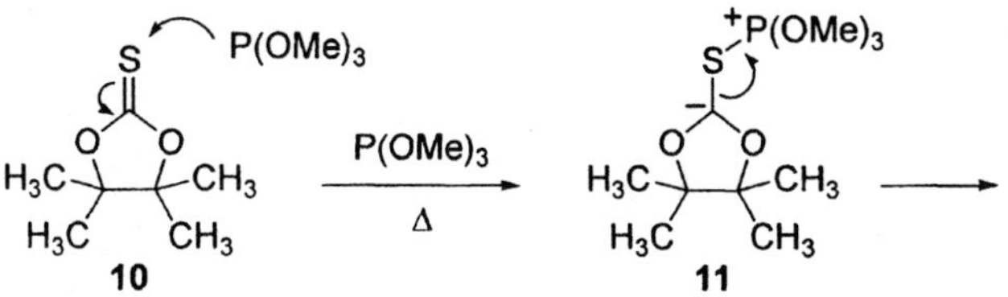

P(OMe)3
10
P(OMe)3
Δ
+P(OMe)3
11
12
85%
13

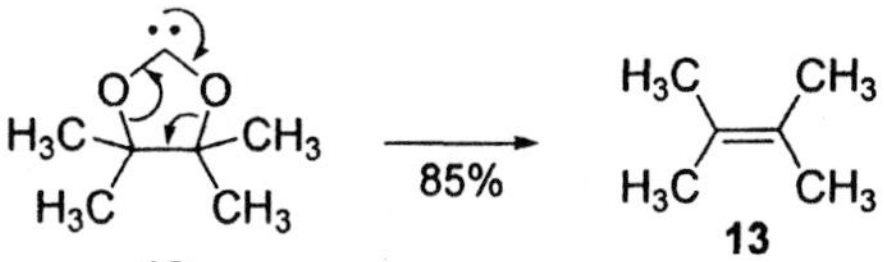

However, evidence opposing the intermediacy of a carbene species in the Corey–Winter olefin synthesis also exists.[3,7] Thermal decomposition of hydrazone salt **14**, which is proposed to proceed *via* carbene **12**, leads to a mixture of products **13**, **15**, and **16**.[7] Thus, an alternative mechanism has been proposed for the Corey–Winter reaction that invokes a phosphorus ylide. In this mechanistic scenario, initial reaction of thionocarbonate **10** with trimethylphosphite affords zwitterion **11**. Cyclization generates **17**, and nucleophilic attack on this intermediate by trimethylphosphite then gives ylide **18**. Breakdown of **18** affords olefin **13**, trimethylphosphite, and carbon dioxide. It is also possible that ylide **18** is generated directly *via* nucleophilic attack on carbene **12** by phosphite. Evidence for an ylide intermediate has been obtained from trapping experiments with trithiocarbonates and benzaldehyde.[3]

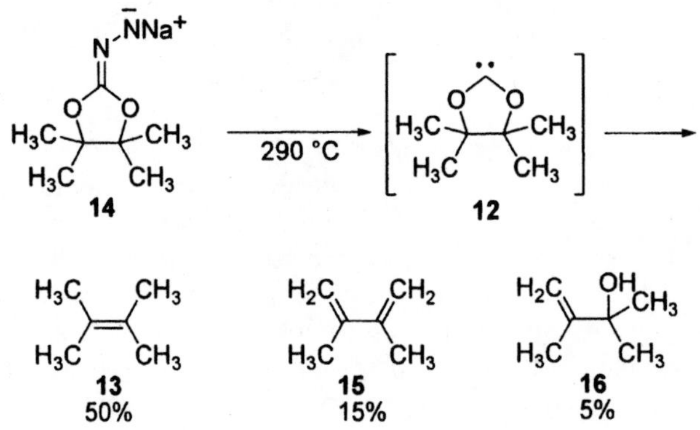

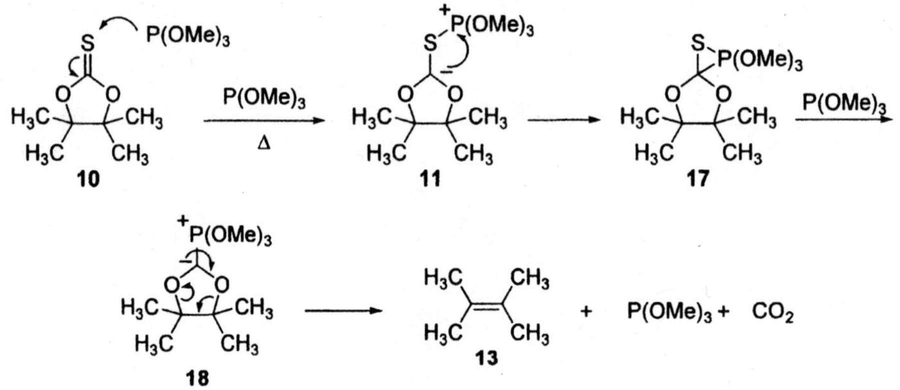

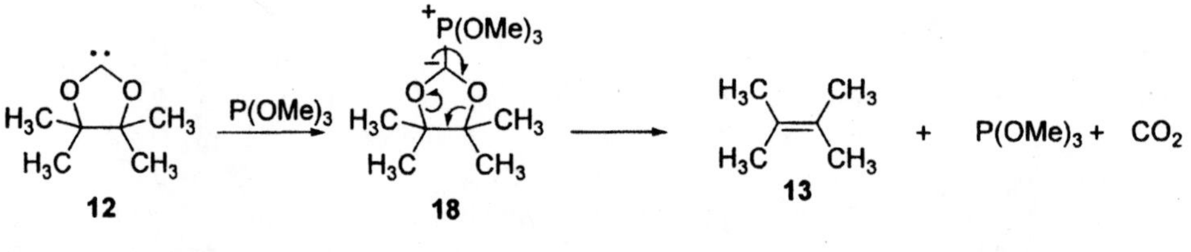

4.3.4 *Variations and Improvements*

There have been several modifications to the Corey–Winter reaction, the most significant of which was published by the Corey group in 1982.[9] Corey and Hopkins reported that by using a diazaphospholidine in place of trimethylphosphite, thionocarbonates could be converted to the corresponding olefins at significantly lower temperatures. Thus, thionocarbonate **20** was prepared from diol **19** in 93% yield, and subsequent treatment of **20** with diazaphospholidine **21** afforded olefin **22** in 88% yield. Several complex examples were disclosed in which olefins were generated in 70–94% yield.[9]

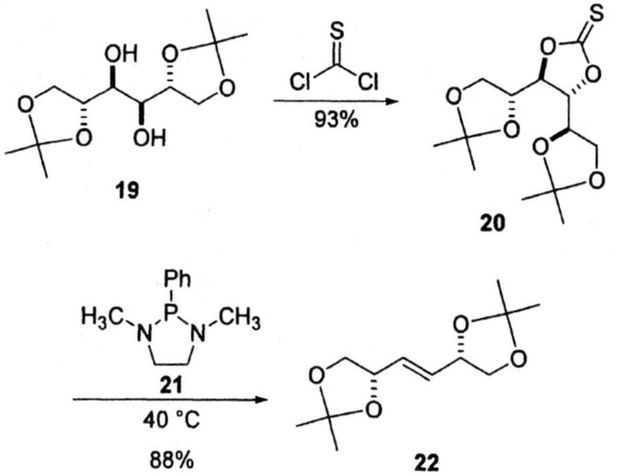

It has also been shown that olefins can be prepared from thionocarbonates *via* the corresponding iodothiocarbonates.[10,11] For example, treatment of thionocarbonate **23** with methyl iodide affords iodothionocarbonate **24** in 95% yield. Exposure of **24** to *n*-BuLi then gives olefin **25** in 99% yield. Several examples employing these reaction conditions are reported with yields of 72–96% overall.[10] This olefin synthesis does require three steps and only terminal olefin products are reported. Finally, it has also been shown that olefins can be generated from the corresponding cyclic sulfates using either magnesium iodide[12] or triphenylphosphine/iodine.[13]

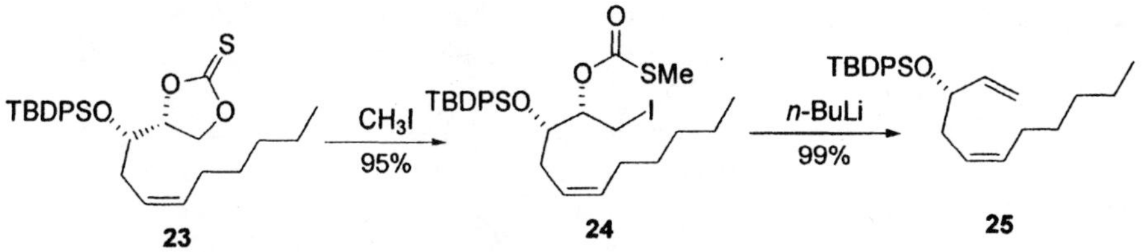

4.3.5 *Synthetic Utility*

The Corey–Winter reaction has found use in complex natural product synthesis. In one example, treatment of thionocarbonate **26** with diazaphospholidine **21** afforded diene **27** in 82% yield. This diene was ultimately used in a total synthesis of (–)-taxol.[14]

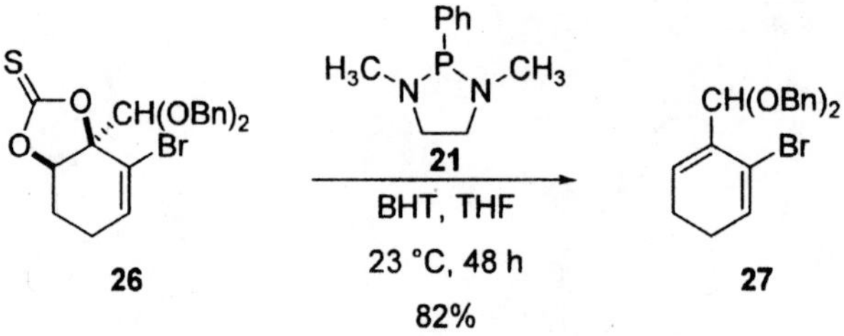

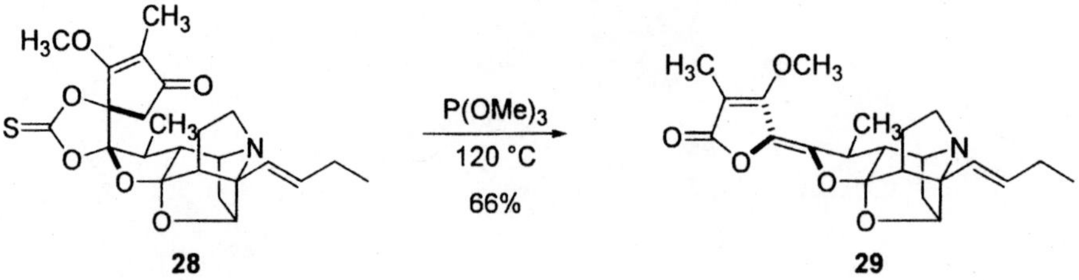

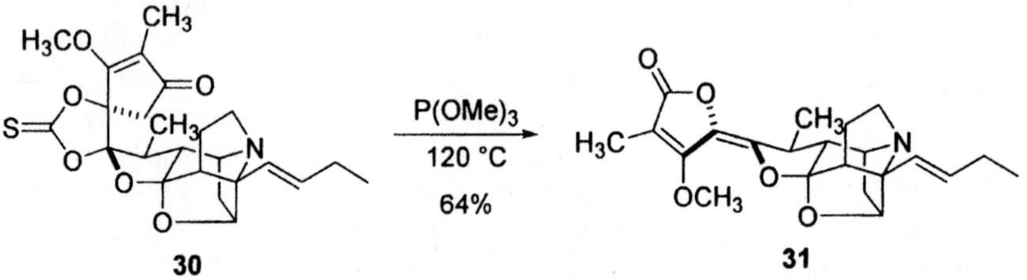

Overman and co-workers used the Corey–Winter reaction in total syntheses of two complex alkaloids. Treatment of thionocarbonate **28** with trimethylphosphite gave didehydrostemofoline **29** in 66% yield. Under the same conditions, isodidehydrostemofoline **31** was obtained from thionocarbonate **30** in 64% yield.[15]

The Corey–Winter reaction has also been used in a formal synthesis of the AIDS drug (–)-abacavir (**35**).[16] Thionocarbonate **33** was prepared from diol **32** in 78% yield. Then, treatment of **33** with phospholidine **21** afforded olefin **34** in 65% yield. The preparation of **34** constitutes a formal synthesis of (–)-abacavir (**35**).

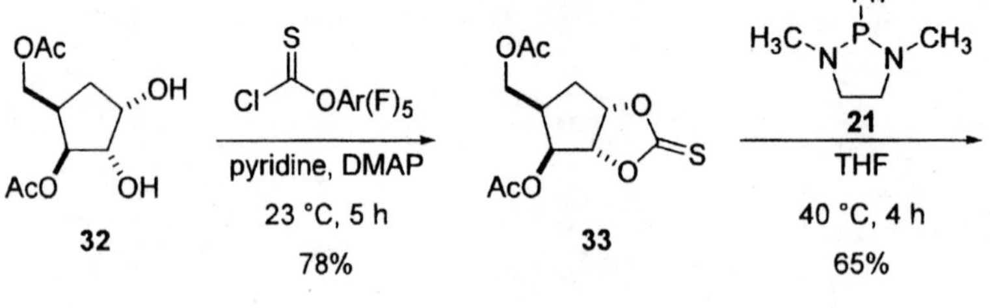

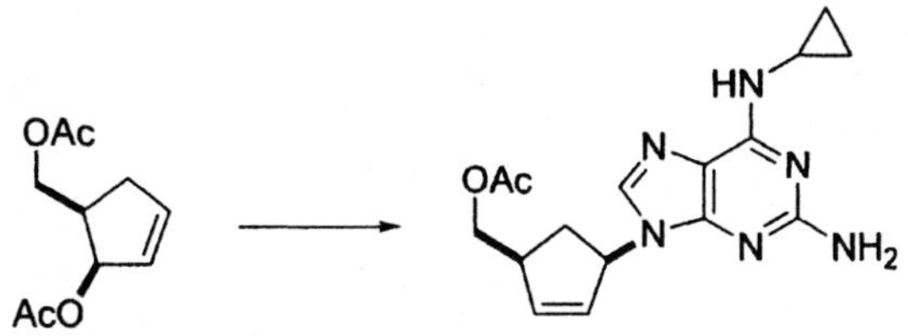

34 (–)-abacavir (35)

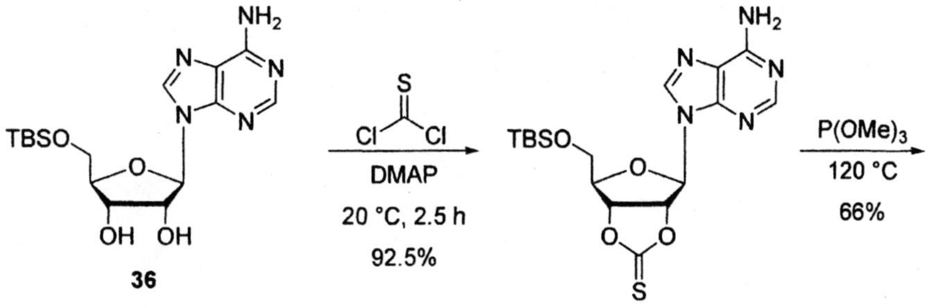

36 37

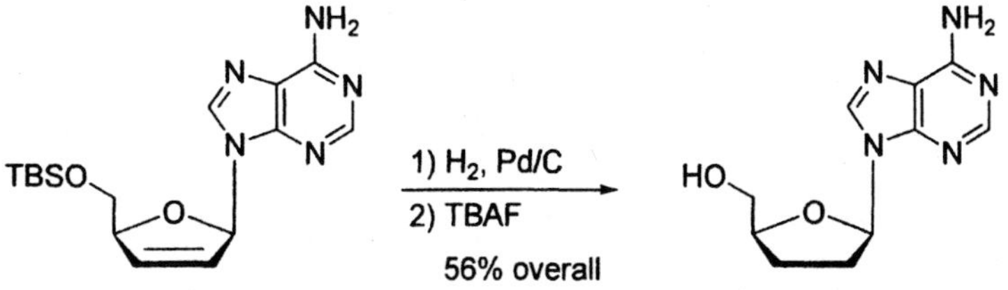

38 39

The Corey–Winter olefination was used on kilogram scale in the synthesis of 2′,3′-dideoxyadenosine (**39**).[17] Exposure of diol **36** to thiophosgene gave thionocarbonate **37** in 92.5% yield, and subsequent treatment of **37** with $P(OMe)_3$ generated olefin **38** in 60.5% yield. Olefin **38** was then converted in two steps and 56% yield to 2′,3′-dideoxyadenosine **39**. Remarkably, the adenosine amino substituent did not require protection during this synthetic sequence. Overall, 905 g of **39** was prepared *via* this route.

The Corey–Winter olefin synthesis has been performed on β-lactam derivatives.[18] This olefination was employed in cases where traditional Wittig-type olefination of the corresponding aldehyde led to disappointing results. In one example, treatment of **40** with carbonyldiimidazole and exposure of the resulting thionocarbonate to $P(OMe)_3$ afforded olefin **41** in 81% yield.

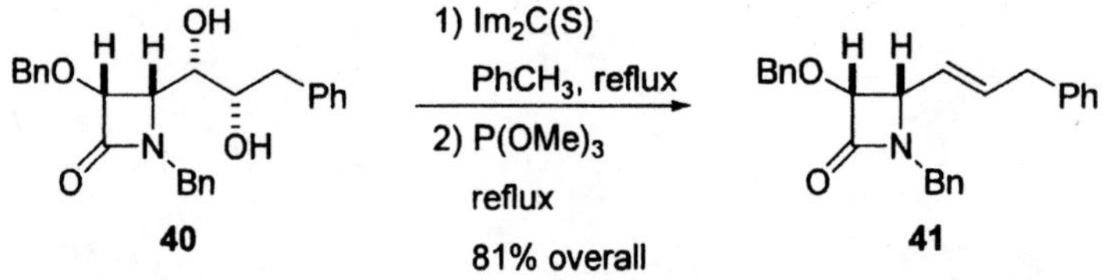

1) $Im_2C(S)$, $PhCH_3$, reflux; 2) $P(OMe)_3$, reflux; 81% overall — 40 → 41

Finally, Paterson and Schlapbach employed the Corey–Winter olefination to generate a trisubstituted olefin in their studies toward a total synthesis of discodermolide.[19] In this example, diols **42** and **43** were treated with thiocarbonyldiimidazole to give the corresponding thionocarbonates. Exposure of this mixture of thionocarbonates to phospholidine **21** at 50 °C generated olefin **44** in 81% yield.

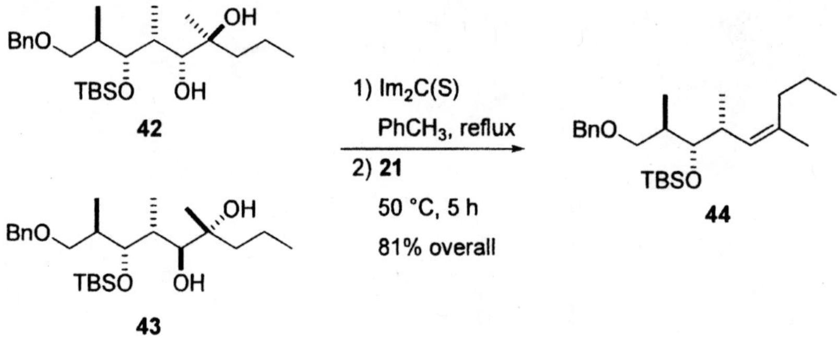

1) $Im_2C(S)$, $PhCH_3$, reflux; 2) **21**, 50 °C, 5 h; 81% overall — 42 + 43 → 44

4.3.6 Experimental

$P(OMe)_3$, 120 °C, 66% — 28 → 29

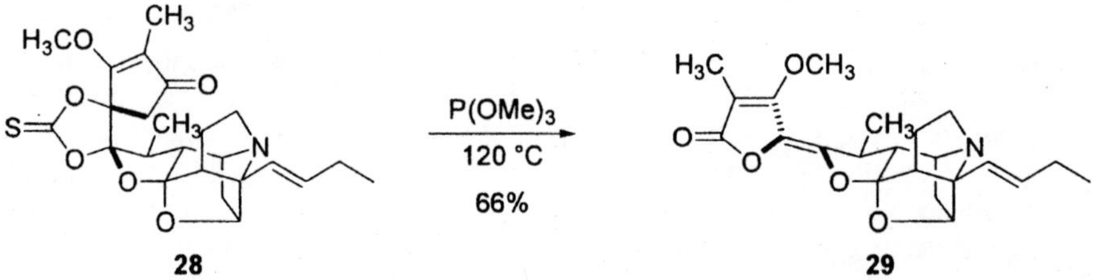

Didehydrostemofoline (29)[15]

A solution of thiocarbonate **28** (14.0 mg, 30.3 μmol; prepared from the corresponding diol and thiophosgene) and $P(OMe)_3$ (1.0 mL, 8.48 mmol) was heated to 120 °C in a sealed tube for 12 h. After cooling to room temperature, the mixture was diluted with EtOAc (10 mL) and washed with saturated aqueous $NaHCO_3$ (4.0 mL). The phases were separated and the aqueous layer was extracted with EtOAc (3 × 20 mL). The combined organic layers were dried (Na_2SO_4) and the solvents and $P(OMe)_3$ were removed *in vacuo*. The ^1H NMR spectrum of the crude product showed the desired product **28**. Chromatography of this residue (50:50 EtOAc:hexanes → EtOAc) gave 7.7 mg (20 μmol, 66%) of pure didehydrostemofoline **29**.

19 → ($Cl_2C{=}S$, 93%) → **20**

21 (40 °C, 88%) → **22**

***trans*-3,4-Didehydro-3,4-dideoxy-1,2,5,6-di-*O*-isopropylidene-*D*-threo-hexitol** (22):[9] To a stirred solution of 262 mg (1.0 mmol) of diol **19** and 293 mg (2.4 mmol) of 4-DMAP in 4.0 mL of dry methylene chloride at 0 °C under argon was added 108 μL (162 mg, 1.2 mmol) of 85% thiophosgene in carbon tetrachloride and the contents were stirred for 1.0 h at 0 °C. Silica gel (2.0 g, Merck) was added and the mixture was allowed to warm to 25 °C. After removal of the methylene chloride *in vacuo*, the remaining solid was loaded onto a column of 6.0 g of silica gel and eluted with 50% ethyl acetate in hexane. Concentration *in vacuo* afforded thionocarbonate **20** (284 mg, 93%) as a colorless solid: mp = 152–156 °C, TLC R_f 0.51 (50% ethyl acetate in hexane), IR ($CHCl_3$) 1320 cm^{-1} (C=S), ^1H NMR ($CDCl_3$, 80 MHz) δ 1.34 (6H, s), 1.46 (6H, s), 4.15 (6H, m), 4.66 (2H, m).

A suspension of thionocarbonate **20** (164 mg, 0.54 mmol) in 0.29 mL (310 mg, 1.6 mmol) of **21** under argon was stirred at 40 °C for 20 h. After cooling to 25 °C, the contents were directly chromatographed on a column of 20 g of silica gel (elution with 5% ether in methylene chloride) to afford 108 mg (88%) of pure *trans* **22** as a colorless solid: mp 80–81 °C, TLC R_f 0.55 (17% ether in methylene chloride), ^1H NMR ($CDCl_3$, 80 MHz) δ 1.35 (s, 6H), 1.39 (6H, s), 3.55 (t, J = 7.7 Hz, 2 H), 4.05 (dd, J = 8.0, 6.2 Hz, 2 H), 4.45 (m, 2H), 5.77 (m, 2H), α_D^{20} +56.7° (c 3.2, $CHCl_3$).

Note: The modified Corey–Winter reaction, using **21**, can also be run without solvent.[16]

4.3.7 References

1. Corey, E. J.; Winter, R. A. E. *J. Am. Chem. Soc.* **1963**, *85*, 2678.
2. Corey, E. J.; Carey, F. A.; Winter, R. A. E. *J. Am. Chem. Soc.* **1965**, *87*, 934.
3. [R] Block, E. *Org. React.* **1984**, *30*, 457.
4. Lemal, D. M.; Gosselink, E. P.; Ault, A. *Tetrahedron Lett.* **1964**, *5*, 579.
5. Lemal, D. M.; Lovald, R. A.; Harrington, R. W. *Tetrahedron Lett.* **1965**, *6*, 2779.
6. Horton, D.; Tindall, C. G. Jr. *J. Org. Chem.* **1970**, *35*, 3558.
7. Borden, W. T.; Concannon, P. W.; Phillips, D. I. *Tetrahedron Lett.* **1973**, *14*, 3161.
8. [R] Bianchi, G.; De Micheli, C.; Gandolfi, R. *Angew. Chem. Int. Ed.* **1979**, *18*, 721.
9. Corey, E. J.; Hopkins, B. *Tetrahedron Lett.* **1982**, *23*, 1979.
10. Adiyaman, M.; Jung, Y.-J., Kim, S.; Saha, G.; Powell, W. S.; FitzGerald, G. A.; Rokach, J. *Tetrahedron Lett.* **1999**, *40*, 4019.
11. Adiyaman, M.; Khanapure, S. P.; Hwang, S. W.; Rokach, J. *Tetrahedron Lett.* **1995**, *36*, 7367.
12. Jang, D. O.; Joo, Y. H. *Synth. Commun.* **1998**, *28*, 871.
13. Jang, D. O.; Joo, Y. H.; Cho, D. H. *Synth. Commun.* **1997**, *27*, 2379.
14. Kusama, H.; Hara, R.; Kawahara, S.; Nishimori, T.; Kashima, H.; Nakamura, N.; Morihira, K. Kuwajima, I. *J. Am. Chem. Soc.* **2000**, *122*, 3811.
15. Brüggermann, M.; McDonald, A. I.; Overman, L. E.; Rosen, M. D.; Schwink, L.; Scott, J. P. *J. Am. Chem. Soc.* **2003**, *125*, 15284.
16. Freiría, M.; Whitehead, A. J.; Motherwell, W. B. *Synthesis* **2005**, 3079.
17. Carr, R. L. K.; Donovan, T. A., Jr.; Sharma, M. N.; Vizine, C. D.; Wovkulich, M. J. *Org. Prep. Proced. Int.* **1990**, *22*, 245.
18. Palomo, C.; Oiarbide, M.; Landa, A.; Esnal, A.; Linden, A. *J. Org. Chem.* **2001**, *66*, 4180.
19. Paterson, I.; Schlapbach, A. *Synlett* **1995**, 498.

Dustin J. Mergott

4.4 Perkin Reaction

4.4.1 Description

The Perkin reaction,[1] which is normally applicable only to aromatic aldehydes, is usually effected by heating a mixture of the aldehyde, an acid anhydride, and a weak base to form cinnamic acids.[2]

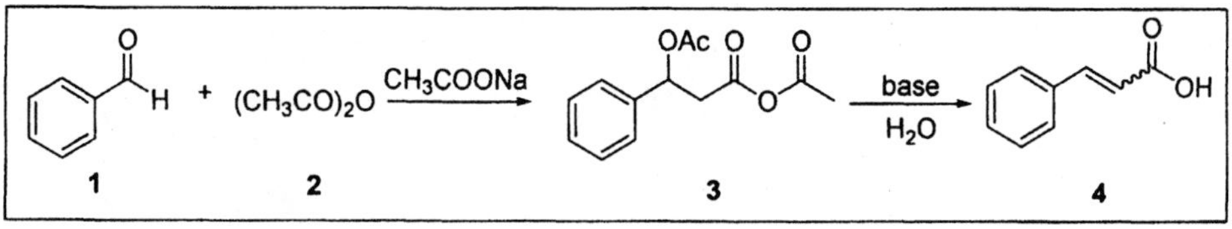
H
+ $(CH_3CO)_2O$
CH_3COONa
OAc O O
base
H_2O
OH
1
2
3
4

4.4.2 Historical Perspective

In 1868, W. H. Perkin[1] described a synthesis of coumarin by heating the sodium salt of salicylaldehyde with acetic anhydride. Further study of this reaction led to the development of a new general method for preparing cinnamic acid and its analogs, which became known as the Perkin reaction.[3]

4.4.3 Mechanism

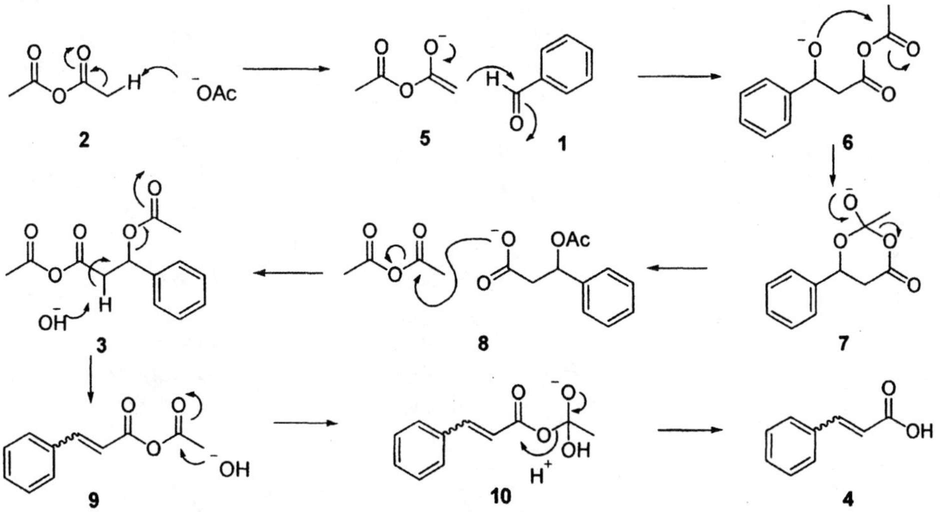
OAc
2
5
1
6
3
8
7
9
10
4
OH
H

The Perkin reaction is believed to proceed by forming an enolate anion **5** from the acid anhydride **2**, which subsequently reacts with the aldehyde **1** to form an alkoxide **6**. The ensuing intramolecular acylation affords β-acyloxy derivative **8**, which undergoes acylation and elimination to form the cinnamic acid ester **9**. Hydrolysis of ester **9** affords the cinnamic acid **4**. The kinetically favored stereoisomer formed in the elimination step has the carboxyl group *trans* to the larger group at the beta carbon atom. When tertiary amines are used as catalysts, the first stage of the proposed mechanism involves the formation of a ketene **14**, which reacts with 4-nitrobenzaldehyde (**15**) yielding a β-lactone **16**, which cleaves to the unsaturated acid **17**.[4]

$$\underset{\mathbf{2}}{(CH_3CO)_2O} + \underset{\mathbf{11}}{NEt_3} \longrightarrow \underset{\mathbf{12}}{CH_3CO\overset{+}{N}Et_3} + \underset{\mathbf{13}}{CH_3COO^-} \longrightarrow \underset{\mathbf{14}}{CH_2=C=O}$$

15 → (CH₂ = C =O) → 16 → 17

4.4.4 *Variations*

Simple aliphatic aldehydes and ketones cannot be used as the carbonyl component in the Perkin reaction. Knoevenagel[5] reported the successful condensation of aldehydes and ketones with malonic acid in the presence of ammonia or amines. The most satisfactory method uses pyridine as a catalyst and is known as the Doebner modification.[6] Thus, acetaldehyde **18** reacts with malonic acid (**19**) in the presence of a pyridine catalyst to afford the acid **20** in 60% yield.

$$\underset{\mathbf{18}}{CH_3CHO} + \underset{\mathbf{19}}{CH_2(CO_2H)_2} \xrightarrow[20\ ^{\circ}C,\ 60\%]{pyr.} \underset{\mathbf{20}}{CH_3CH{=}CHCO_2H}$$

Although acetic anhydride is often used in the Perkin reaction as the acid anhydride component, other acid anhydrides can be used in the Perkin reaction as well. Oglialoro reported that sodium phenylacetate **21** and acetic anhydride give α-phenylcinnamic acid **22**. The Oglialoro modification generates a mixed anhydride *in situ* and is a convenient preparative method that obviates the necessity of using the arylacetic anhydride as a starting material.[2]

$$\underset{\mathbf{1}}{PhCHO} + \underset{\mathbf{21}}{PhCH_2CO_2Na} \xrightarrow[165\ ^{\circ}C,\ 60\%]{(CH_3CO)_2O} \underset{\mathbf{22}}{PhCH{=}C(Ph)CO_2H}$$

Fittig (Fittig synthesis) discovered that aromatic and aliphatic aldehydes (i.e., **1**) react readily with sodium succinate **23** and acetic anhydride at 100 °C, to give γ-phenyl and γ-alkyl paraconic acid (i.e., **24**) in satisfactory yield. The acid loses carbon dioxide upon heating to form a β,γ-unsaturated acid **25**.[2]

-CHO + CO₂Na / CO₂Na — $(CH_3CO)_2O$, 100 °C, 55% → (lactone) CO_2H —Δ→ CO_2H

1 **23** **24** **25**

4.4.5 Synthetic Utility

In general, the Perkin reaction is limited to aromatic aldehydes. The activity of substituted benzaldehydes in the Perkin reaction is similar to the trends observed in other reactions involving the carbonyl group. A halogen (**28**) or nitro (**34**) group in any position increases the rate of reaction and the yield; a methyl group (**26**) in any position decreases the rate and yield, and this effect falls off in the order: *ortho* > *meta* > *para*.[2] A methoxy group in the ortho position (**30**) has a small favorable influence, but in the para position (**32**) it has a definite unfavorable effect on the rate and yield.

+ $(CH_3CO)_2O$ — CH_3COONa, 180°C →

X	Yield %
H (**1**)	45 (**4**)
4-methyl (**26**)	33 (**27**)
4-chloro (**28**)	52 (**29**)
2-methoxy (**30**)	55 (**31**)
4-methoxy (**32**)	30 (**33**)
4-Nitro (**34**)	82 (**35**)

The Perkin reaction has been carried out on aldehydes with aromatic rings other than phenyl, including biphenyl, naphthalene, furan, and indole series using tripolyphosphate (TPP) as the catalyst. However, the Knoevenagel modification is more useful in these cases.[7,8]

RCHO (36) + $CNCH_2CO_2Et$ (37) —TPP, 75–80 °C→ $RCH{=}C(CN)CO_2Et$ (38)

R = 2-furyl (CHO), 3-indolyl (CHO), 1-naphthyl (CHO)

Yield 91% 85% 88%

Cinnamaldehyde (**39**), which is a vinylog of benzaldehyde, gives excellent yields of β-styrylacrylic acid (**40**) under Perkin reaction conditions[3] or Knoevenagel conditions.[9]

39 + $(CH_3CO)_2O$ —CH_3CO_2Na, 165 °C, 60%→ **40**

39 + $CH_2(COOH)_2$ —Py., reflux, 12 h, 56%→ **40**

Brady *et al.* demonstrated that the intramolecular reaction of ketoacid **41** with sodium acetate in the presence of acetic anhydride forms the benzofuran **42**.[10]

41 —NaOAc, Ac_2O, 74%→ **42**

The Perkin reaction has been used to synthesize trifluoromethyl-containing building blocks that are very useful in the pharmaceutical and agrochemical industries. For example, 2,2,2-trifluoro-1-(4-methylphenyl)ethanone (**43**) reacts with acetic anhydride, when heated in the presence of sodium acetate, to produce 3-polyfluoroalkyl-*E*-cinnamic acid **44**.[11]

43 —NaOAc, Ac_2O, Δ, 93%→ **44**

It has been reported that microwave irradiation shortens the reaction time of the Perkin reaction by 60-fold over classical heating. Cesium salts (acetate, carbonate, fluoride) with a small amount of pyridine were found to be the best catalysts.[12] Thus, *p*-hydroxybenzaldehyde (**45**) reacted with acetic anhydride using cesium acetate as a catalyst to provide the α, β-unsaturated acid **46** in 85% yield.

HO 45 → CsOAc/py. Ac_2O, Δ 85% → HO H CO_2H 46

4.4.6 *Experimental*

1 + $(CH_3CO)_2O$ 2 → CH_3COOK → base H_2O → OH 4

Preparation of cinnamic acid (4)[2]

A mixture of 21 g (0.2 mol) of freshly distilled benzaldehyde (**1**), 30 g (0.3 mol) of 95% acetic anhydride (**2**), and 12 g (0.12 mol) of freshly fused potassium acetate is refluxed in an oil bath at 170–175°C for five hours, using an air-cooled condenser. The hot reaction mixture is poured into about 1200 mL of warm water, part of which is used to rinse the reaction flask. Any unreacted benzaldehyde is removed by steam distillation. The residual liquid is cooled slightly, 3–4 g of decolorizing carbon is added, and the mixture is boiled gently for 5–10 minutes. The decolorizing carbon is removed by filtration and the clean filtrate is heated to boiling. Concentrated hydrochloric acid (12–14 mL) is added carefully, and the hot solution is cooled rapidly while stirring vigorously. After the cinnamic acid (**4**) has completely crystallized, the crystals are collected by vacuum filtration, washed with several small portions of water, and then dried. The acid obtained in this way melts at 131.5–132 °C and is pure enough for most purpose. The yield is 16–18 g (55–60%)

4.4.7 *References*

1. Perkin, W. H. *J. Chem. Soc.* **1868**, *21*, 53.
2. [R] Johnson, J. R. *Org. React.* **1942**, *1*, 210.
3. Perkin, W. H. *J. Chem. Soc.* **1877**, *31*, 388.
4. Kinastowski, S.; Nowacki, A. *Tetrahedron Lett.* **1982**, *23*, 3723.
5. Knoevenagel *Ber.* **1898**, *31*, 2598.
6. Doebner *Ber.* **1900**, *33*, 2140.

7. Yadav, J. S.; Reddy, B. V. S.; Basak, A. K.; Visali, B.; Narsaiah, A. V.; *et al. Eur. J. Org. Chem.* **2004**, 546.
8. Mitsch, A.; Wißner, P.; Silber, K.; Haebel, P.; Sattler, I.; *et al. Bioorg. Med. Chem.* **2004**, *12*, 4585.
9. Venkatasamy, R.; Faas, L.; Young, A. R.; Raman, A.; Hider, R. C. *Bioorg. Med. Chem.* **2004**, *12*, 1905.
10. Brady, W. T.; Gu, Y. Q. *J. Heterocyclic Chem.* **1988**, *25*, 969.
11. Sevenard, D. V. *Tetrahedron Lett.* **2003**, *44*, 7119.
12. Veverková, E.; Pacherová, E. *Chem. Papers* **1999**, *53*, 257.

Jin Li

4.5 Perkow Reaction

4.5.1 Description

The Perkow reaction has been used to generate a wide range of vinyl phosphates. Reaction of trialkyl phosphites with α-halogeno-ketones provide vinyl phosphates.[1–5] Alkenes are produced *via* reduction of the dialkyl enol phosphate using either sodium or lithium in liquid ammonia.

$R_3O{-}P(=O)(OR_3){-}OR_3$ + α-halo ketone ⟶ $(R^1O)(R^2O)P(=O){-}O{-}$vinyl + $R^3{-}X$

X = Cl, Br, I

1 **2** **3**

4.5.2 Historical Perspective

In 1952 Perkow discovered a new type of rearrangement in which trialkyl phosphites react with α-aldehyde.[6,7] This "anomalous Arbuzov reaction," which resulted in the formation of dialkylvinylphosphates, subsequently became known as the "Perkow reaction."[1] The reaction of trialkyl phosphites with alkyl halides (Arbuzov reaction) proceeded to yield dialkoxy alkyl phosphonates.[8] The alkyl halide undergoes a nucleophilic displacement to yield the nonisolable alkyl trialkoxy phophonium halide.[8] The halide can then attack an alkoxy group to give phosphonate ester and RX. An increase in temperature favors the Arbuzov reaction leading to phosphonates (S_N2 displacement).

$$P(OR)_3 + R^1X \longrightarrow [R^1P(OR)_3]^+ X^- \longrightarrow R^1{-}P(=O)(OR){-}OR + RX$$

4.5.3 Mechanism

Many different mechanisms have been proposed for the Perkow reaction.[2–4] It involves nucleophilic attack of the phosphite at the carbonyl carbon and affords a zwitterionic intermediate **5** which rearranges to form a cationic species **6** that subsequently dealkylates to give the corresponding vinyl phosphate **7**. The conversion proceeds *via* a Michaelis–Arbuzov cleavage of an alkoxy group by halide ion as shown.

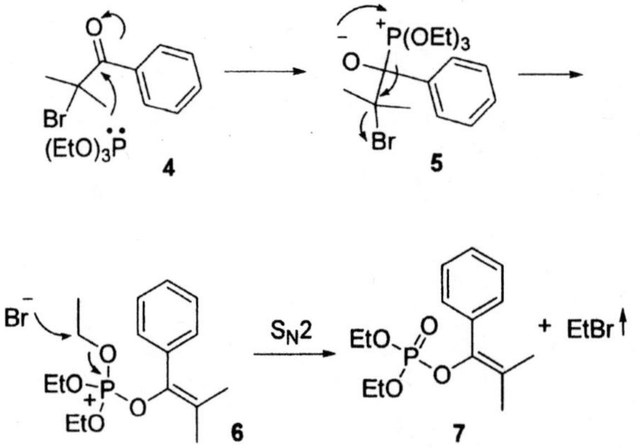

Borowitz[3,9–11] has proposed several facts about the Perkow reaction. The rate determining reversible step involves the addition of the phosphite to carbonyl and an anti-elimination step. He has also reported the determination of relative stereochemistry of *E* and *Z* isomers of vinyl phosphate arising from the Perkow reaction of α-haloketones with trialkyl phosphites.[12] Halide ion loss is not the rate determining step in the Perkow reaction. The steric hindrance of the carbonyl group presumably precludes enol phosphate formation. At least one aliphatic alkoxy group is necessary for the Perkow reaction. Triphenylphosphines are unreactive toward chloral, whereas monoalkyl diarylphosphite yields enol phosphate esters readily indicating that electronic release from the alkoxy group strongly assists the elimination of the halide ion.[2] The reactivity of various carbonyl compounds increases with the affinity of nucleophiles. Aldehydes react more readily than ketones while α-haloesters are unreactive. A change in the halogen atom chlorine to iodine decreases the yield of enol phosphate although the overall reactivity is in the usual order I > Br > Cl.

4.5.4 Variations and Improvements

The Perkow reaction includes the reactions of phosphites $P(OR)_3$, phophonites $P(OR)_2R$, and phosphinites $ORPR_2$ with the spectrum of mono-, di-, and tri- α-halo, α-OMes, and α-OTos carbonyl compounds. Chloral reacts with trimethyl phosphite exothermally at room temperature, but most ketones require the reaction mixture to be heated, usually to 85–110 °C. The reactions are best carried out in a solvent, which can be ether, alcohol, or other non-reactive species. Halogen reactivity is stated to be I > Br > Cl, at least for those cases where carbonyl addition is rate determining. The most commonly used trialkyl phosphites are triethylphosphite (TEP) and trimethyl phosphite (TMP), which is somewhat less reactive.[13] In phosphites containing one or two phenoxy groups, the alkyl group is replaced. Trialkyl phosphites having different alkyl groups usually eliminate the smallest group as the halide. Chloral and bromal react most readily and the reactivity decreases as the number of halogen atoms on the α-carbon atom decreases. Ketones react less readily than aldehydes and esters.[14]

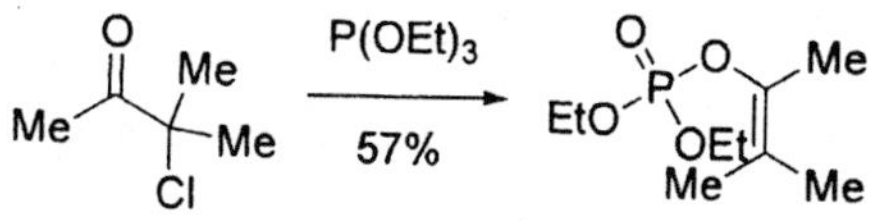
P(OEt)3
57%

8 **9**

A mixture of *E* and *Z* isomers of enol phosphates (**11**) are produced from 2,2-dibromo-1-phenyl-ethanone (**10**) in moderate yields.[13,15–18]

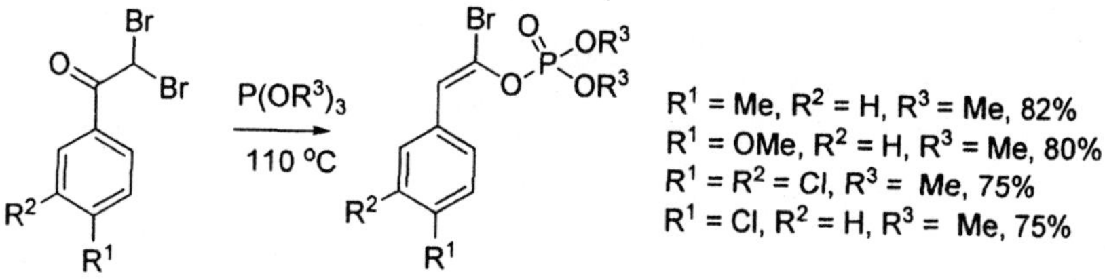
P(OR3)3
110 °C
R1 = Me, R2 = H, R3 = Me, 82%
R1 = OMe, R2 = H, R3 = Me, 80%
R1 = R2 = Cl, R3 = Me, 75%
R1 = Cl, R2 = H, R3 = Me, 75%

10 **11**

Diethylphosphite and thiophosphite anions react in a modified Perkow reaction to yield the enol phosphite and thiophosphates (**13**). $(EtO)_2POK$ is prepared from diethylphosphite and *t*-BuOK (X = O, 72%, X = S, 42%).[19]

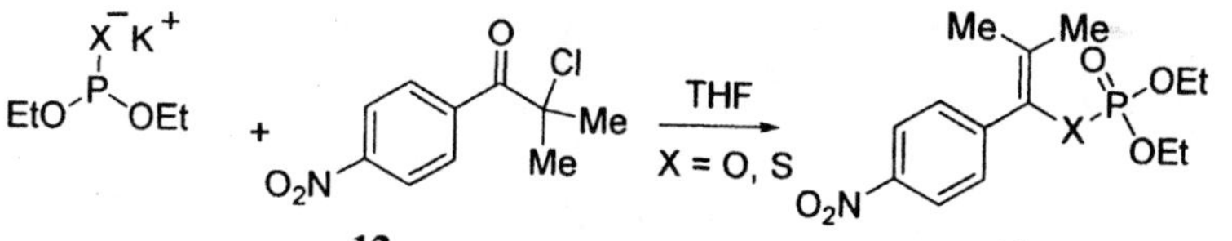
THF
X = O, S

12 **13**

The phosphorylation of **14** with $(EtO)_2P(O)Cl$, $(EtO)_2PCl$, $(EtO)_3P$, $MeP(O)Cl_2$, and $P(NEt_2)_3$ has been studied.[20] Thus, treating **14** with $(EtO)_2PCl$ in Et_2O in the presence of Et_3N at room temperature for 8 h gave **15** in 79%.

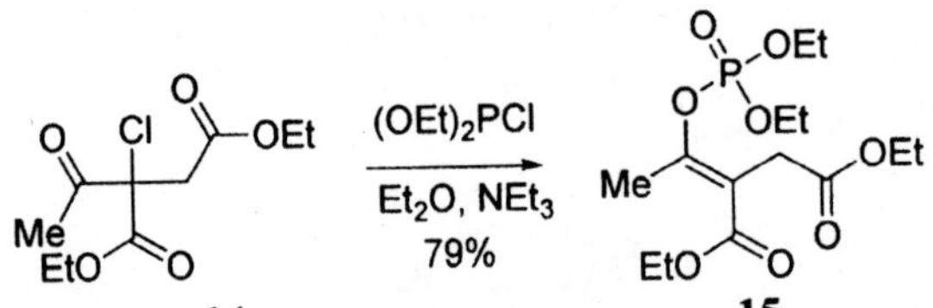

Perkow reaction of ketone (**16**) with $P(OEt)_3$ and $P(OEt)Ph_2$ afforded the vinyl phosphate **17** and vinyl phosphinate **18**, respectively.[21]

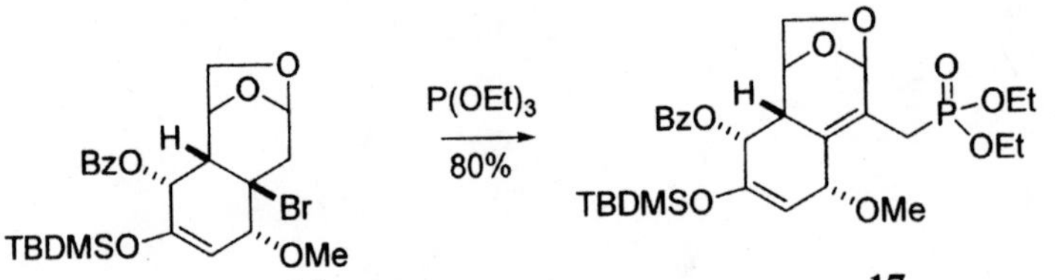

Tris(trimethylsilyl)phosphites, analogous to trialkylphosphites, undergo Perkow reaction with α-halo carbonyl compounds to give vinyl phosphates.[22a] Reaction of α-chloroketone **19** with trimethylphosphite produces dimethyl enol phosphate **20**, which is further converted to bis[trimethylsilyl] esters **21** using Me_3SiBr.[22b] **21** is also produced directly from α-chloroketone **19** on reacting with $P(SiMe_3)_3$.

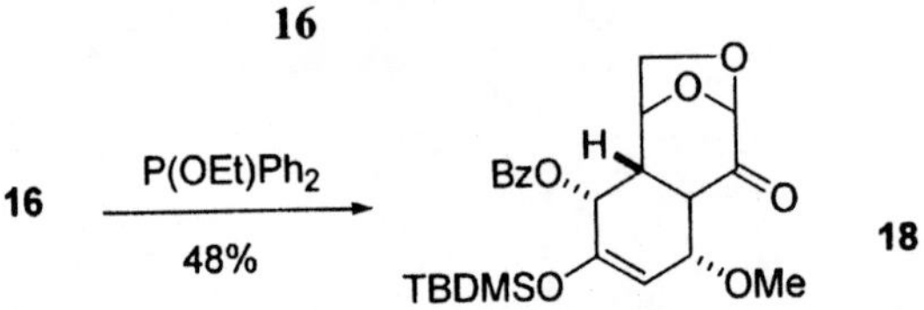

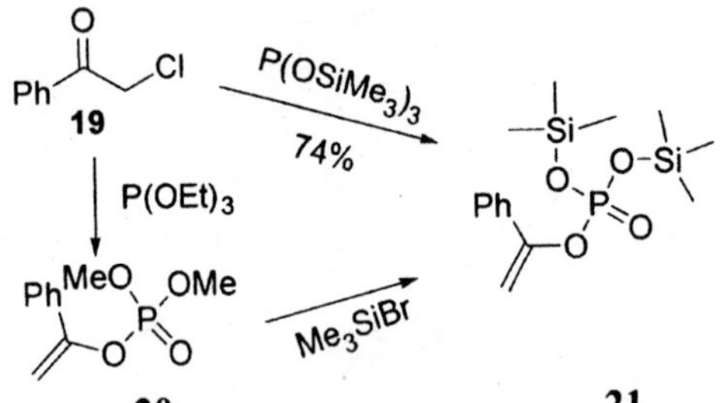

Phosphoenolpyruvates **26** are conveniently prepared from pyruvic acid and dimethyl-trimethylsilyl phosphite **24** through the Perkow reaction.[23,24]

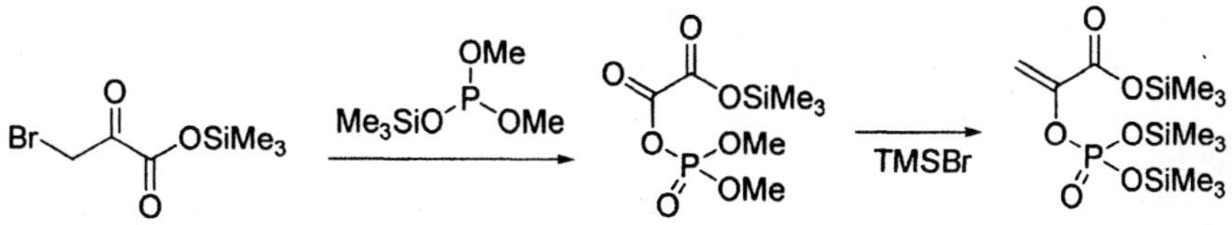

Enol adenosine 5′-phosphate derivatives (**28**, **29**, **30**) are formed by the Perkow reaction of α-chloromethyl ketone with an in situ-generated adenosine-5′-yl bis(trimethylsilyl)phosphite (**27**). Among these reactions, the use of α-chloro-2′,4′-difluoroacetophenone gave predominantly the enol phosphate derivative over the Arbuzov reaction product and the carbonyl adduct, which were formed as by-products.[25]

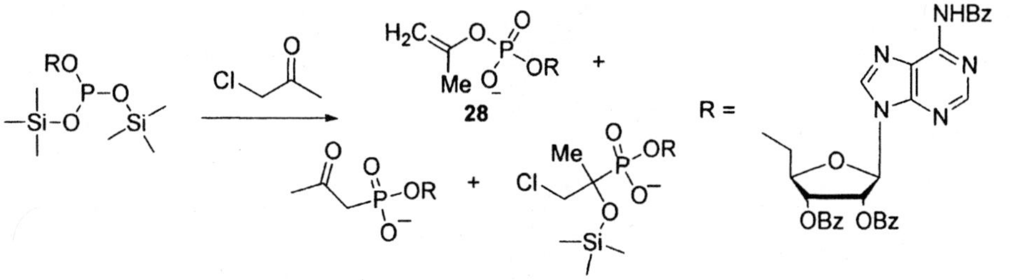

Sal'keeva *et al.* studied the reactions of ethyl and *t*-butyl phosphorodiamidites with chloroacetone, bromoacetone, and α-chloroethyl acetate. The reaction pathway is determined by the structure of the intermediate quasiphosphonium compound responsible for the formation of the Arbuzov products as well as for the occurrence of the anomalous reaction yielding vinyl phosphate (Perkow product).[26] The yield of the Arbuzov product increases from chloroacetone to bromoacetone, because the anions of low basicity are more rapidly ionized, which facilitates the formation of the ionic form of the intermediate. Also *t*-butyl phosphorodiamidite forms *via* Arbuzov reaction as the major product, because the second stage of the S_N1 reaction is accelerated.[26]

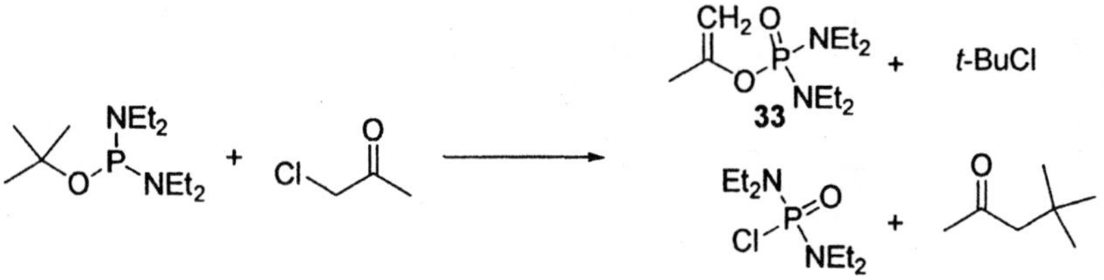

Reaction of benzodioxaphosphorinone **(35)** with dibromoketoester by selective ring cleavage gives the enolate of vinyl phosphate **36**, the latter cyclized upon distillation to give cyclic vinyl phosphate 37. A CNDO/2 study of $H_2P(O)OC(CF_3)$=CHCOMe as a model for the product vinyl phosphates is carried out and they are formed predominantly in the thermodynamically preferred *Z*-configuration. Processes for cyclization of these vinyl phosphates into 2-vinylalkoxy-2-oxo-4H-1,3,2-benzodioxaphosphorin-4-ones were examined (**37**).[27,28]

60 - 70 °C

35 **36**

37

3-Chloro-2-oxopentanoates (**38**) are transformed into phophoryloxyalkanoate **39** by means of standard Perkow reaction.[29]

PhH, reflux, 2h, 91%

E/Z = 30/70

38 **39**

Triethylphosphite readily reacted with chloral to give diethyl-2,2-dichlorovinylphosphate.[30,31]

$P(OEt)_3$

Di-(methyl/ethyl)/isopropyl)indolylphosphite **40** reacted with chloral to provide vinylphosphites **41–43.**[32]

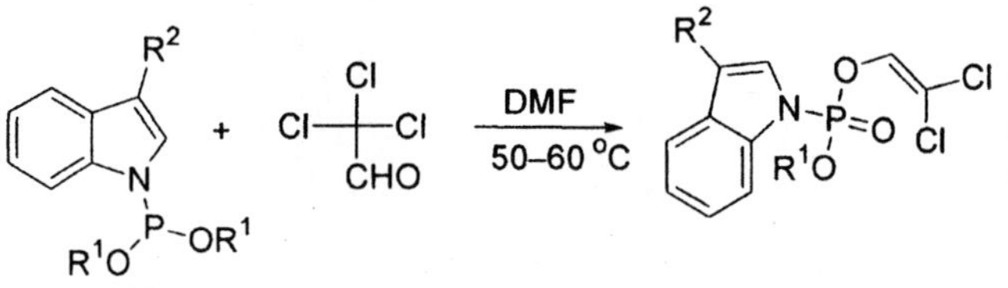

41 R^1 = Me, 57%, R^2 = H
42 R1= *i*-Pr, 69%, R^2 = H
43 R^1 = Et, R^2 = CH_2NMe_2, 73%

Oxazaphospholidine **45** reacts with α-haloacetophenone **45** to give a mixture of vinylphosphonamide diastereomers **47**. The product ratios depend on the nature of the halides.[33]

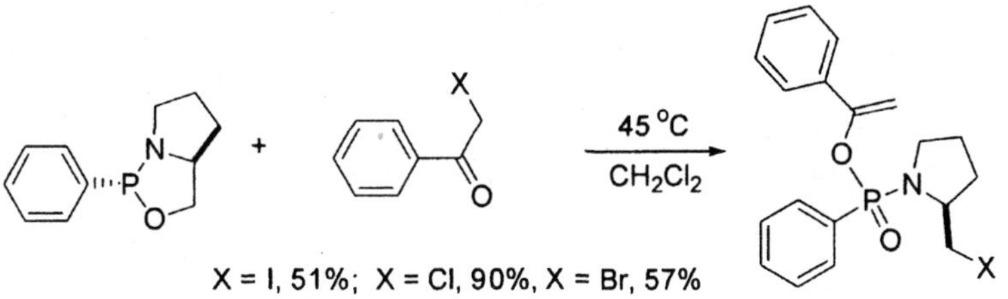

X = I, 51%; X = Cl, 90%, X = Br, 57%

4.5.5 Synthetic Utility

4.5.5.1 Rearrangement to β-ketophosphonates

Diethyl enol phosphate **47** rearranges to β-ketophosphonates **49** upon treatment with LDA.[34] This reaction probably proceeds *via* cleavage with LDA to give an enolate phosphonium ion pair **48** which then interacts to give the anion under equilibrium conditions.[34] The reaction mixture is quenched with acetic acid. The enol phosphate **47** is prepared by Perkow reactions from the corresponding α-halocyclohexanone.[34]

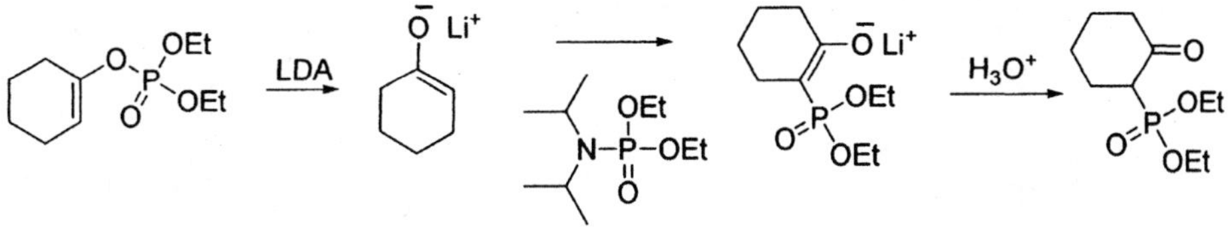

Perkow reaction of bromoketone **50** produces vinyl phosphate **51,** which is stable and used for further reactions.[21]

50 **51**

Treatment of α-chloro-α-acyl-lactone **52** with $P(OEt)_3$ affords the enol phosphate **56**, whereas reaction with $NaP(O)(OEt)_2$ gives **53**.[35] **53** is treated with alcoholic KOH to give the open chain compound **54**, which is further cyclized on acidification to the phostone **55**.[35] **53** is treated with KOH/ether and the product acidified with HCl–$CHCl_3$ to give **55**. Treatment of **55** with PCl_5 produces the corresponding acid chloride, which is converted with p-$MeC_6H_4NH_2$ into characteristic p-toluidides.[35]

52 **53** **54** **55**

56

Dialkyl enol phosphate of lactone **58** undergoes rearrangement to β-ketophosphonate **59** upon treatment with LDA at –100 °C. The enol phosphate of lactone **57** is made by (1) treatment of lactones with LDA to give the lithium enolate, (2) *O*-phosphorylation of the enolate with dialkyl chlorophosphoridate in THF–HMPA under kinetic conditions, or (3) a one step conversion without isolation of the intermediate enol phosphate (**58**).[36] Similarly, several enol phosphates of acyclic esters are rearranged to β-phosphonate esters as shown.[36]

57 58 59

4.5.5.2 Cleavage of enolates with C-alkylation

Enol phosphate **60**, enol phosphinate **61**, and enol phosphonate **62** react with MeLi or BuLi to give a lithium enolate, which is then monoalkylated to **63** in 62–78% yield with various primary and secondary alkyl halides.[11] Some starting ketone is formed and some polyalkylation occurs, the latter especially with MeI.[11] Little or no dialkylation occurs with alkyl halides larger than ethyl. Enol phosphates give more starting ketone than enol phosphinates, perhaps because of a side reaction involving displacement of the –OR group from phosphate by the alkyllithium.[11] Phenyllithium or phenylmagnesium bromide gives a poorer alkylation yield as well as the undesired by-product biphenyl. The method of forming enolates shown competes with the cleavage of enol trimethylsilyl ethers and the direct formation of enolates from ketones with LDA.[37,38]

R^1Li, R^1 = Me, Bu; R^2X; R^2 = Me, Et, *i*-Pr, Bu

60, Y = Z = OEt
61, Y = Z= Ph
62, Y = Ph; Z=OR

63

3-Chloro-2-(diethoxyphosphoryl)oxy-1-propene (**64**) is used by Welch in a "one-pot" annelation of cyclohexanone to provide **66**.[39] Chloropropenylphophonate (**64**) is prepared from 1,3-dichloroacetone and trimethyl phosphite.[39]

LDA; **64**, $Pd(PPh_3)_4$, THF; 1. 10% HCl, 2. 5% KOH

65 **66**

Other conversions lead to *cis*-jasmone and to methylenomycin (**67** and **68**).[40]

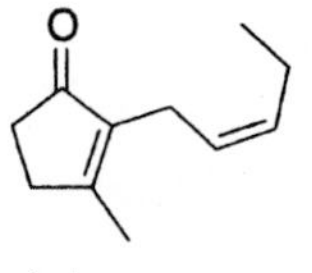

cis-jasmone
67

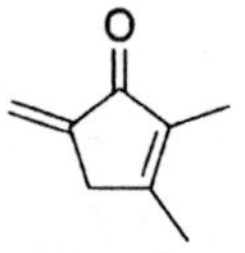

methylenomycin
68

4.5.5.3 Reductive removal of phosphate to alkenes

Enone **69** is converted to the less substituted alkene **71** *via* the intermediacy of the enol phosphate **70.**[41] The sequence shown involves reduction with Li/NH_3 (liquid) followed by phosphorylation of the resulting lithium enolates with diethyl phosphorochloridate to give **70,** which is then treated with Li/$EtNH_2$/*t*-BuOH. Application of such methodology to steroid conversion has been reported by Ireland and Pfister[41] and Kanata *et al.*[42]

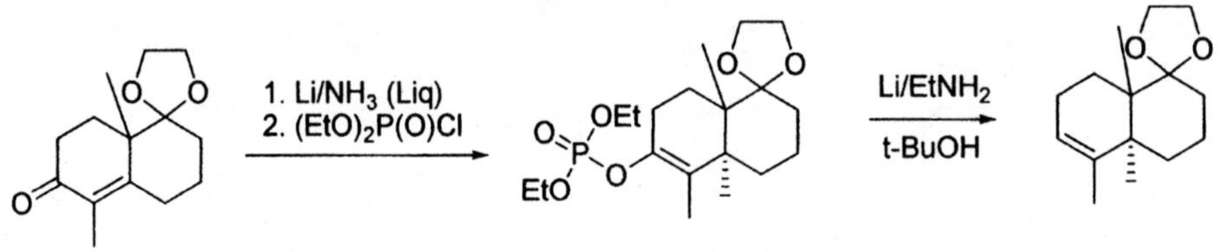

1. Li/NH_3 (Liq)
2. $(EtO)_2P(O)Cl$

Li/$EtNH_2$
t-BuOH

69 **70** **71**

Corey and Wright have utilized reductive conversion of a vinyl diethyl phosphate (**73**) to an alkene **74** in a synthesis of colneleic ester *via* the enol esters **72.**[43]

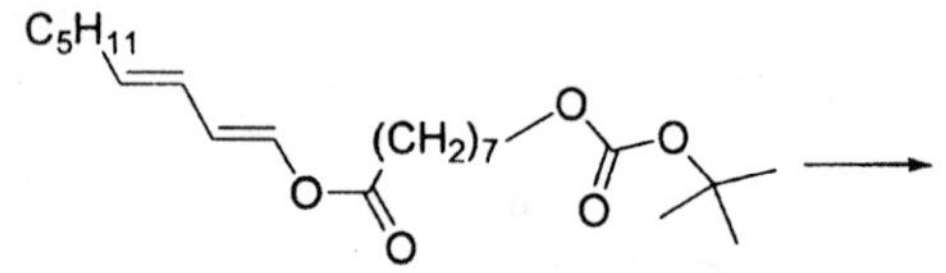

72

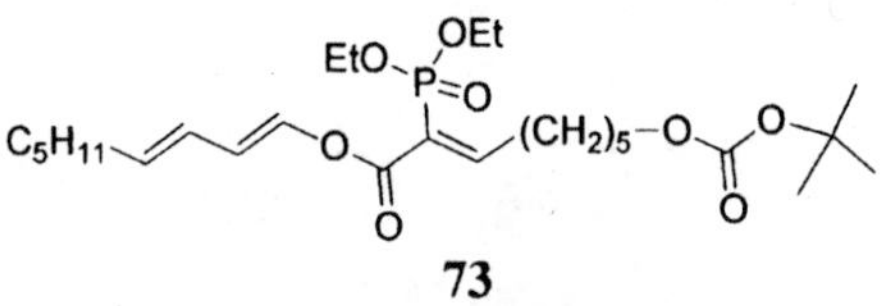

73

$AlEt_3$, $Pd(PPh_3)_4$

74

4.5.5.4 *Synthesis of a-trifluoromethylated ketones*

A series of 1-aryl-3,3,3-trifluoro-1-ethanones **77** have been synthesized from the reaction of $FO_2SCF_2CO_2Me$ with β-bromoenol phosphate **76** in the presence of CuI in moderate yields. Enol phosphates are versatile intermediates in organic synthesis.[17]

$P(OMe)_3$; $FO_2SCF_2CO_2Me$, CuI/DMF, 80 °C, 55-65%

R = Me, OMe, Cl, Br

75 **76** **77**

4.5.5.5 *Double Perkow reaction*

The reaction of 1,4-dibromo-2,3-butanedione **78** with two moles of trimethyl phosphite produces the 2,3-diphosphate **79** *via* a double Perkow reaction.[44]

2 $P(OEt)_2$

78 **79**

Corey *et al.* have used the enol phosphates **80** to synthesize the limonoid intermediate **81** and limonoid **82**.[45] They have also developed a general preparative scheme *via* enol phosphates for the synthesis of limonoids, which are biosynthetically derived from triterpenes.[45]

$Hg(OCOCF_3)_2$, CH_3NO_2

80 **81**

Limonoid

82

4.5.5.6 *Ring enlargement*

The reaction of 2-*tert*-butoxy-5,6-benzo-1,3,2-dioxaphosphinin-4-one (**83**) with chloral involves expansion of the six-membered hetero ring to form 2-hydroxy-3-trichloromethyl-6,7-benzo-1,4,2-λ^5-dioxaphosphepin-5-one-2-oxide (**85**) as a single diastereomer. The product was chlorinated with thionyl chloride to obtain (*RS,SR*)-2-chloro-3-trichloromethyl-6,7-benzo-1,4,2 λ^5-dioxaphosphepin-5-one-2-oxide (**86**).[46]

CCl_3CHO ; 80 °C, PhH ; -CH=CMe$_2$

83 **84** **85**

$SOCl_2$

86

4.5.5.7 *Pharmaceuticals*

Enol phosphates are used in the syntheses of various members of the carbapenam family of antibiotics. The azabicycloheptanedione **87** is enol-phosphorylated to **88,** which reacts with thiol **89** to yield an intermediate ester. The ester is then hydrogenolyzed to yield carbapenam **90.**[47] Similar enol phosphates have been converted to useful intermediates for the synthesis of related carbapenam antibiotics.[48,49]

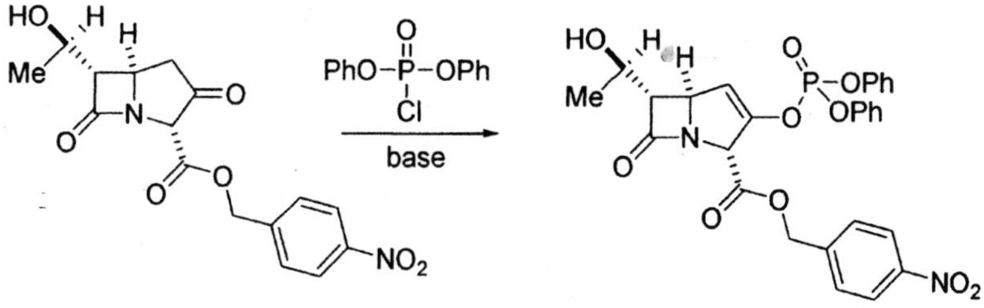

87 88

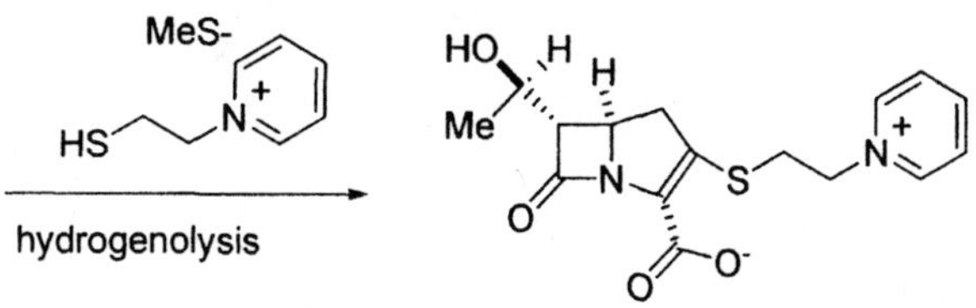

MeS-
HS
hydrogenolysis

89 90

4.5.5.8 Agrochemicals

Tetrachlorvinphos (**91**, stirophos, Gardona®) has been shown to have the *Z* configuration by X-ray analysis as previously mentioned.[50] The insecticide methylbromfenvinphos (**93**) was found to be the *Z*-stereoisomer by single crystal X-ray analysis.[51] Related insecticides bromfenvinphos (**92**) and chlorfenvinphos (**94**) have been synthesized.[52,53]

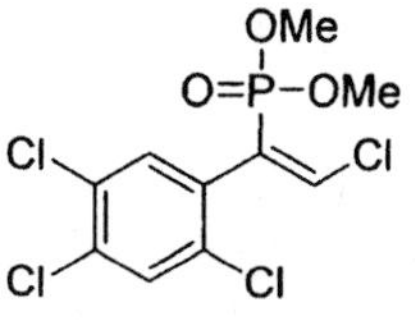

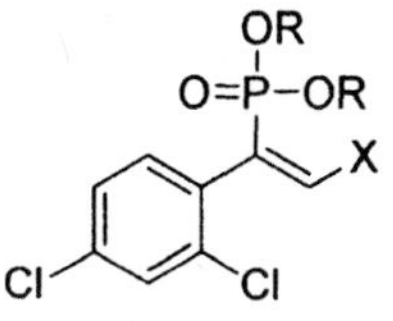

92 Bromfenvinphos, R = Et; X = Br
93 Methylbromfenvinphos, R = Me; X =Br
94 Chlorfenvinphos, R = Et; X = Cl

Tetrachlorvinphos
91

Mevinphos (**96**, Phosdrin®) is prepared from methyl 3-chloroacetoacetate in reaction with trimethyl phosphite. Mevinphos (Phosdrin®) is a mixture of *Z* and *E* stereoisomers in which the *Z* isomer was found to be 100 times more active.[54] In contrast to mevinphos, *E* and *Z* isomers of bromfenvinphos (**92**) and chlorfenvinphos (**94**) show no significant difference for a number of biological properties for insecticides.[55] The less toxic Clodrin (**98**) is synthesized from the 1-phenylethoxy ester of 3-chloracetoacetate **97** which is used for control of ectoparasites in horses, cattle, sheep, and swine.[54]

P(OMe)$_3$

Mevinphos

95 **96**

P(OMe)$_3$

Clodrin

97 **98**

Phosphamidon (**99**, 2-chloro-2-ethylcarbamyl-1-methyl-vinyl dimethyl phosphate) has a broad spectrum of activity against biting and sucking pests and spider mites.[54] A cyclic enol phosphate is prepared from **100**. They are heptenophos (**101**, Hostaquick®), a systematic insecticide, and acaricide (anti-tick and anti-mite agent) with a short residual effect.[54]

Phosphamidon **99** **100** P(OMe)$_3$ Heptenophos **101**

Zwierzak *et al.* prepared enol phosphates containing the α-carboalkoxy and α-carbamoyl groups (**103**), and they showed fungicidal activity.[56]

P(OMe)$_3$

Ar = -C_6H_4-4-*n*-Pr
= -C_6H_4-4-*i*-Pr
= -C_6H_4-4-Br
=-C_6H_3-2,4-di-Cl

Z = -CO_2R^1, -$CON(R^2)_2$

102 **103**

Malinowski and Kroczynski have concluded that the 2-carboalkoxy group in a series of dimethyl and diethyl enol phosphates **104** enhance insecticidal activity.[57] The most effective insecticide and acaricide in their series is **105**.[57]

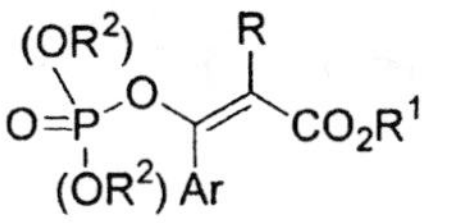

Ar = -C_6H_3-2,4-di-Cl
= -C_6H_3-2,5-di-Cl
= -C_6H_4-4-Cl
R^2 = Me
R^2 = Et

104 **105**

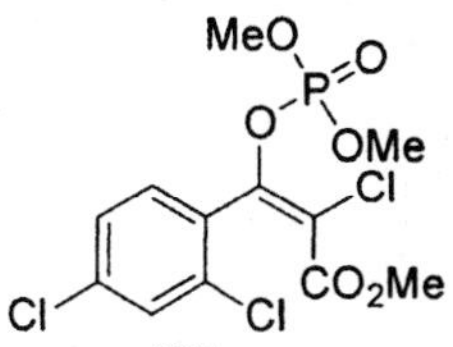

Two series of cyano-substituted ethyl thiopropyl enol phosphate insecticides (**107** and **108**) are synthesized by D'Silva *via* phosphorylation of a precursor ketone **106**.[58] Some of the compounds are effective against the two-spotted mite.[58]

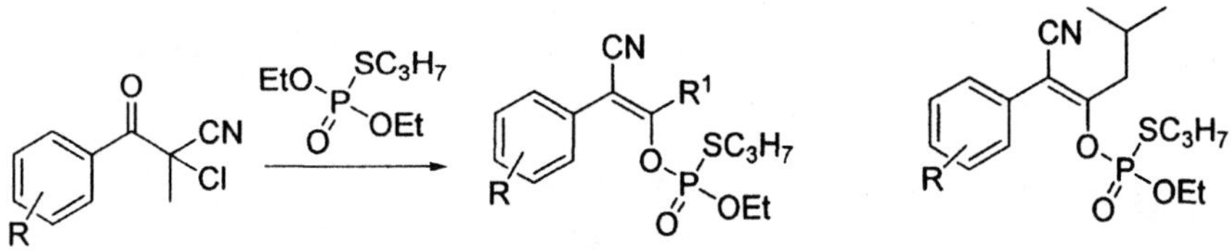

R = H, alkyl, thioalkyl, halo trihalomethyl

106 **107** **108**

4.5.6 Experimental

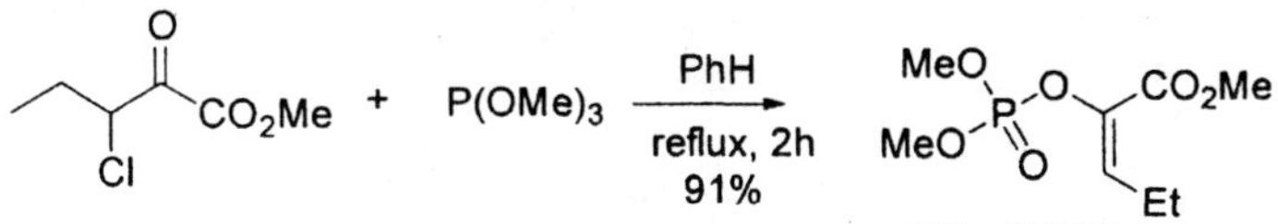

PhH, reflux, 2h, 91%

E/Z = 30/70

38 **39**

A mixture of methyl 3-chloro-2-oxopentanoate (**38**) (0.02 mol) and trimethylphosphite (0.026 mol) in benzene (920 mL) was heated under reflux for 2 hours. The resultant mixture was cooled to room temperature and the solvent was evaporated to yield the crude methyl 2-(dimethoxyphosphoryloxy)-2-pentenoate (**39**) in 91% yield as a mixture of isomers (*E/Z* = 30/70), which was purified by vacuum distillation. Both the structure and ratio of the isomers were unambiguously established by using ^{1}H and ^{31}P NMR spectroscopy. Configurational assignments for (*E*) and (*Z*) were made mainly on the basis of $^{4}J_{POCCH}$, which revealed characteristic larger values for (*E*) than for (*Z*) isomers.[29]

4.5.7 References

1 [R] Borowitz, G. B.; Borowitz, I. J. *Handbook of Organophosphosphorus Chemistry*; Marcel Dekker, Inc., New York; **1992**, 115.
2 Chopard, P. A.; Clark, V. M.; Hudson, R. F.; Kirby, A. J. *Tetrahedron* **1965**, *21*, 1961.
3 Sekine, M.; Okimoto, K.; Yamada, K.; Hata, T. *J. Org. Chem.* **1981**, *46*, 2097.
4 Sekine, M.; Nakajima, M. Hata, T. *J. Org. Chem.* **1981**, *46*, 4030.
5 [R] Li, J. J. *Name Reactions*, 2nd ed.; Springer; **2003**, p. 307.
6 Perkow, W.; Ullerich, K.; Meyer, F. *Naturwissenschaften* **1952**, 39, 353.
7 Whetstone, R. R; Harman, D. (Shell Development Co.,) USP 2,765,331 (**1956**).
8 Kharasch, M. S.; Bengelsdorf, I. S. *J. Org. Chem.* **1955**, *20*, 1356.
9 Borowitz, I. J.; Anschel, M.; Firstenberg, S. *J. Org. Chem.* **1967**, *32*, 1723.
10 Borowitz, I. J.; Firstenberg, S.; Borowitz, G. B.; Schuessler, D. *J. Am. Chem. Soc.* **1972**, 94, 1623.
11 Borowitz, I. J.; Casper, E. W. R., Crouch, R. K.; Yee, K. C. *J. Org. Chem.* **1972**, *37*, 3873.
12 Borowitz, I. J.; Yee, K. C.; Crouch, R. K. *J. Org. Chem.* **1973**, *38*, 1713.
13 Gaydou, E. M. *Bull. Soc. Chim. Fr.* **1973**, 2279.
14 Allen, J. F.; Johnson, O. H. *J. Am. Chem. Soc.*, **1955**, *77*, 2871
15 Cooper, D. J.; Owen, L. N. *J. Chem. Soc.* **1966**, 533.
16 Gololobov, Yu. G.; Kolpdka, T. V.; Oganesyam, A. S.; Chemega, A. N.; Antipin, M. Yu.; *J. Gen. Chem. USSR*, **1986**, *56*, 1512.
17 Huang, X.; He, Y.; Ding, Y. *Phosphorus, Sulfur Silicon Relat. Elem.* **2001**, *174*, 201.
18 Ding, Y.; Huang X. *Heteroatm. Chem.* **2003**, *14*, 304.
19 Russell, G. A.; Ros, F. *J. Am. Chem. Soc.* **1985**, *107*, 2506.
20 Lysenko, V. P.; Gololobov, Yu. G.; Boldeskul, I. E.; Rozhkova, Z. Z. *Ukra. Khim. Zh.* (Russian Edition), **1982**, *48*, 1063.
21 Yamamoto, N.; Isobe, M. *Tetrahedron* **1993**, *49*, 6581.
22 (a) Sekine, M.; Okimoto, K.; Hata, T. *J. Am. Chem. Soc.* **1978**, *100*, 1001. (b) Hata, T.; Yamada, K.; Futatsugi, T.; Sekine, M. *Synthesis* **1979**, *3*, 189.
23 Sekine, M.; Futatsugi, T.; Yamada, K.; Hata, T. *Tetrahedron Lett.* **1980**, *21*, 371.
24 Sekine, M.; Futatsugi, T.; Yamada, K.; Hata, T. *J. Chem. Soc. Perkin Trans. 1* **1982**, 2509.
25 Moriguchi, T. ; Okada, K.; Seio, K.; Sekine, M. *Lett. Org. Chem.* **2004**, *1*, 246.
26 Sal'keeva, L. K.; Nurmaganbetova, M.T.; Kurmanaliev, O. S.; Gazizov, T. K. *Russ. J. Gen. Chem.*, **2002**, *72*, 1760.
27 Mironov, V. F.; Bobrov, M. B.; Mavleev, R. A.; Aminova, R. M.; Konovalova, I. V.; Pashkevich, K. I.; Chernov, P. P. *Zhu. Obs. Khim.*, **1993**, *63*, 797.
28 Mironov, V. F.; Burnaeva, L. A.; Bobrov, M. E.; Pashkevich, K. I.; Konovalova, I. V.; Aminova, R. M., *Zhu. Obsh. Khim.* **1994**, *64*, 1362.
29 Janecki, T.; Bodalski, R. *Heteroatm. Chem.* **2000**, *11*, 115.
30 Perkow, W. *Ber. Dtsch. Chem. Ges.* **1954**, 87, 755.
31 Cramer, F. *Angew. Chemie* **1960**, *72*, 236.
32 Musabekova, Z. R.; Gurevitch, P. A.; Gazizov, T. Kh.; Arkhipov, V. A.; Lukonin, D. E., *Russ. J. Gen. Chem.* **1996**, *66*, 774.
33 Faure, B; Pardigon, O.; Buono, G. *Tetrahedron* **1997**, *53*, 11577.
34 Calogeropoulou, T.; Hammond, G. B.; Weimer, D. F. *J. Org. Chem.* **1987**, *52*, 4185.
35 Korte, F.; Rocchling, H. *Tetrahedron Lett.* **1964**, 2099.
36 Jackson, J. A.; Hammond, G. B.; Weimer, D. F. *J. Org. Chem.* **1989**, *54*, 4750.
37 [R] House, H. O.; *Modern Synthetic Reactions*, (2nd ed.); Benjamin: Menlo Park, California; 1972, p. 546.

38 [R] Carey, F. A.; Sundberg, R. J. *Advanced Organic Chemistry* 2nd ed. part B; Plenum Press: New York; **1983**, p. 10.
39 Welch, S. C.; Assercq, J.-M.; Loh, J.-P. *Tetrahedron Lett.* **1986**, *27*, 1115.
40 Welsch, S. C.; Assercq, J. M.; Loh, J. P.; Glase, S. A. *J. Org. Chem.* **1987**, *52*, 1440.
41 Ireland, R. E.; Pfister, G. *Tetrahedron Lett.* **1969**, 2145.
42 Karmata, S.; Haga, N.; Matsui, T.; Nagata, W. *Synthesis* **1986**, 588.
43 Corey, E. J.; Wright, W. *J. Org. Chem.* **1990**, *55*, 1670.
44 Honig, M.L.; Sheer, M.L.; *J. Org. Chem.* **1973**, *38*, 3434.
45 Corey, E. J.; Reid, J. G.; Myers, A. G.; Hahl, R.W. *J. Am. Chem. Soc.* **1987**, *109*, 918.
46 Konovalova, I. V.; Burnaeva, L. M.; Mironov, V. F.; Gubaidullin, A. T.; Dobrynin A. B.; Litvinov, I. A.; Romanov, S.V.; Zyablikova, T. A.; Yashagina, O. V. *Russ. J. Gen. Chem.* **2002**, *72*, 1186.
47 Dextraze, P. (Bristol Myers Co.), DE 3408347 (**1984**).
48 Fuentes, L. M.; (Texaco), USP 4,845,251 (**1989**).
49 Grabowski, E.; Hughes, D. L.; (Merck Co.), USP 4,894,450 (**1990**).
50 Rohrbaugh, W. J.; Jacobson, R. A. *J. Agri. Food Chem.* **1978**, *26*, 1120.
51 Galdecki, Z.; Glowka, M.; Zwierzak, A. *Phosphorus, Sulfur Silicon Relat. Elem.* 1979, *5*, 299.
52 Zwierzak, A.; Sledzinski, B.; Cieslak, L. *Proc. Instit. Indust. Org. Chem. Tech. Uni. Lodz.* **1974**, *6*, 115.
53 Zwierzak, A.; Sledzinski, B.; Cieslak, L. *Proc. Instit. Indust. Org. Chem. Tech. Uni. Lodz*, **1974**, *6*, 85.
54 Fest, C.; Schmidt, K. J. *Chemistry of Pesticides* (ed. Buchel, K. H. & Trans. Holmwood, G.); **1983**, 48.
55 Kolodynski, J.; Ulaczewski, S.; Witek, S.; *Rocz. Nauk Rol. Ser. E*, **1983**, *10*, 210.
56 Bodalaki, R.; Zwierzak, A., Wroblewski, A.; Markowicz, W.; Koszuk, J.; Kluba, M.; Brylikowska-Piotrowicz, J. *J. Environ. Sci. Health*, **1983**, B18, 579.
57 Malinowski, J.; Kroczynski, J. *Pestycydy (Warsaw)* **1980**, *3*, 17.
58 D'Silva, T. D. J. (Union Carbide), USP 4,469,688 (**1984**).

Marudai Balasubramanian

4.6 Ramberg–Bäcklund Reaction

4.6.1 Description

The Ramberg–Bäcklund reaction is the base-mediated conversion of α-halosulfone **1** to olefin **1** with extrusion of SO_2.[1–8] It is also often referred to as the Ramberg–Bäcklund rearrangement. The thermodynamically stable *E*-olefin is often obtained when a weak base (e.g., NaH) is used whereas a strong base often results in the *Z*-olefin.

X
R S R1
O O
1
Base
R1
R
2
+ O=S=O↑

4.6.2 Historical Perspective

During WWII, Swedish chemist Ludwig Ramberg (1874–1940) at the Organic Chemistry laboratories of the University of Uppsala and his student Birger Bäcklund (1908–1997) discovered the base-mediated conversion of α-halosulfones to regio-defined alkenes. They published the work in a little known journal *Arkiv För Kemi, Mineralogi Och Geologi (Archives for Chemistry, Mineralogy and Geology)* with the title "The Reactions of Some Monohalogen Derivatives of Diethyl Sulfone" in 1940.[9] That ground-breaking work turned out to be Dr. Ramberg's last contribution to chemistry shortly before his death. Over the last six decades, organic chemists have devoted much effort in improving the synthetic utility of the reaction. Many dedicated a great part of their careers and some even committed their whole careers to this field. Largely thanks to their efforts, the Ramberg–Bäcklund reaction is now widely used in the organic synthesis as evidenced by the large number of referenced reviews 1–8.

4.6.3 Mechanism

:B
X H
R S R1
O O
1
R
X S R1
O O
3
S_Ni
R R1
S
O O
4
SO_2
extrusion
R1
R 2
+ O=S=O↑

thiirane dioxide intermediate

The mechanism of the Ramberg–Bäcklund reaction is rather straightforward. When α-halosulfone **1** is treated with a strong base, deprotonation rapidly takes place to give α-anion **3**, which undergoes a backside displacement (intramolecular nucleophilic substitution, S_Ni) to provide thiirane dioxide **4** (also known as episulfone) as the key intermediate.[10] The S_Ni reaction with loss of halide is the rate-limiting step. Finally, the unstable **4** releases sulfur dioxide and the ring strain to deliver alkene **2**.

Thiirane dioxide **4** as the reaction intermediate was unambiguously confirmed after it was isolated from the Ramberg–Bäcklund reaction. In 1989, the Taylor group at the University of York isolated thiirane dioxide **6** as the major product when they treated α-iodosulfone **5** with 1.2 equivalents of potassium *t*-butoxide at low temperature (–78 to 0 °C).[11,12] Not surprisingly, thiirane dioxide **6** was readily converted to alkene **7** (in 88% yield) in the presence of potassium *t*-butoxide at warmer temperature (–20 °C to room temperature).

KO*t*-Bu, THF, –78 °C

5 → **6**, 69% + **7**, 28%

4.6.4 Variations and Improvements

In 1969, Meyers *et al.* at Southern Illinois University published a significant improvement that was synthetically more convenient and useful than the original Ramberg–Bäcklund conditions.[13] The Meyers modification involves *in situ* halogenation of the sulfone, which, without isolation of the α-halosulfone, is converted to the alkene directly. When dibenzyl sulfone **8** was treated with KOH in the presence of a mixture of CCl_4 and *t*-BuOH as solvent, the *E*-olefin **9** was obtained exclusively. Unfortunately, the convenience of the Meyers modification is sometimes offset by sidereactions due to polychlorination.

aq. KOH, CCl_4, *t*-BuOH, 60–80 °C, quant.

8 → **9**

Although the Meyers modification has been widely used to conveniently carry out the Ramberg–Bäcklund reaction, Chan and co-workers at The Chinese University of Hong Kong further modified the one-pot halogenative Ramberg–Bäcklund reaction in

1994.[14] The trick of the Chan modification is the use of alumina-supported potassium hydroxide to increase the activated surface. Therefore, unactivated sulfone **10** was treated with KOH/Al_2O_3 and CF_2Br_2 in *t*-BuOH at reflux to provide alkene **11**. For another un-activated sulfone **12**, the reaction needed to be heated to 50–80 °C and consequently CF_2Br_2 was no longer suitable because of its low boiling point (bp = 22–23 °C). Franck *et al.* replaced CF_2Br_2 with $BrCF_2CF_2Br$ (bp = 47 °C) and obtained alkene **13** in 60% yield.[15] Since the Chan modification is further improvement of the Meyers modification, it is now often known as the Meyers–Chan modification.

KOH/Al_2O_3, CF_2Br_2, reflux, 90%: **10** → **11**

KOH/Al_2O_3, $BrCF_2CF_2Br$, reflux, 60%: **12** → **13**

The third halogenative Ramberg–Bäcklund reaction is the Vedejs modification.[16] He treated substrate **14** with NaH and C_2Cl_6 in DME at room temperature and obtained alkene **15** in 1:1 ratio of *E*/*Z*. However, the Vedejs modification, albeit mild, is limited to sulfones in which the α-position bears a single hydrogen, an ester, and another substituent as exemplified by transformation **1617** .[16] It does not work for simple, un-activated sulfones.

NaH, $Cl_3C\text{-}CCl_3$, DME, rt, 54%: **14** → **15**, *Z* : *E* = 1:1

NaH, $Cl_3C\text{-}CCl_3$, DME, rt, 3 h, 75%: **16** → **17**

The fourth modification, discovered by Hendrickson in 1985,[17,18] uses triflinate (trifluoromethanesulfinate) in place of the halogen as the leaving group as depicted by **1819** .[18] Similarly, Matsuyama used *p*-toluenesulfinate as the leaving group as represented by transformation **20→21**.[19] Since preparation of the substrate itself is lengthy, it seems to be a quite convoluted way of making simple alkenes, which may explain why there has been only one application by Fuchs in the literature thus far.[20]

1.2 eq. KO*t*-Bu, THF
0 °C, 1 h, 83%
18 **19**

NaH, KH
DMSO, rt, 48 h
60%
20 **21**

Numerous variants of the Ramberg–Bäcklund reaction exist. The most prevalent are the vinylogous and Michael-induced Ramberg–Bäcklund reactions. Decarboxylative, epoxy-, and tandem Diels–Alder Ramberg–Bäcklund reactions are also found in the literature.

The substrate for the vinylogous Ramberg–Bäcklund reaction has an additional adjacent double bond to participate in the rearrangement, which gives rise to a diene. Block at SUNY Albany carried out many vinylogous Ramberg–Bäcklund reactions.[21–23] In one case, irradiation of 1-octene and bromomethanesulfonyl bromide (**22**) in a Pyrex tube afforded an adduct, which upon treatment with triethylamine or 1,5-diazabicyclo[4,3,0]-non-5ene (DBN) gave bromomethyl 1-octenyl sulfone **23** with the major isomer as *E*. Exposure of **23** to potassium *t*-butoxide produced 1,3-nonadiene (**24**) in 59% yield after distillation.[23]

n-C_4H_9 → **22**, *hv*, −20 °C, 30 min, CH_2Cl_2, 94% → Et_3N, CH_2Cl_2, 0 °C, 15 min, 97%

23 → KO*t*-Bu, 7:1 HO*t*-Bu/THF, −20 °C, 1 h, 59% → n-C_4H_9 **24**, *Z*: *E* = 83:17

In 1977, de Waard at the University of Amsterdam conceived and prosecuted a novel approach that he dubbed "the Michael-induced Ramberg–Bäcklund (MIRB) olefin synthesis."[24,25] The reaction of dienyl sulfone **25** with one equivalent of sodium phenylsulfinate produced the Michael-induced Ramberg–Bäcklund product **26** in 40% yield.

$NaSO_2Ph$
d_6-DMSO, 1 h, 44%
25
26, *E*:*Z* = 2:1

Taylor and Evans extended the substrate for the MIRB to other nucleophiles including methoxide, thiol, and amines in place of de Waard's sulfinate.[26] For instance, Michael addition of benzylamine to bromo-vinylsulfone **27** gave the adduct as bromo-sulfone **28**, which upon treatment with a base led to the Ramberg–Bäcklund reaction product **29** in 83% yield.

$BnNH_2$, PhH
rt, 95%
27
28

KO*t*-Bu, THF
rt, 83%
29, *E* :*Z* = 95:5

Decarboxylative Ramberg–Bäcklund reaction is relatively rare. Sulfonyl ester **30** was subjected to the Myers's conditions to produce olefin **32**, likely from intermediate **31**. It was established that decarboxylation of **30** did not take place in the absence of CCl_4.[27]

KOH, CCl_4
t-BuOH, heat
n h, 61%
30
31
32

Epoxy-Ramberg–Bäcklund reaction is also uncommon. In place of iodide, sulfonate, or sulfinate leaving groups, Taylor and his colleagues chose epoxide as the leaving group in the Ramberg–Bäcklund reaction.[28] As shown below, exposure of epoxy-sulfone **33** to 2 equivalents LiO*t*-Bu in THF at room temperature delivered allylic alcohol **34** in 69% yield with excellent stereoselectivity.

2 eq. LiO*t*-Bu
THF, rt, 69%
33
34, *E*:*Z* = 92:2

Recently, a tandem Diels–Alder Ramberg–Bäcklund reaction was developed by Block and co-workers.[29] Using chloromethanesulfonyl-ethene (**35**) as the dienophile and cyclohexa-1,3-diene as the enophile, the Diels–Alder cycloaddition produced adduct **36**. A subsequent Ramberg–Bäcklund reaction then converted **36** to diene **37**.[29]

35
100–130 °C, toluene
12–16 h, 97%
36
KO*t*-Bu, THF
rt, 51%
37

4.6.5 *Synthetic Utility*

The Ramberg–Bäcklund reaction is now widely utilized in organic synthesis. It has been applied in synthesizing cyclobutenes, cyclopentenes, cyclohexenes, phenanthrenes, large ring systems, dienes, enynes, polyenes, alkynes, and many natural products.

4.6.5.1 *Cyclobutenes*

Paquette was the first to make cyclobutenes by taking advantage of the Ramberg–Bäcklund reaction.[30,31] Treatment of sulfone **38** with potassium *t*-butoxide led to cyclobutene **39** in 61% yield after distillation regardless of the configuration of the chlorine atom on the substrate (**38**). The ease with which cyclobutene ring formation occurred in this example was particularly noteworthy. In the same fashion, cyclobutene **42** was obtained from the Ramberg–Bäcklund reaction of either epimer **40** or **41**.[32]

KO*t*-Bu, THF
–78 °C, 61%
38
39

KO*t*-Bu, THF
−75 to 0 °C, 5 h

40 → **42**, 35–49% ← **41**

Epoxypropelladiene **45**, a condensed ring system, was prepared by treatment of dichloro-sulfone **44** with potassium *t*-butoxide in THF in 22% yield, whereas intermediate **44** was made from chlorination of disulfide **43** followed by oxidation.[33] Although there is no foreseeable utility of the molecule **45**, it shows that chemists can make almost anything if we so desire.

NCS, then AcOOH, 31%; KO*t*-Bu, THF, −15 °C, 22%

43 → **44** → **45**

4.6.5.2 Cyclopentenes, cyclopentenones, and cyclopentadienes

In 1968, E. J. Corey and his student Eric Block conducted a Ramberg–Bäcklund reaction of bromo-sulfone **46** and obtained $\Delta^{1,5}$-bicyclo[3,3,0]octane (**47**) in 81% yield.[34] (Interestingly, this work planted the seed for Block's life-long fascination with the Ramberg–Bäcklund reaction in his independent career.) Also in 1968, Paquette carried out the same reaction of the chloro-sulfone analog of **46** in refluxing aqueous KOH solution for 48 hours and produced **47** in 75% yield with 25% of recovered starting material.[35] Similarly, typical Ramberg–Bäcklund conditions converted bromosulfone **48** to olefin **49** in 71% yield.[36]

NaH, *t*-BuCH$_2$OH, MeO(CH$_2$CH$_2$)$_4$Me, 70 °C, 1 h, 81%

46 → **47**

KO*t*-Bu, THF, −15 °C to rt, 71%

48 → **49**

Cyclopentenone **51** was easily prepared by treatment of halosulfone **50** with sodium acetate, a weak base.[37] On the other hand, masked cyclopentenone **53** was synthesized from sulfone **52** using the Myers modification.[19]

NHBoc
O
X S
O_2
50
NaOAc, reflux, 10 h
THF-H_2O-MeOH
NHBoc
O
51
48% for X = Cl
62% for X = Br

O O
R
S
O_2
52
KO*t*-Bu, CCl_4, HO*t*-Bu
50 °C, 20 h, 50–84%
O O
R
53

There are not many examples of making cyclopentadienes using the Ramberg–Bäcklund reaction, possibly because of concurrent side reactions. In a rare case, treatment of chlorosulfone hexene **54** with *one* equivalent of *n*-BuLi led to cyclopentadiene **55** after exposure to dilute acid.[38]

Ph S Cl
O_2
54
1. 1 eq. *n*-BuLi, THF, –78 °C
2. dilute acid, 30%
Ph
55

4.6.5.3 Cyclohexenes and phenanthrene

Cyclohexenes can be prepared using the Ramberg–Bäcklund reaction of the seven-membered sulfones. Sulfone **56** was converted to highly oxygenated cyclohexene **57** using the Meyers modification.[39] Not only were the hydroxyl groups tolerated, but a combinatorial library was also prepared—a testimony of how robust and reproducible the reaction is. Moreover, when the diol on **56** was replaced with a di-azide, the Ramberg–Bäcklund worked just as well to produce the corresponding di-azo-hexene (see transformation **116→117** in Section 4.6.6).[40] Similarly, cyclohexadiene **59** was synthesized from sulfone **58** applying the Meyers–Chan modification.[41]

KOH, CCl_4/H_2O
HO*t*-Bu, rt, 1 h, 60%

56 → **57**

KOH/Al_2O_3, CF_2CBr_2
HO*t*-Bu/THF (3:1), 78%

58 → **59**

A rather delicate Ramberg–Bäcklund reaction was carried out by Rigby *et al.* in their total synthesis of (+)-estradiol.[42] As depicted below, sulfone **60** was deprotonated with potassium *t*-butoxide at –105 °C followed by treatment with *N*-iodosuccinamide (NIS) to give the iodo-sulfone, which was further treated with potassium *t*-butoxide at –105 °C to deliver the benzo-fused tetracycle **61** in 60% yield along with some over-oxidized product **62** (by excess NIS).

1. KO*t*-Bu, –105 °C
2. NIS (excess)
3. KO*t*-Bu, –105 °C

60 → **61**, 60% + **62**

Instead of halides as leaving groups, Hendrickson used triflinate (–SO_2CF_3) as the leaving group to carry out the Ramberg–Bäcklund reaction (the Hendrickson modification). For instance, treatment of sulfonyl-triflinate **63** with *n*-BuLi provided cyclohexene **64** in 70% yield.[43]

O_2 S SO_2CF_3 — n-BuLi, THF; –78 to –20 °C; 70% →

63 **64**

Early in his academic career in 1964, Paquette carried out the Ramberg–Bäcklund reaction of chloro-sulfone **65** to make phenanthrene **66**.[44] He discovered that conformational requirements of α-sulfonyl carbanion prolonged the reaction time (5 days) in some cases. In 1986, applying the Meyers modification, Staab *et al.* converted sulfone **67** to phenanthrene **68**.[45] The two chlorine atoms on **68** were the result of over-chlorination where tetrachloro-sulfone was generated *in situ*.

S O_2 Cl — 2 N NaOH; dioxane, 92% →

65 **66**

Me_2N NMe_2 S O_2 — KOH, CCl_4/HO*t*-Bu (1:1); 0 to 20 °C, 1 h, 53% → Me_2N NMe_2 Cl Cl

67 **68**

4.6.5.4 Large ring systems

In the same fashion as transformation **58**→**59**, 2-phenyl-cyclohepta-1,4-diene (**70**) was easily obtained from eight-membered sulfone **69** using the Meyers–Chan modification.[41] The more elaborate cycloheptene **72** was prepared from sulfide **71**,[20] with benzenesulfinate as the leaving group. In addition, eight-membered oxocene **74** was prepared from sulfide **73** in 52% yield.[46] Similarly, the nine-membered oxonene was synthesized using the same Ramberg–Bäcklund olefination process in 48% yield.[47]

Ph O S=O — KOH/Al_2O_3, CF_2CBr_2; HO*t*-Bu/THF (3:1), 80% → Ph

69 **70**

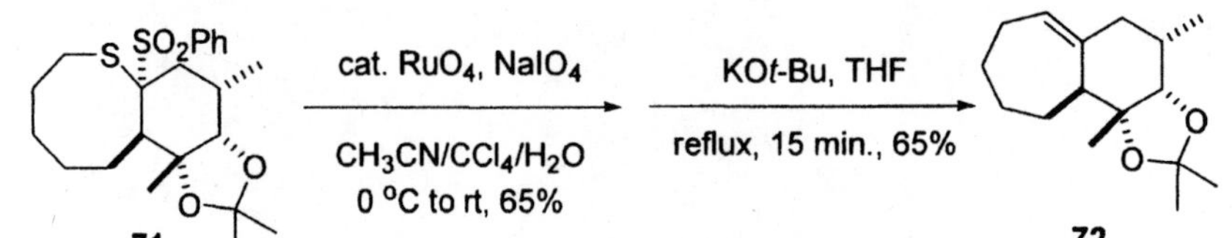

To prepare simple monocyclic conjugated enediynes as substrates for the Bergman cyclization, Nicolaou took advantage of the Ramberg–Bäcklund reaction.[48,49] Chloro-sulfone **76** was prepared from cyclic sulfide **75**, which in turn was assembled by treating the bis-propargyl bromide with Na_2S. Simple treatment of **76** with potassium *t*-butoxide led to enediyne **76** in yields ranging from 24% to 62%.

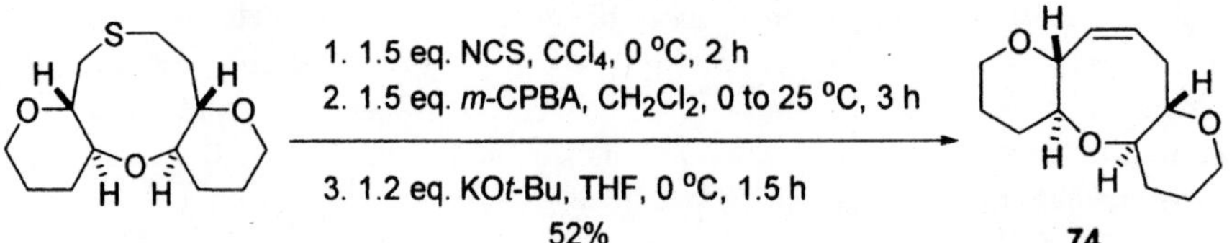

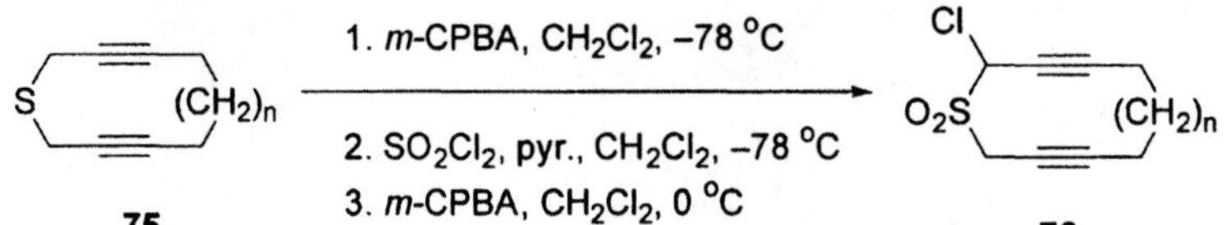

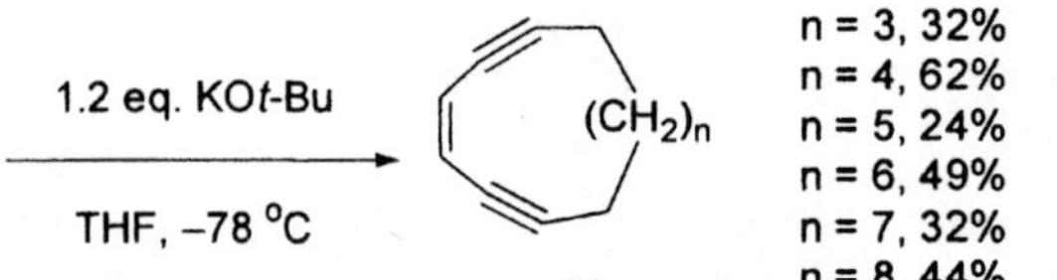

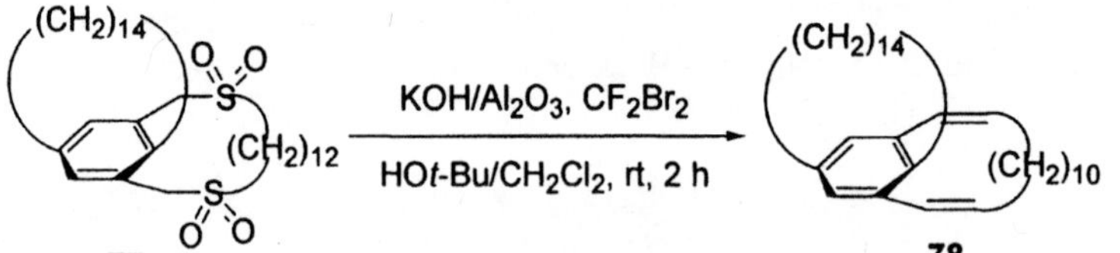

Chan *et al.* prepared many cyclophanes using (what else?) the Chan modification.[50,51] In one instance, the Ramberg–Bäcklund reaction of bis-sulfone 77

produced bis-olefin **78**, which was then hydrogenated to the corresponding cyclophane in 28% yield for the two steps.[51]

Large ring heterocycles have also been synthesized using the Ramberg–Bäcklund reaction. Preparation of [2.2]-(2,6)pyridinopha-1-ene (**80**) from bromo-sulfone **79** is one example,[52] and *trans*-olefin **83** from sulfide **81** is another *via* the intermediacy of chloro-sulfone **82**.[53]

NaOH, aq. dioxane

rt, 2 h, 43%

79 **80**

1. NCS, CCl_4, reflux

2. *m*-CPBA, CH_2Cl_2, rt
58%

KO*t*-Bu, DMSO

rt, 97%

81 **82** **83**

If one thought that the Ramberg–Bäcklund olefination always worked for the synthesis of carbocycles, he certainly would be mistaken. Paquette encountered his first challenge when he first began his exploits with the synthesis of phenanthrene (transformations 6566) in 1964.[44] In 2005, another Ramberg–Bäcklund reaction challenged him again.[54] Utilizing the Myers modification, sulfone **84** was heated for 2 days at 50–80 °C and no desired olefin **85** was observed. Only the starting material was recovered, possibly because the configuration and/or the conformation of **84** did not allow the formation of the episulfone intermediate.

KOH, CCl_4

HO*t*-Bu, 50–80 °C
no reaction

84 **85**

4.6.5.5 *Dienes, eneynes, and polyenes*

Dienes are straightforward to prepare using the Ramberg–Bäcklund reaction. The only requirement is that a double bond is already present on the substrate. We saw earlier examples such as dienes **24, 26, 55,** and **70.** Moreover, Block's bromomethane-sulfonyl bromide (**22**) was a powerful reagent for generating substrates to make dienes, eneynes, and polyenes. As shown below, addition of **22** to olefin **86** under photolysis gave the adduct which underwent an elimination of HBr under the influence of Et_3N to provide bromosulfone **87**. Treatment of **87** with potassium *t*-butoxide then delivered diene **88**.[23] In the same vein, cyclohexenes **89** and **91** were conveniently converted to dienes **90** and **92**, respectively.

86 → Br–CH$_2$–SO$_2$Br (**22**), *hv*, –20 °C, 30 min, CH_2Cl_2 → Et_3N, CH_2Cl_2, 0 °C, 15 min

87 → KO*t*-Bu, 7:1 HO*t*-Bu/THF, –20 °C, 1 h, 74% for 3 steps → **88**

89 → **22** → Et_3N → KO*t*-Bu, 41% → **90**

91 → **22** → Et_3N → KO*t*-Bu, 51% → **92**

Block's building block **22** is also amenable to prepare eneynes. Photo-induced addition of **22** to alkyne **93** gave the adduct **94**, which was converted to eneyne **95** with the aid of potassium *t*-butoxide.[23] The same sequence was applied to terminal alkyne **96** to afford eneyne **97**.

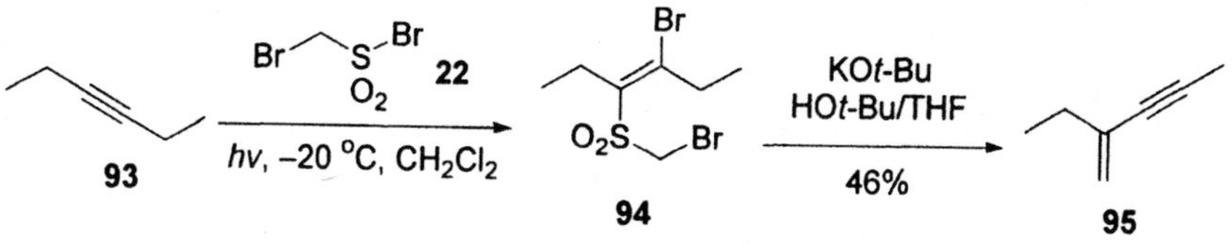

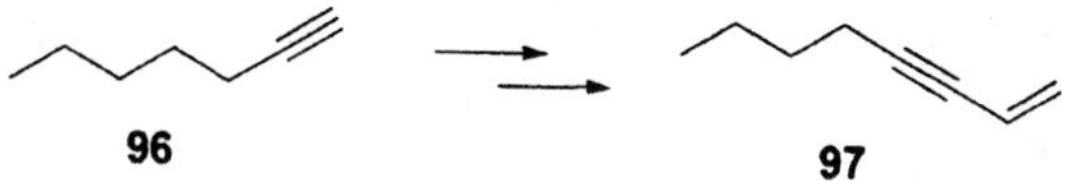

As far as polyenes are concerned, the Ramberg–Bäcklund reaction is an extremely powerful tool in putting the double bonds together. Herein just two examples are listed. Cao applied the Chan modification on a group of allylic dienylic sulfones and achieved the stereoselective synthesis of substituted all-*trans*-1,3,5,7-octatetraenes as exemplified by transformations **9899** .[55] Polyene **101** was produced from sulfone **100** using the Chan modification as well.[56]

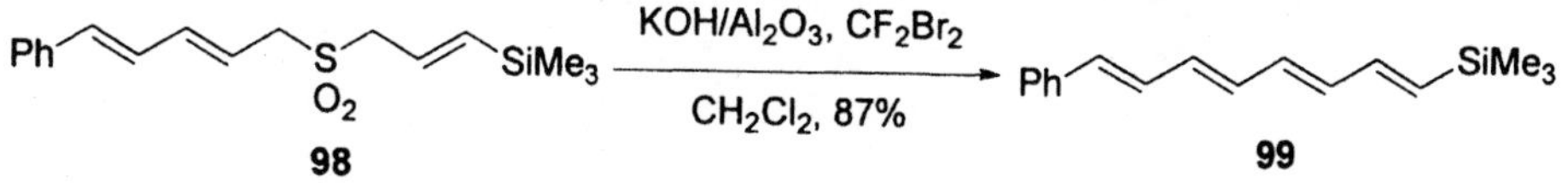

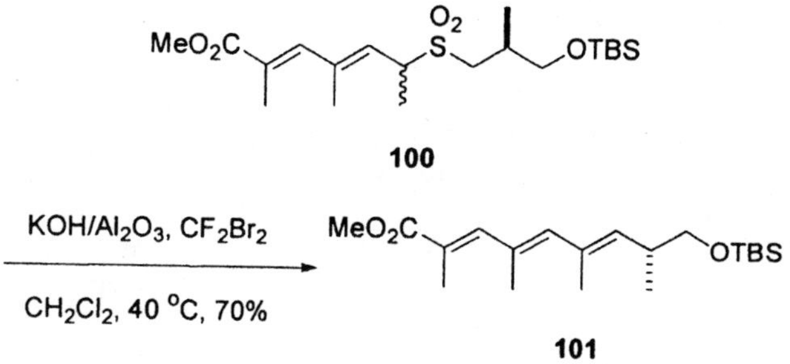

4.6.5.6 Natural products

One measure of a methodology's utility is its applications to the total synthesis of natural products. Since the sulfone substrates can often be assembled convergently, the Ramberg–Bäcklund reaction has been used in the total synthesis of many natural products.

In a formal total synthesis of steroid brassindole, sulfide **102** was used as the substrate of the Ramberg–Bäcklund reaction.[57] Chlorination of **102** followed by oxidation with *m*-CPBA, and treatment of the resulting chloro-sulfone with potassium *t*-butoxide, provided symmetrical olefin **103**.

TBDPSO S OTBDPS

102

1. 1 eq. NCS, CCl4, 90 °C, 15 min
2. 3 eq. *m*-CPBA, 6 eq. $NaHCO_3$, CH_2Cl_2, rt, 3 h
3. 4 eq. KO*t*-Bu, THF, –78 °C, 2 h, 78%

TBDPSO OTBDPS

103

KOH/Al_2O_3, CF_2Br_2

t-BuOH, CH_2Cl_2, 5 °C to rt

26–84%

104 **105**

In a synthesis of *exo*-olefinated deoxoartemisinin, the Chan modification was applied to convert sulfone **104** to *exo*-olefin **105**.[58]

Hart assembled olefin **109** convergently from benzyl bromide **106** by taking advantage of the Ramberg–Bäcklund reaction.[59] As depicted below, the S_N2 displacement of benzyl bromide **106** with thiol **107** led to sulfide **108**. Oxidation of **108** to the sulfone, followed by the Myers modification delivered alkene **109**, an intermediate for the synthesis of *C*-aryl glycosides related to chrysomycins.

DBU, PhH, 83%

106 **107** **108**

1. *m*-CPBA, 94%

2. KOH, CCl_4, *t*-BuOH, 69%

OMOM OTBS OMe OMe OTs

109

Trost left the key operation at the end of his route to (+)-solamin, a natural product with a range of activities such as cytotoxicity, antimalarial and immunosuppressant activities. Applying the Myers modification, sulfone **110** was converted to butenolide diol **111** in good yield after desillylation.[60]

TMSO O_2S O TMSO

110

1. *t*-BuOK, *t*-BuOH CCl_4, rt, 65%

2. TsOH, H_2O, EtOH rt, 95%

HO O HO

111

The trophy probably should go to Boeckman for his exploit in the total synthesis of (+)-eremantholide using the Ramberg–Bäcklund reaction as the final key operation.[61] Treatment of di-halide **112** with TMS_2S/NaOMe led to the required 10-membered-ring sulfide **113** in 45–50% yield. After that, the Ramberg–Bäcklund reaction was employed to effect the crucial ring contraction of **113**. Sequential treatment of **113** with 6 N HCl, oxone, and Amberlyst-15 resin afforded the sulfone enol ether, which was exposed to LiHMDS and Cl_3CCCl_3 to provide a single diastereomeric chloro-sulfone **114** in 57% yield. Treatment of **114** with potassium *t*-butoxide in DME/HMPA at 70 °C led to an 82% yield of the olefin, which was treated with 6 N HCl/THF to deliver (+)-eremantholide (**115**) in 85% yield.

1 eq. TMS_2S, 2 eq. NaOMe

THF (3×10^{-2} M), 0 °C, 4 h

45–50%

112 → 113

1. 6 N HCl/THF (1:10, v/v), rt, 4 h
2. 4 eq. oxone, $MeOH/H_2O$, rt, 6 h
3. Amberlyst-15, 3 Å MS, CH_2Cl_2, rt, 4 h
4. 1.1 eq. LiHMDS, THF, –78 °C, 10 min then C_2Cl_6, 57%

114

1. 2.2 eq. KO*t*-Bu, 10 eq. HMPA DME, 70 °C, 5 min, 82%
2. 6 N HCl/THF (1:10, v/v), rt, 4 h, 85%

115

4.6.6 *Experimental*

A quick perusal of current literature revealed that the Chan modification of the Ramberg–Bäcklund reaction seems to be more prevalent than the Myers modification in the latest publications.

4.6.6.1 *The Chan modification*

98 → 99

KOH/Al_2O_3, CF_2Br_2

CH_2Cl_2, 87%

Preparation of (1E,3E,5E,7E)-1-phenyl-8-trimethylsilyl-1,3,5,7-octatetraene (99)[55]
The sulfone **98** (1 mmol) was added to a stirred suspension of alumina-supported KOH (10 mmol of KOH) in CF_2Br_2/CH_2Cl_2 (1:10 10 mL) at 0 °C. The mixture was then stirred for 2 h. The reaction mixture was filtered through a pad of Celite and the filtered cake was washed thoroughly with CH_2Cl_2. The filtrate was concentrated *in vacuo* to give the crude octatetraene **99**. Flash chromatography over silica gel (hexane) afforded the 1,3,5,7-octatetraene **99** as a yellow oil (87%, Found: C, 80.02; H, 8.53. $C_{17}H_{22}Si$ requires C, 80.25, H, 8.71).

4.6.6.2 The Myers modification

KOH, CCl_4; HO*t*-Bu/H_2O; rt, 2 h, 90%

116 → **117**

Preparation of (+)-(1*S*,2*R*,3*R*,4*R*)-2,3-diazido-1,4-dimethoxycyclohex-5-ene (117)[40]

To 0.20 g (0.70 mmol) of **116**, CCl_4 (2.9 mL), *t*-BuOH (1.9 mL) and H_2O (0.3 mL) were added under N_2. After complete dissolution of **116**, 1.94 g of finely powdered KOH were added and a vigorous stirring was continued for 2 h. Finally, H_2O was added and the reaction mixture was extracted with CH_2Cl_2 giving 0.14 g (0.60 mmol, 90%) of **117** as yellow oil. It was purified by flash chromatography (SiO_2; CH_2Cl_2/light petroleum 3:1).

4.6.7 References

1. [R] Bordwell, F. G. *Organosulfur Chemistry* **1967**, 271–284.
2. [R] Paquette, L. A. *Acc. Chem. Res.* **1968**, *1*, 209–216.
3. [R] Paquette, L. A. *Org. React.* **1977**, *25*, 1–71.
4. [R] Magnus, P. D. *Tetrahedron* **1977**, *33*, 2019–2045.
5. [R] Clough, J. M. The Ramberg–Bäcklund reaction in *Comprehensive Organic Synthesis;* Trost, B. M.; Fleming, I., Eds.; Pergamon, **1991**, *Vol. 3*, 861–886.
6. [R] Taylor, R. J. K. *Chem. Commun.* **1999**, 217–227.
7. [R] Braverman, S.; Cherkinsky, M.; Raj, P. *Sulfur Reports* **1999**, *22*, 49–84.
8. [R] Taylor, R. J. K.; Casy, G. *Org. React.* **2003**, *62*, 357–475.
9. Ramberg, L.; Bäcklund, B. *Arkiv. Kemi, Mineral. Geol.* **1940**, *13A(27)*, 1–50.
10. Fischer, N. H. *Synthesis* **1973**, 393.
11. Sutherland, A. G.; Taylor, R. J. K. *Tetrahedron Lett.* **1989**, *30*, 3269.
12. Jeffrey, S. M.; Sutherland, A. G.; Pyke, S. M.; Powell, A. K.; Taylor, R. J. K. *J. Chem. Soc., Perkin Trans. 1* **1993**, 2317.
13. Meyers, C. Y.; Malte, A. M.; Mattews, W. S. *J. Am. Chem. Soc.* **1969**, *91*, 7510.
14. Chan, T.-L.; Fong, S.; Li, Y.; Man, T.-O.; Poon, C.-D. *J. Chem. Soc., Chem. Commun.* **1994**, 1771.
15. Yang, G.; Franck, R. W.; Byun, H.-S.; Bittman, R.; Samadder, P.; Arthur, G. *Org. Lett.* **1999**, *1*, 2149.
16. Vedejs, E.; Singer, S. P. *J. Org. Chem.* **1978**, *43*, 4884.
17. Hendrickson, J. B.; Palumbo, P. S. *J. Org. Chem.* **1985**, *50*, 2110.
18. Hendrickson, J. B.; Boudreaux, G. J.; Palumbo, P. S. *J. Am. Chem. Soc.* **1986**, *108*, 2358.

19. Matsuyama, H.; Miyazawa, Y.; Takei, Y.; Kobayashi, M. *J. Org. Chem.* **1987**, *52*, 1703.
20. Scarpetti, D.; Fuchs, P. L. *J. Am. Chem. Soc.* **1990**, *112*, 8084.
21. Block, E.; Aslam, M. *J. Am. Chem. Soc.* **1983**, *105*, 6164.
22. Block, E.; Aslam, M.; Eswarakrishnan, V.; Wall, A. *J. Am. Chem. Soc.* **1983**, *105*, 6165.
23. Block, E.; Aslam, M.; Eswarakishnan, V. Eswarakrishnan, V.; Gebreyes, K.; Hutchinson, J.; Iyer, R.; Laffitte, J. A.; Wall, A. *J. Am. Chem. Soc.* **1986**, *108*, 4568.
24. Chen, T. B. R. A.; Burger, J. J.; de Waard, E. R. *Tetrahedron Lett.* **1977**, 4527.
25. Burger, J. J.; Chen, T. B. R. A.; de Waard, E. R.; Huisman, H. O. *Tetrahedron* **1981**, *37*, 417.
26. Evans, P.; Taylor, R. J. K. *Synlett* **1997**, 1043.
27. Wladislaw, B.; Marzorati, L.; Russo, V. F. T.; Zaim, M. H.; Di Vitta, C. *Tetrahedron Lett.* **1995**, *36*, 8367.
28. Evans, P.; Johnson, P.; Taylor, R. J. K. *Eur. J. Org. Chem.* **2006**, 1740.
29. Block, E.; Jeon, H. R.; Zhang, S.-Z.; Dikarev, E. V. *Org. Lett.* **2004**, *6*, 437.
30. Paquette, L. A.; Philips, J. C.; Wingard, R. E., Jr. *J. Am. Chem. Soc.* **1971**, *93*, 4516.
31. Paquette, L. A.; Trova, M. P. *J. Am. Chem. Soc.* **1988**, *110*, 8197.
32. Becker, K. B.; Labhart, M. P. *Helv. Chim. Acta* **1977**, *66*, 1090.
33. Weinges, K.; Baake, H. *Chem. Ber.* **1977**, *110*, 1601.
34. Corey, E. J.; Block, E. *J. Org. Chem.* **1969**, *34*, 1233.
35. Paquette, L. A.; Houser, R. W. *J. Am. Chem. Soc.* **1969**, *91*, 3870.
36. Becker, K. B.; Labhart, M. P. *Helv. Chim. Acta* **1983**, *66*, 1090.
37. Gamble, M. P.; Giblin, G. M.; Montana, J. G.; O'Brien, P.; Ockendon, T. P.; Taylor, R. J. K. *Tetrahedron Lett.* **1996**, *37*, 7457.
38. Burger, J. J.; Chen, T. B. R. A.; de Waard, E. R. Huisman, H. O. *Hetereocycles* **1980**, *14*, 1739.
39. Ceré, V.; Minzoni, M.; Pollicino, S.; Ricci, A.; Gasparrini, F.; Ciogli, A.; D'Acquarica, I. *J. Comb. Chem.* **2006**, *8*, 74.
40. Arcelli, A.; Ceré, V.; Peri, F.; Pollicino, S.; Ricci, A. *Tetrahedron* **2001**, *57*, 3439.
41. Yao, Q. *Org. Lett.* **2002**, *4*, 427.
42. Rigby, J. H.; Warshakoon, N. C.; Payen, A. J. *J. Am. Chem. Soc.* **1999**, *121*, 8237.
43. Hendrickson, J. B.; Boudreaux, G. J.; Palumbo, P. S. *Tetrahedron Lett.* **1984**, *25*, 4617.
44. Paquette, L. A. *J. Am. Chem. Soc.* **1964**, *86*, 4085.
45. Saupe, T.; Krieger, C.; Staab, H. A. *Angew. Chem., Int. Ed. Engl.* **1986**, *25*, 451.
46. Alvarez, E.; Diaz, M. T.; Hanxing, L.; Martin, J. D. *J. Am. Chem. Soc.* **1995**, *117*, 1437.
47. Alvarez, E.; Delgado, M.; Diaz, M. T.; Hanxing, L.; Perez, R.; Martin, J. D. *Tetrahedron Lett.* **1996**, *37*, 2865.
48. Nicolaou, K. C.; Ogawa, Y.; Zuccarello, G.; Schweiger, E. J.; Kumazawa, T. *J. Am. Chem. Soc.* **1988**, *110*, 4866.
49. Nicolaou, K. C.; Zuccarello, G.; Riemer, C.; Estevez, V. A.; Dai, W. M. *J. Am. Chem. Soc.* **1992**, *114*, 7360.
50. Chan, T.-L.; Hung, C.-W.; Man, T.-O.; Leung, M.-k. *J. Chem. Soc., Chem. Commun.* **1994**, 1971.
51. Wei, C.; Mo, K.-F.; Chan, T.-L. *J. Org. Chem.* **2003**, *68*, 2948.
52. Cooke, M. P. *J. Org. Chem.* **1981**, *46*, 1747.
53. MaGee, D. I.; Beck, E. J. *Can. J. Chem.* **2000**, *78*, 1060.
54. Paquette, L. A.; Liu, Z.; Efremov, I. *J. Org. Chem.* **2005**, *70*, 514.
55. Cao, X.-P. *Tetrahedron* **2002**, *58*, 1301.
56. Bouchez, L. C.; Vogel, P. *Chem. Eur. J.* **2005**, *11*, 4609.
57. Schmittberger, T.; Uguen, D. *Tetrahedron Lett.* **1997**, *38*, 2837.
58. Oh, S.; Jeong, I. H.; Shin, W.-S.; Lee, S. *Bioorg. Med. Chem. Lett.* **2004**, *14*, 3683.
59. Hart, D. J.; Merriman, G. H.; Young, G. J. *Tetrahedron* **1996**, *52*, 14437.
60. Trost, B. M.; Shi, Z. *J. Am. Chem. Soc.* **1994**, *116*, 7459.
61. Boeckman, R. K., Jr.; Yoon, S. K.; Heckendorn, D. K. *J. Am. Chem. Soc.* **1991**, *113*, 9682.

Jie Jack Li

4.7 Shapiro Reaction

4.7.1 Description

The Shapiro reaction [1] is the conversion of arylsulfonylhydrazones to alkenes *via* treatment with alkyllithium reagents or alkali metal amides.[2,3,4,5] The reactions proceed *via* alkenyllithium intermediates, which can be transformed to the alkene upon protonation or can be captured with other electrophilic reagents. For example, the tosylhydrazone **1** derived from 2-methylcyclohexanone was converted to 3-methylcyclohexene (**2**) in 98% yield upon treatment with two equivalents of butyllithium.[1]

NNHTs
1. 2 equiv *n*-BuLi
2. H_2O
98%
1 **2**

4.7.2 Historical Perspective

The base-mediated conversion of arylsulfonylhydrazones to alkenes was first observed by Bamford and Stevens in 1952.[6] In a representative transformation, tosylhydrazone **3** was converted to cyclohexene (**4**) in quantitative yield after refluxing for 90 minutes in an alkoxide solution derived from dissolution of sodium in ethylene glycol. These reactions are believed to proceed *via* intermediate diazo compounds, which are transformed to alkenes by thermal elimination processes.

TsHN
N
Na
ethylene glycol
197 °C
100%
3 **4**

In 1967 a significant extension of this chemistry was developed by Shapiro, who reported that treatment of sulfonylhydrazones with methyllithium or *n*-butyllithium at room temperature leads to the formation of vinyllithium reagents.[1] Subsequent studies by Shapiro and others led to widespread use of this transformation.

4.7.3 Mechanism

The mechanism of the Shapiro reaction is believed to involve initial deprotonation of the NH proton from tosylhydrazone **5** to generate **6**, which undergoes a second deprotonation adjacent to the hydrazone group to afford dianion **7**. Elimination of lithium *p*-toluenesulfinate from 7 provides **8**, which yields vinyllithium species **9** upon loss of N_2.[2-4] Evidence in support of this mechanism has been obtained through deuterium labeling experiments that indicate one α-hydrogen (deuterium) atom is lost in the transformation,[1] and through trapping of the intermediate dianion with various electrophiles.[2,7]

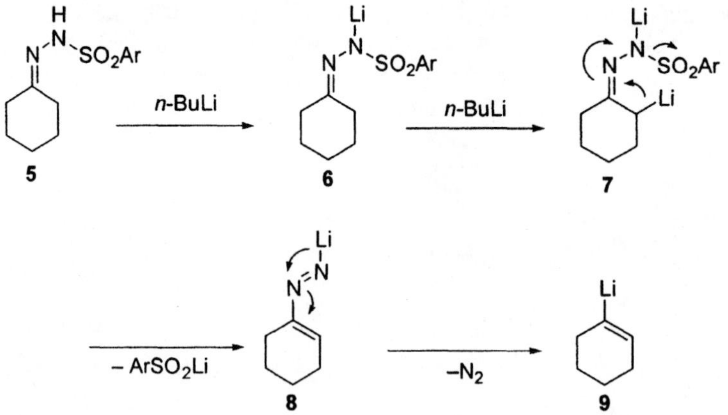

4.7.4 Regiochemistry and Alkene Stereochemistry

In most cases the regiochemistry of the Shapiro reaction is controlled by steric factors, with the second metalation occurring at the less-substituted α-carbon atom of the hydrazone.[2-4] For example, α-methyl ketone **10** was transformed to disubstituted alkene **11** in 69% yield as a single regioisomer *via* conversion to a tosylhydrazone followed by treatment with *n*-BuLi.[8] However, the regiochemical outcome of Shapiro reactions involving α,β-unsaturated hydrazones is difficult to predict in the absence of data obtained from reactions of structurally related substrates, as the second metalation can occur at either the α′- or γ-position.

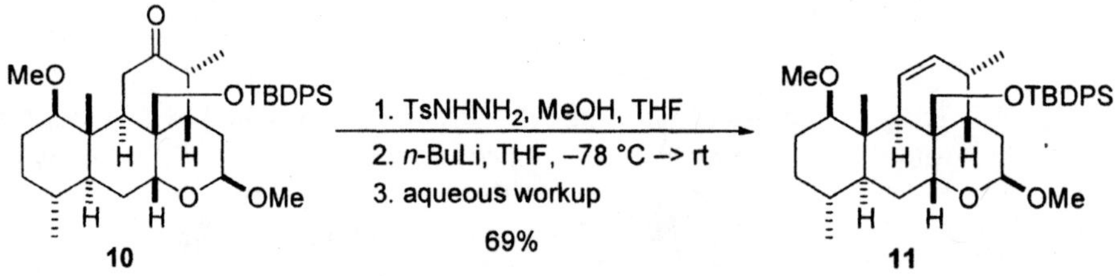

Most Shapiro reactions of acyclic sulfonylhydrazones or cyclic sulfonylhydrazones derived from larger rings proceed with selectivity for the formation of the *E*-vinyllithium reagent. For example, the trisylhydrazone generated from 4-heptanone (**12**)

was converted exclusively to *E*-alkene **14** through sequential treatment with *n*-BuLi/TMEDA followed by chlorotrimethylsilane.[9] This selectivity is likely due to chelation between the alkyllithium base and the metalated hydrazine nitrogen atom, which facilitates deprotonation on the adjacent side of the molecule with the bulky substituent position away from the hydrazine (e.g., **13**). However, branching at the hydrazone α-carbon typically leads to diminished selectivities.[11]

$NNHSO_2Ph$ **12** — 1. *n*-BuLi/TMEDA; 2. TMS–Cl; 97% → [**13**] → **14**

4.7.5 *Variations and Improvements*

The conditions originally employed for the Shapiro reaction involved treatment of the sulfonylhydrazone derivative with an alkyllithium reagent in hexane or ether solvent. Although these conditions are quite effective for the conversion of sulfonylhydrazones to alkenes (e.g., **1**→**2**), efforts to capture the intermediate vinyllithium reagent with electrophiles other than H^+ are often met with limited success due to competing deprotonation of the solvent or the sulfonyl aryl group by the basic vinyllithium species. For example, treatment of **15a** with >2 equiv of *n*-BuLi in hexane followed by quenching with D_2O provided **16** in quantitative yield but with only ~10% deuterium incorporation. A solution to this problem was developed independently by Shapiro and Bond that employs TMEDA (tetramethylethylenediamine) as an additive for Shapiro reactions.[7,10] As shown below, use of TMEDA (4.0 equiv) as a cosolvent led to the conversion of **15b** to **16** in quantitative yield with 95% deuterium incorporation.

15a: Ar = Ts
15b: Ar = Ph

1. *n*-BuLi, Hexanes, **Additive**; 2. D_2O → **16**

Additive = none: 100% yield, 10% D-incorporation (from **15a**)
Additive = TMEDA: 100% yield, 95% D-incorporation (from **15b**)

As mentioned above, one significant problem with the use of phenyl- or tosylhydrazones in the Shapiro reaction is competing deprotonation of an *ortho*-position on the aromatic ring. This side reaction often leads to diminished yields and/or the requirement for ≥3 equivalents of the alkyllithium base. This problem has been addressed through the use of trisylhydrazones (trisyl = 2,4,6-triisopropylphenyl), which do not contain aromatic protons that are easily metalated.[11] For example, trisylhydrazone **15c** was converted to α,β-unsaturated aldehyde **17** in 61% yield *via* treatment with *s*-BuLi followed by quenching with DMF,[11] whereas the analogous reaction of tosylhydrazone **15a** conducted with added TMEDA provided a very low yield (10%) of **17**.[12]

$NNHSO_2Ar$ 1. *n*-BuLi, Hexanes, Additive 2. DMF → CHO

15a: Ar = Ts
15c: Ar =Tris

17

From **15a**: 10% yield (TMEDA additive)
From **15c**: 61% yield (no additive)

The two modifications described above have allowed for the efficient capture of vinyllithium intermediates generated in Shapiro reactions with a wide variety of electrophiles. For example, the alkyllithium reagent **20** prepared from treatment of trisylhydrazone **18** with *n*-BuLi was effectively trapped with benzaldehyde (62% yield, **21**), 1-bromobutane (58% yield, **22**), and bromine from 1,2-dibromoethane (43% yield, **23**).[11] Similarly, formation of **20** from tosylhydrazone **19** followed by trapping with CO_2 afforded **24** (52% yield). The reaction of intermediate **20** (generated from **19**) with cyclohexenone provided **25**, the product of 1,2-addition, in 61% yield.[13]

TrisNH–N=C(CH₃)C_6H_{13} **18** —*n*-BuLi→ **20**
TsNH–N=C(CH₃)C_6H_{13} **19** —*n*-BuLi, TMEDA→ **20**

20 (Li, C_6H_{13}):
PhCHO, 62% → **21** (Ph, OH, C_6H_{13})
BuBr, 58% → **22** (Bu, C_6H_{13})
$(BrCH_2)_2$, 43% → **23** (Br, C_6H_{13})
CO_2, 52% → **24** (CO_2H, C_6H_{13})
cyclohexenone, 61% → **25** (OH, C_6H_{13})

The Shapiro reaction has also been used to convert sulfonylhydrazones into alkenylstannanes or alkenylboronic acids, which have been employed in Pd-catalyzed Stille or Suzuki coupling reactions.[14,15] For example, Keay has described a one-pot procedure for the conversion of trisylhydrazone **26** to alkene **27** *via* sequential treatment with *n*-BuLi followed by B(O*i*-Pr)$_3$. Exchange of the hexane solvent for toluene followed by addition of a palladium catalyst, Na_2CO_3, and 3-bromonitrobenzene afforded **27** in 55% yield.

=N NHTris **26**

1. *n*-BuLi, TMEDA, hexanes
2. B(O*i*-Pr)$_3$
3. Remove hexanes, add toluene
4. cat. Pd(OAc)$_2$, PPh$_3$ Na_2CO_3, 3-bromonitrobenzene

55%

27 NO_2

The dianion generated through treatment of a sulfonylhydrazone can be alkylated at the α-position to yield more highly substituted derivatives,[7] which undergo Shapiro reaction upon treatment with an additional equivalent of base.[16] This protocol allows for the construction of alkenes that derive from deprotonation at the more hindered position, as the monoanion is configurationally stable at low temperature and the deprotonation occurs on the methylene adjacent to the sulfonyl group. For example, treatment of **28** with *s*-BuLi followed by addition of pentyl iodide generates intermediate **29**, which is converted to alcohol **30** in 52% overall yield through a second deprotonation followed by Shapiro reaction.[16] A variety of electrophiles can be used in the second step, including iodine and alkyl iodides.[17,18]

N NHTris **28**

1. *s*-BuLi
2. $C_5H_{11}I$

N $\bar{N}$Tris **29** C_5H_{11}

1. *s*-BuLi
2. HCHO
52%

OH **30** C_5H_{11}

The α-alkylation of sulfonylhydrazone dianions with disulfides followed by Shapiro reaction has been used to effect the 1,2-transposition of carbonyl groups.[19,20] As shown below, treatment of tosylhydrazone **31** with *n*-BuLi/TMEDA followed by addition of dimethyl disulfide and deprotonation with an additional equivalent of *n*-BuLi provided vinylsulfide **32**.[19] Exposure of this compound to mercuric chloride in hot aqueous acetonitrile provided ketone **33** in 75% overall yield.

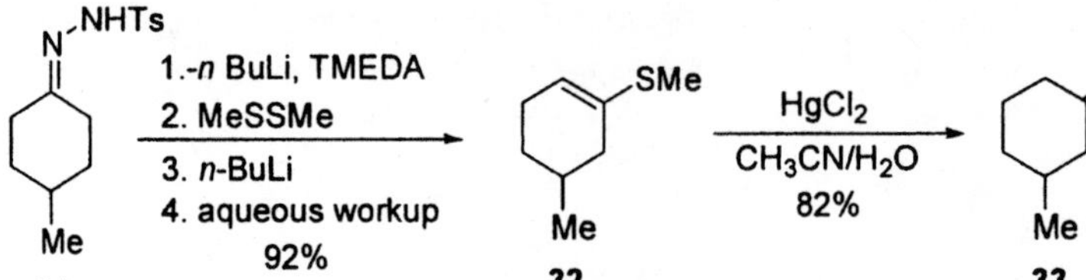

As noted above, the classical Shapiro reaction requires the use of ≥2 equiv of base to effect the conversion of an arylsulfonylhydrazone to an alkene. However, in recent work Yamamoto has described a catalytic version of the Shapiro reaction that employs aziridinyl hydrazones and is effective with only 5–30 mol% of added base.[21] As shown above, treatment of hydrazone **34** with 5 mol% of LDA affords a 93% isolated yield of **35** with >99:1 stereoselectivity. This reaction is believed to proceed *via* metalation of **34** with LDA to yield **36**. This intermediate decomposes with loss of styrene and nitrogen to provide vinyllithium reagent **37**, which is protonated by diisopropylamine, regenerating the LDA base. Unfortunately this method is not amenable to the capture of the intermediate vinyllithium species with electrophiles.

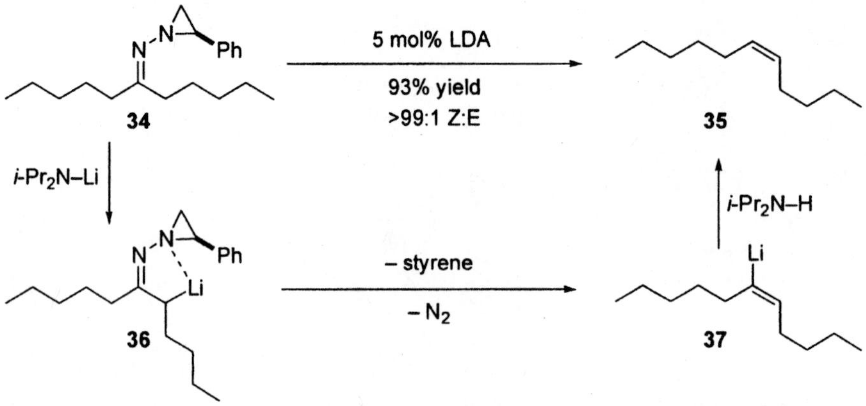

4.7.6 Synthetic Utility

The Shapiro reaction has frequently been employed in the construction of complex molecules. A few recent applications of this method to the synthesis of natural products are described below.

The conversion of ketone **38** to alkene **40** in the context of Winterfeldt's synthesis of (–)-myltaylenol was accomplished in two steps using classical conditions for the Shapiro reaction.[22] The tosylhydrazone **39** was generated cleanly from **38** in near quantitative (98%) yield. Treatment of **39** with excess *n*-BuLi provided **40** in 90% yield.

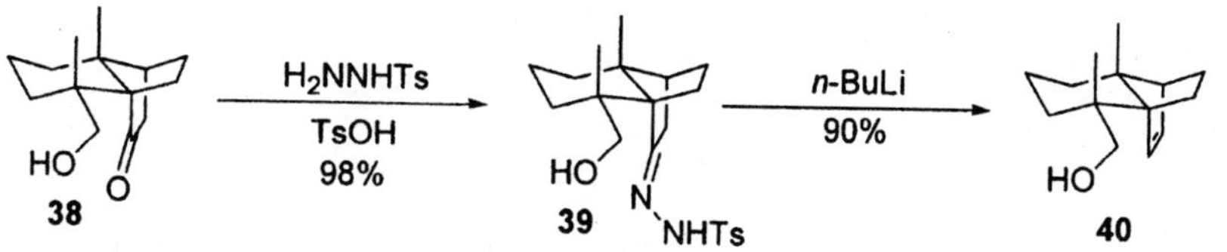

A synthetic route that provides rapid access to the tetracyclic core of ingenol was recently described by Cha in which a Shapiro reaction is used for the construction of a key C–C bond.[23] As shown below, treatment of trisylhydrazone **41** with two equivalents of *t*-BuLi followed by addition of hindered ketone **42** provided allylic alcohol **43** in 75–88% yield. The carbonyl addition occurred with high diastereoselectivity for approach of the nucleophile from the less hindered face. Alcohol **43** was converted to the tetracyclic core of the natural product (**44**) *via* a 7-step sequence.

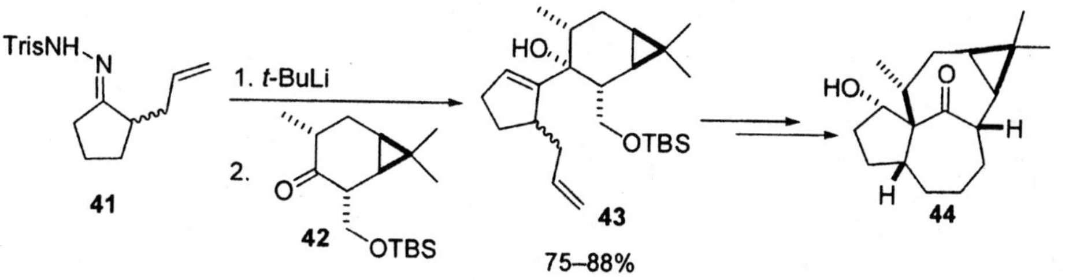

The Shapiro reaction has also been employed as a key step in Sorensen's synthesis of (–)-hispidospermidin.[24] As shown below, Shapiro reaction of **45** followed by transmetalation to Mg and trapping with pyruvate derivative **46** provided **47** in 55% yield as a single diastereomer.

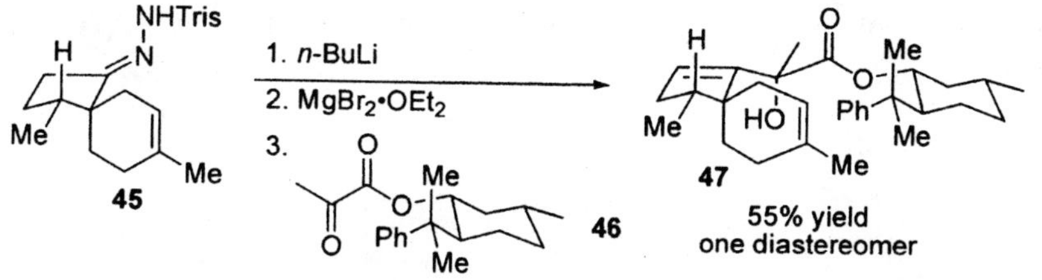

Several approaches to the synthesis of taxol and the taxane skeleton have employed Shapiro reactions.[25] An interesting route to the taxol A-ring that illustrates the utility of dianion functionalization prior to alkene generation was recently described by Koskinen.[25e] As shown below, treatment of tosylhydrazone **48** with 2.2 equiv of *n*-BuLi followed by sequential addition of methyl iodide, another 4 equiv of *n*-BuLi, and paraformaldehyde provided allylic alcohol **49** in 62% yield.

TsNH-N= ... **48** → 1.-*n* BuLi; 2. MeI; 3.-*n* BuLi; 4. paraformaldehyde; 62% → HO ... **49**

4.7.7 *Experimental*

N-NHTris **50** → 1. *n*-BuLi, TMEDA; 2. DMF → O=CH **51**

3-Methylcyclohexene-2-carboxaldehyde (51)[11]
A solution of *s*-BuLi in hexanes (48.7 mL, 53.6 mmol, 1.1 m in hexanes) was slowly added to a solution of **50** (10.0 g, 25.5 mmol) in hexanes/TMEDA (100 mL, 50:50 v:v) at –78 °C. The resulting mixture was then warmed to 0 °C and stirred until nitrogen evolution had ceased. Neat *N,N*-dimethylformamide (2.05 g, 28.0 mmol) was added and the mixture was warmed to room temperature and stirred for 1 h. The resulting mixture was poured into water, transferred to a separatory funnel, and the layers were separated. The aqueous phase was extracted with ether, the combined organic phases were extracted with water until neutral (to remove excess TMEDA), and then dried over anhydrous $MgSO_4$, filtered, and concentrated in vacuo. The crude product was purified by distillation to afford 1.99 g (63%) of the title compound as a colorless oil, bp 88–90 °C (26 Torr).

4.7.8 *References*

1 Shapiro, R. H.; Heath, M. J. *J. Am. Chem. Soc.* **1967**, *89*, 5734–5735.
2 [R] Shapiro, R. H. *Org. React.* **1976**, *23*, 405–507.
3 [R] Adlington, R. M.; Barrett, A. G. M. *Acc. Chem. Res.* **1983**, *16*, 55–59.
4 [R] Chamberlin, A. R.; Bloom, S. H. *Org. React.* **1990**, *39*, 1–83.
5 Leroux, F.; Schlosser, M.; Zohar, E.; Marek, I. in *The Chemistry of Organolithium Compounds*, Rappoport, Z.; Marek, I., Eds., John Wiley and Sons: Hoboken, NJ, 2004, pp. 435–493.
6 Bamford, W. R.; Stevens, T. S. *J. Chem. Soc.* **1952**, 4735–4740.
7 Shapiro, R. H.; Lipton, M. F.; Kolonko, K. J.; Buswell, R. L.; Capuano, L. A. *Tetrahedron Lett.* **1975**, 1811–1814.
8 Grieco, P. A.; Collins, J. L.; Moher, E. D.; Fleck, T. J.; Gross, R. S. *J. Am. Chem. Soc.* **1993**, *115*, 6078–6093.

9 Paquette, L. A.; Fristad, W. E.; Dime, D. S.; Bailey, T. R. *J. Org. Chem.* **1980**, *45*, 3017–3028.
10 Stemke, J. E.; Bond, F. T. *Tetrahedron Lett.* **1975**, 1815–1818.
11 Chamberlin, A. R.; Stemke, J. E.; Bond, F. T. *J. Org. Chem.* **1978**, *43*, 147–154.
12 Traas, P. C.; Boelens, H.; Takken, H. J. *Tetrahedron Lett.* **1976**, 2287–2288.
13 Stemke, J. E.; Chamberlin, A. R.; Bond, F. T. *Tetrahedron Lett.* **1976**, 2947–2950.
14 Nordvall, G.; Sundquist, S.; Nilvebrant, L.; Hacksell, U. *Bioorg. Med. Chem. Lett.* **1994**, *4*, 2837–2840.
15 Passafaro, M. S.; Keay, B. A. *Tetrahedron Lett.* **1996**, *37*, 429–432.
16 Chamberlin, A. R.; Bond, F. T. *Synthesis* **1979**, 44–45.
17 Corey, E. J.; Roberts, B. E. *Tetrahedron Lett.* **1997**, *38*, 8919–8920.
18 Corey, E. J.; Lee, J.; Roberts, B. E. *Tetrahedron Lett.* **1997**, *38*, 8915–8918.
19 Nakai, T.; Mimura, T. *Tetrahedron Lett.* **1979**, 531–534.
20 Kano, S.; Yokomatsu, T.; Ono, T.; Hibino, S.; Shibuya, S. *Chem. Commun.* **1978**, 414–415.
21 Maruoka, K.; Oishi, M.; Yamamoto, H. *J. Am. Chem. Soc.* **1996**, *118*, 2289–2290.
22 Doye, S.; Hotopp, T.; Wartchow, R.; Winterfeldt, E. *Chem. Eur. J.* **1998**, *4*, 1480–1488.
23 Epstein, O. L.; Cha, J. K. *Angew. Chem. Int. Ed.* **2005**, *44*, 121–123.
24 Tamiya, J.; Sorensen, E. J. *Tetrahedron* **2003**, *59*, 6921–6932.
25 (a) Nicolaou, K. C.; Yang, Z.; Sorensen, E. J.; Nakada, M. *Chem. Commun.* **2003**, 1024–1026. (b) Kress, M. H.; Ruel, R.; Miller, W. H.; Kishi, Y. *Tetrahedron Lett.* **1993**, *34*, 5999–6002. (c) Wang, Z.; Warder, S. E.; Perrier, H.; Grimm, E. L.; Bernstiein, M. A. *J. Org. Chem.* **1993**, *58*, 2931–2932. (d) Nicolaoou, K. C.; Claiborne, C. F.; Paulvannan, K.; Postema, M. H. D.; Guy, R. K. *Chem. Eur. J.* **1997**, *3*, 399–409. (e) Bourgeois, D.; Lallemand, J.-Y.; Pancrazi, A.; Prunet, J. *Synlett* **1999**, 1555–1558. (e) Tormakangas, O. P.; Toivola, R. J.; Karvinen, E. K.; Koskinen, A. M. P. *Tetrahedron* **2002**, *58*, 2175–2181. (f) Roy, O.; Pattenden, G.; Pryde, D. C.; Wilson, C. *Tetrahedron* **2003**, *59*, 5115–5151.

John P. Wolfe

4.8 Zaitsev Elimination

4.8.1 Description

The Zaitsev elimination[1] may also be known by the spelling of Saytzeff. The Zaitsev elimination reaction is the formation of an alkene by the dehydrohalogenation of an alkyl halide or the dehydration of an alcohol. Zaitsev's rule states that the alkene formed in greatest amount is the one that corresponds to removal of the hydrogen from the β-carbon having the fewest hydrogen substituents. Another way to state it that is in elimination reactions, the major reaction product is the alkene with the more highly substituted (more stable) double bond. This most-substituted alkene is also the most stable. The original report was about the regioselectivity of the β-elimination reactions of alkylhalides and was expanded to include the β-elimination from alcohols.

H_3C, CH_3, H_3C, Br, H, H
– HBr
H_3C, CH_3, H, CH_3
+
H_3C, H, CH_3CH_2, H
Zaitsev product
1
2
3

4.8.2 Historical Perspective

The Zaitsev elimination, first reported in 1875 by the Russian chemist Alexander Zaitsev (1841-1910), was published in *Justus Liebigs Annalen der Chemie* and the German spelling of his name, Saytzeff, was used.[1] Several historical accounts of Alexander Zaitsev have been reported.[2] The olefin forming E2 elimination has been reviewed.[3]

Zaitsev first studied chemistry at the University of Kazan under Aleksandr Butlerov and earned his diploma degree in 1862; he studied in Western Europe under Hermann Kolbe and Charles Wurtz.[2] Zaitsev earned his Ph.D. degree in 1866 from the University of Leipzig under Kolbe.[2]

He returned to the University of Kazan and published his paper on the formation of alkenes from alcohols in *Justus Liebigs Annalen der Chemie* in 1875.[1] It is in this paper that the generalization that became known as Zaitsev's rule was first published, and led to his lasting fame in sophomore organic chemistry. "The alkene formed in greatest amount is the one that corresponds to removal of the hydrogen from the β-carbon having the fewest hydrogen substituents."[1] The dehydrohalogenation of alkylhalides to form olefins was reported by Zaitsev in 1875; the reaction was latter expanded to include β-elimination of alkyl halides and alcohols to form alkenes.[2]

4.8.3 Mechanism

The β-hydride elimination reactions can be classified into three distinct mechanisms which are the E2, E1, and E1cb mechanisms.[3]

The E2 mechanism involves a bimolecular transition state in which removal of a proton that is β to the leaving group is concerted with the departure of the leaving group.[3]

B: H X trans antiperiplanar H X

4 5 6

Alternatively, if the dihedral angle is zero degrees (eclipsed X and H) the elimination can also take place.[3]

B: H X H X

7 8

The eclipsed conformation is 3.0–3.3 kcal/mol higher in E, so the elimination takes place primarily through the *trans* antiperiplanar arrangement.

The E1 mechanism is the unimolecular ionization of the reactant and elimination is completed by the removal of the b proton. When A and E are large groups the major product will have the *trans* orientation.[3]

A D E H B A D B E

9 10

The E1cb like the E1 mechanism involves two steps but the order is reversed. Proton abstraction precedes expulsion of the leaving group.[3]

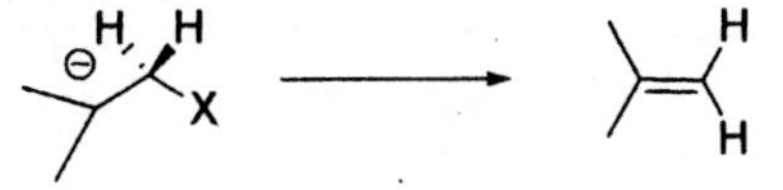

11 12

An example of the Zaitsev elimination is the formation of pentene from 2-bromopentane. A mixture of pentene isomers are produced.[4,5]

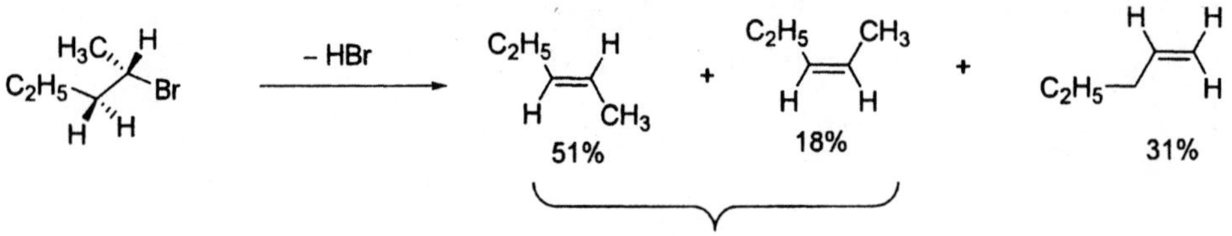

Zaitsev products

13 14 15 16

The *syn*-elimination transition states are very much destabilized relative to the *anti*-elimination transition. In the *anti*-elimination the conformation that leads to the *trans* product is more stable that the *cis* by approximately 1 kcal/mol and the *trans* product is formed to a greater extent.

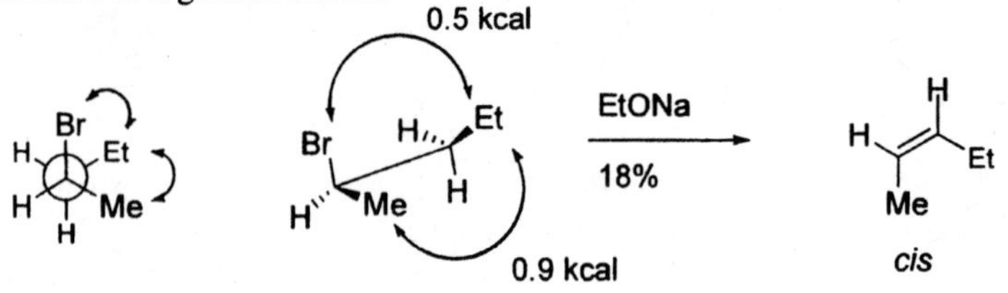

ΔE = 0.9 kcal/mol 17 18

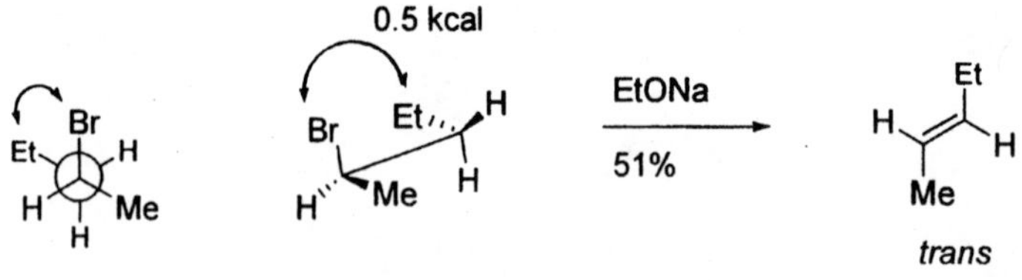

19 20

The least substituted product or *anti*-Zaitsev product is also formed. The mechanism is shown below.[3]

Br H H H Me Et
Br Et H H Me H
21
EtONa
51%
Et H H H H H
22

Br H H H H H Et H or Br H H H H Et H H

4.8.4 *Variations and Improvements*

The Zaitsev elimination reaction is commonly used in organic synthesis and since it has been known for over 130 years many examples are known in the literature.

The formation of olefins from acyclic alcohols or alkyl halides often results in a mixture of olefin stereoisomers. When cyclic halides or alcohols are substrates a more controlled mixture of products is observed.

The dehydrohalogenation of alkyl halides is a very general reaction and can be accomplished with chlorides, bromides, and iodides. Fluorides generally undergo a Hoffman elimination. The order of leaving group reactivity is I > Br > Cl > F. Hot alcoholic KOH is the most frequently used base[6], though stronger bases (OR^-, NH_2^-, *etc.*) can be used[7] or weaker bases (amines) may be used[8]. The bicyclic amidines (DBU or DBN) are good reagents for difficult cases.[9,10] Dehydrohalogenation using the non-ionic phosaphazine bases is even faster.[11] Phase transfer catalysts with ^-OH have been used.[12] As previously mentioned certain weak bases in dipolar aprotic solvents are effective reagents for dehydrohalogenation. Among the most often used for synthetic purposes are LiCl, LiBr, or Li_2CO_3 in DMF.[13] Dehydrohalogenation has also been effected by heating of the alkyl halide in HMPA with no other reagent present.[14]

Dehydration of alcohols can be accomplished in several ways.[6] Both sulphuric acid and phosphoric acid are common regents, but in many cases these reagents lead to rearrangement or ether formation. If the alcohol can be evaporated, vapour phase elimination over Al_2O_3 is an excellent method since side reactions are greatly reduced. Many other dehydrating agents have been used on occasion: P_2O_5, I_2, $ZnCl_2$, Ph_3BiBr_2/I_2, BF_3-etherate, DMSO, $KHSO_4$, anhydrous $CuSO_4$, and phthalic anhydride among others.

Secondary and tertiary alcohols can also be dehydrated without rearrangement simply by refluxing in HMPA. Nearly all of these reagents follow Zaitsev's rule.

4.8.5 Synthetic Utility

4.8.5.1 Dehydrohalogenation of acyclic halides

2-Pentene can be prepared by the action of alcoholic potash on 2-iodopentane[15] or 2-bromopentane[16] by dehydration of 2-pentanol[17] or 3-pentanol[18] by pyrolysis of the acetate of 3-pentanol.[19] The potassium ethoxide mediated dehydrobromination of 2-bromopentane (**23**) yields a mixture of products. The 1-pentene (**26**) is formed in 31% while the *trans* isomer (**24**) and *cis* isomer (**25**) of 2-pentene are formed in 51% and 18%, respectively.

H_3C, H, CH_3CH_2, Br, H, H — 1 M KOEt, 80 °C → CH_3CH_2, H, H, CH_3 + H, H, CH_3CH_2, CH_3 + H, H, C_3H_7, H

23 **24** **25** **26**

4.8.5.2 Synthesis of strained ring systems

In 1994, Ito *et al.* utilized a Zaitsev elimination in the synthesis of **28**.[20] The reaction utilized the base KOH to accomplish a double elimination of bromide from **27** to form diene **28**.

F_3C, O, Br, Br — KOH, EtOH, H_2O → OH, O

27 **28**

4.8.5.3 Synthesis of valuable intermediates

Fuchs recently reported an economical and environmentally friendly syntheses of 2-(phenylsulfonyl)-1,3-cyclohexadiene (**30**) and 2-(phenylsulfonyl)-1,3-cycloheptadienes.[21] These compounds are valuable intermediates to several groups of researchers. Dehydrobromination of intermediate **29** was accomplished by Hunig's base and 5 mol% Pd catalyst in acetonitrile to yield 2-(phenylsulfonyl)-1,3-cyclohexadiene (**30**).

SO_2Ph / Br — *i*-Pr_2NEt, MeCN; $Pd(PPh_3)_4$, 100% — SO_2Ph

29 → **30**

Practical synthesis of enantiopure spiro[4.4]nonane C-(2′-deoxy) ribonucleosides was accomplished using a $LiBr/Li_2CO_3$ mediated dehydrobromination of intermediate **31**.[22]

TBS, H, O, Br — LiBr, Li_2CO_3; DMF, 120 °C; 80% — TBSO, H, O

31 → **32**

4.8.5.4 Natural product synthesis

Tadano *et al.* utilized a Zaitsev elimination in the synthesis of (±)-mycoepoxydiene.[23] The vicinal dibromo adduct **33**, as a diastereomeric mixture to potassium *t*-butoxide, provided a 9-oxabicyclo [4.2.1]nona-2,4-diene derivative **34** in a 58% yield.

Br, Br, O, OTBDPS, Me — *t*-BuOK, *t*-BuOH; 75 °C, 58% — O, OTBDPS, Me

33 → **34**

Corey *et al.* utilized a Zaitsev elimination in the synthesis of (±)-mycoepoxydiene.[24] The vicinal dibromo adduct **35**, as a diastereomeric mixture to potassium *t*-butoxide, provided a 9-oxabicyclo [4.2.1]nona-2,4-diene derivative **36** in a 35% yield.

Br, CO_2CH_3, CH_3 — DBU; benzene reflux; 2 h; 35% — CO_2CH_3, CH_3

92–94% *ee*

35 → **36**

In 2005 Nicolaou *et al.* in the synthesis of thiostrepton utilized reaction of **37** and DBU at ambient temperature to form allylic epoxide **38** in 96% yield.[25]

DBU
rt, 96%

37 **38**

4.8.6 *Experimental*

t-BuOK, DMSO
88%

39 **40**

3,3-Dimethylcyclobut-1-ene (40)[26]
1-Bromo-3,3-dimethylcyclobutane (**39**) (49.3 g, 302 mmol) was added in one portion to a stirred solution of *t*-BuOK (52 g, 463 mmol) in anhydrous DMSO (460 mL), kept at 20 °C (cold water bath) under nitrogen. The mixture was stirred at ambient temperature for 20 min, the volatile material was bulb to-bulb-distilled (T bath 50 °C/20 Torr) into a –78 °C (dry ice/acetone bath) cold, pointed-bottom Schlenk flask. After the distillation was complete, the crude product was transferred by cannula from the cooled collecting flask into a separating funnel, washed with water (2 × 20 mL), brine (10 mL), and placed in a storage bottle containing a small amount of BHT (ca. 10 mg) and molecular sieves (4 Å) (ca. 1 g). The colorless liquid weighed 21.9 g (88%). It was pure 3,3-dimethylcyclobut-1-ene (**40**) according to its ^{1}H NMR spectrum.

4.8.7 *References*

1. Saytzeff, A. *Ann. Chem.* **1875**, *179*, 296–301.
2. Lewis, D. E. *Bull. Hist. Chem.* **1995**, *17*, 21–30.
3. (a) [R] Bartsch, R. A.; Zavada, J. *Chem. Rev.* **1980**, *80*, 453–494. (b) Boger D. L. *Modern Organic Synthesis Lecture Notes*; TSRI: La Jolla, **1999**, 26–27.
4. Dhar, M. L.; Hughes, E. D.; Ingold, C. K. *J. Chem. Soc.* **1948**, 2058–2065.
5. Brown, H. C.; Moritani, I. *J. Am. Chem. Soc.* **1956**, *78*, 2203–2210.
6. [R] Larock, R.C. *Comprehensive Organic Transformations;* VCH: New York, **1999**, 256–258.
7. Truscheit, E.; Eiter, K. *Liebigs Ann. Chem.* **1962**, *658*, 65–72.
8. Oediger, H.; Moehler F. *Angew. Chem. Int. Ed. Engl.* **1967**, *6*, 76.
9. Oediger, H.; Moller, F.; Eiter, F. *Synthesis* **1972**, 591–598.
10. Schwesinger, R.; Schlemper, H. *Angew. Chem. Int. Ed. Engl.* **1987**, *26*, 1167–1172.

11. Kimura, Y.; Regen, S. L. *J. Org. Chem.* **1983**, 48, 195–198.
12. [R] Fieser, L. F.; Fieser, M. *Reagents for Organic Syntheses*, vol. 1; Wiley; Hoboken, NJ, **1967**, 606.
13. Hanna, R. *Tetrahedron Lett.* **1968**, *9*, 2105–2106.
14. Matsubara, S.; Matsuda, H.; Hamatani, T.; Schlosser, M. *Tetrahedron* **1988**, *44*, 2855–2863.
15. Fieser, L. F.; Fieser, M. *Advanced Organic Chemistry*, Reinhold, Hoboken, NJ, **1961**, 140.
16. Wagner, G.; Saytzev, A. *Ann.Chem.* **1875**, *175*, 373–378.
17. (a) Lucas, H. J.; Moyse, H. W. *J. Am. Chem. Soc.* **1925**, *47*, 1462–1469; (b) Sherrill, M. L.; Otto, B.; Pickett, L. W. *J. Am. Chem. Soc.* **1929**, *51*, 3023–3033.
18. Tissier, A. *Bull. Soc. Chim.* **1893**, *9*, 100.; Sherrill, M. L. Baldwin, C.; Haas, D. *J. Am. Chem. Soc.* **1929**, *51*, 3034–3037.
19. Leendertse, J. J.; Tulleners, A. J.; Waterman, H. I. *Rec. Trav. Chim.* **1934**, *53*, 715–724.
20. Ito, K.; Kawaji, H.; Nitta, M. *Tetrahedron Lett.* **1994**, *35*, 2561–2564.
21. Meyers, D. J.; Fuchs, P. L. *J. Org. Chem.* **2002**, *1*, 200–204.
22. Hartung, R.; Paquette, L. A. *J. Org. Chem.* **2005**, *5*, 1597–1604.
23. Takao, K.; Watanabe, G.; Yasui, T.; Tadano, K. *Org. Lett.* **2002**, 17, 2941–2944.
24. Corey, E. J.; Guzman-Perez, A.; Lazerwith, S. E. *J. Am. Chem. Soc;* **1997**, *119*; 11769–11776.
25. Nicolaou, K. C.; Safina, B. S.; Zak, M.; Lee, S. H.; Nevalainen, M.; Bella, M.; Estrada, A. A.; Funke, C.; Zecri, F. J.; Bulat, S. *J. Am. Chem. Soc.* **2005** *31*, 11159–11175.
26. Larsen, N.W.; Pedersen, T. *J. Mol. Spectrosc.* **1994**, *166*, 372–382.

Timothy J. Hagen

Chapter 5 Amine Synthesis 423

5.1 Fukuyama Amine Synthesis

5.1.1 Desciption

The Fukuyama amine synthesis represents a general protecting group strategy for the formation of a wide variety of secondary amines (**5**) including differentially protected primary amines.[1,2] It utilizes a nitro sulfonamide (**3a–c**), often abbreviated as Ns or nosyl, to protect a primary amine **1**. This sulfonamide is easily formed by reaction between any primary amine **1** and sulfonyl chloride **2a–c** to give **3a–c**. Alternatively, the parent (unsubstituted) sulfonamide can be alkylated with any alkyl halide to give **3a–c**. This Ns protected amine (**3a–c**) cleanly undergoes alkylation with any alcohol activated by DEAD (diethylazodicarboxylate) and PPh_3 to give **4a–c**.

$R{-}NH_2$ (**1**) + SO_2Cl (X, Y) → pyr. *or* lutidine → R–NH**Ns** (**3a–c**)
2a: X=NO_2, Y=H
2b: X=H, Y=NO_2
2c: X,Y=NO_2

3a–c → R'–OH, DEAD/PPh_3 *or* R'-X, K_2CO_3, DMF → R–N(Ns)–R' (**4a–c**)

4a–c → $HSCH_2CO_2H$ or PhSH with base (**2a–c**) *or n*-$PrNH_2$ (**2c**) → R–NH–R' (**5**)

The major advantage of this method is apparent during nosylate removal. Thiols, such as thiophenol or $HSCH_2CO_2H$, remove the nosyl group in the presence of a base.[3] Using $HSCH_2CO_2H$ and Et_3N to deprotect the nosyl group is advantageous in that the nitrophenylthioacetic acid that is generated may be easily removed with an aqueous base wash during workup. The nosyl protecting group is ideal for secondary amine synthesis in that it is orthogonal to most other common protecting groups. It is also quite stable to acidic (10 equiv HCl in MeOH at 60 °C for 4 hours) and basic (10 equiv NaOH in MeOH at 60 °C for 4 hours) conditions. In general, 2,4-dinitro-benzenesulfonamide (**4c**) is much more easily installed and removed, as even excess *n*-$PrNH_2$ may serve to deprotect the secondary amine. Mono-nitrobenzenesulfonamides (**4a–b**) typically require PhSH and K_2CO_3 in DMF at room temperature for removal. From a commercial perspective, the mono-nitrobenzenesolfonamides are less expensive and more readily available on scale.

5.1.2 Historical Perspective

The ubiquitous role of amines in both nature and in a vast variety of biologically important synthetic molecules gives this functionality a place of special prominence and interest in organic chemistry. Therefore, much effort and attention has been exerted toward developing methods for the selective synthesis of primary, secondary, and tertiary amines.[4] In particular, the selective synthesis of secondary amines and orthogonally protected primary amines is quite important as these are often featured as valuable synthetic intermediates.[5]

Outlined below is a brief summary of these efforts along with their limitations. The traditional method of amine synthesis is simple alkylation of a primary amine with either an alkyl halide or activated alcohol (LG = leaving group). Although this method appears simple, the formation of over-alkylated products, such as tertiary and quaternary amines limits this method's utility to the synthesis of either "bulky" secondary amine products, or electronically deactivated ones. Another well established methodology is reductive amination of an aldehyde. While this procedure is quite efficient and general, it can in certain instances suffer from the formation of unwanted tertiary amines if the secondary amine product is sufficiently nucleophilic to undergo reaction with another equivalent of aldehyde. The metal hydride reductant employed may also not be compatible with other functionalities present in the molecule, such as electron deficient olefins and esters. Amide reduction is another option for forming secondary amines, but it too suffers from very harsh reducing conditions ($LiAlH_4$, *i*-Bu_2AlH, or borane). A modern, palladium-catalyzed means of forming aryl amines has been intensively investigated by the groups of Buchwald and Hartwig.[6] Palladium complexes catalyze the reaction between aryl sulfonates and halides with a wide variety of amines to give substituted anilines under exceedingly mild conditions. This procedure is limited to sp^2 carbon atoms in the electrophile and is therefore of limited use in forming purely aliphatic secondary amines.

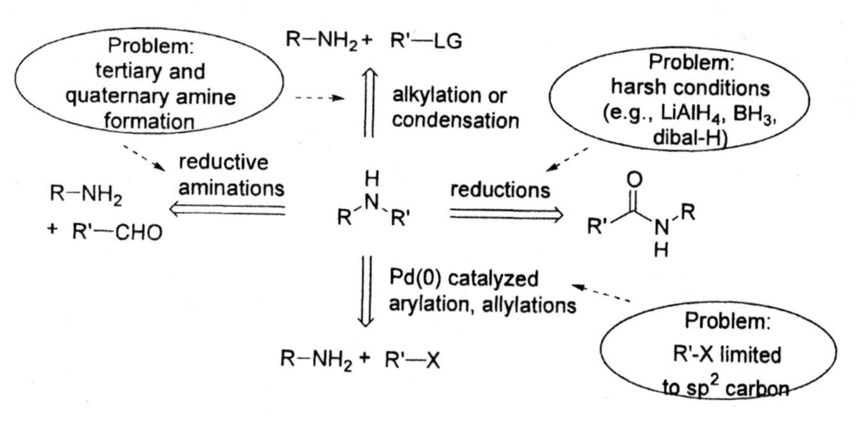

Given the limitations described above, a selective and reliable secondary amine synthesis with broad applicability and functional group tolerance requires a protecting group strategy. This approach would first introduce a protecting group on a primary amine or treat a protected ammonia derivative with an electrophile. This protected primary amine would then react with another electrophile, such as an alkyl halide or activated alcohol, giving a protected secondary amine which can be easily deprotected. Central to this method is the appropriate choice of protecting group. It should prevent over-alkylation but still allow the amine to be nucleophilic enough to react with one equivalent of the electrophile. Ideally, it should be orthogonal to common amine protecting groups like carbamates, benzyl, and allyl as this would provide access to differentially protected primary amines and allow for other sensitive protecting groups in the molecule to survive.

This strategy was first realized by S. M. Weinreb and co-workers when they showed that a variety tosyl (tosyl = *p*-toluenesulfonyl) protected primary amines **6** could react with alkyl alcohols under Mitsunobu conditions to give the tosylated secondary amine **7**.[7] In addition, they showed that BocNH(Ts) amines (**6**) may also react to give the differentially protected primary amine **7**. They also demonstrated ring closures using the Mitsunobu conditions on an *N*-tosyl amino alcohols. The drawback to this reported methodology is that the tosyl group is removed under strong reducing conditions with sodium naphthalenide to give **5**.[8]

S. M. Weinreb et. al.

An improvement with regard to protecting group was later made by T. Tsunoda and co-workers.[9] They employed a trifluoroacetamide protecting group (**8**) and showed that it too can be alkylated with an alcohol **9** under modified Mitsunobu conditions with TMAD (*N,N,N',N'*-tetramethylazodicarboxamide) and *n*-Bu_3P. The trifluoroacetamide group of **10** is easily removed with 1 M NaOH (aq.) in methanol at ambient temperatures to give amine **11**. The more basic Mitsunobu conditions are required as the pK_a of the trifluoroacetamide is ~ 13.6 versus the more acidic tosylated analog (TsNHMe) with a pK_a of ~ 11.7.

T. Tsunoda *et al.*

F_3C N Ph H **8** — **9** OH, TMAD/*n*-Bu_3P → F_3C N Ph **10** — 1M NaOH, MeOH → HN Ph **11**

5.1.3 Mechanism

Based on the aforementioned examples, it is the facile removal of the nosyl protecting group developed by Fukuyama that makes this methodology useful. Treatment of a nosyl protected amine with a variety of thiols, or other nucleophiles, gives the Meisenheimer complex **12**. Upon loss of SO_2, this generates the desired secondary amine **5** along with a thioether **13a–c** which must be removed.

R' N R SO_2 HO_2CCH_2S X Y **12** — $-SO_2$ → R N(H) R' **5** + SCH_2CO_2H X Y **13a–c**

5.1.4 Synthetic Utility

5.1.4.1 Preparation of secondary and differentially protected primary amines

This protocol appears to be fairly general for the reaction of any nosylated amine and alcohol.[2] Fukuyama showed that this methodology could be used to generate differentially protected primary amines by reaction with various alkyl and aryl alcohols under the standard Mitsunobu conditions.[10] This generates PMB, Ph, Bn, Boc, Alloc, and Cbz protected amines **5** after removal of the Ns group with a thiol. As with most Mitsunobu reactions, this alcohol (**14**) displacement by the nosylate goes with complete inversion of configuration regardless of the substituents on the nosylate. Chiral alcohol **15** is cleanly inverted to give protected amines **16** without any loss of enantiopurity.

R'–OH **14** — R–NHNs, DEAD, PPh_3 → RSH → R' N(H) R **5**

R' = alkyl, aryl
R = PMB, Ph, Bn, Boc
Alloc, or Cbz

EtO_2C–OH (15) → R–NHNs, DEAD, PPh_3 → RSH → EtO_2C–NHR (16)

R = PMB, Boc, Alloc, or Cbz

Furthermore, Fukuyama has shown that the order of deprotection of these differentially protected primary amines is irrelevant. The nosylate **4** may be removed first with a thiol followed by acidic removal or hydrogenolysis of the second protecting group giving **17**. The reverse order works equally well with acidic or hydrogenative removal of the first protecting group of **4** followed by denosylation with thiol to give **17**.

R'N(R)Ns (4) → RSH → R'NHR (5) → Acid Or H_2/Pd(0) → R'NH_2 (17)

R = Boc, Alloc, or Cbz

R'N(R)Ns (4) → Acid Or H_2/Pd(0) → R'NHNs (18) → RSH → R'NH_2 (17)

5.1.4.2 Preparation of nitrogen heterocycles

The Fukuyama amine synthesis can also be performed on nosylated amino alcohols in an intramolecular sense and thus represents a powerful means of macrocyclization.[11] His group has shown that halo alcohols **19** can react with sulfonamide **20** under Mitsunobu conditions to give nosylated amino halide **21**. This is then cleanly alkylated with Cs_2CO_3 giving Ns-protected nitrogen macrocycles **22**. Interestingly, the order of reaction with halo alcohols **19** can be reversed by first reaction with NsNHBoc (**23**) and K_2CO_3. Removal of the Boc moiety then gives nosylated amino alcohols **24**. The ring closure is then accomplished with DEAD/PPh_3 giving **22**.

19 (for n = 1–3) → 20 Ns–NH_2, DEAD, PPh_3 (67–74%) → 21 → Cs_2CO_3, n-Bu_4NI (62–66%) → 22

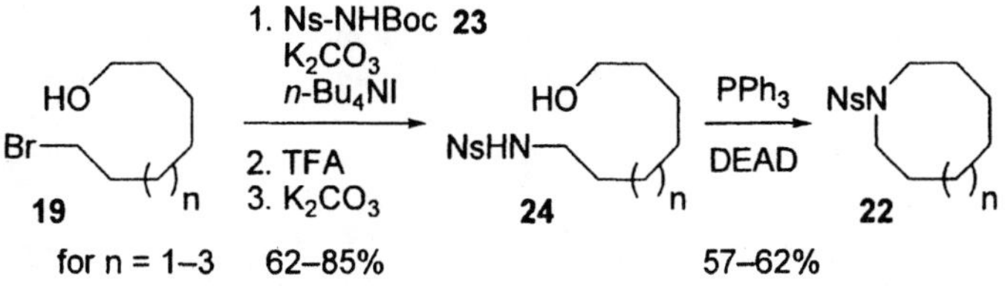

5.1.4.3 Application to total synthesis

The Fukuyama amine synthesis has seen wide application in the context of natural product synthesis. Complex polyamine natural products that highlight the orthogonal nature of the nosyl protecting group, such as lipogrammistin-A[12] and various spider toxins,[13] have been efficiently synthesized. This protocol has also been used in the context of medicinal chemistry,[14] glucosylamines,[15] and blasticidin amino acids.[16]

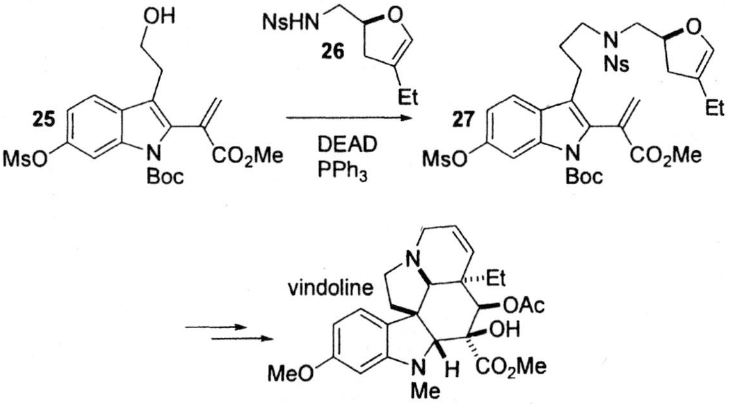

The most challenging application of this nosyl protecting group strategy has been with densely functionalized polycyclic alkaloid natural products. These include FR900482,[17] flustramines,[18] lentiginosine,[19] queuine,[20] aspidosperma alkaloids,[21] and aspidophytine.[22] Perhaps the most impressive achievement in total synthesis using the nosylate protecting group is Fukuyama's synthesis of vinblastine, a dimeric alkaloid with potent anti-cancer properties.[23] Vinblastine is composed of two major alkaloid fragments, vindoline and a catharanthine derivative. Fukuyama uses his nosylate methodology to couple indole alcohol **25** with nosyl amine **26** under Mitsunobu conditions to give fragment **27** which is then taken on to vindoline. Despite the complexity of vindoline, the nosylate is easily removed with pyrrolidine and this triggers a rearrangement and intramolecular Diels–Alder reaction to give the polycyclic tertiary amine of vindoline. The other half of vinblastine is made by alkylating $NsNH_2$ **20** with indole alcohol **28** to give fragment **29**. This nosyl amino epoxide undergoes macrocyclization with epoxide ring opening by simple base treatment to give 11-member ring **30**. Fragment **30** is later coupled with vindoline to give vinblastine using an efficient electrophilic aromatic substitution.

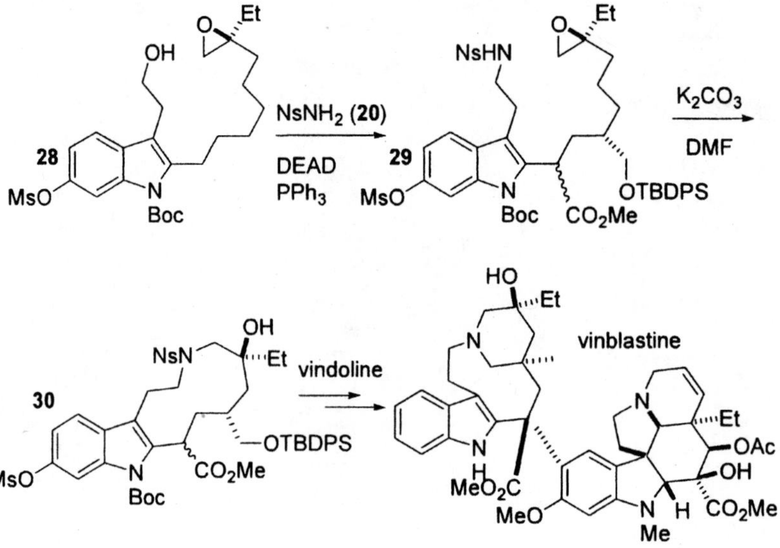

5.1.4.4 Application to solid phase synthesis

Due to the stability of the nosyl protecting group and its orthogonal nature, it has been rapidly adopted to solid phase synthesis where the mild deprotection conditions in the final stage of a linear synthesis provide Ns with a distinct advantage over other common amine protecting groups. Solid phase has proved to be quite useful for the synthesis of Ns-protected polyamine toxins where final purification can be a challenge due to the highly polar nature of polyamines.[24] The Fukuyama amine synthesis has also been applied to the solid phase synthesis of *N,N*-disubstituted α-amino acids where the mild Ns removal conditions provide access to any type of desired substitution on the nitrogen of the amino acid.[25] Solid phase synthesis of diverse chemical libraries which incorporate the nosyl protecting group strategy have also been reported.[26]

5.1.4.5 Limitations and side products

Despite the broad scope of the Fukuyama amine synthesis, there are possible complications that may arise. Common to any Mitsunobu reaction, there is the issue of purification that is often required in order to remove the copious amounts of $Ph_3P{=}O$ that are generated. This can be difficult, but modifications to the phosphine base employed can mitigate this problem (*vida infra*). The Mitsunobu alkylation also oftentimes generates the product **34** resulting from addition of the hydrazine that is generated from DEAD to the activated alcohol **32**.[7a] This can be a problem if the alcohol is of high value, or the limiting reagent. Once again, minor modifications to both the base phosphine base and the diazo compound can diminish this side product.

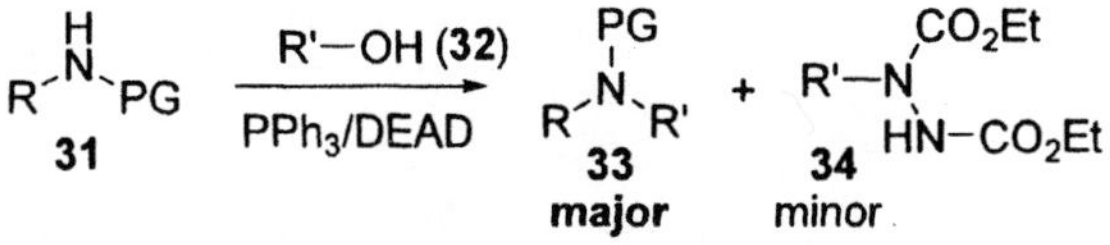

As depicted below, alkylation of nosyl amines, such as **36b**, can also suffer from competing intramolecular reactions of the Ph_3P/DEAD activated alcohol.[27] In this instance, rather than giving the expected amino piperidine product **37**, alcohol **35** undergoes an intramolecular reaction to form a highly strained aziridine intermediate that then is opened by nosyl amine **36b** to give the observed pyrrolidine **38**.

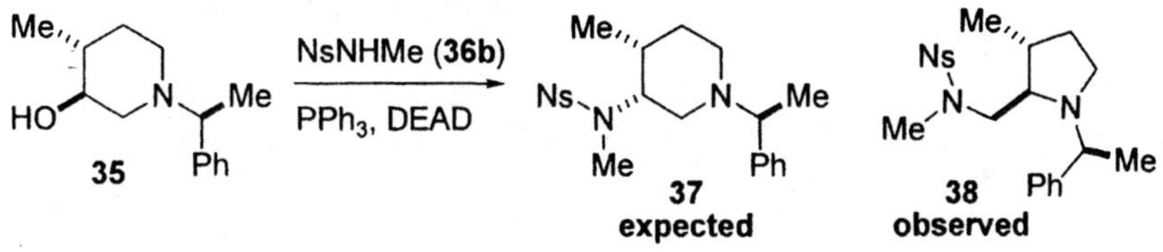

Problems may also occur during the installation of the nosyl moiety during the reaction of a primary amine **1** and nosyl chloride **2c**.[2b] A cyclic Meisenheimer complex **39** is formed followed by loss of SO_2 to give nitro aniline **40** as a minor product during nosylation. None of this side product is observed with the less reactive mono-nitro sulfonates (**2a–b**).

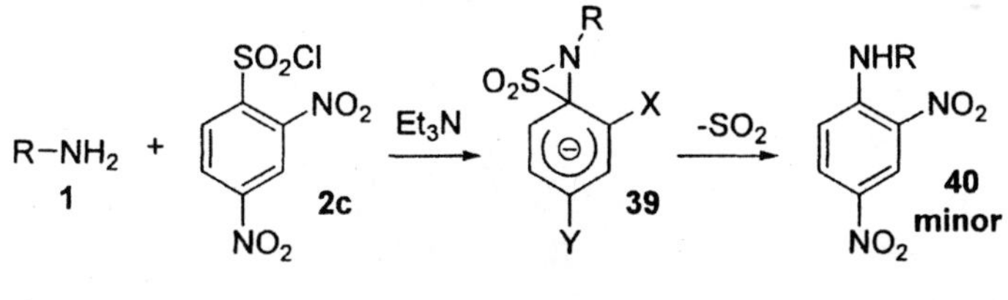

Complications may also arise during the thiol deprotection of the nosyl group.[28] It was observed that during the *p*-Ns deprotection of a Taxol intermediate, 9% of an impurity was generated which results from thiol substitution at the nitro center of nosylate **4b**. A study of this phenomenon revealed that this is a general problem with *p*-nosylate **4b** deprotection using *i*-Pr_2NH and PhSH in DMF with varying amounts of **41**

being formed depending upon the secondary amine substrate. This problem was not observed with *o*-nosylated amines (**4a**).

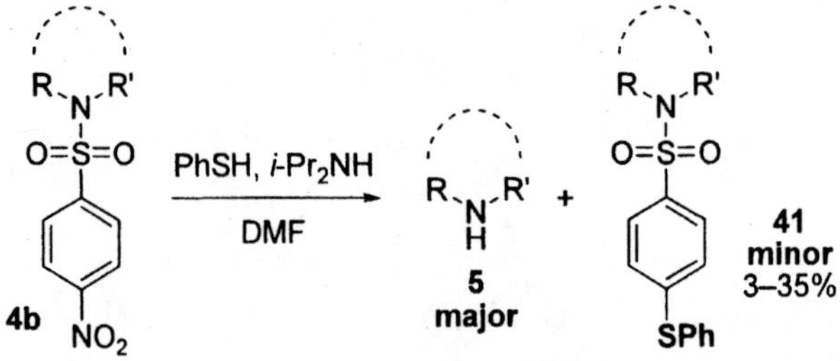

5.1.5 *Variations*

5.1.5.1 Phosphine oxide removal

Other researchers have made improvements to and expanded the scope of the original Fukuyama amine synthesis as described above. For example, it has been reported that a combination of $PyPh_2P$ and DTBAD (di-*tert*-butylazodicarboxylate) activates a variety of primary and some secondary alcohols **42** to react with $NsNH_2$ (**20**) giving nosyl amines **43** with consistently excellent yields.[29] Importantly, the phosphine oxide and DTBAD by-products that are generated are easily removed from **44** as the DTBAD decomposes to isobutene, nitrogen, and CO_2 upon acid quench. The $PyPh_2P$=O is also readily removed by an aqueous acid wash.

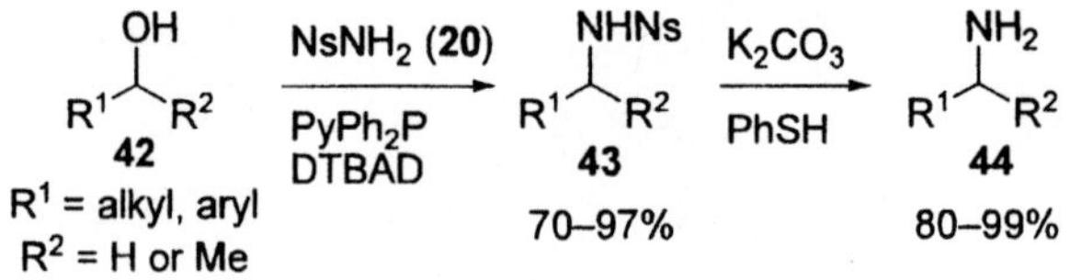

5.1.5.2 Regioselectivity

The regioselectivity of the Fukuyama amine synthesis has also been examined. As shown, the NsCl (1 equiv) preferentially reacts with the primary amine of **45** despite the presence of a rather nucleophilic secondary nitrogen of the pyrrolidine to give **46** as the sole product.[30]

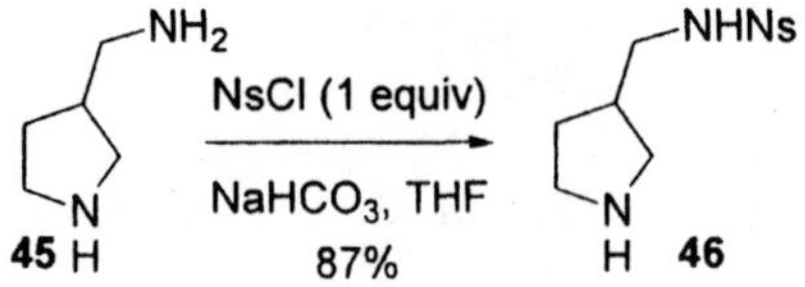

5.1.5.3 Chemoselectivity

The Mitsunobu alkylation conditions of **48** also exhibit high chemoselectivity when subjected to reaction with diol **47**.[31] In this example, judicious choice of phosphine and diazo compound dictate which alcohol is activated. Using *n*-Bu_3P in combination with TMAD (*N,N,N',N'*-tetramethylazodicarboxamide) gives primarily reaction with the primary alcohol yielding **49**. Whereas, Me_3P and ADDP (1,1'-(azodicarbonyl)-dipiperidide) allow for reaction at the secondary alcohol with another equivalent of **48** giving a fully protected di-amine **50**.

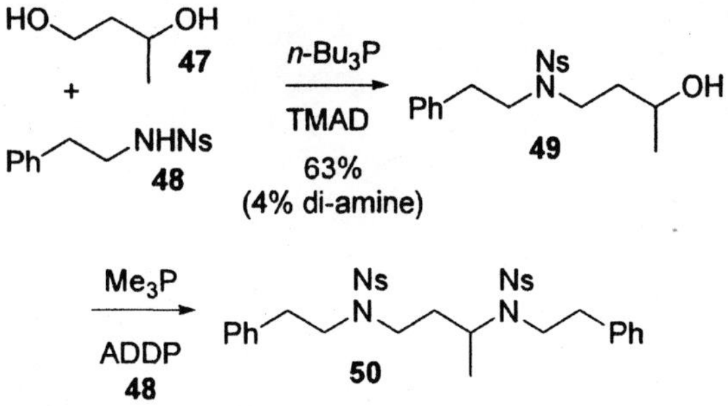

Even compounds containing both a free amine and reactive phenol can be selectively nosylated under the right conditions.[32] Tyrosine **51** when treated with NsCl using lyophilized Na_2CO_3 in a 8:1 mixture of THF/DMF gives predominantly *N*-Ns product **52** with only minor amounts of *N,O*-di-Ns product **53**. This was found to be very dependent upon solvent and base choice as other solvent–base combinations gave mixtures of products.

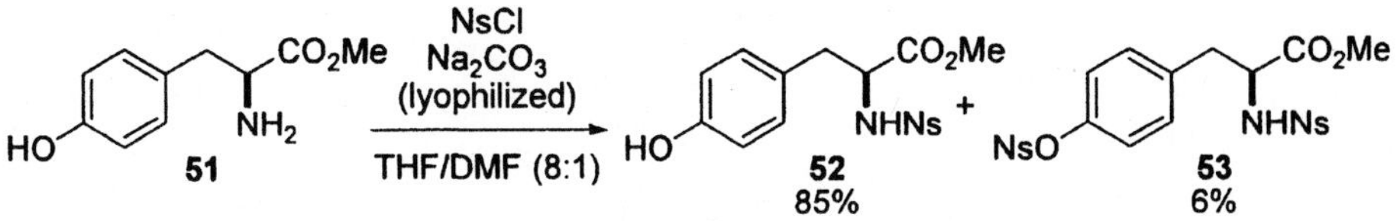

5.1.5.4 Preparation of protected hydroxylamines

The Fukuyama amine synthesis can also be applied to protected hydroxylamines and hydroxamates.[33] Various protected hydroxylamines **54** undergo clean reaction with NsCl giving **55**. This *N,O*-protected hydroxylamine **55** may then react with a variety of alkyl halides to give fully protected hydroxylamines **56**.

54 $RO{-}NH_2$ $\xrightarrow[NsCl]{Et_3N}$ $RO{-}NHNs$ (55) $\xrightarrow[Cs_2CO_3]{R'{-}Br}$ 56

R = Bn, allyl, t-Bu, THP, 2,4-$(MeO)_2Bn$ — 51–80% — >90%

5.1.5.5 *Nosyl deprotection and functionalization*

An interesting deprotective-acylation strategy for the nosyl moiety has also been reported.[34] This modification made use of the Meisenheimer complex **12** that is generated when the nosyl group is treated with a nucleophile. In this literature example, nucleophilic acylating agents are utilized rather then the standard thiol nucleophile, which results in a wide variety of acylated amines being generated. For example, treatment of nosyl amine **57** with a hydroxamic acid results in a non-symmetric urea **58**. Treatment of **57** with a dithiocarbamic acid gives thioureas (**59**). Subjecting **57** to thioacids gives amides **60**, whereas treatment with a dithioacid gives the expected thioamides **61**.

5.1.6 *Experimental*

62 + 2a $\xrightarrow[2.\ Ph(CH_2)_3Br,\ K_2CO_3,\ DMF]{1.\ NEt_3,\ CH_2Cl_2}$

63 $\xrightarrow[CH_3CN]{PhSH,\ KOH}$ 64

N-(4-Methoxy-benzyl)-2-nitro-N-(3-phenyl-propyl)-benzenesulfonamide (63)[2c]

A solution containing 6.81 g of 4-methoxybenzylamine (**62**, 49.6 mmol), 100 mL of CH_2Cl_2, and 6.93 mL of triethylamine (49.6 mmol) was cooled to 0 °C in an ice-water bath. To this was then added 10.0 g of 2-nitrobenzenesulfonyl chloride (**2a**, 45.1 mmol) over a period of 5 min. The mixture was then allowed to warm to room temperature, stirred for 15 min, and quenched with 100 mL 1 N HCl. The aqueous layer was then extracted with CH_2Cl_2 (2 × 100 mL), and the combined organic layers were washed with 50 mL brine, dried with $MgSO_4$, filtered, and concentrated under reduced pressure to give 14.2 g (98% yield) of the crude sulfonamide. This crude sulfonamide was then recrystallized from 500 mL of 1:1 EtOAc:hexanes to give 13.15 g (91% yield) of pure N-(4-methoxybenzyl)-2-nitrobenzenesulfonamide; mp 123 °C; IR (neat) 3312, 2941, 1543, 1511, 1363, 1337, 1243, 1160 cm^{-1}; ^1H NMR ($CDCl_3$, 400 MHz) δ 3.76 (s, 3H), 4.25 (d, 2H, J = 6.2 Hz), 5.63 (br, t, 1H, J = 6.2 Hz), 6.75 (d, 2H, J = 8.5 Hz), 7.13 (d, 2H, J = 8.5 Hz), 7.63–8.03 (m, 4H); ^{13}C NMR ($CDCl_3$, 100 MHz) δ 47.4, 55.3, 114.0, 125.2, 127.7, 129.2, 131.1, 132.7, 133.3, 134.0, 159.3.

A flask was then charged with 10.0 g of N-(4-methoxybenzyl)-2-nitrobenzenesulfonamide (31.0 mmol), 12.9 g of powdered K_2CO_3 (93.1 mmol), and 40 mL of anhydrous DMF. To this stirred mixture was added 5.19 mL (34.1 mmol) of 3-phenylpropyl bromide over a period of 5 min. This mixture was then heated to 60 °C and allowed to stir for 70 min. After cooling to room temperature, the reaction was diluted with 250 mL of water, and extracted with ether (3 × 250 mL). The combined organics were washed with 100 mL brine, dried with $MgSO_4$, filtered, and concentrated under reduced pressure to give a pale, yellow oil which may then be purified by silica gel column chromatography (eluted with EtOAc/hexanes, gradient of EtOAc from 10–40%). This provided 13.5 g (99% yield) of title compound **63** as a thick, pale yellow oil.

(4-Methoxy-benzyl)-(3-phenyl-propyl)-amine (64)[2c]

A solution containing 7.82 mL of thiophenol (76.5 mmol) and 20 mL of CH_3CN was cooled in an ice-water bath to 0 °C. To this solution was added 10.9 M NaOH (7.02 mL, 76.5 mmol) over a 10 min period. After an additional 5 min, the cooling bath is removed and 13.5 g of N-(4-methoxy-benzyl)-2-nitro-N-(3-phenyl-propyl)-benzenesulfonamide (**63**, 30.6 mmol) in 20 mL CH_3CN was added over 20 min. The reaction was then heated to 50 °C for 40 min. After cooling to room temperature, the mixture was diluted with 80 mL water, and extracted with CH_2Cl_2 (3 × 80 mL). The combined organic extracts were washed with 80 mL brine, dried over $MgSO_4$, filtered, and concentrated under reduced pressure. The crude material was then purified by silica gel column chromatography (eluted with CH_2Cl_2/MeOH, gradient of MeOH from 0–2% then 95:2.5:2.5 = CH_2Cl_2:MeOH:i-$PrNH_2$) to give 7.81 g of **64** along with its hydrochloride salt. This oil was then dissolved in 120 mL of CH_2Cl_2 and washed with 1 M NaOH (2 × 80 mL), 40 mL brine, dried over $MgSO_4$, filtered, and concentrated under reduced pressure. Bulb-to-bulb distillation (0.25 torr with 150 °C oven) provided 7.08 g (91% yield) of title amine **64**.

5.1.7 References

1. [R] Kan, T.; Fukuyama, T. *Chem. Commun.* **2004**, 353.
2. (a) Fukuyama, T.; Jow, C. -K.; Cheung, M. *Tetrahedron Lett.* **1995**, *36*, 6373. (b) Fukuyama, T.; Cheung, M.; Jow, C. -K.; Hidai, Y.; Kan, T. *Tetrahedron Lett.* **1997**, *38*, 5831. (c) Kurosawa, W.; Kan, T.; Fukuyama, T. *Org. Syn.* **2003**, *79*, 186. (d) Kobayashi, Y.; Fukuyama, T. *J. Het. Chem.* **1998**, *35*, 1043. For a Ns-deprotection study, see: (e) Nihei, K.; Kato, M. J.; Yamane, T.; Palma, M. S.; Konno, K. *Synlett.* **2001**, 1167. for alternative Ns- alkylation conditions, see: (f) Albanese, D.; Landini, D.; Lupi, V.; Penso, M. *Eur. J. Org. Chem.* **2000**, 1443.
3. For an alternative, fluorous deprotection strategy, see: Christensen, C.; Clausen, R. P.; Begtrup, M.; Kristensen, J. L. *Tetrahedron Lett.* **2004**, *45*, 7991.
4. [R] (a) Brown, B. R. *The Organic Chemistry of Aliphatic Nitrogen Compounds*; Oxford University: New York, 1994. [R] (b) Mitsunobu, O. *Comprehensive Organic Synthesis*; Trost, B. M., Fleming, I., Eds.; Pergamon: Oxford, 1991; Vol. 6, p 65.
5. [R] Salvatore, R. N.; Yoon, C. H.; Jung, K. W. *Tetrahedron* **2001**, *57*, 7785.
6. [R] (a) Jiang, L., Buchwald, S. L. *Metal-Catalyzed Cross-Coupling Reactions (2nd Edn.)*; De Meijere, A., Diederich, F., Eds.; Wiley-VCH: Weinheim, 2004; pp 699–760. [R] (b) Yang, B. H.; Buchwald, S. L. *J. Org. Met. Chem.* **1999**, *576*, 125. [R] (c) Wolfe, J. P.; Wagaw, S.; Marcoux, J. F.; Buchwald, S. L. *Acc. Chem. Res.* **1998**, *31*, 805. [R] (d) Hartwig, J. F. *Modern Arene Chemistry*; Astruc, D., Ed.; Wiley-VCH: Weinheim, 2002; pp 107–168. [R] (e) Hartwig, J. F. *Handbook of Organopalladium Chemistry for Organic Synthesis*; Negishi, E., Ed.; John Wiley & Sons, Inc.: Hoboken, 2002; pp 1051–1096. [R] (f) Hartwig, J. F. *Angew. Chem. Int. Ed., Engl.* **1998**, *37*, 2046; [R] (g) Hartwig, J. F. *Synlett.* **1997**, 329.
7. (a) Henry, J. R.; Marcin, L. R.; McIntosh, M. C.; Scola, P. M.; Harris, D.; Weinreb, S. M. *Tetrahedron Lett.* **1989**, *30*, 5709. For the use of other sulfonamides, see: (b) Grehn, L.; Ragnarsson, U. *J. Org. Chem.* **2002**, *67*, 6557. (c) Vedejs, E.; Kongkittingam, C. *J. Org. Chem.* **2001**, *66*, 7355.
8. For tosyl removal with *i*-PrMgCl and catalytic Ni(II), see: Milburn, R. R.; Snieckus, V. *Angew. Chem. Int. Ed., Engl.* **2004**, *43*, 892.
9. Tsunoda, T.; Otsuka, J.; Yamaiya, Y.; Ito, S. *Chem. Lett.* **1994**, 539.
10. Fukuyama, T.; Cheung, M.; Kan, T. *Synlett.* **1999**, 1301.
11. (a) Kan, T.; Kobayashi, H.; Fukuyama, T. *Synlett.* **2002**, 697. (b) Kan, T.; Fujiwara, A.; Kobayashi, H.; Fukuyama, T. *Tetrahedron* **2002**, *58*, 6267.
12. See ref. 11b, and: (a) Fujiwara, A.; Kan, T.; Fukuyama, T. *Synlett.* **2000**, 1667. (b) Kan, T.; Fujiwara, A.; Kobayashi, H.; Fukuyama, T. *Tetrahedron* **2002**, *58*, 6267. For other spermidines, see: (c) Amssoms, K.; Augustyns, K.; Yamani, A.; Zhang, M.; Haemers, A. *Synth. Commun.* **2002**, *32*, 319. (d) Scheiper, B.; Glorius, F.; Leitner, A.; Fürstner, A. *Proc. Natl. Acad. Sci.U. S. A.* **2004**, *101*, 11960. For additional polyamines, see: (e) Heinrich, M. R.; Kaskman, Y.; Spiteller, P.; Steglich, W. *Tetrahedron* **2001**, *57*, 9973. (f) Heinrich, M. R.; Steglich, W.; Banwell, M. G.; Kashman, Y. *Tetrahedron* **2003**, *59*, 9239.
13. (a) Hidai, Y.; Kan, T.; Fukuyama, T. *Tetrahedron Lett.* **1999**, *40*, 4711. (b) Hidai, Y.; Kan, T.; Fukuyama, T. *Chem. Pharm. Bull.* **2000**, *48*, 1570. (c) Nihei, K.; Kato, M. J.; Yamane, T.; Palma, M. S.; Konno, K. *Bioorg. Med. Chem. Lett.* **2002**, *12*, 299.
14. Kessler, A.; Faure, H.; Petrel, C.; Ruat, M.; Dauban, P.; Dodd, R. H. *Bioorg. Med. Chem. Lett.* **2004**, *14*, 3345.
15. Turner, J. J.; Wilschut, N.; Overkleeft, H. S.; Klaffke, W.; van der Marel, G. A.; van Boom, J. H. *Tetrahedron Lett.* **1999**, *40*, 7039.
16. (a) Ichikawa, Y.; Ohbayashi, M.; Hirata, K.; Nichizawa, R.; Isobe, M. *Synlett.* **2001**, 1763. (b) Ichikawa, Y.; Hirata, K.; Ohbayashi, M.; Isobe, M. *Chem. Eur. J.* **2004**, *10*, 3241.
17. (a) Suzuki, M.; Kambe, M.; Tokuyama, H.; Fukuyama, T. *J. Org. Chem.* **2004**, *69*, 2831. (b) Kambe, M.; Suzuki, M.; Tokuyama, H.; Fukuyama, T. *Org. Lett.* **2001**, *3*, 2575.
18. Fuchs, J. R.; Funk, R. L. *Org. Lett.* **2005**, *7*, 677.
19. Ichikawa, Y.; Ito, T.; Isobe, M. *Chem. Eur. J.* **2005**, *11*, 1949.

20. Barnett, C. J.; Grubb, L. M. *Tetrahedron* **2000**, *56*, 9221.
21. (a) Kobayashi, S.; Peng, G.; Fukuyama, T. *Tetrahedron Lett.* **1999**, *40*, 1519. (b) Hilton, S. T.; Ho, T. C. T.; Pljevaljcic, G.; Schulte, M.; Jones, K. *Chem. Commun.* **2001**, 209.
22. (a) Sumi, S.; Matsumoto, K.; Tokuyama, H.; Fukuyama, T. *Org. Lett.* **2003**, *5*, 1891. (b) Sumi, S.; Matsumoto, K.; Tokuyama, H.; Fukuyama, T. *Tetrahedron* **2003**, *59*, 8571.
23. (a) Kobayashi, S.; Ueda, T.; Fukuyama, T. *Synlett.* **2000**, 883. (b) Yokoshima, S.; Ueda, T.; Kobayashi, S.; Sato, A.; Kuboyama, T.; Tokuyama, H.; Fukuyama, T. *J. Am. Chem. Soc.* **2002**, *124*, 2137. (c) Yokoshima, S.; Ueda, T.; Kobayashi, S.; Sato, A.; Kuboyama, T.; Tokuyama, H.; Fukuyama, T. *Pure Appl. Chem.* **2003**, *75*, 29. (d) Kuboyama, T.; Yokoshima, S.; Tokuyama, H.; Fukuyama, T. *Proc. Natl. Acad. Sci. U. S. A.* **2004**, *101*, 11966. (e) Schneider, C. *Angew. Chem. Int. Ed., Engl.* **2002**, *41*, 4217.
24. (a) Chhabra, S. R.; Khan, A. N.; Bycroft, B. W. *Tetrahedron Lett.* **2000**, *41*, 1099. (b) Strømgaard, K.; Andersen, K.; Ruhland, T.; Krogsgaard-Larsen, P.; Jaroszewski, J. W. *Synthesis* **2001**, 877. (c) Kan, T.; Kobayashi, H.; Fukuyama, T. *Synlett.* **2002**, 1338.
25. (a) Yang, L.; Chiu, K. *Tetrahedron Lett.* **1997**, *42*, 7307. (b) Combs, A. P.; Rafalski, M. *J. Comb. Chem.* **2000**, *2*, 29. (c) Lin, X.; Dorr, H.; Nuss, J. M. *Tetrahedron Lett.* **2000**, *41*, 3309.
26. (a) Piscopio, A. D.; Miller, J. F.; Koch, K. *Tetrahedron Lett.* **1998**, *39*, 2667. (b) Kan, T.; Tominari, Y.; Natsugari, H.; Tomita, T.; Iwatsubo, T.; Fukuyama, T. *Chem. Commun.* **2003**, 2244. (c) Arya, P.; Wei, C. -Q.; Barnes, M. L.; Daroszewska, M. *J. Comb. Chem.* **2004**, *6*, 65.
27. See ref. 2: Cai, W.; Colony, J. L.; Frost, H.; Hudspeth, J. P.; Kendall, P. M.; Krishnan, A. M.; Makowski, T.; Mazur, D. J.; Phillips, J.; Ripin, D. H. B.; Ruggeri, S. G.; Stearns, J. F.; White, T. D. *Org. Process. Res. Dev.* **2005**, *9*, 51.
28. Wuts, P. G. M.; Northuis, J. M. *Tetrahedron Lett.* **1998**, *39*, 3889.
29. Guisado, C.; Waterhouse, J. E.; Price, W. S.; Jorgensen, M. R.; Miller, A. D. *Org. Biomol. Chem.* **2005**, *3*, 1049.
30. Favre-Réguillon, A.; Segat-Dioury, F.; Nait-Bouda, L.; Cosma, C.; Siaugue, J. -M.; Foos, J.; Guy, A. *Synlett.* **2000**, 868.
31. Olsen, C. A.; Jørgensen, M. R.; Witt, M.; Mellor, I. R.; Usherwood, P. N. R.; Jaroszewski, J. W.; Franzyk, H. *Eur. J. Org. Chem.* **2003**, 3288.
32. Penso, M.; Albanese, D.; Landini, D.; Lupi, V.; Tricarico, G. *Eur. J. Org. Chem.* **2003**, 4513.
33. Reddy, P. A.; Schall, O. F.; Wheatly, J. R.; Rosik, L. O.; McClurg, J. P.; Marshall, G. R.; Slomczynska, U. *Synthesis* **2001**, 1086.
34. (a) Messeri, T.; Sternbach, D. D.; Tomkinson, N. C. O. *Tetrahedron Lett.* **1998**, *39*, 1669. (b) Messeri, T.; Sternbach, D. D.; Tomkinson, N. C. O. *Tetrahedron Lett.* **1998**, *39*, 1673.

Jacob M. Janey

5.2 Gabriel Synthesis

5.2.1 Description

The Gabriel synthesis involves the formation of primary amines from alkyl halides by using potassium phthalimide, thus preventing over-alkylation.

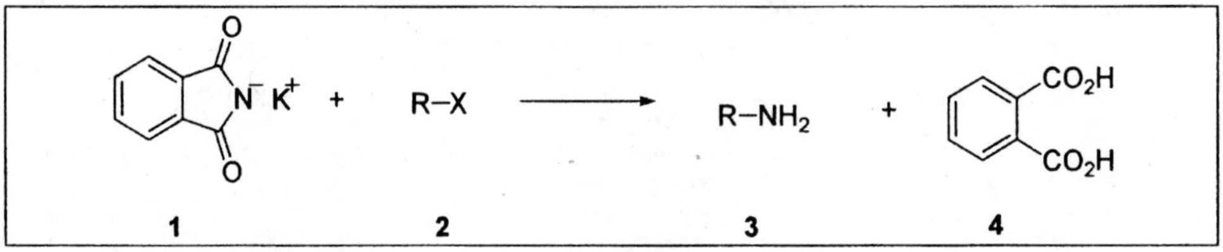

5.2.2 Historical Perspective

The Gabriel synthesis was established in the late 1880s by the German chemist Sigmund Gabriel.[1-3] It is a classical method for forming primary amines and involves alkylation of potassium phthalimide followed by cleavage of the phthaloyl protecting group. Since its instigation as a reliable amination method, the experimental procedure has undergone several improvements, in both the alkylation and phthalimide cleavage steps. The most important of these have been to change the somewhat harsh conditions originally employed to more facile, milder reaction conditions. Principally, novel aminating agents in place of phthalimide, and alternative conditions for cleavage, for example, hydrazinolysis, have been investigated to this effect.

5.2.3 Mechanism

Potassium phthalimide **1** is an "NH_2" synthon which, after alkylation, is no longer nucleophilic and therefore does not react any further. The alkylated phthalimide is cleaved by reaction with base leading to the requisite amine **3** and phthalic acid **4** as a by-product.

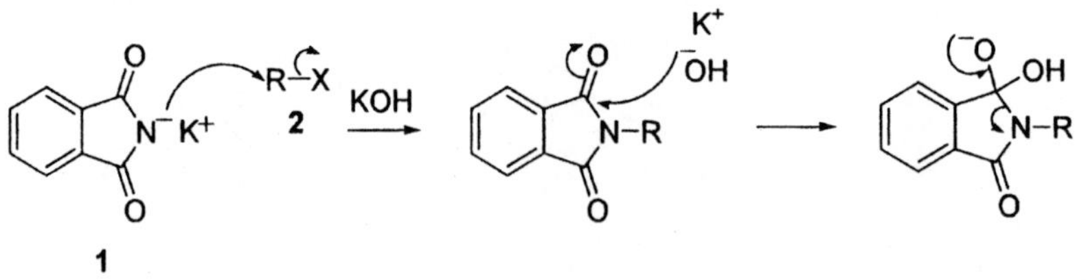

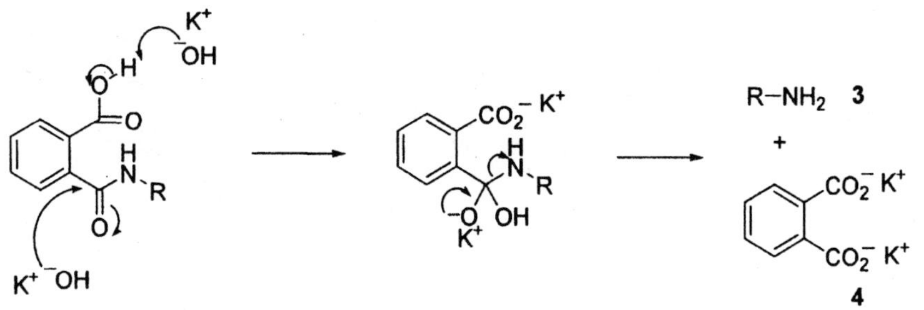

5.2.4 Variations and Improvements

The Gabriel synthesis has been successful in amine synthesis where other methods have failed thus making it, even after a century, one of the most useful preparations of amines available.[4,5] One such amine, 2,6-difluorobenzyl amine **8** (2,6-DFBA), has been used in the synthesis of various medicines; however, its preparation has often been lengthy.[6a–d] A more straightforward synthesis from 2,6-difluorobenzyl chloride **6** using the Gabriel method has been described.[7] A phase-transfer catalyst was used to obtain the *N*-(2,6-difluorobenzyl) phthalimide **7** in a 95% yield. Hydrolysis with aqueous KOH and the use of Dean–Stark apparatus gave 2,6-DFBA which was dried with solid KOH thus avoiding repetitive extraction of the amine with ether.

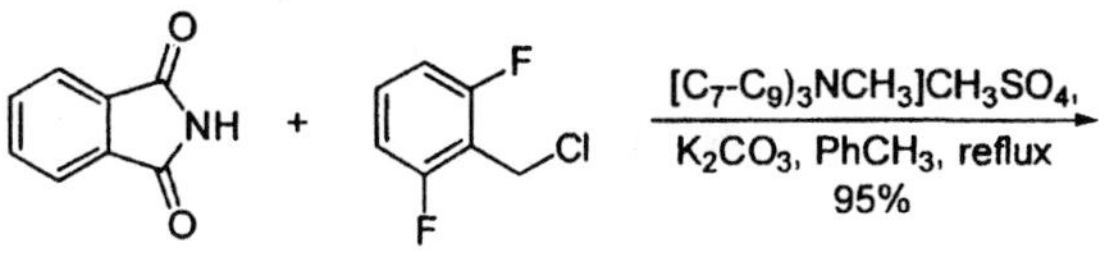

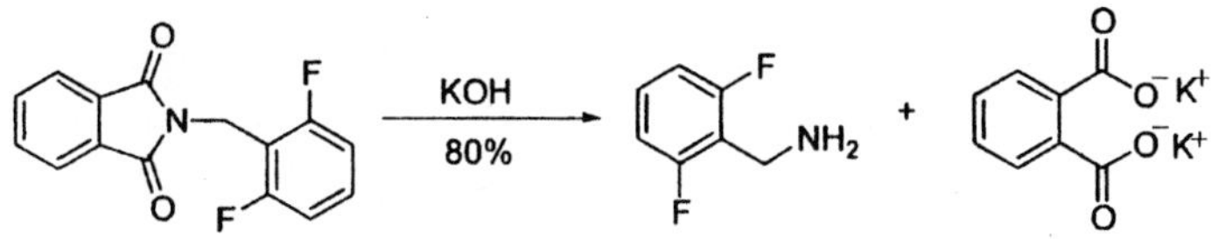

Alternative methods to remove the phthaloyl group have been developed over the years as the classical method is often deemed too harsh, sometimes leading to degradation of other functionalities present. An example is the two-stage one-pot method using sodium borohydride followed by acetic acid reported by Ganem and co-workers.[8] Previously, $NaBH_4$ reduction of phthalimides has been described but these examples were highly solvent and concentration and workup dependent.[9a–c] Ganem and co-workers found that the reduction of phthalimides with $NaBH_4$ proceeded smoothly in high yields in a 6:1 2-propanol:H_2O solvent system. The *o*-hydroxymethyl benzamides **12** could then be cyclised with aqueous acetic acid to form a lactone with concomitant release of the desired primary amine.

NaBH$_4$; H$^+$, pH 5, 80 °C; + RNH$_2$

10 **11** **12** **13** **14**

R groups included long chain alkyls, benzyl group, and several *L*-amino acids. Gratifying, the phthalimide structures of the α-amino acids were deprotected without loss of optical activity, which may otherwise be expected from an acid reflux step.

A commonly used modification of the Gabriel synthesis is the Ing–Manske procedure.[10] Instead of removal of the phthaloyl group with aqueous base or acid, the Ing–Manske modification utilizes hydrazinolysis. It has been used in the synthesis of optically active *α*-amino acids **17**, as illustrated by Guifa and Lingchong.[11]

TBAB, THF, 70 °C, 4 h

1 **15** **16**

$NH_2H_2.H_2O$, EtOH, reflux, 2 h; overall yields 17-47%

17

R = CH_3, C_2H_5, nC_3H_7, nC_4H_9
R' = C_2H_5, bornyl

As well as amines, the Ing–Manske modification has also been used to synthesise *O*-substituted hydroxylamines, for example, **20**.[12] The hydroxyamines thus obtained were subsequently condensed with steroids to form *O*-alkyloximes.

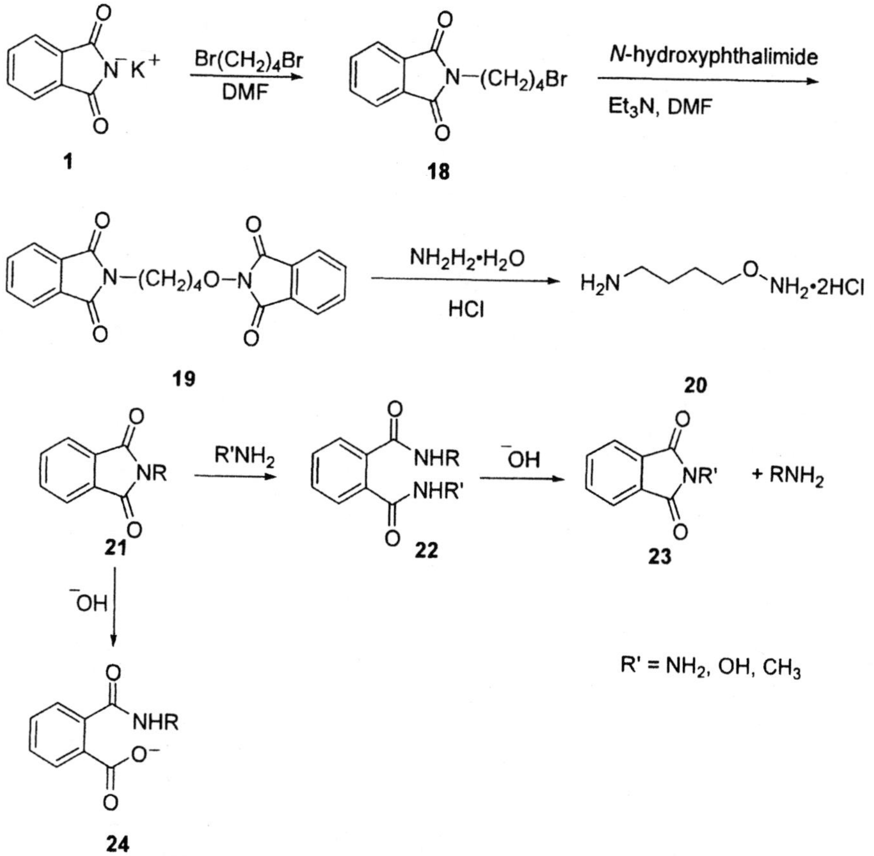

Several kinetic studies on the second step of the Gabriel synthesis, namely, phthalimide cleavage, have been carried out; as a result improved procedures were developed.[13,14] In particular, kinetic studies carried out by Ariffin's group on the hydrazinolysis of phthalimides led to the conclusion that increasing the pH of the reaction mixture *after* complete consumption of the starting phthalimide **21** to form **22** will result in increased rates of formation of RNH_2.[15] The general scheme is outlined. If the base, sodium hydroxide, is added at the start of the reaction, the rate of formation of **24** increases, thus decreasing the yield of RNH_2 through a competing pathway.

Cleavage of the phthaloyl group to give the primary amine in the second step of the Gabriel mechanism usually requires rigorous conditions which are not always compatible with the functionalities already present in the compound. To overcome this, many reagents have been put forth as alternatives to the phthalimide in the Gabriel reaction.[16] An example is sodium diformylamide **25**; the *N,N*-diformylalkylamines **26** thus formed can be hydrolysed under mild conditions.[17]

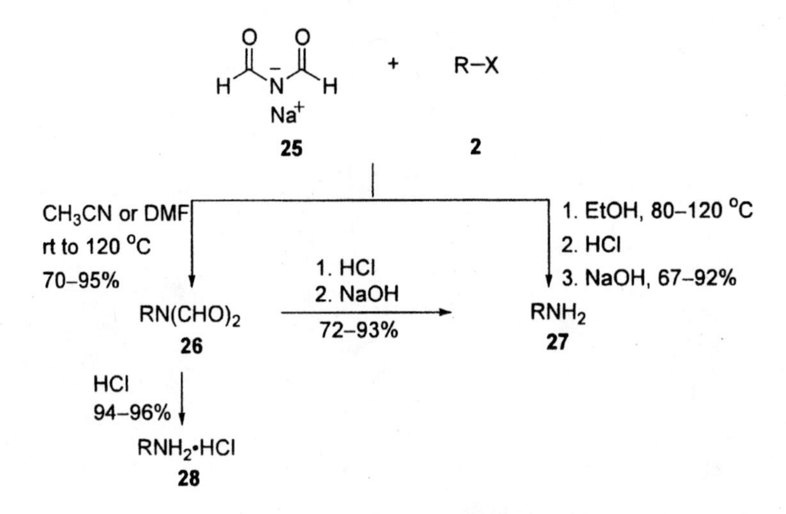

The synthesis can be carried out in a number of ways. A one-pot reaction in ethanol followed by hydrolysis with hydrochloric acid and sodium hydroxide successively gives rise to the primary amines in yields between 67% to 92%. Alternatively, reaction in anhydrous solvents gives the *N,N*-diformylalkylamines **26** which can be hydrolysed as previously, or simply treated with hydrochloric acid to give the corresponding hydrochloride salts. R groups included alkyl chains, allyl, benzyl, bromobenzyl, esters, and ketones.

Similarly, in their synthesis of γ-aminobutyric acid analogs, Allan and co-workers use a modified Gabriel reagent in order to avoid using basic conditions, which would have led to the ring opening of the tetronic acid ether.[18] Consequently, the potassium salt of bis(*t*-butoxycarbonyl)amine **33** was utilised to give the protected amine **30**. Amine **31** was then unmasked, converted to the salt, and the tetronic acid deprotected in one step using HBr-acetic acid at room temperature. The bis(*t*-butoxycarbonyl)amine potassium salt was first developed by Carpino as an amine equivalent.[19] A convenient preparation has been described by Grehn and Ragnarsson from commercially available starting materials, and is outlined below.[20]

t-BuO; Br; $(t\text{-}BuOCO)_2NK^+$; 71%; t-BuO; $N(OCOt\text{-}Bu)_2$; HBr-AcOH; 15%; HO; $\overset{+}{N}H_3Br^-$

29 **30** **31**

H, NH_2; Boc_2O (2.2 eq.), DMAP (0.1 eq.), CH_3CN, rt, 6 h; $[H(C{=}O)N(Boc)_2]$; DEAEA (1.2 eq.), CH_3CN, rt, 15 h; 80%; HN(Boc)Boc

32 **33**

DEAEA = 2-diethylaminoethylamine

An additional advantage of the di-*t*-butyl imidocarbonate is selective removal of one Boc protecting group, leaving a primary Boc-protected amine. Allylic halides and mesylates when treated with the imidocarbonate and K_2CO_3 in butanone or diglyme lead to the *N,N*-Boc-protected amines **35**.[21] Subsequent reaction with TFA (1.5–2 eq.) gave the Boc-protected amines **36**. Propargylic and homoallylic amines could also be obtained with this method; in these cases using Cs_2CO_3 instead of K_2CO_3 gave better yields. Some representative examples are shown. Regiochemistry and alkene geometry remained intact indicating that displacement of the leaving group by the iodide catalyst, and subsequently the imidodicarbonate, occurred *via* an S_N2 reaction.

X, R_1, R_4, R_2, R_3; $HN(Boc)_2$, LiI; Boc–N–Boc; TFA (1.5–2 eq.), CH_2Cl_2; Boc–NH

34 **35** **36**

Br 91%; Cl 80%; OMs 94%

Zwierzak and co-workers have reported on the use of a phosphoramidate as an alternative to the phthalimide in the Gabriel synthesis.[22a–c] However, several drawbacks to their method led to revisions in the original method. In particular, the use of toxic benzene as a solvent, deprotection by gaseous HCl in benzene, and low yields were discouraging. The modified procedure reaction involves nucleophilic amination by diethyl *N*-sodio-*N*-(*t*-butoxycarbonyl)phosphoramidate **37** with tetrabutylammonium bromide as a catalyst in acetonitrile.[23] Carrying out the reactions in tetrahydrofuran led to lower yields of the ammonium tosylates **39**. The phosphoramidates **38** could be deprotected to the ammonium tosylates without purification. R groups include alkyl chains, cyclohexyl, benzyl, and alkyne containing chains, although the reaction is limited to primary organic halides; *N*-alkylation of secondary alkyl bromides led solely to recovery of starting materials.

$(EtO)_2P(O)N(Na)C(O)Ot\text{-}Bu$ **37** —[RBr, Bu_4NBr (10 mol%); MeCN, reflux, 3 h]→ $(EtO)_2P(O)N(R)C(O)Ot\text{-}Bu$ **38** —[pTsOH·H_2O, EtOH, reflux, 6 h, 44–80%]→ $R{-}NH_3^+TsO^-$ **39**

$(EtO)_2P(O)NH_2$ **40** —[oxalyl chloride; CCl_4, 0 °C]→ $(EtO)_2P(O)N{=}C{=}O$ **41** —[t-BuOH; CCl_4, rt]→ $(EtO)_2P(O)NHC(O)Ot\text{-}Bu$ **42** —[$NaOCH_3$; CH_3OH, rt]→ **37**

$(EtO)_2P(O)NHSi(CH_3)_3$ **43**

Preparation of the phosphoramidate **37** is outlined.[22a] Zwierzak also describes the use of diethyl *N*-(trimethylsilyl)-phosphoramidate **43** as a Gabriel reagent.[24] In a similar method to the above, *N*-alkylphosphoramidates could be obtained using this reagent in good yields.

In their synthesis of a Pirmenol metabolite, Goel and Purchase found that the phthalimide derivative **44** did not undergo addition reactions with organolithium reagents.[25] Alternative Gabriel reagents such as phosphoramidates or sulfonamides have also been found to interfere with organolithium reagents. Ultimately, dibenzylamine was employed which did not impede the desired 2-pyridyllithium addition. The benzyl groups were then conveniently removed under mild conditions by catalytic transfer hydrogenolysis (CAT)[26] using ammonium formate in methanol and 10% Pd/C as a catalyst. Protection of the ketone as the cyclic ketal avoided by-products.

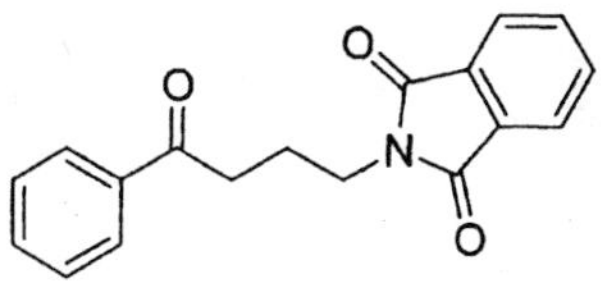

44

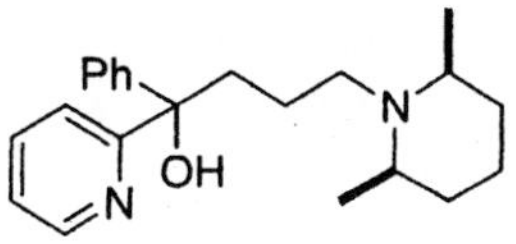

Pirmenol

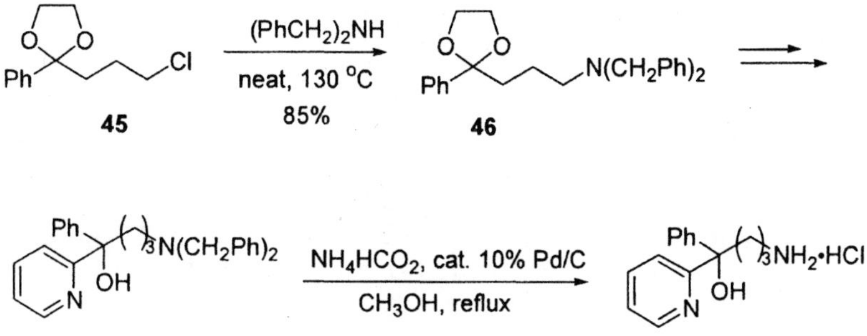

45 → 46: $(PhCH_2)_2NH$, neat, 130 °C, 85%

47 → 48: NH_4HCO_2, cat. 10% Pd/C, CH_3OH, reflux

The conditions to couple the phthalimide group with primary halides have also been modified to obtain the best yields. The most common method by which this step has been altered is by utilizing Mitsunobu conditions.[27a,b] In a recent example, the conversion of diol **49** to the corresponding diamine was attempted using several methods. Employing the Mitsunobu conditions resulted in the best yields.[28] Thus, reaction of diol **49** with phthalimide in the presence of PPh_3 and DEAD gave diimide **50**. Treatment with hydrazine followed by hydrogenation gave diamine **51**. The absolute configuration of the norbornene backbone, confirmed by X-ray analysis, was unchanged by the Mitosunobu and Gabriel-type amine synthesis.

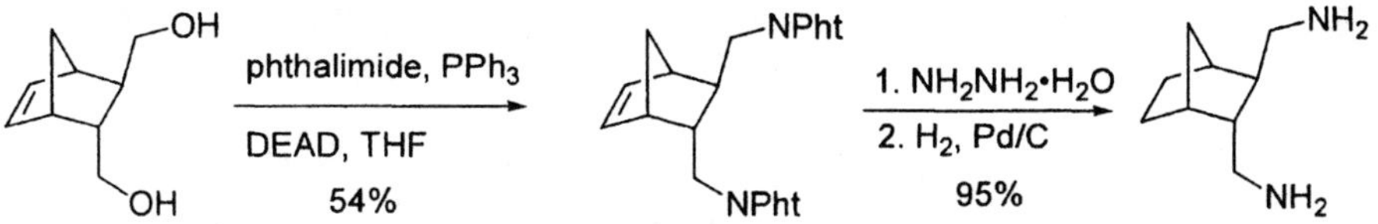

49 → 50: phthalimide, PPh_3, DEAD, THF, 54%

50 → 51: 1. $NH_2NH_2{\bullet}H_2O$ 2. H_2, Pd/C, 95%

Allylic amines **54** have also been synthesised using the Mitosunobu protocol. Sen and Roach describe the formation of such amines from allylic alcohols.[29] The phthaloyl group can be removed cleanly by hydrazine at room temperature, while for more bulky groups (e.g., nerol, farnesol) phthaloyl cleavage was better effected with methylamine. Deprotection with methylamine was found to be exceptionally mild, and in substrates containing several double bonds, geometric integrity was preserved. In addition the products were obtained in high purity thus eliminating the need for further purification.

DIAD, PPh_3, phthalimide, THF, rt, 4 h, 76–98%; NH_2H_2 or CH_3NH_2, CH_3OH, 76–88%

52 53 54

5.2.5 *Synthetic Utility*

The preparation of a phthalimide-containing resin to be used under Mitsunobu conditions means that primary amines can be generated cleanly after cleavage by hydrazinolysis.[30] With this goal in mind, the resin linker **57** was obtained in > 90% yield *via* the synthesis outlined below. An example of the resin in use is shown with N^6-benzyladenosine **58**, where the less hindered hydroxyl group was successfully converted to the amine **59** in 85%.

1. $(NH_4)_2CO_3$ 2. TrCl, DIPEA, CH_3CN

55 56

1. Aminomethyl resin PyBOP, DIPEA, DMA 2. Ac_2O, DIPEA, CH_2Cl_2 3. TFA, reflux > 90%

57

1. resin X, PPh_3, DEAD, THF 2. NH_2H_2, EtOH, CH_2Cl_2 85%

58 59

The formation of aminals *via* the Gabriel reaction can arise on changing the solvent from toluene to dimethylformamide, in the presence of sodium salts of *N*-heterocycles.[31] In this unique example, **63** could be prepared in useful quantities in a more practical technique than with previous methods which required high heat and/or pressure conditions.[32a,b] The required phthalimide derivatives **62** was also formed but in low yields.

DMF, 70–100 °C, 6–14 h

60 **61** **62** 18% **63** 5–20% **64** 56%

63 **65** **66**

67 **68** **69** → **64**

The mechanism for formation of **64** was proposed as follows. The *N*-vinyl-phthalimide **63** can undergo nucleophilic addition with the sodium salt of imidazole present in the reaction mixture to form the ring-opened phthalimide **65**, which then tautomerizes to the imine **66**. A second nucleophilic addition takes place, leading to ring closure with loss of the nitrogen heterocycle to give aminal **64**. A number of *N*-containing heterocycles gave rise to these aminal derivatives.

Arguably one of the most valuable exploitations of an amination reaction is in the formation of amino acids. Wu and Zhang have employed chiral reverse micelles as asymmetric microenvironments for enantioselective synthesis of 2-phthalimides-esters **72**.[33] Hydrazinolysis and hydrolysis of the esters then lead to α-amino acids. It was found that asymmetric induction varied according to the reaction temperature, the alkyl chain of the surfactants, and the structure of the surfactants.

N^-K^+ + Br–OEt → THF, reflux
1 **70** **71**

RX, KOH, chiral reverse micelle → **72** → 1. hydrazinolysis 2. hydrolysis, 40–60%, *ee* 21–59% → **73**

RX: CH_3I, $CH_3CH_2CH_2I$, i-C_3H_7Br, $PhCH_2Br$

Although the *ee*'s and yields were not particularly high, optically active amino acids were successfully synthesized and, perhaps more importantly, the reaction scheme illustrates the versatility of the Gabriel amination, well over a hundred years after its conception.

5.2.6 *Experimental*

74 → **75** → **76**

Dimethyl α,δ-diphthalimidoadipate (75)[34]
A mixture of dimethyl α,δ-dibromoadipate (69 g, 0.21 mol) and potassium phthalimide in dimethylformamide (260 mL) was heated at 90 °C for 40 min. The reaction mixture was then cooled, diluted with CCl_4 (300 mL) and H_2O (1200 mL) and extracted with CCl_4 (2 × 100 mL). The combined organic extracts were washed successively with aqueous NaOH (200 mL, 0.1 M) and H_2O (200 mL), then dried over sodium sulfate. After concentration under reduced pressure, the addition of ether (300 mL) induced a rapid crystallisation. Recrystallisation with EtOAc gave the phthalimide derivative **75** (87 g, 90%).

α,δ-Diaminoadipic acid (76)
A. By Acid Hydrolysis
A mixture of dimethyl α,δ-diphthalimidoadipate **76** (50 g), hydrobromic acid (100 mL, 40%), and glacial acetic acid (100 mL) was heated under reflux until a clear solution resulted (10 days). On cooling, most of the phthalic acid crystallized. After filtration, the filtrate and water washes were concentrated under reduced pressure practically to dryness. The residue was dissolved in H_2O (100 mL), filtered, and neutralized with concentration ammonia. After crystallization at 0 °C for 12 h, α,δ-diaminoadipic acid (17.3 g, 91.2%) was obtained.

B. By Treatment with Hydrazine
A mixture of dimethyl α,δ-diphthalimidoadipate **75** (4.64 g, 0.01 mol), CH_3OH (50 mL), and aqueous NH_2NH_2 (1.2 mL, 0.02 mol, 85%) was heated under reflux for 1 h. After cooling, H_2O (25 mL) was added, the methanol removed under reduced pressure. Concentrated HCl (25 mL) was added to the residual aqueous suspension and the mixture heated under reflux for 1 h. After cooling to 0 °C, crystalline phthalhydrazide was removed by filtration. The filtrate was then concentrated under reduced pressure to remove the hydrochloride acid and the moist residue was dissolved in H_2O (50 mL). A small amount of insoluble matter was removed by filtration and the clear filtrate was neutralized with 2 M NaOH. After cooling at 0 °C for 12 h, α,δ-diaminoadipic acid **76** (1.4 g, 79.5%) was obtained.

5.2.7 References

1. Gabriel, S. *Ber. Dtsch. Chem. Ges.* **1887**, *20*, 2224.
2. Gabriel S. *Ber. Dtsch. Chem. Ges.* **1908**, *41*, 1132.
3. Colman, J.; Albert, A. *Ber. Dtsch. Chem. Ges.* **1926**, *59A*, 7.
4. [R] Gibson, M. S.; Bradshaw, R. W. *Angew. Chem.* **1968**, *80*, 986,
5. Gibson, M. S.; Bradshaw, R. W. *Angew. Chem. Int. Ed. Engl.* **1968**, *7*, 919
6. (a) Roe, A. M.; Burton, R. A.; Wiley, G. L.; Baines, M. W.; Rasmussen, A. C. *J. Med. Chem.* **1968**, *11*, 814; (b) U.S. 4,107,326 in CA, 1979, 90, P103685q.; (c) Kelley, J. L.; Krochmal, M. P.; Lin, J. A.; McRean, E. W.; Soroko, F. E.; *J. Med. Chem.* **1988**, *31*. 1005; (d) Jpn. Kokai Tokkoyo Koho, JP 0753, 476 [9553,476] in CA, **1995**, 122, 314267x.
7. Vasilevskaya, T. N.; Yakovleva, O. D.; Kobrin, V. S. *Synthetic Commum.* **1998**, *28(2)*, 201.

8. Osby, J. O.; Martin, M. G.; Ganem, B. *Tetrahedron Lett.* **1984,** *25(20)*, 2093.
9. (a) Hubert, J. C.; Wijnberg, J. P. B. A.; Speckamp, W. N. *Tetrahedron* **1975**, *31*, 1437; (b) Horii, Z. –I.; Iwata, C.; Tamura, Y. *J. Org. Chem.* **1961**, *26*, 2273; (c) Uhle, F. C. *J. Org. Chem.* **1961**, *26*, 2273.
10. Ing, H. R.; Manske, R. H. F. *J. Chem. Soc.* **1926**, 2349.
11. Guifa, S.; Lingchong, Y. *Synthetic Commun.* **1993**, *23(9)*, 1229.
12. Mikola, H.; Hanninen E. *Bioconjugate Chem.* **1992**, *3*, 182.
13. Khan. M. N. *J. Chem. Soc. Perkin Trans. 2* **1988**, 213.
14. Khan, M. N. *J. Org. Chem.* **1995**, *60*, 4563. Khan, M. N. *J. Org. Chem.* **1996**, *61*, 8063.
15. Ariffin A.; Khan, M. N.; Lan, L. C.; May, F. Y.; Yun, C. S. *Synthetic Commun.* **2004**, *34(24)*, 4439.
16. [R] Ragnarsson, U.; Grehn, L. *Acc. Chem. Res.* **1991**, *24*, 285.
17. Yinglin, H.; Hongwen, H. *Synthesis* **1990**, 122.
18. Allan, R. D.; Johnstone, G. A. R.; Kazlauskas R.; Tran, H. W. *J. Chem. Soc. Perkin. Trans. 1* **1983**, 2983.
19. Carpino, L. A. *J. Org. Chem.* **1964**, *29*, 2820.
20. Grehn, L.; Ragnarsson, U. *Synthesis* **1987**, 275.
21. van Benthem, R. A. T. M.; Michels, J. J.; Hiemstra, H.; Speckamp, W. N. *Syntheic Lett.* **1994**, 368.
22. (a) Zwierzak, A.; Pilichowska, S. *Synthesis* **1982**, 922; (b) Koziara, A.; Zwierzak, A. *Synthesis* **1992**, 1063; (c) Osowska-Pacewicka, K.; Zwierzak, A. *Synthesis* **1996**, 333.
23. Zwierzak, A. *Synthetic Commun.* **2000**, *30(13)*, 2287.
24. Zwierzak, A. *Synthesis* **1982**, 920.
25. Purchase, C. F.; Goel, O. P. *J. Org. Chem.* **1991**, *56*, 457.
26. [R] Johnstone, R.; Wiley, A. *Chem. Rev.* **1985**, *85*, 129.
27. (a) Mitsunobu, O.; Takizawa, S.; Morimoto, H. *J. Am. Chem. Soc.* **1976**, *98*, 7858; (b) Kolb, M. Danzin, C.; Barth, J.; Claverie, N. *J. Med. Chem.* **1982**, *25*, 550.
28. Tanyeli, C.; Özçubukçu, S. *Tetrahedron Asymmetry* **2003**, *14,* 1167.
29. Sen, S. E.; Roach, S. L. *Synthesis*, **1995**, 756.
30. Aronov, A. M.; Gelb, M. H. *Tetrahedron Lett.* **1998**, 4947.
31. Press, J. B.; Haug, M. F.; Wright, W. B. Jr. *Synthetic Commun.* **1985**, *15(9)*, 837.
32. (a) Furukawa, J.; Onishi, A.; Tsurata, T. *J. Org. Chem.* **1958**, *23*, 672 and references therein; (b) Bayer, E.; Geckeler, K. *Angew. Chem.* **1979**, *91*, 568.
33. Wu, W.; Zhang, Y. *Tetrahedron Asymmetry* **1998**, *9*, 1441.
34. Sheehan, J. C.; Bolhofer, V. A. *J. Am. Chem. Soc.* **1950**, *72*, 2786.

Nadia M. Ahmad

5.3 Leuckart–Wallach Reaction

5.3.1 Description

The Leuckart–Wallach reaction[1,2] is the reductive amination of carbonyl compounds **1** in the presence of excess formic acid (**3**) as a reducing agent. The Eschweiler–Clarke reaction represents a specific example of this reaction, where the carbonyl compound is formaldehyde.

$$R_1C(=O)R_2 \;(\mathbf{1}) \quad R_3CH(NH_2)R_4 \;(\mathbf{2}) \quad HC(=O)OH \;(\mathbf{3}) \longrightarrow R_1CH(NR_3R_4)R_2 \;(\mathbf{4})$$

The advantages in this reaction lie in economy and simplicity; however long reaction times, sealed tube conditions, and high temperatures remain major drawbacks. In addition, formation of the *N*-formyl derivative is frequently seen, and the selective synthesis of a primary amine from ammonia is difficult.[3,4]

5.3.2 Historical Perspective

In 1885, R. Leuckart reported that the reaction of ammonia and primary or secondary amines with ketones (and aldehydes) was possible with ammonium formate, to produce the corresponding primary, secondary, or tertiary amine.[1] This reaction became known as the Leuckart reaction. In 1891, Wallach reported that the use of excess formic acid led to an improvement in yield, and that the reaction could be carried out at lower temperatures.[2] The Leuckart–Wallach reaction was born. Wallach went on to apply this reaction to a wide variety of carbonyl compounds, including alicyclic and terpenoid ketones.[5]

5.3.3 Mechanism

The mechanism of the Leuckart–Wallach reaction has been the subject of much discussion and is generally accepted to proceed *via* the reaction of the amine with the carbonyl functionality to give an α-amino alcohol, which dehydrates to give an iminium ion.[3] The reduction of this iminium is effected by formic acid.[6–8]

Although a radical mechanism has been proposed,[10] more recently a rate-limiting C–H bond cleavage step has been invoked,[11] and it is generally accepted that the formic acid acts as a hydride donor.[9] This is consistent with the observation of production of carbon dioxide and reduction product in equivalent quantities.[10]

2 5 6 7 8 3

α-amino alcohol

iminium ion

$-H^{\oplus}$

$-CO_2$

4 9

$-H^{\oplus}$

5.3.4 Variations

In 1931, Ingersoll *et al.* noted that formamide could act as an intermediate, and consequently ammonium formate could be replaced with formamide in the Leuckart reaction,[12] and in 1968, Bach carried out Leuckart–Wallach reactions on cyclic aliphatic ketones with dimethylformamide in high yields, and with reaction times of only 8–16 h.[13]

HCO_2H H_2O 190°C 8 - 16h

10 11 12 75%

Kitamua *et al.* have identified a useful catalyst for the Leuckart–Wallach reaction that can be used to form primary amines at lower temperatures, and in a higher yield than the standard conditions.[14] The reaction uses 0.5 mol% $[RhCp^*Cl_2]_2$ and a 1:5 substrate:ammonium formate ratio.

O
H ONH$_4$
[RhCp*Cl$_2$]$_2$
70°C, 7h
13 → 14 (NH_2) 92%

In addition, this catalyst has been used to selectively aminate α-keto acids to form α-amino acids.[14]

O
H ONH$_4$
[RhCp*Cl$_2$]$_2$
70°C, 7h
15 → 16 (NH_2, OH) 91%

Martinez *et al.* have carried out several studies into the Leuckart-Wallach reaction on 1-substituted-2-norbornanones.[15,16] The expected products result from 2-norbonanone and (1*R*)-*N*-(3,3-dimethyl-2-oxo-1-norbornyl)acetamine, but the reaction of (1*R*)-*N*-(7,7-dimethyl-2-oxo-1-norbornyl)acetamine gives rise to an unexpected intramolecular transamidation *via* a Wagner–Meerwein shift.[15]

17 (O, $NHCOCH_3$) → 18 (OHCHN, $NHCOCH_3$)
H_2NCHO
HCO_2H
150°C

19 (O, $NHCOCH_3$) → 20 (CH_3COHN, NHCHO)
H_2NCHO
HCO_2H
150°C

5.3.5 Synthetic Utility

The Leuckart–Wallach reaction was originally used in wide scope for the reaction of aliphatic and alicyclic ketones with ammonia and primary and secondary amines.[1–5] In the original cases, however, the only aldehyde attempted was formaldehyde and, to a lesser extent, valeradehyde.[5e] In 1950, DeBenneville and Macartney reported that the Leuckart–Wallach reaction with aliphatic aldehydes and secondary amines is initiated at a lower temperature than in the case of ketones and aromatic aldehydes.[17] A wide range of aliphatic aldehydes were reacted, and the products were isolated in high yields; the reaction temperature was maintained between 50 and 95 °C.

HCO_2H, 60°C; 21 + 22 → 23, 84%

More recently the reaction has been applied widely to the reaction of heteroaromatic carbonyls and heteroaromatic amines. Musumarra successfully reacted thiophene-2-carbaldehyde with pyramid-2-yl amine in a moderate yield, but with a reaction time of only 7 hours.[18]

HCO_2H, reflux, 7h; 24 + 25 → 26, 45%

Recently, the application of microwave conditions to reactions that normally proceed thermally, but with long reaction times, has led to a much-hastened procedure for the Leuckart–Wallach reaction.[19] Loupy *et al.* have carried out the reductive amination of various aromatic ketones using a 1:3:3 ratio of carbonyl:formamide:formic acid under microwave irradiation for 30 minutes to produce the corresponding amines in high yields.[19] In comparative studies the authors report considerable improvements in yield with microwave use.

$HCONH_2$, HCO_2H, microwave, 30 min, 202 °C; 27 → 28 (NHCHO), 99%

Similarly, the use of the microwave to accelerate the Leuckart reaction has been employed in the synthesis of pyrimidines[20] and the stereoselective synthesis of *endo*-tropanamine.[21]

5.3.6 Experimental

$$\text{3,5,5-trimethylhexanal} + \text{Me}_2\text{NH} \xrightarrow{HCO_2H} \text{dimethyl-(3,5,5-trimethylhexyl)amine}$$

Dimethyl-(3,5,5-trimethyl-hexyl)-amine[17]
Dimethylamine was added dropwise to formic acid and heated to 60 °C. One equivalent of 3,5,5-trimethyl hexanal was added dropwise over the course of one hour, and the reaction mixture heated until no further carbon dioxide was evolved (about one hour). The reaction mixture was poured onto dilute HCl and washed with ether. The aqueous layer was basified with aqueous NaOH and extracted with diethyl ether. The combined extracts were concentrated *in vacuo* to yield an orange oil, which was distilled to a colourless oil, 84%.

5.3.7 References

1. Leuckart, R. *Ber. Dtsch. Chem. Ges.* **1885**, *18*, 2341.
2. Wallach, O. *Ber. Dtsch. Chem. Ges.* **1891**, *24*, 3992.
3. Moore, M. L. *Org. React.* **1949**, *5*, 301.
4. Gibson, H. W. *Chem. Rev.* **1969**, *69*, 673.
5. Wallach, O. (a) *Ibid.*, **1891**, *24*, 3992; (b) *Ann.* **1892**, *269*, 347; (c) *Ann.* **1893**, *276*, 306; (d) *Ibid.* **1898**, *300*, 283; (e) *Ann.* **1905**, *343*, 54.
6. Noyce, D. S.; Bachelor, F. W. *J. Am. Chem. Soc.* **1952**, *74*, 4577.
7. Sauers, R. R. *J. Am. Chem. Soc.* **1958**, *80*, 4721.
8. Sauers, R. R.; Leonard, N. J. *J. Am. Chem. Soc.* **1957**, *79*, 6210.
9. Staple, E.; Wagner, E. *J. Org. Chem.* **1949**, *14*, 559.
10. Lukasiewicz, A. *Tetrahedron* **1963**, *19*, 1789.
11. Agwada, V. C.; Awachie, P. I. *Tetrahedron*, **1990**, *46*, 1899.
12. Ingersoll, A. W.; Brown, J. H.; Kim C. K.; Beauchamp, W. D.; Jennings, G. *J. Am. Chem. Soc,.* **1936**, *58*, 1808.
13. Bach, R. D. *J. Org. Chem.*, **1968**, *33*, 1647.
14. Kitamua, M.; Lee, D.; Hayashi, S.; Tanaka, S.; Yoshimura, M. *J. Org. Chem.*, **2002**, *67*, 8685.
15. Martinez, A. G.; Vilar, E. T.; Fraile, A. G.; Martinez-Ruiz, P. M.; San Antonio, R. M.; Alcazar, M. P. M. *Tetrahedron: Assym.* **1999**, *10*, 1499.
16. Martinez, A. G.; Vilar, E. T.; Fraile, A. G.; Martinez-Ruiz, P. M. (a) *Tetrahedron: Assym.* **2001**, *12*, 2153; (b) *Tetrahedron* **2003**, *59*, 1565.
17. DeBenneville, P. L.; Macartney, J. H. *J. Am. Chem. Soc.* **1950**, *72*, 3073.
18. Musumarra, G.; Sergi, C. *Heterocycles*, **1994**, *37*, 1033.
19. Loupy, A.; Monteux, D.; Petit, A.; Aizpurua, J. M.; Dominguez, E.; Palomo, C. *Tetrahedron Lett.* **1996**, *37*, 8177.
20. Tyagarajan, S.; Chakravarty, P.K. *Tetrahedron Lett.* **2005**, *46*, 7889.
21. Allegretti, M.; Berdini, V.; Cesta, C.; Curti, R.; Nicolini, L.; Topai, A. *Tetrahedron Lett.* **2001**, *42*, 4257.

Alice R. E. Brewer

CHAPTER 6 Carboxylic Acid Derivatives Synthesis 457

6.1 Fischer–Speier Esterification

6.1.1 Description

Esterification of acids is achieved by refluxing with excess alcohol in the presence of hydrogen chloride or other acid catalysts.[1]

$$RCO_2H \; (\mathbf{1}) + R_1OH \; (\mathbf{2}) \xrightarrow{H^+} RCO_2R_1 \; (\mathbf{3})$$

6.1.2 Historical Perspective

In 1895, Fischer and Speier reported that carboxylic acids were esterified with alcohol in the presence of acid catalysts (i.e., HCl, H_2SO_4, *etc.*)[2]

$$PhCO_2H \; (\mathbf{4}) + CH_3CH_2OH \; (\mathbf{5}) \xrightarrow[\Delta,\ 2\ d,\ 76\%]{H_2SO_4} PhCO_2CH_2CH_3 \; (\mathbf{6})$$

6.1.3 Mechanism

$$RCO_2H \; (\mathbf{1}) + H^+ \rightleftharpoons RC(=OH^+)OH \; (\mathbf{7}) + R_1OH \; (\mathbf{2})$$

$$\rightleftharpoons RC(OH)_2(\overset{+}{O}HR_1) \; (\mathbf{8}) \underset{}{\overset{-H^+}{\rightleftharpoons}} RC(OH)_2(OR_1) \; (\mathbf{9}) \overset{+H^+}{\rightleftharpoons}$$

$$RC(OH)(OH_2^+)(OR_1) \; (\mathbf{10}) \overset{-H_2O}{\rightleftharpoons} RC(=OH^+)OR_1 \; (\mathbf{11}) \overset{-H^+}{\rightleftharpoons} RCO_2R_1 \; (\mathbf{3})$$

Protonation of the carbonyl oxygen in acid **1** activates it toward nucleophilic attack by the alcohol **2**, furnishing the tetrahedral intermediate **8**. A facilitated proton transfer between the hydroxyl moieties **8–10**, followed by the loss of a water molecule, provides **11**. Finally, the loss of a proton from **11** yields the ester **3**.

6.1.4 Synthetic Utility

Many acid catalysts have been employed in the Fischer–Speier esterification process, such as dry HCl,[3–5] H_2SO_4[6–8] and $MeSO_3H$.[9] Adams[4] reported that esterification of acetonedicarboxylic acid **12** with alcohol **13** (saturated with HCl) provided the diester **14** in about 40% yield.

$$HO_2C\text{-}CH_2\text{-}C(=O)\text{-}CH_2\text{-}CO_2H\ (\mathbf{12}) + EtOH\ (\mathbf{13}) \xrightarrow[\text{45 °C, 20 min, rt, 12 h; 39–43\%}]{HCl} EtO_2C\text{-}CH_2\text{-}C(=O)\text{-}CH_2\text{-}CO_2Et\ (\mathbf{14})$$

It is known that, when equimolar quantities of reactants are used, the equilibrium in the esterification reaction lies quite short of completion. General methods to obviate this and shift the equilibrium in favor of the product include using an excess of alcohol or acid reagents, adding a water scavenger, or simply collecting the desired ester product by distillation. In the esterification of adipic acid **15** with alcohol **13**,[6] the reaction was driven to completion by azeotropic removal of the water produced in the reaction. The ester **16** was obtained in 97% yield.

$$HO_2C(CH_2)_4CO_2H\ (\mathbf{15}) + EtOH\ (\mathbf{13}) \xrightarrow[\text{toluene, }\Delta\text{; 95–97\%}]{H_2SO_4} EtO_2C(CH_2)_4CO_2Et\ (\mathbf{16})$$

Newman has described a method for esterification of certain sterically hindered acids. This method consists of dissolving the acid to be esterified in 100% sulfuric acid then pouring the solution into the desired alcohol.[10]

$$\text{2,4,6-Me}_3C_6H_2\text{-}CO_2H\ (\mathbf{17}) + MeOH\ (\mathbf{18}) \xrightarrow[\text{78\%}]{100\%\ H_2SO_4} \text{2,4,6-Me}_3C_6H_2\text{-}CO_2Me\ (\mathbf{19})$$

Unlike the conventional Fischer esterification process, which does not allow for functionality in the molecule that is sensitive to the strong acidic conditions of the reaction, the boron trifluoride etherate reagent is mild enough to convert functionalized acids cleanly to their esters. Other functional groups present in addition to the carboxyl moiety are not affected. Thus this method offers a general method for the facile esterification of functionalized carboxylic acids.[11–14]

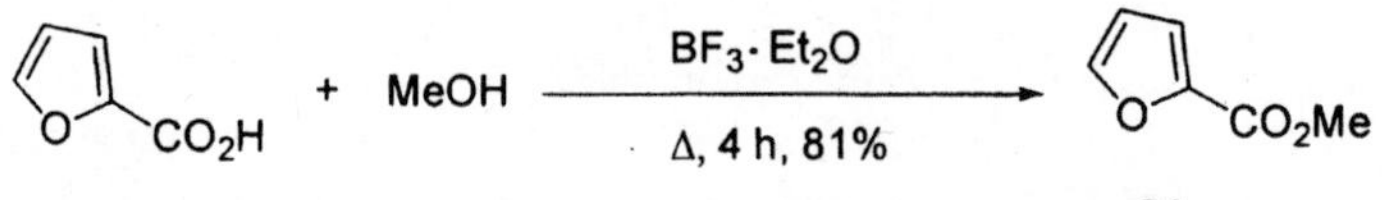

20 + MeOH (18) —($BF_3 \cdot Et_2O$; Δ, 4 h, 81%)→ 21

Polymer protected $AlCl_3$ and organotin dichloride have been used as catalysts as well in esterification.[15,16]

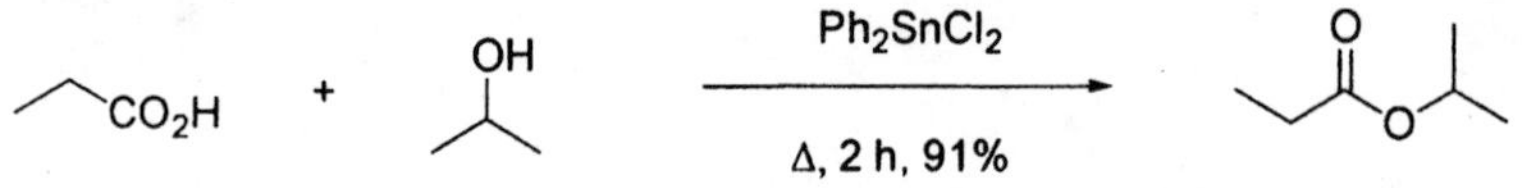

22 + 23 —(Ph_2SnCl_2; Δ, 2 h, 91%)→ 24

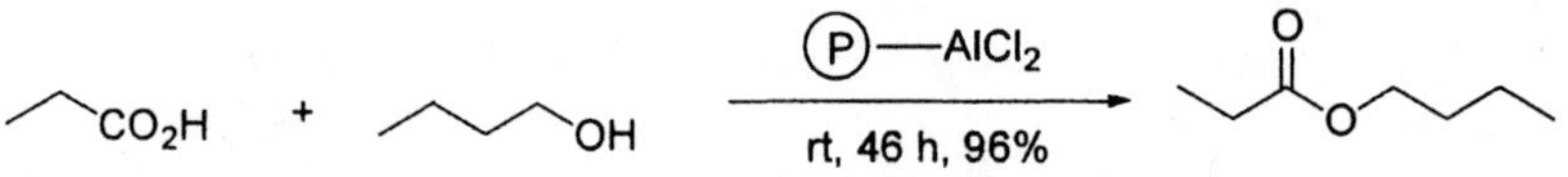

22 + 25 —(Ⓟ—$AlCl_2$; rt, 46 h, 96%)→ 26

6.1.5 *Experimental*

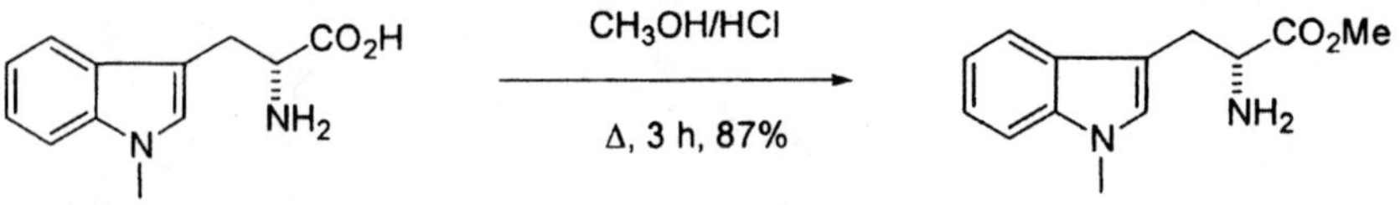

27 —(CH_3OH/HCl; Δ, 3 h, 87%)→ 28

Preparation of *N*-methyltryptophan methyl ester hydrochloride (28)[17]

N-Methyltryptophan (95 g) was dissolved in a freshly saturated solution of methanolic hydrogen chloride (850 mL). The mixture which resulted was heated to reflux for 3.5 h and then allowed to cool to room temperature. The crystalline product which formed upon cooling was collected by filtration to provide *N*-methyltryptophan methyl ester hydrochloride (102.2 g, 87%).

6.1.5 References

1. [R] Euranto, E. K. *The Chemistry of Carboxylic Acid and Esters*; Interscience: New York, 1969; pp 505.
2. Fischer, E.; Speier, A. *Ber.* **1895**, *28*, 3252.
3. Swann, S. J.; Oehler, R.; Buswell, R. J. *Organic Synthesis* **1943**, *Coll. Vol. 2*, 276.
4. Adams, R.; Chiles, H. M. *Organic Synthesis* **1941**, *Coll. Vol. 1*, 237.
5. Lapworth, A.; Baker, W. *Organic Synthesis* **1941**, *Coll. Vol. 1*, 451.
6. Fuson, R. C.; Cleveland, E. A. *Organic Synthesis* **1943**, *Coll. Vol. 2*, 264.
7. Corson, B. B.; Doges, R. A.; Harris, S. A.; Hazen, R. K. *Organic Synthesis* **1941**, *Coll. Vol. 1*, 241.
8. Inglis, J. K. H. *Organic Synthesis* **1941**, *Coll. Vol. 1*, 254.
9. Cassady, J.; Howie, G. A.; Robison, J. M.; Stamos, I. K. *Organic Synthesis* **1990**, *Coll. Vol. 7*, 319.
10. Newman, M. S. *J. Am. Chem. Soc.* **1941**, *63*, 2431.
11. Kadaba, P. K. *Synthesis* **1972**, 628.
12. Kadaba, P. K. *Synthesis* **1971**, 316.
13. Marshall, J. L.; Erickson, K. C.; Folsom, T. K. *Tetrahedron Lett.* **1970**, 4011.
14. Taber, D. F.; Hoermer, R. S. *J. Org. Chem.* **1992**, *57*, 441.
15. Kumar, A. K.; Chattopadhyay, T. *Tetrahedron Lett.* **1987**, *28*, 3713.
16. Blossey, E. C.; Turner, L. M.; Neckers, D. C. *Tetrahedron Lett.* **1973**, 1823.
17. Fu, X. *Ph.D. Thesis, UW-Milwaukee* **1992**.

Jin Li

6.2 Mukaiyama Esterification

6.2.1 *Description*

The formation of esters from carboxylic acids by using *N*-alkyl-2-halopyridinium salts, such as 2-chloro-1-methylpyridine iodide (**1**) and alcohols, was first reported by Mukaiyama and, thus, the reaction is defined as the Mukaiyama esterification. The active ester (**2**) forms esters in good yield under mild conditions when subjected to alcohols.

R_1-CO_2H —(**1**, TEA)→ [**2**] —(R_2OH, TEA)→ $R_1CO_2R_2$

Reagents such as **1** are coined Mukaiyama reagents and have demonstrated a great deal of synthetic versatility by converting carboxylic acids and alcohols into esters, amides, lactones, lactams, thiol esters, acid fluorides, olefins, allenes, carbodiimides, isocyanates, isothiocyanates, and nitriles.[1]

6.2.2 *Historical Perspective*

In 1975, Teruaki Mukaiyama published two papers. One described the synthesis of carboxylic esters,[2] and the other, carboxamides,[3] by using 1-methyl-2-halopyridinium iodide (**1**) as the coupling reagent.

6.2.3 *Mechanism*

1 —(Et_3N, H–O–C(O)–R_1 **3**; $-\overset{+}{N}H(Et)_3\overset{-}{I}$, $-\overset{+}{N}H(Et)_3Cl^-$)→ [**2**] —(Nu-H (**4**); $-\overset{+}{N}H(Et)_3\ \overset{-}{I}$)→ R_1C(O)Nu (**5**) + **6**

When **1** is subjected to the carboxylate anion of **3** (TEA, R_1CO_2H), 2-acyloxy-1-methylpyridinium iodide (**2**) is formed. The hydrogen halide (HCl) is scavenged with triethylamine. A nucleophile (NuH) (**4**) such as an amine or alcohol in the presence of TEA adds to the carbonyl center of **2**, yielding the desired product **5** and the by-product, 1-methyl-2-pyridone (**6**).

6.2.4 *Variations and Improvements*

Although Mukaiyama reagent is commonly referred to as **1**, 2-halo-benzoxazolium and benzothiazolium (**7** and **8**, respectively) species are also referred to as "Mukaiyama reagents."[1]

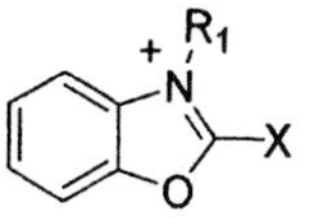

7 Y^- 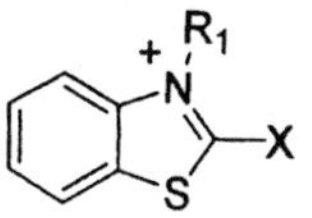 **8** Y^-

R_1 = Methyl, Ethyl, Phenyl

X = Cl, Br, F

Y = I, BF_4, TsO, FSO_3

The reagent 7 (where R^1 = Ethyl, X = Cl, Y = BF_4) has been shown to convert primary and secondary alcohols to their respective chlorides.[4] Note that this transformation proceeds with inversion of the stereocenter.

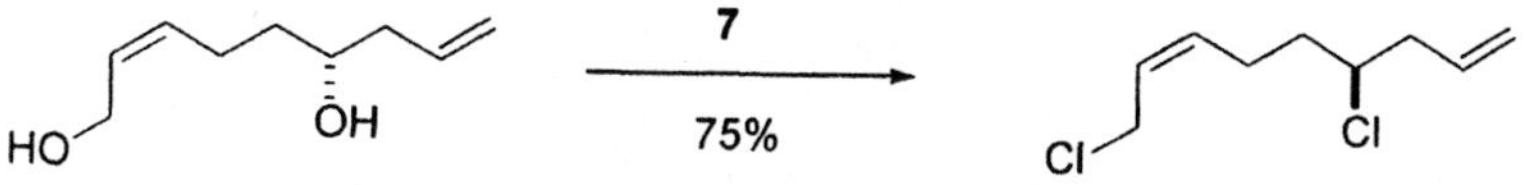

HO OH —**7**, 75%→ Cl Cl

The reagents **9**, **10**, **11**, and **12**, developed by Xu,[5] are very effective coupling reagents for the synthesis of peptides containing sterically hindered amino acids, such as α,α-dialkylated or *N*-methylated amino acids.

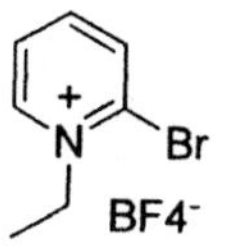

Br BF_4^- F BF_4^- Br $SbCl_6^-$ F $SbCl_6^-$

9 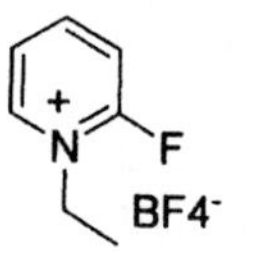 **10** 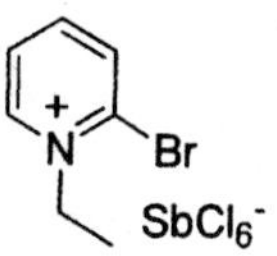 **11** 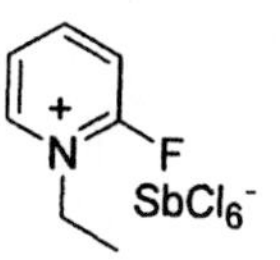 **12**

The main reactive intermediates of the pyridinium salt mediated couplings are the corresponding acyl halides and acyloxypyridinuim salts of the *N*-protected amino acid or peptide.

A polymer supported version of **1** (**13**) was synthesized by Taddei.[6]

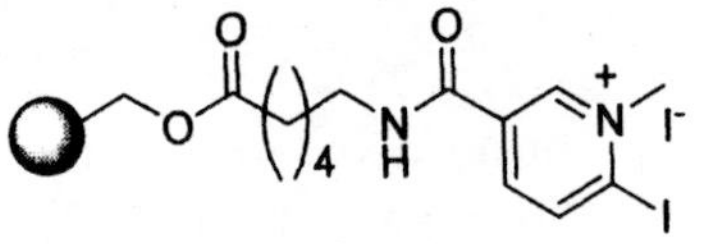

13

The butyl chain serves as a spacer to separate the reactive center of the molecule from the surface of the resin. This resin was used for the synthesis of β-lactams (**14** to **15**).

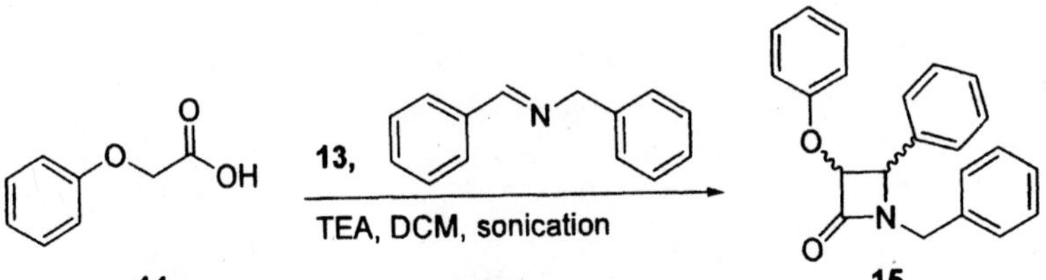

14 → (13, TEA, DCM, sonication; 60%) → **15**

Reagent **13** was also used under microwave conditions to create a library of esters.[7] Another polymer supported version of Mukaiyama's reagent (**16**) was disclosed by Swinnen.[8]

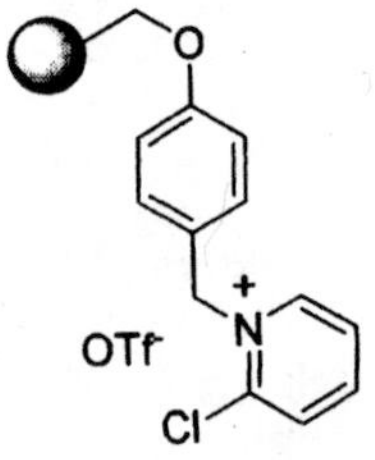

16

Wang resin served as the polymer support. Reagent **16** is a very effective coupling reagent for the synthesis of esters (**17**) and amides (**18**).

16, TEA
DCM, rt
92%
17

16, TEA
DCM, rt
80%
18

A fluorous tagged version of **1** has also been prepared and used as a "separation friendly" Mukaiyama condensation reagent (**19**) for amide formation.[9]

19

6.2.5 *Synthetic Utility*

As mentioned previously, **1** is very useful to form active esters of carboxylic acids which can then form esters or other functional groups under mild conditions. The reagent rarely causes epimerization of chiral centers and is a very effective coupling reagent. These valuable attributes have led many to use this reagent in the total synthesis of natural products. During the total synthesis of balanol, Nicolaou used **1** to couple key fragments together via an ester bond, to provide the fully functional group protected version of Balanol (**20**).[10] No racemization was observed and the coupling was accomplished in good yield.

1 (1.3 eq), DMAP (0.5 eq)

TEA (20 eq), DCM, 25 °C

5h

79%

20

The use of **1** with ω-oxocarboxylic acids **21**, in the presence of *O*-acetyl quinidine (10 mol%, **22**, NR_3) to produce chiral β-lactones (**23**), has been demonstrated by Romo.[11]

When **21** is exposed to **1**, with triethylamine, ketene **24** is produced *in situ*. This strategy is attractive since it does not employ a ketene generator or activated aldehydes to form the ketene intermediate. **22** adds to **24**, producing **25**, which engages the aldehyde in an intramolecular aldol reaction to yield **26**. **26** then spontaneously cyclizes to β-lactone **23** with concomitant regeneration of the chiral catalyst **22**.

Mukiayama's reagent (**1**) has also been used as an effective coupling reagent of acids with resin bound secondary amines (**27**).[12]

The newly formed resin bound amides are liberated from the solid phase by acidic conditions to provide the free amides (**28**) in good yield without further purification. It has also been shown that carboxylic acids containing an aldehydic moiety (i.e., 4-formyl benzoic acid) give the expected amides (no imine formation) in 90% yield and with 95% purity (e.g., where R = phenethyl).

Reagent **1** has been used to provide rapid access to a variety of 3-substituted coumarins (**29** and **30**) in good yield.[13]

22

1, TEA, ACN

CHO

CO_2H

21

25 °C

37-54%

(90-92% ee)

23

$-NR_3$ (22)

CHO

CHO

R_3N

NR_3

24

25

26

27

1, DIEA

R_1-CO_2H

TFA, DCM

28

(42–90%)

HO_2C

1, TEA, ACN

reflux

45%

29

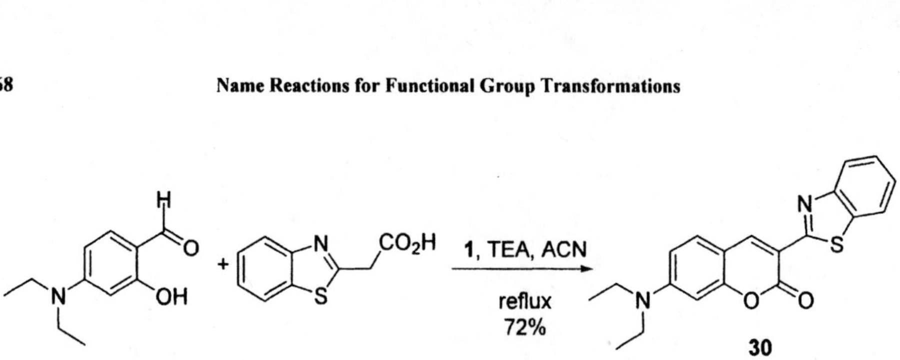

1, TEA, ACN

reflux

72%

30

It was first disclosed by Amin[14] that the synthesis of β-lactams could be accomplished by using **1** as a ketene generator. Jun had modified the procedure by utilizing chiral oxazoldin-2-ones (**31**), **1**, and *trans*-imines (**34**) to conduct stereoselective synthesis of *cis*-β-lactams (**35**).[15]

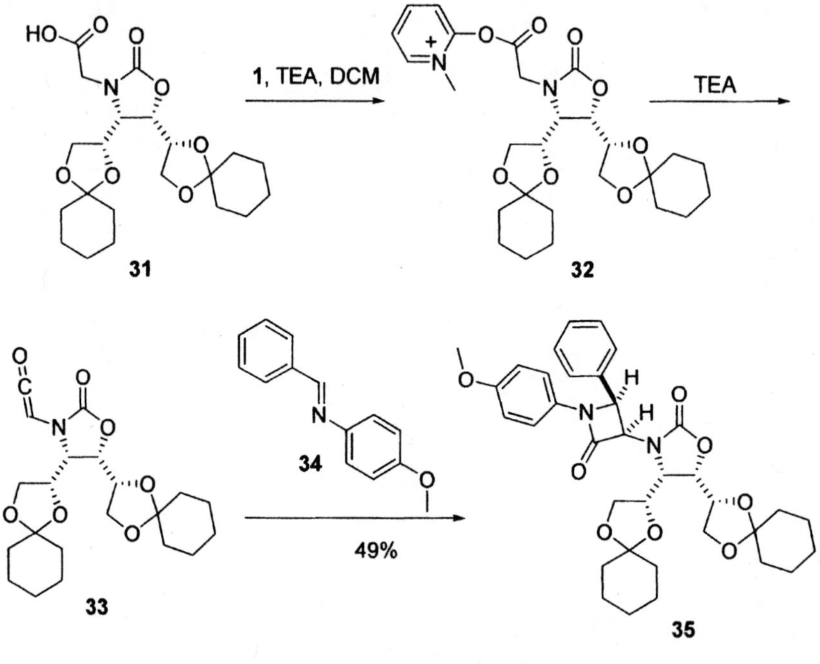

1, TEA, DCM

TEA

31

32

33

34

49%

35

Active ester formation of **31** with **1** (**32**) was followed by *in situ* dehydration to the ketene intermediate **33**. The non-concerted [2 + 2] cycloaddition of **33** with *trans*-imine **34** underwent a symmetry allowed *con* rotation that led to the *cis*-β-lactam **35** in modest yield.

The synthesis of a 400 member library of aryl substituted 2-iminohydantoin tripeptide compounds has been accomplished by subjecting resin bound thioureas (**36**) with 10 equivalents of **1** in the presence of excess triethylamine in DMF for 18 hours.[16]

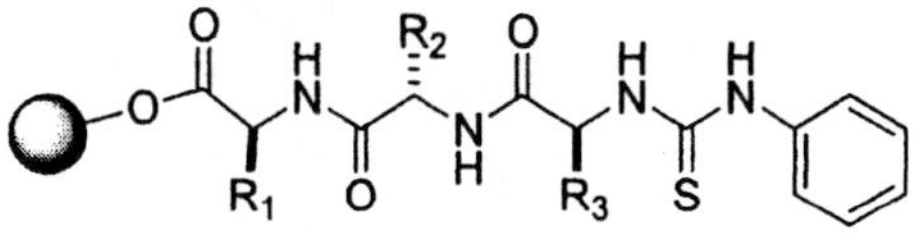

36

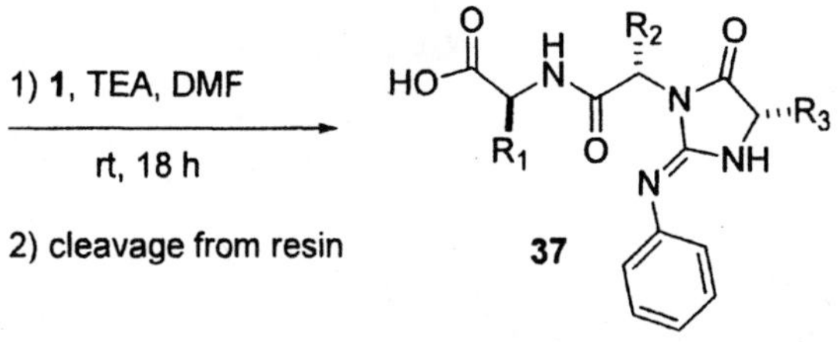

1) 1, TEA, DMF

rt, 18 h

2) cleavage from resin

37

Compound **1** serves as a dehydrothiolation reagent, which is followed by intramolecular cyclization to yield resin bound 2-iminohydantoins. Release of the products (**37**) from solid phase was accomplished with a solution of 20% hexafluoro-2-propanol in dichloromethane.

The use of **1** to form the key intermediate lactam **38** via an intramolecular cyclization has been employed en route to the total synthesis of the alkaolid (±)-pinnaic acid.[17]

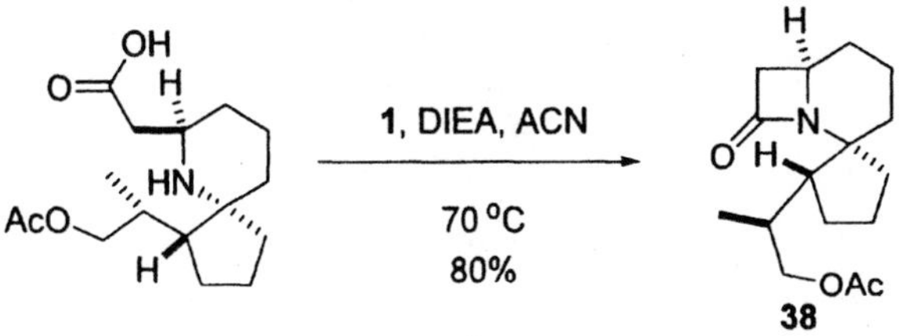

1, DIEA, ACN

70 °C

80%

38

6.2.6 *Experimental*[3]

A general procedure is as follows. To 1-methyl-2-chloropyridinium iodide (1.2 mmol, **1**), was added a mixture of amine (1.0 mmol), carboxylic acid (1.0 mmol), and triethylamine (2.4 mmol) in dichloromethane (10 mL). After being refluxed for 1 hour under argon, the mixture was cooled to room temperature and ether (20 mL) added. The mixture was then washed with 5% aqueous hydrochloric acid solution (v/v, 50 mL) three times followed by water (1 × 50 mL). The organic layer was concentrated under reduced pressure and the carboxamide product was isolated by silica-gel column chromatography.

6.2.7 *References*

1. Mukaiyama, T. *Angew. Chem. Int. Ed. Engl.* **1979**, *18*, 707.
2. Mukaiyama, T.; Usui, M.; Shimada, E.; Saigo, K. *Chem. Lett.* **1975**, 1045.
3. Mukaiyama, T.; Saigo, S.; Bald, E. *Chem.Lett.* **1975**, 1163.
4. Orru, R. V. A.; Groenendaal, B.; Heyst, J. V.; Hunting, M.; Wesseling, C.; Schmitz, R. F.; Mayer, S.; Faber, K. *Pure Appl. Chem.* **2003**, *75*, 259.
5. Li, P.; Xu, J-C. *Tetrahedron* **2000**, *56*, 8119.
6. Donati, D.; Morelli, C.; Porcheddu, A.; Taddei, M. *J. Org. Chem.* **2004**, *69*, 9316.
7. Donati, D.; Morelli, C.; Taddei, M. *Tetrahedron Lett.* **2005**, *46*, 2817.
8. Crosignani, S.; Gonzalez, J.; Swinnen, D. *Org. Lett.* **2004**, *6*, 4579.
9. Nagashima, T.; Lu, Y.; Petro, M. J.; Zhang, W. *Tetrahedron Lett.* **2005**, *46*, 6585.
10. Nicolaou, K. C.; Bunnage, M. E.; Koide, K. *J. Am. Chem. Soc.* **1994**, *116*, 8402.
11. (a) Romo, D.; Cortez, G. S.; Tennyson, R. L. *J. Am. Chem. Soc.* **2001**, *123*, 7145. (b) Schneider, C. *Angew. Chem. Int. Engl. Ed.* **2002**, *41*, 744.
12. Tao, B.; Boykin, D. W. *J. Chem. Research (S)* **2003**, 10.
13. Mashraqui, S. H.; Vashi, D.; Mistry, H. D. *Syn. Commun.* **2004**, *34*, 3129.
14. Amin, S. G.; Glazer, R. D.; Mahas, M. S. *Synthesis* **1979**, 210.
15. Shin, D. G.; Heo, H. J.; Jun, J-G. *Syn. Commun.* **2005**, *35*, 845.
16. Gavrilyuk, J. I.; Evindar, G.; Batey, R. A. *J. Comb. Chem.* **2006**, *8*, 237.
17. Christie, H. S.; Heathcock, C. H. *PNAS* **2004**, *101*, 12079.

Daniel D. Holsworth

6.3 Ritter Reaction

6.3.1 Description

The Ritter reaction occurs when a stable carbenium ion **1** reacts with a nitrile **2** in the presence of an acid. Hydrolysis of the adduct then leads to an amide **3**.[1,2]

N≡—R

1) $H^{\oplus}$

2) H_2O

1 **2** **3**

The carbenium ion may of course arise from many different starting materials, and though the proton source was originally a strong mineral acid, recent modifications[3,4] have tackled this potential limitation. The nitrile may be aromatic or aliphatic. The reaction usually requires moderate (40–50 °C) temperatures (which may occur naturally on reaction), and yields are normally high (80–100%).

6.3.2 Historical Perspective

In 1948, Ritter and his student Minieri reported that the reaction of alkenes **4** with mononitriles **2** in the presence of concentrated sulfuric acid, followed by aqueous workup gave amides **3**.[1] The report was illustrated with many examples, indicating a reaction of high reproducibility and wide applicability. In the paper immediately following this one, Ritter expanded on this report and detailed the mechanism further.[2]

N≡—R

1) $H^{\oplus}$

2) H_2O

4 **2** **3**

6.3.3 Mechanism

The first step in the Ritter reaction is the formation of a carbenium ion **1**. Thus any substrate capable of generating a stable carbenium ion is a suitable starting material. In the case of tertiary alcohols **5**, the mechanism begins with an acid mediated E1 elimination of the hydroxyl to give the requisite carbenium ion **1**. Nucleophilic attack by the nitrogen lone pair on the nitrile **2** leads to the formation of a nitrilium ion intermediate **7** and hydrolysis by simple aqueous workup then produces the amide **3**.

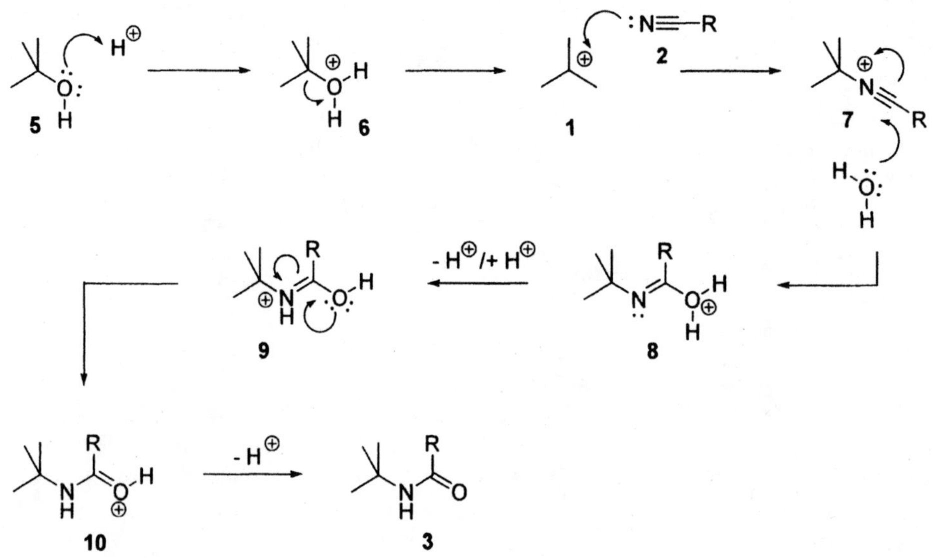

6.3.4 *Variations*
The problems with the Ritter reaction are associated either with the starting material (unstable carbenium ions, hygroscopicity) or with the reaction conditions (strong acid required, competing side reactions). Attention has focused on improving the reaction with respect to these potential limitations.

6.3.4.1 *Weaker acids to promote rearrangements*
A major problem of the Ritter reaction is the requirement of such a strong acid. Recent modifications have tackled this problem in various ways. Vogel[3] reported a modification of the Ritter reaction involving the use of CF_3SO_3H to promote oxa-ring opening and concomitant reaction with acetonitrile to produce the amino-conduritol derivative **16**.

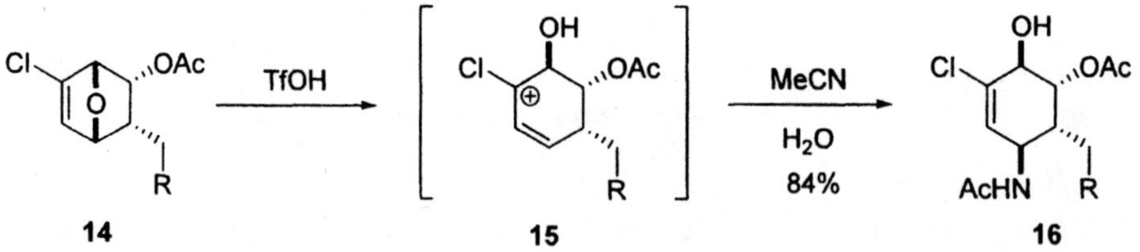

Tillequin[4] implemented an electrogenerated acids (EGA)[6] method to produce a similar product to that seen in the example above. In this method, electrolysis generated a base at the cathode and the area surrounding the anode becomes acidic. The conditions overall are neutral, and yet an acid-mediated process is rendered possible.

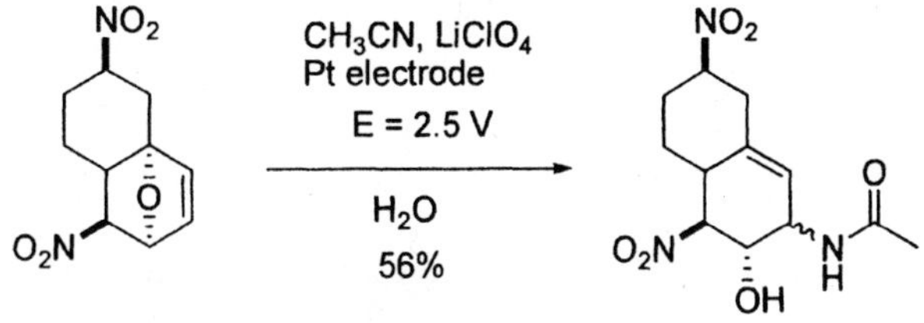

CH_3CN, $LiClO_4$
Pt electrode
E = 2.5 V
H_2O
56%

17 18

6.3.4.2 *Stabilisation of the carbenium ion*

The ease of formation of carbenium ions from substrates such as tertiary alcohols means these are ideal starting materials for the Ritter reaction. However, formation of such an ion at a less substituted centre is much less facile, if not impossible. There has been a large amount of work that focused on the stabilisation or promotion of carbenium ion formation at such centres.

6.3.4.3 *Transition metal stabilisation*

The carbenium ion intermediate can be stabilised with transition metals. However the resulting complex must be unstable enough to allow further reaction. This approach has important implications for the Ritter reaction, allowing the conversion of primary and secondary alcohols to stabilised carbenium species, a transformation that would otherwise be impossible. Several examples of this type of modification of the Ritter reaction exist, and notably in these complexed systems the reaction takes place with complete stereocontrol.

In 1981, Jaouen reported the applicability of this approach to the Ritter reaction of primary and secondary alcohols using $Cr(CO)_3$ catalysis.[7]

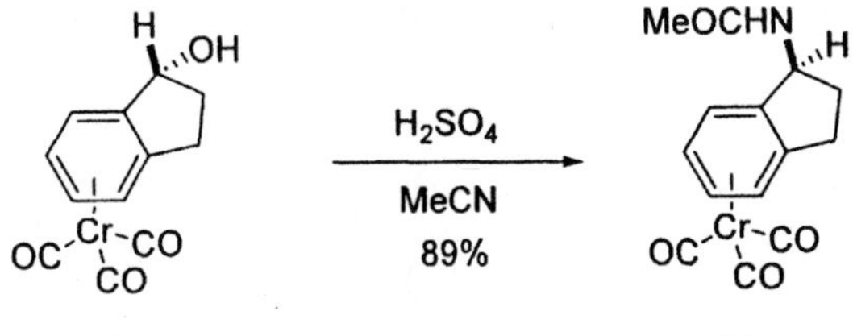

H_2SO_4
MeCN
89%

19 20

Iqbal[8] has similarly reported the efficient Co(III)DMG complex-catalysed conversion of allylic alcohols **21** in the presence of nitriles to the corresponding allylamides **22** and **23**. Usefully, this reaction is carried out in the absence of acid.

OH, R^1, R^2 **21** — Co(III), MeCN → NHAc, R^1, R^2 **22** + NHAc, R^1, R^2 **23**

6.3.4.4 *Heteroatom stabilisation*

Various heteroatoms have been used to facilitate carbenium ion formation at less substituted centres, including halogens,[9–11] nitrogen,[12] selenium,[13] and sulfur.[14] These reactions generally proceed *via* activation of the alkene, through formation of a cyclic cationic intermediate **25**.[15]

R^1 **24** + $E^{\oplus}$ → [R^1, $E^{\oplus}$ **25**] — RCN, H_2O → R^2, HN, O, R^1, E **26** → E, R^1, H, N, R^2, O **27**

In most cases these reactions proceed without stereocontrol, but Toshimitsu has shown that the reaction of β-hydroxy sulfide **28** with aqueous trifluoromethanesulfonic acid gave amide **30**.[16]

OH, C_5F_5, SPh **28** — CF_3SO_3H, MeCN, H_2O, RT → [H, Ph, $S^{\oplus}$, C_5F_5 **29**] — 95%, 97% ee → O, NH, C_5F_5, SPh **30**

6.3.4.5 *Cascade reactions implementing the Ritter reaction*

Recent examples of the Ritter reaction have combined this useful synthetic tool with other reactions and rearrangements to produce complex reaction sequences. For example, Bishop has reported the combination of the Ritter reaction with the Wagner–Meerwein rearrangement, which can result in complex and sometimes unpredictable reaction sequences.[17] The use of the Ritter reaction to trap cationic intermediates in the CAN-mediated dimerization of alkoxystyrenes has recently allowed the synthesis of α-aminotetralin derivatives by Nair.[18]

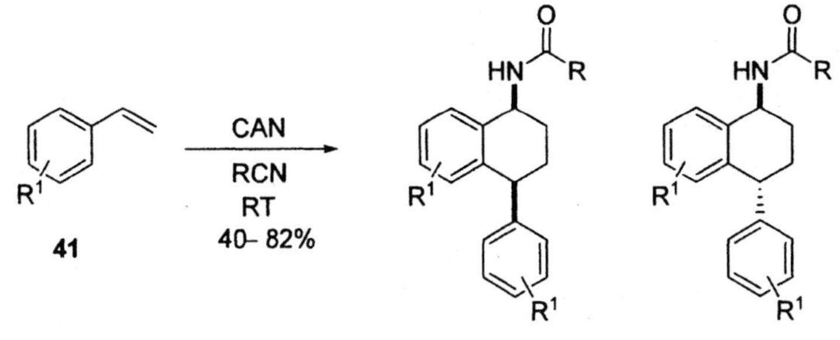

6.3.5 *Synthetic Utility*

Carbenium ions can be produced from numerous different functionalities *via* many different techniques, and consequently the Ritter reaction represents an extremely versatile technique for the formation of amides. For example, we have above illustrated the formation of a stable carbenium ion from alcohols and alkenes, and Reddy has reported the use of an ester as a source of the necessary carbenium ion.[5] In this case, *t*-butyl acetate (**12**) was converted to the corresponding cation with catalytic concentrated sulphuric acid and was reacted with a wide variety of aromatic nitriles **11** with commendable success.

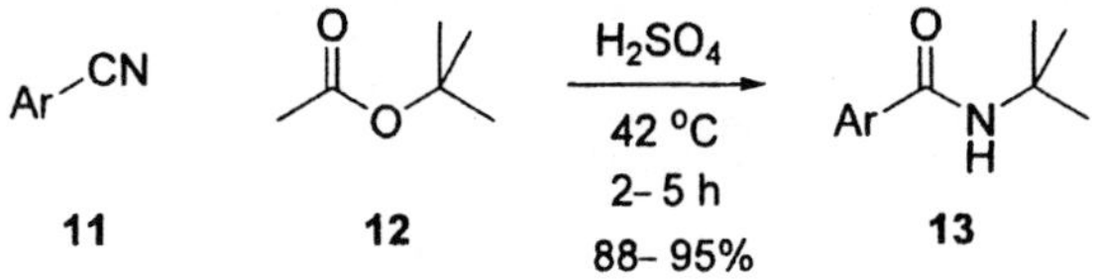

It is also worthwhile to note that the hydrolysis of amides produced in the Ritter reaction leads to amines, and thus the Ritter reaction provides a synthetically useful way to convert tertiary alcohols to tertiary amines.

6.3.6 *Experimental*

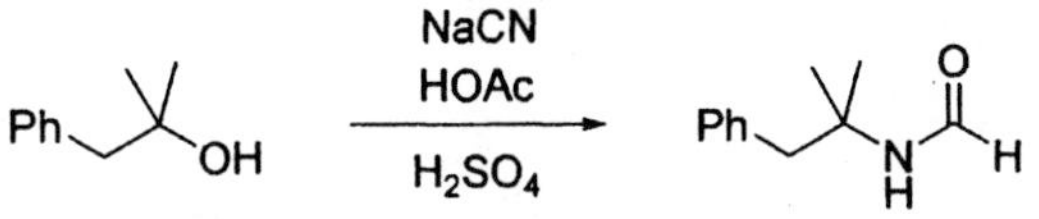

N-Formyl-α,α-dimethyl-β-phenethylamine[2]

To glacial acetic acid (500 mL) is added sodium cyanide (110 g, 2.0 mol) in small portions and the temperature of the mixture is maintained at room temperature. A cooled solution of concentrated sulfuric acid (272 mL, 4.9 mol) in glacial acetic acid (250 mL) is added slowly, again maintaining the temperature at room temperature. α,α-Dimethyl-β-phenethyl alcohol (300 g, 2 mol) is added over 20 minutes, during which time the temperature of the mixture rises slowly to 35–45°. The reaction mixture is stirred for an additional 90 minutes and allowed to stand overnight. The amber-coloured mixture containing some solid sodium sulfate is aerated with nitrogen for 2 hours, poured into 3 L of ice water, and the supernatant oil separated. The aqueous phase is neutralized with sodium carbonate and extracted with dietheyl ether. The ethereal extract is combined with the original oily supernatant, neutralized with sodium carbonate, and dried over anhydrous sodium sulfate. The solvent is removed under reduced pressure, and the residue is distilled to yield 230–248 g (65–70%) of product, bp 137–141° (2 mmHg). Redistillation of the ether-containing fore-run yields up to an additional 14 g of material.

6.3.7 References

1. Ritter, J. J.; Minieri, P. P. *J. Am. Chem. Soc.* **1948**, *70*, 4045.
2. Ritter, J. J.; Kalish, J. *J. Am. Chem. Soc.* **1948**, *70*, 4048.
3. Pasquarello, C.; Picasso, S.; Demange, R.; Malissard, M.; Berger, E. G.; Vogel, P. *J. Org. Chem.* **2000**, *65*, 4256.5.
4. Le Goanvic, D; Lallemond, M.-C.; Tillequin, F.; Martens, T. *Tetrahedron Lett.* **2001**, *42*, 5175.
5. Reddy, K. L. *Tetrahedron Lett.* **2003**, *44*, 1453.
6. Torii, S. *Electroorganic Syntheses, Methods and Applications, Part I: Oxidations*; Kodansha: NCH, **1985**; pp. 315-330.
7. Jaouen, G.; Top, S. *J. Org. Chem.* **1981**, *46*, 78.
8. (a) Mukhopadhyay, M.; Reddy, M. M.; Maikap, G. C.;Iqbal, J. *J. Org. Chem.* **1995**, *60*, 2670. (b) Mukhopadhyay, M.; Iqbal, J. *J. Org. Chem.* **1997**, *62*, 1843.
9. Stavber, S.; Pecan, T.S.; Papez, M.; Zupan, M. *Chem. Commun.* **1996**, 2247.
10. Bellucci, G.; Bianchini, R.; Chiappe, C. *J. Org. Chem.* **1991**, *56*, 3067.
11. Barluenga, J.; Gonzalez, J. M.; Campos, P. J.; Asensio, G. *Tetrahedrom Lett.* **1986**, *27*, 1715.
12. Bloom, A.J.; Fleischmann, M.; Mellor, J. M. *J. Chem. Soc., Perkin Trans. I* **1984**, 2357.
13. (a) Toshimitsu, A.; Aoai, T.; Owada, H.; Uemura, S.; Okano, M. *J. Org. Chem.* **1981**, *46*, 4727. (b) Yoshida, M.; Satoh, N.; Kamigata, N. *Chem. Lett.* **1989**, 1433. (c) Tiecco, M.; Testaferri, L.; Tingoli, M; Bartoli, D. *Tetradedron* **1989**, *45*, 6819. (d) Toshimitsu, A.; Nakano, K.; Mukai, T.; Tamao, K. *J. Am. Chem. Soc.* **1996**, *118*, 2756. (e) Toshimitshu, A.; Ito, M.; Uemura, S. *J. Chem. Soc., Chem. Commun.* **1989**, 530. (f) Toshimitsu, A.; Uemura, S.; Okano, M. *J. Chem. Soc., Chem. Commun.* **1982**, 87. (g) Toshimitsu, A.; Hayashi, G.; Terao, K.; Uemura S. *J. Chem. Soc., Perkin Trans. I* **1986**, 343. (h) Bewick, A.; Coe, D. E.; Fuller, G. B.; Mellor, J. M. *Tetrahedrom Lett.* **1980**, *21*, 3827. (i) Morella, A. M.; Ward, A. D. *Aust. J. Chem.* **1995**, *48*, 445.
14. (a) Bewick, A.; Coe, D. E.; Mellor, J. M.; Owton. W. M. *J. Chem. Soc., Perkin Trans. I* **1985**, 1033. (b) Benati, L.; Montevecchi, P. C. *J. Chem. Soc., Perkin Trans. I* **1987**, 2815.
15. Eastgate, M. D.; Fox, D. J.; Morley, T. J.; Warren, S. *Synthesis* **2002**, *17*, 2124.
16. (a) Toshimitsu, A.; Hirosawa, C.; Tanimoto, S. *Tetrahedron Lett.* **1991**, *32*, 4317. (b) Toshimitsu, A.; Hirosawa, C.; Tamao, K. *Tetrahedron* **1994**, *50*, 8997.
17. Djaidi, D.; Leung, I. S. H.; Bishop, R; Craig, D. C.; Scudder, M. L. *J. Chem. Soc., Perkin 1* **2000**, 2037.
18. Nair, V.; Rajan, F.; Rath, N. P. *Org. Lett.* **2002**, *4*, 1575.

Alice R. E. Brewer

6.4 Strecker Amino Acid Synthesis

6.4.1 *Description*

The Strecker reaction[1] describes the synthesis of α-amino acids from aldehydes (and ketones) with ammonia and cyanide equivalents by hydrolysis of the resultant α-aminonitrile intermediates.

$$\mathrm{RCHO} + \mathrm{NH_3} + \mathrm{HCN} \longrightarrow \mathrm{RCH(NH_2)CN} \xrightarrow{\text{hydrolysis}} \mathrm{RCH(NH_2)CO_2H}$$

This three-component coupling reaction, with its greater than 150 year history, continues to enjoy great utility. It is the most efficient synthesis of non-natural amino acids due to its generality, simplicity, and low cost. These attributes are particularly attractive to process chemists where the cost of goods is a key driver for the scale up and development of chemical matter. Initial studies around this reaction gave rise to racemic versions of the product. However, more recent examples of the Strecker reaction have focused on asymmetric variations to provide optically active α-aminonitriles or α-amino acids.

6.4.2 *Historical Perspective*

Adolph Strecker (1822–1871) first reported[2] his work on the reaction of acetaldehyde and ammonia with hydrogen cyanide in 1850. The intermediate was hydrolyzed to afford alanine, the first synthetically prepared α-amino acid. Its name was derived from the nature of its starting material, aldehyde. While Friedrich Wöhler's (1880–1882) synthesis of urea in 1828 from a solution of potassium cyanate and ammonium chloride led scientists to begin questioning the Vitalism theory,[3] the work of Strecker and others certainly contributed to the death knell of the privileged status that "organic" compounds held at the time. Later, Emil Erlenmeyer (1825–1909) in 1875 showed that the intermediate was probably the α-aminonitrile, which resulted in this reaction becoming a general synthesis of amino acids. In 1880 Johann Tiemann (1848–1899), of the Reimer–Tiemann reaction, altered the synthetic protocol by reversing the order of addition and showed that the cyanohydrin of the carbonyl compound could react with ammonia to produce the same intermediate. Improved yields were obtained with this modification compared to Strecker's original procedure. Zelinskii–Stadnikov further modified the reaction conditions by the use of potassium or sodium cyanide with ammonium salts as replacements for the volatile hydrogen cyanide and ammonia. For aldehydes that exhibited reduced reactivity, the Knoevenagel–Bucherer modification makes use of bisulfite adducts to improve reactivity and provide access to the desired product.

6.4.3 Mechanism

The mechanism of the Strecker reaction has received considerable attention over its life-span.[4] The conversion of a carbonyl compound into an α-amino acid, by this method, requires a two-step process. The first step consists of the three-component condensation of cyanide and ammonia with the carbonyl compound **1** to produce an intermediate, α-aminonitrile **3**. The second step involves the hydrolysis of the nitrile functional group to reveal the latent carboxylic acid **4**. Whereas the second step is fairly straightforward and can be done under basic or acid conditions, the first step is more involved than one may expect. The widely accepted sequence for the first step is the nucleophilic addition of ammonia to the carbonyl carbon to produce the corresponding imine derivative **2**. Once formed, this initial species is captured by the cyanide anion to generate the requisite α-aminonitrile **3**.

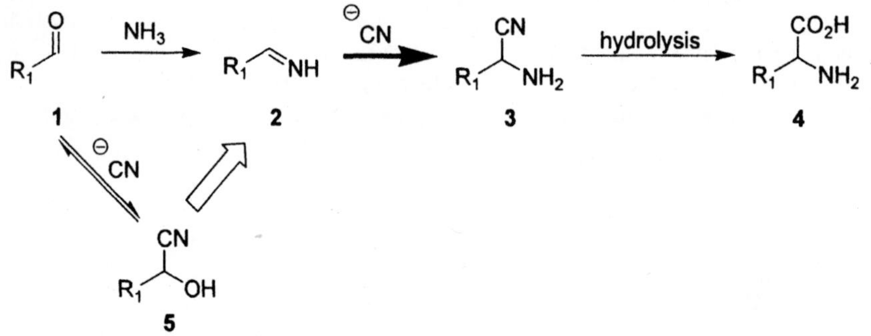

Although this is the generally accepted mechanism for the Strecker reaction, and the reaction sequence eventually must pass through these species (**2** → **3**), the order of addition of the reagents can change the initial course of events. If cyanide is added first or the ammonia and cyanide are added together, a competition for the electrophilic carbonyl carbon is set up. The favored reaction product (kinetic), in the short-term, from this competition is cyanohydrin **5**. However, the relative lack of stability of this functional group results in **5** collapsing back to starting material **1**, which over time ultimately leads to **2**. Once formed, this intermediate is consumed by cyanide and drives the equilibria to α-aminonitrile **3**. One may compare the Strecker reaction to the Mannich reaction. In the Mannich reaction, ammonia is condensed with an aldehyde (such as formaldehyde) and an active methylene compound (such as an enolate), whereas in the Strecker reaction, the active methylene compound is replaced with cyanide anion.

6.4.4 Variations and Improvements or Modifications

The classical Strecker reaction[1–2] was typically conducted using alkaline cyanide in aqueous solution with ammonium salts. While these conditions have proved to be very useful for the preparation of α-amino acids, the reagents and the nature of the solvent places practical limitations on the scope of this reaction.[5] For example, as the complexity of the carbonyl component increases, its water solubility decreases, thus limiting the generality of the Strecker reaction and the diversity of the products produced.

One solution to this limitation was the introduction of cyanide sources that are soluble in organic solvents. The initial illustration used diethyl phosphorocyanidate (DEPC)[6] for the conversion of **6** to α-aminonitrile **7**. This stood in contrast to the classical Strecker reaction conditions which afforded no desired product.

C_8H_{17} H H H O **6** $(EtO)_2P(O)CN$ NH THF 62% C_8H_{17} N NC H H H **7**

Trimethylsilyl cyanide (TMSCN), in conjunction with a Lewis acid catalyst, was found to be of greater practical value compared to DEPC in performing these reactions.[7] Thus, using non-aqueous reaction conditions, carbonyl compounds **8** could be transformed into their corresponding α-aminonitrile derivatives **9**. In these examples, greater reactivity was observed for the carbonyl compound if it contained electron-donating groups rather than electron-withdrawing groups. Additionally, aldehydes gave greater yields than ketones (77–99% compared to 5–50%, respectively).

R_1 =O R_2 **8** TMSCN R_3R_4NH cat. ZnI_2 R_1 CN R_2 N-R_3 R_4 **9**

The introduction of TMSCN has greatly expanded the interest in the Strecker reaction as witnessed by the number of modifications of this reagent's use in forming α-aminonitriles. Various groups have investigated alternative Lewis acid catalysts for this process with the goal of improving various attributes of the reaction. Lewis acids, including $InCl_3$, $BiCl_3$, $Pr(OTf)_3$, and $RuCl_3$, have been reported to have efficacy in catalyzing the reaction of carbonyl **1** and amine **10** to the desired product **11**.[8] The Lewis acids were found to be powerful tools for catalyzing the nucleophilic addition to an imine. The mild reaction conditions facilitated the reaction of alkyl and aryl aldehydes with amines at room temperature in standard organic solvents with good to excellent yields over several hours.

R_1CHO (1) + R_2-NH_2 (10) —TMSCN, Lewis acid→ $R_1CH(CN)NHR_2$ (11)

A recent report[9] described the use of scandium triflate in conjunction with tributyltin cyanide, a more water stable cyanide source compared to TMSCN. These reaction conditions could be carried out in both organic and aqueous solutions. It was observed that rare earth triflates are stable Lewis acids in water. Thus, a variety of aldehydes **1** and amine **12** were converted to α-aminonitriles **13** in excellent yield. The spent reagent could be completely recovered and, along with the scandium reagent, recycled for subsequent use.

R_1CHO (1) + Ph_2CH-NH_2 (12) —$Sc(OTf)_3$, Bu_3SnCN, rt→ $R_1CH(CN)NHCHPh_2$ (13)

79–94% in H_2O
83–94% in MeCN/Toluene

Montmorillonite KSF clay[10] has also been found to catalyze the nitrile addition to aryl imines generated from aldehydes **1**. The mild reaction conditions and excellent yields contribute to the utility of this variation. The clay can be recovered and recycled, which allows this reaction to be classified as green chemistry.

R_1CHO (1) + R_2-NH_2 (10) —TMSCN, KSF–CH_2Cl_2, 85–94%→ $R_1CH(CN)NHR_2$ (11)

Even iodine has been reported[11] to act as a mild Lewis acid for these transformations. It was able to catalyze the formation of α-aminonitriles **11** with greater catalytic activity than the other Lewis acids using reduced catalyst loading to afford the desired product with shorter reaction times and in good to excellent yields.

R_1CHO (1) + R_2-NH_2 (10) —TMSCN, I_2, 72–94%, 1–8 h→ $R_1CH(CN)NHR_2$ (11)

(Bromodimethyl)sulfonium bromide[12] was found to also provide better conversions of **1** and **10** to afford **11**, with increased reaction rates and lower catalyst loading compared to the other Lewis acids.

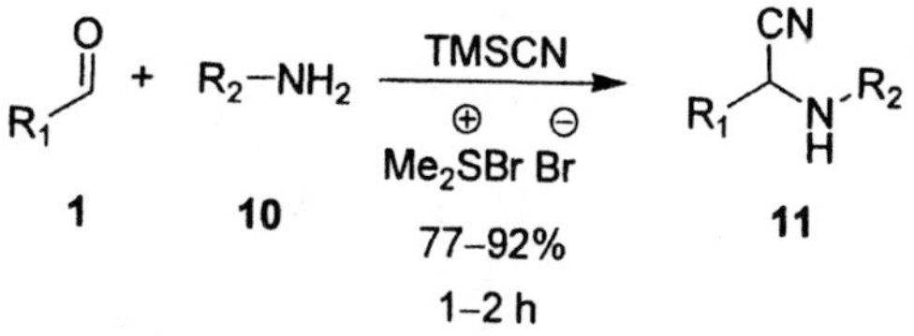

The application of ultrasound[13] to the Strecker reaction was shown to improve the product outcome of this reaction. Treatment of aminoketone **14** using the classical Strecker reaction conditions afforded only the cyanohydrin. The desired product **15** could be produced after 12–13 days of stirring in acetic acid. Ultrasound conditions not only improved the yield of this reaction, but accelerated the reaction times by about 12-fold.

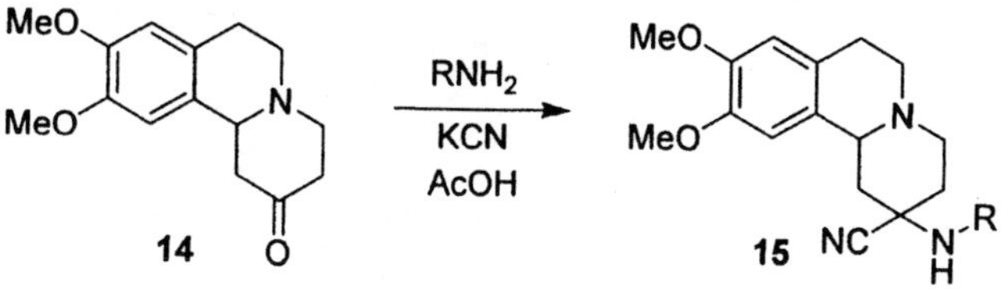

Yield (%)

R	Δ	((((((
R = H	62	100
Bu	60	88
Ph	73	99
Bn	79	99

The use of cyclodextrins[14] has provided the ability to conduct the Strecker reaction with TMSCN in water *via* supramolecular catalysis involving reversible guest–host interactions. Activation of imine **16** by complexation with the hydroxyl groups present in cyclodextrins was found to work best with β-cyclodextrin. This chemically "green" reaction could be applied to ketones as well as aldehydes.

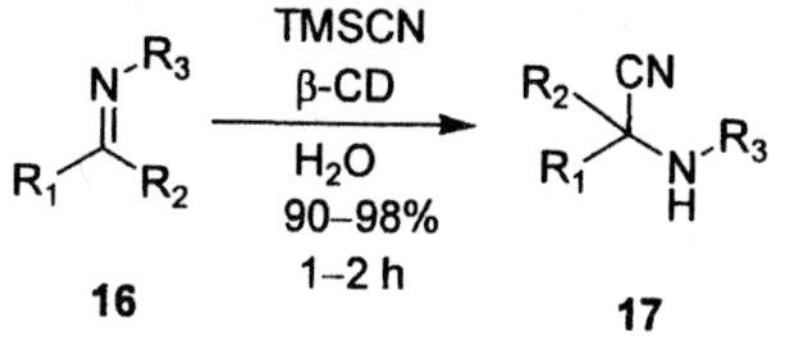

One may consider the Strecker reaction as the prototypical multicomponent reaction.[15] Three reactants, an amine, a carbonyl compound, and a source of cyanide, come together in a single reaction vessel to afford a single product, an α-aminonitrile. Variations on this reaction process include the Bucherer–Bergs, Petasis, Ugi, and amidocarbonylation reactions.

For the Bucherer–Bergs reaction,[16] similar reaction partners to the Strecker reaction are involved. The one major point of divergence is the addition of a "carbon dioxide" source. Thus the final product from this reaction is hydantoin **19**, which could be hydrolyzed to give a Strecker-like product, if so desired. Mechanistically, a reaction sequence analogous to the Strecker reaction can be invoked to generate α-aminonitrile **18** from carbonyl compound **8**. In the presence of the carbon dioxide source, the α-aminonitrile product **18** is trapped and results in the formation of cyclized product **19**.

R_1 R_2 C=O **8** —KCN, $(NH_4)_2CO_3$→ [R_1 R_2 C(CN)(NH_2) **18**] → **19**

The utility of this reaction was exemplified in the preparation of aldose reductase inhibitors for the treatment of diabetes.[17] Exposure of 8-azachromanone **20** to the Bucherer–Bergs reaction conditions afforded the spirohydantoin **21**.

20 —$NaHSO_3$, KCN, $(NH_4)_2CO_3$→ **21**

In its most general form, the boronic acid Mannich or Petasis reaction[18] involves the reaction of boronic acid **22**, a carbonyl compound **8**, and an amine **23** to produce secondary amines **24**. If one uses α-keto acid **25** for the carbonyl component then the corresponding product from the Petasis reaction was α-amino acid **26**. The key mechanistic step was proposed to occur intramolecularly with alkyl migration from intermediate boronate ester **28** formed from aminol **27**.

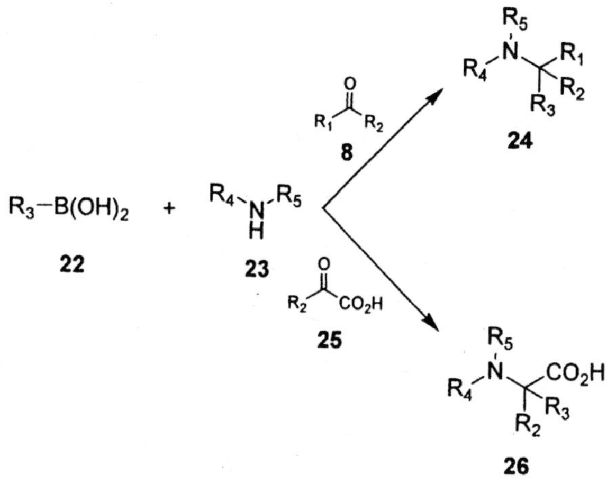

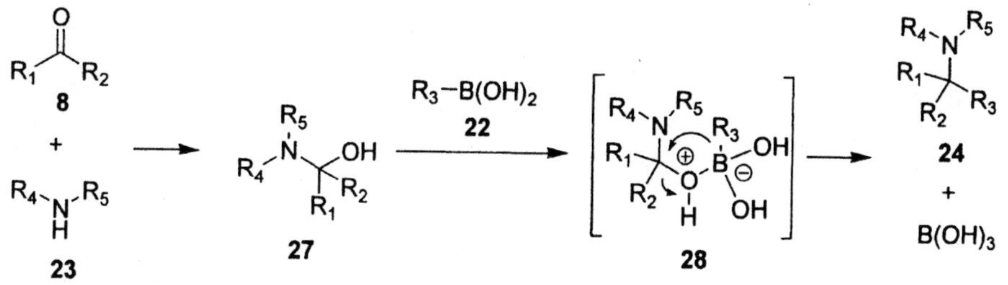

The Ugi reaction[19] is a four-component condensation requiring a carbonyl compound **8**, an amine **29**, a carboxylic acid derivative **30**, and an isocyanide **31**. The product from this mixture is α-aminoacyl amide **32** that again could be hydrolyzed to an α-amino acid. Pharmaceutically valuable thiazoles **33** have been elaborated from intermediates formed by this reaction sequence.[20]

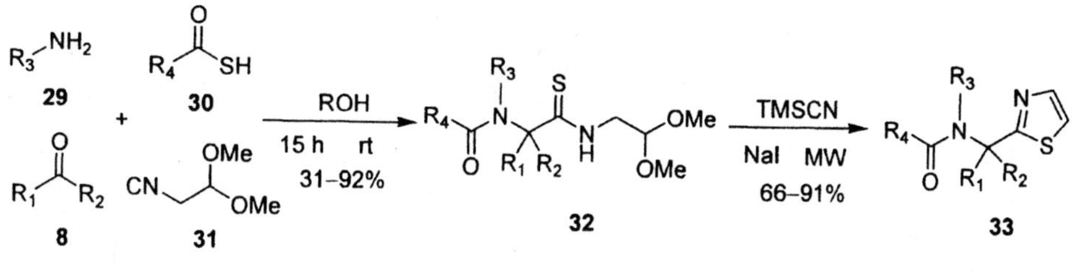

An alternate route to α-amino acids involved the chemistry of amidocarbonylation.[21] Originally communicated by Wakamatsu,[22] Pino[23] confirmed the general utility of this process. The reaction employs the use of a cobalt catalyst to facilitate the condensation of aldehyde **1**, an amine **34**, and carbon monoxide to afford *N*-acyl-α-amino acids **35** in a single step.

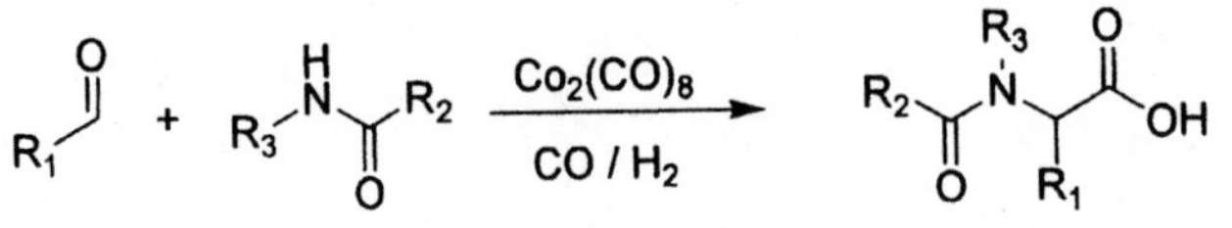

$Co_2(CO)_8$, CO / H_2

1 **34** **35**

The observation that palladium could also catalyze this process enhanced the scope of amidocarbonylation reactions.[24] This method provided ready access to the arylglycine building block **39**, from benzaldehyde **37** and amide **38**, required for the synthesis of the chloropeptins **36** that have been shown to possess anti-HIV activity.

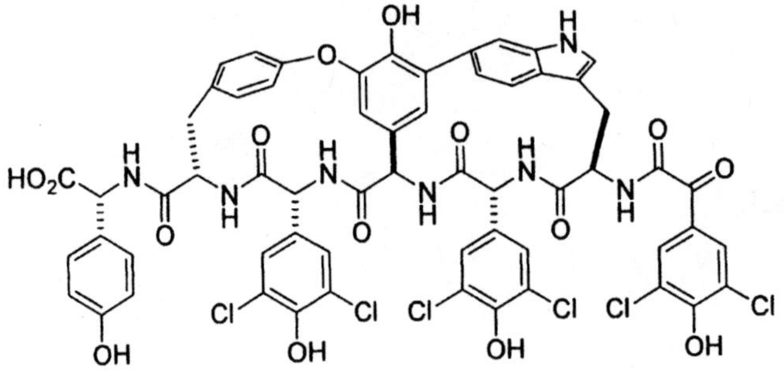

36

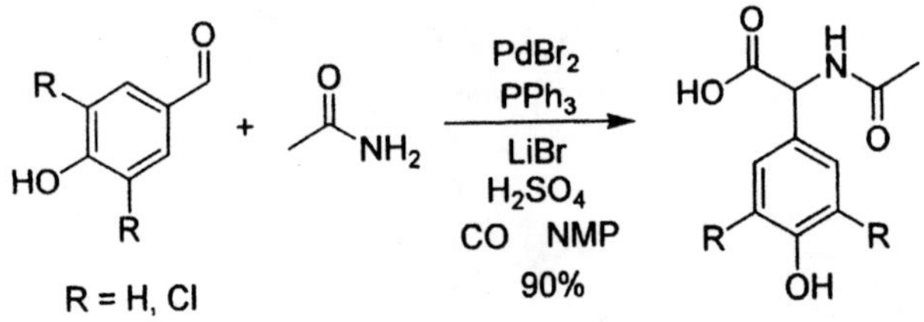

$PdBr_2$, PPh_3, LiBr, H_2SO_4, CO NMP, 90%

R = H, Cl

37 **38** **39**

The Strecker reaction has seen a renaissance in its utility in recent years with the development of methods for preparing α-amino acids **17** asymmetrically. Specifically, it is the asymmetric addition of cyanide to imine **16** which has been the focus of this attention.

16 → 17 ($^{\ominus}$CN)

One could envision the generation of the asymmetric center of α-aminonitrile **17** in one of two possible modes, not considering approaches using resolutions techniques. The first mode is more traditional and employs the use of a chiral auxiliary as part of the imine. The second takes advantage of the great strides made in asymmetric catalysis, whether mediated by transition metals or by the more recent organocatalysts.

The first example of a substrate-controlled asymmetric Strecker reaction was reported by Harada.[25] The reaction was conducted using aldehyde **1** and *D*-(–)-α-methylbenzylamine **40**. Hydrolysis of the α-aminonitrile and hydrogenolysis unmasked the amine to afford α-amino acid **4**.

1 + **40** → **4**: 1. NaCN; 2. HCl; 3. $Pd(OH)_2$, H_2; 9–58%, 22–58% *ee*

Later, Ojima[26] repeated this work but used the improved reaction conditions (*vide supra*) of TMSCN and Lewis acid catalysis. The best results were obtained with $ZnCl_2$ rather than $AlCl_3$ and the enantiomeric excesses were in the range 55–70%.

41 + TMSCN → **4**: 1. $ZnCl_2$; 2. $H^{\oplus}$; 3. H_2 / Pd; 97–100%

Carbohydrates[27] have also been utilized as chiral auxiliaries in these reactions. Tetra-*O*-pivaloyl-β-*D*-galactosylamine **42** was determined to provide optimal levels of chiral induction for the addition of cyanide. Thus, **43** was constructed from the intermediate imine formed from carbonyl component **1**.

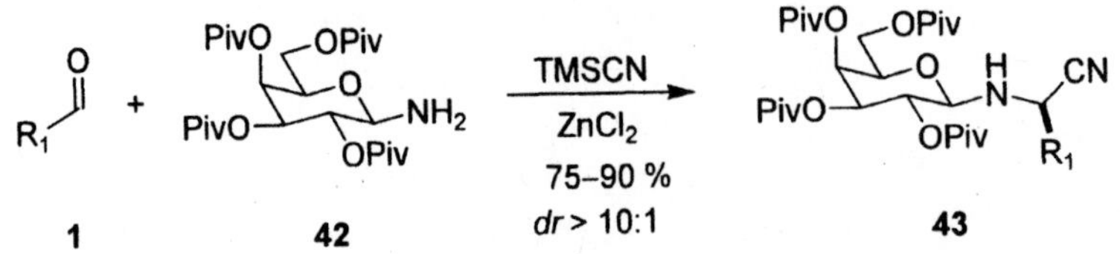

1 **42** **43**

Enantiopure sulfinimines (thiooxime-*S*-oxides) **44** have been reported to facilitate the asymmetric Strecker reaction.[28] Davis found that typical cyanide sources, such as potassium cyanide and TMSCN, did not possess sufficient reactivity for addition to the sulfinimines. Product **46**, however, could be obtained using the more Lewis acidic Et_2AlCN. Not only did the coordination of the aluminum to the oxygen of the sulfinimine activate the imine toward nucleophilic addition, this complexation also facilitated the delivery of the nitrile (see **45**). These results triggered numerous modifications and variations that have enhanced this approach to chiral α-amino acids.

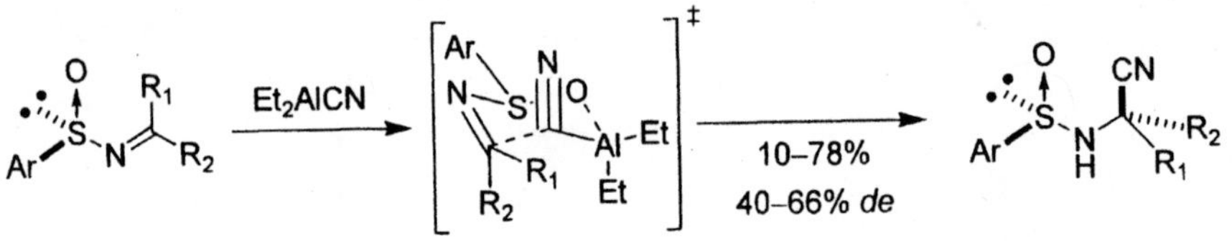

44 **45** **46**

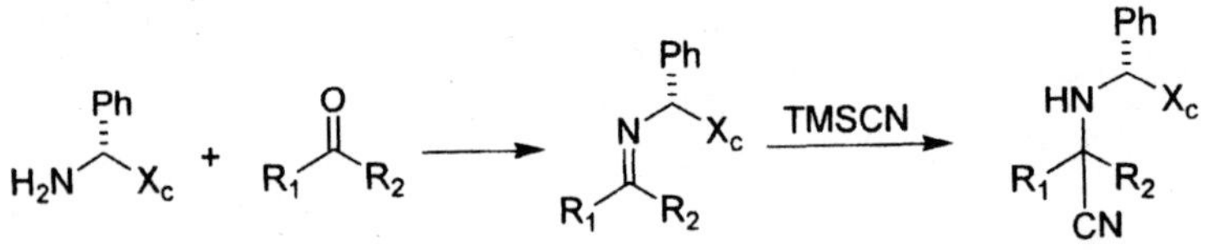

47 **8** **48** **49**

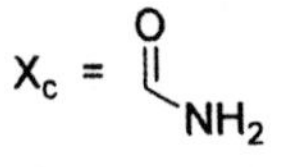

-CH2OH

glycinol 85–91% dr 2:1–25:1

glycinamide 51–93% dr up to 99:1

Amino acid derivatives have also shown efficacy as chiral auxiliaries for substrate-controlled Strecker reactions. Since they can be readily accessed in both enantiomeric forms (*R* and *S*), these types of chiral auxiliaries can serve as a straightforward method of generating optically enriched α-amino acids. To this end, phenylglycinol and phenylglycine amide **47** have been employed in this fashion.[29]

Reaction with carbonyl compound **8** produced optically active imine **48** that could add cyanide to afford the chiral α-aminonitrile **49**.

For the phenylglycinol auxiliary, cyclization of the alcohol moiety onto the intermediate imine limits the utility of this auxiliary. This side reaction could be eliminated by use of phenylglycinamide, which also provided products of higher optical purity.

An interesting variation on the use of chiral amine auxiliaries was reported by Taillades.[30] This group took advantage of the chiral pool and transformed optically active monoterpene **50** into amino-based chiral auxiliary **51**. Exposure of **51** to **8** ultimately gave rise to **52**.

50 ⇒ 51 + 8 —KCN→ 52

42–67%
dr 62:38–79:21

Enders's chiral hydrazones **53** have also been employed to address the problem of asymmetric synthesis.[31] His (*S*)-1-amino-2-methoxymethylpyrrolidine (SAMP) exhibits the ability to control the diastereoselective addition of TMSCN to chiral hydrazones **53** that gave rise to chiral α-aminonitriles **54**.

53 —TMSCN, $TiCl_4$→ 54

75–93%
de 88–91%

The first report of catalyst-controlled Strecker reactions was based on a report by Inoue[32] with the disclosure that cyclic dipeptide **55** could catalyze the asymmetric formation of cyanohydrins **5** from aldehydes **1**. This group showed that the corresponding acyclic version of the catalyst did not possess any catalytic activity. The cyclic catalyst behaved best for aryl and heteroaryl aldehydes.

catalyst (2 mol%)

30–97%

53–97% *ee*

1 5 55

cyclo[(*S*)-phenylalanyl-(*S*)-histidyl]

While this catalyst could not be applied to the Strecker reaction, Lipton was able to modify this species and was the first to report an asymmetrically catalyzed Strecker reaction.[33] He reasoned the imidazole did not possess sufficient basicity and prepared the corresponding guanidine derivative, cyclo[(*S*)-phenylalanyl-(*S*)-norarginyl] **56**.

56

cyclo[(*S*)-phenylalanyl-(*S*)-norarginyl]

catalyst (2 mol%)

70–97%

64–99% *ee*

57 **58**

The transformation of imine **57**, derived from aromatic aldehydes, afforded **58** with good to excellent enantiocontrol. This observation stands in contrast to imines related to **57** prepared from heteroaryl or alkyl aldehydes. In these cases, products related to **58** could be obtained but did so with poor optical purity (10–32% *ee*).

With these results in hand, the Corey labs were able to utilize a readily available chiral C_2-symmetical bifunctional guanidine catalyst **59**.[34] They rationalized the origin of the enantioselectivity to a pre-transition state assembly of the catalyst **59**, the imine **60**, and cyanide. Additionally, DFT modeling studies using the B3LYP method gave rise to two competitive pathways for the catalytic cycle.[34b] Concomitant hydrolysis of the nitrile and deprotection of the amine converted **61** to amino acids **62**.

59

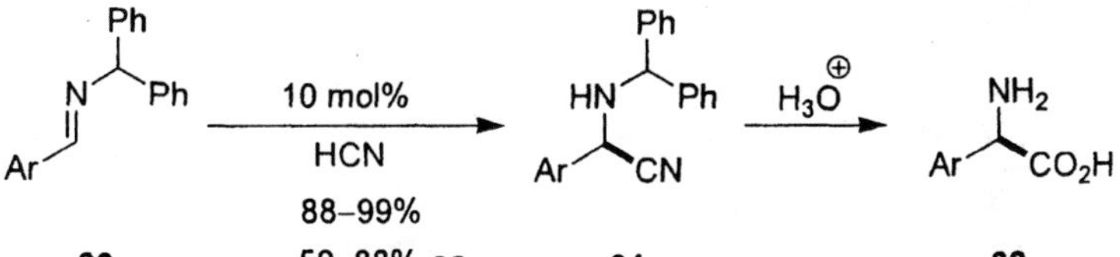

From their extensive mechanistic studies on the asymmetric dihydroxylation reaction, the Corey group extended the utility of these catalysts to the Strecker reaction.[35] Treatment of their dihydroxylation catalyst with trifluoroacetic acid generated a stable crystalline solid **63** that was demonstrated to catalyze the asymmetric Strecker reaction. The *N*-allyl moiety was found to be preferred to benzyl as a protecting group for the nitrogen atom. Thus, when **64** was treated with catalyst **63**, nitrile product **65** could be obtained in excellent yield. A solvent effect on the % *ee* was noted such that CH_2Cl_2 was optimal compared to toluene that may compete with the aldimine for the catalyst's binding site.

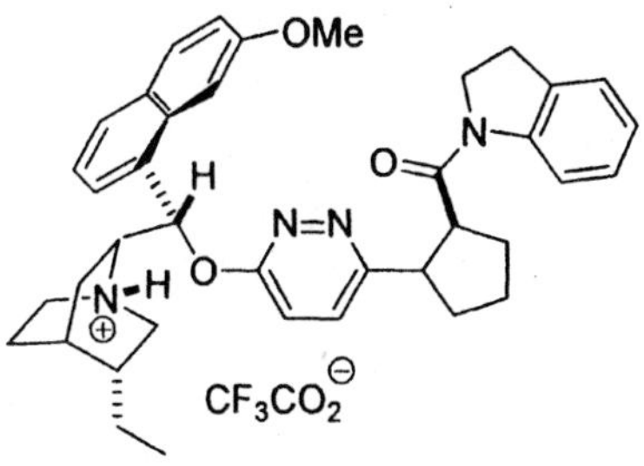

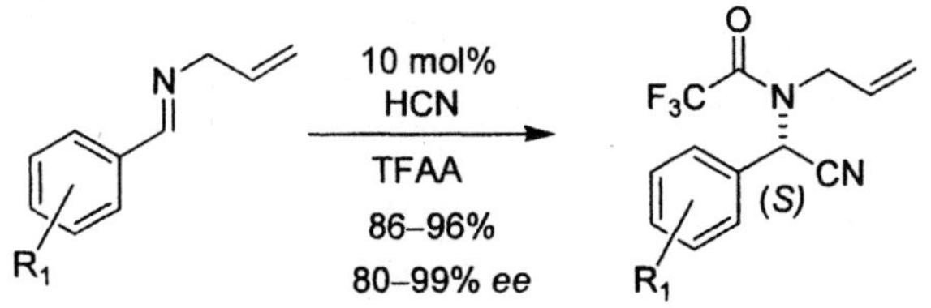

Concomitant with these efforts, Jacobsen was extending the utility of his salen catalysts.[36] Initially, the aluminum-based salen catalyst **68** was found to afford α-aminonitriles **67** from imines **66** with excellent enantioselectivities.

66 → 67 (5 mol% HCN, TFAA, 69–99%, 37–99% *ee*); catalyst 68

Upon further studies that employed a strategy of catalyst optimization *via* combinatorial chemistry methods, it was subsequently found that the metal was not required for catalysis and suitably functionalized chiral Schiff bases, such as **69**, were sufficient.

66 → 67 (2 mol% HCN, TFAA, 65–92%, 70–91% *ee*); catalyst 69

Along similar lines, a titanium complex of a tripeptide Schiff base **72** was found to be an efficient catalyst for the transformation of **70** into **71**.[37]

70 → 71 (10 mol% Ti(O*i*Pr)$_4$, TMSCN, 82–97%, 85–99% *ee*); ligand 72

The great versatility of binol as a chiral ligand has been extended to the Strecker reaction. Shibasaki[38] has developed bifunctional catalysts employing the use of this species. After the reportewd use of an aluminum variation of this catalyst **75** for the asymmetric formation of cyanohydrins, aluminum and gadolinium derivatives of **78** were shown to efficiently catalyze the addition of cyanide to imines.

9 mol%

TMSCN

66–97%

70–95% *ee*

73 **74** **75**

$Gd(OiPr)_3$ (10 mol%)

20 mol%

TMSCN

58–94%

51–98% *ee*

76 **77** **78**

Kobayashi has extended this chemistry and has reported a zirconium version of this binol catalyst.[39] He has described binuclear catalyst **79** in which the optimized version, essential for asymmetric induction, made use of (*R*)-6-bromobinol and (*R*)-3-bromobinol.

79

10 mol%

Bu_3SnCN

55–98%

74–99% *ee*

1. MeI, K_2CO_3

2. H_2O_2 / $HO^{\ominus}$

94%

80 **81**

82 → 83 (CAN, 63%)

Heterobimetallic binol catalyst based on titanium, aluminum, and scandium have also been reported.[40] The results showed that the titanium catalyst was not effective. While the aluminum catalyst was better, it turned out that the scandium derivative was the most efficient.

84 → 86 (catalyst, HCN or TMSCN)

87

M = Ti^{III}	20% *ee*
M = Al^{III}	4–39% *ee*
M = Sc^{III}	20–95%
	45–95% *ee*

6.4.5 Synthetic Utility

Under the assumption that the Earth's primitive atmosphere was "reducing," Miller in 1953,[41] exposed a mixture of methane, ammonia, water, and hydrogen to an electric discharge. This mixture of gases simulated primordial Earth conditions and the electric discharge simulated lightning events proposed to have existed at the time. After running continuously for a week, the resulting gimish was analyzed and glycine **88**, α-alanine **89**, β-alanine **92**, aspartic acid **91**, and α-aminobutyric acid **90** could be isolated. These products could all be rationalized based on the Strecker reaction. This seminal study gave rise to the field of prebiotic chemistry and studies to understand this chemistry continue to be evaluated.[42]

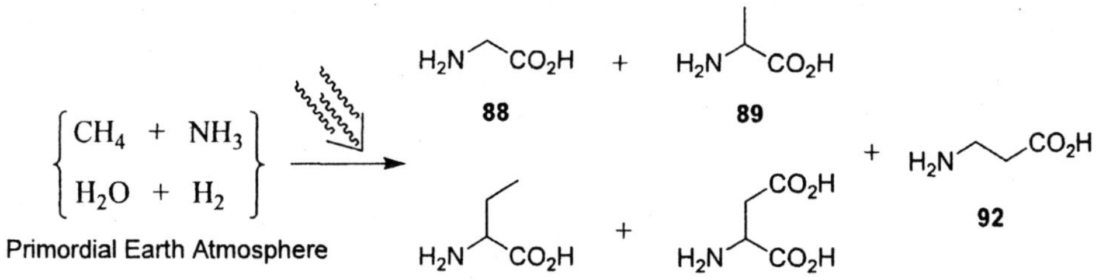

Lactacystin **97** inhibits cell proliferation and induces neurite outgrowth as a result of proteosome-mediated peptidase inhibition. In one approach to this target, the α,α-disubstituted α-amino acid contained in this molecule was assembled using an intramolecular asymmetric Strecker reaction.[43] Esterification of α-hydroxyketone **93** with Boc-*L*-Phe afforded ester **94**. Deprotection of the Boc group and intramolecular imine formation preceeded the addition of cyanide to produce **95**. Ozonolysis and treatment with conc. HCl gave rise to the desired amino acid **96**. This advanced intermediate constituted a formal total synthesis of lactacystin **97**.

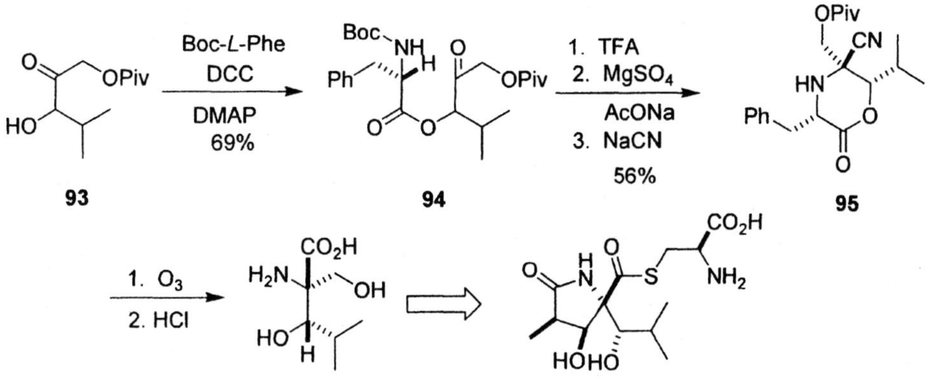

Cyclomarin A **98** exhibits cytotoxicity towards cancer cells and has anti-inflammatory activity. It is a cyclopeptide isolated from an *actinomycete* and has attracted the attention of synthetic chemists due to the number of non-natural amino acids present in this structure. The (2*S*,4*R*)-*D*-hydroxyleucine **103** contained within the target has been prepared by using the Davis chiral auxiliary (*vide supra*).[44] Condensation of an enantiopure *p*-toluenesulfinamide **100** with aldehyde **99** afforded sulfinimine **101**.

Exposure to Et_2AlCN afforded the desired amino nitrile **102**. Deprotection and hydrolysis of the nitrile was accomplished with methanolic HCl to produce **103**.

98

99 + **100** → $Ti(OEt)_4$, 84% → **101** → Et_2AlCN, 92%, 91% *de* →

102 → HCl, MeOH, 86% → **103**

A series of cyclobutane derivatives that possess neurokinin-1 (NK1) receptor antagonist activity has been reported.[45] These compounds could not be prepared utilizing the classical Strecker reaction conditions, as the intermediate α-aminonitrile **105** could not be hydrolyzed. Thus, ketone **104** was exposed to the Bucherer–Bergs reaction conditions to produce the corresponding hydantoins **106** and **107**. The related α-amino acids, **108** and **109**, respectively, could be generated by hydrolysis.

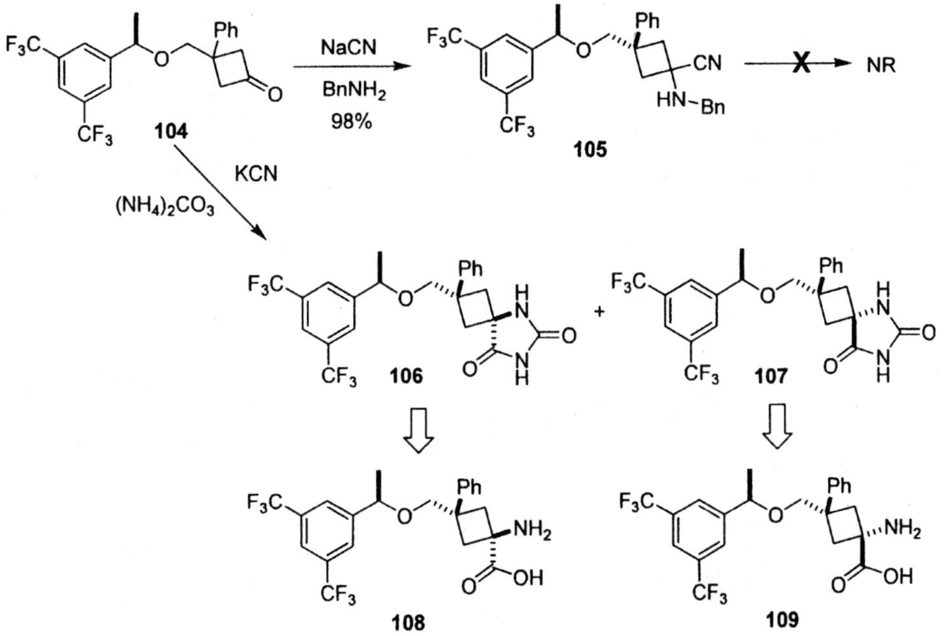

Synthetic use of the α-aminonitrile intermediates from the Strecker reaction have also been investigated.[46] In this manner, it was possible to assemble the core scaffold of the indolizidine alkaloids **112**. Treating **110**, which contained the amine component and a dimethoxyacetal as a latent carbonyl component, with cyanide under acidic conditions initiated an intramolecular cyclization to afford the Strecker reaction intermediate α-aminonitrile **111**. While one could hydrolyze the nitrile functionality to unmask a carboxylic acid, in this case, the nitrile was used to introduce the requisite sidechains needed in the construction of the indolizidine alkaloids **112**.

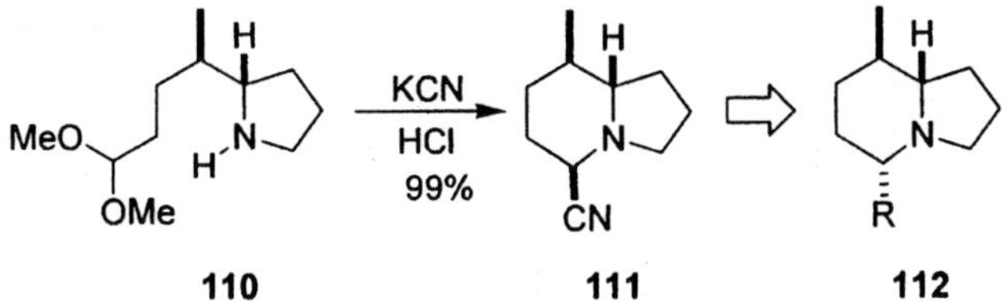

6.4.6 *Experimental*

Hydrogen cyanide (HCN) is highly toxic and must be kept in a well-ventilated hood and handled with protective gloves.

For the classical Strecker reaction conditions, there are multiple references in the literature that one could use to conduct this reaction.[1, 47] The following two examples illustrate the modified Strecker reaction using TMSCN and an asymmetric variation of this reaction.

TMSCN, NH_3
ZnI_2
97%
113 → 114 (NH_2 - HCl)

2-Amino-2-(3-chlorophenyl)acetonitrile hydrochloride (114)[7]
To a mixture of **113** (5.6 g, 40 mmol) and TMSCN (5.0 g, 50 mmol) was added a catalytic amount of ZnI_2. The mixture was stirred for 15 min before adding a saturated solution of ammonia in methanol (30 mL). The reaction was heated to 40 ° C and stirred an additional 2 h before removing the solvent. The crude product was diluted with ether, dried ($MgSO_4$), and filtered. HCl was bubbled through the filtrate and **114** (8.1 g, 97%) precipitated as an off-white solid.

1. HCN
63 (10 mol %)
2. TFAA
98%
99% *ee*
115 → 116 (S)
63 ($CF_3CO_2^{\ominus}$)

(*S*)-*N*-Allyl-*N*-(cyano(*m*-tolyl)methyl)-2,2,2-trifluoroacetamide (116)[35]
To a solution of **115** (0.2 mmol) and **63** (0.02 mmol) in CH_2Cl_2 (1 mL) under nitrogen at −78 °C was added TMSCN (0.4 mmol) and isopropanol (0.4 mmol). The reaction mixture was stirred for 40 h before treating with TFAA. The solvent was removed *in vacuo* after flushing with nitrogen gas that was passed through bleach to remove any remaining HCN. The crude product was chromatographed on silica to afford **116** in 98% yield and > 99% *ee*.

6.4.7 References

1. (a) [R] Block, R. J. *Chem. Rev.* **1946**, *38*, 501–571. (b) [R] Mowry, D. T. *Chem. Rev.* **1947**, *47*, 189–283. (c) [R] Williams, R. M. *Synthesis of Optically Active a-Amino Acids* Pergamon Press, New York, **1989**, 208–229. (d) [R] Shafran, Y. M.; Bakulev, V. A.; Mokrushin, V. S. *Russ. Chem. Rev.* **1989**, *58*, 148–162. (e) [R] Williams, R. M.; Hendrix, J. A. *Chem. Rev.* **1992**, *92*, 889-917. (f) [R] Duthaler, R. O. *Tetrahedron* **1994**, *50*, 1539–1650. (g) [R] Tolman, V. *Amino Acids* **1996**, *11*, 15–36. (h) Kobayashi, S.; Ishitani, H. *Chem. Rev.* **1999**, *99*, 1069-1094. (i) [R] Enders, D.; Shilvock, J. P. *Chem. Soc. Rev.* **2000**, *29*, 359–373. (j) [R] Yet, L. *Angew. Chem. Int. Ed.* **2001**, *40*, 875–877. k) [R] Groger, H. *Chem. Rev.* **2003**, *103*, 2795–2827.
2. (a) Strecker, A. *Liebigs Ann. Chem.* **1850**, *75*, 27–45. b) Strecker, A. *Liebigs Ann. Chem.* **1854**, *79*, 321–335.
3. a) Szabadvary, F. *Chemistry* **1973**, *46*, 12–14. (b) Sourkes, T. L. *Ambix* **2000**, *47*, 37–46. (c) Kauffman, G. B.; Chooljian, S. H. *Chem. Ind.* **2000**, 774–775.
4. (a) Sannie, C. *Bull. Soc. Chim.* **1925**, *37*, 1557–1576. (b) Stewart, T. D.; Li, C-H. *J. Am. Chem. Soc.* **1938**, *60*, 2782–2787. (c) Harris, G. H.; Harriman, B. R.; Wheeler, K.W. *J. Am. Chem. Soc.* **1946**, *68*, 846–848. (d) Sander, E. G.; Jencks, W. P. *J. Am. Chem. Soc.* **1968**, *90*, 6154–6162. (e) Schlesinger, G.; Miller, S. L. *J. Am. Chem. Soc.* **1973**, *95*, 3729–3735. (f) Taillades, J.; Commeyars, A. *Tetrahedron* **1974**, *30*, 127–132. (g) Taillades, J.; Commeyars, A. *Tetrahedron* **1974**, *30*, 2493–2501. (h) Taillades, J.; Commeyars, A. *Tetrahedron* **1974**, *30*, 3407–3414. (i) Commeyras, A.; Taillades, J.; Mion, L.; Bejaud, M. *Informat. Chim.* **1976**, 199–207. (j) Bejaud, M.; Mion, L.; Commeyras, A. *Bull. Soc. Chim. Fr.* **1976**, 233–236. (k) Bejaud, M.; Mion, L.; Commeyras, A. *Bull. Soc. Chim. Fr.* **1976**, 1425–1430. (l) Moutou, G.; Taillades, J.; Benefice-Malouet, S.; Commeyras, A. *J. Phys. Org. Chem.* **1995**, *8*, 721–730. m) Atherton, J. H.; Blacker, J.; Crampton, M. R.; Grosjean, C. *Org. Biomol. Chem.* **2004**, *2*, 2567–2571.
5. Georgiadis, M. P.; Haroutounian, S. A. *Synthesis* **1989**, 616–617.
6. Harusawa, S.; Hamada, Y.; Shioiri, T. *Tetrahedron Lett.* **1979**, 4663–4666.
7. (a) Mai, K.; Patil, G. *Tetrahedron Lett.* **1984**, *25*, 4583–4586. (b) Mai, K.; Patil, G. *Org. Prep. Proced. Int..* **1985**, 17, 183–186.
8. a) Ranu, B. C.; Dey, S. S.; Hajra, A. *Tetrahedron* **2002**, *58*, 2529–2532. b) De, S. K.; Gibbs, R.A. *Tetrahedron Lett.* **2004**, *45*, 7407–7408. (c) De, S. K.; Gibbs, R. A. *Synth. Commun.* **2005**, *35*, 961–966. (d) De, S. K. *Synth. Commun.* **2005**, *35*, 653–656.
9. Kobayashi, S.; Busujima, T.; Nagayama, S. *Chem. Commun.* **1998**, 981–982.
10. Yadav, J. S.; Reddy, B. V. S.; Eeshwaraiah, B.; Srinivas, M. *Tetrahedron Lett.* **2004**, *60*, 1767–1771.
11. Royer, L.; De, S. K.; Gibbs, R. A. *Tetrahedron Lett.* **2005**, *46*, 4595–4597.
12. Das, B.; Rammu, R.; Ravikanth, B.; Reddy, K. R. *Synthesis* **2006**, 1419–1422.
13. Menendez, J. C.; Trigo, G. G.; Sollhuber, M. M. *Tetrahedron Lett.* **1986**, *27*, 3285–3288.
14. Surenda, K.; Krishnaveni, N. S.; Mahesh, A.; Rama Rao, K. *J. Org. Chem.* **2006**, *71*, 2532–2534.
15. (a) Dyker, G. *Angew. Chem. Intl. Ed. Engl.* **1997**, *36*, 1700–1702. (b) Dyker, G. *Org. Synth. Highlights IV* **2000**, 53–57.
16. a) Haroutounian, S. A.; Georgiadis, M. P.; Polissiou, M. G. *J. Heterocyclic Chem.* **1989**, *26*, 1283–1287. b) Meusel, M.; Guetschow, M. *Org. Prep. Proceed. Intl.* **2004**, *36*, 391–443. c) Li, J.-T.; Wang, S.-X.; Chen, G-F.; Li, T.-S. *Curr. Org. Synth.* **2005**, *2*, 415–436.
17. Sarges, R.; Goldstein, S. W.; Welch, W. M.; Swindell, A. C.; Siegel, T. W.; Beyer, T. A. *J. Med. Chem.***1990**, *33*, 1859–1865.
18. (a) Petasis, N. A.; Akritopoulou, I. *Tetrahedron Lett.* **1993**, *34*, 583–586. (b) Petasis, N. A.; Zavialov, I. A. *J. Am. Chem. Soc.* **1997**, *119*, 445–446. (c) Petasis, N. A.; Goodman, A.; Zavialov, I. A. *Tetrahedron* **1997**, *53*, 16463–16470. (d) Klopfenstein, S. R.; Chen, J. J.; Golebiowski, A.; Li, M.; Peng, S. X.; Shao, X. *Tetrahedron Lett.* **2000**, *41*, 4835–4839. (e) McLean, N. J.; Tye, H.; Whittaker, M. *Tetrahedron Lett.* **2004**, *45*, 993–995. (f) Follmann, M.; Graul, F.; Schafer, T.; Kopec, S.; Hamley, P. *Synlett* **2005**, 1009–1011. (g) Southwood, T. J.; Curry, M. C.; Hutton, C. A. *Tetrahedron* **2006**, *62*, 236–242.
19. (a) [R] Domling, A.; Ugi, I. *Angew. Chem. Intl Ed.* **2000**, *39*, 3168–3210. (b)Ugi, I.; Domling, A.; Werner, B. *J. Heterocyclic Chem.* **2000**, *37*, 647–658. (c) Ugi, I. *Pure Appl Chem.* **2001**, *73*, 187–191. (d) Tempest, P. A. *Curr. Opin. Drug Disc. Develop.* **2005**, *8*, 776–788.
20. Kazmaier, U.; Ackermann, S. *Org. Biomol. Chem.* **2005**, *3*, 3184–3187.
21. (a) Magnus, P.; Slater, M. *Tetrahedron Lett.* **1987**, *28*, 2829–2832. (b) Beller, M.; Eckert, M. *Angew. Chem. Intl. Ed.* **2000**, *39*, 1010–1027.
22. Wakamatsu, H.; Uda, J.; Yamakami, N. *Chem. Commun.* **1971**, 1540.
23. Parnaud, J.-J.; Campari, G.; Pino, P. *J. Mol. Catal.* **1979**, *6*, 341–350.

24. (a) Beller, M.; Eckert, M.; Vollmuller, F.; Bogdanovic, S.; Geissler, H. *Angew. Chem. Intl. Ed. Engl.* **1997**, *36*, 1494–1496. (b) Beller, M.; Eckert, M.; Vollmuller, F. *J. Mol. Catal.* **1998**, *135*, 23–33. (c) Beller, M.; Eckert, M.; Holla, E.W. *J. Org. Chem.* **1998**, *63*, 5658–5661. (d) Beller, M.; Moradi, W.A.; Eckert, M.; Neumann, H. *Tetrahedron Lett.* **1999**, *40*, 4523–4526. (e) Beller, M.; Eckert, M.; Moradi, W.A. *Synlett* **1999**, 108–110.

25. (a) Harada, K. *Nature* **1963**, *200*, 1201. (b) Harada, K.; Okawara, T. *J. Org. Chem.* **1973**, *38*, 707–710.

26. Ojima, I.; Inaba, S-I. *Chem. Lett.* **1975**, 737–740. See also: a) Inaba, T.; Fujita, M.; Ogura, K. *J. Org. Chem.* **1991**, *56*, 1274–1279. b) Vincent, S. P.; Schleyer, A.; Wong, C-H. *J. Org. Chem.* **2000**, *65*, 4440–4443. c) Wede, J.; Volk, F-J.; Frahm, A.W. *Tetrahedron: Asymmetry* **2000**, *11*, 3231–3252. d) Meyer, U.; Breitling, E.; Bisel, P.; Frahm, A. W. *Tetrahedron: Asymmetry* **2004**, *15*, 2029–2037.

27. (a) Kunz, H.; Sager, W.; Pfrengle, W.; Laschat, S.; Schanzebach, D. *Chem. Peptides Proteins* **1993**, *5*, 91–98. (b) Kunz, H.; Sager, W. *Angew. Chem. Int. Ed. Engl.* **1987**, *26*, 557–559. (c) [R] Kunz, H.; Ruck, K. *Angew. Chem. Int. Ed. Engl.* **1993**, *32*, 336–358. (d) Knauer, S.; Kranke, B.; Krause, L.; Kunz, H. *Curr. Org. Chem.* **2004**, *8*, 1739–1761. e) Zhou, G.; Zhang, P.; Pan, Y.; Guo, J. *Org. Prep. Proced. Int.* **2005**, *37*, 65–73.

28. (a) Davis, F. A.; Portonovo, P. S.; Reddy, R. E.; Chiu, Y-H. *J. Org. Chem.* **1996**, *61*, 440–441. (b) Davis, F. A.; Qu, J.; Srirajan, V.; Joseph, R.; Titus, D. D. *Heterocycles* **2002**, *58*, 251–258. (c) Wang, H. Zhao, X.; Li, Y.; Lu, L. *Org. Lett.* **2006**, *8*, 1379–1381. (d) Morton, D.; Stockman, R. A. *Tetrahedron* **2006**, *62*, 8869–8905.

29. (a) Chakraborty, T. K.; Reddy, G. V.; Hussain, K. A. *Tetrahedron Lett.* **1991**, *32*, 7597–7600. (b) Chakraborty, T. K.; Hussain, K. A.; Reddy, G. V. *Tetrahedron* **1995**, *51*, 9179–9190. (c) Ma, D.; Tian, H.; Zou, G. *J. Org. Chem.* **1999**, *64*, 120–125. (d) Warmuth, R.; Munsch, T. E.; Stalker, R. A.; Li, B.; Beatty, A. *Tetrahedron* **2001**, *57*, 6383–6397. (e) Boesten, W. H. J.; Seerden, J-P. G.; de Lange, B.; Dielemans, H. J. A.; Elsenberg, H. L. M.; Kaptein, B.; Moody, H. M.; Kellogg, R.M.; Broxterman, Q. B. *Org. Lett.* **2001**, *3*, 1121–1124.

30. Bousquet, C.; Tadros, Z.; Tonnel, J.; Mion, L.; Taillades, J. *Bull. Soc. Chim. Fr.* **1993**, *130*, 513–520.

31. Enders, D.; Moser, M. *Tetrahedron Lett.* **2003**, *44*, 8479–8481.

32. (a) Oku, J-I.; Inue, S. *J.C.S. Chem. Commun.* **1981**, 229–230. (b) Tanaka, K.; Mori, A.; Inue, S. *J. Org. Chem.* **1990**, *55*, 181–185.

33. (a) Iyer, M. S.; Gigstad, K. M.; Namdev, N. D.; Lipton, M. *J. Am. Chem. Soc.* **1996**, *118*, 4910–4911. (b) Iyer, M. S.; Gigstad, K. M.; Namdev, N. D.; Lipton, M. *Amino Acids* **1996**, *11*, 259–268.

34. (a) Corey, E. J.; Grogan, M. J. *Org. Lett.* **1999**, *1*, 157–160. (b) Li, J.; Jiang, W-Y.; Han, K-L.; He, G-Z.; Li, C. *J. Org. Chem.* **2003**, *68*, 8786–8789.

35. Huang, J.; Corey, E. J. *Org. Lett.* **2004**, *6*, 5027–5029.

36. (a) Sigman, M. S.; Jacobsen, E. N. *J. Am. Chem. Soc.* **1998**, *120*, 5315–5316. (b) Sigman, M. S.; Jacobsen, E. N. *J. Am. Chem. Soc.* **1998**, *120*, 4901–4902. (c) Su, J. T.; Vachal, P.; Jacobsen, E. N. *Adv. Synth. Catal.* **2001**, *343*, 197–200.

37. (a) Krueger, C. A.; Kuntz, K. W.; Dzierba, C. D.; Wirschun, W. G.; Gleason, J. D.; Snapper, M. L.; Hoveyda, A. H. *J. Am. Chem. Soc.* **1999**, *121*, 4284–4285. (b) Porter, J. R.; Wirschun, W. G.; Kuntz, K. W.; Snapper, M. L.; Hoveyda, A. H. *J. Am. Chem. Soc.* **2000**, *122*, 2657–2658. (c) Josephsohn, N. S.; Kuntz, K. W.; Snapper, M. L.; Hoveyda, A. H. *J. Am. Chem. Soc.* **2001**, *123*, 1154–1159.

38. (a) Hamashima, Y.; Sawada, D.; Kanai, M.; Shibasaki, M. *J. Am. Chem. Soc.* **1999**, *121*, 2641–2642. (b) Takamura, M.; Hamashima, Y.; Usuda, H.; Kanai, M.; Shibasaki, M. *Angew. Chem. Intl. Ed.* **2000**, *39*, 1650–1652. (c) Masumoto, S.; Usuda, H.; Suzuki, M.; Kanai, M.; Shibasaki, M.; *J. Am. Chem. Soc.* **2003**, *125*, 5634–5635. (d) Kato, N.; Mita, T.; Kanai, M.; Therrien, B.; Kawano, M,; Yamaguchi, K.; Danjo, H.; Sei, Y.; Sato, A.; Furusho, S.; Shibasaki, M. *J. Am. Chem. Soc.* **2006**, *128*, 6788–6769.

39. (a) Ishitani, H.; Komiyama, S.; Kobayashi, S. *Angew. Chem. Intl. Ed.* **1998**, *37*, 3186–3188. (b) Ishitani, H.; Komiyama, S.; Hasegawa, Y.; Kobayashi, S. *J. Am. Chem. Soc.* **2000**, *122*, 762–766. (c) Kobayashi, S.; Ishitani, H. *Chirality* **2000**, *12*, 540–543.

40. (a) Byrne, J. J.; Chavarot, M.; Chavant, P-Y.; Vallee, Y. *Tetrahedron Lett.* **2000**, *41*, 873–876. (b) Chavarot, M.; Byrne, J. J.; Chavant, P. Y.; Vallee, Y. *Tetrahedron: Asymmetry* **2001**, *12*, 1147–1150.

41. (a) Miller, S. L. *Science* **1953**, *117*, 528–530. (b) Miller, S. L. *Biochim. Biophys. Acta* **1957**, *23*, 480–489.

42. (a) [R] Lemmon, R. M. *Chem. Rev.* **1970**, *70*, 95–109. (b) Miller, S. L.; Van Trump, J. E. in *Origin of Life*, D. Reidel Publishing Co. **1981**, 135–141. (c) Miller, S. L. *Chemica Scripta* **1986**, *26B*, 5–11. (d) Schulte, M.; Shock, E. *Origins Life Evol. Biosphere* **1995**, *25*, 161–173. (e) Brandes, J. A.; Hazen, R. M.; Yoder, H. S, Jr.; Cody, G. D. in *Perspectives in Amino Acid and Protein Geochemistry*, Oxford University Press, **2000**,

41–47. (f) Bada, J. L. *Earth Planetary Sci. Lett.* **2004**, *226*, 1–15. (g) Simoneit, B. R. T. *Adv. Space Res.* **2004**, *33*, 88–94. (h) Lerner, N. R.; Cooper, G. W. *Geochim. Cosmochim. Acta* **2005**, *69*, 2901–2906.

43. (a) Moon, S-H.; Ohfune, Y. *J. Am. Chem. Soc.* **1994**, *116*, 7405–7406. (b) Ohfune, Y.; Shinada, T. *Bull. Chem. Soc. Jpn.* **2003**, *76*, 1115–1129.
44. Hansen, D. B.; Starr, M-L.; Tolstoy, N.; Joullie, M. M. *Tetrahedron: Asymmetry* **2005**, *16*, 3623–3627.
45. Wrobleski, M. L.; Reichard, G. A.; Paliwal, S.; Shah, S.; Tsui, H-C.; Duffy, R. A.; Lachowicz, J. E.; Morgan, C. A.; Varty, G. B.; Shih, N-Y. *Bioorg. Med. Chem. Lett.* **2006**, *16*, 3859–3863.
46. Polniaszek, R. P.; Belmont, S. E. *J. Org. Chem.* **1991**, *56*, 4868–4874.
47. (a) Kendall, E. C.; McKenzie, B. F. *Org. Synth., Coll. Vol. 1*, **1941**, p. 21. (b) Clarke, H. T.; Bean, H. J. *Org. Synth., Coll. Vol. 2*, **1943**, p. 29. (c) Steiger, R. E. *Org. Synth., Coll. Vol. 3*, **1955**, pp. 66; 84; 88. (d) Redemann, C. E.; Icke, R. N. *J. Org. Chem.* **1943**, *8*, 159–161.

Paul Galatsis

6.5 Yamada Reactions

6.5.1 Description

Diethyl phosphoryl cyanide or diethyl cyanophosphonate (**3**) activates carboxylic acid (**1**) and serves as a coupling agent with amine (**2**) to provide amide or peptide (**4**).[1-3]

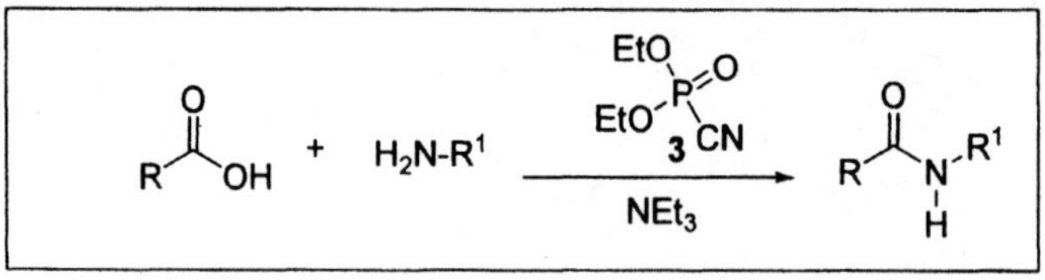

1 2 4

6.5.2 Historical Perspective

Diethyl cyanophosphonate is also known as diethylphosphoryl cyanide, diethoxyphosphoryl cyanide, diethylphosphorocyanidate, diethyl cyanophosphate, and DEPC.[1-3] It can be readily prepared from triethyl phosphite and cyanogen bromide.[2] It is very toxic and corrosive and it must be handled in a fume hood. The reagent is moisture sensitive and should be stored under nitrogen in a refrigerator.

6.5.3 Mechanism

The addition of NEt_3 to carboxylic acid **1** generates carboxylate anion which attacked at phosphorus of diethyl cyanophosphonate and subsequently displaces −CN to produce activated carboxylic acid phosphate ester **5**. Nucleophilic addition of amine **2** to phosphate ester **5** gives the amide **4** and diethylphosphate.[1-3]

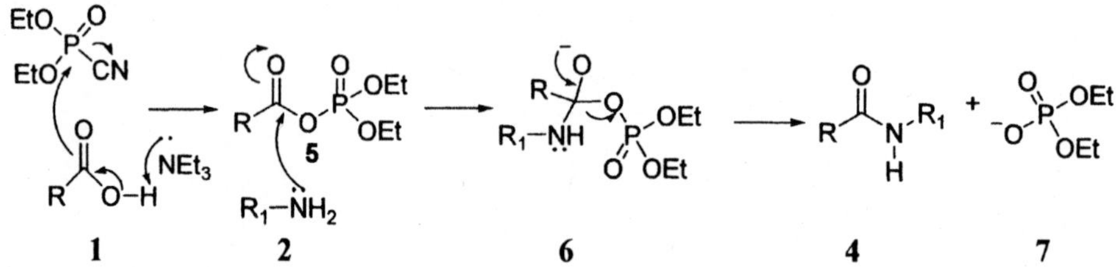

1 2 6 4 7

6.5.4 Variations and Improvements

DEPC serves as a coupling agent for the synthesis of simple amides, esters, α-aminonitriles, and aryl thiocyanates; it is also used as an efficient coupling agent for racemization-free peptide synthesis. Amides of various types can be obtained by the simple mixing of carboxylic acids and amines with DEPC in the presence of triethylamine.[3] Both aromatic and aliphatic acids easily react with aromatic and aliphatic amines.[3] The reaction is rapid, clean, and usually free of contamination from side products. The widely used coupling agent DCC did not give any trace of *N*-cyclohexylbenzamide from benzoic acid and cyclohexylamine, proving the superiority of DEPC to 1,3-dicyclohexylcarbodiimide (DCC).

CO_2H NH_2 O=P(OEt)$_2$CN NEt$_3$, DMF, O HN

8 **9** **10**

DEPC in combination with NEt$_3$ has proved to be a new efficient reagent for the direct *C*-formylation of active methylene compounds with carboxylic acids and also for the *N*-acylation (peptide bond formation), *S*-acylation (thioester formation) and *O*-acylation (esterification).[4, 5] Reaction of DEPC with carboxylic acids **11** in the presence of triethylamine produces transient acyl cyanides, which in the presence of alcohols or thiols results in the formation of the corresponding esters (**12**) or thioesters (**13**).

O R^1 OH DMF R^{2}OH O R^1 OR^2 **12** R^{2}SH O R^1 SR^2

11 **13**

Although esterification yields are not generally satisfactory, the advantage of this method is that the reaction proceeds under mild and almost neutral conditions. The method can be efficiently applied to the synthesis of thioesters in high yields with little racemization, as demonstrated by the formation of the ethylthiol ester of benzoyl-*L*-leucine. Thioesters are directly and conveniently prepared from carboxylic acids and thiols with diphenyl phosphorazide (DPPA) or diethyl phosphorocyanidate (DEPC) in the presence of NEt$_3$ in DMF solution.[6] Both aromatic and aliphatic carboxylic acids easily couple with aromatic and aliphatic thiols, especially when DEPC is a condensing agent. A highly functionalized penicillin thiol ester is obtained by DPPA, but the DEPC proved fruitless in this particular base. DCC is relatively superior to DPPA and DEPC in the coupling of 3-phenylpropionic acid (**14**) with benzenethiol, but far inferior in the formation of *S*-ethyl 3-benzenepropanethionates.[7]

$$\text{14} + \text{15} \xrightarrow[\text{NEt}_3\text{, DMF,}]{\text{O=P(OEt)}_2\text{CN}} \text{16}$$

Homologated hydroxy carboxylic acids **19** are prepared from carboxylic acids **17** *via* hydrolysis of intermediate dicyanophosphates (**18**).[8] In general, homologation is achieved by addition of HCN to aldehyde followed by hydrolysis of the resulting cyanohydrins. This is complementary to the cyanohydrin route but the DEPC method involves fewer steps and gives superior results.[8]

$$\underset{\textbf{17}}{\text{RCO}_2\text{H}} \xrightarrow[\text{Et}_3\text{N, DMF, 23 °C}]{\text{(EtO)}_2\text{P(O)CN}} \underset{\textbf{18}}{\text{R–C(OP(O)(OEt)}_2\text{)(CN)}_2} \xrightarrow{\text{H}_3\text{O}^+} \underset{\textbf{19}}{\text{RCH(OH)CO}_2\text{H}}$$

Peroxycarboxylic acid esters are utilized as initiators for radical polymerization and are interesting intermediates for the decarboxylation of carboxylic acids. They are generally prepared by the acylation of hydro peroxides with acid chlorides, acid anhydrides, or imidazolides in the presence of a base. Condensation of carboxylic acids (**20**) with *t*-butyl-hydroperoxide (**21**) has been smoothly achieved by the use of diethyl phophorocyanidate and NEt_3 under mild conditions giving *t*-butyl peroxycarboxylates (**22**) in good yield.[9]

$$\underset{\textbf{20}}{\text{RCO}_2\text{H}} + \underset{\textbf{21}}{t\text{-BuOOH}} \xrightarrow[\text{Et}_3\text{N}]{\text{DEPC}} \underset{\textbf{22}}{\text{R–C(=O)–O–O–}t\text{-Bu}}$$

6.5.5 *Synthetic Utility*

6.5.5.1 *Amides*

Amides of various types can be obtained by the simple mixing of carboxylic acids and amines with DEPC in the presence of NEt_3. Amide **25** is prepared from 1-phenyl-2-piperidinylethylamine (**24**) and methoxy ethoxy acetic acid (**23**) using DEPC in 90% yield.[10]

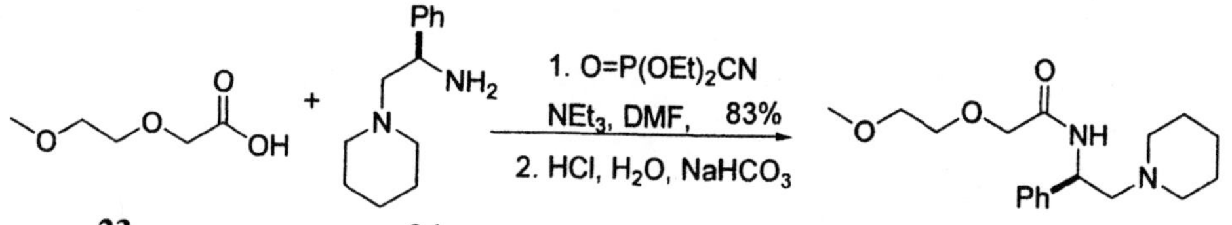

23 24 25

An intermediate amide **27** required for the synthesis of alkaloids suaveoline and norsuaveoline is prepared *via* the coupling of amine **26** and monoethyl malonate in the presence of DEPC.[11]

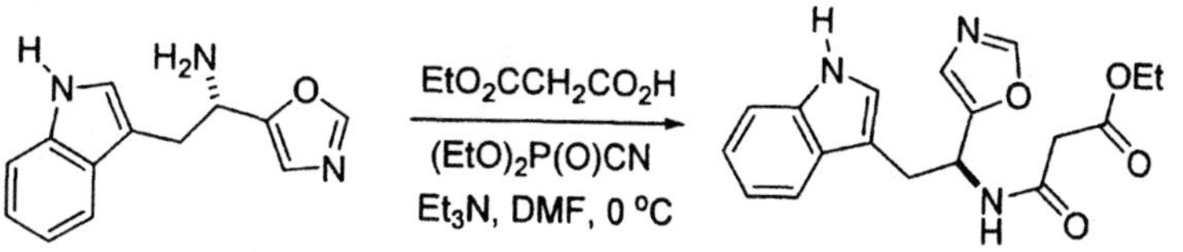

26 27

Amidation of pyrazine 2-amino-3-carboxylic acid (**28**) with *t*-$BuNH_2$ using DEPC provides amide **29** in the presence of NEt_3 but failed with base diazabicycloundecane (DBU).[12] DEPC effectively coupled amino ester **30** with **28** to give amide **31** in 95% yield but the coupling reaction failed with DCC.[13,14] DEPC served as an effective coupling agent for the preparation of amide **33** from methyl pyrrolidine carboxylate (**32**) and pyrazine carboxylic acid **28**.[13]

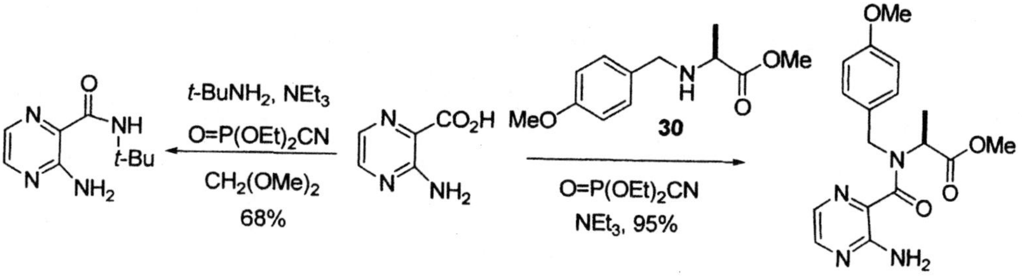

29 28 31

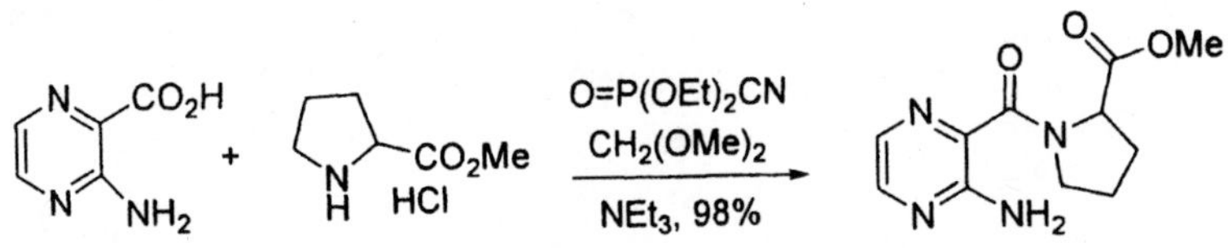

28 32 33

Diamide **36** has been used as a precursor for making a novel phase transfer catalyst.[15] Diamide **36** is prepared from diaminedichloride (**34**) and mandelic acid (**35**) in the presence of DEPC.[15]

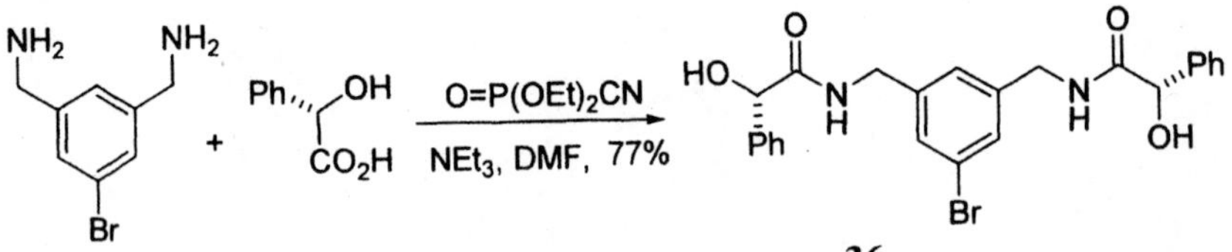

34 35 36

DEPC promoted the condensation reaction of thiophene carboxylic acid (**37**) and hydroxyethanol amine to give amide **38** in 63% yield.[16] Amide **38** is a key intermediate used in the synthesis of thienopyrimidines.

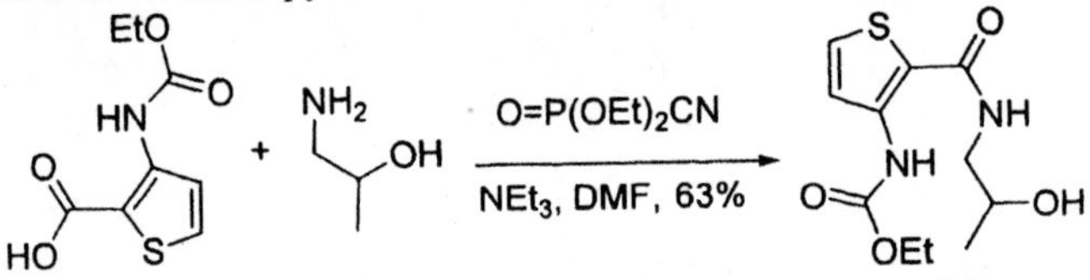

37 38

Unsaturated aryl carboxylic acid **40** is readily coupled with amino ethyl acetate (**39**) to afford amide **41** in 95% yield, which is a key intermediate for the synthesis of a thirteen-membered cyclic peptide skeleton.[17]

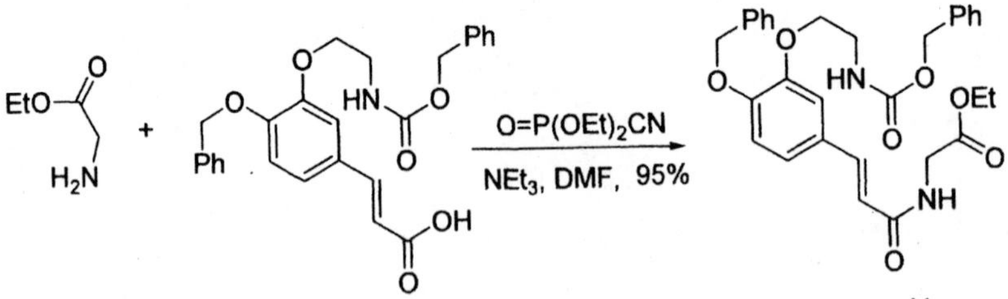

39 40 41

The prodrug (**44**) for suicide gene therapy is prepared *via* condensation of difluorobenzoic acid (**42**) with di-*t*-butyl *L*-glutamate HCl (**43**) in the presence of DEPC.[18]

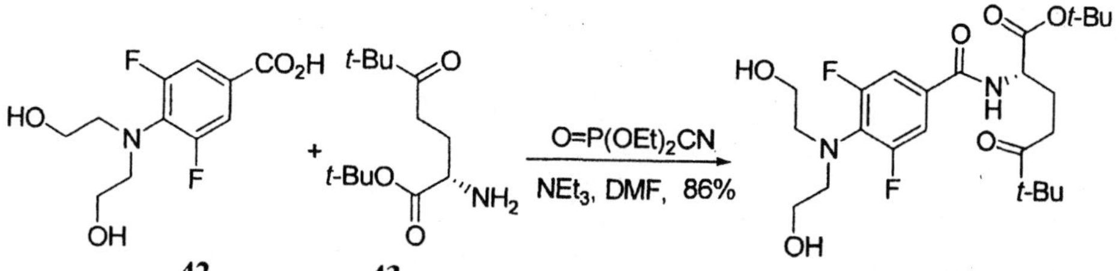

O=P(OEt)$_2$CN
NEt$_3$, DMF, 86%

42 **43** **44**

DEPC is an effective coupling agent for the condensation reaction of benzyloxyethylamine (**45**) and the dimethoxybenzoic acid derivative (**46**), and provides the aryl amide **47** in 77% yield.[19]

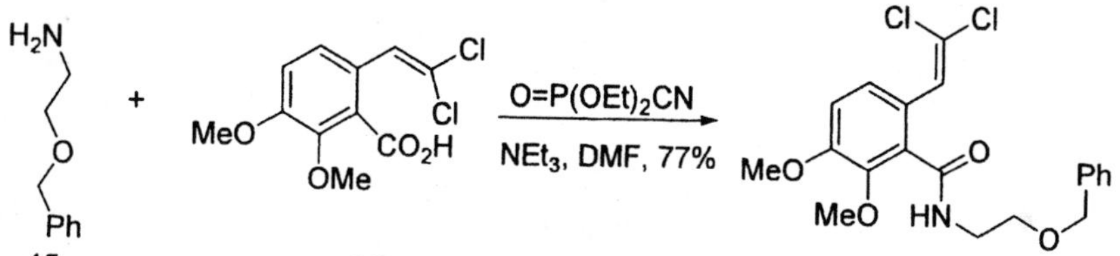

O=P(OEt)$_2$CN
NEt$_3$, DMF, 77%

45 **46** **47**

Total synthesis of the antibiotic globomycin is achieved *via* macrolactonization, from the key intermediate **50**. Coupling of hydroxy acid **48** and amide **49** is mediated by DEPC without protection of the hydroxyl group in **48**.[20]

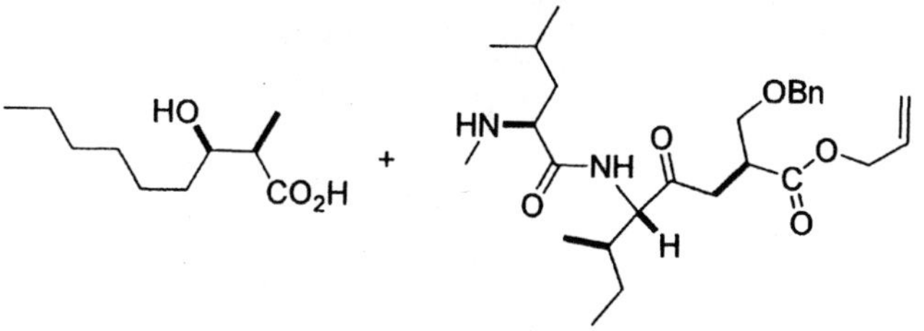

48 **49**

$O{=}P(OEt)_2CN$

NEt_3, DMF, 1 h, rt
99%

50

The primary amine **52** is coupled with benzothiophene carboxylic acid (**51**) to give the desired amide **53**; its antimicrobial activity has been evaluated.[21] Amide **53** shows good antimicrobial (*in vitro*) and inhibitory activity (*in vivo*) against *P. oryzate*.[21]

$O{=}P(OEt)_2CN$

NEt_3, DMF, 60%

51 52 53

Bisamide **56** is prepared by coupling of the diacid **55** with 1-(*S*)-2-(*R*)-1-amino-2-indanol (**54**) using DEPC as an activating agent.[22] The binding data for **56** compare very favorably with the established HIV therapeutic agent indinavir.[22]

$O{=}P(OEt)_2CN$

NEt_3, DMF, 75%

54 55 56

The key intermediate **59** required for the synthesis of sulfobacin A (**59**) is achieved successfully *via* a coupling reaction of aliphatic acid **57** and amino ester **58** using DEPC.[23] Sulfobacin A is biologically active sulfonolipids isolated from the culture broth of *Chryseobacterium sp*. NR 2993 in a soil sample collected from Japan.

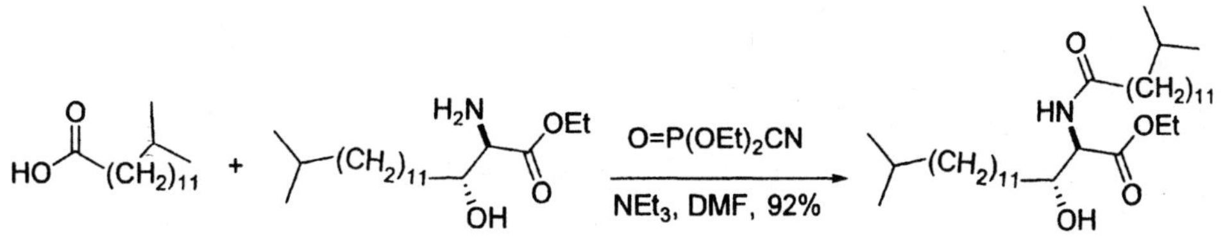

57 + 58 → 59 ($O=P(OEt)_2CN$, NEt_3, DMF, 92%)

Hygomycin A is a fermentation-derived natural product first isolated from *Streptomyces hygrospicus* in 1953. Hygomycin A shows a broad spectrum of activity against gram-positive and gram-negative bacteria. The protected diol of hygomycin A (**62**) is synthesized by coupling of furanoside (**61**) with aminocyclitol (**60**) without epimerization at C-4.[24]

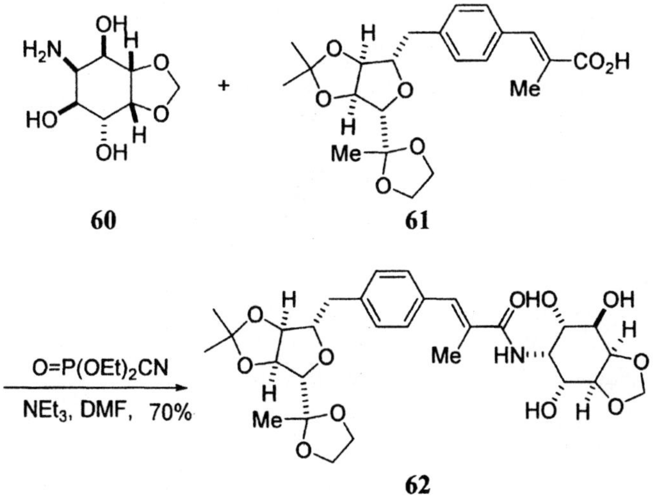

60 + 61 → 62 ($O=P(OEt)_2CN$, NEt_3, DMF, 70%)

The natural product bengamide A (**65**) is synthesized *via* coupling of 3-amino-hexahydro-2-azepinone (**64**) with optically active polyhydroxy alkanyl carboxylic acid (**63**).[25] Bengamide A was first isolated from choristid sponge and shows anti-infective activity.

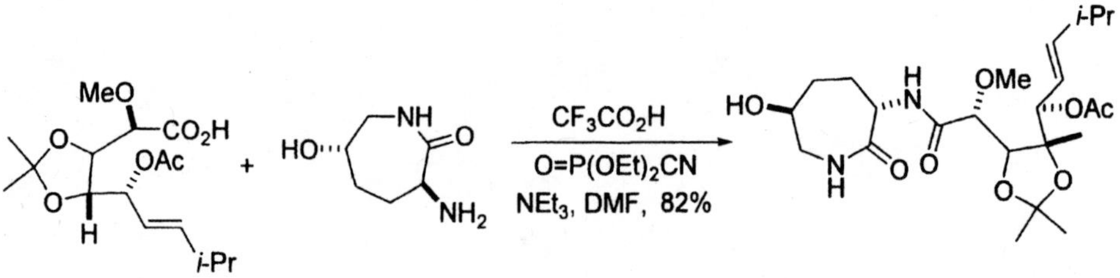

Synthesis of the spermidine alkaloid (–)-(4*R*)-dihydroisomyricoidine involves the intramolecular cyclization of an amino acid **66** in the presence of DEPC which leads to a 13-membered amino lactam **67**.[26]

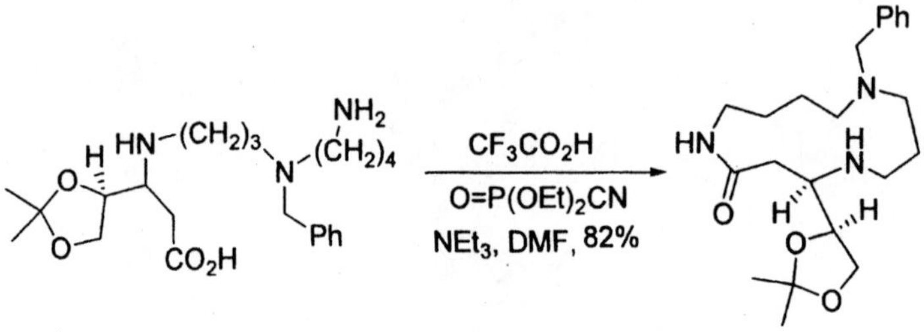

Cyclamenol A is one of the very few noncarbohydrate and nonpeptide natural products that inhibits leukocyte adhesion to endothelial cells. Bis-vinyl iodide (**70**), a precursor for cyclamenol A, is prepared from iodo acid **68** and amine **69**.[27]

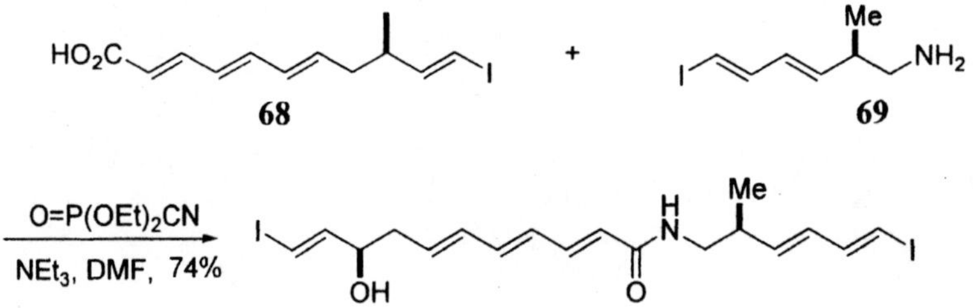

A key component required for the synthesis of an antitumor alkaloid (**73**, 9-demethyl-tubulosine) is obtained in 82% yield *via* amidation of heterocyclic acid **71** and 5-benzyloxy-tryptamine (**72**).[28] 9-Demethyltubulosine increases the survival time of mice infected with leukemia L1210 or P388.

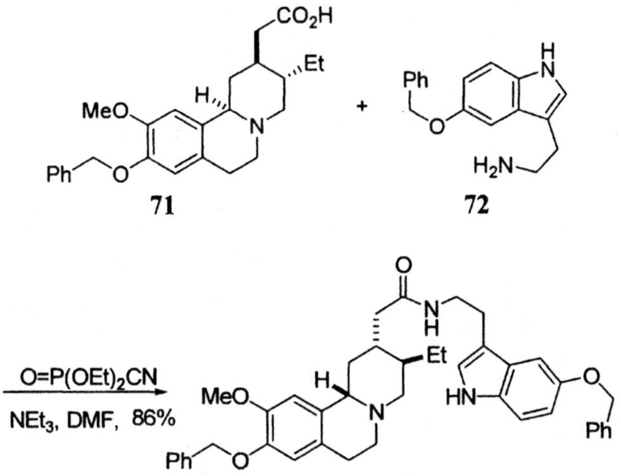

71 72

O=P(OEt)$_2$CN

NEt$_3$, DMF, 86%

73

Acylation of tricyclic amine **74** with furoic acid in the presence of DEPC yields the analog of the α1-blocker prazoicin (**75**, pyrimidobenzoxathiepin derivative).[29] However, **75** did not show any significant α1-blocking action on *in vitro* biological evaluations.

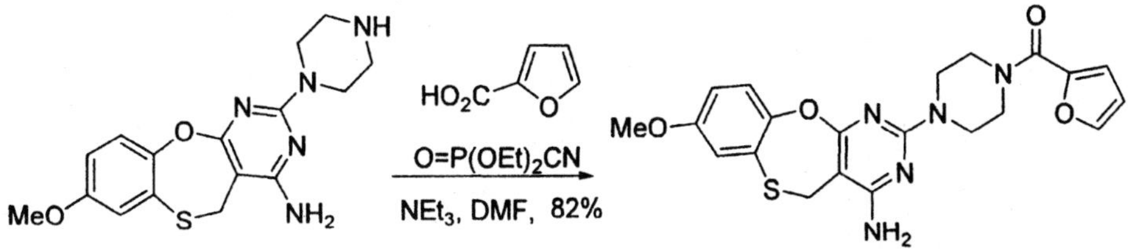

O=P(OEt)$_2$CN

NEt$_3$, DMF, 82%

74 75

A stereoselective synthesis of tilivalline, a metabolite from klebsiella, has been achieved *via* intermediate **78** by coupling 3-hydroxyanthranilic acid (**76**) with pyrrolidine acetal (**77**).[30]

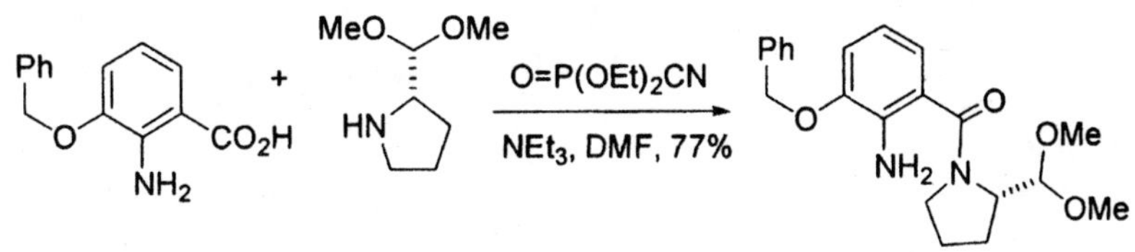

O=P(OEt)$_2$CN

NEt$_3$, DMF, 77%

76 77 78

Dichlorovinylamide (**81**) is a key intermediate used in the synthesis of the alkaloid lennoxamine which is prepared *via* coupling of amine **79** and aryl acid **80** using DEPC.[31] Lennoxamine was isolated from the Chilean barberries *Berberis darwinii* and belongs to a group of isoindolobenzazepine alkaloids.

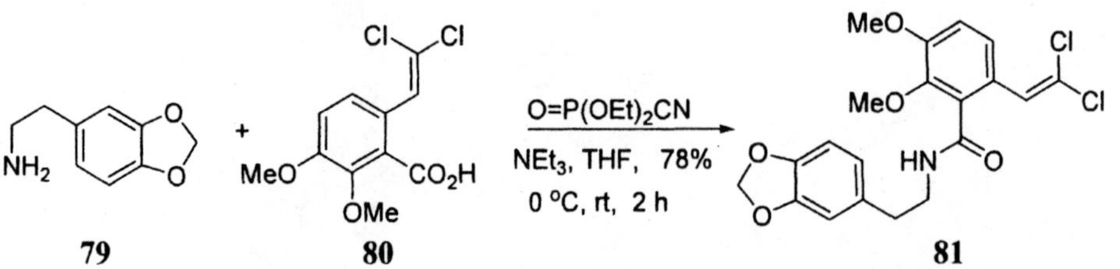

O=P(OEt)$_2$CN, NEt$_3$, THF, 78%, 0 °C, rt, 2 h

79 **80** **81**

A series of cholic amides **84**, **87** and **90** are synthesized from cholic acids **82**, **85** and **88** and α-amino esters **83**, **86** and **89** using DEPC.[32] Condensation reaction of muricholic acid (**82**) with taurine or glycine methyl ester (**83**) provides amide **84**.[33] A novel amide **87** is prepared by coupling of cholic acid **85** to amino acid ester **86**, using DEPC as a coupling agent.[34]

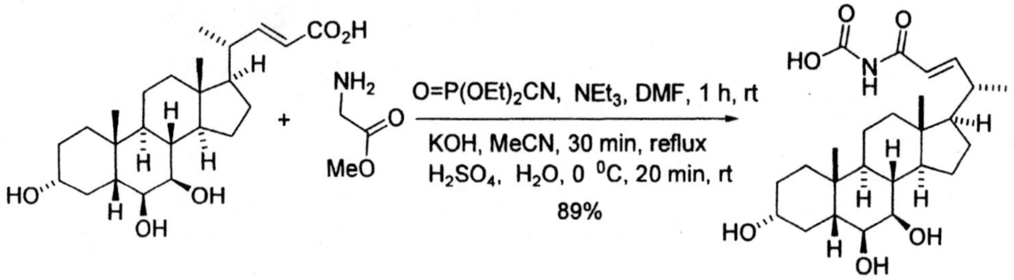

O=P(OEt)$_2$CN, NEt$_3$, DMF, 1 h, rt; KOH, MeCN, 30 min, reflux; H$_2$SO$_4$, H$_2$O, 0 °C, 20 min, rt; 89%

82 **83** **84**

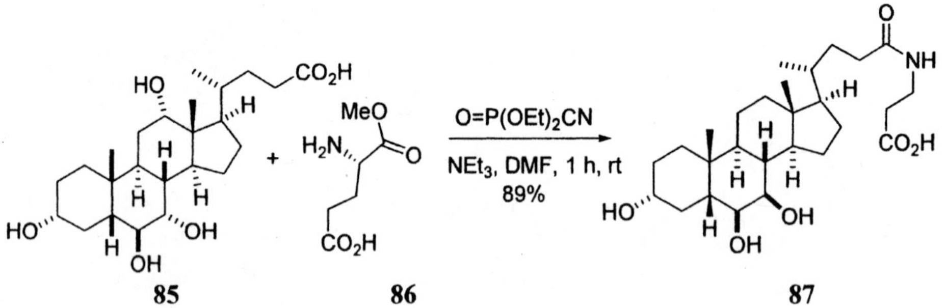

O=P(OEt)$_2$CN, NEt$_3$, DMF, 1 h, rt, 89%

85 **86** **87**

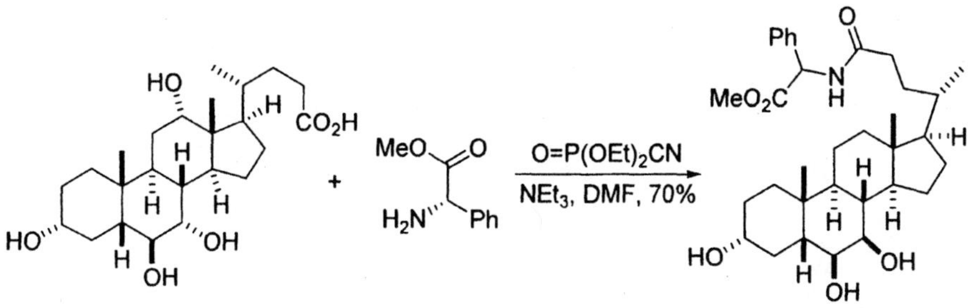

88 89 90

The indole derivative (**92**) is obtained by coupling of tryptamine and unprotected hydroxy acid (**91**) by means of DEPC to give 3-indolylethyl amide in 65% yield.[35] No considerable antibacterial activity against *Staphylococcus aureus* was observed.

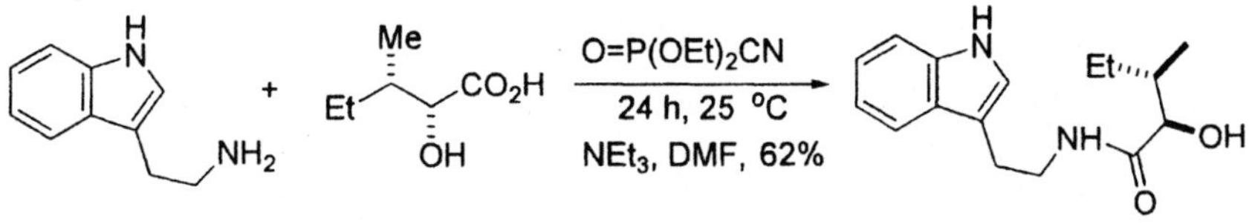

90 91 92

The antiviral marine natural product **93**, [(±)-hennoxazole A] is synthesized by a convergent approach and the final step includes the coupling of oxazole carboxylic acid (**95**) and pyran derivative (**94**) using DEPC.[36,37] Hennoxazole A is active against herpes with peripheral analgesic activity comparable to that of indomethacin.

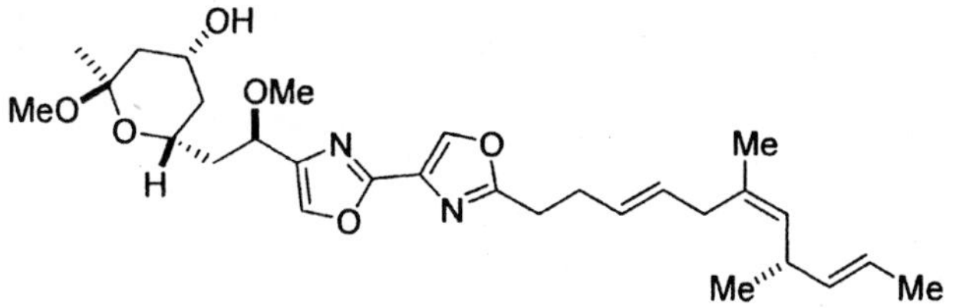

93

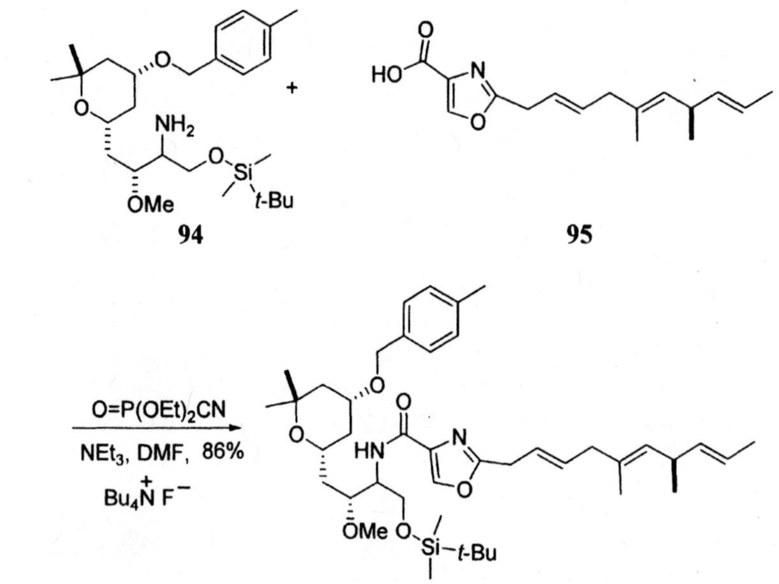

The first macrocyclic steroid derivative intended for use in molecular recognition chemistry has been synthesized by direct cyclodimerization of amide **99**, which in turn was prepared from acid **97** and amine **98** using DEPC.[38]

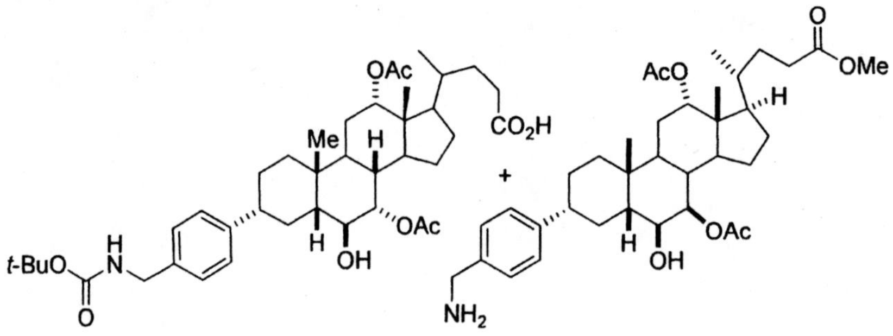

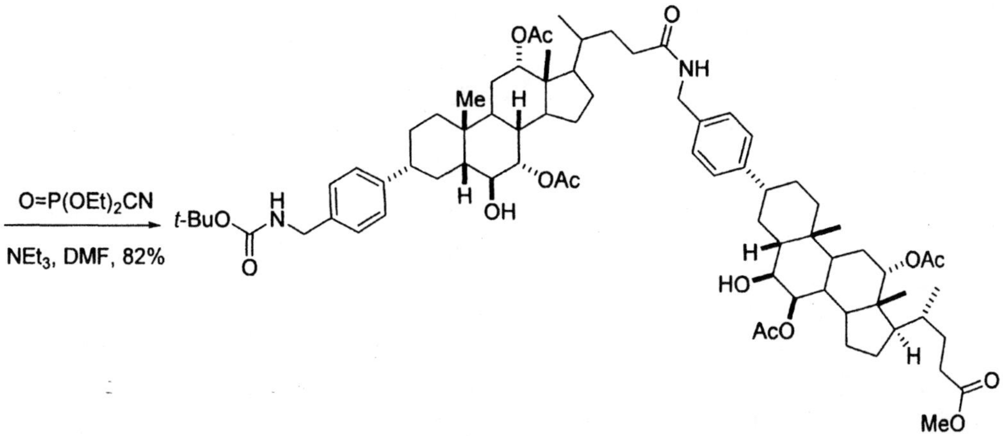

$O=P(OEt)_2CN$

NEt_3, DMF, 82%

99

Coupling of amino cyclohexene derivative **100** with 6-bromopiperonylic acid (**101**) provides amide **102**, which is a key intermediate for the synthesis of phenanthridone alkaloid lycoricidine (**103**).[39]

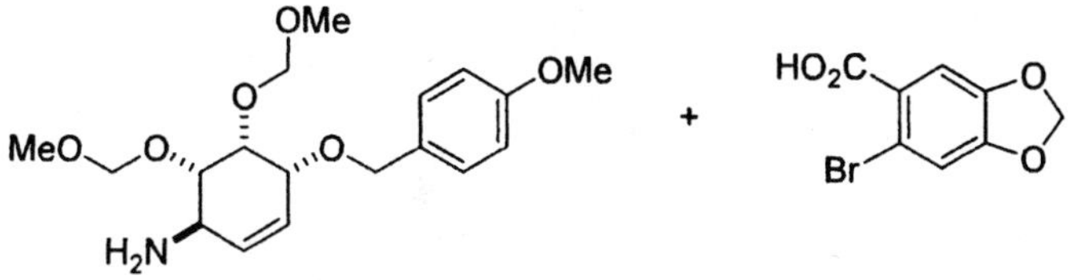

100

101

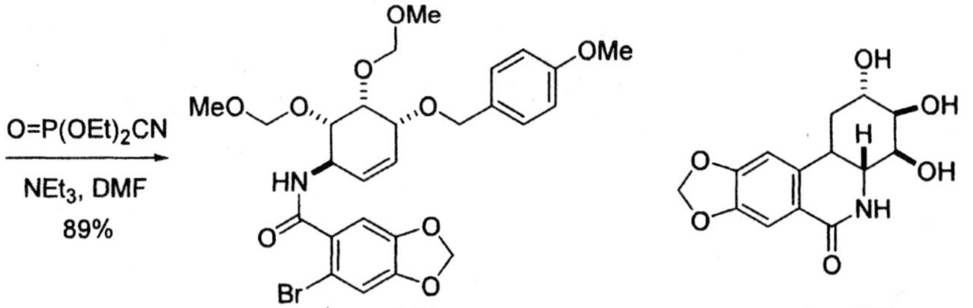

$O=P(OEt)_2CN$

NEt_3, DMF

89%

102

103

Diazonamide A is a potently antimitotic natural product and its penultimate step involves the coupling of macrocyclic amine **104** with (*S*)-α-hydroxy isovaleric acid and DEPC in 90% yield.[40,41]

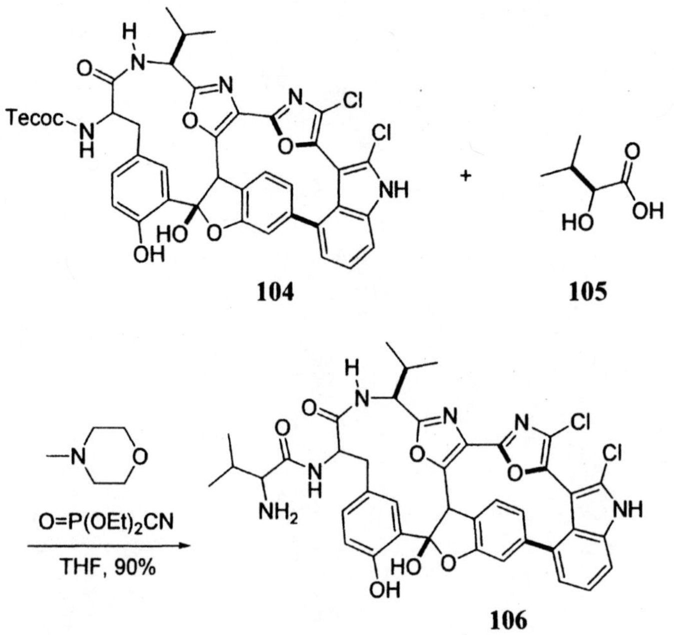

104

105

$O=P(OEt)_2CN$

THF, 90%

106

Semi-synthesis of a photoactivatable redox protein is accomplished from mono porphyrin carboxylic acid (**107**) and α-aminomethyl-2,2-bipyridine (bipy) (**108**) using DEPC.[42]

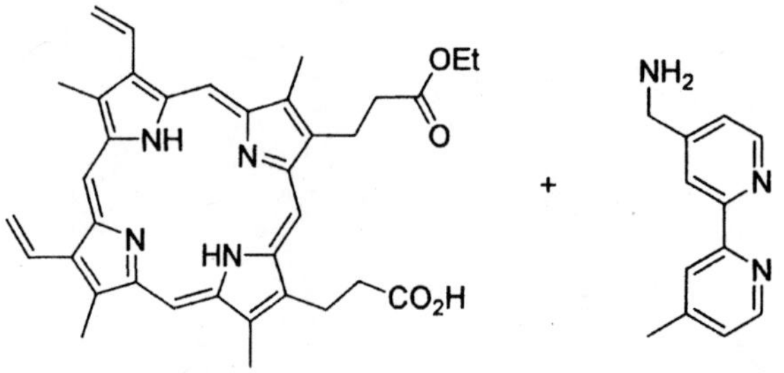

107

108

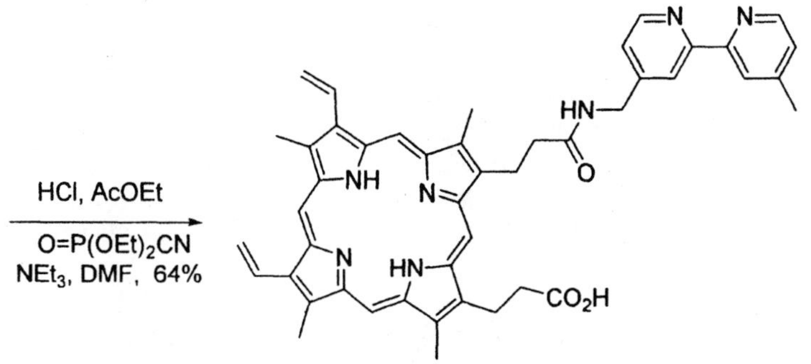

HCl, AcOEt

O=P(OEt)$_2$CN
NEt$_3$, DMF, 64%

109

A hydrophobic vitamin B_{12} derivative **(111)** is synthesized from monocyanocobyrinic acid hexamethyl ester **(110)** using DEPC.[43] The cyanide ion from DEPC is trapped by the cobalt complex as an axial ligand.[43] In this condensation reaction DEPC is a superior coupling agent compared to DCC.[43]

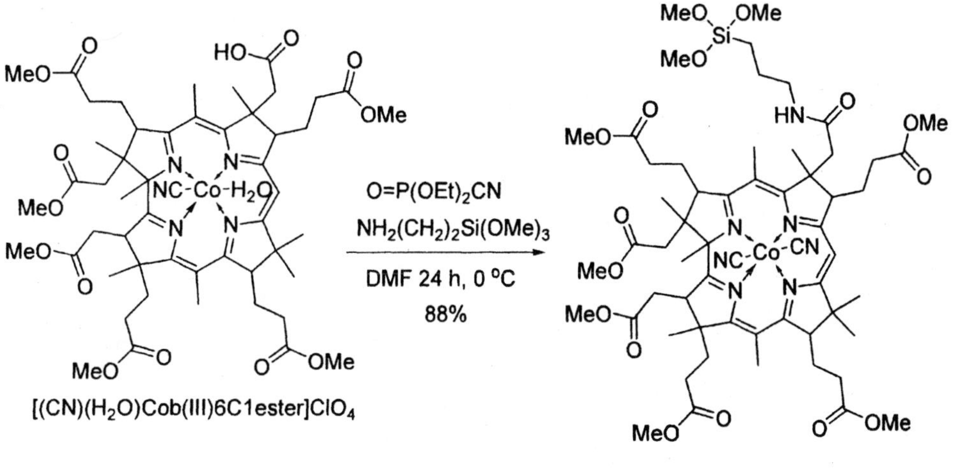

O=P(OEt)$_2$CN
$NH_2(CH_2)_2Si(OMe)_3$
DMF 24 h, 0 °C
88%

[(CN)(H_2O)Cob(III)6C1ester]ClO_4

110 **111**

6.5.5.2 ***Peptides***

There are two known methods for the formation of peptides from carbonyl and amino components. One method involves the use of adducts of phosphorus and tetrahalomethanes as coupling agents.[44] The other involves the use of coupling agents such as DEPC and diphenylphosphoryl azide (DPPA).[45a] Both DPPA and DEPC have been shown to be effective peptide coupling agents in solution and solid-phase peptide synthesis.[45b] Using these reagents, synthesis of porcine motilin, comprising 22 amino acids, is successfully achieved.[45a] Comparisons of the coupling agents such as DPPA and DEPC with DCC are often made in peptide chemistry.[45b,45c] DECP is more useful for more nucleophilic amines containing electron-donating substitutents in an aromatic ring, whereas phosphorus trichloride is effective for less nucleophilic amines. Studies of selected coupling methods for the attachment of amino acid derivatives to *cis*- and *trans*-4-aminocyclohexanecarboxylic acid have shown DEPC to be the reagent of choice as coupling agent.[46]

A study directed toward racemization during peptide coupling of *L*-proline *t*-butyl ester **112** with benzyloxycarbonyl-*L*-phenylalanyl-*L*-valine **113** revealed that, among different coupling agents, DEPC coupling afforded the least racemization.[47]

$O{=}P(OEt)_2CN$, NEt_3, DMF, 69%

112 **113** **114**

Under mild conditions, DEPC promotes the coupling of *N*-protected amino acid (**115**) with *N*-substituted amino piperidine **116** to afford substituted piperidine amide **117**.[48] Amide **117** shows *N*-type calcium channel antagonism and it is endowed with potent action on pain and neuropathic pain.

$O{=}P(OEt)_2CN$, Et_3N, DMF, 0 °C, 12 h, rt, 70%

115 **116** **117**

The synthesis of an asymmetric glycocluster polypeptide required a component **120** which has been prepared by coupling of the amino sugar **118** with β-alanine derivative **119** using DEPC.[49]

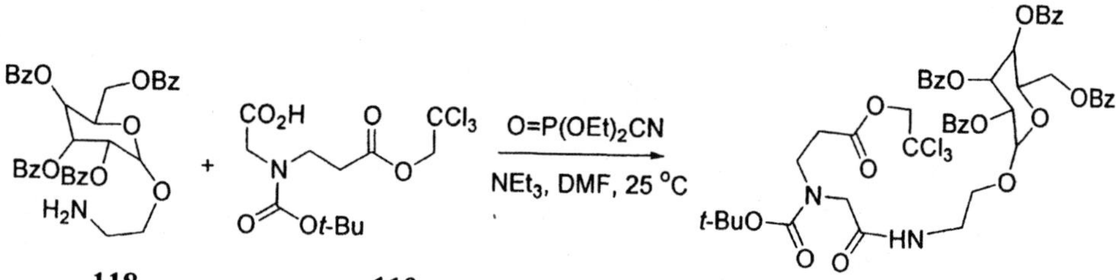

Dolastatin 10 (**123**), a potent antineoplastic and tubulin inhibitory substance, was isolated from an Indian Ocean sea hare *Dolabella auricularia.* Dolastatin 10 (**123**) is prepared by the coupling of Dov-Val-dil (**121**) with Dap-Doe (**122**) using DEPC in 97% yield.[50–52]

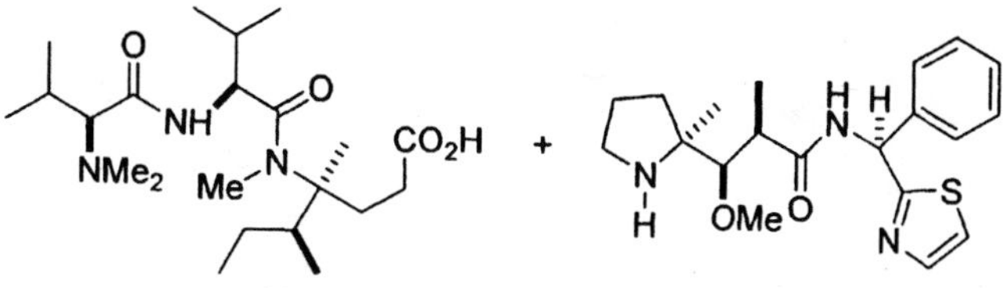

Coupling of *N*-protected amine **124** and thiazole carboxylic acid (**125**) is accomplished using the coupling agent DEPC to provide intermediate **126**, which is further cyclized to give dolastatin 3 (**127**), a unique cyclic pentapeptide containing three common amino acids, Pro, Leu, and Val, and two unusual heterocyclic amino acids, the achiral (Gly)Thz and the chiral (Gln)Thz.[53]

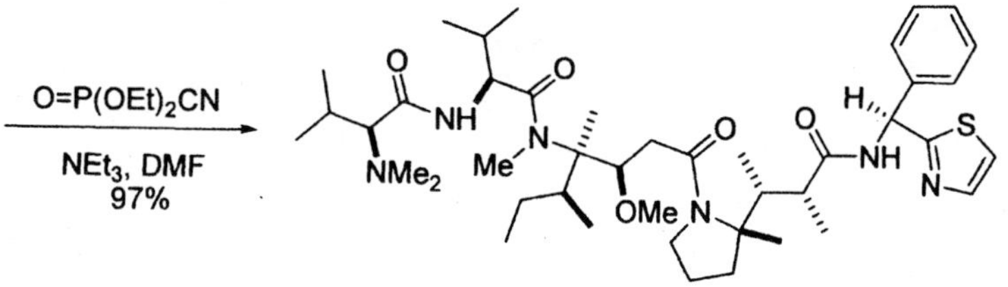

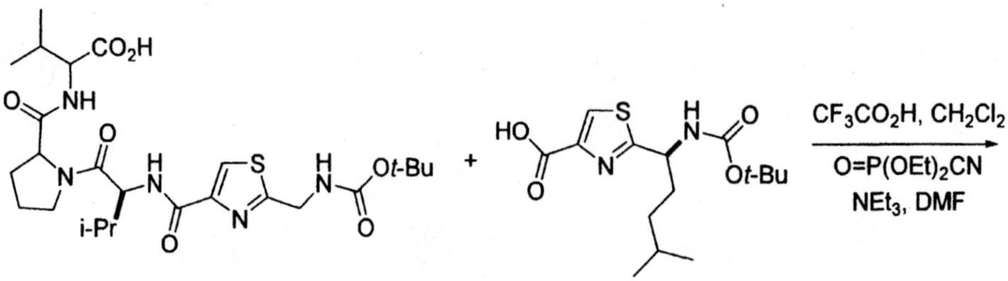

CF_3CO_2H, CH_2Cl_2
O=P(OEt)$_2$CN
NEt_3, DMF

124 125

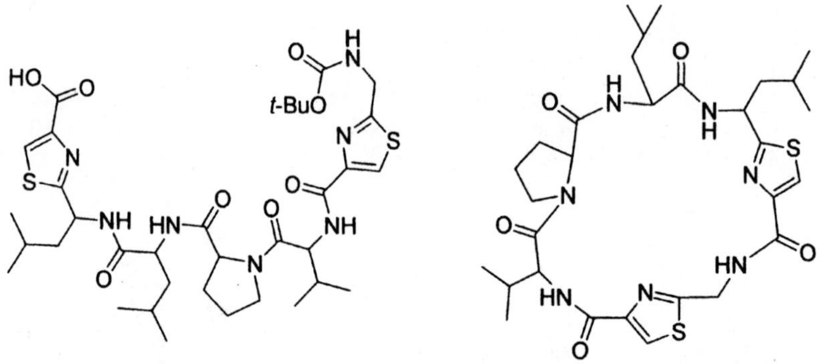

126 127, Dolastatin 3

β-Keto acid **128** and dipeptide **129** are successfully coupled in the presence of DEPC to afford dolastatin 18 (**130,** Dhex-(*S*)-Leu-*N*-Me-(*R*)-Phe-Doe) in 60% yield.[52] The structure of cell growth inhibitor dolastatin 18 is confirmed by X-ray crystal studies.

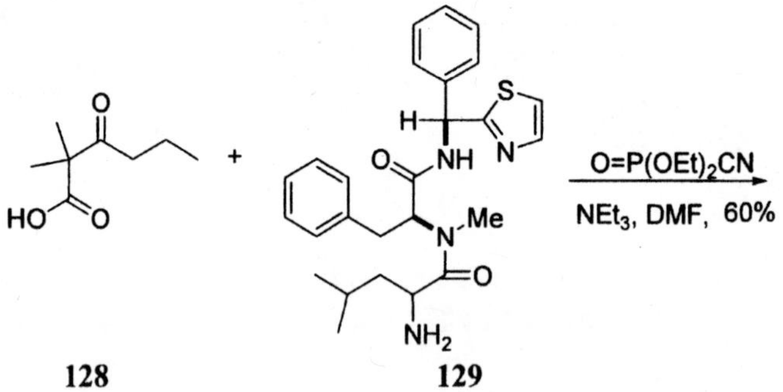

O=P(OEt)$_2$CN
NEt_3, DMF, 60%

128 129

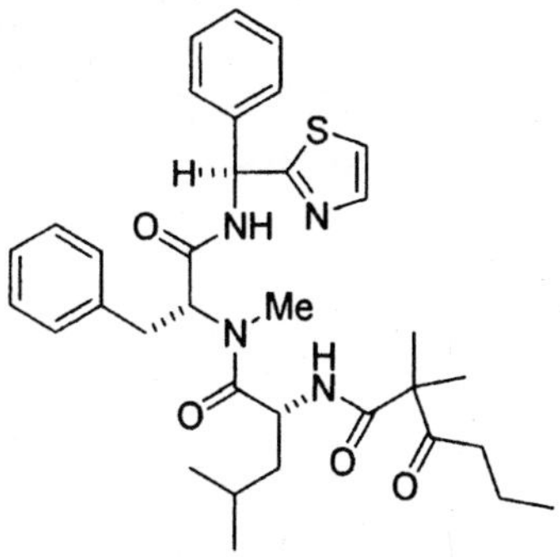

130, dolastatin 18

Antibiotics

Boc-(*S*)-2-aminobutanoic acid (**132**) is coupled with (*S*)-serine methyl ester HCl (**131**) using DEPC to afford the dipeptide **133**.[54] Dipeptide **133** is a building block for the synthesis of a cyclic peptide keramamide B.

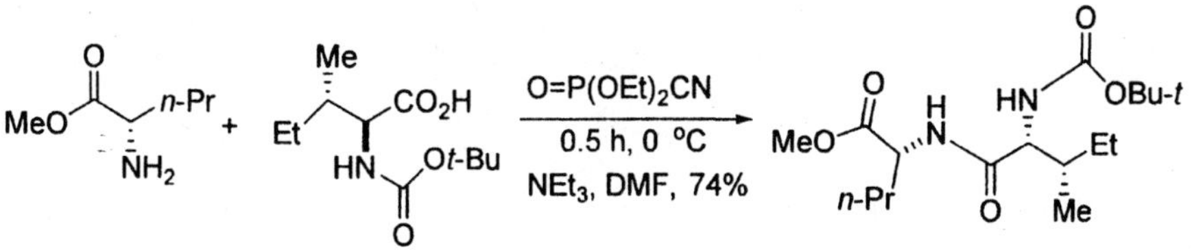

O=P(OEt)$_2$CN, 0.5 h, 0 °C, NEt$_3$, DMF, 74%

131 **132** **133**

The synthesis of the penicillin bulgecin (**136**) hybrid is achieved from benzhydryl 6-aminopenicillanate (**134**) *via* coupling with pyrrolidine carboxylic acid ester **135** using DEPC.[55]

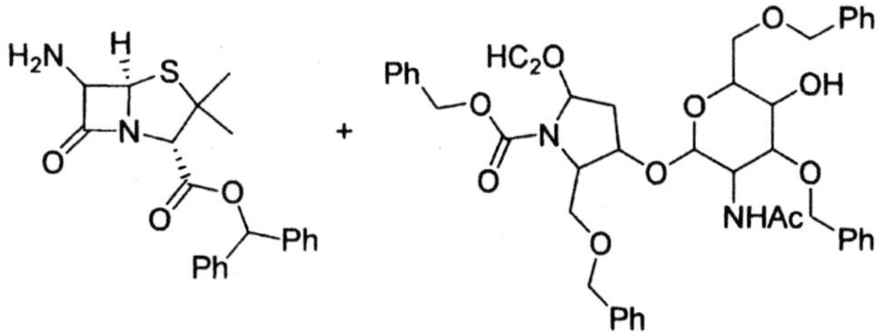

134 **135**

$O{=}P(OEt)_2CN$
NEt_3, DMF, 70%

136

N-Carbobenzoxyampicillin phenacyl ester (**139**) is conveniently prepared by the condensation reaction of *N*-carbobenzyloxy-*D*-phenylglycine (**138**) with phenacyl 6-amino-penicillanate **137**.[3]

$O{=}P(OEt)_2CN$
NEt_3, DMF, 61%

137 **138** **139**

Anticancer agents

4-(3-Substituted-2-quinoxalinylamino)phenylacetyl-*L*-glutamate (**142**) is prepared *via* coupling of acid **141** and diethyl glutamate (**140**) and has been evaluated for anticancer activity.[56]

$O{=}P(OEt)_2CN$
NEt_3, DMF, 2 h, 73%

140 **141** **142**

The methotrexate (MTX) analog (**145**) is prepared by condensing pyridopyrimidone carboxylic acid (**143**) with glutamic acid dimethyl ester (**144**).[57] Testing of amide **145** in vitro against CCRF-CEM leukemia cells revealed that it is inactive.

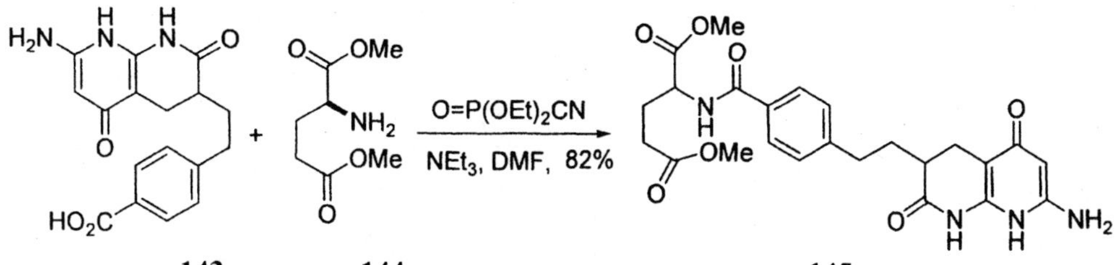

A key intermeditate **148** for the synthesis of cleavable prodrug, carboxypeptidase A is prepared *via* coupling of *N*-Boc-*L*-Orn-*L*-Phe *t*-Bu ester **147** and 4-amino-4-deoxy-1-formylpteroic acid (**146**).[58]

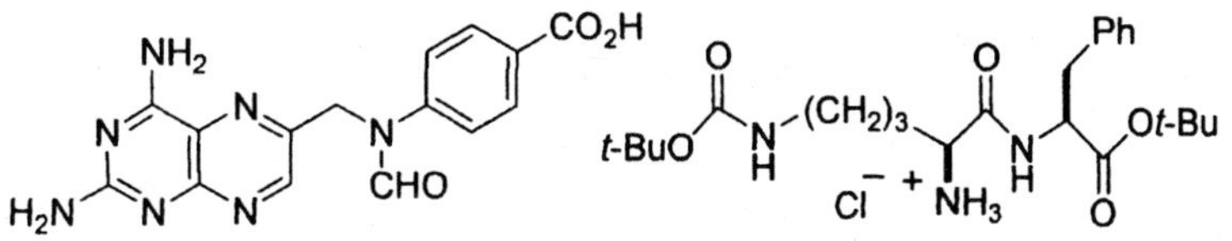

DEPC activates the coupling reaction of amido carboxylic acid **149** and tetracyclic amine **150** to produce dipeptide **151** in 88% yield.[59]

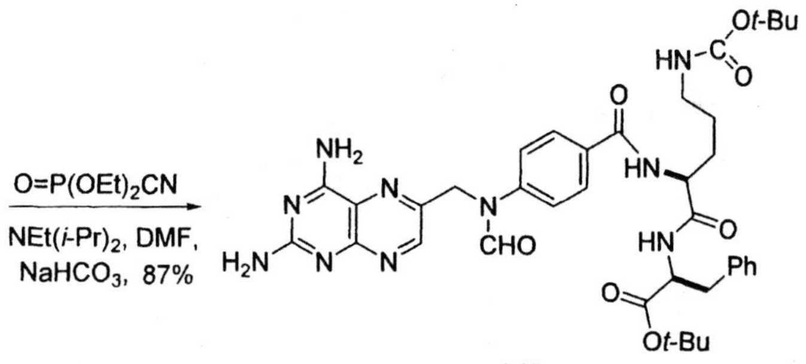

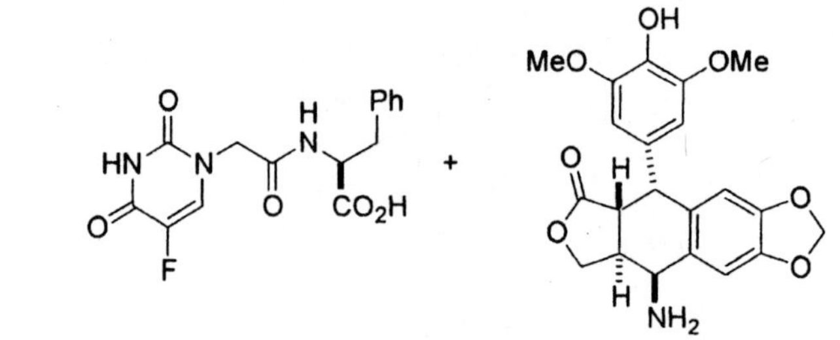

149 + 150

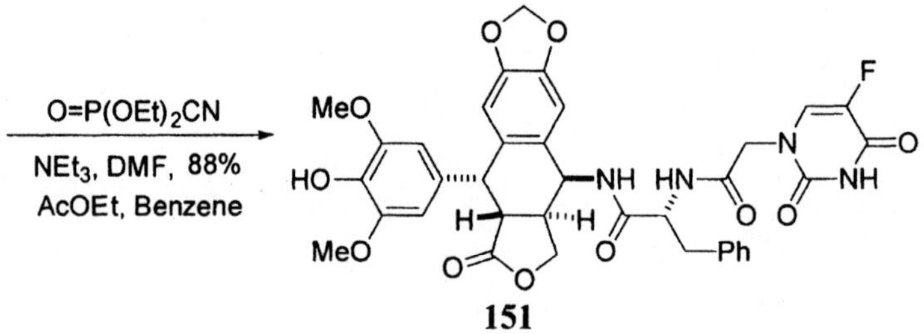

$O{=}P(OEt)_2CN$

NEt_3, DMF, 88%

AcOEt, Benzene

151

Antitumor agents

Synthesis of an analog of the antitumor drug methotrexate (**154**) is accomplished *via* condensation of the amino diester (**152**) and 4-amino-4-deoxy-*N*-methylpteronic acid (**153**) in the presence of DEPC.[60,61]

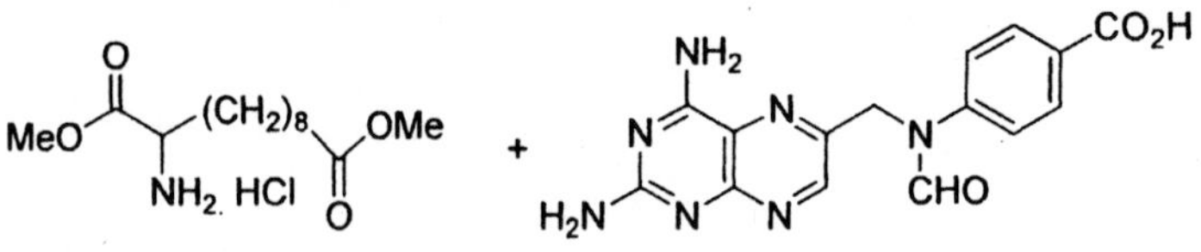

152 + 153

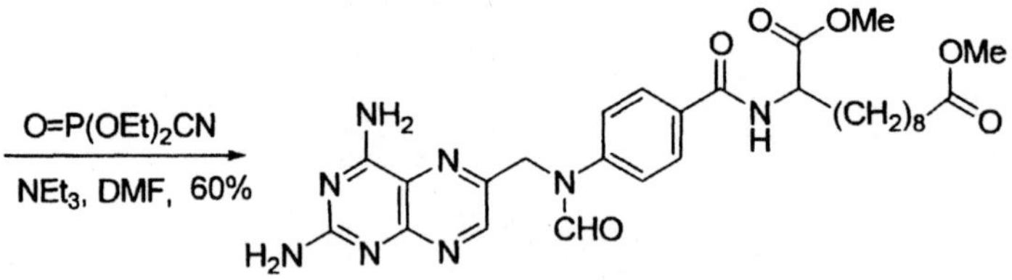

$O{=}P(OEt)_2CN$

NEt_3, DMF, 60%

154

Preparation of the peptolide lyngbyabellin A (**155**) involves the synthesis of a thiazole derivative intermediate (**158**) *via* coupling of Boc-glycine **157** with thiazole amine **156** using DEPC.[62] Lyngbyabellin A was isolated from the marine cyanobacterium *Lyngbya majuscule* collected at Apra Harbour, Guam. It exhibits attractive cytotoxic properties against the human cancer cell lines and shown to be a potent disrupter of the cellular microfilament network.

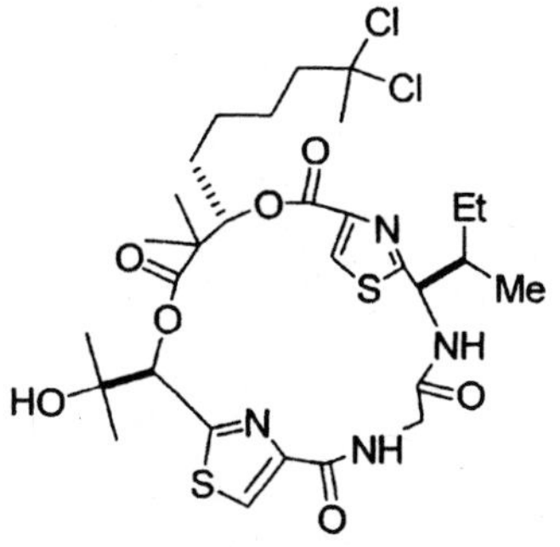

155, lyngbyabellin A

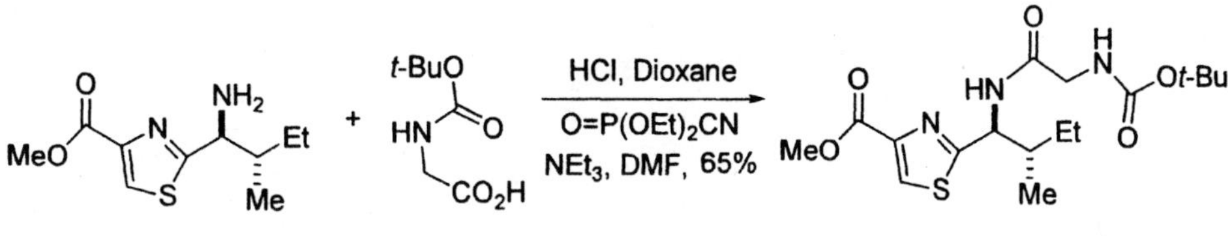

HCl, Dioxane

O=P(OEt)$_2$CN

NEt$_3$, DMF, 65%

156 **157** **158**

Thrombin inhibitors

The synthesis of the macrocyclic thrombin inhibitor cyclotheonamide B (**161**) is achieved *via* coupling of amine **159** and ornithine containing tetrapeptide acid **160**.[63] Cyclotheonamide B, isolated from a marine sponge *Theonella,* strongly inhibit various proteinase, particularly thrombin, and is the first macrocyclic thrombin inhibitor having two novel amino acid residues: a vinylogous tyrosine and α-keto homolog of arginine.

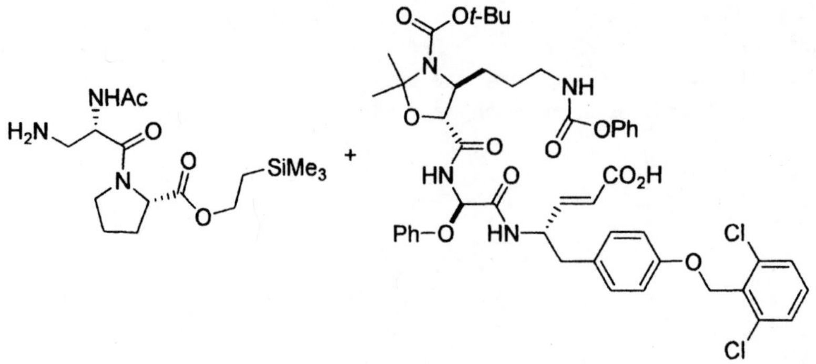

159 + 160

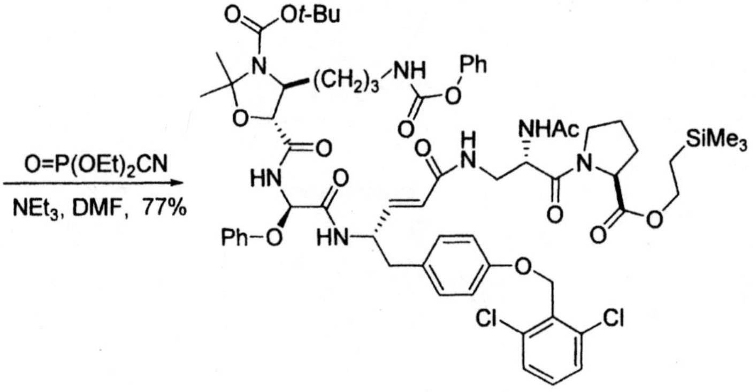

$O=P(OEt)_2CN$

NEt_3, DMF, 77%

161

Nazumamide A (**164**), a thrombin inhibitory linear tetrapeptide of marine origin, is effectively constructed using DEPC.[64] Stepwise incorporation of each amino acid residue is efficiently carried out by use of DEPC as a coupling agent.

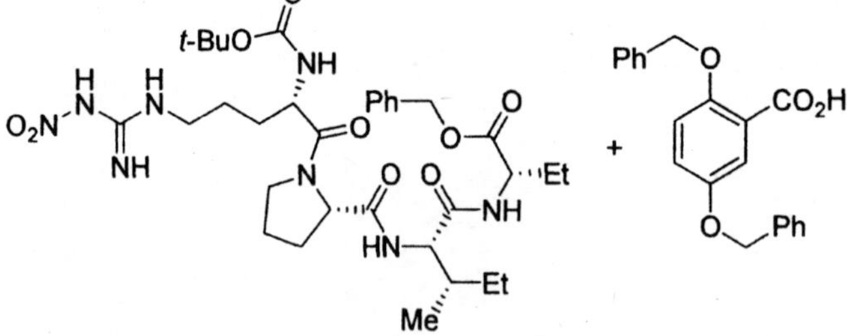

162 + 163

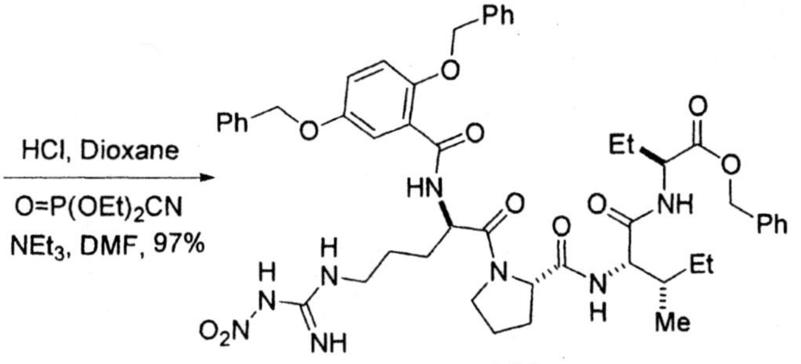

HCl, Dioxane

O=P(OEt)$_2$CN

NEt$_3$, DMF, 97%

164

Cytotoxic depsipeptides

The intermediate **167** required for the total synthesis of doliculide, a potent cytotoxic cyclodipeptide, is achieved *via* coupling of amine **165** and acid **166** using DEPC.[65] Another key component amide **169**, required for the synthesis of doliculide, is also prepared from acid **168** using DEPC.[66]

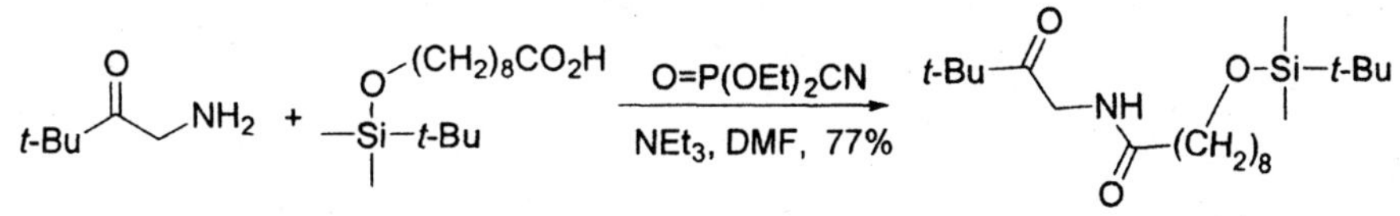

O=P(OEt)$_2$CN

NEt$_3$, DMF, 77%

165 **166** **167**

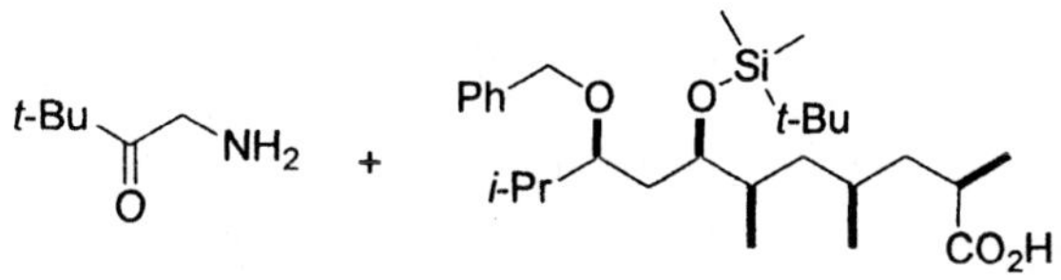

168

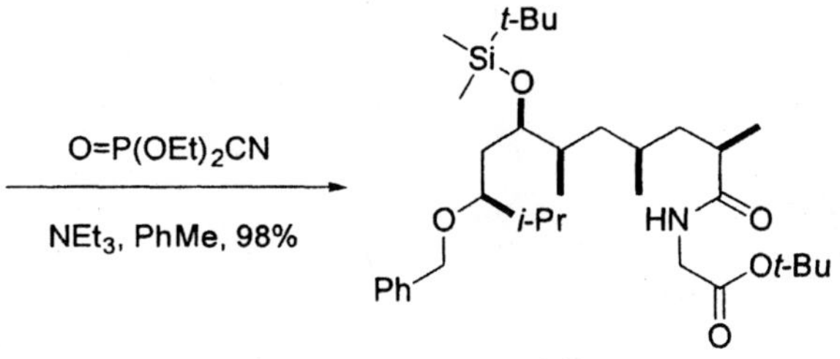

O=P(OEt)$_2$CN

NEt$_3$, PhMe, 98%

169

Synthesis of arenastatin A (**172**), an extremely potent cytotoxic depsipeptide, involves the coupling of β-alanine 2-(trimethylsilyl)ethyl ester (**170**) and the *D,O*-methyltyrosine derivative (**171**) using DEPC.[67,68]

O=P(OEt)$_2$CN

NEt$_3$, DMF, 82%

170 **171** **172**

Didemnins were isolated from a Caribbean tunicate *Trididemnum solidum*. These structurally unique cyclic depsipeptides have quite interesting cytotoxic, antiviral, and immunosuppressive activities. The synthesis of the didemnins A, B, and C (**173–175**) involves the key component **178**, which is prepared from acid **176** and amine **177** using the coupling agent DEPC.[69]

173, Didemnin A: R = H

174, Didemnin B: R =

175, Didemnin C: R =

O=P(OEt)$_2$CN

NEt$_3$, DMF, 83%

176 **177** **178**

Renin inhibitors

The structurally modified renin inhibitor **181** is prepared by the coupling of amine (**179**) and the histidine derivative (**180**) using DEPC.[70a] Dipeptide is evaluated for its susceptibility to cleavage by the serine protease chymotrypsin.

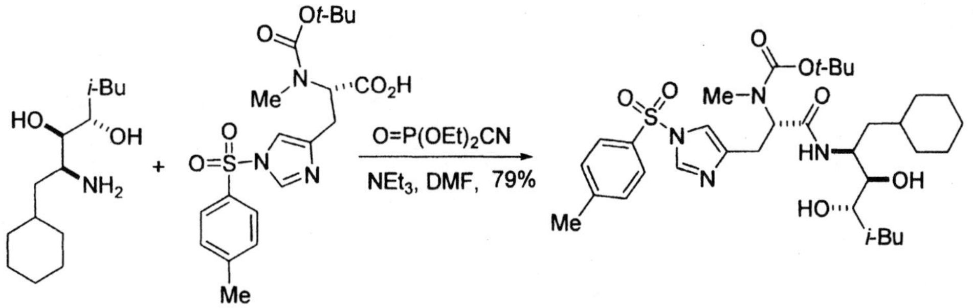

O=P(OEt)$_2$CN

NEt$_3$, DMF, 79%

179 **180** **181**

Renin inhibitory peptide **184** is prepared from methyl butyl amine **182** and γ-protected hydroxy acid **183** using DEPC.[70b]

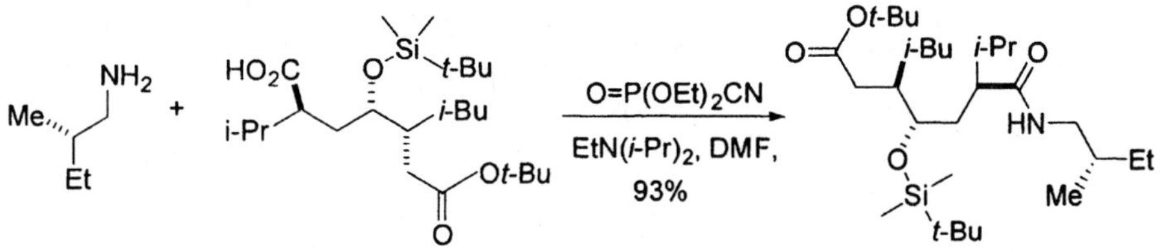

O=P(OEt)$_2$CN

EtN(i-Pr)$_2$, DMF, 93%

182 **183** **184**

Linear renin inhibitory peptide **187** is synthesized by coupling of pyridine derivative **185** and substituted pyrrolidone carboxylic acid **186** using DEPC.[71]

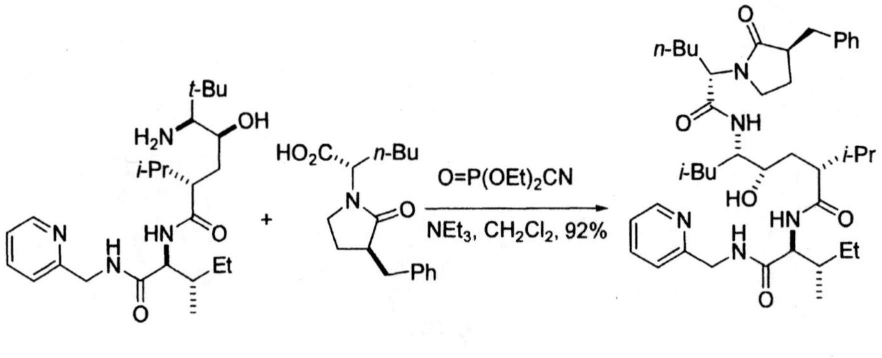

185 186 187

Renin inhibitor containing pseudopeptide component **190** is prepared *via* DEPC coupling reaction of acid **188** and amine **189**.[72]

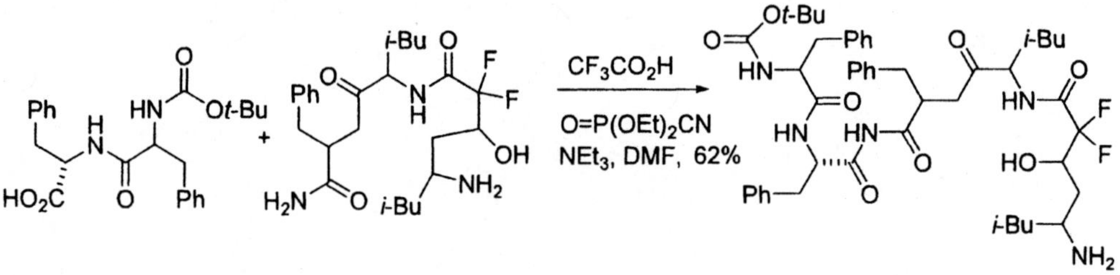

188 189 190

The synthesis of microginin, an angiotension converting enzyme inhibitory pentapeptide, involves the multiple steps wherein DEPC is used as coupling agent.[73] A key component **193** required for the synthesis of microginin is obtained from *N*-protected amino acid (**191**) and amine **192**.

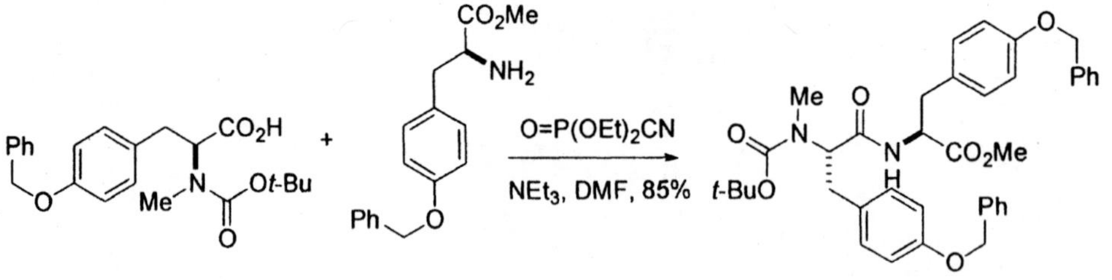

191 192 193

The *N*-Boc pentapeptide Phe-Phe-Phe-difluorostatone-Leu-Phe-NH_2 has been found to be a potent inhibitor of human renin which is obtained from a building block **196.** Peptide **196** is prepared from amine **194** and amino acid **195** using DEPC.[74]

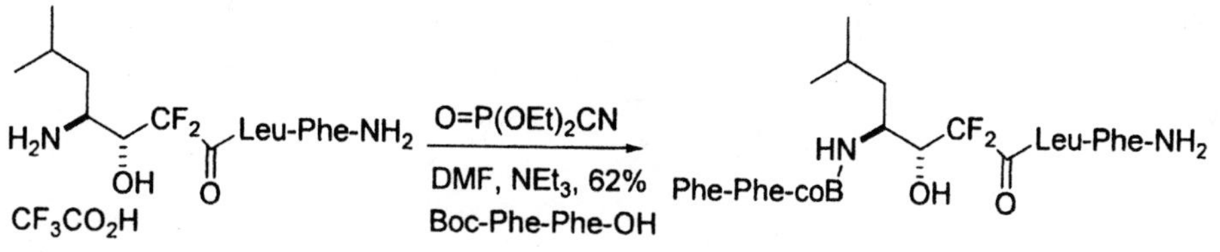
H2N
CF2
Leu-Phe-NH2
OH
O
CF3CO2H
O=P(OEt)2CN
DMF, NEt3, 62%
Boc-Phe-Phe-OH
HN
Phe-Phe-coB
CF2
Leu-Phe-NH2
OH
O

194 **195** **196**

Angiotensinogen analogs such as renin inhibitory peptide requires a key intermediate **199**, which is prepared from imidazole carboxylic acid **197** and amine **198** in the presence of DEPC.[75]

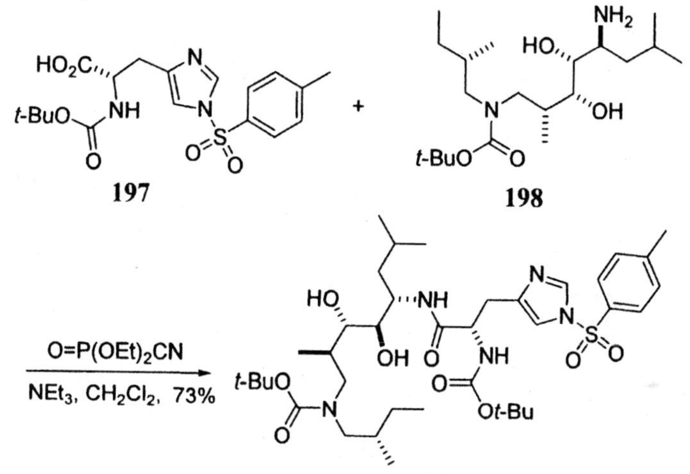
HO2C
t-BuO
NH
O
N
N
O=S
O
+
NH2
HO
N
OH
t-BuO
O
O=P(OEt)2CN
NEt3, CH2Cl2, 73%
HO
NH
OH
O
NH
O=
Ot-Bu
N
N
S
O
O
t-BuO
N
O

197 **198** **199**

Antifungal agents

Amide **202** is a key intermediate for the synthesis of various antifungal compounds (TAK–456 and TAK–457). Amide **202** is prepared *via* the coupling reaction of substituted triazole amine **200** and glycine derivative of tetrazole **201** using DEPC.[76]

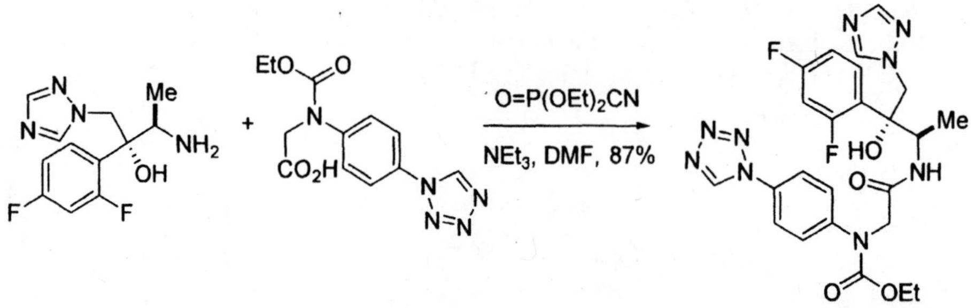

O=P(OEt)$_2$CN
NEt$_3$, DMF, 87%

200 201 202

Geodiamolide A (**203**) and jaspamide (**204**) are cytotoxic antifungal cyclic depsipeptides of marine sponge origin. The key component **207** required for their synthesis is achieved *via* coupling of acid **206** with amine **205** using DEPC.[77]

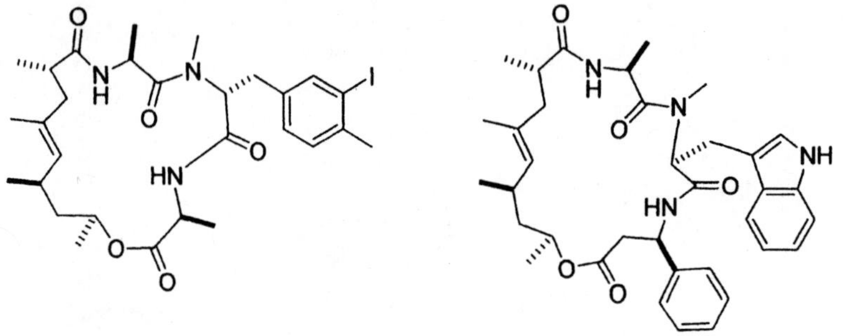

203 204

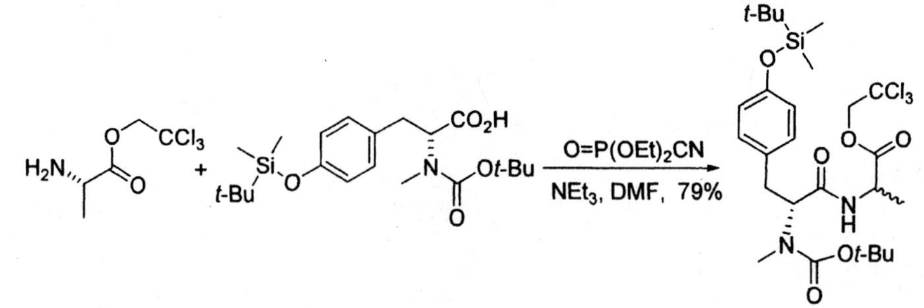

O=P(OEt)$_2$CN
NEt$_3$, DMF, 79%

205 206 207

Folic acid analogues

A series of 5,8-dideaza analogs of folic acid such as isofolic acid aminopterine have been evaluated as inhibitors of thymidylate synthase. Coupling of 10-thia-5,8-dideaza-pteroic acid (**208**) with di-*t*-Bu-glutamate (**209**) using DEPC affords amide (**210**).[78]

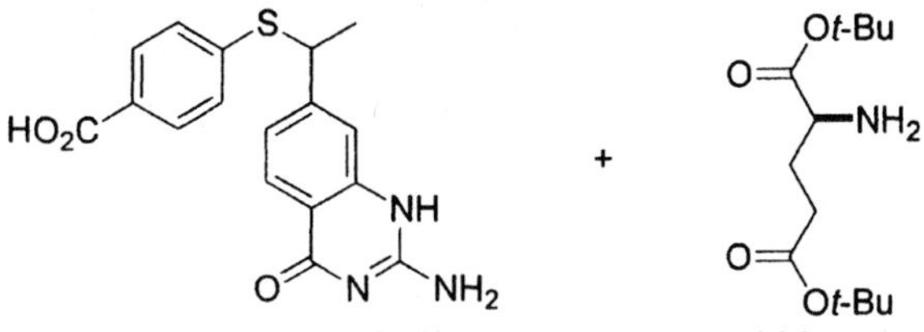

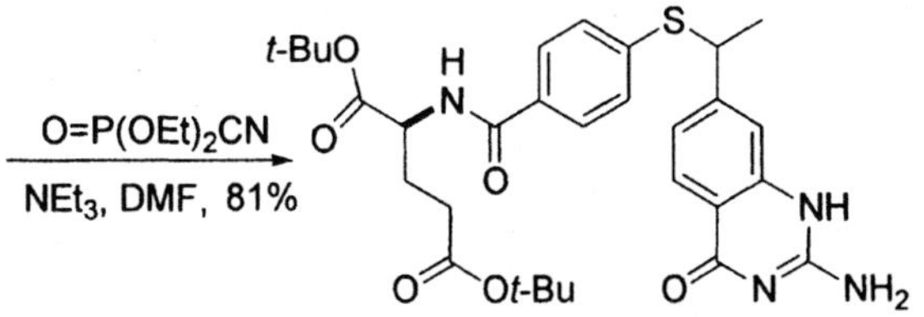

The quinazoline analog of folic acid, **212** is synthesized from substituted diaminoquinazoline (**211**) and amino diester **209** using DEPC.[79] Several quinazoline analogs of folic acid bearing Cl or Me at C-5 are evaluated as inhibitors of the four human gastrointestinal adenocarcinoma cell lines *in vitro*.

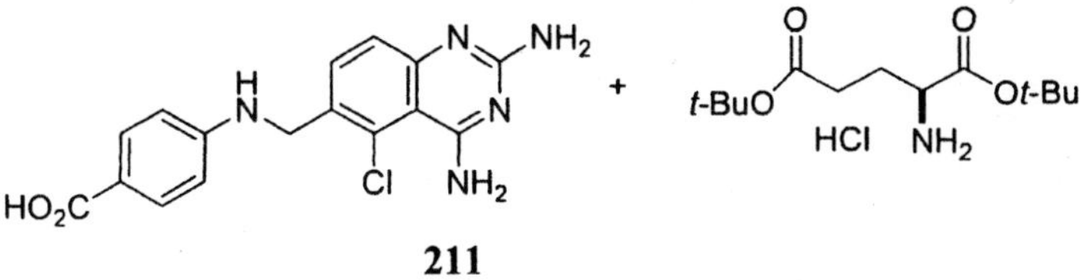

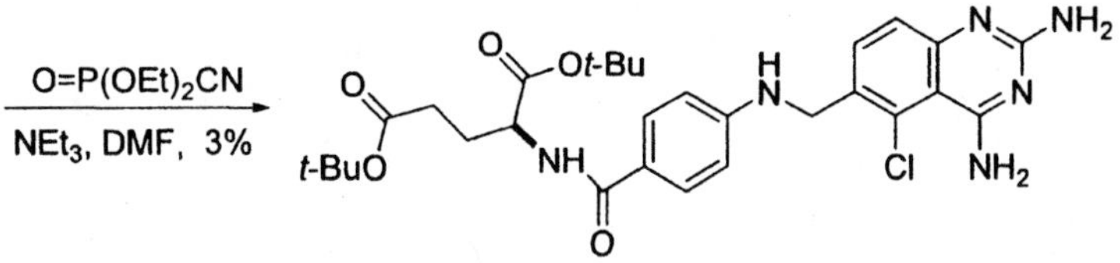

5,8-Dideaza analogs of folic acid (**214**) are prepared *via* coupling of dideazapteroic acid (**213**) with di-*t*-Bu-*L*-glutamate using DEPC.[80] 2-Desamino-5,8-dideazaisofolic acid is found to be 4–6 fold cytotoxic toward L1210 leukemia cells than their 2-NH_2 counterparts and to be poor inhibitors of mammalian thymidylate synthase.

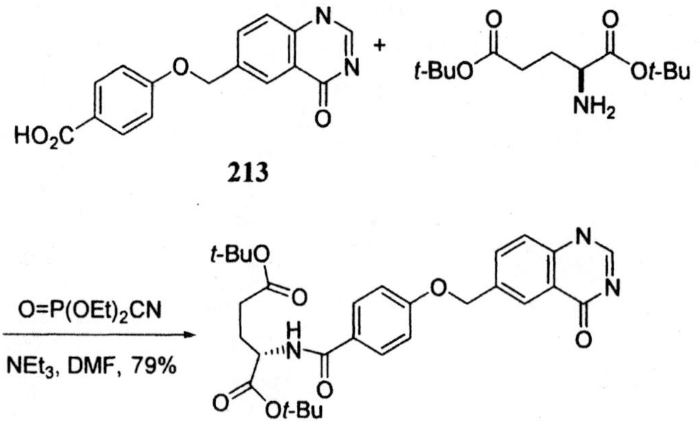
HO2C
O
N
N
O
+
t-BuO
O
O
Ot-Bu
NH2
O=P(OEt)2CN
NEt3, DMF, 79%
t-BuO
O
H
N
O
O
Ot-Bu

213

214

Condensation of the substituted quinazoline carboxylic acid (**215**) with the polyglutamyl derivative (216) using DEPC as coupling agent provides quinazoline antifolates (217).[81]

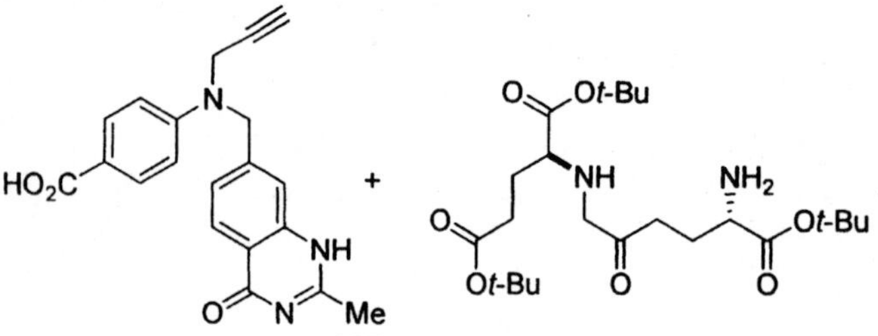
HO2C
N
NH
O
N
Me
+
O
Ot-Bu
NH
NH2
O
Ot-Bu
O
O
Ot-Bu
O

215 **216**

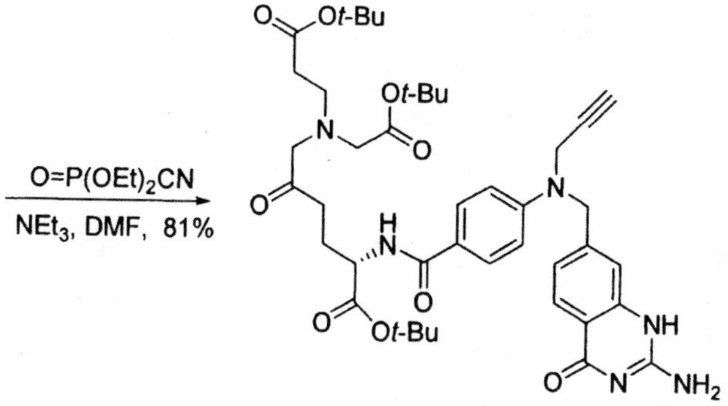

O=P(OEt)$_2$CN
NEt$_3$, DMF, 81%

217

Tripeptide **220** is prepared from amino pyrrole **218** and pyrrole carboxylic acid **219** using DEPC and NEt_3.[82]

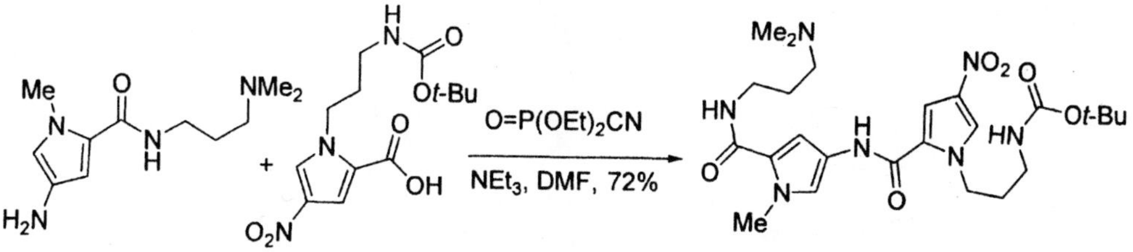

O=P(OEt)$_2$CN
NEt$_3$, DMF, 72%

218 **219** **220**

Synthesis of the bicyclic dodecapeptide theonellamide F is achieved *via* cyclic peptide **224** which in turn is prepared from building block **223**. The key intermediate **223** is prepared by coupling of tetrapeptide **221** and a β-alanine derivative (**222**).[83]

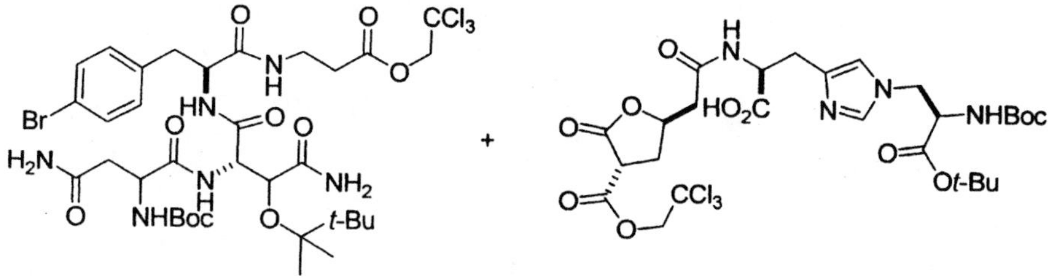

221 **222**

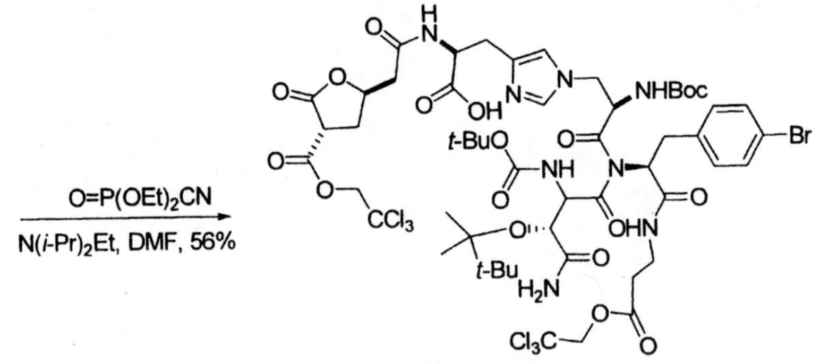

O=P(OEt)$_2$CN

N(i-Pr)$_2$Et, DMF, 56%

223

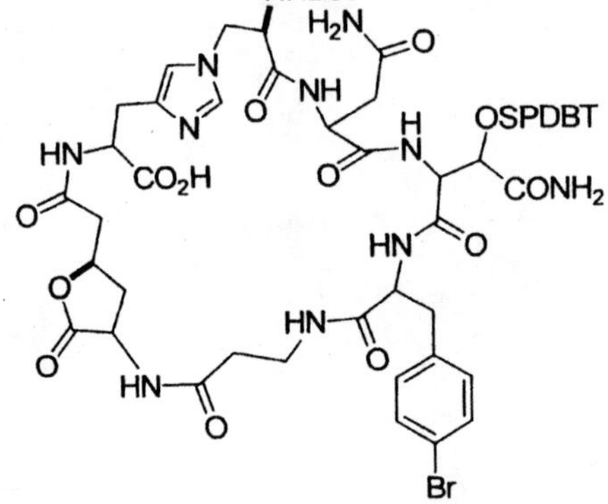

224

The pentapeptides avellanins A and B (**228**) are isolated from fungus *Hamigera avellanea*, which cause remarkable increase in blood pressure in rats and mice. They are synthesized in a stepwise manner from dipeptide **225** and acid **226** using DEPC.[84]

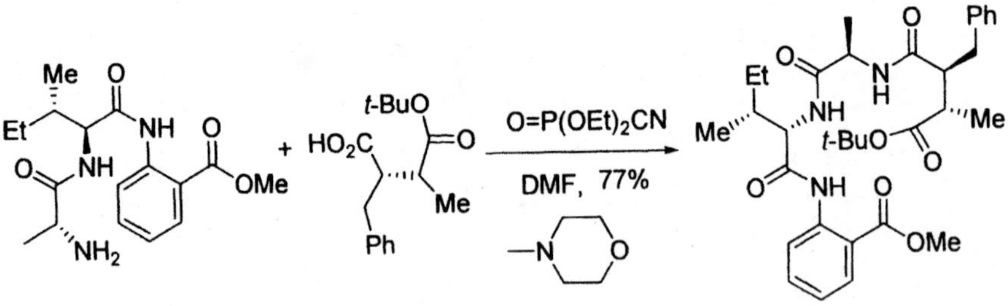

O=P(OEt)$_2$CN

DMF, 77%

225 **226** **227**

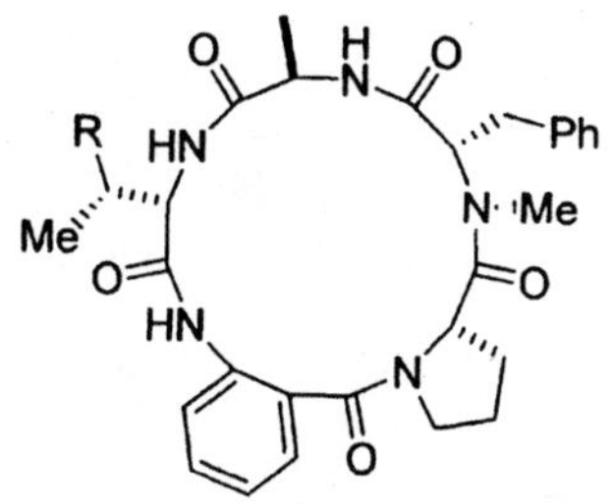

228 avellanin A (R = Et), avellanin B (R = Me)

The mammalian folypolyglutamate synthetase inhibitor **230** is prepared from 2-desamino-10-thiapteroic acid (**229**) *via* coupling with glutamate using DEPC.[85]

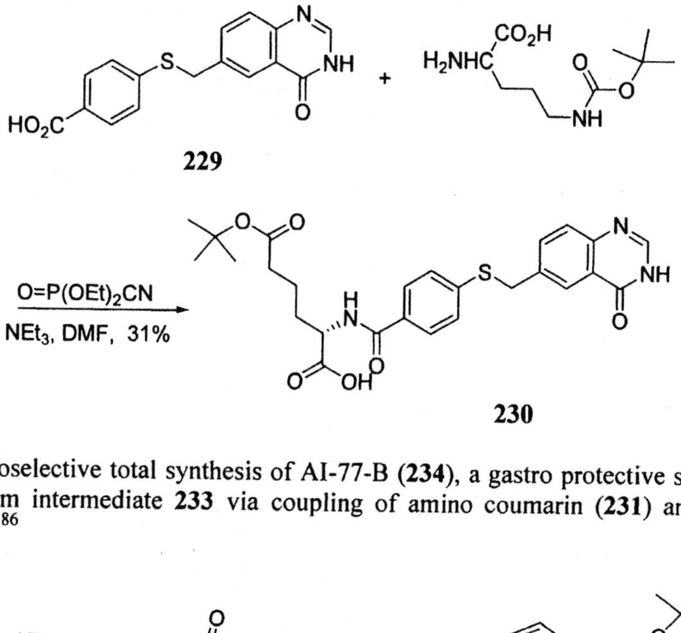

229

$O{=}P(OEt)_2CN$

NEt_3, DMF, 31%

230

Stereoselective total synthesis of AI-77-B (**234**), a gastro protective substance, is achieved from intermediate **233** via coupling of amino coumarin (**231**) and acid **232** using DEPC.[86]

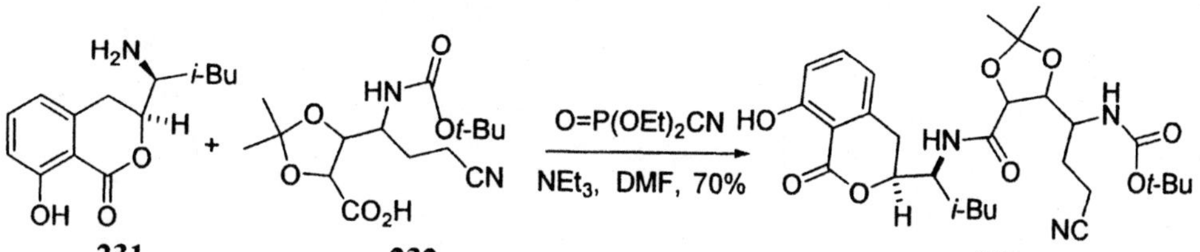

$O{=}P(OEt)_2CN$

NEt_3, DMF, 70%

231 **232** **233**

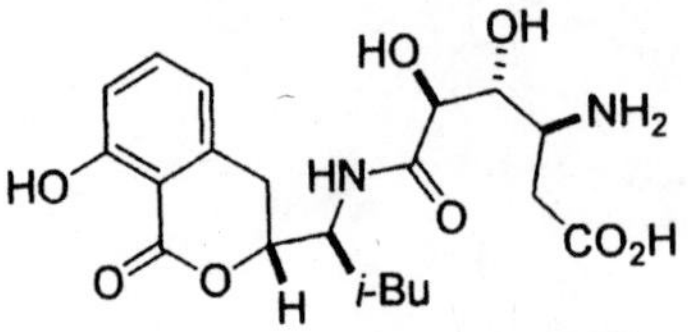

234 (AI-77-B)

The oxazole derivative **237** is a key component for the synthesis of cell permeable phosphatase inhibitors calyculins A and B and it is prepared *via* coupling of alkylamino oxazole **235** and *N*-protected acid **236**.[87]

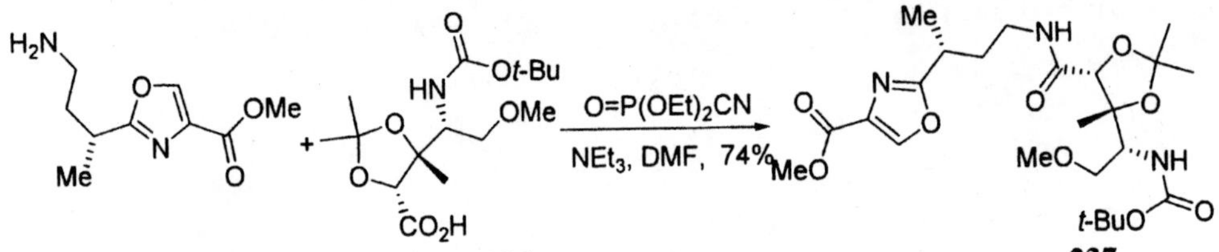

$O=P(OEt)_2CN$, NEt_3, DMF, 74%

235 **236** **237**

The pyrrolopyrimidine antifolate masked ornithine analog **240** is prepared from methyl esters of lysine **238** and substituted benzoic acid **239** using DEPC.[88]

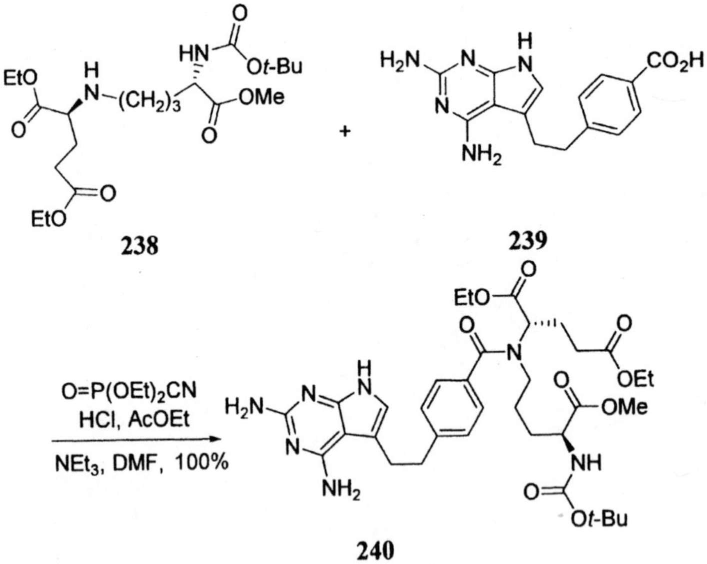

238 **239**

$O=P(OEt)_2CN$, HCl, AcOEt, NEt_3, DMF, 100%

240

Condensation of the hydroxy acid (**241**) with glycine *t*-butyl ester hydrochloride (**242**) smoothly proceeds in the presence of DEPC and NEt_3 to give the amide (**243**).[89] Amide **243** is used as an intermediate for the synthesis of topostin B567 (**244**) and topostin D654 (**245**); both have proved to be novel inhibitors of mammalian DNA topoisomerase.[89]

HCl, AcOEt

$O{=}P(OEt)_2CN$

NEt_3, DMF, 96%

t-BuO, $NH_2{\cdot}HCl$

241 **242** **243**

244 **245**

Tyrosine kinase inhibitors are prepared from key intermediate **247** *via* condensation of tryptophan (**246**) and benzylamine using DEPC.[90]

EtO_2CCF_3

$PhCH_2NH_2$

$O{=}P(OEt)_2CN$

NEt_3, DMF, 80%

246 **247**

The dipeptide **250** is prepared from amino adenine **248** and Boc-*D*-Ala **249** using DEPC as the coupling agent.[91] The amino acid carrying the adenine in the side chain has been employed for the synthesis of the naturally occurring dipeptide NK374200, a novel insecticide agent isolated from the culture of the fungus *Talaromyces sp.*[91]

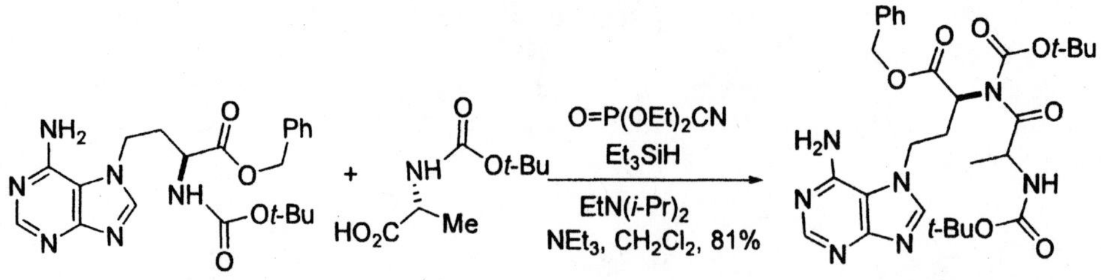

248 249 250

Chromogenic and fluorogenic peptidase substrate **252** is prepared from α-amino acid **251** and amine using DEPC.[92] H-Gly-O*t*-Bu is condensed with Boc-*L*-glu(OMe)-OH using DEPC.[92]

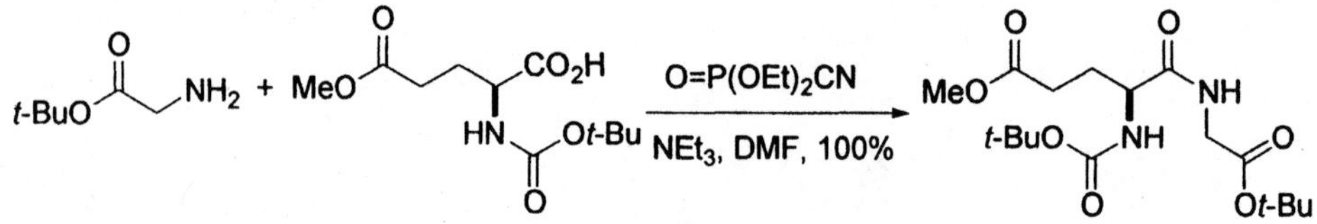

251 252

Synthesis of the cyclic hexapeptide echinocandin D (**256**) involves a key intermediate tetrapeptide **255** which is prepared from **253** and *N*-linoleyl-*N*-Boc ornithine (**254**) using DEPC.[93] Echinocandin D, isolated from *Aspergillus ruglosus*, is a member of a family of lipopeptides possessing high antifungal activity.

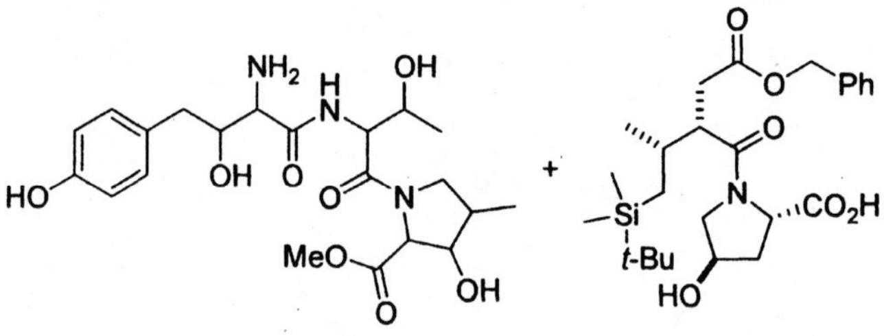

253 254

O=P(OEt)$_2$CN

NEt$_3$, DMF, 86%

255

256

6.5.6 *Experimental*

28 + 32 (HCl) → 33

O=P(OEt)$_2$CN, CH$_2$(OMe)$_2$, NEt$_3$, 98%

Methyl (2*S*)-*N*-(2-aminopyrazinecarbonyl)pyrrolidine-2-carboxylate (33)[13]

To a mixture of 3-aminopyrazine-2-carboxylic acid (**22**) (714 mg, 5.13 mmol) and *L*-proline methyl ester hydrochloride (**32**) (1020 mg, 6.16 mmol, 1.2 equiv) in dry DME (30.0 mL) was added dropwise DEPC (93 %, 1.00 mL, 6.16 mmol, 1.2 equiv) and triethylamine (1.70 mL, 12.2 mmol, 2.4 equiv), respectively, at 0 °C. The resultant solution was stirred at 0 °C for 1 h and at 40 °C for 1 h under nitrogen. The mixture was diluted with ethyl acetate (500 mL) and washed with water (2 × 50 mL), saturated sodium hydrogen carbonate solution (2 × 50 mL), water (2 × 50 mL), and saturated sodium chloride solution (50 mL) successively. The combined organic layer was dried ($MgSO_4$) and evaporated under reduced pressure to afford the crude product, which was purified on a silica gel column chromatography using A and H (1:2, v/v) as an eluent to give amide derivative **33** (oil, 1259 mg, 5.03 mmol, 98%). A mixture of *syn* and *anti* rotamers (57:43); yield 98 %; *Rf* = 0.18 (A:H 1:1); oil; $[\alpha]_D^{25} = -52.3°$ (*c* 0.68, $CHCl_3$).

HO O O H S NH N O Me N O NH2 + O=P(OEt)2CN NEt3, DMF, 60%

128 **129**

H S O NH N Me N H N O O O

130, dolastatin 18

Dolastatin 18 (2, Dhex-*L*-Leu-*N*-Me-*D*-Phe-Doe 130)[52]

A solution of 2,2-dimethyl-3-oxohexanoic acid (**128**, 30 mg) in DCM (3 mL) was added to a stirred solution of tripeptide **129** (92 mg) in DCM (2 mL), and the mixture was cooled to 0 °C under argon. Triethylamine (1 mL) and DEPC (0.05 mL) were added, and the mixture was stirred for 2.5 h, warming to room temperature. Removal of solvent yielded a creamy oil that was fractionated by flash chromatography (5:1 hexane-acetone) to give dolastatin 18 (**130**, 40.5 mg; 0.065 mmol). Further purification on SephadexLH-20 [hexane-toluene-methanol (3:1:1)] gave a colorless solid that crystallized from acetone-hexane in a cluster of needles: mp 108–109 °C; $[\alpha]_D^+$ 44 (*c* 0.05 CH_2Cl_2).

$[(CN)(H_2O)Cob(III)6C1ester]ClO_4$

$O{=}P(OEt)_2CN$
$NH_2(CH_2)_2Si(OMe)_3$
DMF 24 h, 0 °C
88%

110 **111**

Vitamin B_{12} derivative (111)[43]

A hydrophobic vitamin B_{12} derivative (**111**) was synthesized from monocyanocobyrinic acid hexamethyl ester (**110**) $[(CN)(H_2O)Cob(III)6C1ester]ClO_4$. In 10 mL of dry DMF, 50 mg (0.047 mmol) of $[(CN)(H_2O)Cob(III)6C1ester]ClO_4$ was dissolved and the brown solution was cooled to 0 °C using an ice-bath under nitrogen atmosphere. To the solution was added 15 mg (0.092 mmol) of diethylphosphoryl cyanide (DEPC). After stirring for 30 min at 0 °C, 18.1 mg (0.08 mmol) of 3-(trimethoxysilyl)propylamine and 16 mL (0.096 mmol) of triethylamine were added and further stirred for 6 h at 0 °C, then 15 h at room temperature under nitrogen atmosphere. To the resulting purple solution was added 60 mL of dichloromethane and washed with water (3 × 60 mL). After drying over anhydrous Na_2SO_4, the dichloromethane extract was concentrated to dryness. The product was reprecipitated from benzene upon addition of hexane to afford a purple powder. It is noteworthy that DEPC is a superior reagent for this condensation reaction compared to other reagents, such as *N,N*-dicyclohexylcarbodiimide. The product was characterized by UV–VIS, NMR, IR, and HR mass (FAB) spectroscopies. Yield 88%.

6.5.7 References

1 [R] Li, J. J. *Name Reactions*, 2nd ed., Springer-Verlag: New York, 2003, p447.
2 Shioiri, T.; Yokoyama, Y.; Kasai, Y.; Yamada, S.-I. *Tetrahedron,* **1976**, *32*, 2211.
3 Yamada, S.-I.; Kasai, Y.; Shioiri, T. *Tetrahedron Lett.* **1973**, *14*, 1595.
4 Shioiri, T.; Hamada, Y. *J. Org. Chem.* **1978**, *43*, 3631.
5 Kato, N.; Hamada, Y.; Shioiri, T. *Chem. Pharm. Bull.* **1984**, *32*, 3323.
6 Shioiri, T.; Ninomiya, K.; Yamada, S.-I. *J. Am. Chem. Soc.* **1972**, *94*, 6203.
7 Yokoyama, Y.; Shioiri, T.; Yamada, S.-I. *Chem. Pharm. Bull.* **1977**, *25*, 2423.
8 Mizuno, M.; Shioiri, T. *Tetrahedron Lett.* **1998**, *39*, 9209.
9 Hamada, Y.; Mizuno, A.; Ohno, T.; Shioiri, T. *Chem. Pharm. Bull.* **1984**, *32*, 3683.
10 Shirai, R.; Aoki, K.; Sato, D.; Kim, H.-D.; Murakata, M.; Yatsuro, T.; Koga, K. *Chem. Pharm. Bull.* **1994**, *42*, 690.
11 Ohba, M.; Natsutani, I.; Sakuma, T. *Tetrahedron Lett.* **2004**, *45*, 6471.
12 Okawa, T.; Kawase, M.; Eguchi, S.; Kakehi, A.; Shiro, M. *J. Chem. Soc. Perkin. Trans. 1*, **1998**, 2277.
13 Okawa, T.; Eguchi, S. *Tetrahedron* **1998**, *54*, 5853.
14 Okawa, T.; Eguchi, S. *Tetrahedron Lett.* **1996**, *37*, 81.
15 Manabe, K. *Tetrahedron* **1998**, *54*, 14465.
16 Sugiyama, M.; Sakamoto, T.; Tabata, K.; Endo, K.; Ito, K.; Kobayashi, M.; Fukumi, H. *Chem. Pharm. Bull.* **1989**, *37*, 2091.
17 Mutoh, R.; Shirai, R.; Koiso, Y.; Iwasaki, S. *Heterocycles* **1995**, *41*, 9.
18 Friedlos, F.; Davies, L.; Scanlon, I.; Ogilvie, L. M.; Martin, J.; Stribbling, S. M.; Spooner, R. A.; Niculescu-Duvaz, I.; Marais, R.; Springer, C. J. *Cancer Res.* **2002**, *62*, 1724.
19 Koseki, Y.; Nagasaka, T. *Chem. Pharm. Bull.* **1995**, *43*, 1604.
20 Kiho, T.; Nakayam, M.; Kogen, H. *Tetrahedron*, **2003**, *59*, 1685.
21 Kasemura, K.; Yamamoto, A.; Nakano, T.; Murakami, H.; Fujihara, Y.; Nomura, M. *Bokin Bobai* **2001**, *29*, 689.
22 Chen, C.-A; Sieburth, S. M.; Glekas, A.; Hewitt, G. W.; Trainor, G. L.; Erickson-Viitanen, S.; Garber, S. S.; Cordova, B.; Jeffry, S.; Klabe, R. M. *Chem. Bio.* **2001**, *8*, 1161.
23 Shioiri, T.; Irako, N. *Tetrahedron* **2000**, *56*, 9129.
24 Trost, B.M.; Dudash, J.; Dirat, O. *Chem. Eur. J.* **2002**, *8*, 259.
25 Chida, N.; Tobe, T.; Okada, S.; Ogawa, S. *J. Chem. Soc. Chem. Commun.* **1992**, 1064.
26 Horni, A.; Linden, A.; Hesse, M. *Helv. Chim. Acta.* **1998**, *81*, 1303.
27 Nazare, M; Waldmann, H. *Chem. Euro. J.* **2001**, *7*, 3363.
28 Fujii, T.; Ohba, M.; Hatakeyama, H. *Chem. Pharm. Bull.* **1987**, *35*, 2355.
29 Sugihara, H; Mabuchi, H; Kawamatsu, Y. *Chem. Pharm. Bull.* **1987**, *35*, 1919.
30 Mori, S.; Aoyama, T.; Shioiri, T. *Tetrahedron Lett.* **1986**, *27*, 6111.
31 Koseki, Y.; Katsura, S.; Kusnano, S.; Sakata, H.; Sato, H.; Monzene, Y.; Nagaska, T. *Heterocycles* **2003**, *59*, 527.
32 Willemen, H. M.; Vermonden, T.; Koudijs, A.; Marcelis, A. T. M.; Sudhölter, E. J. R. *Colloids and Surfaces, A: Physicochemical and Engineering Aspects,* **2003**, *218*, 59.
33 Kakiyama, G.; Iida, T.; Yoshimoto, A.; Goto, T.; Mano, N.; Goto, J.; Nambara, T.; Hagey, L. R.; Hofmann, A. F. *J. Lipid Res.*, **2004**, *45*, 567.
34 Willemenn, H. M.; Vermonden, T.; Marcelis, A. T. M. M.; Sudhölter, E. J. R. *Eur. J. Org.Chem.* **2001**, *66*, 2329.
35 Paik, S.; Park, M.K.; Jhun, S. H.; Park, H. K.; Lee, C. S.; Cho, B. R.; Byun, H. S.; Choe, S. B.; Suh, S. *Bull. Korean Chem. Soc.* **2003**, *24*, 623.
36 Yokokawa, F.; Asano, T.; Shioiri, T. *Tetrahedron* **2001**, *57*, 6311.
37 Yokokawa, F.; Asano, T.; Shioiri, T. *Org. Letters* **2000**, *2*, 4169.
38 Bonar-Law, R.; Davis, A. P. *Tetrahedron* **1993**, *49*, 9829.
39 Chida, N.; Ohtsuka, M.; Ogawa, S. *Tetrahedron Lett.* **1991**, *32*, 4525.
40 Burgett, A. W. G.; Li, Q.; Wei, Q.; Harran, P. G. *Angew. Chem. Int. Ed.* **2003**, *42*, 4961.
41 Wipf, P.; Methot, J.-L. *Org. Lett.* **2001**, *3*, 1261.
42 Hamachi, I.; Tanaka, S.; Tsukiji, S; Shinkai, S.; Oishi, S. *Inorg. Chem.* **1998**, *37*, 4380.
43 Shimakoshi, H.; Tokunaga, M.; Kuroiwa, K.; Kimizuka, N.; Hisaeda, Y. *Chem. Commun.* **2004**, *50*.
44 Yamada, S.-I; Takeuchi, Y. *Tetrahedron Lett.* **1971**, *39*, 3595.

45 (a) Hamada, Y.; Rishi, S.; Shioiri, T.; Yamada, S. *Chem. Pharm. Bull.* **1977**, *25*, 224.
(b) Han, S.-Y.; Kin, Y.-A., *Tetrahedron* **2004**, *60*, 2447. (c) Yamada, S.-I; Shioiri, T; Ikota, N.; Tachbana, S. *J. Am. Chem. Soc.* **1975**, *97*, 7174.
46 Chen, W.-Y.; Olsen, R. K. *J. Org. Chem.* **1975**, *40*, 350.
47 Takuma, S.; Hamada, Y.; Shioiri, T. *Chem. Phar. Bull.* **1982**, *30*, 3147.
48 Teodori, E.; Baldi, E.; Dei, S.; Gualtieri, F.; Romanelli, M. N.; Scapecchi, S.; Bellucci, C.; Ghelardini, C.; Matucci, R. *J. Med. Chem.* **2004**, *47*, 6070.
49 Sato, K.; Hada, N.; Takeda, T. *Tetrahedron Lett.* **2003**, *44*, 9331.
50 Pettit, G. R.; Srirangam, J. K.; Sing, S. B.; Williams, M. D.; Herald D. L.; Barkoczy, J.; Kantoci, D.; Hogon, F. *J. Chem. Soc., Perkin Trans. 1* **1996**, 859.
51 Shioiri, T.; Hayashi, K.; Hamada, Y.; *Tetrahedron* **1993**, *43*, 1913.
52 Pettit, G. R.; Hogan, F.; Herald, D. L. *J. Org. Chem.* **2004**, *69*, 4019.
53 Holzapfel, C.; van Zyl, W. J.; Roos, M. *Tetrahedron* **1990**, *46*, 649.
54 Shioiri, T.; Hughes, R. J. *Heterocycles* **2003**, *61*, 23.
55 Barrett, A. G.M.; Pilipauskas, D. *J. Org. Chem.* **1991**, *56*, 2787.
56 Piras, S.; Loriga, M.; Paglietti, G. *Farmaco* **2002**, *57*, 1.
57 Borrell, J.; Teixido, J.; Matallana, J. L.; Martĭnez-Teipel, B.; Colominas, C.; Costa, M.; Balcells, M.; Schuler, E.; Castillo, M. *J. Med. Chem.* **2001**, *34*, 2366.
58 Wright, J. E.; Rosowsky, A. *Bioorg. Med. Chem.* **2002**, *10*, 493.
59 Zhang, F.-M.; Tian, X. *Huaxue Xuebao* **2002**, *60*, 720.
60 Rosowsky A.; Bader, H.; Kohler, W.; Freisheim, J. H.; Moran, R. G. *J. Med. Chem.* **1988**, *31*, 1338.
61 Rosowsky, A.; Freisheim, J. H.; Bader, H.; Forsch, R. A.; Susten, S. S.; Cucchi, C. A.; Frei, E. *J. Med. Chem.* **1985**, *28*, 660.
62 Yokokawa, F.; Sameshima, H.; Shioiri, T. *Tetrahedron Lett.* **2001**, *42*, 4171
63 Deng, J.; Hamada, Y.; Shioiri, T. *Tetrahedron Lett.* **1996**, *37*, 2261.
64 Hayashi, K.; Hamada, Y.; Shioiri, T. *Tetrahedron Lett.* **1992**, *33*, 5075.
65 Ishiwata, H.; Sone, H.; Kigoshi, H.; Yamada, K. *Tetrahedron* **1994**, *50*, 12853.
66 Ishiwata, H.; Sone, H.; Kigoshi, H.; Yamada, K. *J. Org. Chem.* **1994**, *59*, 4712.
67 Kobayashi, M.; Kurosu, M; Wang, W.; Kitagawa, I. *Chem. Pharm. Bull.* **1994**, *42*, 2394.
68 Kobayashi, M.; Kurosu, M.; Ohyabu, N.; Wang, W.; Fujii, S.; Kitagawa, I. *Chem. Pharm. Bull.* **1994**, *42*, 2196.
69 Hamada, Y.; Kondo, Y.; Shibata, M.; Shioiri, T. *J. Am. Chem. Soc.* **1989**, *111*, 669.
70 (a) Plattner, J. J.; Marcote, P. A.; Kleinert, H. D.; Stein, H. H.; Greer, H. S.; Bolis, G.; Fung, A. K. L.; Bopp, B. A.; Luly, J. R.; Sham, H. L.; Kempf, D. J.; Rossenberg, S. H.; Dellaria, J. F.; Merits, B. D. I.; Perum, T. J. *J. Med. Chem.* **1988**,*3 1*, 2277.;
(b) Thaisrivongos, S.; Mao, B.; Pals, D.T.; Turner, S. R.; Kroll, L. T. *J. Med. Chem.* **1990**, *33*, 1337.
71 Thaisrivongos, S.; Pals, D.T.; Turner, S. R.; Kroll, L. T. *J. Med. Chem.* **1988**,*3 1*, 1369.
72 Tenbrink, R. E.; Pal, D. T.; Harris, D. W.; Johnson, G. A. *J. Med. Chem.* **1988**, *31*, 671.
73 Matsuura, F.; Hamada, Y.; Shioiri, T. *Tetrahedron* **1994**, *50*, 11303.
74 Fearon, K.; Spaltenstein, A.; Hopkins, P. B.; Gelb, M. H. *J. Med. Chem.* **1987**, *30*, 1617.
75 Thaisrivongos, S.; Pals, D. T.; Kroll, L. T.; Turner, S. R.; Han, F.-S. *J. Med. Chem.* **1987**, *30*, 976.
76 Ichikawa, T.; Yamada, M.; Yamaguchi, M.; Kitazaki, T.; Matsushita, Y.; Higashikawa, K.; Itoh, K. *Chem. Pharm. Bull.* **2001**, *49*, 1110.
77 Imaeda, T.; Hamada, Y.; Shioiri, T. *Tetrahedron Lett.* **1994**, *35*, 591
78 Hynes, J. B.; Patil, S. A.; Tomažįč, A.; Kumar, A.; Pathak, A.; Tan, X.; Xianqiang, L.; Ratnam, M.; Delcamp, T. J.; Friesheim, J. H. *J. Med. Chem.* **1988**, *31*, 449.
79 Hynes, J.B.; Kumar, A.; Tomažiĉ, A.; Washtien, W.L. *J. Med. Chem.* **1987**, *30*, 1515.
80 Hynes, J. B.; Patil, S. A.; Hagen, R. L.; Cole, A.; Kohler, W.; Freisheim, J. H. *J. Med. Chem.* **1989**, *32*, 852.
81 Bisset, M. F. G.; Pawelczak, K.; Jackman, A. L.; Calvert, A. H.; Hughes, L. R. *J. Med. Chem.* **1992**, *35*, 859.
82 He, G-X.; Browne, A.; Groppe, B. J. C.; Blasko, A.; Mei, H-Y.; Bruice, T. C. *J. Am. Chem. Soc.* **1993**, *115*, 7061.
83 Tohdo, K.; Hamada, Y.; Shioiri, T. *Synlett* **1994**, 247.
84 Nako, K.; Hamada, Y.; Shioiri, T. *Chem. Pharm. Bull.* **1989**, *37*, 930.
85 Patil, S. A.; Shane, B.; Freisheim, J. H.; Singh, K. S.; Hynes, J. B. *J. Med. Chem.* **1989**, *28*, 1559.
86 Hamada, Y.; Kawai, A.; Kohono, Y.; Hara, O.; Shioiri, T. *J. Am. Chem. Soc.* **1989**, *111*, 1524.

87 Smith, A. B.; Friestad, G. F.; Barbosa, J.; Bertonesque, E.; Duan, J. J.-W.; Hull, K. G.; Iwashima, M.; Qiu, Y.; Spoors, P.; Salvatore, B. A. *J. Am. Chem. Soc.* **1999**, *121*, 10478.
88 Itoh, F.; Yoshioka, Y.; Yukishige, S.; Yoshida, S.; Ootsu, K.; Akimoto, H.; *Chem. Pharm. Bull.* **2000**, *48*, 1270.
89 Shioiri, T.; Terao, Y.; Irako, N.; Aoyama, T. *Tetrahedron* **1998**, *54*, 15701.
90 Showalter, H. D. H.; Sercel, A. D.; Leja, B.M.; Wolfangel, C. D.; Ambroso, L. A.; Elliott, W. L.; Fry, D. W.; Kraker, A. J.; Howard, C. T. Howard, Lu, G. H. Moore, C. W.; Nelson, J. M.; Roberts, B. J.; Vincent, P. W.; Denny, W. A.; Thompson, A. M. *J. Med Chem.* **1997**, *40*, 413.
91 Ciapetti, P.; Soccolini, F.; Taddei, M. *Tetrahedron* **1997**, *53*, 1167.
92 Shioiri, T.; Murata, M.; Hamada, Y. *Chem. Pharm. Bull.* **1987**, *35*, 2698.
93 Evans, D.; Weber, A. E. *J. Am. Chem. Soc.* **1987**, *109*, 7151.

Marudai Balasubramanian

6.6 Yamaguchi Esterification

6.6.1 Description

The Yamaguchi esterification utilises 2,4,6-trichlorobenzoyl chloride, also known as the Yamaguchi reagent, to form esters from carboxylic acids and alcohols.

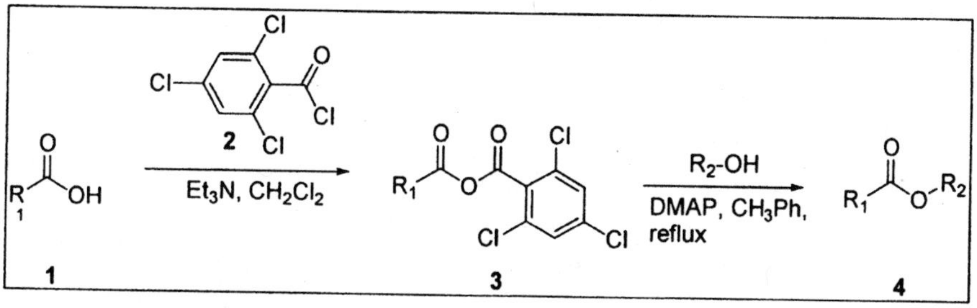

2

Et_3N, CH_2Cl_2

1

3

R_2-OH

DMAP, CH_3Ph, reflux

4

6.6.2 Historical Perspective

Prior to the esterification pioneered by Yamaguchi and co-workers few examples existed for the preparation of large ring lactone systems.[1–4] Yamaguchi recognised that to overcome the unfavourable entropic factors leading to the formation of polymers the ring closure of such systems would need to be rapid. The method also needed to be mild so as to be applicable to complex natural systems without adversely affecting other functionalities present.

Yamaguchi and co-workers then screened a number of acid chlorides for the formation of the mixed anhydride with two features in mind: (1) the component should be a good leaving group, and (2) the carbonyl group of the component should be sterically hindered from nucleophilic attack.[5,6] Eventually, 2,4,6-trichlorobenzoyl chloride was deemed to be the best in terms of rate of reaction and yield.

6.6.3 Mechanism

The carboxylic acid coupling partner is initially deprotonated by triethylamine typically, forming the carboxylate ion which then reacts with the acid chloride to give a mixed anhydride. This forms the first part of the two-step procedure. The acylating agent dimethylaminopyridine (DMAP) is then added, which reacts with the mixed anhydride to form an acyl pyridinium salt with the elimination of an aromatic carboxylate anion. The second coupling partner, the alcohol, then reacts with the acyl pyridinium salt to give the desired ester.

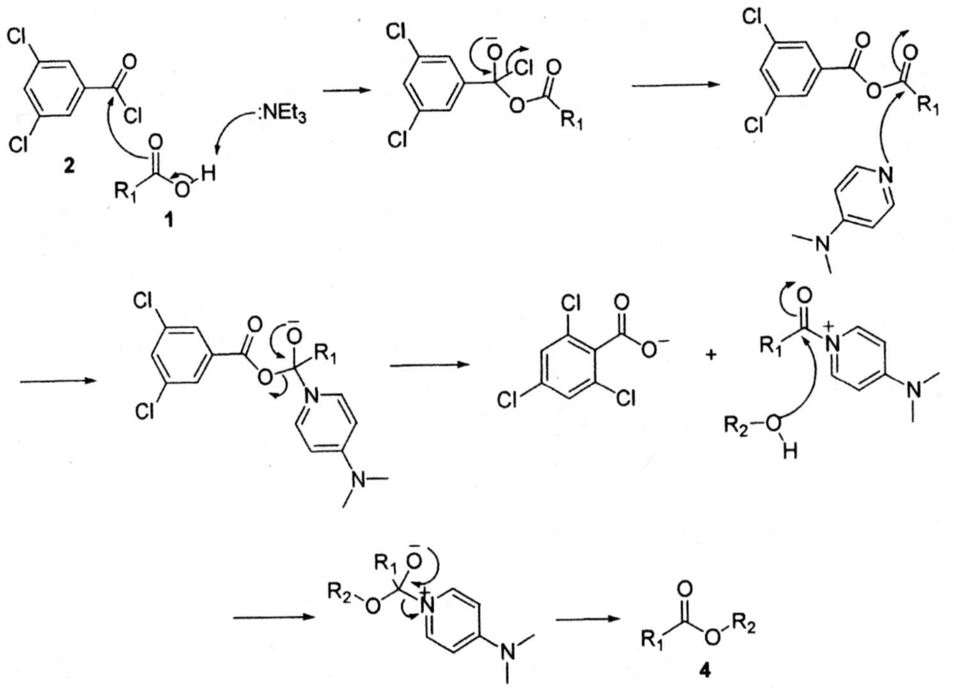

6.6.4 *Variations and Improvements*

Variations in the Yamaguchi protocol include changing the concentration of DMAP and altering the dilution of the reaction mixture. Arguably the most significant variation to the Yamaguchi procedure is the Yonemitsu modification. In their synthesis of 9-dihydroerythronolide A, Yonemitsu *et al.* were able to affect coupling to obtain the macrolide structure by modified Yamaguchi reaction only.[7] Other methods such as Corey's pyridothio ester formation were unsuccessful.[8–10] In this case, the modification involved considerably increasing the equivalents of DMAP.

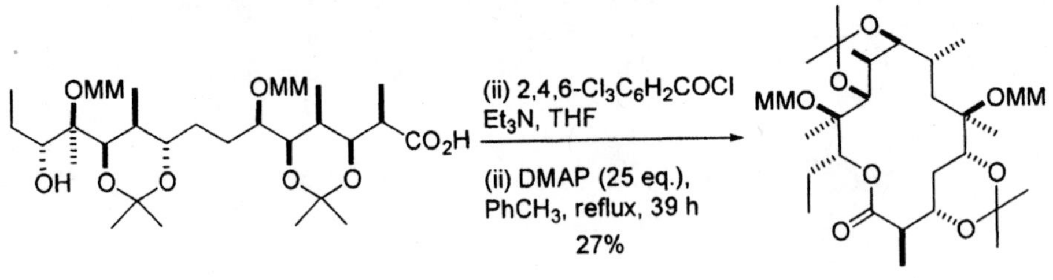

(ii) 2,4,6-$Cl_3C_6H_2COCl$, Et_3N, THF

(ii) DMAP (25 eq.), $PhCH_3$, reflux, 39 h

27%

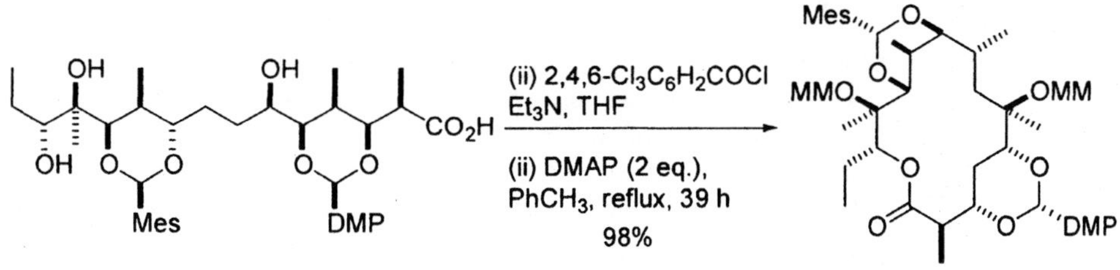

(ii) $2,4,6\text{-}Cl_3C_6H_2COCl$
Et_3N, THF
(ii) DMAP (2 eq.),
$PhCH_3$, reflux, 39 h
98%

7 8

The yield was still low; however, the authors then reported the total synthesis of Erythonolide A by the macrocyclisation of a slightly different substrate, **7**.[11] The excellent yield was attributed to a suitable conformation of **7** and a more efficient formation of the acylpyridinium salt. A more detailed study on the macrocyclisation of such substrates by employing different conditions discusses different coupling reagents and varying equivalents of triethylamine.[12] It is demonstrated that the Yamaguchi reaction can be carried as a one-pot procedure.

A recent example employing the Yonemitsu modification is demonstrated in Paterson's total synthesis of (–)-Baconipyrone C.[13] Initially, the Keck modification of the Steglich esterification was attempted for the coupling of the two fragments to give the complete backbone of the natural product. This however, resulted in the epimerisation of a remote stereocentre, giving an unseparable mixture of diastereoisomers. The Yonemitsu–Yamaguchi procedure gave similar results. However, by modifying the Yonemitsu–Yamaguchi protocol further, by warming the mixture from –78 to 0 °C over 10 min, then rapid quenching with $NaHCO_3$ solution, the desired ester was obtained with less than 10% epimerisation detected.

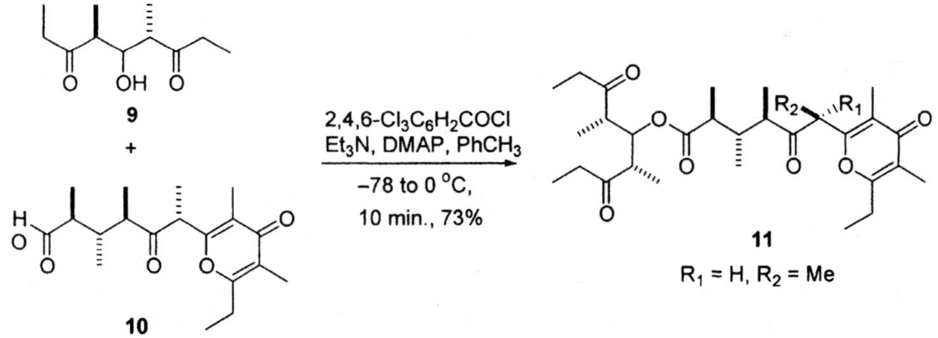

9 + 10

$2,4,6\text{-}Cl_3C_6H_2COCl$
Et_3N, DMAP, $PhCH_3$
–78 to 0 °C,
10 min., 73%

11
R_1 = H, R_2 = Me

Shiina and co-workers report the result of coupling reactions carried out by using benzoic anhydride and its derivatives in place of the Yamaguchi coupling reagent.[14] These can provide useful alternatives to the conventional Yamaguchi conditions. In addition, DMAP can be used as both the base and the activator for the dehydration condensation step, thus nullifying the need for a separate base such as triethylamine. Consequently, the coupling can be carried out in one step. The example illustrated shows the use of 2-methyl-6-nitrobenzoic anhydride (MNBA, **13**) in the synthesis of eight-membered-ring lactones.

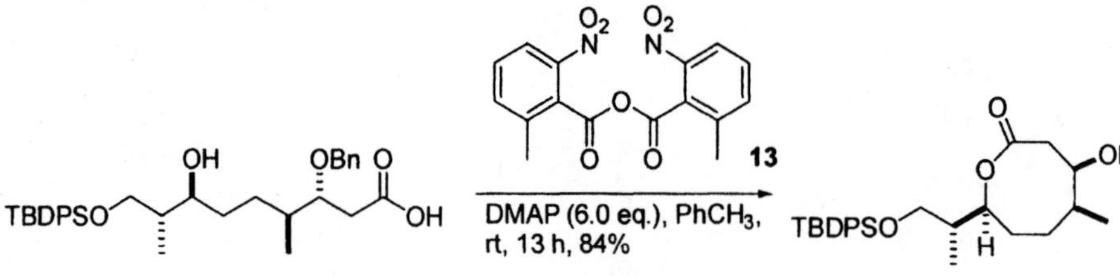

TBDPSO OH OBn O OH **12** → **13**, DMAP (6.0 eq.), $PhCH_3$, rt, 13 h, 84% → TBDPSO O O OBn H **14**

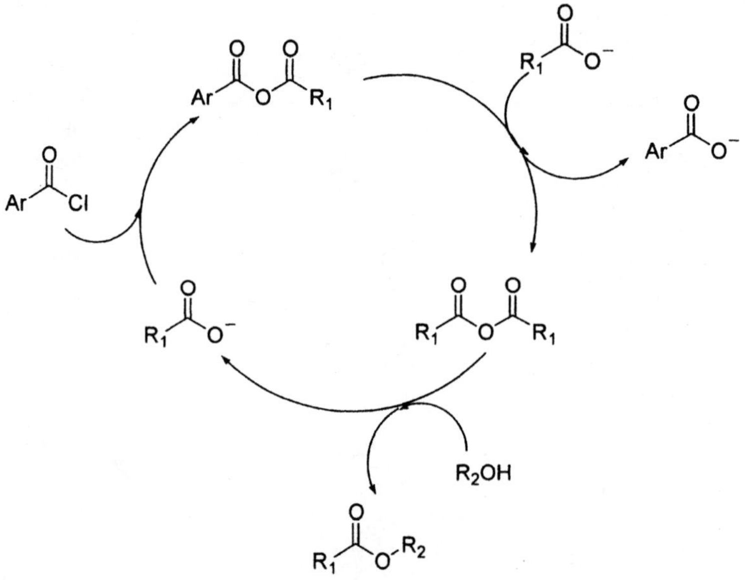

In a similar fashion, Dhimitruka and SantaLucia argue that the use of 2,4,6-trichlorobenzoyl chloride is not always necessary and that other inexpensive acid chlorides can be employed.[15] This is attractive especially when dealing with scale-up reactions. The authors also postulate the formation of a symmetric aliphatic anhydride to explain the regioselectivity of the Yamaguchi esterification. The mechanism assumes that aliphatic carboxylates are better nucleophiles than aromatic carboxylates and alcohols, and is illustrated below. Thus, since the mechanism is based on competing reactivities, there is no need for a two-step method or excess DMAP.

Variations in the Yamaguchi protocol also involve differing the order of addition. For example, in the synthetic studies toward bafilomycin A, the C1-C11 and C12-C17 fragments were coupled by first premixing the acid **15** with DMAP (2 eq.) in toluene at room temperature, then adding triethylamine (4 eq.) and 2,4,6-trichlorobenzoyl chloride (2 eq.) successively, then finally adding the alcohol **16** in toluene.[16] An excellent yield was obtained.

OCH_3 CO_2H OTES **15** + OH nBu_3Sn ODMT **16** → 89% → OCH_3 O OTES O n-Bu_3Sn ODMT **17**

6.6.5 Synthetic Utility

Due to the prevalence of many macrolide natural products, Yamaguchi's esterification has been used extensively in the total syntheses of such compounds. Increasingly, Yamaguchi's method has succeeded where other notable esterification conditions have failed. A recent example involves the coupling of two monomers to form the clavosolide A dimer.[17] The Corey–Nicolaou protocol mentioned previously gave only a 30% conversion whereas the Yamaguchi esterification resulted in a clean reaction with a 66% yield.

H H OH O O BnO OH **18** → Yamaguchi reagents, 125 °C, 6 h, 66% → H H H O O O O O BnO O H H H **19**

6.6.6 Experimental

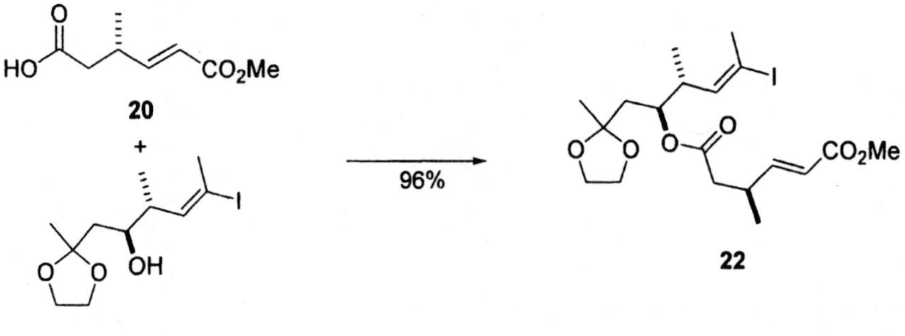

4-Methyl-hex-2-enedioic acid, 6-[4-iodo-2-methyl-1-(2-methyl-[1,3]dioxolan-2-yl-methyl)-pent-3-enyl]ester 1-methyl ester (22)[18]

To a solution of carboxylic acid **20** (74 mg, 0.432 mmol) in toluene (4 mL) was added triethylamine (0.180 mL, 1.29 mmol). 2,4,6-Trichlorobenzoyl chloride (68 mg, 0.432 mmol) was introduced and the resulting mixture was stirred for 1 h at ambient temperature. A solution of alcohol **21** (128 mg, 0.392 mmol) and DMAP (48 mg, 0.392 mmol) in toluene (4 mL) was added and the reaction mixture was allowed to stir for 1 h. Evaporation of the solvent followed by flash column chromatography of the residue (hexanes/EtOAc 4:1 + 1% Et_3N) provided ester **22** as a colorless oil (200 mg, 96%).

6.6.7 References

1. [R] Nicolaou, K. C. *Tetrahedron* **1977**, *33*, 683.
2. Back, T. G. *Tetrahedron*, **1977**, *33*, 3041.
3. Mukaiyama, T. *Chem. Lett.* **1977**, 441.
4. [R] Masamune, S., Bates, G. S. *Angew. Chem. Int. Ed. Engl.* **1977**, *16*, 585.
5. Inanaga, J.; Hirata, K.; Sakki, H.; Katsuki, T.; Yamaguchi, M. *Bull. Chem. Soc. Jpn.* **1979**, *52 (7)*, 1989.
6. Yamaguchi, M.; Hirao, I. *Tetrahedron Lett.* **1983**, *23*, 885.
7. Tone, H.; Nishi, T.; Oikawa, Y.; Hikota, M.; Yonemitsu, O. *Tetrahedron Lett.* **1987**, *28(39)*, 4569.
8. Corey, E. J.; Nicolaou, K. C. *J. Am. Chem. Soc.* **1974**, *96*, 5614.
9. Corey, E. J.; Brunelle, D. J.; Stork, P. J. *Tetrahedron Lett.* **1976**, 3405.
10. Woodward, R. B. *J. Am. Chem. Soc.* **1981**, *103*, 3213.
11. Hikota, M.; Tone, H.; Horita, K.; Yonemitsu, O. *J. Org. Chem.* **1990**, *55*, 7.
12. Hikota, M.; Sakurai, Y.; Horita, K.; Yonemitsu, O. *Tetrahedron Lett.* **1990**, *31(44)*, 6367.
13. Paterson, I.; Chen, D. Y.-K.; Acena, J. L.; Franklin, A. S. *Org. Lett.* **2000**, *2(11)*, 1513.
14. Shiina, I.; Kubota, M.; Oshiumi, H.; Hashizume, M. *J. Org. Chem.* **2004**, *69*, 1822.
15. Dhimitruka, I.; SantaLucia, J., Jr. *Org. Lett.* **2006**, *8(1)*, 47.
16. Quéron, E.; Lett, R. *Tetrehedron Lett.* **2004**, *45*, 4533.
17. Smith, A. B. III.; Simov, V. *Org. Lett.* **2006**, *8(15)*, 3315.
18. Furstner, A.; Kattnig, E.; Lepage, O. *J. Am. Chem. Soc.* **2006**, *128*, 9194.

Nadia M. Ahmad

CHAPTER 7 Miscellaneous Functional Group Manipulations 551

7.1 Balz–Schiemann Reaction

7.1.1 *Description*

The Balz–Schiemann (BS) reaction (often called the Schiemann reaction) describes the preparation of an aromatic diazonium fluoborate **1** followed by its thermal decomposition to afford the corresponding aryl fluoride **2**.[1–4]

$$ArNH_2 \xrightarrow[2.\ HBF_4]{1.\ HNO_2} \underset{\mathbf{1}}{ArN_2BF_4} \xrightarrow{heat} \underset{\mathbf{2}}{ArF} + N_2 + BF_3$$

The two-step BS reaction has several variations in both the diazonization step to afford **1** and the decomposition step to give the aryl fluoride **2**.

7.1.2. *Historical Perspective*

In their landmark publication, Balz and Schiemann[5] reported the syntheses of fluorobenzene, 4-fluorotoluene, 2,4-dimethylfluorobenzene, 1-fluoronaphthalene, and 4,4′-difluorodiphenyl. Diazotization of the appropriate aromatic amine with nitrous acid ($NaNO_2$, HCl), followed by the addition of fluoroboric acid gave the corresponding aryl diazonium salts (62–67% yields). Thermal decomposition of the dry aryl diazonium salts gave the aryl fluoride (97–100% yields). Only 2-, 3-, and 4-nitrophenyl diazonium fluoroborates, which were prepared in 74–100% yields, failed to yield the expected nitrofluorobenzenes upon thermolysis.

An interesting account of the life of Günther Schiemann on the occasion of his 100th birthday was published in 1999.[6]

7.1.3 *Mechanism*

$$NaNO_2 + H^+ \longrightarrow \underset{\mathbf{3}}{HNO_2} \xrightarrow[-H_2O]{H^+} \underset{\mathbf{4}}{\overset{+}{NO}}$$

$$ArNH_2 + NO^+ \longrightarrow Ar{-}\overset{+}{N}H_2{-}N{=}O \rightleftharpoons Ar{-}NH{-}N{=}O \overset{H^+}{\rightleftharpoons}$$

$$Ar{-}NH{-}N{=}\overset{+}{O}H \longleftrightarrow Ar{-}\overset{+}{N}H{=}N{-}OH \overset{-H^+}{\rightleftharpoons} Ar{-}N{=}N{-}OH$$

$$\overset{H^+}{\rightleftharpoons} Ar{-}N{=}N{-}\overset{+}{O}H_2 \xrightarrow{-H_2O} Ar{-}\overset{+}{N}{\equiv}N$$

The mechanism of the diazonization step is well established[7] and involves the sequence shown on the previous page.

Nitrous acid (**3**), which is generated from sodium nitrite and aqueous acid, undergoes dehydration under the acidic conditions to form nitrosyl cation **4**. Reaction of **4** with the aryl amine eventually affords the aryl diazonium salt after tautomerization and dehydration. The addition of fluoroboric acid sets the stage for the second step leading to aryl fluoride.

There is strong evidence that the mechanism of the decomposition of the aryl diazonium fluoroborates involves an aryl cation intermediate that rapidly collapses to the aryl fluoride and boron trifluoride by reaction of the aryl cation with tetrafluoroborate.[8–12] Aryl radicals do not appear to be involved. For example, whereas iodoacetic acid is an effective aryl radical trap in several hydrodediazonization reactions, it does not interrupt the normal (ionic) pathway involved in the Schiemann reaction.[11b]

$$\underset{\mathbf{1}}{Ar{-}\overset{+}{N}{\equiv}N\ \ BF_4^-} \rightleftharpoons \underset{\mathbf{5}}{Ar^+\ N_2\ BF_4^-} \xrightarrow{-N_2} \underset{\mathbf{6}}{Ar^+BF_4^-} \longrightarrow Ar{-}F + BF_3$$

Theoretical calculations support the intermediacy of a singlet aryl cation **6** preceded by an ion–molecule pair (**5**).[13]

7.1.4 Variations and Improvements

Variations and improvements in both steps are known. For example, the simple expedient of using tetrahydrofuran as the solvent to prepare diazonium tetrafluoroborates greatly improves the yields in several cases.[14]

The preparation and use of diazonium fluorosilicates is superior in a few cases to diazonium fluoroborates.[15] For example, the synthesis of methyl 5-fluoronicotinate **7** was only successful using the diazonium fluorosilicate method.[15]

H$_2$N–(pyridine)–CO$_2$Me $\xrightarrow[\text{EtOH}]{H_2SiF_6}$ $\xrightarrow[\text{EtONO}_2]{\text{HOAc}}$ $\overset{+}{N_2}$–(pyridine)–CO$_2$Me $\xrightarrow{\text{heat}}$ F–(pyridine)–CO$_2$Me (**7**)

More prevalent than diazonium fluorosilicates in so-called modified Balz–Schiemann reactions have been diazonium hexafluorophosphates.[16] In some cases these latter diazonium salts give higher yields of aryl fluorides. Some examples are shown below, with the normal Balz–Schiemann reaction yields using diazonium tetrafluoroborates given in parentheses.[16a]

R	**9**, %	**10**, %
2-CO_2H	79 (46)	78 (19)
4-CO_2H	77 (84)	64 (0)
2-Br	97 (50)	77 (81)
4-Br	100 (64)	79 (75)
4-OH	77 (0)	20 (–)
4-OCH_3	100 (85)	70 (67)
4-NO_2	100 (100)	63 (58)
4-CH_3	97 (90)	71 (70)

The combination of boron trifluoride etherate and *tert*-butyl nitrite,[17] in place of the conventional conditions ($NaNO_2$, HCl, HBF_4), followed by decomposition of the diazonium salt in chlorobenzene afforded 5-fluorobenzo[*c*]phenanthrene (**11**).[17b] Nitrosonium tetrafluoroborate has been used to diazotize 8-amino-1-tetralone (**12**) *en route* to 8-fluoro-1-tetralone (**13**).[18]

Another modification of the BS reaction is to employ a triazene and HF as the source of fluoride.[19] For example, aminotamoxifen (**14**) was transformed into fluorotamoxifen (**16**) *via* triazene **15** using this milder method.[20]

Ph Et
$Me_2N(CH_2)_2O$ NH_2
14
1. $NaNO_2$, HCl
2. piperidine (N-H)
69%
Ph Et
$Me_2N(CH_2)_2O$ N=N N
15
40% HF
35%
Ph Et
$Me_2N(CH_2)_2O$ F
16

Other deaminative fluorination techniques include the use of pyridinium poly(hydrogen fluoride)-sodium nitrite[21] and silicon tetrafluoride-*tert*-butyl nitrite.[22]

Several variations for the decomposition of aryl diazonium tetrafluoroborates are available. One of the most widely used is a photodecomposition method first described by two groups in 1971.[23,24] For example, 4,5-difluoroimidazole (**17**) was prepared using this method.[25]

NHBoc F
1. HBF_4
2. $NaNO_2$, H_2O
$\overset{+}{N_2}\ \overset{-}{BF_4}$ F
$h\nu$
0 °C
36%
F F
17
(+ F 8%)

Other photoinduced fluoro-dediazonizations have been described, and in some cases excellent yields of aryl fluorides are obtained as shown for **18** and **19**.[26] In these examples the aryl diazonium salts were prepared using 3-methylbutyl nitrite in boron trifluoride etherate and hydrogen fluoride (1:1).[27]

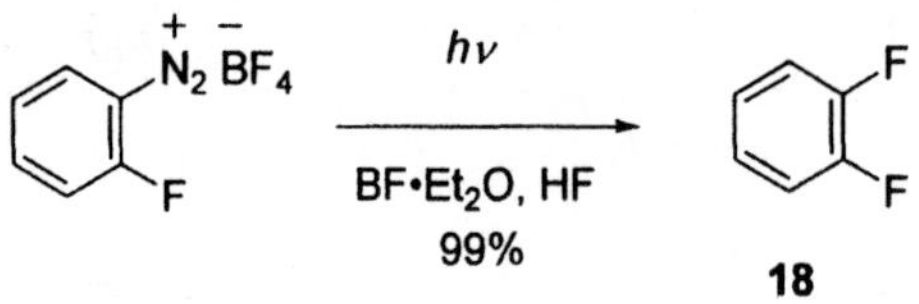

Several catalysts are known to facilitate the decomposition of aryl diazonium tetrafluoroborates, including the combination of crown ethers and copper powder[28] and the employment of ionic liquids.[29] An example of each is depicted below for the synthesis of **20** and **21**. It should be noted that the former method leads in most cases to *reduction* of the aryl diazonium salt and not to introduction of fluorine.

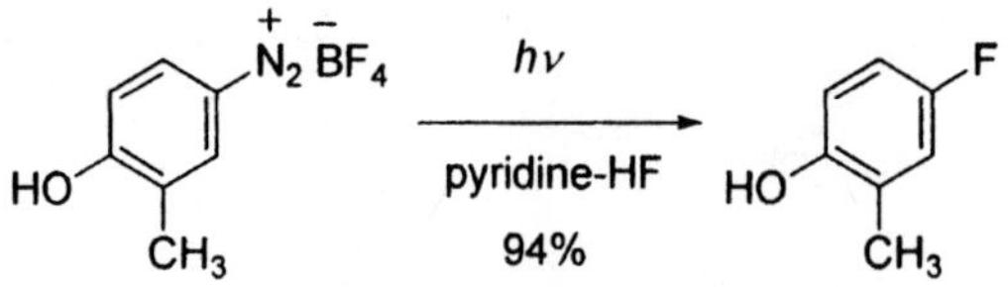

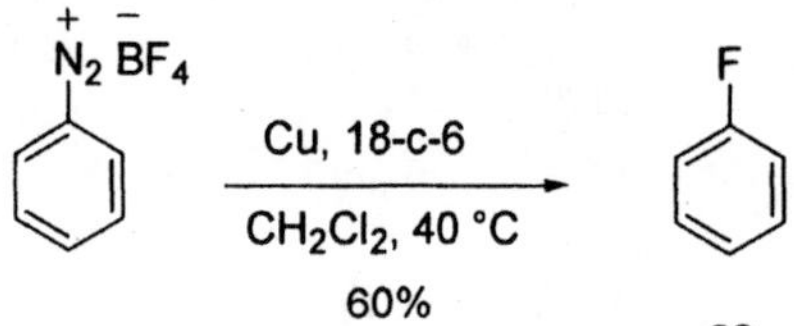

The incorporation of radiolabelled fluoride-18 employs a clever modified BS reaction using the decomposition of aryl diazonium tetrachloroborate **22** in the presence of $^{18}F^-$, as shown.[30]

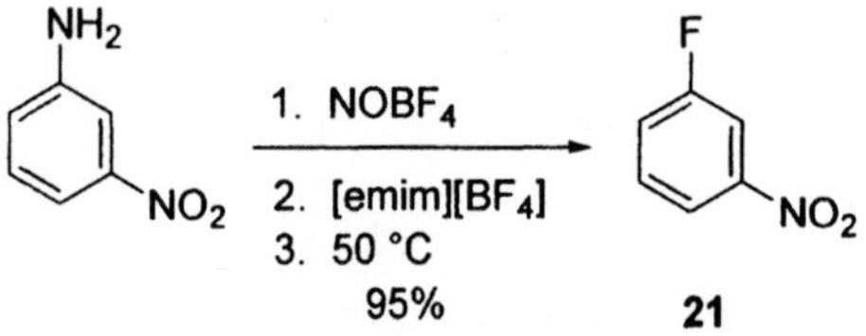

NH_2 / CH_3 — i-amylONO, HCl, EtOH, 83% → $\overset{+}{N_2}\overset{-}{Cl}$ / CH_3 — BCl_3, $CHCl_3$, hexane, 94% →

$\overset{+}{N_2}\ \overset{-}{BCl_4}$ / CH_3 **22** — n-Bu_4NOH, $^{18}F^-$, heptane reflux → ^{18}F / CH_3 19%

7.1.5 Synthetic Utility

Me, NO_2, H_2N, N — 1. Ac_2O 2. H_2, Pd/C 3. $NaNO_2$ HBF_4 83% → Me, $\overset{+}{N_2}\ \overset{-}{BF_4}$, AcHN, N — toluene, 110 °C, 47% → Me, F, AcHN, N (ref. 31h)

NH_2, N, N, HO, O, N, N, NH_2, OH — HF, HBF_4, aq THF, $NaNO_2$, 43% → NH_2, N, N, HO, O, N, N, F, OH (ref. 35f)

NH_2, MeO_2C, S — $NaNO_2$, HCl, HPF_6, 79% → $\overset{+}{N_2}\ \overset{-}{PF_6}$, MeO_2C, S — 180-200 °C, sand, 42% → F, MeO_2C, S (ref. 38c)

NH_2, N, N, H — $NaNO_2$, aq HBF_4 → F, N, N, H 40% + OH, N, N, H 52% (ref. 42)

As we have seen to some extent in the previous section, the range of aryl and heteroaryl fluorides that are available from the BS reaction is truly enormous. Noteworthy are the several BS syntheses of fluorinated heterocycles such as pyridines,[31] quinolines,[32] isoquinolines,[33] naphthyridines,[34] purines,[35] benzoquinolines,[36] benzocinnolines,[37] thiophenes,[38] pyrazoles,[39] selenazoles,[40] benzimidazoles,[41] and a pyrrolo[2,3-*b*]pyridine.[42] Some representative recent examples are depicted here.

A wide variety of fluorinated aromatic and polyaromatic systems are accessible *via* the BS synthesis, such as **11**, **13**, **16**, and **18–21** presented earlier. Other examples are fluorinated mesitylenes,[43] several polyfluorinated benzenes,[44], 4-chloro-2-fluoronitrobenzene,[45] bromofluorotoluenes,[46] 7-fluoroindanone,[18] fluorinated polycyclic aromatic hydrocarbons,[17,47] 4-tritylfluorobenzene,[48] fluorinated diaryl thioethers,[49] fluorophenols,[50] fluoronorepinephrines,[51] fluorinated estrogens,[19,52] and an unusual metacyclophane.[53] A selection of these compounds is shown here.

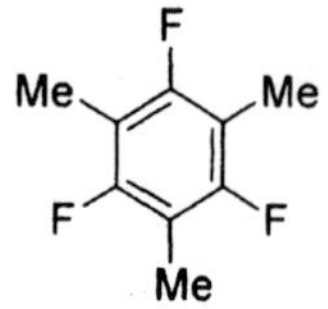

ref. 43 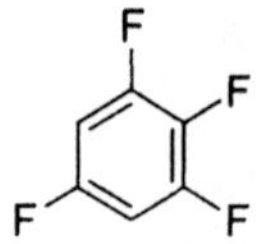 ref. 44d 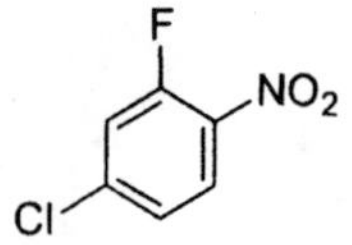 ref. 45 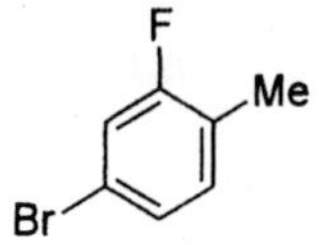 ref. 46

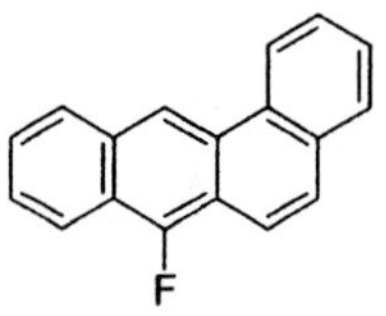

ref. 47a 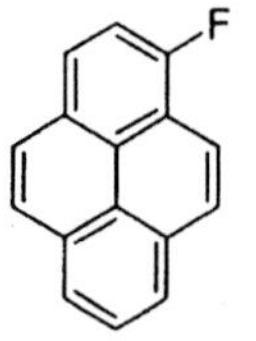 ref. 47c 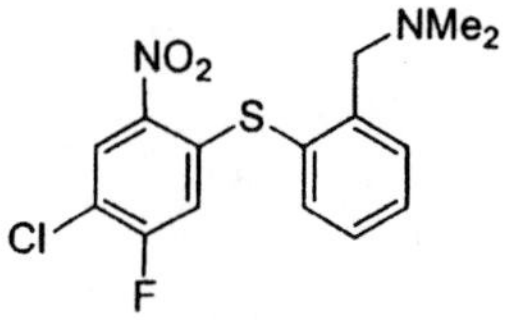 ref. 49 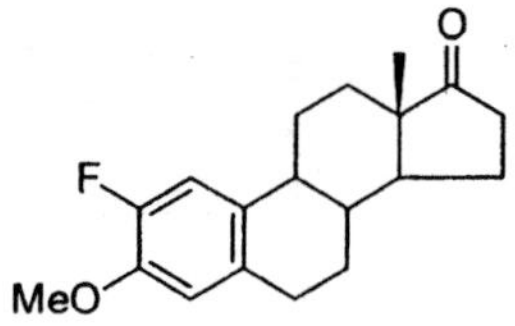 ref. 52a

The BS reaction is an excellent method for the introduction of radioactive fluorine-18 into molecules of biological interest. Examples include ^{18}F-labeled haloperidol (**23**),[54] fluconazole (**24**),[55] 5-fluoro-*L*-dopa (**25**),[56] amino acids,[57a] and 5-fluorouracil.[57b]

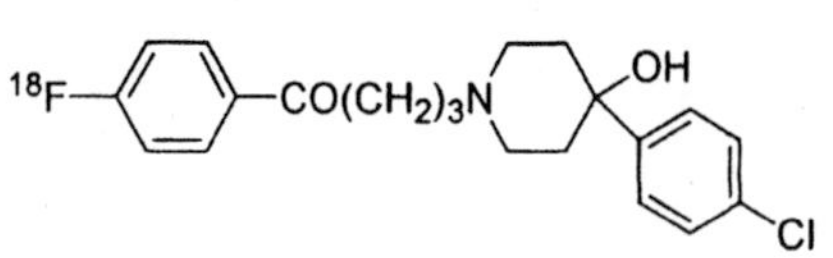

23[53] 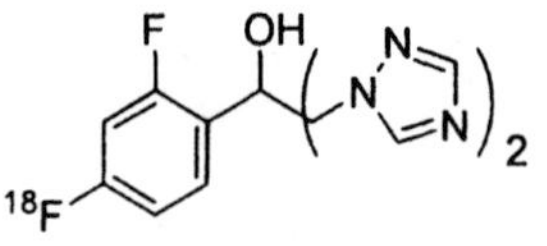 **24**[54] 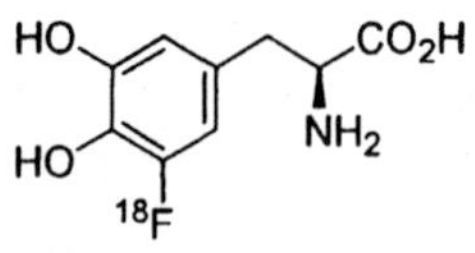 **25**[55]

Some specific reactions are illustrated below for the syntheses of 2,3,6-trifluorochlorobenzene (**26**),[44e] 7-fluorobenz[*a*]anthracene (**27**),[47a] and 4-tritylfluorobenzene (**28**).[48]

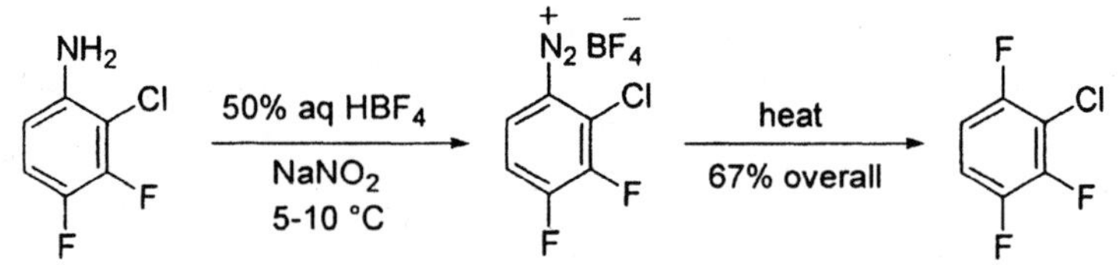
NH2
Cl
F
F
50% aq HBF4
NaNO2
5-10 °C
N2 BF4
heat
67% overall

26[44e]

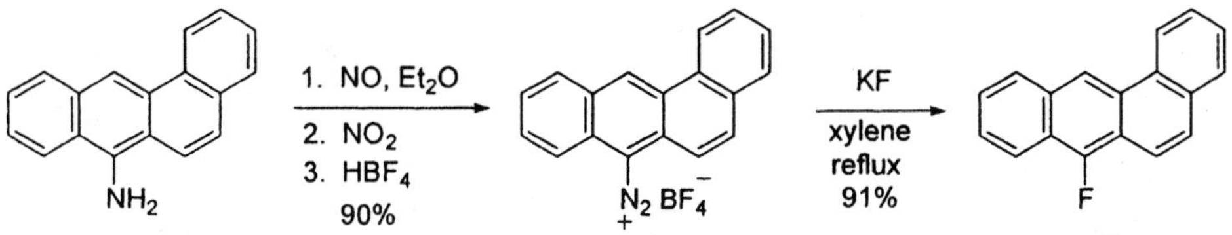
NH2
1. NO, Et2O
2. NO2
3. HBF4
90%
N2 BF4
KF
xylene
reflux
91%
F

27[47a]

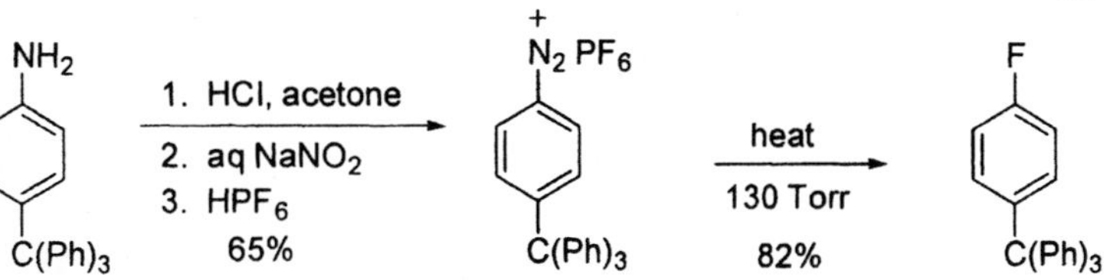
NH2
C(Ph)3
1. HCl, acetone
2. aq NaNO2
3. HPF6
65%
N2 PF6
heat
130 Torr
82%
F

28[48]

7.1.6 Experimental

The reader is referred to the synthesis of 4,4′-difluorobiphenyl,[58] 4-fluorobenzoic acid,[59] fluorobenzene,[60] and 1-bromo-2-fluorobenzene,[61] published in *Organic Syntheses*, and to the syntheses of 3-fluorotoluene, 3-nitrofluorobenzene, 4-fluorobromobenzene, 4-fluoroanisole, 2-fluoronaphthalene, 3-fluoropyridine, and 4,4′-difluorobiphenyl published in *Organic Reactions*.[1]

Due to the potential explosive nature of aryl diazonium salts, the reader is urged to consult the original literature before attempting any BS synthesis. For example, a great difference exists in the stability of the diazonium tetrafluoroborates derived from the isomeric aminopyridines, aminoquinolines, and aminoisoquinolines. Thus, whereas the diazonium tetrafluoroborate from 2-aminoquinoline decomposes as it forms at room temperature, the same diazonium salt from 3-aminoquinoline is stable enough to be isolated and dried, decomposing only at 95 °C.[32]

2,6-Dichloro-3-fluoropyridine[31f]

To a stirred solution of 3-amino-2,6-dichloropyridine (16.3 g, 0.1 mol) in tetrafluoroboric acid (42%, 300 mL) was added dropwise an aqueous saturated solution of sodium nitrite (6.9 g, 0.1 mol) at 5 °C. The resulting precipitate was filtered off, washed successively with cold water and Et_2O, and dried under reduced pressure (3 Torr) below 80 °C, giving 14.8 g (57%) of 2,6-dichloro-3-pyridinediazonium tetrafluoroborate, mp 167–169 °C; IR (KBr) 2260 cm^{-1}. Anal. Calcd for $C_5H_2BCl_2F_4N_3$: C, 22.94; H, 0.76; Cl, 27.09; F, 29.03; N, 16.05. Found: C, 22.91; H, 0.77; Cl, 27.13; F, 29.24; N, 16.14. A mixture containing the diazonium salt (13.1 g, 0.05 mol) and anhydrous $MgSO_4$ (13.1 g) was heated at 170–200 °C under reduced pressure (4–10 Torr). The product distilled and/or sublimed during the reaction course and was collected under cooling with dry ice/acetone, and taken up in $CHCl_3$. The organic solution was washed with 1 N NaOH, dried (K_2CO_3), and concentrated to dryness. The residue was crystallized from hexane-Et_2O (1:1) to give 5.6 g (67%) of the very sublimable product, mp 44–46 °C; ^1H NMR ($CDCl_3$) δ 7.47 (1H, dd, J = 8.5, 7.0 Hz), 7.27 (1H, dd, J = 8.5, 3.3 Hz).

4-Chloro-2-fluoronitrobenzene[45]

Sodium nitrite (3.4 g, 49 mmol) in water (10 mL) was added dropwise to a stirred solution of 5-chloro-2-nitroaniline (8.0 g, 46 mmol) in concentrated hydrochloric acid (280 mL) cooled at 5–10 °C over 60 min. After further stirring at the same temperature for 60 min, hexafluorophosphoric acid (40%, 33.8 g) was added and the obtained mixture was stirred at 0–5 °C for 2 h. The precipitate was collected by filtration, washed sequentially with cold water, ethanol, and ether, and air dried to give a slightly yellow solid (12.4 g). The dry diazonium salt (**Caution**: explosive at high temperature) was added in portions to dichlorobenzene at about 150 °C with stirring over a period of 20 min. Decomposition of the salt proceeded smoothly and the same temperature was maintained for a further 1 h. After cooling to room temperature, the reaction mixture was poured into Na_2CO_3 solution and partitioned. The organic layer was washed with water, dried (Na_2SO_4), and distilled at reduced pressure with a Vigreux column. The distillate of 133–136 °C/30 Torr was collected to give the product (4.1 g, 50%), mp 46–48 °C; ^1H NMR ($CDCl_3$): δ 8.05 (1H, dd, J = 8.7, 8.2 Hz), 7.34–7.26 (2H, m). Anal. Calcd for $C_6H_3ClFNO_2$: C, 41.05; H, 1.72; N, 7.98. Found: C, 41.11; H, 1.74; N, 7.92.

8-Fluoro-1-tetralone (13)[18]

8-Amino-1-tetralone (0.62 g, 3.85 mmol) in dry acetone (10 mL) was added to a mixture of $NOBF_4$ (0.59 g, 5.1 mmol) in acetone (10 mL) at –20 °C. After 45 min, more nitrosonium tetrafluoroborate (~0.67 g, 5.73 mmol) was added to the mixture. The reaction was continued (30 min) until TLC analysis showed no starting material. The mixture was poured into anhydrous $CHCl_3$ (50 mL) and stirred for 30 min. The mixture was dried ($MgSO_4$) and the solvent was removed under vacuum. The solids were added portion-wise to a solution of toluene at reflux. Heating was continued for 15 min whereupon the mixture was cooled to room temperature and filtered through Celite. The solids were flushed with $CHCl_3$ and the concentrated residue was purified on a column of silica gel. The product **13** was eluted with 60% EtOAc:hexane to give a yellow oil, 0.26 g (41%); ^{1}H NMR ($CDCl_3$) δ 7.39 (1H, dt, J = 7.8, 4.8 Hz), 7.03 (1H, d, J = 7.5 Hz), 6.96 (1H, dd, J = 8.7, 11.4 Hz), 2.95 (2H, t, J = 5.7 Hz), 2.63 (2H, t, J = 6.3 Hz), 2.13–2.05 (2H, m). Anal. Calcd for $C_{10}H_9FO$: C, 73.16; H, 5.53. Found: C, 73.08; H, 5.53.

7.1.7 References

1. [R] Roe, A. *Org. React.* **1949**, *5*, 193.
2. [R] Suschitsky, H. *Advances in Fluorine Chemistry*, Eds. Stacey, M.; Tatlow, J. C.; Sharpe, A. G., Vol. 4, Buttersworth, Washington, 1965, p 1.
3. [R] Pavlath, A. E.; Leffler, A. J. *Aromatic Fluorine Compounds*, Reinhold, New York, 1965.
4. [R] *Methoden der Organischen Chemie (Houben-Weyl)*, Vol. 5, Part 3, Halogen Compounds – Fluorine and Chlorine; Ed. Miller, E., Georg Thieme Verlag, Stuttgart, 1962.
5. Balz, G.; Schiemann, G. *Chem. Ber.* **1927**, *60*, 1186.
6. Cornils, B.; Baerns, M. *Nachr. Chem. Tech. Lab.* **1999**, *47*, 1316.
7. [R] Gould, E. S. *Mechanism and Structure in Organic Chemistry*, Holt, Rinehart and Winston, New York, 1959, pp 174–175.
8. Olah, G. A.; Tolgyesi, W. S. *J. Org. Chem.* **1961**, *26*, 2053, and references to the earlier Russian literature.
9. (a) Swain, C. G.; Sheats, J. E.; Gorenstein, D. G.; Harbison, K. G.; Rogers, R. J. *Tetrahedron Lett.* **1974**, *15*, 2973. (b) Swain, C. G.; Sheats, J. E.; Harbison, K. G. *J. Am. Chem. Soc.* **1975**, *97*, 783. (c) Swain, C. G.; Rogers, R. J. *J. Am. Chem. Soc.* **1975**, *97*, 799.
10. (a) Szele, I.; Zollinger, H. *J. Am. Chem. Soc.* **1979**, *100*, 2811. (b) Hashida, Y.; Landells, R. G. M.; Lewis, G. E.; Szele, I.; Zollinger, H. *J. Am. Chem. Soc.* **1978**, *100*, 2816. (c) Mauer, W.; Szele, I.; Zollinger, H. *Helv. Chim. Acta* **1979**, *62*, 1079. (d) Ravenscroft, M. D.; Zollinger, H. *Helv. Chim. Acta* **1988**, *71*, 507. (e) Ravenscroft, M. D.; Skrabal, P.; Weiss, B.; Zollinger, H. *Helv. Chim. Acta* **1988**, *71*, 515.
11. (a) Becker, H. G. O.; Israel, G. *J. Prakt. Chem.* **1979**, *321*, 579. (b) Wassmundt, F. W.; Kiesman, W. F. *J. Org. Chem.* **1997**, *62*, 8304. (c) Filippi, A.; Lilla, G.; Occhiucci, G.; Sparapani, C.; Ursini, O.; Speranza, M. *J. Org. Chem.* **1995**, *60*, 1250. (d) Steenken, S.; Ashokkumar, M.; Maruthamuthu, P.; McClelland, R. A. *J. Am. Chem. Soc.* **1998**, *120*, 11925. (e) For an alternative (free-radical) mechanism, see Yan, D. *Acta Chim. Sinica* **1989**, 422.
12. (a) Canning, P. S. J.; McCrudden, K.; Maskill, H.; Sexton, B. *Chem. Commun.* **1998**, 1971. (b) Canning, P. S. J.; McCrudden, K.; Maskill, H.; Sexton, B. *J. Chem. Soc., Perkin Trans. 2* **1999**, 2735. (c) Canning, P. S. J.; Maskill, H.; McCrudden, K.; Sexton, B. *Bull. Chem. Soc. Jpn.* **2002**, *75*, 789. These papers are excellent sources for additional references.
13. (a) Hrusák, J.; Schröder, D.; Iwata, S. *J. Chem. Phys.* **1997**, *106*, 7541. (b) Nicolaides, A.; Smith, D. M.; Jensen, F.; Radom, L. *J. Am. Chem. Soc.* **1997**, *119*, 8083. (c) Aschi, M.; Harvey, J. N. *J. Chem. Soc., Perkin Trans. 2* **1999**, 1059.
14. (a) Fletcher, T. L.; Namkung, M. J. *Chem. Ind.* **1961**, 179. (b) Brill, E. *J. Chem. Soc. (C)* **1966**, 748.
15. (a) Cheek, P. H.; Wiley, R. H.; Roe, A. *J. Am. Chem. Soc.* **1949**, *71*, 1863. (b) Hawkins, G. F.; Roe, A. *J. Org. Chem.* **1949**, *14*, 328.
16. (a) Rutherford, K. G.; Redmond, W.; Rigamonti, C. S. B., J. *J. Org. Chem.* **1961**, *26*, 5149. (b) Setliff, F. L. *Ark. Acad. Sci. Proc.* **1968**, *22*, 88.
17. (a) Doyle, M. P.; Bryker, W. J. *J. Org. Chem.* **1979**, *44*, 1572. (b) Mirsadeghi, S.; Prasad, G. K. B.; Whittaker, N.; Thakker, D. R. *J. Org. Chem.* **1989**, *54*, 3091.
18. Nguyen, P.; Corpuz, E.; Heidelbaugh, T. M.; Chow, K.; Garst, M. E. *J. Org. Chem.* **2003**, *68*, 10195.
19. (a) Rosenfeld, M. N.; Widdowson, D. A. *J. Chem. Soc., Chem. Comm.* **1979**, 914. (b) Ng, J. S.; Katzenellenbogen, J. A.; Kilbourn, M. R. *J. Org. Chem.* **1981**, *46*, 2520.
20. Shani, J.; Gazit, A.; Livshitz, T.; Biran, S. *J. Med. Chem.* **1985**, *28*, 1504.
21. (a) Olah, G. A.; Welch, J. T.; Vankar, Y. D.; Nojima, M.; Kerekes, I.; Olah, J. A. *J. Org. Chem.* **1979**, *44*, 3872. (b) Yoneda, N.; Fukuhara, T. *Tetrahedron* **1996**, *52*, 23.

22. (a) Tamura, M.; Shibakami, M.; Sekiya, A. *Eur. J. Org. Chem.* **1998**, 725. (b) Tamura, M.; Shibakami, M.; Kurosawa, S.; Arimura, T.; Sekiya, A. *J. Fluorine Chem.* **1996**, *78*, 95.
23. Petterson, R. C.; DiMaggio, III, A.; Hebert, A. L.; Haley, T. J.; Mykytha, J. P.; Sarkar, I. M. *J. Org. Chem.* **1971**, *36*, 631.
24. (a) Kirk, K. L.; Cohen, L. A. *J. Am. Chem. Soc.* **1971**, *93*, 3060. (b) Kirk, K. L.; Cohen, L. A. *J. Am. Chem. Soc.* **1973**, *95*, 4619.
25. Dolensky, B.; Takeuchi, Y.; Cohen, L. A.; Kirk, K. L. *J. Fluorine Chem.* **2001**, *107*, 147.
26. Sawaguchi, M.; Fukuhara, T.; Yoneda, N. *J. Fluorine Chem.* **1999**, *97*, 127.
27. (a) Anderson, L. C.; Roedel, M. J. *J. Am. Chem. Soc.* **1945**, *67*, 955. (b) Shinhama, K.; Aki, S.; Furuta, T.; Minamikawa, J. *Synth. Commun.* **1993**, *23*, 1577.
28. Hartman, G. D.; Biffar, S. E. *J. Org. Chem.* **1977**, *42*, 1468.
29. Laali, K. K.; Gettwert, V. J. *J. Fluorine Chem.* **2001**, *107*, 31.
30. Knöchel, A.; Zwernemann, O. *Appl. Radiat. Isot.* **1991**, *42*, 1077.
31. (a) Roe, A.; Hawkins, G. F. *J. Am. Chem. Soc.* **1947**, *69*, 2443. (b) Finger, G. C.; Starr, L. D.; Roe, A.; Link, W. J. *J. Org. Chem.* **1962**, *27*, 3965. (c) Link, W. J.; Borne, R. F.; Setliff, F. L. *J. Heterocycl. Chem.* **1967**, *4*, 641. (d) Setliff, F. L. *Org. Prep. Proc. Int.* **1971**, *3*, 217. (e) Desai, P. B. *J. Chem. Soc., Perkin Trans. 1* **1973**, 1865. (f) Matsumoto, J.; Miyamoto, T.; Minamida, A.; Nishimura, Y.; Egawa, H.; Nishimura, H. *J. Heterocycl. Chem.* **1984**, *21*, 673. (g) Sanchez, J. P.; Gogliotti, R. D. *J. Heterocycl. Chem.* **1993**, *30*, 855. (h) Nakayama, K.; Kuru, N.; Ohtsuka, M.; Yokomizo, Y.; Sakamoto, A.; Kawato, H.; Yoshida, K.; Ohta, T.; Hoshino, K.; Akimoto, K.; Itoh, J.; Ishida, H.; Cho, A.; Palme, M. H.; Zhang, J. Z.; Lee, V. J.; Watkins, W. J. *Bioorg. Med. Chem. Lett.* **2004**, *14*, 2493.
32. Roe, A.; Hawkins, G. F. *J. Am. Chem. Soc.* **1949**, *71*, 1785.
33. (a) Roe, A.; Teague, C. E., Jr. *J. Am. Chem. Soc.* **1951**, *73*, 687. (b) Zára-Kaczián, E.; Deák, G.; György, L. *Acta Chim. Hung.* **1989**, *126*, 573.
34. (a) Matsumoto, J.; Miyamoto, T.; Minamida, A.; Nishimura, Y.; Egawa, H.; Nishimura, H. *J. Med. Chem.* **1984**, *27*, 292. (b) Nishimura, Y.; Matsumoto, J. *J. Med. Chem.* **1987**, *30*, 1622.
35. (a) Montgomery, J. A.; Hewson, K. *J. Heterocycl. Chem.* **1967**, *4*, 463. (b) Montgomery, J.A.; Hewson, K. *J. Org. Chem.* **1969**, *34*, 1396. (c) Ikehara, M.; Yamada, S. *Chem. Pharm. Bull.* **1971**, *19*, 104. (d) Montgomery, J. A.; Clayton, S. D.; Shortnacy, A. T. *J. Heterocycl. Chem.* **1979**, *16*, 157. (e) Montgomery, J. A.; Shortnacy, A. T.; Secrist, III, J. A. *J. Med. Chem.* **1983**, *26*, 1483. (f) Barai, V. N.; Zinchenko, A. I.; Eroshevskaya, L. A.; Zhernosek, E. V.; De Clercq, E.; Mikhailopulo, I. A. *Helv. Chim. Acta* **2002**, *85*, 1893. (g) Prekupec, S.; Svedruzic, D.; Gazivoda, T.; Mrvos-Sermek, D.; Nagl, A.; Grdisa, M.; Pavelic, K.; Balzarini, J.; De Clercq, E.; Folkers, G.; Scapozza, L.; Mintas, M.; Raic-Malic, S. *J. Med. Chem.* **2003**, *46*, 5763. (h) Barai, V. N.; Zinchenko, A. I.; Eroshevskaya, L. A.; Zhernosek, E. V.; Balzarini, J.; De Clercq, E.; Mikhailopulo, I. A. *Nucleosides & Nucleotides* **2003**, *22*, 751. (i) Hardcastle, I. R.; Arris, C. E.; Bentley, J.; Boyle, F. T.; Chen, Y.; Curtin, N. J.; Endicott, J. A.; Gibson, A. E.; Golding, B. T.; Griffin, R. J.; Jewsbury, P.; Menyerol, J.; Mesguiche, V.; Newell, D. R.; Noble, M. E. M.; Pratt, D. J.; Wang, L.-Z.; Whitfield, H. J. *J. Med. Chem.* **2004**, *47*, 3710.
36. Saeki, K.; Tomomitsu, M.; Kawazoe, Y.; Momota, K.; Kimoto, H. *Chem. Pharm. Bull.* **1996**, *44*, 2254.
37. Kilic, E.; Tuzun, C. *Synth. Commun.* **1992**, *22*, 545.
38. (a) Corral, C.; Lasso, A.; Lissavetzky, J.; Alvarez-Insúa, A. S.; Valdeolmillos, A. M. *Heterocycles* **1985**, *23*, 1431. (b) Kobarfard, F.; Kauffman, J. M.; Boyko, W. J. *J. Heterocyclic Chem.* **1999**, *36*, 1247. (c) Kiryanov, A. A.; Seed, A. J.; Sampson, P. *Tetrahedron Lett.* **2001**, *42*, 8797. (d) Dvornikova, E.; Bechcicka, M.; Kamienska-Trela, K.; Krowczynski, A. *J. Fluorine Chem.* **2003**, *124*, 159.
39. (a) Fabra, F.; Vilarrasa, J.; Coll, J. *J. Heterocycl. Chem.* **1978**, *15*, 1447. (b) Fabra, F.; Fos, E.; Vilarrasa, J. *Tetrahedron Lett.* **1979**, *20*, 3179.
40. Archer, S.; McGarry, R. *J. Heterocycl. Chem.* **1982**, *19*, 1245.
41. Fisher, E. C.; Joullié, M. M. *J. Org. Chem.* **1958**, *23*, 1944.
42. Thibault, C.; L'Heureux, A.; Bhide, R. S.; Ruel, R. *Org. Lett.* **2003**, *5*, 5023.
43. Finger, G. C.; Reed, F. H.; Maynert, E. W.; Weiner, A. M. *J. Am. Chem. Soc.* **1951**, *73*, 149.
44. (a) Finger, G. C.; Reed, F. H. *J. Am. Chem. Soc.* **1944**, *66*, 1972. (b) Finger, G. C.; Reed, F. H.; Burness, D. M.; Fort, D. M.; Blough, R. R. *J. Am. Chem. Soc.* **1951**, *73*, 145. (c) Finger, G. C.; Reed, F. H.; Oesterling, R. E. *J. Am. Chem. Soc.* **1951**, *73*, 152. (d) Finger, G. C.; Reed, F. H.; Finnerty, J. L. *J. Am. Chem. Soc.* **1951**, *73*, 153. (e) Finger, G. C.; Shiley, R. H.; Dickerson, D. R. *J. Fluorine Chem.* **1974**, *4*, 111.
45. Xu, Z.-Y.; Du, X.-H.; Xu, X.-S.; Ni, Y.-B. *J. Chem. Res.* **2004**, 496.

46. Zou, X.; Qiu, Z. *J. Fluorine Chem.* **2002**, *116*, 173.
47. (a) Newman, M. S.; Lilje, K. C. *J. Org. Chem.* **1979**, *44*, 1347. (b) Luthe, G.; Brinkman, U. A. Th. *The Analyst* **2000**, *125*, 1699. (c) Luthe, G.; Scharp, J.; Brinkman, U. A. Th.; Gooijer, C. *Anal. Chim. Acta* **2001**, *429*, 49. (d) Luthe, G.; Broeders, J.; Brinkman, U. A. Th. *J. Chromat. A* **2001**, *933*, 27.
48. Sweeny, Jr., A.; Blair, R. D.; Chang, A.; Clark, L. *J. Fluorine Chem.* **1975**, *6*, 389.
49. Oya, S.; Choi, S. R.; Coenen, H.; Kung, H. F. *J. Med. Chem.* **2002**, *45*, 4716.
50. Lindenstruth, A. F.; Fellman, J. H.; VanderWerf, C. A. *J. Am. Chem. Soc.* **1950**, *72*, 1886.
51. Kirk, K. L.; Cantacuzene, D.; Nimitkitpaisan, Y.; McCulloh, D.; Padgett, W. L.; Daly, J. W.; Creveling, C. R. *J. Med. Chem.* **1979**, *22*, 1493.
52. (a) Utne, T.; Jobson, R. B.; Babson, R. D. *J. Org. Chem.* **1968**, *33*, 2469. (b) Heiman, D. F.; Senderoff, S. G.; Katzenellenbogen, J. A.; Neeley, R. J. *J. Med. Chem.* **1980**, *23*, 994. (c) Horwitz, J. P.; Iyer, V. K.; Vardhan, H. B.; Corombos, J.; Brooks, S. C. *J. Med. Chem.* **1986**, *29*, 692.
53. Tsuge, A.; Moriguchi, T.; Mataka, S.; Tashiro, M. *J. Chem. Research (S)* **1995**, 460.
54. Kook, C. S.; Reed, M. F.; Digenis, G. A. *J. Med. Chem.* **1975**, *18*, 533.
55. Livni, E.; Fischman, A. J.; Ray, S.; Sinclair, I.; Elmaleh, D. R.; Alpert, N. M.; Weiss, S.; Correia, J. A.; Webb, D.; et al. *Nucl. Med. Biol.* **1992**, *19*, 191.
56. Argentini, M.; Wiese, C.; Weinreich, R. *J. Fluorine Chem.* **1994**, *68*, 141.
57. (a) Clark, J. C.; Goulding, R. W.; Roman, M.; Palmer, A. J. *Radiochem. Radioanal. Lett.* **1973**, *1412*, 101. (b) Fowler, J. S.; Finn, R. D.; Lambrecht, R. M.; Wolf, A. P. *J. Nucl. Med.* **1973**, *14*, 63.
58. Schiemann, G.; Winkelmüller, W. *Org. Syn. Coll. Vol. II* **1943**, 188.
59. Schiemann, G.; Winkelmüller, W. *Org. Syn. Coll. Vol. II* **1943**, 299.
60. Flood, D. T. *Org. Syn. Coll. Vol. II* **1943**, 295.
61. Rutherford, K. G.; Redmond, W. *Org. Syn. Coll. Vol. V* **1973**, 133.

Gordon W. Gribble

7.2 Buchwald–Hartwig Amination

7.2.1 Desciption

The Buchwald–Hartwig amination is an exceedingly general method for generating any type of aromatic amine from an aryl halide or aryl sulfonates.[1,2] The key feature of this methodology is the use of catalytic palladium modulated by various electron-rich ligands. Strong bases, such as sodium *tert*-butoxide, are essential for catalyst turnover.

$$R^1\text{-}C_6H_4\text{-}X + HNR^2R^3 \xrightarrow[\text{NaO}t\text{-Bu}]{\text{cat. } L_nPd(0)} R^1\text{-}C_6H_4\text{-}NR^2R^3$$

$X = I, Br, Cl, OSO_2R$

The most noteworthy feature of this palladium-catalyzed amination is the wide tolerance of nitrogen nucleophiles. Depending upon the proper choice of ligand for palladium, most primary and secondary amines of all types are well received. This reaction even works for much weaker nitrogen nucleophiles, such as amides, carbamates, sulfonamides, imines, oximes, and hydrazines. As for the aromatic component, a new generation of ligands for this reaction has even allowed for room temperature reactions of aryl chlorides and further expanded the synthetic utility. Given the relatively mild reaction conditions and broad scope, the Buchwald–Hartwig amination is one of the premier methods for constructing the nitrogen–aryl bond.

7.2.2 Historical Perspective

Aromatic amines are prevalent in a wide variety of natural products and are ubiquitous in therapeutic agents and other synthetic, bioactive molecules. As such, a number of classical methods exist for creating this bond. Among the most common are S_NAr reactions[3] and copper catalyzed Goldberg–Ullmann type reactions.[4] Although these reactions offer much literature precedence, they can often be limited by harsh reaction conditions, poor functional group tolerance, narrow substrate scope, or undesired side products. A modern alternative is the analogous Ni(0) catalyzed process, but this protocol often requires careful catalyst handling under an inert atmosphere.[5]

The first paper that demonstrated the feasibility of a palladium-catalyzed amination was a report by T. Migita and co-workers.[6] They demonstrated that a tri-*ortho*-tolyl phosphine palladium chloride catalyst will catalyze the cross-coupling of *N,N*-diethylamino-tributyltin with bromobenzene. The process can be thought of as an amino-Stille cross-coupling.

1983 T. Migata *et al.*

$$n\text{-}Bu_3SnNEt_2 + ArBr \xrightarrow[\text{tol., 100 °C, 3 h; 16–81\% yield}]{\text{10 mol\% }[(o\text{-tol})_3P]_2PdCl_2} ArNEt_2 + n\text{-}Bu_3SnBr$$

A stoichiometric, intramolecular variant of the Buchwald–Hartwig amination was reported by D. L. Boger and J. S. Panek as early as 1984 in the synthesis of lavendamycin.[7] They showed that 1.2 equiv of $Pd(PPh_3)_4$ could effect the intramolecular cyclization of an amino bromide **1** to the desired heterocycle **2**. The absence of an external base to sequester the generated HBr rendered this reaction super-stoichiometric in palladium.

1984 D. L. Boger and J. S. Panek

MeO_2C, N, CO_2Me, H_2N, Me, Br, **1** → [$Pd(PPh_3)_4$ (1.2 equiv); THF, 80 °C, 21 h; 84% yield] → MeO_2C, N, CO_2Me, HN, Me, **2**

Contemporaneously with each other, S. L. Buchwald and J. F. Hartwig were both examining the mechanism and working to optimize the original amino-stannane coupling reported by T. Migita.[8] Hartwig was able to isolate and characterize by X-ray crystallography a catalytically competent Pd-amine intermediate (*trans*-$[(o\text{-tolyl})_3P](NHEt_2)PdCl_2$. Buchwald and A. S. Guram were able to improve the orginal Migita procedure by developing a way to generate the amino-stannane *in situ*. This procedure worked well for the coupling of secondary amines and anilines with aryl bromides.

1994 S. L. Buchwald and A. S. Guram

$$n\text{-}Bu_3SnNEt_2 + HNR^2R^3 \xrightarrow[-HNEt_2]{\text{80 °C, Ar purge}} n\text{-}Bu_3Sn{-}NR^2R^3 \xrightarrow[\text{tol., 105 °C, 3 h; 55–88\% yield}]{\text{ArBr, 1-2.5 mol\% }[(o\text{-tol})_3P]_2PdCl_2} Ar{-}NR^2R^3$$

1995 S. L. Buchwald, A. S. Guram, and R. A. Rennels
1995 J. F. Hartwig and J. Louie

cat. $[(o\text{-tol})_3P]_2PdCl_2$
NaO*t*-Bu (Buchwald)
Or
LiHMDS (Hartwig)

Buchwald was able to eliminate toxic tin by first attempting the cross-coupling with $B(NMe_2)_3$ instead.[9] Key to this "amino-Suzuki" alternative was the use of NaO*t*-Bu as a stoichiometric base. Subsequently, they found that boron was not necessary and simple amines would readily couple. At the same time, J. F. Hartwig and J. Louie reached the same conclusions. Based upon their examination of the catalytic cycle of the Migita reaction, Hartwig reasoned that amines could cross-couple in a catalytic sense if a base was present to turn over the palladium by deprotonating the Pd-amine intermediate giving an intermediate palladium amido species. They used $LiN(SiMe_3)_2$ (LiHMDS) to accomplish this purpose, thereby generating LiBr and $HN(SiMe_3)_2$.

7.2.3 *Mechanism*

Pd(BINAP)$_2$ catalyzed

$$\frac{-d[ArX]}{dt} = \frac{k_1 k_2}{k_{-1}[L]}[ArX][Pd]$$

The generally accepted catalytic cycle for the Buchwald–Hartwig amination mirrors that of other palladium catalyzed cross-coupling reactions.[10,11,12] There is an oxidative addition (**A** to **B**), followed by an exchange on palladium (**B** to **C**), and finally a reductive elimination (**C** to **D** and **A**). The main difference involves the exchange step. In a Suzuki, or Stille, reaction this step proceeds through a discrete transmetallation event, whereas the Buchwald–Hartwig amination requires stoichiometric NaO*t*-Bu with which to deprotonate the Pd-amine complex giving **C**, thereby also forming NaBr and *t*-BuOH.

The Pd/BINAP catalyzed Buchwald–Hartwig amination has been extensively studied using reaction calorimetry to determine the kinetics along with isolation and characterization of some of the proposed intermediates.[11a] Using this and several other lines of evidence, a general rate equation can be derived for this particular catalyst system. It's instructive to note that for this transformation, the rate depends upon dissociation of BINAP from $(BINAP)_2Pd$ to form the active Pd(BINAP) catalyst, and the rate of oxidative addition to ArX. The amine and base do not play a significant role in the overall kinetics.

For some combinations of ligands and palladium, amines containing α-protons are prone to β-hydride elimination (**C** to **F**), giving an imine (**E**) and de-halogenated arene (**G**) as the major side products. For the first generation of ligands, this was a particular limitation. Chelating phosphines (second generation) and the latest third generation ligands have eliminated this side reaction as a major concern, although some difficult amine/haloarene combinations still suffer from this pathway.

7.2.4 *Synthetic Utility*

7.2.4.1 First generation ligands: mono-dentate phosphines

The early examples of the Buchwald–Hartwig amination typically utilize $(o\text{-tol})_3P$ as the ligand for palladium with NaO*t*-Bu as the stoichiometric base.[9, 13] The reaction is typically conducted at 100 °C in toluene. For aryl bromides, electron rich (EDG = electron donating group) and electron neutral aromatics substituted at either the *para* or *meta* positions work quite well, whereas electron withdrawing groups (EWG = electron withdrawing group) give lower yields. The amines are primarily limited to secondary acyclic (aliphatic and aromatic) and cyclic. Primary amines can participate with some aryl bromides, but they tend to give high levels of de-halogenated arene and imine side products.

Br, R^1, + $HN(R^2)(R^3)$ → 2-5 mol% $[(o\text{-tol})_3P]_2PdCl_2$; NaO*t*-Bu, tol. 100 °C, 2–4 h; 67–89% yield → R^1-aryl-$N(R^2)R^3$

R^1 = EWG or EDG
amine = 2° cyclic or acyclic

Aryl iodides may also couple with a variety of amines using the aforementioned reaction conditions and phosphine ligand.[14] In general, the reaction is limited to secondary cyclic and acyclic amines. Primary amines give very poor yields unless the aryl iodide contains an *ortho* methyl group.

0.5 mol% $Pd_2(dba)_3$, 2 mol% $P(o\text{-tol})_3$; NaO*t*-Bu, dioxane, 65–100 °C, 2–24 h, 18–79% yield

R^1 = EWG or EDG
amine = 2° cyclic or acyclic
amine = 1° aliphatic: low yield, unless R^1 *ortho*

Interestingly, the intramolecular variant of this transformation works best with palladium tetrakis as the catalyst.[9a,13] A combination of NaO*t*-Bu and K_2CO_3 serves as the base, although Et_3N as solvent may also be used.

N(H)Bn; I or Br; n = 1–3; cat. $Pd(PPh_3)_4$; NaO*t*-Bu/K_2CO_3 or Et_3N, tol., 22–100 °C; 83–96% yield; N Bn

Evidence of an undesired β-elimination pathway can be demonstrated when enantioenriched α-phenyl pyrrolidine is subjected to the standard reaction conditions. Although this amine couples in 60% yield with bromobenzene, the resulting product is racemic. This indicates that the palladium-amido intermediate is in equilibrium with an imine.

Ph + ArBr; 98% *ee*; $Pd_2(dba)_3$, $P(o\text{-tol})_3$; NaO*t*-Bu, tol., 100 °C, 60% yield; Ph, N Ar, 0% *ee*

N-Aryl piperazines can be synthesized using this catalyst and unprotected piperazines.[15] Key to this protocol is the use of excess piperazine to favor mono-arylation versus the di-arylated product. Moderate yields of the mono-arylated product can be obtained with a variety of aryl bromides. The major side product is the de-halogenated arene caused by the β-elimination pathway.

3 mol% $[(o\text{-tol})_3P]_2PdCl_2$

NaO*t*-Bu, tol.
100 °C, 2–4 h
10–57% yield

1.2–4 equiv

+ di-arylation
+ de-halogenation

These amine coupling conditions have been adapted to the solid phase by a group at Merck in order to prepare a library of substituted anilines.[16] The aryl bromide is attached to the solid phase through an amide linkage. *Para* and *meta* substituted solid phase aryl bromides were found to couple efficiently with secondary amines. This methodology was later extended to primary aliphatic amines as well using the second generation of phosphine ligands.

7.2.4.2 Second generation ligands: chelating phosphines

A major advance and expansion in reaction scope was made when Buchwald and Hartwig reported that common chelating phosphines, in particular, BINAP and DPPF, allowed for the general arylation of primary aliphatic amines. Buchwald found that *rac*-BINAP with palladium could efficiently catalyze the coupling of a variety of aryl bromides and primary amines of all types, aliphatic, aromatic, allyl, and benzyl.[17] Important to this reaction is the pre-mixing of $Pd_2(dba)_3$ and BINAP (1.5/1 = BINAP/Pd). Primary amines branched at the alpha position required more forcing reaction conditions (higher temperature, more catalyst, and longer times). Secondary cyclic amines and substituted anilines gave high yields, but secondary acyclic aliphatic amines proved to be difficult substrates.

0.25 mol% $Pd_2(dba)_3$
0.75 mol% *rac*-BINAP

NaO*t*-Bu, tol.
80 °C, 2–39 h
66–98% yield

R^1 = EWG or EDG
amine = 2° cyclic or acyclic: **low yields with acyclic aliphatic**
amine = 1° aliphatic or aromatic: high yields

BINAP

Although these conditions are quite general for a wide variety of aryl bromides, the use of a strong base (NaO*t*-Bu) can be incompatible with base sensitive, enolizable, substituents (R^1). To address this limitation, Buchwald was able to replace NaO*t*-Bu with Cs_2CO_3. This allowed for aryl bromides substituted with an ester, nitrile, nitro, ketone, or aldehyde to couple cleanly with primary amines and cyclic secondary amines.

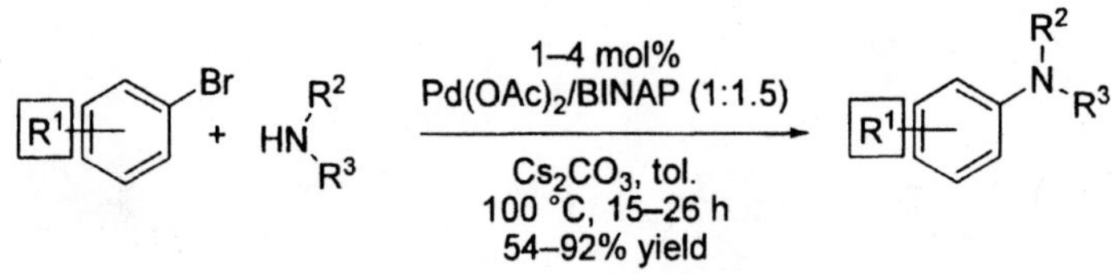

R^1 = Base sensitive: NO_2, CO_2R, CN, C(O)R, CHO
amine = 2° cyclic or acyclic: **low yields with acyclic aliphatic**
amine = 1° aliphatic or aromatic

Aryl iodides may also couple with primary amines using BINAP or tol-BINAP as a ligand.[18] Importantly, the more reactive aryl iodides undergo this cross coupling at lower temperatures than with the first generation catalyst (22–40 °C in THF). As with aryl bromides, this catalyst system does not perform well with acyclic aliphatic secondary amines, but appears quite general for cyclic secondary and primary amines.

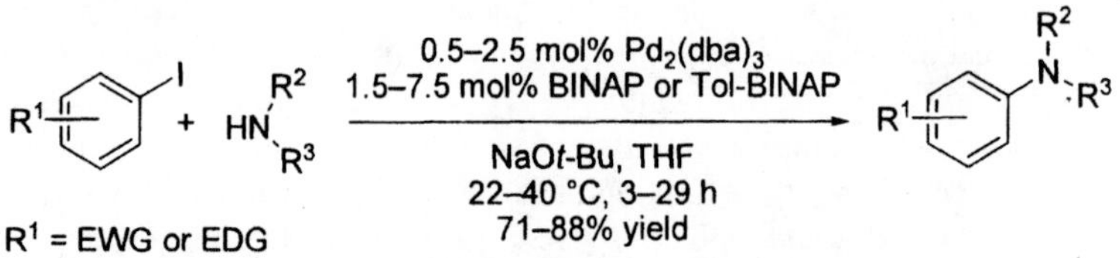

R^1 = EWG or EDG
amine = 2° cyclic or acyclic: **low yields with acyclic aliphatic**
amine = 1° aliphatic or aromatic

The versatility of this catalyst was demonstrated by Loeppky who showed that cyclopropylamine reacts well with many aryl bromides.[19] This proved to be a great improvement over a traditional multi-step reductive amination methodology. As before, this catalyst proved robust enough to work well on the solid phase for the cross-coupling of primary amines to immobilized aryl bromides.[20]

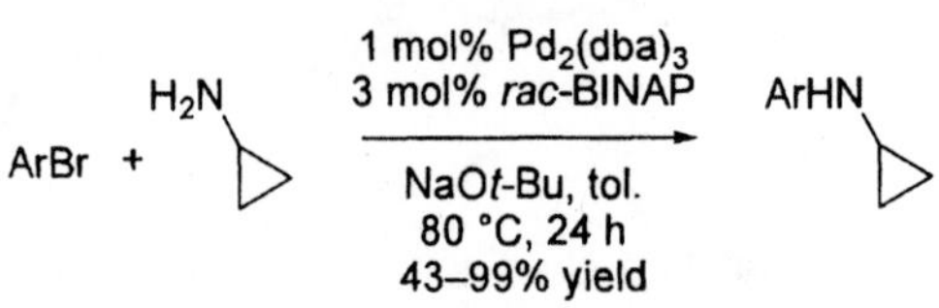

Hartwig's group has developed chelating ferrocene ligands for this amination.[21] They demonstrated that readily available DPPF serves as a competent ligand for the palladium catalyzed amination of primary aliphatic and aromatic amines. The coupling is conducted in THF and works equally well for both aryl iodides and bromides. Secondary amines were not extensively reported.

5 mol% (DPPF)$PdCl_2$
15 mol% DPPF
NaO*t*-Bu, THF
100 °C (sealed), 3 h
80–96% yield
(11 examples)

X = Br or I
R^1 = EWG or EDG
amine = 2° acyclic (one example)
amine = 1° aliphatic or aromatic

DPPF

Buchwald and co-workers examined a P,O-chelating ferrocene ligand, *rac*-PPF-OMe.[22] This ligand is general for secondary acyclic and cyclic amines of all types. Primary amines were not reported with this ligand. The aryl bromides work best with either electron donating or electron withdrawing R^1. Electron neutral aryl bromides tend to give lower yields of the *N*-arylated product.

0.25–2.5 mol% $Pd_2(dba)_3$
0.75–7.5 *rac*-PPF-OMe
NaO*t*-Bu, tol.
80 °C, 4–12 h
52–93% yield

R^1 = EWG or EDG: **lower yields with neutral aromatic**
R^1 = neutral aromatic, or base sensitive: used Cs_2CO_3
amine = 2° cyclic or acyclic: aromatic or aliphatic

rac-PPF-OMe

all >99% ee

Ar—Br
Pd(II)salt/L
L = BINAP or DPPF
NaO*t*-Bu, tol.
70–100 °C

40–86% yield
94–99%ee

43–63% yield
96%ee

72–82% yield
>99%ee

As mentioned previously, the first generation (o-tol)$_3$P ligand gave partial to complete racemization for alpha-chiral amines.[23] The exception is the intramolecular variant in which the enantiopurity of the starting amine is maintained. As shown, the use of chelating phosphines (BINAP or DPPF) limits epimerization through the β-elimination pathway for secondary and primary chiral amines.

Buchwald has also shown that the easily synthesized DPEphos ligand is an ideal catalyst for the arylation of anilines with aryl bromides.[24] Even di-*ortho* substituted aniline and aryl bromide can be efficiently coupled. Furthermore, Buchwald has reported that readily available, and cheaper, $PdCl_2$ may serve as the Pd source rather than $Pd(OAc)_2$.

0.05–5 mol% $Pd(OAc)_2$ or $PdCl_2$
0.075–7.5 mol% DPEphos
NaO*t*-Bu or Cs_2CO_3, tol.
80–100 °C, 2–21 h
80–96% yield

R^1 = EDG or neutral aromatic
R^2 = EDG or EWG, also with ortho substitution

DPEphos

Others have explored modifications of BINAP and DPPF. Boche and co-workers used a water soluble sulfonated BINAP type ligand in order to conduct this amination in a water/MeOH mixture using NaOH as the base.[25] Beletskaya and co-workers have examined the effects of mono-oxide DPPF type ligands on the amination.[26]

A useful advance in ligand design was made by van Leeuwen who reports that Xantphos serves as an excellent ligand for the arylation of anilines.[27] Sterically encumbered aryl bromide **3** cross-couple quite easily with **4** giving aniline **5** using Xantphos as a ligand.

0.5 mol% $Pd(OAc)_2$
0.75 mol% Xantphos
NaO*t*-Bu, tol.
80 °C, 68 h
97% yield

Xantphos

The Beletskaya group has extensively examined the mono- versus di-arylation of diamines.[28] Using excess diamine **6**, they have shown that even an aryl dibromide can be mono-aminated with minimal (8%) double amination. An aryl iodo bromide was shown to mono-aminate at the iodide position, leaving the bromide intact. Other permutations upon this theme were also reported with varying degrees of success.

2–4 mol% $Pd_2(dba)_3$
3–6 mol% DPPF
NaO*t*-Bu, dioxane
100 °C, 3–12 h

6
NH_2 NH_2
3 equiv

X = Y = Br: 71-93% yield (8% double amination)
X = I and Y = Br: 72% yield
X = Br and Y = Cl: 75% yield (15% double amination)
X = Y = I: 57% yield

Beletskaya has also published an example of bis-amination using two different amines and a dihaloquinoline **7**.[29] The reaction is catalyzed by a modified version of DPPF (**L1**). Addition of 1 equiv of morpholine followed by piperidine gives diaminoquinoline **8**, where the more reactive aryl bromide position reacts first with the morpholine. The process can be reversed by adding piperidine and then morpholine to give **9**. They also report a sequential amination followed by Suzuki cross-coupling in which the same reactivity pattern is obeyed using **L1** as a ligand for palladium.

2 mol% $PdCl_2$(**L1**)
NaO*t*-Bu
dioxane

8 85%
9 78%

L1 Ar = *o*-*i*-Pr phenyl

The palladium catalyzed amination of aryl triflates has proved to be a more challenging problem. Aryl triflates are less favored to undergo oxidative addition as compared to aryl bromides and iodides. Another, more daunting, challenge with aryl triflates is their proclivity to react with NaO*t*-Bu at sulfur to give a sodium phenoxide as the major by-product. Buchwald's laboratory reported that with fairly electron rich (EDG) and neutral aryl triflates, they could use NaO*t*-Bu as a base with minor amounts of phenol by-product being generated.[30] Later, they developed a more general solution by replacing NaO*t*-Bu with Cs_2CO_3. Under these reaction conditions, with BINAP, both electron rich and electron poor aryl triflates gave good to excellent yields of the expected substituted anilines with minimal phenol side products. Secondary cyclic and primary amines were among the best substrates, as was seen in other Pd(BINAP) catalyzed reactions. This methodology has also been used to aminate a 2-triflatropone with an array of functionalized anilines.[31]

cat. $Pd(OAc)_2$/BINAP (P/Pd =1.5/1); Cs_2CO_3, THF, dioxane, or tol, 65–100 °C, 4–16 h, 46–92% yield

main side product

R^1 = EDG, EWG, or base sensitive
amine = 2° cyclic or acyclic (benzylic)
amine = 1° aliphatic or aromatic

Hartwig's solution to aryl triflate amination was by slow addition of the aryl triflate to a solution of Pd(DPPF) catalyst, amine, and NaO*t*-Bu.[32] This procedure works well for primary amines and secondary cyclic amines, but yields can be lower if the aryl triflate contains R^1 = Me or is base sensitive.

slow addition

1.5 mol% $Pd_2(dba)_3$, 3–4 mol% DPPF; NaO*t*-Bu, tol., 80–100 °C, 2–5 h, 42–97% yield

R^1 = EDG or EWG
amine = 2° cyclic
amine = 1° aliphatic or aniline

7.2.4.3 *Third generation ligands: bulky phosphines and* N*-heterocyclic carbenes*

The preceding examples demonstrate that the Buchwald–Hartwig amination is a versatile strategy for the amination of aryl bromides, iodides, and triflates. The use of chelating phosphines (BINAP and DPPF) has allowed for the efficient arylation of most primary amines and cyclic secondary amines. The original Migita catalyst, $[(o\text{-tol})_3P]_2PdCl_2$, was found to be adequate for many secondary acyclic amines. An improvement to these previously described catalyst systems would be to lower reaction temperatures, tolerate a more diverse range of haloarenes (such as aryl chlorides and heterocycles), and lower catalyst loadings. To this end, bulky phosphines, such as *t*-Bu_3P, have been shown to catalyze many other palladium catalyzed cross-couplings at ambient temperatures with low loadings of palladium.

The first to demonstrate t-Bu_3P for the Buchwald–Hartwig amination was Koie and co-workers.[33] They reported the mono-arylation of unprotected piperazine by using 6 equivalents with aryl bromides. A ligand screen revealed that t-Bu_3P in a 4:1 ratio with Pd gave the arylated piperazine in high yields. The reaction still required high temperature (120 °C) and *ortho*-substituted aryl bromides gave poor yields. They did report one example coupling chlorobenzene with piperazine, but no isolated yield was recorded.

cat. $Pd_2(dba)_3$ or $Pd(OAc)_2$
(t-Bu)$_3$P (P/Pd = 4/1)
NaOt-Bu, o-xylene
120 °C, 3–4 h
22–94% yield
6 equiv

R^1 = EDG or EWG: poor yields for o-OMe and o-F

Koie expanded this methodology to the coupling between *N*-phenyl anilines and di-bromobenzenes.[34] With the exception of 1,2-dibromobenzene, these reactions all gave the expected di-aminated products. They also showed that tribromobenzenes and even aryl chlorides could be efficiently aminated under these conditions.

0.05 mol% $Pd(OAc)_2$
(t-Bu)$_3$P (P/Pd = 4/1)
NaOt-Bu, o-xylene
120 °C, 3–4 h
77–92% yield

Tribromides also give high yield
No reaction for 1,2-dibromobenzene

Hartwig was able to show that ligand stoichiometry with palladium has a profound influence on catalyst activity.[35] Decreasing the amount of t-Bu_3P such that P/Pd = 0.8/1 gives a catalyst which allows for room temperature amination of aryl bromides with most secondary amines (cyclic and acyclic) and anilines. They reported that the arylation of primary aliphatic amines with unhindered aryl halides was not amenable to these conditions. They did find that various azoles could be arylated using this catalyst at elevated temperatures.

R^1-ArBr + HNR^2R^3 → (1–2 mol% $Pd(dba)_2$, $(t\text{-}Bu)_3P$ (P/Pd = **0.8/1**), NaO*t*-Bu, tol., **22 °C**, 1–6 h, 81–99% yield) → R^1-Ar-NR^2R^3

R^1 = EDG or EWG
amine = 2° cyclic or acyclic: aromatic, aliphatic, or azoles
amine = 1° anilines: **no aliphatic**

Buchwald and co-workers have developed a library of bulky, electron rich monodentate phosphine ligands based on a bi-aryl backbone.[36] The other two substitutents on the phosphine are usually either cyclohexyl or *t*-Bu moieties. The unique feature of this bi-aryl substituent is that it is quite easy to tune both the steric bulk and electronics of the ligand *via* substitution on the *ortho* aryl ring (e.g., see ligand **L3**). They began by screening for the room temperature amination of aryl chloride. Initially, **L2** and **L3** were found to be capable ligands for this amination, but the preparation of these ligands required a lengthy multi-step synthesis. They subsequently found that **L4** and **L5** were excellent ligands for the palladium catalyzed amination of aryl chloride, bromides, iodides, and triflates. Using a slight excess of **L4** to palladium (**L4**/Pd = 1.5 /1), Buchwald reports that many aryl bromides react at room temperature with a diverse set of cyclic and acyclic secondary amines. Some aryl bromides containing base sensitive functionality required higher temperature (80 °C) and K_3PO_4 rather than NaO*t*-Bu. Primary aliphatic amines were not reported, but primary anilines, allyl amines, and benzyl amines were found to undergo efficient arylation.

L2 (PCy_2, PCy_2) **L3** (PCy_2, Me_2N) **L4** ($P(t\text{-}Bu)_2$) **L5** (PCy_2)

R^1-ArBr + HNR^2R^3 → (1–5 mol% $Pd(OAc)_2$, **L4** (P/Pd = 1.5/1), NaO*t*-Bu, tol., 22–80 °C, 20 h, 52–96% yield) → R^1-Ar-NR^2R^3

R^1 = EDG or EWG
amine = 2° cyclic or acyclic: aromatic, aliphatic, or azoles
amine = 1° anilines, allyl, or benzyl: no aliphatic

Aryl iodides also could react with amines of all type at room temperature when NaO*t*-Bu was used as base with this class of ligands.[37] For most examples, either **L3** or **L5** was utilized, although Xantphos was found to be a useful ligand for some select examples. Base sensitive aryl iodides could be effectively aminated by using Cs_2CO_3 instead, although higher temperatures were generally required. Buchwald has also shown that this class of bi-aryl phosphines can be attached to the solid phase through an ether linkage on the *ortho* aryl group.[38] This allows for simple purification of the aniline products without phosphine contamination, along with catalyst recycling without adding additional palladium.

0.5 mol% $Pd(OAc)_2$
2 mol% **L3, L5,** or Xantphos
NaO*t*-Bu or Cs_2CO_3
dioxane/*t*-BuOH or THF
22–120 °C, 2–20 h
70–100% yield

R^1 = EDG or EWG: Cs_2CO_3 if base sensitive
amine = 2° cyclic or acyclic: aromatic or aliphatic
amine = 1° amines: anilines or aliphatic

This ligand class is quite useful for the synthesis of unsymmetrical alkyl diaryl amines.[39] Typically, a primary aliphatic amine is first arylated using the Pd/BINAP catalyst system. The second arylation is then carried out with **L6**, Xantphos, or PPF-OMe (with Cs_2CO_3 if a base sensitive aryl bromide). Ligand choice for the second arylation depends upon the electronic nature of the *N*-alkylaniline and Ar^2Br as outlined in the table. They were unable to couple 2-bromobenzaldehyde, 4-(4′-bromophenyl)-2-butanone, or 4′-bromoacetophenone with *N*-alkylanilines (alkyl > Me).

R–NH₂ → R–NH–Ar¹ → R–N(Ar¹)(Ar²)

Step 1: Ar^1–Br, 1 mol% $Pd(OAc)_2$, BINAP (P/Pd = 1.5/1), NaO*t*-Bu, tol., 80 °C

Step 2: Ar^2–Br, 1 mol% $Pd(OAc)_2$ or $Pd_2(dba)_3$, **Ligand** (P/Pd = 1.5/1), NaO*t*-Bu or Cs_2CO_3, tol., 80 °C

R = alkyl

L6: 2-(dimethylamino)-2′-(diphenylphosphino)biphenyl (Me_2N, PPh_2)

Ar^2Br \ $Ar^1(R)NH$	e- poor	e- rich
e- poor	Xantphos	**L6**
e- rich	N/A	**L6**

PPF-OMe with Cs_2CO_3 if base sensitive

It has also been reported that 1,2-dibromobenzene can be sequentially aminated with two different amines.[40] As before, the first amination of an aryl bromide with an alkyl amine is catalyzed by Pd/BINAP. The second amination of the product bromoaniline can then be catalyzed by either Pd/BINAP or Pd/**L3** with a variety of cyclic secondary or primary amines.

H_2N–CH(CH$_3$)Ph; 4 mol% $Pd_2(dba)_3$; 8 mol% BINAP; NaO*t*-Bu, tol. 135 °C, sealed

RNH_2; 4 mol% $Pd_2(dba)_3$; 8 mol% **L3** or BINAP; NaO*t*-Bu, tol. or DME 80–135 °C, sealed

R = 2° cyclic or 1° aliphatic and aromatic

The Buchwald ligands can also be applied to the catalytic amination of a variety of aryl sulfonates. Aryl triflates containing diverse functionalities (except for R^1 = Cl or Br) are coupled with most amines using Pd/**L4**.[36] Electron rich or neutral aryl nonaflates (ArONf = $ArOSO_2(CF_2)_3$-CF_3) are aminated using **L3** or **L4** as a ligand for palladium.[41] Electron deficient and base sensitive ArONf can be aminated using the Xantphos ligand. Interestingly, ArONf where R^1 = Cl or Br can be selectively aminated using either **L3** or BINAP. Milder tertiary amine bases such as DBU (1,8-diazabicyclo[5.4.0]undec-7-ene) or MTBD (7-methyl-1,5,7-triazabicyclo[4.4.0]dec-5-ene) may replace NaO*t*-Bu for the amination of aryl nonaflates under microwave conditions with **L7** (XPhos).

R^1–C_6H_4–OSO_2R + HNR^2R^3 → R^1–C_6H_4–NR^2R^3; cat. $Pd(OAc)_2$ or $Pd_2(dba)_3$, **Ligand**; NaO*t*-Bu or K_3PO_4, 22–105 °C

amines = 2° cyclic or acyclic: aromatic or aliphatic
amines = 1° aromatic or aliphatic

L7 (XPhos): PCy_2, *i*-Pr, *i*-Pr, *i*-Pr

	R^1			
	EDG	EWG	base sensitive	Cl/Br
-OTf	**L4**	**L4**	**L4**	N/A
-ONf	**L4/L3**	Xantphos	Xantphos	**L3**/BINAP
-OTs*	**L7**	**L7**	**L7**	**L7**

* tol./*t*-BuOH using Cs_2CO_3, K_2CO_3, or KOH/H_2O

Aryl bezenesulfonates and aryl tosylates (ArOTs) are cheap and easily prepared compared to their triflate counterparts. Buchwald reports that these useful sulfonates can be catalytically aminated using **L7** (XPhos) as a ligand.[42] A variety of base and solvent combinations using *t*-BuOH was found to be necessary depending upon the nature of the aryl tosylate and amine. In all of the aforementioned examples, a sterically and electronically diverse library of bi-aryl phosphine ligands allowed Buchwald and co-workers to specifically optimize reaction conditions for most ArX/amine combinations.

A variation upon this theme has been reported by A. S. Guram and co-workers.[43] They introduced an oxygen containing moiety at the *ortho* position of the aryl phosphine ligand to create a series of P,O-chelating ligands. Extensive screening revealed **L8** was an ideal ligand for the amination of aryl bromides and iodides with most secondary amines and primary anilines. As with Buchwald's bi-aryl phosphines and *t*-Bu_3P, primary aliphatic amines were poor substrates. For aryl chlorides, **L9** proved to be best as it could efficiently couple aryl chlorides with all classes of amines.

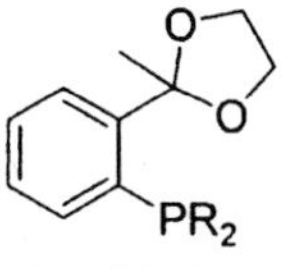

PR_2

R = Ph (**L8**)
R = Cy (**L9**)

L8: ArBr and ArI with 2° amines (acyclic and cyclic) also with 1° anilines, no aliphatic

L9: ArCl with 2° amines (acyclic and cyclic) also with aliphatic and aromatic 1° amines

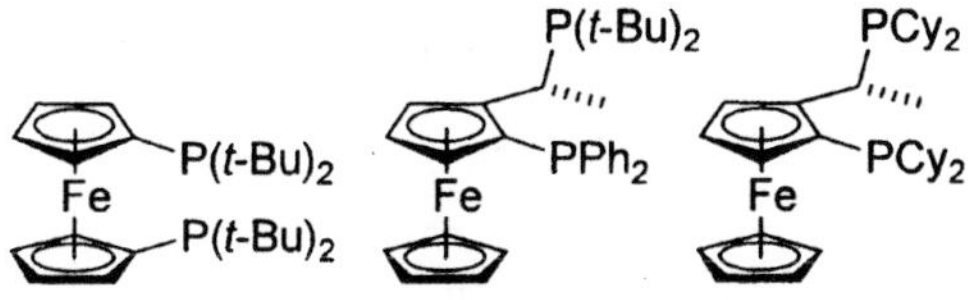

P(*t*-Bu)$_2$ PCy$_2$ P(*t*-Bu)$_2$ PPh$_2$ PCy$_2$ Fe Fe Fe P(*t*-Bu)$_2$

L10 **L11** **L12**

ArI, ArBr, ArCl, ArOTs
+
2° cyclic amines (poor for acyclic)
1° aliphatic and anilines

Hartwig has found that bulky ferrocene ligands are quite versatile ligands for the amination of aryl chlorides, bromides, iodides, and tosylates at room temperature in many cases.[44] Ligands **L10, L11,** and **L12** serve as highly activating ligands for palladium and allow for the coupling of primary aliphatic amines with any ArX. As with second generation chelating phosphines, secondary acyclic aliphatic amines are poor substrates.

Hartwig was able to address the limitations of the aforementioned ferrocene ligands by using Q-Phos.[45] This is the first bulky mono-dentate phosphine ligand which was shown to be effective for primary aliphatic amines as well as many secondary cyclic and acyclic amines and exhibits perhaps the broadest reaction scope to date. Notable limitations are the couplings of acyclic aliphatic secondary amines with *ortho* aryl halides, or electron rich aryl halides. These problematic substrate combinations tend to give de-halogenated arene as the major product.

Q-phos

$P(t\text{-Bu})_2$ Fe Ph Ph Ph Ph Ph

General for ArBr and ArCl with all amines

R X + $HNBu_2$

poor yields

J. G. Verkade has synthesized a series of bicyclic triamino phosphines that are potent σ-donors to palladium, yet can be modified to incorporate a variety of steric encumberance.[46] It was found that ligand **L13** could promote the catalytic amination of both aryl chlorides and bromides with secondary cyclic amines and anilines. **L13** was not as effective for aliphatic secondary acyclic amines and primary aliphatic amines. The scope was improved with ligand **L14**, but as with other third generation ligands, primary aliphatic amines were prone to giving de-halogenated arene.

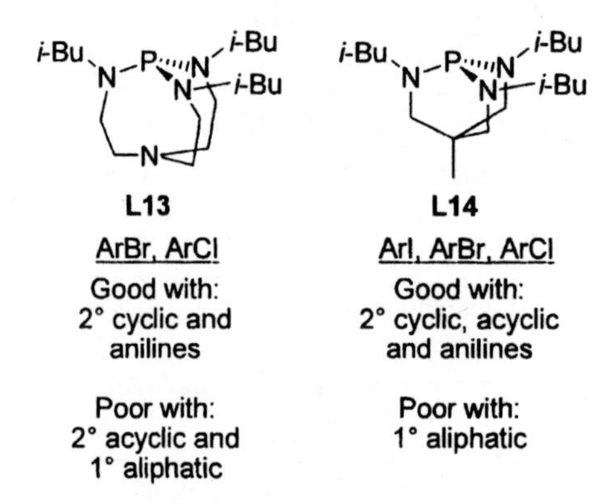

Verkade was able to aminate aryl bromides and chlorides where R^1 contains an acidic proton (phenol, amide, or ketone) by simply using 2.4 equiv of LiHMDS rather than the usual 1.4 equiv. NaO*t*-Bu.[46a] This procedure works well for cyclic secondary amines and anilines of all types, but gives poor yields for primary aliphatic and secondary acyclic amines.

Br or Cl + HNR^2R^3 → cat. $Pd(OAc)_2$, cat. **L13** (L/Pd = 2:1), LiHMDS, THF/tol., 80 °C, 24 h, 52–95% yield

R^1 = Contains acidic proton (e.g., phenol, amide, ketone, *etc.*)
amine = 2° cyclic: **no alipahtic acyclic**
amine = 1° anilines: **no aliphatic**

Verkade's P,N-ligand **L13** also catalyzes the Heck reaction of aryl iodides. It was thus shown that the Buchwald–Hartwig amination could be appended to a Heck reaction using Pd/**L13**.[47] The operation is performed by first adding an aryl iodide to the catalyst solution and an aminostyrene **10** at 60 °C. This gives an intermediate *N*-aryl aminostyrene **11** to which 2 equiv of a second aryl iodide is added at 110 °C. This gives a second arylation of the aminostyrene as well as the Heck arylation of the olefin (**12**).

H_2N **10** → Ar^1—I, 2 mol% $Pd_2(dba)_3$, 4 mol% **L13**, NaO*t*-Bu, tol., 60 °C, 3 h, 78% yield → **11** → Ar^2—I (2 equiv), 110 °C, 16 h, 79% yield → **12**

S. P. Nolan's laboratory and others have published extensively on the use of *N*-heterocyclic carbenes (NHC) as ligands for a variety of palladium catalyzed transformations.[48,49] The key feature of *N*-heterocyclic carbenes is their strong σ-donor/π-acceptor character. These electronic properties act to both activate palladium towards oxidative addition and stabilize the Pd(0) oxidation state. The modular nature of this ligand motif allows for both steric and electronic modification. Early results from Nolan showed that *ortho*-substituted aryl groups on nitrogen provided the requisite steric bulk, and saturation of the imidazole backbone gave a more activating ligand.[48b,c] The reported nomenclature is such that **SIPr** = **Saturated Imidazol** ***i*-Pr$_2$Ph** and **cin** = cinnamyl. The most recent reports have shown that the palladium pre-catalyst can have a large effect on reactivity.[48a] After screening a number of combinations, **(SIPr)Pd(cin)Cl** emerged as the most active catalyst with the broadest substrate scope for the Buchwald–Hartwig amination. Almost all the reported aminations of aryl chlorides and bromides were complete in less than 90 min at room temperature. Once again, primary aliphatic amines were not reported, but all other secondary and primary aromatic amines were arylated in high yield. Many of these NHC type ligands are available as their HCl salt from commercial vendors and these can be used in combination with $Pd_2(dba)_3$.

1 mol% **(SIPr)Pd(cin)Cl**
KO*t*-Bu, DME
22 °C, 1–90 min
83–99% yield

X = Br or Cl
R^1 = Me or Cl
amine = 2° cyclic or acyclic: aliphatic or aromatic
amine = 1° aniline or benzyl: **no aliphatic**

(SIPr)Pd(cin)Cl

M. L. Trudell and Cheng synthesized a chelating bis-NHC ligand (**L15**) that was optimized specifically for the synthesis of *N*-heteroaryl-7-azabicyclo[2.2.1]heptanes.[50] They demonstrated that this ligand could aminate various pyridyl bromides and chlorides to give the desired product in low to moderate yields.

4 mol% $Pd_2(dba)_3$
4 mol% **L15**
NaO*t*-Bu, dioxane
100 °C, 1–2 d
36–71% yield

L15
Ar = 2,4,6-triisopropylphenyl

From the preceding discussion of third generation ligands as well as the chelating phosphine ligands, most ArX/amine combinations can be coupled given the correct choice of ligand and reaction conditions. Beller and co-workers have reported a general ligand screening approach for specific examples of Buchwald–Hartwig aminations.[51a] Hartwig has used fluorescence resonance energy transfer (FRET) to design a high throughput assay for amine arylation which screened a fairly diverse library of known palladium ligands.[51b] These screens show that many of the aforementioned ligands are quite comparable, with the specific limitations mentioned throughout with regard to amines. From a practical perspective, ligand choice is often dictated by the specific nature of ArX and amine (as outlined above and summarized later), as well as ligand availability, cost, and convenience.

7.2.4.4 Coupling of aryl chlorides

The amination of aryl chlorides is desirable due to their prevalence and lower costs as compared to aryl bromides and iodides. Aryl chlorides are less prone to oxidative addition to palladium(0) and therefore require more activating ligands. An early example of ArCl amination was reported by M. Beller and co-workers.[52] They found a palladacycle derived from $Pd(OAc)_2$ and $(o\text{-tol})_3P$, for example, *trans*-di(μ-acetato)-bis[*o*-(di-*o*-tolylphosphino)benzyl]dipalladium(II). This catalyst allows for the amination of electron deficient aryl chlorides (**13**) with piperidine in the presence of catalytic LiBr and KO*t*-Bu as the stoichiometric base. The high temperature required (135 °C) also gives rise to a minor amount of the *meta* aminated product **15**, likely due to aryne formation. It is unclear from this example whether a true oxidative addition to the aryl chloride is occurring, or whether the *para* product **14** also results from aryne formation, as a 1:1 mixture of products **14** and **15** also results without palladacycle present.

F_3C **13** (Cl) + NH → [1 mol% palladacycle, 20 mol% LiBr; KO*t*-Bu, tol., 135 °C, 1 d] → **14** + **15**

with Pd catalyst: 7:1 (74% yield)
no Pd catalyst: 1:1 (79% yield)

Bulky phosphines, such as $(t\text{-Bu})_3P$, were shown to provide a palladium catalyst which is able to aminate aryl chlorides at room temperature with secondary amines of all type along with anilines.[35,53,54,55] Primary amines give very low yields , except if coupling an *o*-tolyl aryl chloride. In addition, Hartwig has shown that NaOH may be used as a base under biphasic conditions (toluene/aq. NaOH) with a phase transfer catalyst ($C_{16}H_{33}NMe_3Br$).

M. Beller has also developed a similar bulky mono-dentate phosphine ligand, di(1-adamantyl)-*n*-butylphosphine, for the coupling of aryl chlorides.[56] Although $(Ad)_2P(n\text{-Bu})$ allows for the arylation of bulky anilines and secondary amines, as well as $t\text{-BuNH}_2$, the reaction temperatures are significantly higher (100–120 °C). In addition, electron neutral or rich aryl chlorides couple best, whereas electron deficient aryl chlorides were not extensively reported.

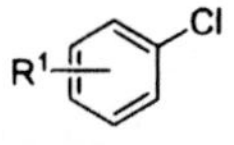

	$(t\text{-Bu})_3P$[a]	$(Ad)_2P(n\text{-Bu})$[c]	L3	L4[f]	L9	L16 L17	Q-phos[f]	L18[g]	L13[g]	(SIPr)Pd(cin)Cl[i]
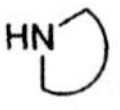 HN (cyclic)	EDG EWG	EDG	EDG EWG	EDG EWG	EDG EWG	EDG EWG	EDG EWG	n/a	EDG EWG	EDG EWG
HNR_2	EDG EWG	EDG	EDG	EDG	EDG EWG	n/a	EDG	n/a	n/a[h]	EDG
HN(R)Ar	EDG EWG	EDG EWG	EDG	EDG EWG	EDG	EDG EWG	EDG EWG	n/a	EDG EWG	EDG
H_2NR	n/a[b]	n/a[d]	n/a[e]	n/a[e]	*o*-EDG	EDG	EDG EWG	EDG EWG	n/a[h]	EDG

[a] Conducted at 22-70 °C with either NaO*t*-Bu, or KOH/H_2O/tol. with cat. $C_{16}H_{33}NMe_3Br$ at 90 °C. [b] Not broadly applicable, limited to *o*-tolyl chloride. [c] 120 °C. [d] Not broadly applicable, limited to $t\text{-BuNH}_2$ or $(Ad)NH_2$. [e] Not broadly applicable, limited to $BnNH_2$ or hexylamine with *o*-tolyl chloride. [f] 22-80 °C. Used K_3PO_4 in DME if R^1 is base sensitive. [g] Used LiHMDS if R^1 = acidic. [h] <70% yields for EWG and EDG. [i] 22 °C in DME

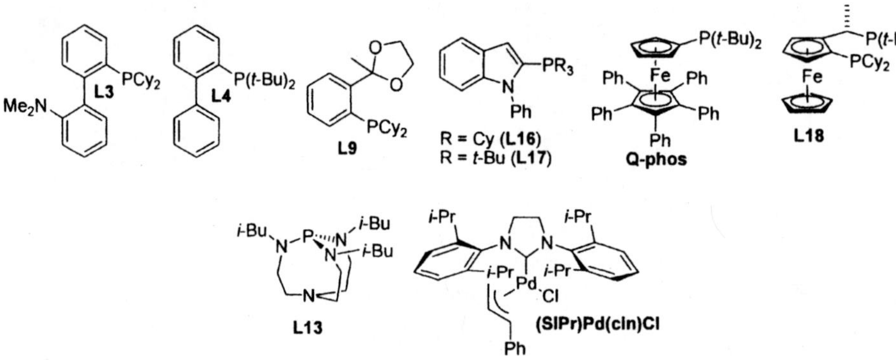

The Buchwald series of bulky phosphines containing the tunable bi-aryl moiety have also been shown to allow for the efficient amination of aryl chlorides as well.[57] In particular, Pd/**L3** catalyzed the reaction between secondary cyclic amines and most aryl chlorides at 80 °C. Secondary acyclic amines and anilines tend to couple best with neutral to electron rich aryl chlorides. Primary aliphatic amines are quite limited to either the amination of *o*-tolyl chlorides, or more activated primary amines such as benzyl amine. A better ligand for this transformation was reported to be **L4** as the amination could be conducted at room temperature in many cases.[58] Base sensitive R^1 required the use of K_3PO_4 rather than NaO*t*-Bu. Most amines work well with this catalyst system, except for unactivated primary aliphatic amines (e.g., *n*-hexyl amine). Allyl amine, hydrazines, imines, benzyl amine, and $(EtO)_2CH_2NH_2$ all couple well with aryl chlorides under the standard reaction conditions. A thermally stable palladacycle of **L4** may also serve to catalyze the amination of aryl chlorides.

The P,O-ligand developed by A. S. Guram (**L9**) is a versatile ligand for the amination of aryl chlorides as most classes of amines work well.[43] Primary amines have a more limited scope and tend to couple best with electron neutral to rich aryl chlorides, especially those with *ortho* substitution.

M. Beller reports an improved phosphine ligand for aryl chloride amination based on *N*-phenyl indoles, **L16** or **L17**.[59] This class of ligands allows for the catalytic amination of aryl chlorides with primary aliphatic amines and anilines. Secondary cyclic amines also couple, but secondary acyclic amines were not reported.

As stated previously, Q-phos shows perhaps the broadest scope for the Buchwald–Hartwig amination.[45] Aryl chlorides are efficiently aminated by all amine classes. The only coupling which gave poor yields was between *ortho* substituted aryl chlorides and acyclic secondary aliphatic amines. Typically, reaction temperatures of 70–100 °C are required and base sensitive aryl chlorides can be aminated by replacing NaO*t*-Bu with K_3PO_4.

Chelating ferrocene phosphine **L18** was reported by Hartwig to efficiently catalyze the amination of most aryl chlorides with any type of primary aliphatic amine, imine, or hydrazine at 80–100 °C with NaO*t*-Bu in DME.[60] Base sensitive aryl chlorides, or those containing acidic protons, may be aminated using LiHMDS as the stoichiometric base. Impressively, catalyst loadings as low as 0.005 mol% can be used.

The Verkade P,N-ligand **L13** catalyzes the amination of aryl chlorides at 100 °C with a variety of secondary cyclic amines and anilines.[46a,b,d,61] Aliphatic primary and secondary amines give the expected aminated products in lower yields (< 70%) and do not aminate electron rich aryl chlorides.

Hartwig initially demonstrated that **SIPr** carbene ligand with $Pd(dba)_2$ is an excellent catalyst for the amination of ArCl with secondary cyclic amines, secondary acyclic amines, and anilines.[62] Primary amines were low yielding, except with benzyl amine. Nolan's catalyst, **(SIPr)Pd(cin)Cl**, gave room temperature amination of aryl chlorides in DME with all secondary amines and anilines.[48] Primary aliphatic amines were reported to work well starting with $Pd_2(dba)_3$ and **IPr**-HCl using KO*t*-Bu in dioxane. Typically, electron neutral and electron rich aryl chlorides are the best substrates.

7.2.4.5 Coupling of aromatic heterocycles

The palladium-catalyzed amination of heteroaromatic halides, or the arylation of heteroaromatic containing amines, is complicated by the Lewis basicity of the heteroatom. The heterocycle may thus act as a ligand for palladium and cause catalyst inhibition and/or poisoning. R. Dommisse demonstrated that by using excess K_2CO_3 with the Pd/BINAP or **L5** catalyst system, a series of pyridyl chlorides could be efficiently aminated by either anilines or aminopyridines.[63]

cat. Pd $(OAc)_2$
cat. BINAP or **L5**
K_2CO_3 (**20 equiv**)
110 °C, tol., 1–2 d
41–93% yield
also 3,5-dichloro
X = C or N

Other previously described catalysts were found to be tolerant of pyridyl halides as coupling partners. M. Beller's adamantyl phosphines were able to catalyze the coupling of protected piperazines and various chloropyridines in order to prepare a small library of bioactive compounds.[64] Buchwald and Wagaw report that either Pd/BINAP or dppp (1,3-bis(diphenylphosphino)propane) was a good catalyst for coupling pyridyl bromides and chlorides with all classes of amines using the standard NaO*t*-Bu conditions.[65] Buchwald ligand **L4** was also broadly applicable to aminating pyridyl chlorides with most amines.[36] Lower yields and di-arylation were a problem for aliphatic primary amines using **L4**. Microwaves with ligand **L5** and Pd were also shown to couple 2 and 3 chloropyridines with anilines.[66]

Beller's improved catalyst system using ligands **L16** or **L17** could efficiently aminate chloropyridines with only secondary cyclic amines and anilines.[59] The ferrocene ligand **L18,** as reported by Hartwig, provides a catalyst which allows for smooth amination of 2 and 3 chloropyridines with primary aliphatic amines, anilines, and a variety of weak nitrogen nucleophiles.[60] Verkade briefly described the use of ligand **L13** for the amination of chloropyridines with secondary cyclic amines and anilines.[46d] Nolan demonstrated that NHC ligand **L19** was ideal for the cross-coupling of halopyridines with cyclic secondary amines and anilines.[48b]

$(Ad)_2P(n\text{-}Bu)$
NBn
HN

BINAP
amines
2° cyclic
2° acyclic
anilines
1° aliphatic
X = Cl or Br

L4
amines
2° cyclic
2° acyclic
anilines
1° aliphatic

L16/L17
amines
2° cyclic
anilines

L18
amines
anilines
1° aliphatic
imine
hydrazine
amides

L13
amines
2° cyclic
anilines

L19
amines
2° cyclic
anilines
X = Cl or Br

Ar = 2,6-*i*-Pr
L19

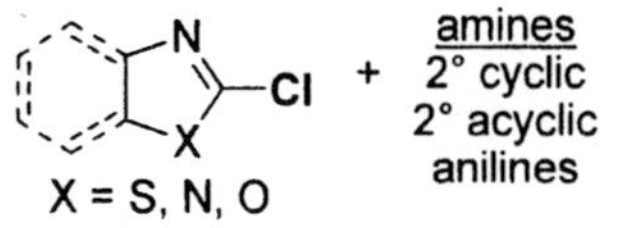

Buchwald's XPhos ligand (**L7**) works well to catalyze the amination of 5-bromopyrimidine with anilines and secondary cyclic amines.[67] Buchwald also found that ligands **L20** and **L21** were the best ligand for the palladium-catalyzed amination of benzothiazoles with secondary amines and anilines. Hartwig reports that $(t\text{-Bu})_3P$ was excellent for the catalytic amination of 2-chloroheteroaryl compounds with secondary amines and anilines.[35,68] As with other examples using bulky mono-dentate phosphines, primary aliphatic amines were not reported.

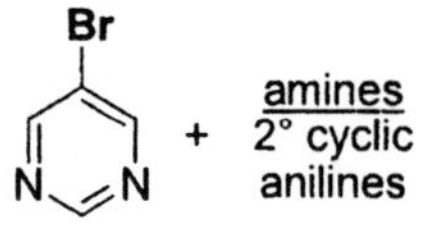

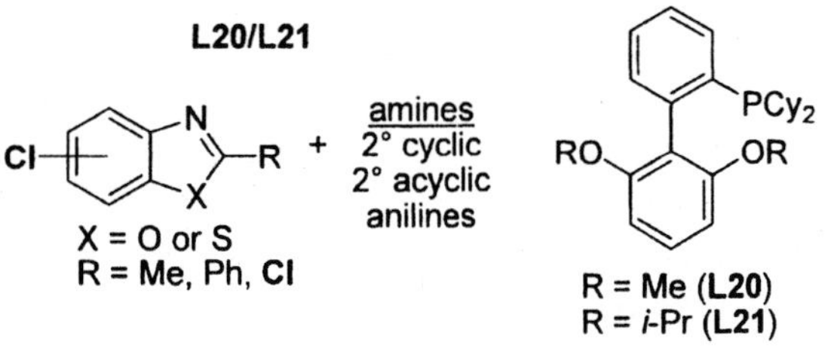

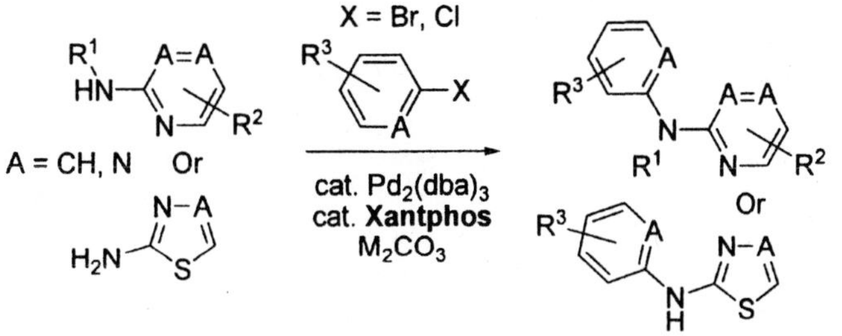

The Merck process research group found that Xantphos was the best ligand for the palladium-catalyzed cross-coupling of aminoheterocycles and aryl bromides, or halopyridines.[69] Depending upon the reaction partners, Cs_2CO_3, Na_2CO_3, K_3PO_4, or Na*Ot*-Bu was used as the base.

An interesting double amination of 2-chloro-3-iodo-pyridine (**16**) with 2-aminopyridine (**17**) catalyzed by Pd/BINAP or Pd/Xantphos, which gives dipyridoimidazoles (**18**), was described by B. U. W. Maes and co-workers. 2-Aminopyrazine was also found to be a capable coupling partner.[70] The mechanism is most likely an initial amination at the 3-iodo position followed by an intramolecular amination at the 2-chloro position.

cat. $Pd(OAc)_2$
cat. BINAP or
Xantphos
Cs_2CO_3, tol.
110 °C, 17 h
96% yield

16 17 18

The bulky mono-dentate phosphine ligands $(t\text{-Bu})_3P$[68, 71] and **L20/L21**[67] also catalyze the amination of halothiophenes and halofurans with secondary cyclic amines or anilines. Di-halothiophenes can also be di-aminated under these conditions. Pd/BINAP is a good catalyst choice for the amination of electron deficient halothiophenes with primary aliphatic amines and anilines.[72]

$(t\text{-Bu})_3P$ — amines: 2° cyclic, anilines; X = Br, Cl; A = S, O; 2-halo heterocycle give low yields

L20/L21 — amines: 2° cyclic, anilines; X = Br, Cl

BINAP — EWG; amines: anilines, 1° aliphatic; X = Br, Cl, OTf

7.2.4.6 Applications

The Buchwald–Hartwig amination has been widely adopted for the synthesis of natural products, and especially by medicinal chemists for building libraries of bioactive compounds. The broad substrate scope afforded by readily available ligands ensures that most aromatic amines can be accessed by this methodology. In the realm of natural products, the early stoichiometric example reported by Boger and Panek for the synthesis of lavendamycin demonstrates the utility of this transformation.[7] In a similar manner, B. B. Snider used an intramolecular Buchwald–Hartwig amination of **19** toward the synthesis of (–)-Asperlicin and (–)-Asperlicin C.[73] It is interesting to note that the original $(o\text{-tol})_3P$ ligand often performs best for intramolecular cyclization reactions.

$Pd_2(dba)_3$, $P(o\text{-tolyl})_3$
K_2CO_3, tol.
83% yield

19 NHCbz, i-Bu → 20 NCbz, i-Bu → (–)-Asperlicin, (–)-Asperlicin C

S. P. Nolan and co-workers described the use of a Pd/NHC complex for an intramolecular amination/cyclization with an aryl chloride (**21**). They used this methodology as an approach to the *Cryptocarya* alkaloids.[74]

MeO BnO NH Cl 21

(IMes)Pd(allyl)Cl
6 mol%

NaO*t*-Bu, DME
80 °C, 14 h
99% yield

MeO BnO N 22 O O

cryptaustoline
cryptowoline

A general route to phenazines (**27**) was reported by T. Kamikawa in which both an intermolecular and an intramolecular Pd/BINAP catalyzed amination was utilized.[75] The intermolecular reaction gave the expected product **26** in quantitative yield, whereas the intramolecular step was much less efficient (50% yield **27**), giving undesired furan **28** as a side product.

CO_2Me Br NO_2 23 + H_2N O OMe Ph 24

cat. $Pd(OAc)_2$
cat. BINAP

Cs_2CO_3, tol.
100 °C, 99% yield

CO_2Me H N NO_2 O OMe Ph 25

CO_2Me H N Br NH_2 O OMe Ph 26

cat. $Pd(OAc)_2$
cat. BINAP

Cs_2CO_3, tol.
100 °C, 50% yield

CO_2Me N N O OMe Ph 27 3 : 1 CO_2Me N N O Ph 28

Buchwald was able to exploit his XPhos (**L7**) ligand for the catalytic di-amination of 4,5-dichlorophthalic acid methyl ester (**29**).[76] This protocol can be used for the rapid synthesis of a small library of 4,5-dianilinophthalimides for the treatment of Alzheimer's disease.

Cl Cl CO_2Me CO_2Me 29 + NH_2 R 30

0.5 mol% $Pd_2(dba)_3$
1.1 mol% **L7**

K_3PO_4, tol.
110 °C, 1 d
84–92% yield

MeO_2C CO_2Me 31 Ar–NH HN–Ar

Many other examples from the literature have used the Buchwald–Hartwig amination to rapidly access bioactive anilines. S. Morita formed aripiprazole **32** and its metabolite by distinguishing between an aryl chloride and bromide during the amination catalyzed by Pd/BINAP.[77] An efficient synthesis of naturally occurring α-carboline skeleton **33** utilized an intramolecular amination with Pd/BINAP.[78] BINAP was also used to synthesize norastemizole (**34**) by catalyzing the amination of a 2-chlorobenzimidazole with primary amines.[79] Other benzimidazoles (**35**)[80] and imidazolpyridines (**36**)[81] have been functionalized *via* palladium-catalyzed amination. A β-carboline (**37**) was aminated with benzylamine using catalytic Pd/BINAP.[82] Other carbolines (**38**) have also been formed using an intramolecular amination.[83] Fairly complex C8-adenosine adducts (**39**) were synthesized using Pd/BINAP-catalyzed amination,[84] as well as the synthesis of C8-deoxyguanosine adduct (**40**) of the food mutagen IQ.[85]

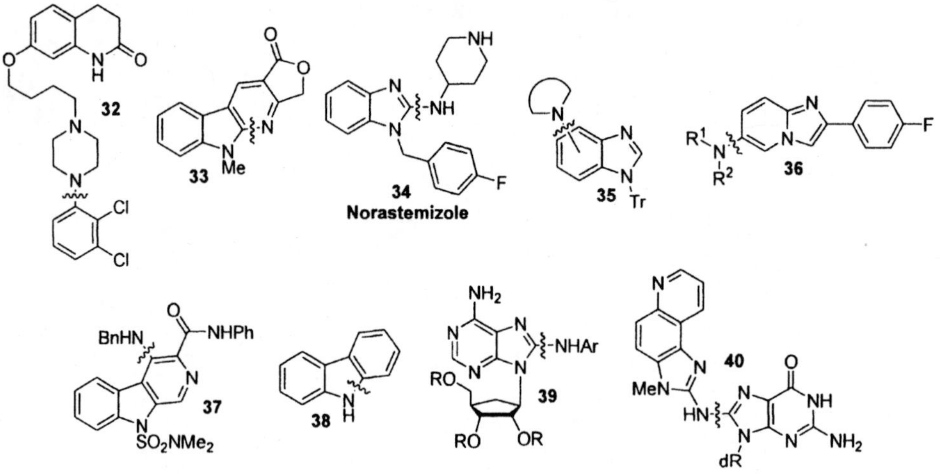

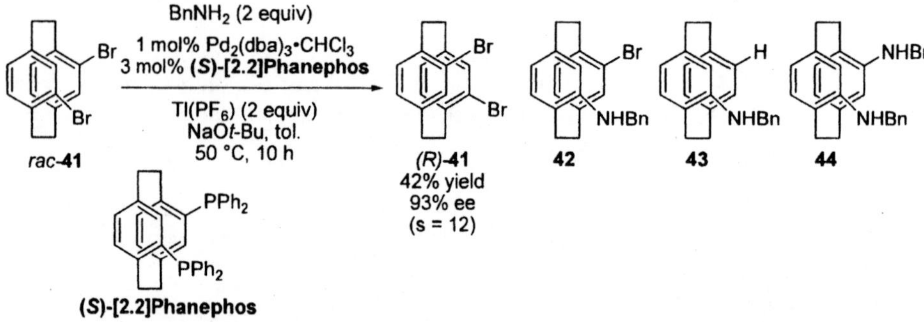

The Merck process group was able to use the Buchwald–Hartwig amination as a means to resolve *rac*-4,12-dibromo[2.2]paracyclophane **41**.[86] They report that (*S*)-[2.2]Phanephos was able to perform a catalytic, asymmetric amination of *rac*-4,12-dibromo[2.2]paracyclophane **41**. This kinetic resolution selectively aminated (**42**), de-halogenated (**43**), or di-aminated (**44**) the (*S*)-4,12-dibromo[2.2]paracyclophane **41**, leaving the (*R*)-4,12-dibromo[2.2]paracyclophane **41** in 42% yield (93% *ee*, s = 12). Enantiopure (*R*)-4,12-dibromo[2.2]paracyclophane **41** may then be used to synthesize (*R*)-[2.2]Phanephos.

I. P. Beletskaya's group has extensively examined the synthesis of macrocycles *via* the Buchwald–Hartwig amination by the reaction of long chain diamines with di-halo arenes.[87] High dilution is often required as dimers and other oligiomers are major side products. Yields are moderate at best (20–60% yield of **45** or **46**), but this methodology allows access to aza-crown ethers and modified macrocycles of lithocholic acid (**47**).[88]

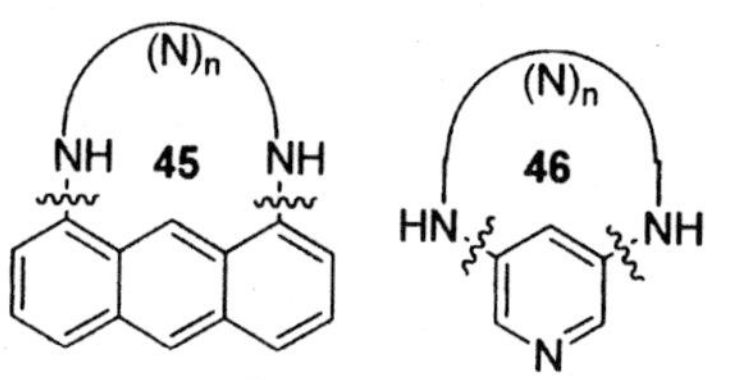

(N)n NH 45 NH
(N)n 46 HN NH N
+ linear and cyclic oligiomers
H H H H O O HN X 47 N H

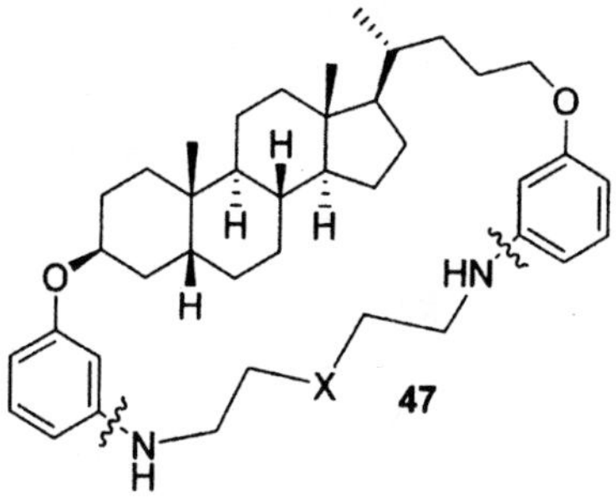

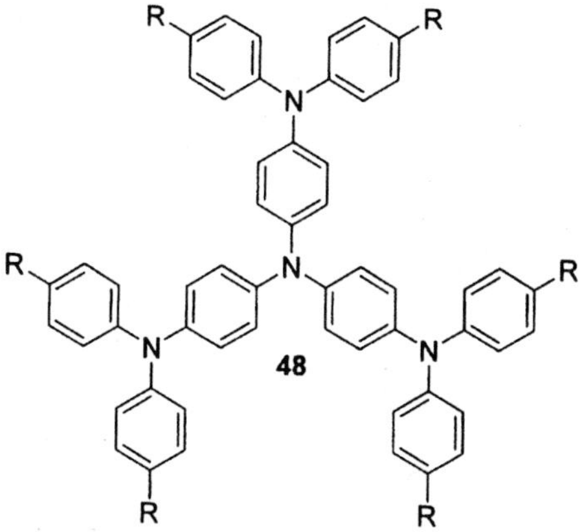

R R N N R N R N 48 R R

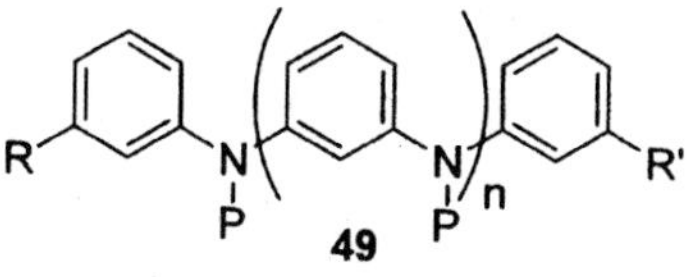

R N P N P n R' 49

Both Buchwald and Hartwig have explored the use of palladium-catalyzed amination for the synthesis of dendrimers and polyanilines. Dendrimers, such as **48**, can be easily assembled through sequential arylation of aminoanilines.[89] These dendrimers are useful in creating metal-organic frameworks and often can exhibit interesting electronic effects, such as diradicals, and as optical materials. Polymeric anilines **49** are of intense interest due to their unique electrical conducting properties. Using the

Buchwald–Hartwig amination, a wide variety of controlled length polyanilines may be synthesized with tailored properties.[90]

7.2.4.7 *Ligand summary*

Proper ligand selection is key to a successful Buchwald–Hartwig amination. It may also provide a starting point for screening. Many of the Buchwald bi-aryl phosphine ligands are commercially available and can serve as a useful screening tool due to their wide ranging substrate tolerance and versatility. Outlined in the table below is an approximate guide for ligand selection based on the preceding examples. Many ligands can service multiple classes of substrates, but particular emphasis has been placed on readily available ligands.

Conditions:
1.4 equiv NaO*t*-Bu
toluene, 22–110 °C (Use Cs_2CO_3 if R is base sensitive)

	ArI	ArBr	ArCl	ArOTf[a]
HN (cyclic amine)	**L3/L5** Xantphos	**L4** $(t\text{-Bu})_3P$	**L4** $(t\text{-Bu})_3P$	**L4**
HNR_2	**L3/L5** Xantphos	**L4** $(t\text{-Bu})_3P$	$(t\text{-Bu})_3P$	**L4**
HN(R)Ar	BINAP	BINAP $(t\text{-Bu})_3P$	**L4** $(t\text{-Bu})_3P$	BINAP DPPF
H_2NR	BINAP DPPF	BINAP DPPF	**L18** Q-phos	BINAP DPPF

[a] Use Cs_2CO_3 or slow addition of triflate

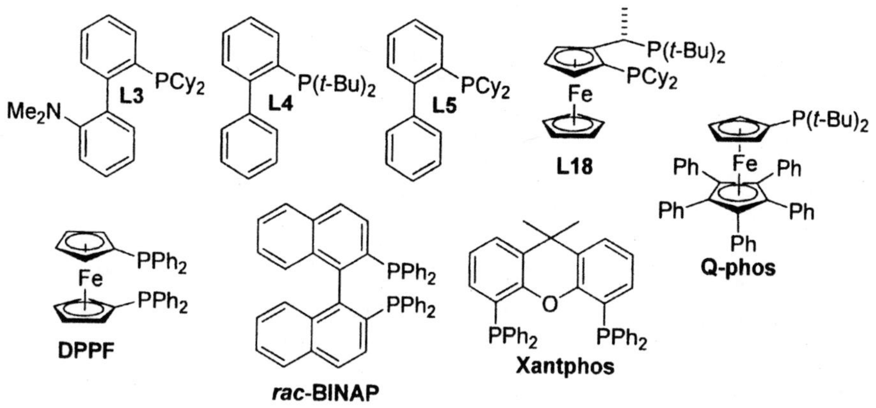

L3 L4 L5 L18 Q-phos

DPPF rac-BINAP Xantphos

7.2.5 *Variations*

7.2.5.1 Ammonia equivalents

The use of ammonia as a nitrogen source for the palladium-catalyzed amination would be advantageous as it provides access to anilines from any aryl halide. Buchwald's initial strategy to achieve this was to use commercially available benzophenone imine as an ammonia surrogate.[91] Despite the weak basicity of the imino nitrogen, the imine was found to undergo arylation with a variety of aryl chlorides, bromides, iodides, and triflates using either ligand **L4** or BINAP. Mild acidic hydrolysis, or hydrogenolysis, liberates the free aniline.

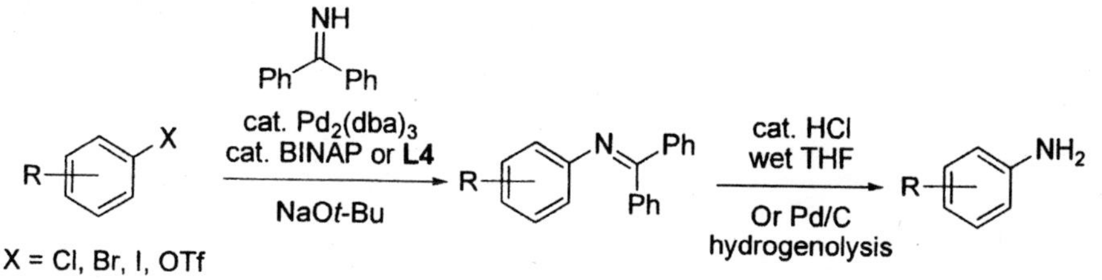

cat. $Pd_2(dba)_3$, cat. BINAP or **L4**, NaO*t*-Bu; cat. HCl, wet THF, Or Pd/C hydrogenolysis

X = Cl, Br, I, OTf

D. G. Putnam described di- or mono-allylamine as a masked ammonia equivalent.[92] The amination proceeds well with aryl bromides or halopyridines using either (DPPF)$PdCl_2$ or $[(o\text{-tol})_3P]_2PdCl_2$. Deprotection was accomplished using standard Pd/C hydrogenolysis conditions.

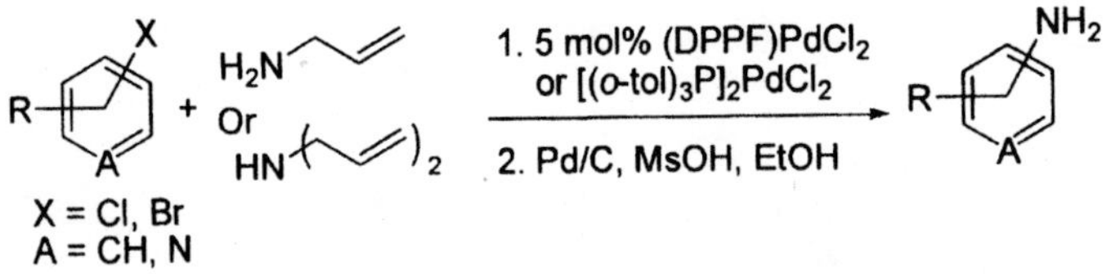

Nitrogen gas may serve as an ammonia equivalent by first capturing nitrogen with TiX_4/Li/TMSCl, which gives a $L_nTi\text{-}N(SiMe_3)_2$ complex **50**.[93] (DPPF)Pd or (BINAP)Pd catalyze the amination of aryl bromides and triflates with this titanium amide complex in low yields after an acidic workup. Di-arylation is the major byproduct.

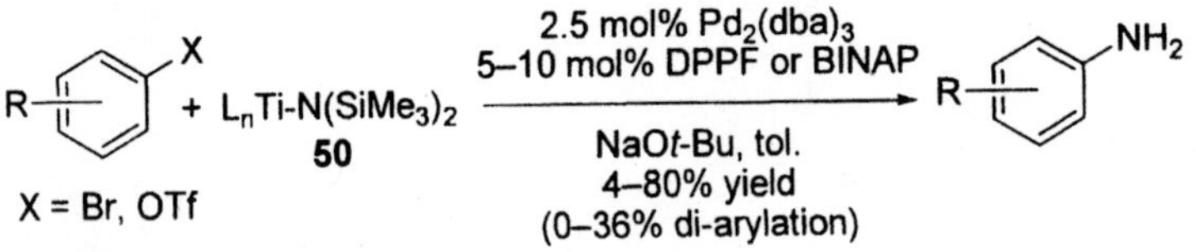

Hartwig was able to show that LiHMDS could serve as both an ammonia equivalent and as the stoichiometric base for the amination reaction catalyzed by $Pd(t\text{-}Bu)_3P$.[94] The reaction conditions tolerate many aryl bromides and chlorides, but due to the size of LiHMDS, *ortho*-substituted haloarenes are not acceptable. As before, acidic workup reveals the aniline by cleavage of the N–Si bond.

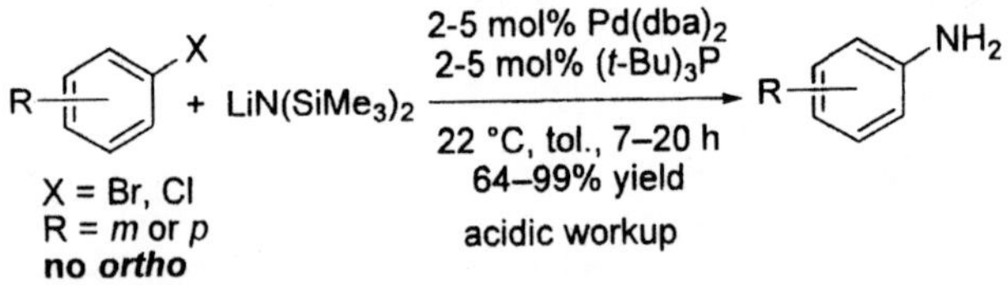

Buchwald and Huang addressed the aforementioned limitations of using LiHMDS by employing a smaller silyl amine, namely, Ph_3SiNH_2.[95] Aryl bromides and chlorides were easily aminated using ligand **L5**. They also report that $LiNH_2$ provides di- and tri-arylated ammonia after acidic hydrolysis using ligand **L4**.

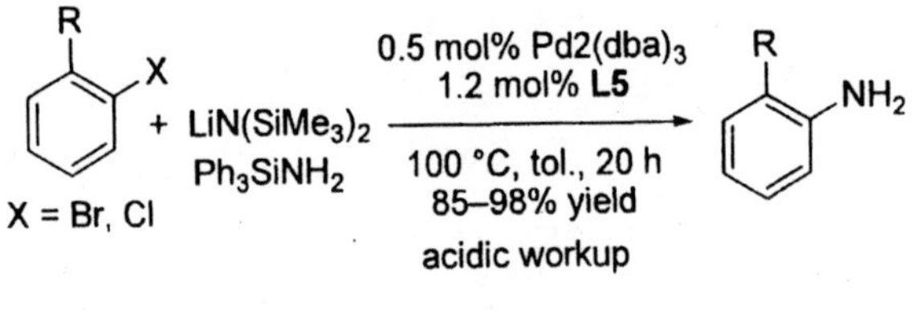

Hartwig next described a milder basic ammonia equivalent, $Zn(HMDS)_2$, as both the ammonia equivalent and as base.[96] Key to this transformation is the LiCl additive. In this manner, aryl bromides, chlorides, and triflates may be transformed to their aniline analogs after acidic hydrolysis. As with LiHMDS, *ortho*-substitution is not tolerated. The advantage of $Zn(HMDS)_2$ is that the attenuated basicity allows for base sensitive nitro, nitrile, ester, and ketone functionalities to be tolerated.

R–C6H4–X + $ZnN(SiMe_3)_2$ → (1-3 mol% $Pd(dba)_2$, 1-3 mol% $(t\text{-Bu})_3P$, LiCl (0.6 equiv), 50–90 °C, THF, 1–20 h, 80–97% yield, acidic workup) R–C6H4–NH_2

X = Br, Cl, OTf
R = *m* or *p*
no *ortho*

A major advance in ammonia arylation was made by the Hartwig group.[97] They report that ammonia itself could be arylated using ferrocene **L18** as ligand. By keeping the reaction dilute (0.05 M) under a pressurized ammonia atmosphere (80 psi), di-arylation could be greatly suppressed in the coupling with various aryl halides and triflates. As a convenient alternative, they also found $LiNH_2$ (10 equiv) at 0.05 M haloarene in DME gives the desired unsubstituted aniline using **L18**.

R–C6H4–X + NH_3 or $LiNH_2$ → (1 mol% (**L18**)$PdCl_2$, 90 °C, DME) R–C6H4–NH_2

X = Br, Cl, OTf, I
R = *m, p, o*

7.2.5.2 Indoles: synthesis and functionalization

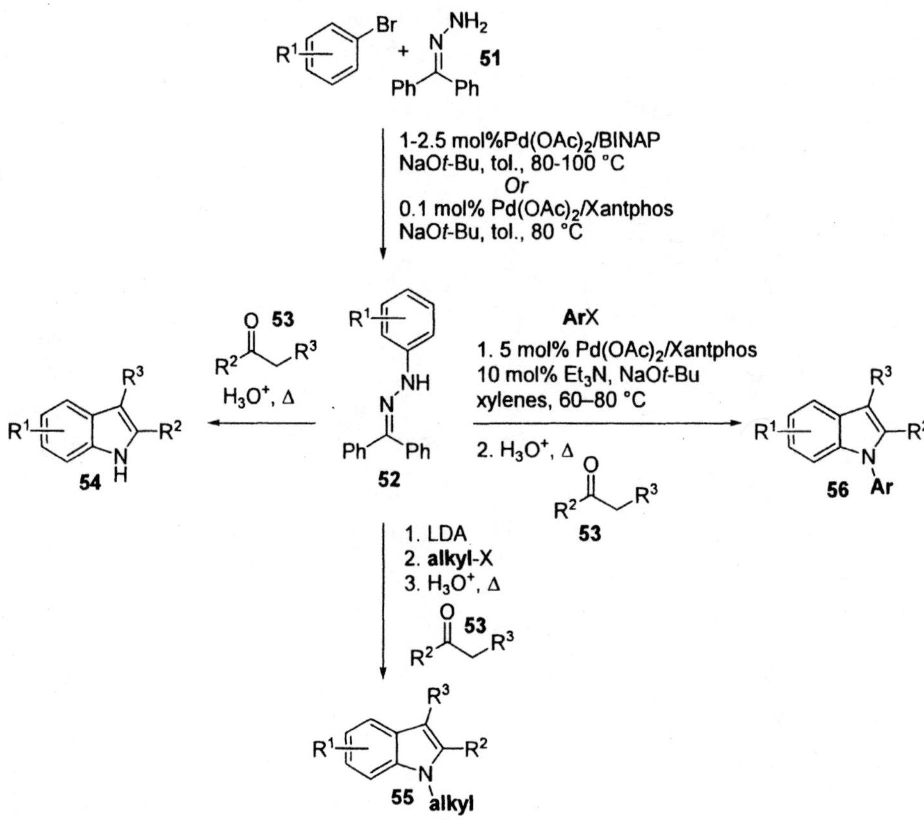

The rich biological activity of functionalized indoles has made their synthesis an area of constant interest and attention. Buchwald's laboratory has reported a useful and general synthesis using palladium-catalyzed amination methodology in conjunction with a Fischer indole synthesis to provide an array of functionalized indoles.[98] They found that benzophenone hydrazone **51** is efficiently arylated with a variety of aryl bromides using either BINAP or Xantphos as a ligand. These stable *N*-aryl hydrazones **52** can then be treated with an enolizable ketone **53** under mild acidic conditions to give the Fischer indole product **54**. The reaction proceeds by an initial hydrolysis of the benzophenone hydrazone, followed by condensation with the added ketone. The intermediate enolizable hydrazone thus formed then undergoes a Fischer cyclization to the indole **54**.

These *N*-aryl benzophenone hydrazones **52** can also be functionalized by treatment with LDA followed by alklation of nitrogen. The aforementioned Fischer cyclization conditions with an enolizable ketone **53** gives *N*-alkyl indoles **55**. *N*-Aryl indoles **56** can be accessed by Buchwald–Hartwig amination of the *N*-aryl benzophenone hydrazone **52** followed by Fischer cyclization. Using these strategies, a diverse range of substituted indoles can be easily synthesized.

A less direct strategy for indole synthesis (**58**) is the intramolecular Buchwald–Hartwig amination of an enamine **57** catalyzed by (DPPF)$PdCl_2$.[99] It is necessary to use the *Z*-enamine isomer **57** that was set by a Horner–Wadsworth–Emmons reaction.

57 → (DPPF)$PdCl_2$, KOAc/DMF, 90° C → 58

Buchwald has also demonstrated that indoles can be *N*-arylated with aryl bromides, chlorides, iodides, and triflates using a variety of ligands based upon his bi-aryl phosphine scaffold.[100] The exact choice of ligand depends upon the coupling partners and can be optimized for each individual case.

X = I, Br, Cl, OTf; cat. $Pd_2(dba)_3$, **L3,L4,L5,L22,L23,L24** or $(t\text{-Bu})_3P$; NaO*t*Bu, tol., 80–100 °C, 43–95% yield

L22 **L23** **L24**

7.2.5.3 Coupling of amides, carbamates, hydrazines, and sulfoximes

As many previous examples have alluded to, weak nitrogen nucleophiles such as imines, hydrazones, amides, carbamates, sulfonamides, and sulfoximes may also be arylated under the standard Buchwald–Hartwig conditions. As Buchwald's indole synthesis demonstrated, hydrazones (**52**) are arylated with many haloarenes using either BINAP[98] or DPPF[101] as a ligand for palladium. In addition, *t*-butylcarbazate (**59**)[102] and di-*t*-butylcarbazate (**60**)[103] may be arylated with various aryl bromides catalyzed by (DPPF)$PdCl_2$.

52 BINAP or DPPF

59 DPPF

60 DPPF

Intramolecular amination of hydrazines **61** can be catalyzed by Pd/DPPF.[104] This gives the cyclized product indazoles **62** in excellent yields. *N*-Aryl hydrazines and unprotected hydrazines work well, whereas *N*-triphenylphosphonium bromide hydrazines are more difficult substrates.

61 → **62**: cat. $Pd(OAc)_2$, cat. DPPF; NaO*t*-Bu, tol., 90 °C, 15 h, 40–90% yield

R^1 = H, F, OMe
R^2 = H, $P^+Ph_3Br^-$, Ar
R^3 = H, Ar

Weaker hydrazine derived nucleophiles may also participate in the amination reaction as has been reported by H. Katayama.[105] They found that *N*-acyl hydrazines **63** underwent a clean intramolecular cyclization to give the desired indol[1,2-*b*]indazole derivatives **64** using DPEphos as the ligand.

63 → **64**: cat. $Pd(OAc)_2$, cat. DPEphos; Cs_2CO_3, tol., 100–110 °C, 18 h, 84–96% yield

An intramolecular aryl guanidinylation was reported by R. A. Batey and G. Evindar.[106] They found palladium tetrakis worked well for this cyclization, but catalytic CuI and 1,10-phenanthroline exhibited broader scope and lower loadings.

65 → **66**: 10 mol% $(PPh_3)_4Pd$; Cs_2CO_3, DME, 80 °C, 16 h, 63–93% yield

C. Bolm has found that the sulfoximes **67** can be *N*-arylated using catalytic BINAP/Pd with Cs_2CO_3 as base.[107] Aryl bromides with a variety of substitution were suitable for this transformation and gave the expected *N*-aryl sulfoximes **68** in high yield.

5 mol% $Pd(OAc)_2$, BINAP or Tol-BINAP; Cs_2CO_3, tol., 110 °C; 67–94% yield

R^2 = Ph, *p*-tol, Me

Sulfoxime arylation may be coupled with an intramolecular aldol-type condensation giving benzothiazines.[108] After arylation of the sulfoxime **67** with an *ortho*-bromobenzaldehyde using catalytic Pd/BINAP, the product spontaneously condenses under the reaction conditions to give the cyclized product **69**. Also, the enantiopurity of the starting sulfoximes **67** is maintained throughout the reaction course.

5 mol% $Pd(OAc)_2$, 7.5 mol% BINAP; Cs_2CO_3, tol., 110 °C, 40 h; 73–81% yield

Analogous to the original Migita cross-coupling of aminostannanes, *N*-silyl imines **70** may also be arylated using palladium catalysis. J. Barluenga and co-workers found that using Buchwald's ligand **L3** *N*-TMS imines **70** could be arylated (**71**).[109] They propose that the base, rather then capturing HX from the Pd-amine intermediate, attacks silicon as a nucleophile to give *t*-BuOTMS and NaX and thereby drives the transmetallation event. A variety of aryl bromides couple with *N*-TMS imine **70** using catalytic BINAP or **L3**. The advantage of *N*-TMS imines is their stability as compared to their non-silylated counterparts. This cross-coupling may also be conducted in super-critical CO_2 as well.

$Pd_2(dba)_3$, BINAP or **L3**; Na*Ot*-Bu, tol., 90 °C; 76–97% yield

R^2 = Ar or cinnamyl

Secondary amides and carbamates are very good substrates for the palladium-catalyzed arylation. Buchwald reports the intramolecular cyclization of amides and common carbamate protected nitrogens on to aryl bromides to give 5–7 member rings.[9a,13,110] Depending upon the susbstrate, DPEphos, Xantphos, or *rac*-MOP were found to be excellent ligands for palladium. Cs_2CO_3 and K_2CO_3 in toluene at 100 °C were the best reaction conditions.

N(H)Bn, Br, O — Or — N(H)R, Br

R = Ac, Cbz, Boc
n = 1-3

$Pd(OAc)_2$
DPEphos, Xantphos
or *rac*-MOP

Cs_2CO_3, tol.
100 °C
79–94% yield

=O, N, Bn — Or — N, R

PPh$_2$
OMe

***rac*-MOP**

The intermolecular arylation of amides catalyzed was first demonstrated by William Shakespeare for a series of 4–7 member lactams.[111a] More complete studies by Buchwald and co-workers[111b,c] report that aryl iodides, bromides, and triflates all undergo cross-coupling with lactams, primary and secondary amides, carbamates, and sulfonamides.[111d] Xantphos with Cs_2CO_3 in dioxane were the optimal reaction conditions. Oxazolidinones and ureas were also found to be arylated under these conditions. Furthermore, they found that Xantphos could be *trans*-chelating to palladium based on an X-ray structure of an isolated intermediate in the catalytic cycle. Others have described a microwave mediated cross-coupling of aryl chlorides and sulfonamides using ligand **L3**.

X = I, Br, OTF
R^1 = EWG, EDG
low yields if enolizable

Xantphos
Cs_2CO_3/dioxane

+

n = 1–4

R^2 = H, alkyl
R^3 = H, Me, Ph

Et–NH–C(O)–OEt

R^2–NH–SO_2p-Tol
R^2 = H, Me

A. Ghosh developed an efficient protocol for the arylation of oxazolidinones **72** which uses Buchwald ligands **L4**.[112] *Meta* and *ortho* methoxy aryl chlorides gave low yields in the cross-coupling, whereas most other aryl chlorides gave excellent yields of the *N*-aryl oxazolidinone **73**, even with base sensitive functionality present on the aryl chloride.

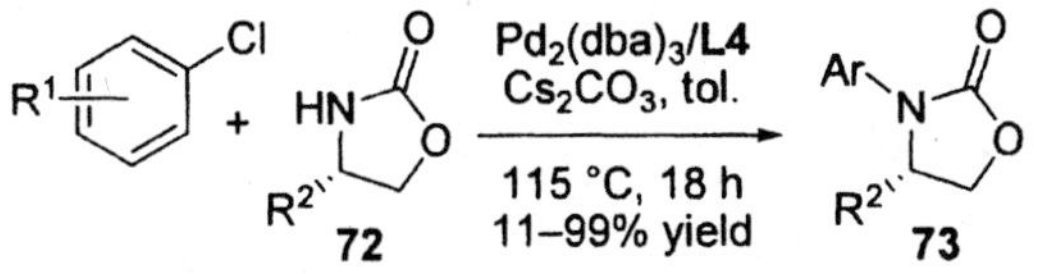

S. D. Edmondson described arylation of vinylogous amides **75** with aryl bromides (**74**) and chlorides catalyzed by Pd/**L3**.[113] This arylation was then coupled with a Heck cyclization by adding additional $Pd_2(dba)_3$, giving an indole **76**.

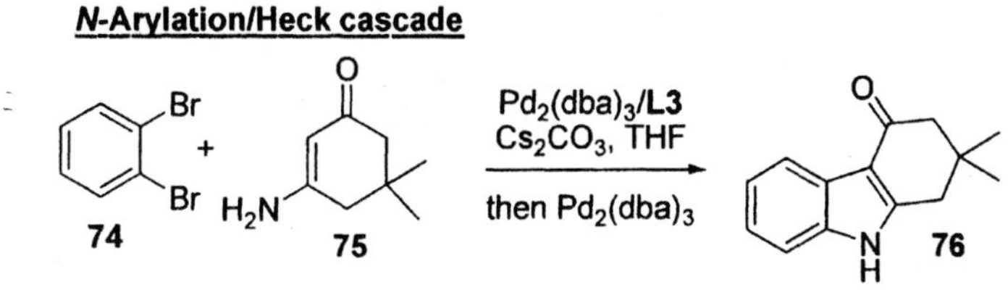

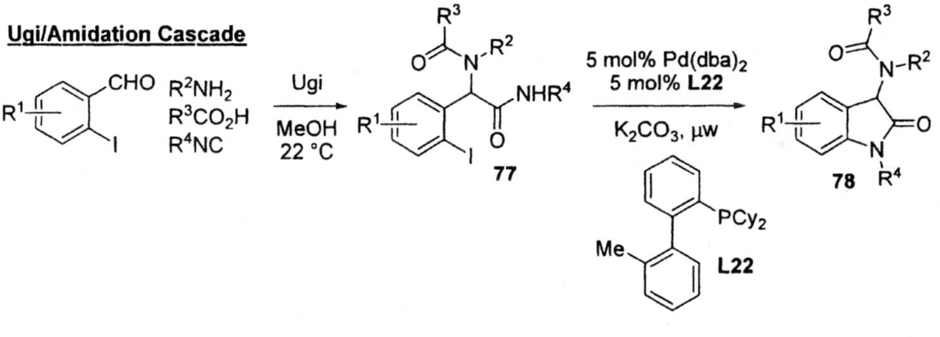

Another reaction cascade appears in a report by J. Zhu and co-workers.[114] They performed a 4-component Ugi reaction followed by a Buchwald–Hartwig amination in a one-pot sequence. 2-Iodobenzaldehyde, amide, acid, and isonitrile are stirred in methanol to give an aryl iodide **77** containing a secondary and tertiary amide. The secondary amide then cyclizes onto the iodide by adding $Pd(dba)_2$, Buchwald ligand **L22**, and K_2CO_3 in toluene/acetonitrile. Microwave heating gave a rapid cyclization (1.5 h) to the desired oxindole product **78**.

7.2.5.4 Coupling of amines with alkenes

The Buchwald–Hartwig amination is not limited to aryl halides and sulfonates; vinyl halides and sulfonates may participate as well. An early example of this process was shown in an intramolecular sense for a carbapenam synthesis in the laboratory of M. Mori.[115] Pd/DPEphos catalyzed the cyclization of a vinyl bromide **79** giving β-lactam **80** in 97% yield.

OTBS H NH Br OTBS O **79** — $Pd(OAc)_2$, DPEphos; Cs_2CO_3, tol., 80 °C, 1.5 h, 97% yield → OTBS H N O **80** TBSO

A study of the intermolecular vinylation of azoles (pyrrole, carbazole, and indoles) and phenothiazines was published by I. P. Beletskaya and co-workers.[116] This vinylation is catalyzed by Pd(*t*-Bu)$_3$P and utilizes the lithium amide of various azoles. Various vinyl bromides efficiently undergo the cross-coupling with full retention of their configuration (*E* or *Z*).

R¹ N Li + Br R² / Br Ph / Br Ph — 4 mol% $Pd(dba)_2$, 8 mol% $(t\text{-Bu})_3P$; tol./DME, 50–100 °C, 1–10 h, 30–99% yield → R¹ N R² / R¹ N Ph / R¹ N Ph

A variety of secondary cyclic and acyclic amines may also couple with vinyl bromides catalyzed by Pd/BINAP or **L13.**[117] This serves as a useful alternative for enamine synthesis. The vinylation of primary amines gives enamines which immediately tataumerize to the corresponding imine. Competition experiments reveal that vinyl bromides react in preference to aryl bromides. Vinyl chlorides are also aminated to give enamines or amino dienes[118] as well. These may then serve as useful nucleophiles or dienes for Diels–Alder reactions.

Vinyl Bromides
Pd(II)/BINAP (or **L13**)
NaO*t*-Bu, tol.

(imine)

Vinyl Chlorides
$Pd_2(dba)_3$/**L3**
NaO*t*-Bu, tol.

The palladium-catalyzed coupling of vinyl triflates with amides, carbamates, and sulfonamides has been reported by researchers at Merck.[119] They identified Xantphos as the best ligand for this reaction. The product enamides serve as valuable synthetic intermediates as they can undergo asymmetric hydrogenation to give enantiopure amines.

3 mol% $Pd_2(dba)_3$
9 mol% Xantphos
Cs_2CO_3, dioxane
50 °C, 8–14 h
10–97% yield

M. C. Willis has described a tandem reaction process where pyrrolidine is coupled with vinyl bromides (**81**) to give an enamine.[120] In the same reaction vessel, an alkylidene malonate **82** is then added at 80 °C to give the Michael addition product **83**.

Vinylation/Michael Addition

1. $Pd(OAc)_2$/BINAP
NaO*t*-Bu, tol.
2. **82**

M. Lautens has also reported a tandem process in which a palladium-catalyzed amine vinylation is coupled with an inter- or intramolecular Heck reaction.[121] This gives a 2-vinyl indole **85**. In many examples, no phosphine ligand is required for palladium as long as 1 equiv of Me_4NCl is present.

Vinylation/Heck Cascade

84 → 85: 4 mol% $Pd(OAc)_2$, Me_4NCl (1 equiv) **Or** 8 mol% $(o\text{-tol})_3P$; $K_3PO_4 \cdot H_2O/Et_3N$, tol., 110 °C

7.2.6 *Experimental*

4-bromo-3-methylanisole + *n*-HexNH$_2$ (1.2 equiv.) → **86**: 0.25 mol% $Pd_2(dba)_3$, 0.75 mol% *rac*-BINAP; NaO*t*-Bu (1.4 equiv), tol., 80 °C, 18–23 h, 94% yield

_N_-Hexyl-2-methyl-4-methoxyaniline (86)[17c]

A flame dried 250 mL, round-bottomed flask fitted with a magnetic stirbar is charged with tris(dibenzylidenacetone)dipalladium (114 mg, 0.125 mmol, 0.0025 equiv), *rac*-BINAP (233 mg, 0.375 mmol, 0.0075 equiv.), and sodium *tert*-butoxide (6.73 g, 70.0 mmol, 1.4 equiv). After purging with argon for 5 minutes, freshly distilled toluene is added (50 mL) while being stirred. Next, the 4-bromo-3-methylanisole (10.0 g, 50.0 mmol, 1 equiv), *n*-hexylamine (distilled, 7.9 mL, 60.0 mmol, 1.2 equiv), and toluene (distilled, 50 mL) are added. The dark red mixture is placed in an 80 °C oil bath with stirring until all the aryl bromide is consumed as judged by GC analysis (18–23 hours). After cooling to room temperature, the mixture is transferred to a separatory funnel. The reaction flask is rinsed with diethyl ether (2 × 50 mL), brine (100 mL), water (20 mL), and again with diethyl ether (50 mL). All of the rinses are also transferred to the separatory funnel and the layers are then mixed and separated. The aquesous layer is extracted with diethyl ether (50 mL), and the combined organics are dried over $MgSO_4$. The mixture is filtered and the $MgSO_4$ is washed with 50 mL diethyl ether. Concentration *in vacuo* gave a crude brown oil which can then be purified by distillation (bulb-to-bulb, bp 92 °C at 0.001 Torr) to give 10.35 g (94%) of the title compound (**86**) as a pale yellow oil.

7.2.7. *References*

1. [R] (a) Buchwald, S. L.; Mauger, C., Mignani, G.; Scholz, U. *Adv. Synth. Catal.* **2006**, *348*, 23. [R] (b) Schlummer, B.; Scholz, U. *Adv. Synth. Catal.* **2004**, *346*, 1599. [R] (c) Jiang, L.; Buchwald, S. L. *Metal-Catalyzed Cross-Coupling Reactions (2nd Edition)*; De Meijere, A.,

Diederich, F., Eds.; Wiley-VCH: Weinheim, 2004; pp 699–760. [R] (d) Hartwig, J. F. *Modern Arene Chemistry*; Astruc, D., Ed.; Wiley-VCH: Weinheim, 2002; pp 107–168. [R] (e) Hartwig, J. F. *Handbook of Organopalladium Chemistry for Organic Synthesis*; Negishi, E., Ed.; John Wiley & Sons, Inc.: Hoboken, 2002; pp 1051–1096. [R] (f) Fu, G. C., Littke, A. F. *Angew. Chem. Int. Ed., Engl.* **2002**, *41*, 4176. [R] (g) Murahashi, S. *J. Org. Met. Chem.* **2002**, *653*, 27. [R] (h) Prim, D., Campagne, J.-M., Joseph, D., Andrioletti, B. *Tetrahedron* **2002**, *58*, 2041. [R] (i) Yang, B. H.; Buchwald, S. L. *J. Org. Met. Chem.* **1999**, *576*, 125. [R] (j) Hartwig, J. F. *Pure Appl. Chem.* **1999**, *71*, 141. [R] (k) Belfield, A. J., Brown, G. R., Foubister, A. J. *Tetrahedron* **1999**, *55*, 11399.

2. Past reviews: [R] (a) Wolfe, J. P., Wagaw, S., Marcoux, J.-F., Buchwald, S. L. *Acc. Chem. Res.* **1998**, *31*, 805. [R] (b) Hartwig, J. *Acc. Chem Res.* **1998**, *31*, 852. [R] (c) Hartwig, J. F. *Angew. Chem. Int. Ed., Engl.* **1998**, *37*, 2046. [R] (d) Frost, C. G., Mendonca, P. *J. Chem. Soc., Perkin Trans. 1*, **1998**, 2615. [R] (e) Baranano, D., Mann, G., Hartwig, J. F. *Curr. Org. Chem.* **1997**, *1*, 287. [R] (f) Hartwig, J. F. *Synlett.* **1997**, 329.
3. Rossi, R., de Rossi, R. H. *Aromatic Substitution by the $S_{RN}1$ Mechanism, Vol. 178*, American Chemical Society, Washington, DC, **1983**.
4. (a) Rao, H., Fu, H., Jiang, Y., Zhao, Y. *J. Org. Chem.* **2005**, *70*, 8107. (b) Hu, T., Li, C. *Org. Lett.* **2005**, *7*, 2035. (c) Ran, C., Dai, Q., Harvey, R. G. *J. Org. Chem.* **2005**, *70*, 3724. (d) Antilla, J. C., Baskin, J. M., Barder, T. E., Buchwald, S. L. *J. Org. Chem.* **2004**, *69*, 5578. (e) Okano, K., Tokuyama, H., Fukuyama, T. *Org. Lett.* **2003**, *5*, 4987. (f) Kwong, F. Y., Buchwald, S. L. *Org. Lett.* **2003**, *5*, 793. (g) Antilla, J. C., Buchwald, S. L. *Org. Lett.* **2001**, *3*, 2077.
5. (a) Kelly, R. A. III, Scott, N. M., Diez-Gonzales, S., Stevens, E. D., Nolan, S. P. *Organomettalics* **2005**, *24*, 3442. (b) Chen, C., Yang, L.-M., *Org. Lett.* **2005**, *7*, 2209. (c) Omar-Amrani, R., Thomas, A., Brenner, E., Schneider, R., Fort, Y. *Org. Lett.* **2003**, *5*, 2311. (d) Desmarets, C., Schneider, R., Fort, Y. *J. Org. Chem.* **2002**, *67*, 3029.
6. Kosugi, M., Kameyama, M., Migita, T. *Chem. Lett.* **1983**, 927.
7. (a) Boger, D. L., Panek, J. S. *Tetrahedron Lett.* **1984**, *25*, 3175. (b) Boger, D. L., Duff, S. R., Panek, J. S., Yasuda, M. *J. Org. Chem.* **1985**, *50*, 5782. (c) Boger, D. L., Duff, S. R., Panek, J. S., Yasuda, M. *J. Org. Chem.* **1985**, *50*, 5790.
8. (a) Paul, F., Patt, J., Hartwig, J. F. *J. Am. Chem. Soc.* **1994**, *116*, 5969. (b) Guram, A., Buchwald, S. L. *J. Am. Chem. Soc.* **1994**, *116*, 7901.
9. (a) Guram, A. S., Rennels, R. A., Buchwald, S. L. *Angew. Chem. Int. Ed., Engl.* **1995**, *34*, 1348. (b) Louie, J., Hartwig, J. F. *Tetrahedron Lett.* **1995**, *36*, 3609.
10. First generation ligand mechanism: (a) Widenhoefer, R. A., Buchwald, S. L. *Organometallics* **1996**, *15*, 3534. (b) Widenhoefer, R. A., Buchwald, S. L. *Organometallics* **1996**, *15*, 2755. (c) Widenhoefer, R. A., Zhong, A., Buchwald, S. L. *Organometallics* **1996**, *15*, 2745. (d) Louie, J., Hartwig, J. F. *Angew. Chem. Int. Ed., Engl.* **1996**, *35*, 2359. (e) Louie, J., Paul, F., Hartwig, J. F. *Organometallics* **1996**, *15*, 2794. (f) Hartwig, J. F., Richards, S., Baranano, D., Paul, F. *J. Am. Chem. Soc.* **1996**, *118*, 3626. (g) Hartwig, J. F., Paul, F. *J. Am. Chem. Soc.* **1995**, *117*, 5373.
11. Second generation (chelating phosphines) mechanism: (a) Shekhar, S., Ryberg, P., Hartwig, J. F., Mathew, J. S., Blackmond, D. G., Strieter, E. R., Buchwald, S. L. *J. Am. Chem. Soc.* **2006**, *128*, 3584. (b) Shekhar, S., Ryberg, P., Hartwig, J. F. *Org. Lett.* **2006**, *8*, 851. (c) Klingensmith, L. M., Strieter, E. R., Barder, T. E., Buchwald, S. L. *Organometallics* **2006**, *25*, 82. (d) Hooper, M. W., Hartwig, J. F. *Organometallics* **2003**, *22*, 3394. (e) Yamashita, M., Vicario, J. V. C., Hartwig, J. F. *J. Am. Chem. Soc.* **2003**, *125*, 16347. (f) Singh, U. K., Strieter, E. R., Blackmond, D. G., Buchwald, S. L. *J. Am. Chem. Soc.* **2002**, *124*, 14104. (g) Alcazar-Roman, L. M., Hartwig, J. F. *Organometallics* **2002**, *21*, 491. (h) Alcazar-Roman, L. M., Hartwig, J. F., Rheingold, A. L., Liable-Sands, L. M., Guzei, I. A. *J. Am. Chem. Soc.* **2000**, *122*, 4618. (i) Driver, M. S., Hartwig, J. F. *J. Am. Chem. Soc.* **1997**, *119*, 8232.
12. Third generation (bulky phosphines) mechanism: (a) Strieter, E. R., Buchwald, S. L. *Angew. Chem. Int. Ed., Engl.* **2006**, *45*, 925. (b) Yamashita, M., Hartwig, J. F. *J. Am. Chem. Soc.* **2004**,

126, 5344. (c) Strieter, E. R., Blackmond, D. G., Buchwald, S. L. *J. Am. Chem. Soc.* **2003**, *125*, 13978. (d) Alcazar-Roman, L. M., Hartwig, J. F. *J. Am. Chem. Soc.* **2001**, *123*, 12905.
13. Wolfe, J. P., Rennels, R. A., Buchwald, S. L. *Tetrahedron* **1996**, *52*, 7525.
14. Wolfe, J. P., Buchwald, S. L. *J. Org. Chem.* **1996**, *61*, 1133.
15. Zhao, S.-H., Miller, A. K., Berger, J., Flippin, L. A. *Tetrahedron Lett.* **1996**, *37*, 4463.
16. Willoughby, C. A., Chapman, K. T. *Tetrahedron Lett.* **1996**, *37*, 7181.
17. (a) Wolfe, J. P., Buchwald, S. L. *J. Org. Chem.* **2000**, *65*, 1144. (b) Wolfe, J. P., Wagaw, S., Buchwald, S. L. *J. Am. Chem. Soc.* **1996**, *118*, 7215. (c) Wolfe, J. P., Buchwald, S. L. *Org. Syn.* **2000**, *78*, 23.
18. Wolfe, J. P., Buchwald, S. L. *J. Org. Chem.* **1997**, *62*, 6066.
19. Cui, W., Loeppky, R. N. *Tetrahedron* **2001**, *57*, 2953.
20. Ward, Y. D., Farina, V. *Tetrahedron Lett.* **1996**, *37*, 6993.
21. Driver, M. S., Hartwig, J. F. *J. Am. Chem. Soc.* **1996**, *118*, 7217.
22. (a) Marcoux, J.-F., Wagaw, S., Buchwald, S. L. *J. Org. Chem.* **1997**, *62*, 1568. (b) Wolfe, J. P., Buchwald, S. L. *Tetrahedron Lett.* **1997**, *38*, 6359.
23. Wagaw, S., Rennels, R. A., Buchwald, S. L. *J. Am. Chem. Soc.* **1997**, *119*, 8451.
24. (a) Sadighi, J. P., Harris, M. C., Buchwald, S. L. *Tetrahedron Lett.* **1998**, *39*, 5327. (b) Zhang, X.-X., Harris, M. C., Sadighi, J. P., Buchwald, S. L. *Can. J. Chem.* **2001**, *79*, 1799.
25. Wullner, G., Jansch, H., Kannenberg, S., Schubert, F., Boche, G. *Chem. Commun.* **1998**, 1509.
26. Gusev, O. V., Peganova, T. A., Kalsin, A. M., Vologdin, N. V., Petrovskii, P. V., Lyssenko, K. A., Tsvetkov, A. V., Beletskaya, I. P. *J. Organomet. Chem.* **2005**, *690*, 1710.
27. (a) Guari, Y., van Es, D. S., Reek, J. N. H., Kamer, P. C. J., van Leeuwen, P. W. N. M. *Tetrahedron Lett.* **1999**, 3789. (b) Guari, Y., van Strijdonck, G. P. F., Boele, M. D. K., Reek, J. N. H., Kamer, P. C. J., van Leeuwen, P. W. N. M. *Chem. Eur. J.* **2001**, *7*, 475.
28. (a) Beletskaya, I. P., Bessmertnykh, A. G., Guilard, R. *Synlett.* **1999**, 1459. (b) Beletskaya, I. P., Bessmertnykh, A. G., Guilard, R. *Tetrahedron Lett.* **1999**, *40*, 6393.
29. Beletskaya, I. P., Tsvetkov, A. V., Tsvetkov, P. V., Latyshev, G. V., Lukashev, N. V. *Russ. Chem. Bull., Int. Ed.* **2005**, *54*, 215.
30. (a) Ahman, J., Buchwald, S. L. *Tetrahedron Lett.* 1997, 38, 6363. (b) Wolfe, J. P., Buchwald, S. L. *J. Org. Chem.* **1997**, *62*, 1264.
31. Hicks, F. A., Brookhart, M. *Org. Lett.* **2000**, *2*, 219.
32. Louie, J., Driver, M. S., Hamann, B. C., Hartwig, J. F. *J. Org. Chem.* **1997**, *62*, 1268.
33. Nishiyama, M., Yamamoto, T., Koie, Y. *Tetrahedron Lett.* **1998**, *39*, 617.
34. Yamamoto, T., Nishiyama, M., Koie, Y. *Tetrahedron Lett.* **1998**, *39*, 2367.
35. Hartwig, J. F., Kawatsura, M., Hauck, S. I., Shaughnessy, K. H., Alcazar-Roman, L. M. *J. Org. Chem.* **1999**, *64*, 5575.
36. Wolfe, J. P., Tomori, H., Sadighi, J. P., Yin, J., Buchwald, S. L. *J. Org. Chem.* **2000**, *65*, 1158.
37. Ali, M. H., Buchwald, S. L. *J. Org. Chem.* **2001**, *66*, 2560.
38. Parrish, C. A., Buchwald, S. L. *J. Org. Chem.* **2001**, *66*, 3820.
39. Harris, M. C., Geis, O., Buchwald, S. L. *J. Org. Chem.* **1999**, *64*, 6019.
40. Rivas, F. M., Riaz, U., Diver, S. T. *Tetrahedron: Asymm.* **2000**, *11*, 1703.
41. (a) Anderson, K. W., Mendez-Perez, M., Priego, J., Buchwald, S. L. *J. Org. Chem.* **2003**, *68*, 9563. (b) Tundel, R. E., Anderson, K. W., Buchwald, S. L. *J. Org. Chem.* **2006**, *71*, 430.
42. Huang, X., Anderson, K. W., Zim, D., Jiang, L., Klapars, A., Buchwald, S. L. *J. Am. Chem. Soc.* **2003**, *125*, 6653.
43. (a) Bei, X., Uno, T., Norris, J., Turner, H. W., Weinberg, W. H., Guram, A. S. *Organometallics* **1999**, *18*, 1840. (b) Bei, X., Guram, A. S., Turner, H. W., Weinberg, W. H. *Tetrahedron Lett.* **1999**, *40*, 1237.
44. (a) Hamann, B. C., Hartwig, J. F. *J. Am. Chem. Soc.* **1998**, *120*, 7369. (b) Roy, A. H., Hartwig, J. F. *J. Am. Chem. Soc.* **2003**, *125*, 8704.

45. Kataoka, N., Shelby, Q., Stambuli, J. P., Hartwig, J. F. *J. Org. Chem.* **2002**, *67*, 5553.
46. (a) Urgaonkar, S., Verkade, J. G. *Adv. Synth. Catal.* **2004**, *346*, 611. (b) Urgaonkar, S., Xu, J.-H. Verkade, J. G. *J. Org. Chem.* **2003**, *68*, 8416. (c) Urgaonkar, S., Nagarajan, M., Verkade, J. G. *J. Org. Chem.* **2003**, *68*, 452. (d) Urgaonkar, S., Nagarajan, M., Verkade, J. G. *Org. Lett.* **2003**, *5*, 815.
47. (a) Nandakumar, M. V., Verkade, J. G. *Tetrahedron* **2005**, *61*, 9775. (b) Nandakumar, M. V., Verkade, J. G. *Angew. Chem. Int. Ed., Engl.* **2005**, *44*, 3115.
48. (a) Marion, N., Navarro, O., Mei, J., Stevens, E. D., Scott, N. M., Nolan, S. P. *J. Am. Chem. Soc.* **2006**, *128*, 4101. (b) Grasa, G. A., Viciu, M. S., Huang, J., Nolan, S. P. *J. Org. Chem.* **2001**, *66*, 7729. (c) Huang, J., Grasa, G., Nolan, S. P. *Org. Lett.* **1999**, *1*, 1307.
49. (a) Nielsen, D. J., Cavell, K. J., Viciu, M. S., Nolan, S. P., Skelton, B. W., White, A. H. *J. Organomet. Chem.* **2005**, *690*, 6133. (b) Viciu, M. S., Germaneau, R. F., Navarro-Fernandez, O., Stevens, E. D., Nolan, S. P. *Organometallics* **2002**, *21*, 5470. (c) Viciu, M. S., Navarro, O., Germaneau, R. F., Kelly III, R. A., Sommer, W., Marion, N., Stevens, E. D., Cavallo, L., Nolan, S. P. *Organometallics* **2004**, *23*, 1629. [R] (d) Hillier, A., Grasa, G. A., Viciu, M. S., Lee, H. M., Yang, C., Nolan, S. P. *J. Organomet. Chem.* **2002**, *653*, 69. (e) Viciu, M. S., Kissling, R. M., Stevens, E. D., Nolan, S. P. *Org. Lett.* **2002**, *4*, 2229. (f) Titcomb, L. R., Caddick, S., Cloke, F. G. N., Wilson, D. J., McKerrecher, D. *Chem. Commun.* **2001**, 1388. (g) Caddick, S., Cloke, F. G. N., Clentsmith, G. K. B., Hitchcock, P. B., McKerrecher, D., Titcomb, L. R., Williams, M. R. V. *J. Organomet. Chem.* **2001**, *617-618*, 635.
50. Cheng, J., Trudell, M. L. *Org. Lett.* **2001**, *3*, 1371.
51. (a) Frisch, A. C., Zapf, A., Briel, O., Kayser, B., Shaikh, N., Beller, M. *J. Mol. Cat. A* **2004**, *214*, 231. (b) Stauffer, S. R., Hartwig, J. F. *J. Am. Chem. Soc.* **2003**, *125*, 6977.
52. Beller, M., Riermeier, T. H., Reisinger, C.-P., Herrmann, W. A. *Tetrahedron Lett.* **1997**, *38*, 2073.
53. (a) Kuwano, R., Utsunomiya, M., Hartwig, J. F. *J. Org. Chem.* **2002**, *67*, 6479. (b) Stambuli, J. P., Kuwano, R., Hartwig, J. F. *Angew. Chem. Int. Ed., Engl.* **2002**, *41*, 4746.
54. Bedford, R. B., Blake, M. E. *Adv. Synth. Catal.* **2003**, *345*, 1107.
55. Li, G. Y. *Angew. Chem. Int. Ed., Engl.* **2001**, *40*, 1513.
56. (a) Ehrentraut, A., Zapf, A., Beller, M. *J. Mol. Cat. A* **2002**, *182–183*, 515. (b) Tewari, A., Hein, M., Zapf, A., Beller, M. *Tetrahedron* **2005**, *61*, 9705.
57. (a) Old, D. W., Wolfe, J. P., Buchwald, S. L. *J. Am. Chem. Soc.* **1998**, *120*, 9722. (b) Schnyder, A., Indolese, A. F., Studer, M., Blaser, H.-U. *Angew. Chem. Int. Ed., Engl.* **2002**, *41*, 3668.
58. (a) Wolfe, J. P., Buchwald, S. L. *Angew. Chem. Int. Ed., Engl.* **1999**, *38*, 2413. (b) Zim, D., Buchwald, S. L. *Org. Lett.* **2003**, *5*, 2413.
59. (a) Rataboul, F., Zapf, A., Jackstell, R., Harkal, S., Riermeier, T., Monsees, A., Dingerdissen, U., Beller, M. *Chem. Eur. J.* **2004**, *10*, 2983. (b) Harkal, S., Rataboul, F., Zapf, A., Fuhrmann, C., Riermeier, T., Monsees, A., Beller, M. *Adv. Synth. Catal.* **2004**, *346*, 1742.
60. Shen, Q., Shekhar, S., Stambuli, J. P., Hartwig, J. F. *Angew. Chem. Int. Ed., Engl.* **2005**, *44*, 1371.
61. Urgaonkar, S., Verkade, J. G. *J. Org. Chem.* **2004**, *69*, 9135.
62. Stauffer, S. R., Lee, S., Stambuli, J. P., Hauck, S. I., Hartwig, J. F. *Org. Lett.* **2000**, *2*, 1423.
63. Jonckers, T. H. M., Maes, B. U. W., Lemiere, G. L. F., Dommisse, R. *Tetrahedron* **2001**, *57*, 7027.
64. Michalik, D., Kumar, K., Zapf, A., Tillack, A., Arlt, M., Heinrich, T., Beller, M. *Tetrahedron Lett.* **2004**, *45*, 2057.
65. Wagaw, S., Buchwald, S. L. *J. Org. Chem.* **1996**, *61*, 7240.
66. Maes, B. U. W., Loones, K. T. J., Lemiere, G. L. F., Dommisse, R. A. *Synlett.* **2003**, 1822.
67. Charles, M. D., Schultz, P., Buchwald, S. L. *Org. Lett.* **2005**, *7*, 3965.
68. Hooper, M. W., Utsunomiya, M., Hartwig, J. F. *J. Org. Chem.* **2003**, *68*, 2861.
69. Yin, J., Zhao, M., Huffman, M. A., McNamara, J. M. *Org. Lett.* **2002**, *4*, 3481.

70. Loones, K. T. J., Maes, B. U. W., Dommisse, R. A., Lemiere, G. L. F. *Chem. Commun.* **2004**, 2466.
71. (a) Watanabe, M., Yamamoto, T., Nishiyama, M. *Chem Commun.* **2000**, 133. (b) Ogawa, K., Radke, K. R., Rothstein, S. D., Rasmussen, S. C. *J. Org. Chem.* **2001**, *66*, 9067.
72. Luker, T. J., Beaton, H. G., Whiting, M., Mete, A., Cheshire, D. R. *Tetrahedron Lett.* **2000**, *41*, 7731.
73. He, F., Foxman, B. M., Snider, B. B. *J. Am. Chem. Soc.* **1998**, *120*, 6417.
74. Cammerer, S. S., Viciu, M. S., Stevens, E. D., Nolan, S. P. *Synlett* **2003**, 1871.
75. Emoto, T., Kubosaki, N., Yamagiwa, Y., Kamikawa, T. *Tetrahedron Lett.* **2000**, *41*, 355.
76. Hennessy, E. J., Buchwald, S. L. *J. Org. Chem.* **2005**, *70*, 7371.
77. Morita, S., Kitano, K., Matsubara, J., Ohtani, T., Kawano, Y., Otsubo, K., Uchida, M. *Tetrahedron* **1998**, *54*, 4811.
78. Abouabdellah, A., Dodd, R. H. *Tetrahedron Lett.* **1998**, *39*, 2119.
79. (a) Hong, Y., Senanayake, C. H., Xiang, T., Vandenbossche, C. P., Tanoury, G. J., Bakale, R. P., Wald, S. A. *Tetrahedron Lett.* **1998**, *39*, 3121. (b) Hong, Y., Tanoury, G. J., Wilkinson, H. S., Bakale, R. P., Wald, S. A., Senanayake, C. H. *Tetrahedron Lett.* **1997**, *38*, 5607.
80. Lopez-Rodriguez, M., Benhamu, B., Ayala, D., Rominguera, J. L., Murcia, M., Ramos, J. A., Viso, A. *Tetrahedron* **2000**, *56*, 3245.
81. Enguehard, C., Allouchi, H., Gueiffier, A., Buchwald, S. L. *J. Org. Chem.* **2003**, *68*, 4367.
82. Batch, A., Dodd, R. H. *J. Org. Chem.* **1998**, *63*, 872.
83. Iwaki, T., Yasuhara, A., Sakamoto, T. *J. Chem. Soc., Perkin Trans. 1* **1999**, 1505.
84. Schoffers, E., Olsen, P. D., Means, J. C. *Org. Lett.* **2001**, *3*, 4221.
85. Wang, Z., Rizzo, C. J. *Org. Lett.* **2001**, *3*, 565.
86. Rossen, K., Pye, P., Maliakal, A., Volante, R. P. *J. Org. Chem.* **1997**, *62*, 6462.
87. (a) Beletskaya, I. P., Averin, A. D., Bessmertnykh, A. G., Guilard, R. *Tetrahedron Lett.* **2001**, *42*, 4987. (b) Beletskaya, I. P., Averin, A. D., Bessmertnykh, A. G., Guilard, R. *Tetrahedron Lett.* **2001**, *42*, 4983. (c) Beletskaya, I. P., Bessmertnykh, A. G., Averin, A. D., Denat, F., Guilard, R. *Eur. J. Org. Chem.* **2005**, 281. (d) Averin, A. D., Ulanovskaya, O. A., Fedotenko, I. A., Borisenko, A. A., Serebryakova, M. V., Beletskaya, I. P. *Helv. Chim. Acta* **2005**, *88*, 1983. (e) Averin, A. D., Ulanovskaya, O. A., Borisenko, A. A., Serebryakova, M. V., Beletskaya, I. P. *Tetrahedron Lett.* **2006**, *47*, 2691. (f) Zhang, X.-X., Buchwald, S. L. *J. Org. Chem.* **2000**, *65*, 8027.
88. Averin, A. D., Ranyuk, E. R., Lukashev, N. V., Beletskaya, I. P. *Chem. Eur. J.* **2005**, *11*, 7030.
89. (a) Louie, J., Hartwig, J. F. *J. Am. Chem. Soc.* **1997**, *119*, 11695. (b) Chae, H. K., Eddaoudi, M., Kim, J., Hauck, S. I., Hartwig, J. F., O'Keeffe, M., Yaghi, O. M. *J. Am. Chem. Soc.* **2001**, *123*, 11482. (c) Hauck, S. I., Lakshmi, K. V., Hartwig, J. F. *Org. Lett.* **1999**, *1*, 2057. (d) Harris, M. C., Buchwald, S. L. *J. Org. Chem.* **2000**, *65*, 5327.
90. (a) Watanabe, M., Nishiyama, M., Yamamoto, T., Koie, Y. *Tetrahedron Lett.* **2000**, *41*, 481. (b) Goodson, F. E., Hartwig, J. F. *Macromolecules* **1998**, *31*, 1700. (c) Goodson, F. E., Hauck, S. I., Hartwig, J. F. *J. Am. Chem. Soc.* **1999**, *121*, 7527. (d) Singer, R. A., Sadighi, J. P., Buchwald, S. L. *J. Am. Chem. Soc.* **1998**, *120*, 213. (e) Sadighi, J. P., Singer, R. A., Buchwald, S. L. *J. Am. Chem. Soc.* **1998**, *120*, 4960. (f) Zhang, X.-X., Sadighi, J. P., Mackewitz, T. W., Buchwald, S. L. *J. Am. Chem. Soc.* **2000**, *122*, 7606.
91. (a) Wolfe, J. P., Ahman, J., Sadighi, J. P., Singer, R. A., Buchwald, S. L., *Tetrahedron Lett.* **1997**, *38*, 6367. (b) Singer, R. A., Buchwald, S. L. *Tetrahedron Lett.* **1999**, *40*, 1095.
92. Jaime-Figueroa, S., Liu, Y., Muchowski, J. M., Putnam, D. G. *Tetrahedron Lett.* **1998**, *39*, 1313.
93. Hori, K., Mori, M. *J. Am. Chem. Soc.* **1998**, *120*, 7651.
94. Lee, S., Jorgensen, M., Hartwig, J. F. *Org. Lett.* **2001**, *3*, 2729.
95. Huang, X., Buchwald, S. L. *Org. Lett.* **2001**, *3*, 3417.
96. Lee, D.-Y., Hartwig, J. F. *Org. Lett.* **2005**, *7*, 1169.

97. Shen, Q., Hartwig, J. F. *J. Am. Chem. Soc.* **2006**, *128*, 10028.
98. (a) Wagaw, S., Yang, B. H., Buchwald, S. L. *J. Am. Chem. Soc.* **1999**, *121*, 10251. (b) Wagaw, S., Yang, B. H., Buchwald, S. L. *J. Am. Chem. Soc.* **1998**, *120*, 6621. (c) Ortner, B., Waibel, R., Gmeiner, P. *Angew. Chem. Int. Ed., Engl.* **2001**, *40*, 1283.
99. Brown, J. A. *Tetrahedron Lett.* **2000**, *41*, 1623.
100. Old, D. W., Harris, M. C., Buchwald, S. L. *Org. Lett.* **2000**, *2*, 1403.
101. Hartwig, J. F. *Angew. Chem. Int. Ed., Engl.* **1998**, *37*, 2090.
102. Wang, Z., Skerlj, R. T., Bridger, G. J. *Tetrahedron Lett.* **1999**, *40*, 3543.
103. Arterburn, J. B., Rao, K. V., Ramdas, R., Dible, B. R. *Org. Lett.* **2001**, *3*, 1351.
104. (a) Song, J. J., Yee, N. K. *Org. Lett.* **2000**, *2*, 519. (b) Song, J. J., Yee, N. K. *Tetrahedron Lett.* **2001**, *42*, 2937.
105. Zhu, Y., Kiryu, Y., Katayama, H. *Tetrahedron Lett.* **2002**, *43*, 3577.
106. Evindar, G., Batey, R. A. *Org. Lett.* **2003**, *5*, 133.
107. (a) Bolm, C., Hildebrand, J. P. *J. Org. Chem.* **2000**, *65*, 169. (b) Cho, G. Y., Okamura, H., Bolm, C. *J. Org. Chem.* **2005**, *70*, 2346.
108. Harmata, M., Pavri, N. *Angew. Chem. Int. Ed., Engl.* **1999**, *38*, 2419.
109. (a) Barluenga, J., Aznar, F., Valdes, C. *Angew. Chem. Int. Ed., Engl.* **2004**, *43*, 343. (b) Smith, C. J., Early, T. R., Holmes, A. B., Shute, R. E. *Chem. Commun.* **2004**, 1976.
110. Yang, B. H., Buchwald, S. L. *Org. Lett.* **1999**, *1*, 35.
111. (a) Shakespeare, W. C. *Tetrahedron Lett.* **1999**, *40*, 2035. (b) Yin, J., Buchwald, S. L. *J. Am. Chem. Soc.* **2002**, *124*, 6043. (c) Yin, J., Buchwald, S. L. *Org. Lett.* **2000**, *2*, 1101. (d) Burton, G., Cao, P., Li, G., Rivero, R. *Org. Lett.* **2003**, *5*, 4373.
112. (a) Ghosh, A., Sieser, J. E., Riou, M., Cai, W., Rivera-Ruiz, L. *Org. Lett.* **2003**, *5*, 2207. (b) Madar, D. J., Kopecka, H., Pireh, D., Pease, J., Pliushchev, M., Sciotti, R. J., Wiedeman, P. E., Djuric, S. W. *Tetrahedron Lett.* **2001**, *42*, 3681.
113. Edmondson, S. D., Mastracchio, A., Parmee, E. R. *Org. Lett.* **2000**, *2*, 1109.
114. Bonnaterre, F., Bois-Choussy, M., Zhu, J. *Org. Lett.* **2006**, *8*, 4351.
115. Kozawa, Y., Mori, M. *Tetrahedron Lett.* **2002**, *43*, 111.
116. Lebedev, A. Y., Izmer, V. V., Kazyul'kin, D. N., Beletskaya, I. P., Voskoboynikov, A. Z. *Org. Lett.* **2002**, *4*, 623.
117. (a) Barluenga, J., Fernandez, M. A., Aznar, F., Valdes, C. *Chem. Eur. J.* **2004**, *10*, 494. (b) Reddy, C., Reddy, V., Urgaonkar, S., Verkade, J. G. *Org. Lett.* **2005**, *7*, 4427.
118. Barluenga, J., Fernandez, M. A., Aznar, F., Valdes, C. *Chem. Commun.* **2004**, 1400.
119. (a) Wallace, D. J., Klauber, D. J., Chen, C., Volante, R. P. *Org. Lett.* **2003**, *5*, 4749. (b) Willis, M. C., Brace, G. N. *Tetrahedron Lett.* **2002**, *43*, 9085.
120. Willis, M. C., Chauhan, J., Whittingham, W. G. *Org. Biomol. Chem.* **2005**, *3*, 3094.
121. Fayol, A., Fang, Y.-Q., Lautens, M. *Org. Lett.* **2006**, *8*, 4203.

Jacob M. Janey

7.3 Haloform Reaction

7.3.1 Description

The Haloform reaction (also known as the Lieben iodoform reaction or Lieben haloform reaction) describes the two-step conversion of a methyl ketone **1** to a carboxylic acid *via* threefold halogenation to the trihaloketone **2** followed by base-cleavage to the carboxylate **3** and haloform **4**. Acidification affords the carboxylic acid **5**.[1-7]

$$\underset{\mathbf{1}}{R-\overset{O}{\overset{\|}{C}}-CH_3} \xrightarrow[HO^-]{3X_2} \underset{\mathbf{2}}{R-\overset{O}{\overset{\|}{C}}-CX_3} \xrightarrow{HO^-} \underset{\mathbf{3}}{R-\overset{O}{\overset{\|}{C}}-\bar{O}} + \underset{\mathbf{4}}{HCX_3} \xrightarrow{H^+} \underset{\mathbf{5}}{R-\overset{O}{\overset{\|}{C}}-OH}$$

R = alkyl, aryl, H; X = Cl, Br, I

An important variation is that compounds such as methyl carbinols also undergo the Haloform reaction since they are typically oxidized to the corresponding methyl ketone under the reaction conditions. This reaction is particularly well known to undergraduate organic chemistry students as the "Iodoform Test" in qualitative organic chemistry.[1,3,7,8]

7.3.2 Historical Perspective

Given the historic availability of simple ketones and the halogens chlorine, bromine, and iodine, it is not surprising that the Haloform reaction has a long and rich history, summarized exhaustively by Fuson and Bull[1] and by Parkin.[2] Interestingly, the initial discoveries of the Haloform reaction all involved the treatment of ethanol with iodine, chlorine, or bromine in the presence of alkali to afford the respective haloform following *in situ* oxidation of ethanol to acetaldehyde.[9] Indeed, chloroform was used extensively in the U.S. Civil War as a battlefield anesthetic during amputations.[10] The now well-known "Iodoform Test" was inaugurated in 1870 by Lieben.[3] As will be presented later, nature produces enormous quantities of chloroform and bromoform, most probably involving the Haloform reaction.

7.3.3 Mechanism

$$\underset{\mathbf{1}}{R{-}C(=O){-}CH_3} \xrightleftharpoons{HO^-} \underset{\mathbf{6}}{R{-}C(O^-){=}CH_2} \xrightarrow{X_2} \underset{\mathbf{7}}{R{-}C(=O){-}CH_2X} \xrightleftharpoons{HO^-} R{-}C(O^-){=}CHX \xrightarrow{X_2}$$

$$R{-}C(=O){-}CHX_2 \xrightleftharpoons{HO^-} R{-}C(O^-){=}CX_2 \xrightarrow{X_2} \underset{\mathbf{8}}{R{-}C(=O){-}CX_3} \xrightarrow[2.\ H^+]{1.\ HO^-} \underset{\mathbf{5}}{R{-}C(=O){-}OH} + \underset{\mathbf{4}}{HCX_3}$$

The mechanism of the Haloform reaction involves base-catalyzed generation of an enolate **6** followed by *in situ* halogenation to afford an α-haloketone **7**. Two subsequent repetitions afford trihaloketone **8**, which then is susceptible to cleavage by alkali to furnish carboxylic acid **5** and haloform **4**, after acidification.

Early mechanistic studies of the Haloform reaction are summarized by Fuson and Bull.[1] In support of the stepwise pathway, examples of the isolation of all three haloketones are known, and several cases of stable trihaloketones have been described. For example, haloketones **9–11** have been isolated from the base-induced halogenation of the appropriate ketone depending on the reaction time.[11,12]

9: 2,4,6-trichlorophenyl $C(=O)CH_2Cl$ **10**: 2,4,6-tribromo-3-carboxyphenyl $C(=O)CH_2Br$ **11**: 2,4,6-tribromo-3-carboxyphenyl $C(=O)CHBr_2$ **12**: 2,4,6-tribromo-3-carboxyphenyl $C(=O)CBr_3$

Furthermore, in a series of papers Fuson and co-workers have shown that base-cleavage of the trihalo ketone is strongly retarded by adjacent bulky groups.[11–13] For example, α,α,α-tribromo-2,4,6-trichloroacetophenone is unaffected by extended treatment with cold solutions of hydroxide.[11] Interestingly, 1,1,1-trichloropropanone is found in chlorine-treated drinking water and is presumed to be a precursor of chloroform that is ubiquitous in such water.[14]

Several reported kinetic studies shed light on the various steps in the Haloform reaction. Bartlett showed in 1934 that the reaction of acetone with strong alkaline solutions of hypobromite and hypoiodite involves enolization of acetone as the rate-determining step, but the rate of the same reaction with alkaline hypochlorite involves as the slow step reaction of acetone enol with hypochlorite.[15] Likewise, the mechanism of the iodoform reaction and its optimization for quantitative analysis have received much scrutiny[16] including studies of acetone-1-C-14[17] and deuterated acetone[18] for isotope effects in the Haloform reaction. The latter study concludes that the rate-determining step is proton abstraction by base. Guthrie and Cossar have described an exhaustive kinetics examination of the chlorination of acetone and the subsequent cleavage to chloroform and acetate.[19] A subsequent paper addresses the alkaline cleavage of trihaloaceto-phenones, which shows this relative rate order: $PhCOCF_3$ (1.0), $PhCOCCl_3$ (5.3×10^{10}), $PhCOCBr_3$ (2.2×10^{13}).[20] The cleavage of ring-substituted 2,2,2-trichloroacetophenones has been reported,[21] and a theoretical examination of the base-promoted cleavage of 4-alkyl-4-hydroxy-3,3-difluoromethyl trifluoromethyl ketones was described.[22]

7.3.4 Variations and Improvements

Only a few departures from the basic Haloform reaction conditions (Section *7.3.5*) have been developed. Both sodium bromite[23] and benzyltrimethylammonium tribromide[24] in aqueous sodium hydroxide are convenient alternatives to the use of bromine in the Haloform reaction (e.g., **13** to **14**,[23] **15** to **16**[24]). Both reagents also effect conversion of methyl carbinols to carboxylic acids.

$$\underset{\mathbf{13}}{R-\overset{O}{\overset{\|}{C}}-CH_3} \xrightarrow[\text{aq. NaOH}]{NaBrO_2,\ NaBr} \left[R-\overset{O}{\overset{\|}{C}}-\overset{-}{O}\overset{+}{Na} \right] \xrightarrow[80\text{-}96\%]{H_3O^+} \underset{\mathbf{14}}{R-\overset{O}{\overset{\|}{C}}-OH}$$

R = *i*-Bu, *t*-Bu, *n*-Pen, Ph, 4-Tol, 4-ClPh, 4-NO_2Ph, 2-Naphthyl

$$\underset{\mathbf{15}}{\text{2-thienyl-}COCH_3} \xrightarrow[\text{2. } H_3O^+ \quad 97\%]{\text{1. } PhCH_2N(CH_3)_3Br \text{, aq. NaOH}} \underset{\mathbf{16}}{\text{2-thienyl-}CO_2H}$$

The use of benzyltrimethylammonium tribromide is particularly effective for the synthesis of aromatic and heteroaromatic carboxylic acids as shown in Table 1.

The Haloform reaction is catalyzed by cyclodextrins in what the authors label as "inverse phase transfer catalysis,"[25] but the synthetic utility of this variation remains to be seen. An alternative to the use of halogen is a nitroarene catalyzed oxidation of acetophenone with sodium percarbonate or sodium perborate.[26] However, the yields of substituted benzoic acids furnished by this method are mediocre (23–73%) in comparison to the conventional Haloform conditions. Likewise, the Haloform reaction of acetone with iodine in liquid ammonia is without synthetic merit (8–12%).[27]

An important variation of the Haloform reaction is the synthesis of esters instead of carboxylic acids, and several groups have reported this reaction, but Arnold and co-workers seem to be the discoverers.[28] Thus, they isolated 5-carbomethoxy-8-

methoxytetralin (**18**) from 5-acetyl-8-methoxytetralin (**17**) using either aqueous methanolic hypochloride (80% yield) or hypobromite (83% yield).

$$\textbf{17}\ (\text{COCH}_3,\ \text{OMe}) \xrightarrow[\text{aq. MeOH, KOH, K}_2\text{CO}_3\ \ 80\%]{\text{CaOCl}} \textbf{18}\ (\text{CO}_2\text{Me},\ \text{OMe})$$

This result is not surprising since it was shown earlier that trichloromethyl ketones react more rapidly with methanol (alcohols) than with water under alkaline conditions.[29] The ethanolic cleavage of a trichloromethyl ketone has been used to prepare ethyl 3,3-diethoxypropanoate,[30] and aryl trifluoromethyl ketones can be cleaved to give carboxylic acids or esters.[31] A similar reaction of 3-trifluoroacetylindole (**19**) with lithium dialkylamides affords good yields of the corresponding indole-3-carboxamides (**20**).[32]

$$\textbf{19}\ (\text{COCF}_3) \xrightarrow[\text{2. H}_2\text{O}\ \ 74\text{–}88\%]{\text{1. LiNR}_2\text{ (5 eq), 0 °C, THF}} \textbf{20}\ (\text{CONR}_2)$$

An electrocatalytic Haloform reaction to convert methyl ketones into esters has been described by two groups,[33,34] and the Haloform reaction has been extended to non-methyl ketones,[35] but this is considered to be outside the scope of the present article. Likewise, a Haloform reaction of ß-sulfonylmethyl ketones affords α-halosulfones and acetate,[36] and various quinones yield iodoform upon treatment with alkaline iodine.[37]

Table 1. Comparison of $NaBrO_2$-NaBr and $PhCH_2N(CH_3)_3Br$ in Haloform reactions from the corresponding methyl ketone

	Method, % yield	
Product	$NaBrO_2$-NaBr[23]	$PhCH_2N(CH_3)_3Br$[24]
CO_2H	94	95
CO_2H, Me	96	99
CO_2H, Cl	90	99
CO_2H, O_2N	87	88
CO_2H, MeO	—	92
CO_2H	85	99
CO_2H	—	70
MeO, CO_2H, OMe	—	99
S, CO_2H	40	97
O, CO_2H	—	95

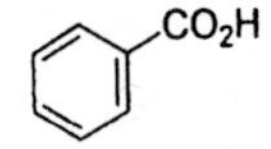

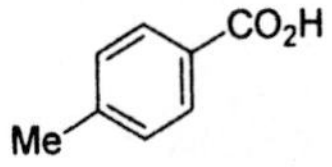

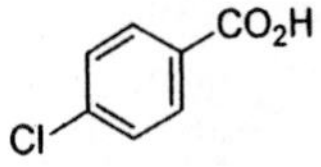

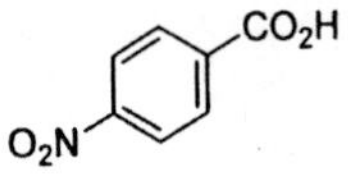

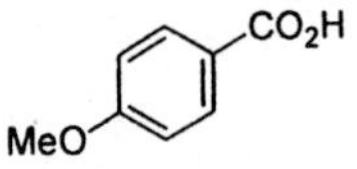

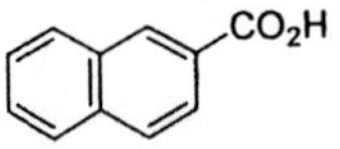

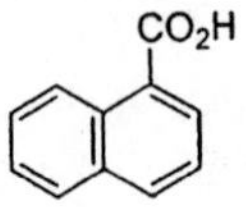

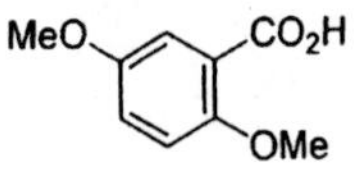

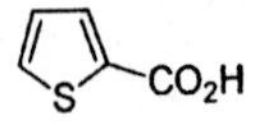

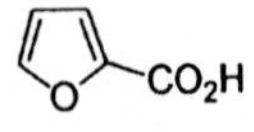

7.3.5 Synthetic Utility

Despite its age, the Haloform reaction remains a unique and powerful transformation of methyl ketones to carboxylic acids. The range of substrates is truly enormous. In addition to the examples shown below, the Haloform reaction has been employed to synthesize these carboxylic acids (from the corresponding methyl ketones): 3-acrylamido-3-methylbutanoic acid (34%),[38] kainic acid analogs,[39] triterpenoids,[40] 3-carboxytropolone (83%),[41] 7-carboxy-2-methoxytropone (33%),[41] 2-arylpropanoic acids (95%),[42] dimethylbenzoic acids,[43] tetrahydro-naphthalenes (70–100%),[44] *N*-(4-carboxyphenylethyl)acetamide (44–75%),[45] pyridone-substituted benzoic acids,[46] epicucurbic acid (89%),[47] 4-(pyren-6-yl)butanoic acid (79%),[48] 3-(benzo[*a*]pyren-6-yl)propanoic acid (56%),[48] and $^{13}C_{12}$-benzoic acid (90%).[49] In the following examples, a solution of hypohalite is prepared initially followed by reaction with the substrate.

Br_2, NaOH
82%
CO_2H
(ref. 50)

O
CH_3
Br_2, NaOH
aq. dioxane
rt, 2 h
98%
CO_2H
(ref. 51)

OMe O
Me
MeO_2C H
Br_2, aq. KOH
dioxane
rt to reflux
86%
OMe
CO_2H
MeO_2C H
(ref. 52)

O
H
NaOBr
aq. NaOH
15 °C to rt
91% (as methyl ester)
CO_2H
H
(ref. 53)

Br_2, NaOH
aq. dioxane, 5 °C
56%
(ref. 54)

Br_2, NaOH
aq. THF, 0 °C to rt
90%
(ref. 55)

Br_2, NaOH
aq. dioxane, 0 °C
100%
(ref. 56)

Noteworthy in the above reactions is the chemoselectivity and the absence of side reactions involving potentially reactive sites (i.e., electron-rich alkenes and aromatic rings and esters). However, on occasion such side reactions do occur (**21**→**22** and **23**),[57] although in this case temperature and hypochlorite concentration control minimize the formation of **23**.

NaOCl
aq. NaOH
21 **22** **23**
(ref. 57)

The synthetic utility of the Haloform reaction is revealed further by a synthesis of isotopically pure (99.2%) $CDCl_3$ by the reaction of trichloroacetophenone with NaOD in D_2O.[58] The alkoxide-cleavage of trichloromethyl ketones has been employed to synthesize dihydropyridine esters (e.g., **24**) in which ester exchange does not occur.[59]

NaOR^1, R^{1}OH, THF, rt, 70-95% → **24** (ref. 59a)

R^1 = Me, *i*-Pr; R^2 = 2-CO_2Et, 3-CO_2Et, 2-CF_3, 2-Cl

Treatment of trichloroacetylpyrroles (and trichloroacetylimidazoles) with amines provides access to polyamides and distamycin analogs (e.g., **25**).[60]

NaH, THF, 79% → **25** (ref. 60d)

A Haloform-type reaction occurs between trichloromethyl-substituted 1,3-diketones and hydrazine or *N*-methylhydroxylamine to afford hydroxypyrazoles (e.g., **26**) and 2-methyl-3-isoxazolones, respectively.[61]

N_2H_4, MeCN, rt, 97% → **26** (ref. 61)

The aqueous base cleavage of a trifluoromethyl group from a lumazine hydrate is known,[62] and the rupture of ring D of 3-benzyl-*O*-estrone with hypoiodite to afford the corresponding tricyclic diacid (75% yield) has been reported.[63] Indeed, over the years the iodoform version of the Haloform reaction has undergone extensive improvements[64] and has seen applications in quantitative analysis.[65]

7.3.6 Biological Haloform Reaction

No discussion of the Haloform reaction is complete without mention of Nature's role in the production of haloform, especially chloroform and bromoform.[66] It is estimated that 90% of atmospheric chloroform has a biogenic origin while only 10% is anthropogenic.[67]

For example, termites produce up to 15% of the global chloroform emissions,[68] most likely by a Haloform reaction. Halogenated ketones are ubiquitous in the oceans and a few examples are shown (**27–30**).[66] The antibacterial tribromoacetamide is present in an Okinawan *Wrangelia* alga,[69] and myriad other halogenated natural products can be envisioned to produce haloform.[66]

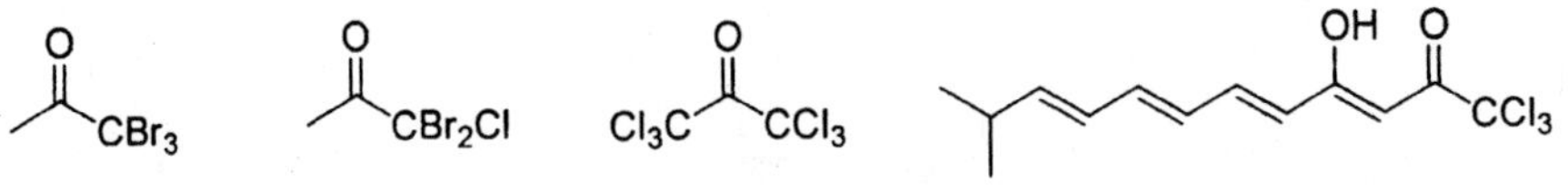

A second biogenic source of chloroform is the chloroperoxidase-catalyzed degradation of humic acid leading ultimately to chloroform and chlorinated acetic acids.[66] Moreover, this pathway seems to operate when drinking water (containing organic matter) is disinfected with chlorine.[70–72] This halogenation mechanism is supported by the isolation of $^{13}CHCl_3$ and $^{13}CHBr_3$ when 1,3-dihydroxybenzene-2-^{13}C (**31**) is treated with aqueous chlorine and bromine, respectively.[73]

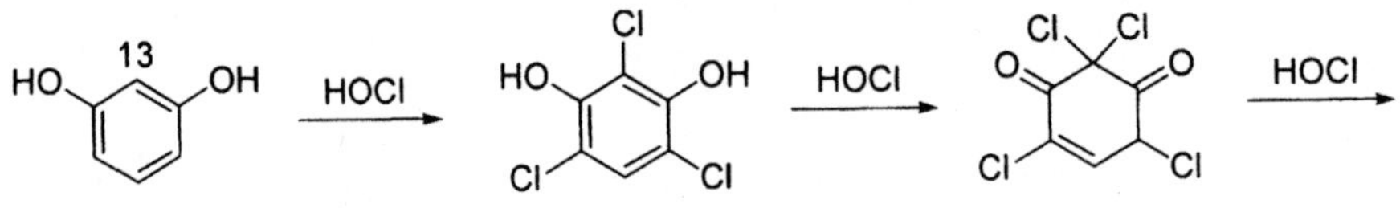

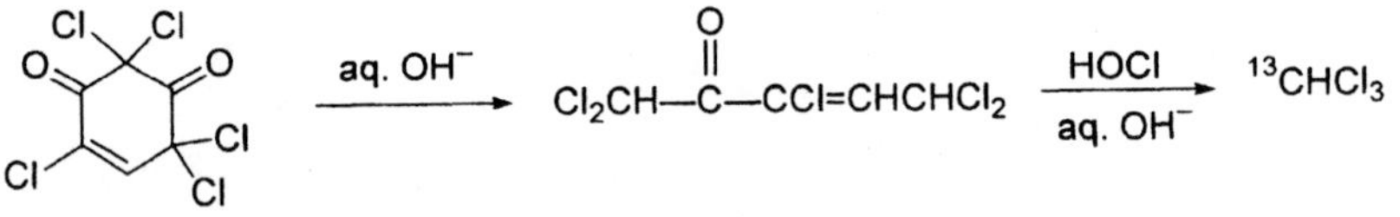

7.3.7 *Experimental*

The reader is referred to the synthesis of trimethylacetic acid,[74] 2-naphthoic acid,[75] 3,3-dimethylacrylic acid,[76] 3,3-dimethylglutaric acid,[77] 3β-acetoxy-5-androstene-17β-carboxylic acid,[78] ethyl pyrrole-2-carboxylate,[79] and bicyclo[1.1.1]pentane-1,3-dicarboxylic acid[80] published in *Organic Syntheses.*

Cinnamic acid[23]

Benzylideneacetone (1.0 g, 6.8 mmol) is added to a solution of sodium bromite (94.7% purity, 3.1 g, 21.7 mmol), sodium bromide (2.3 g, 22.4 mmol), and sodium hydroxide (0.3 g, 7.5 mmol) in water (10 mL) at room temperature, and then the mixture is stirred for 6 h. A brown precipitate obtained is filtered from the reaction mixture and washed with chloroform (20 mL). The resultant colorless crystals are acidified with hydrochloric acid (20 wt%, 30 ml) and extracted with chloroform. The chloroform solution is dried with magnesium sulfate and evaporated *in vacuo* to give cinnamic acid as colorless crystals; yield: 0.71 g (70%); mp 132–134 °C (from water).

Benzoic acid[24]

Benzyltrimethylammonium tribromide (6.04 g, 15.5 mmol) was added to a solution of acetophenone (0.60 g, 5 mmol) and NaOH (1.2 g, 30 mmol) in water (50 mL), and the mixture was stirred at room temperature for 20 min. The orange color of the solution immediately faded and bromoform formed during the period. To the mixture was added 5% aq. solution of $NaHSO_3$ (10 mL) and bromoform was filtered off. The filtrate was made acidic with 6 N HCl and was extracted with ether (40 mL × 4). The ether layer was dried over $MgSO_4$, filtered and evaporated at reduced pressure to give benzoic acid; yield: 0.58 g (95%); mp 122–123 °C.

2-Carboxytropolone[41]

To a stirred mixture of 40% sodium hydroxide solution (1 mL) and sodium hypochlorite solution, prepared from chlorine gas and sodium hydroxide (1.84 g) in ice water (15 mL), was added a solution of 3-acetyltropolone (820 mg, 5 mmol) in 6% NaOH solution (3 mL) with cooling in an ice water bath. The reaction mixture was stirred for 1 hr at 15 °C. The mixture was added to a solution of sodium hydrogensulfite (500 mg) in water (2 mL) to remove an excess of sodium hypochlorite. The mixture was carefully made slightly acidic with concentrated hydrochloric acid. The precipitate was collected and dissolved in hot water, the solution was made acidic with concentrated hydrochloric acid to give crystals which were recrystallized from methanol to give 3-carboxytropolone as yellow needles in a yield of 690 mg (85%), mp 217–218 °C; IR (KBr): ν_{max} 3212 (OH), 1718 (COOH), 1585 cm^{-1} (C=O); ^1H NMR (deuteriochloroform): δ 7.05–7.8 (3H, m), 8.15 (1H, d, J = 10 Hz, H-4), 10.3 (2H, br, OH).

6,7-Dimethoxynaphthalene-2-carboxylic acid[55]

To a stirred solution of NaOH (34.89 g, 870 mmol) in 120 ml of H_2O was added Br_2 (42.6 g, 266.2 mmol) over 30 min at 0 °C. To this mixture was added a solution of 2-acetyl-6,7-dimethoxynaphthalene (20 g, 87 mmol) in 60 mL of THF slowly over 30 min at 0 °C. The mixture was warmed to room temperature and stirred for 8 h. The organic phase was separated in a separatory funnel and dispatched. To the aqueous solution were added ice (200 g) and 20% $NaHSO_3$ soln. (100 mL). After acidification of the aq. soln. with 37% HCl (pH $\leq$ 3), the solidified acid was filtered with suction and dried at 60 °C (18.2 g, 90%) to give white crystals, mp 238–240 °C; ^1H NMR (200 MHz, (D_6)-DMSO): 12.65 (br, s, 1H), 8.42 (br, s, 1H), 7.79 (m, 2H), 7.48 (s, 1H), 7.36 (s, 1H), 3.90 (s, 3H), 3.89 (s, 3H). ^{13}C NMR (50 MHz (D_6)-DMSO): 169.5, 152.9, 151.6, 133.2, 130.6, 129.8, 128.1, 127.8, 125.3, 109.5, 108.1, 57.4 (2 C).

7.3.8 *References*

1. [R] Fuson, R. C.; Bull, B. A. *Chem. Rev.* **1935**, *15*, 275.
2. [R] Parkin, H. *Mendel Bull.* **1936**, *9*, 3.
3. [R] Lieben, A. *Ann.* (Suppl.) **1870**, 218.
4. [R] Chakrabartty, S. K. in Trahanovsky, W. S. *Oxidation in Organic Chemistry, Part C*, Academic Press, New York, 1978, pp 348–370.
5. [R] House, H. O. *Modern Synthetic Reactions*, 2nd Ed., W. A. Benjamin, Menlo Park, CA, 1972, pp 459-478.
6. [R] Mundy, B. P.; Ellerd, M. G.; Favaloro, Jr., F. G. *Name Reactions and Reagents in Organic Synthesis*, 2nd Ed., Wiley-Interscience, New Jersey, 2005, pp 392-393.
7. [R] Seelye, R. N.; Turney, T. A. *J. Chem. Ed.* **1959**, *36*, 572.
8. (a) Cheronis, N. D.; Entrikin, J. B.; Hodnett, E. M. *Semimicro Qualitative Organic Analysis*, 3rd Ed., Interscience Publishers, New York, 1965, p 361. (b) Shriner, R. L.; Hermann, C. K. F.; Morrill, T. C.; Curtin, D. Y.; Fuson, R. C. *The Systematic Identification of Organic Compounds*, 8th Ed., John Wiley & Sons, Hoboken, NJ, 2004, pp 273-276.
9. (a) Serullas, M. *Ann. Chim.* **1822**, *20*, 165; **1823**, *22*, 172, 222; **1824**, *25*, 311. (b) Guthrie, S. *Am. J. Sci. and Arts* **1832**, *21*, 64. (c) Soubeiran, E. *Ann. Chim.* **1831**, *48*, 131. (d) Dumas, J.-B. A. *Ann. Chim.* **1834**, *56*, 113, 122. (e) Liebig, J. *Ann. Physik. Chem.* **1831**, *23*, 444.
10. Bollet, A. J. *Civil War Medicine — Challenges and Triumphs*, Galen, Arizona, 2002.
11. Fuson, R. C.; Bertetti, J. W.; Ross, W. E. *J. Am. Chem. Soc.* **1932**, *54*, 4380.
12. Bull, B. A.; Ross, W. E.; Fuson, R. C. *J. Am. Chem. Soc.* **1935**, *57*, 764.
13. (a) Gray, A. R.; Walker, J. T.; Fuson, R. C. *J. Am. Chem. Soc.* **1931**, *53*, 3494. (b) Fuson, R. C.; Farlow, M. W.; Stehman, C. J. *J. Am. Chem. Soc.* **1931**, *53*, 4097. (c) Fuson, R. C.; Lewis, P. H.; Du Puis, R. N. *J. Am. Chem. Soc.* **1932**, *54*, 1114. (d) Fisher, C. H.; Snyder, H. R.; Fuson, R. C. *J. Am. Chem. Soc.* **1932**, *54*, 3665. (e) Bull, B. A.; Fuson, R. C. *J. Am. Chem. Soc.* **1933**, *55*, 3424. (f) Johnson, R.; Fuson, R. C. *J. Am. Chem. Soc.* **1934**, *56*, 1417.
14. Suffet, I. H.; Brenner, L.; Silver, B. *Environ. Sci. Technol.* **1976**, *10*, 1273.
15. Bartlett, P. D. *J. Am. Chem. Soc.* **1934**, *56*, 967.
16. (a) Morgan, K. J.; Bardwell, J.; Cullis, C. F. *J. Chem. Soc.* **1950**, 3190. (b) Cullis, C. F.; Hashmi, M. H. *J. Chem. Soc.* **1956**, 2512.
17. Ropp, G. A.; Bonner, W. A.; Clark, M. T.; Raaen, V. F. *J. Am. Chem. Soc.* **1954**, *76*, 1710.
18. Pocker, Y. *Chem. Ind.* **1959**, 1383.
19. Guthrie, J. P.; Cossar, J. *Can J. Chem.* **1986**, *64*, 1250.
20. Guthrie, J. P.; Cossar, J. *Can J. Chem.* **1990**, *68*, 1640.
21. Zucco, C.; Lima, C. F.; Rezende, M. C.; Vianna, J. F.; Nome, F. *J. Org. Chem.* **1987**, *52*, 5356.
22. Olivella, S.; Solé, A.; Jiménez, O.; Bosch, M. P.; Guerrero, A. *J. Am. Chem. Soc.* **2005**, *127*, 2620.
23. Kajigaeshi, S.; Nakagawa, T.; Nagasaki, N.; Fujisaki, S. *Synthesis* **1985**, 674.
24. Kajigaeshi, S.; Kakinami, T.; Yamaguchi, T.; Uesugi, T.; Okamoto, T. *Chem. Express* **1989**, *4*, 177.
25. Trotta, F.; Cantamessa, D.; Zanetti, M. *J. Incl. Phenom. Macrocycl. Chem.* **2000**, *37*, 83.
26. Bjørsvik, H.-R.; Merinero, J. A. V.; Ligøuori, L. *Tetrahedron Lett.* **2004**, *45*, 8615.
27. Vaughn, T. H.; Nieuwland, J. A. *J. Am. Chem. Soc.* **1932**, *54*, 787.
28. Arnold, R. T.; Buckles, R.; Stoltenberg, J. *J. Am. Chem. Soc.* **1944**, *66*, 208.
29. Houben, J.; Fischer, W. *Ber.* **1931**, *64*, 240.
30. Tietze, L. F.; Voss, E.; Hartfiel, U. *Org. Syn.* **1990**, *69*, 238.
31. Delgado, A.; Clardy, J. *Tetrahedron Lett.* **1992**, *33*, 2789.
32. Hassinger, H. L.; Soll, R. M.; Gribble, G. W. *Tetrahedron Lett.* **1998**, *39*, 3095.
33. (a) Nikishin, G. I.; Elinson, M. N.; Makhova, I. V. *Angew. Chem. Int. Ed. Engl.* **1988**, *27*, 1716. (b) Nikishin, G. I.; Elinson, M. N.; Makhova, I. V. *Tetrahedron* **1991**, *47*, 895.
34. Matsubara, Y.; Fujimoto, K.; Maekawa, H.; Nishiguchi, I. *J. Jpn. Oil Chem. Soc.* **1996**, *45*, 21.
35. (a) Levine, R.; Stephens, J. R. *J. Am. Chem. Soc.* **1950**, *72*, 1642. (b) Rothenberg, G.; Sasson, Y. *Tetrahedron* **1996**, *52*, 13641.
36. Buttero, P. D.; Maiorana, S. *Gazz. Chim. Ital.* **1973**, *103*, 809.
37. Booth, H.; Saunders, B. C. *Chem. Ind.* **1950**, 824.
38. Hoke, D. I.; Robins, R. D. *J. Poly. Sci.* **1972**, *10*, 3311.
39. Goldberg, O.; Luini, A.; Teichberg, V. I. *Tetrahedron Lett.* **1980**, *21*, 2355.

40. Kikuchi, T.; Kadota, S.; Suehara, H.; Namba, T. *Tetrahedron Lett.* **1981**, *22*, 465.
41. Zhang, L.-C.; Imafuku, K. *J. Heterocycl. Chem.* **1991**, *28*, 717.
42. Ferrayoli, C. G.; Palacios, S. M.; Alonso, R. A. *J. Chem. Soc., Perkin Trans. 1* **1995**, 1635.
43. Neudeck, H. K. *Monatsh. Chem.* **1996**, *127*, 185.
44. (a) Parlow, J. J.; Mahoney, M. D. *Pestic. Sci.* **1995**, *44*, 137. (b) Parlow, J. J.; Mahoney, M. D. *Pestic. Sci.* **1996**, *46*, 227.
45. Hanano, T.; Adachi, K.; Aoki, Y.; Morimoto, H.; Naka, Y.; Hisadome, M.; Fukuda, T.; Sumichika, H. *Bioorg. Med. Chem. Lett.* **2000**, *10*, 881.
46. Hartmann, R. W.; Reichert, M. *Arch. Pharm. Pharm. Med. Chem.* **2000**, *333*, 145.
47. Hailes, H. C.; Isaac, B.; Javaid, M. H. *Tetrahedron* **2001**, *57*, 10329.
48. Dyker, G.; Kadzimirsz, D. *Eur. J. Org. Chem.* **2003**, 3167.
49. Breuer, S. W. *J. Labelled Cpd. Radiopharm.* **2000**, *43*, 283.
50. Ohloff, G.; Vial, C.; Wolf, H. R.; Job, K.; Jégou, E.; Polonsky, J.; Lederer, E. *Helv. Chim. Acta* **1980**, *63*, 1932.
51. Musa, A.; Sridharan, B.; Lee, H.; Mattern, D. L. *J. Org. Chem.* **1996**, *61*, 5481.
52. Matsumoto, T.; Takeda, Y.; Usui, S.; Imai, S. *Chem. Pharm. Bull.* **1996**, *44*, 530.
53. Nussbaumer, C.; Fráter, G.; Kraft, P. *Helv. Chim. Acta* **1999**, *82*, 1016.
54. Storm, J. P.; Andersson, C.-M. *J. Org. Chem.* **2000**, *65*, 5264.
55. Göksu, S.; Kazaz, C.; Sütbeyaz, Y.; Secen, H. *Helv. Chim. Acta* **2003**, *86*, 3310.
56. Larionov, O. V.; Kozhushkov, S. I.; de Meijere, A. *Synthesis* **2005**, 158.
57. Bjørsvik, H.-R.; Norman, K. *Org. Process Res. Dev.* **1999**, *3*, 341.
58. Boyer, W. M.; Bernstein, R. B.; Brown, T. L.; Dibeler, V. H. *J. Am. Chem. Soc.* **1951**, *73*, 770.
59. (a) Bennasar, M.-L.; Roca, T.; Monerris, M.; Juan, C.; Bosch, J. *Tetrahedron* **2002**, *58*, 8099. (b) Bennasar, M.-L.; Zulaica, E.; Alonso, Y.; Bosch, J. *Tetrahedron: Asymmetry* **2003**, *14*, 469.
60. (a) Xiao, J.; Yuan, G.; Huang, W.; Chan, A. S. C.; Lee, D. K.-L. *J. Org. Chem.* **2000**, *65*, 5506. (b) Xiao, J.-H.; Yuan, G.; Huang, W.-Q.; Du, W.-H.; Wang, B.-H.; Li, Z.-F. *Chin. J. Chem.* **2001**, *19*, 116. (c) Xiao, J.; Yuan, G.; Huang, W.; Lum, J. H. K.; Lee, D. K.-L.; Chan, A. S. C. *J. Chin. Chem. Soc.* **2001**, *48*, 929. (d) Yuan, G.; Xiao, J.; Huang, W.; Tang, F.; Zhou, Y. *Arch. Pharm. Res.* **2002**, *25*, 585.
61. Flores, A. F. C.; Zanatta, N.; Rosa, A.; Brondani, S.; Martins, M. A. P. *Tetrahedron Lett.* **2002**, *43*, 5005.
62. Scheuring, J.; Kugelbrey, K.; Weinkauf, S.; Cushman, M.; Bacher, A.; Fischer, M. *J. Org. Chem.* **2001**, *66*, 3811.
63. Fischer, D. S.; Woo, L. W. L.; Mahon, M. F.; Purohit, A.; Reed, M. J.; Potter, B. V. L. *Bioorg. Med. Chem.* **2003**, *11*, 1685.
64. (a) Fuson, R. C.; Tullock, C. W. *J. Am. Chem. Soc.* **1934**, *56*, 1638. (b) Schaeffer, H. F. *J. Chem. Ed.* **1942**, *19*, 15. (c) Gillis, B. T. *J. Org. Chem.* **1959**, *24*, 1027. (d) Kallianos, A. G.; Mold, J. D. *Anal. Chem.* **1962**, *34*, 1174.
65. (a) Doré, M.; Goichon, J. *Water Res.* **1980**, *14*, 657. (b) Sastry, C. S. P.; Naidu, P. Y. *Talanta* **1998**, *46*, 1357.
66. [R] Gribble, G. W. *Prog. Chem. Org. Nat. Prod.* **1996**, *68*, 1.
67. Laturnus, F.; Haselmann, K. F.; Borch, T.; Grøn, C. *Biogeochem.* **2002**, *60*, 121.
68. Khalil, M. A. K.; Rasmussen, R. A.; French, J. R. J.; Holt, J. A. *J. Geophys. Res.* **1990**, *95*, 3619.
69. Kigoshi, H.; Ichino, T.; Takada, N.; Suenaga, K.; Yamada, A.; Yamada, K.; Uemura, D. *Chem. Lett.* **2004**, *33*, 98.
70. (a) Boyce, S. D.; Hornig, J. F. *Environ. Sci. Technol.* **1983**, *17*, 202. (b) Boyce, S. D.; Hornig, J. F. *Water Res.* **1983**, *17*, 685.
71. Cheh, A. M. *Mutation Res.* **1986**, *169*, 1.
72. (a) Walter, B.; Ballschmiter, K. *Fresenius J. Anal. Chem.* **1991**, *341*, 564. (b) Walter, B.; Ballschmiter, K. *Fresenius J. Anal. Chem.* **1992**, *342*, 827. (c) Urhahn, T.; Ballschmiter, K. *Chemosphere* **1998**, *37*, 1017.
73. Boyce, S. D.; Barefoot, A. C.; Hornig, J. F. *J. Labelled Cpd. Radiopharm.* **1983**, *20*, 243.
74. Sandborn, L. T.; Bousquet, E. W. *Org. Syn. Coll. Vol. I* **1932**, 526.
75. Newman, M. S.; Holmes, H. L. *Org. Syn. Coll. Vol. II* **1943**, 428.
76. Smith, L. I.; Prichard, W. W.; Spillane, L. J. *Org. Syn. Coll. Vol. III* **1955**, 302.
77. Smith, W. T.; McLeod, G. L. *Org. Syn. Coll. Vol. IV* **1963**, 345.
78. Staunton, J.; Eisenbraun, E. J. *Org. Syn. Coll. Vol. V* **1973**, **8**.
79. Bailey, D. M.; Johnson, R. E.; Albertson, N. F. *Org. Syn. Coll. Vol. VI* **1988**, 618.

80. Levin, M. D.; Kaszynski, P.; Michl, J. *Org. Syn.* **1999**, *77*, 249.

Gordon W. Gribble

7.4 Hunsdiecker reaction

7.4.1 Description

The Hunsdiecker reaction is the reaction of silver carboxylate **1** with bromine to give bromide **2** at elevated temperature.[1–4] The reaction also works for making chlorides and iodides. It is also known as the Hunsdiecker–Borodin reaction.

$$\underset{\mathbf{1}}{RCO_2Ag} \xrightarrow[\Delta]{Br_2} \underset{\mathbf{2}}{R{-}Br} + CO_2 + AgBr$$

The Hunsdiecker reaction is closely related to the Simonini reaction.[5] As a matter of fact, the first half of the Simonini reaction *is* the Hunsdiecker reaction. Therefore, treatment of one equivalent of silver carboxylate **1** with iodine gives alkyl iodide **3**. In the presence of another equivalent of silver carboxylate **1**, an S_N2 displacement reaction takes place, affording ester **4**. Since there are so many practical ways to make esters, the Simonini reaction is no longer widely used.

$$\underset{\mathbf{1}}{RCO_2Ag} \xrightarrow{I_2} \underset{\mathbf{3}}{R{-}I} \xrightarrow[\mathbf{1}]{RCO_2Ag} \underset{\mathbf{4}}{RCO_2R}$$

7.4.2 Historical Perspective

Aleksandr Porfirevič Borodin (1833–1887), the illegitimate son of an elderly prince, was born in St. Petersburg. He studied chemistry under Nikolai Nikolaevie Zinin (1812–1880) and succeeded him as Professor of Chemistry at the St. Petersburg Academy of Medicine and Surgery in 1864. Borodin was also an accomplished composer and kept a piano outside his laboratory. He is now best known for his musical masterpiece, opera Prince Egor. Borodin prepared methyl bromide from silver acetate and bromine in 1861,[6] but another eighty years elapsed before Cläre and Heinz Hunsdiecker converted Borodin's synthesis into a general method[7,8]—the Hunsdiecker or Hunsdiecker–Borodin reaction in 1939. Cläre Hunsdiecker was born in 1903 and educated in Cologne. She developed the bromination of silver carboxylate alongside her husband, Heinz.

7.4.3 Mechanism

The first step of the Hunsdiecker reaction is quite straightforward. The reaction between silver carboxylate **1** and bromine gives rise to insoluble silver bromide along with acyl hypobromite **3**. The unstable acyl hypobromite **3** undergoes a homolytic cleavage of the O–Br bond to provide carboxyl radical **4**. Carboxyl radical **4** then decomposes *via* radical decarboxylation to release carbon dioxide and alkyl radical **5**, which subsequently reacts with another molecule of acyl hypobromite **3** to deliver alkyl bromide **2**, along with regeneration of carboxyl radical **4**.[9–12] Because of the radical pathway, chirality is often lost for the chiral carbon atom immediately adjacent to the carboxylic acid.

X–X, 1 ⟶ AgX + 3 —homolytic cleavage⟶

X· + 4 ⟶ CO_2↑ + R· (5) —3⟶ R–X (2) + 4

7.4.4 Variations and Improvements

In addition to Ag, many other metals (Hg, Pb, Tl, and Mn) can be used in the Hunsdiecker reaction.[13–20] The Cristol–Firth modification is a one-pot reaction using excess red HgO and one equivalent of halogen.[13] The advantage of the Cristol–Firth modification is that the mercury carboxylate does not need to be isolated and purified, unlike the Hunsdiecker reaction using the silver carboxylate. Therefore, as depicted by transformations **6**→**7**[14] and **8**→**9**,[15] the Cristol–Firth modification is operationally more straightforward and more widely used than the classical Hunsdiecker reaction using silver carboxylate **1**.

Cl–C₄H₆–CO_2H (**6**) —HgO, Br_2, Δ / CCl_4, dark, 35–46%⟶ Cl–C₄H₆–Br (**7**)

Cl–C₆H₄–CO_2H (**8**) —HgO, Br_2 / *hv*, 80%⟶ Cl–C₆H₄–Br (**9**)

Another valuable alternative to the Hunsdiecker reaction is the Kochi modification,[3] which employs a Pb(IV) reagent. It is a one-carbon oxidative degradation of carboxylic acids and is suitable for synthesis of secondary and tertiary chlorides as exemplified by the transformation of carboxylic acid **10** to cyclobutyl chloride (**11**) mediated by lead tetra-acetate (LTA).[16] Further improvement of the Kochi modification employs LTA under milder conditions where *N*-chlorosuccinimide (NCS) is used as chlorinating agent and a mixture of DMF and HOAc as solvent.[17]

$Pb(OAc)_4$, LiCl, 100%: **10** → **11** + CO_2 + $LiPb(OAc)_3$ + HOAc

In addition to mercury and lead, thallium(I) carboxylates also work for the Hunsdiecker reaction.[18,19] For instance, treatment of thallium(I) carboxylate **12** with bromine provided bromide **13** in 86% yield.[18] In one occasion, $Mn(OAc)_2$-catalyzed Hunsdiecker reaction provided a facile entry to α-(dibromomethyl)-benzenemethanol.[20]

CO_2H → ($Tl(I)_2OEt$, petroleum ether, quant.) → **12** ($CO_2Tl(I)$) → (Br_2, CCl_4, reflux, 4 h, 86%) → **13** (Br)

The third modification of the Hunsdiecker reaction is the so-called Suárez modification,[21] where the steroidal acids were treated with hypervalent iodine reagent in CCl_4 to prepare steroidal chloride. The Suárez modification also works for bromination using iodobenzene diacetate, bromine, and CH_2Br_2 as the solvent under irradiation as exemplified by transformations **14**→**15**.[22] Unlike many variations described before, the Suárez modification tolerates a variety of functional groups.

14 (CO_2H) → ($PhI(OAc)_2$, Br_2, CH_2Br_2, 2 × 200 W, tungsten lamps, 4 h, 66%) → **15** (Br)

The fourth modification of the Hunsdiecker reaction, pioneered by Barton, is the use of *t*-butyl hypoiodide.[23] Thus, acid **16** was treated with *t*-butyl hypoiodide to give the acyl hypoiodite, which underwent white-light photolysis at room temperature to give iodide **17**. The reaction works for primary, secondary, and tertiary acids.

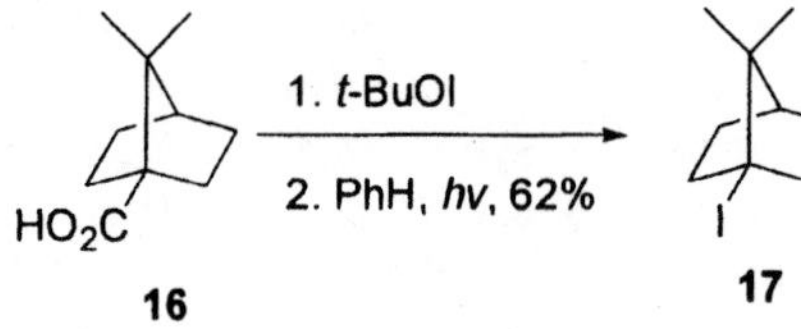

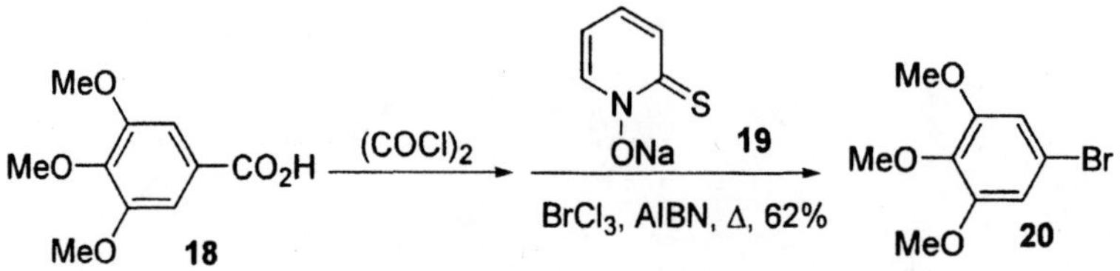

The last, but certainly not the least, is the Barton modification to the Hunsdiecker reaction.[24–26] It involves decomposition of thiohydroxamate esters in halogen donor solvents such as CCl_4, $BrCCl_3$, CHI_3, or CH_2I_2 promoted by a source of radical initiation, which could be radical initiator (e.g., **18**→**20**),[24] thermal (e.g., **21**→**22**),[25] or photolytic[26] conditions. The Barton modification is highly compatible with most functional groups. For example, under photolytic conditions, acid **23** was converted to acid chloride **24**, which, without isolation, was treated with the sodium salt of *H*-hydroxypyridine-2-thione (**19**) with bromotrichloromethane as solvent to give alkyl bromide **25** in 90% yield.[26]

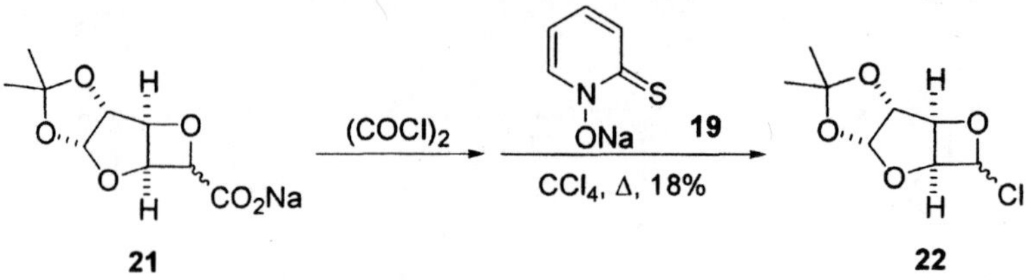

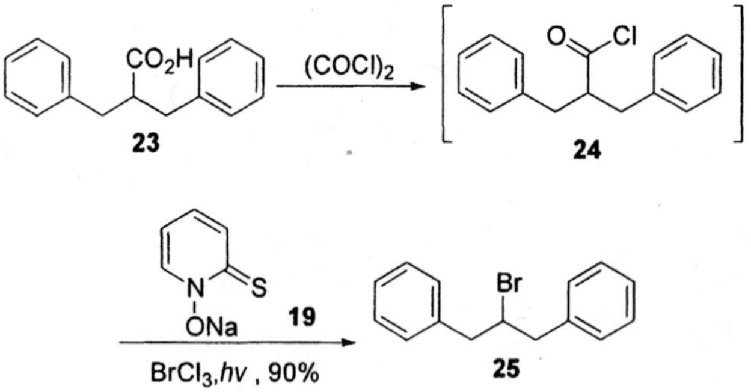

7.4.5 Synthetic Utility

The classical Hunsdiecker conditions using Ag, and modifications using metals such as Hg, Tl, Pb, and Mn(II), are not very synthetically useful because of the use of toxic metals, requirement of high temperature, and poor yields. As a consequence, many variants of "greener" chemistry have been developed to replace heavy metals. In addition to Barton's radical approaches, Roy *et al.* developed a metal-free Hunsdiecker reaction where the acid was treated with *N*-bromosuccinamide (NBS) and a catalytic amount of LiOAc[27] or the phase transfer catalyst (PTC) tetrabutylammonium trifluoroacetate (TBATFA).[28–30] As shown below, cinnamic acid **26** was converted to β-bromostyrene **27** in almost quantitative yield.[28] The authors also found that a mixture of 93:7 MeCN/H_2O was also a good solvent for the metal-free Hunsdiecker reaction.[29] In place of TBATFA, another phase transfer catalyst "Select flur" was found to be an efficient catalyst for the metal-free Hunsdiecker reaction as well (e.g., **29**→**30**).[31]

CO2H; NBS, n-$Bu_4N^+CF_3CO_2^-$; $ClCH_2CH_2Cl$, 96%; MeO **26** → Br; MeO **27**

CO2H; Cl, N+, $2\ \bar{B}F_4$, +N, F "Select fluor"; KBr, CH_3CN, 82%; HO **29** → Br; HO **30**

The aforementioned reaction was extended to a nitro-Hunsdiecker reaction to make nitrostyrenes and nitroarenes from unsaturated carboxylic acids such as **31**→**32**.[32] This is a rare example of nitro-olefin preparation.

CO2H; HNO_3, cat. AIBN; CH_3CN, 50 °C, 29 h, 62%; O, O **31** → NO_2; O, O **32**

In an effort to develop a green Hunsdiecker reaction, synthesis of β-bromostyrenes was accomplished from the reaction of α,β-unsaturated aromatic carboxylic acids with KBr and H_2O_2 catalyzed by $Na_2MoO_4 \cdot 2H_2O$ in aqueous medium.[33]

$Na_2MoO_4 \cdot 2H_2O$, H_2O_2; KBr, H_2O, 65–85%

R = H, 4-OMe, 4-Cl, 2-Me, or 2-Cl

33 → **34**

Recently, it was found that microwave-induced Hunsdiecker reaction afforded stereoselective synthesis of (*E*)-β-arylvinyl halides as exemplified by transformation **35**→**36**.[34,35]

NBS, LiOAc; MeCN-H_2O (93:7); MW 2 min. 68%

35 → **36**

7.4.6 *Experimental*

HgO, Br_2, Δ; CCl_4, dark, 35–46%

6 → **7**

Preparation of 1-bromo-3-chlorocyclobutane[14]

In a 1-L, three-necked, round-bottomed flask, wrapped with aluminum foil to exclude light, and equipped with a mechanical stirrer, a reflux condenser, and an addition funnel, is suspended 37 g (0.17 mole) of red mercury(II) oxide in 330 mL of carbon tetrachloride. To the flask is added 30.0 g. (0.227 mole) of 3-chlorocyclobutanecarboxylic acid (**6**). With stirring, the mixture is heated to reflux before a solution of 40 g (0.25 mole) of bromine in 180 mL of carbon tetrachloride is added dropwise, but as fast as possible (4–7 minutes) without loss of bromine from the condenser. After a short induction period, carbon dioxide is evolved at a rate of 150–200 bubbles per minute. The solution is allowed to reflux for 25–30 minutes, until the rate of carbon dioxide evolution slows to about 5 bubbles per minute. The mixture is cooled in an ice bath, and the precipitate is removed by filtration. The residue is washed with carbon tetrachloride, and the filtrates are combined. The solvent is removed by distillation using a modified Claisen distillation apparatus with a 6-cm Vigreux column; vacuum distillation of the residual oil gives 13–17 g (35–46%) of 1-bromo-3-chlorocyclobutane (7), bp 67–72 °C (45 mm.).

7.4.7 References

1. [R] Johnson, R. G.; Ingham, R. K. *Chem. Rev.* **1956**, *56*, 219–269.
2. [R] Wilson, C. V. *Org. React.* **1957**, *9*, 341–381.
3. [R] Sheldon, R. A.; Kochi, J. K. *Org. React.* **1972**, *19*, 326–387.
4. [R] Crich, D. In *Comprehensive Organic Synthesis;* Trost, B. M.; Steven, V. L., Eds.; Pergamon, **1991**, *Vol. 7*, 723–734.
5. Simonini, A. *Monatsh. Chem.* **1892**, *13*, 320.
6. Borodin, A. *Justus Liebigs Ann. Chem.* **1861**, *119*, 121.
7. Hunsdiecker, H.; Hunsdiecker, C.; Vogt, E. U.S. Patent 2,176,181 **(1939)**.
8. Hunsdiecker, H.; Hunsdiecker, C. *Ber. Dtsch. Chem. Ges.* **1942**, *75B*, 291.
9. Cristol, S. J.; Douglass, J. R.; Firth, W. C., Jr.; Krall, R. E. *J. Am. Chem. Soc.* **1961**, *82*, 1829.
10. Jenning, P. W.; Ziebarth, T. D. *J. Org. Chem.* **1969**, *34*, 3216.
11. Bunce, N. J.; Urban, L. O. *Can. J. Chem.* **1971**, *49*, 821.
12. Cason, J.; Walba, D. M. *J. Org. Chem.* **1972**, *37*, 669.
13. Cristol, S. J.; Firth, W. C., Jr. *J. Org. Chem.* **1961**, *26*, 280.
14. Lampman, G. M.; Aumiller, J. C. *Org, Synth.* **1988**, *Coll. Vol. 6*, 179.
15. Myers, A. I.; Fleming, M. P. *J. Org. Chem.* **1979**, *44*, 3405.
16. Kochi, J. K. *J. Am. Chem. Soc.* **1965**, *87*, 2500.
17. Becher, K. B.; Geisel, M.; Grob, C. A.; Kuhnen, F. *Synthesis* **1973**, 493.
18. McKillop, A.; Bromley, D.; Taylor, E. C. *J. Org. Chem.* **1969**, *34*, 1172.
19. Cambie, R. C.; Hayward, R. C.; Jurlina, J. L.; Rutedge, P. S.; Woodgate, P. D. *J. Chem. Soc., Perkin Trans. 1* **1981**, 2608.
20. Chowdhury, S.; Roy, S. *Tetrahedron Lett.* **1996**, *37*, 2623.
21. Concepcion, J. I.; Francisco, C. G.; Freire, R.; Hernadez, R.; Salazar, J. A.; Suárez, E. *J. Org. Chem.* **1986**, *51*, 402.
22. Camps, P.; Lukach, A. E.; Pujol, X.; Vazquez, S. *Tetrahedron* **2000**, *56*, 2703.
23. Barton, D. H. R.; Faro, H. P.; Serebryakov, E. P.; Woolsey, N. F. *J. Chem. Soc.* **1965**, 2438.
24. Barton, D. H. R.; Lacher, B.; Zard, S. Z. *Tetrahedron* **1987**, *43*, 4321.
25. Fleet, G. W.; Son, J. C.; Peach, J. M.; Hamor, T. A. *Tetrahedron Lett.* **1988**, *29*, 1449.
26. Barton, D. H. R.; Crich, D.; Motherwell, W. B. *Tetrahedron* **1985**, *41*, 3901.
27. Chowdhury, S.; Roy, S. *J. Org. Chem.* **1997**, *62*, 199.
28. Naskar, D.; Chowdhury, S.; Roy, S. *Tetrahedron Lett.* **1998**, *39*, 699.
29. Naskar, D.; Roy, S. *Tetrahedron* **2000**, *56*, 1369.
30. Das, J. P.; Roy, S. *J. Org. Chem.* **2002**, *67*, 7861.
31. Das, J. P.; Ye, C.; Shreeve, J. M. *J. Org. Chem.* **2004**, *69*, 8561.
32. Das, J. P.; Sinha, P.; Roy, S. *Org. Lett.* **2002**, *4*, 3055.
33. Sinha, J.; Layek, S.; Mandal, G. C.; Bhattacharjee, M. *Chem. Commun.* **2001**, 1916.
34. Kuang, C.; Yang, Q.; Senboku, H.; Tokuda, M. *Synthesis*, **2005**, 1319.
35. Kuang, C.; Senboku, H.; Tokuda, M. *Synlett* **2005**, 1439.

Jie Jack Li

7.5 Japp–Klingemann Hydrazone Synthesis

7.5.1 Description

The reaction between a compound with an active methinyl carbon **1** and a diazonium salt **2** is assumed to involve the formation of an unstable azo compound that undergoes hydrolytic cleavage to form a hydrazone **3**.

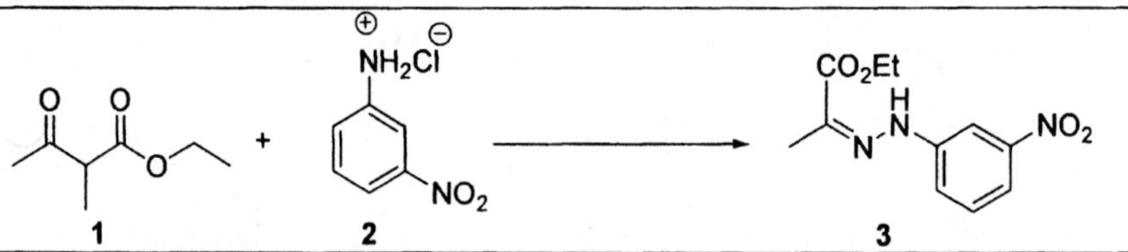

7.5.2 Historical Perspective

In 1887, in an attempt to prepare the azo ester **4** by coupling benzenediazonium chloride **2** with ethyl 2-methylacetoacetate **1**, Japp and Klingemann obtained a product which was recognized as the phenylhydrazone of ethyl pyruvate **3**.[1–4]

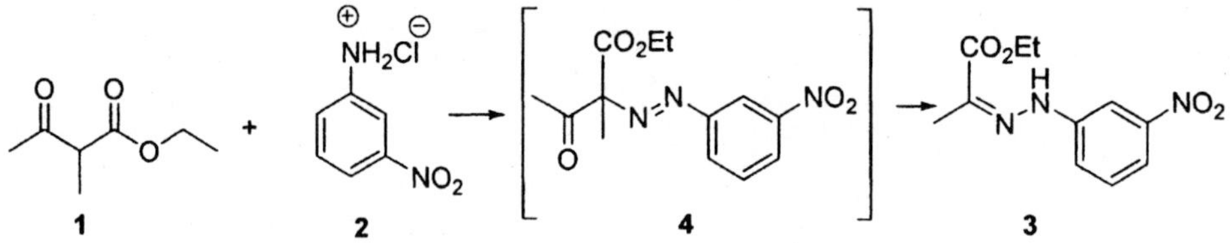

7.5.3 Mechanism

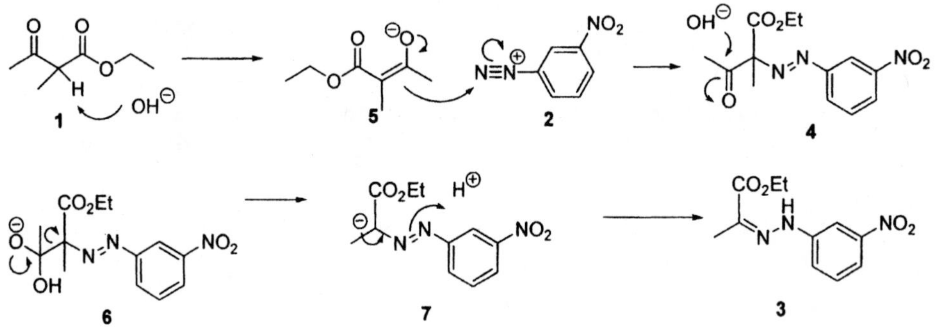

Deprotonation of acetoacetate **1** with base provides the active methinyl carbanion **5**, which couples with diazonium salt **2** to form the intermediate **4**. The acetyl group is then cleaved by base to form **7**, which upon protonation yields the hydrazone **3**.[5]

7.5.4 Variations

Japp and Klingemann found that benzenediazonium salt **2** coupled with ethyl methylacetoacetate **1** in alkaline solution to give the phenylhydrazone of ethyl pyruvate **3** with elimination of acetic acid. However, if the ketonic ester was first hydrolyzed with alkali and then coupled, the product was monophenylhydrazone **10**, with carbon dioxide being eliminated rather than acetic acid.[6]

$C_6H_5N_2^{\oplus}$

8 **9** **10**

The requirement for the occurrence of the Japp–Klingemann reaction is the presence of a hydrogen atom of sufficient activity to permit the coupling with the diazonium salt. A methinyl group in the α-position of a pyridyl acetic acid **11** is reactive enough to participate in the Japp–Klingemann process.[7]

0 °C, NaOAc, 82%

11 **12** **13**

0 °C, NaOAc, 7%

14 **12** **15**

When the Japp–Klingemann reaction is applied to a cyclic β-keto ester **16**, the ring is opened in the second stage of the process to yield **18**.[8] As in the reactions of acyclic β-keto esters, the reaction takes the decarboxylation course if the ester is saponified before the coupling. Thus the monophenylhydrazone of cyclohexane-1,2-dione **20** is obtained from ethyl cyclohexanone-2-carboxylate **19**.[9]

CO_2Et =O + NH_2Cl 0 °C, KOH / 50% → HO_2C NH N CO_2Et

16 17 18

O CO_2H + NH_2Cl 0 °C, NaOAc / EtOH/H_2O → O NNHPh

19 17 20

7.5.5 Synthetic Utility

The Japp–Klingemann reaction has been used extensively in the synthesis of indoles. 5-Methoxy-1*H*-indole-2-carboxylic acid **25** has been prepared through a Japp–Klingemann–Fischer indole process in 81% overall yield from *p*-anisidine **21**.[10]

α-Amino acids **27** have been prepared by reductive cleavage of phenylhydrazones **26** of α-keto acids and esters **1**. These hydrazones can be readily prepared by Japp–Klingemann reaction conditions. Various reductive conditions have been employed including: $Pd/C/H_2$, $Pd(OH)_2/H_2$, PtO/H_2, $Zn/HgCl_2/EtOH$, $Zn/CaCl_2/EtOH$, and Zn/AcOH.[6,11]

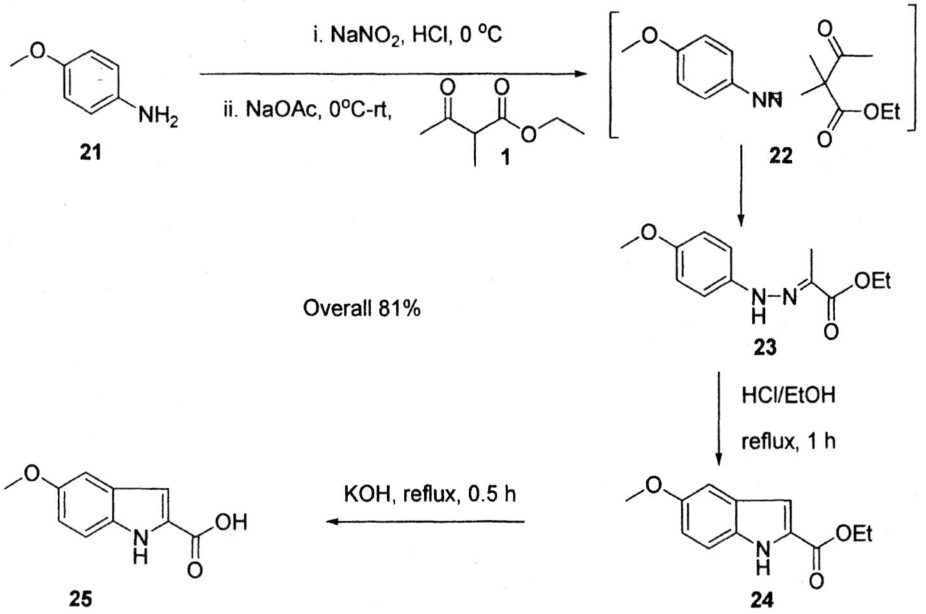

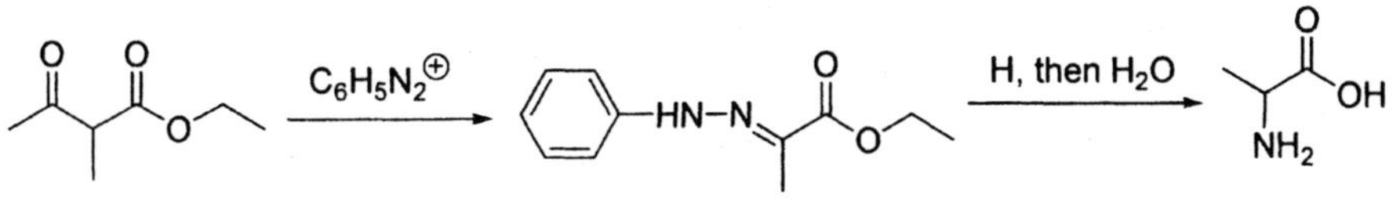

7.5.6 Experimental

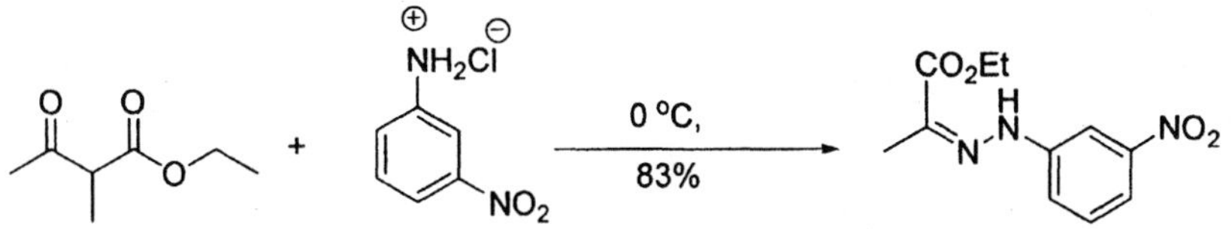

Preparation of ethyl pyruvate *o*-nitrophenylhydrazone (3)[5]

To an ice-cold solution of 20.5 g (0.14 mol) of ethyl 2-methylacetoacetate (**1**) in 150 mL of ethanol is added 51 mL of 50% aqueous potassium hydroxide. This mixture is then diluted with 300 mL of ice water; and the cold diazonium salt **2** solution, prepared from

20.0 g (0.14 mol) of *o*-nitroaniline, 60 mL of concentrated hydrochloride acid, 90 mL of water, and 10.5 g of the sodium nitrite, is rapidly run in with stirring. Stirring is continued for 5 minutes. The separated ethyl pyruvate *o*-nitrophenylhydrazone (**3**) is then collected by filtration and recrystallized from ethanol (30 g, 83%). mp 106 °C.

7.5.7 References

1. Japp, F. R.; Klingemann, F. *Ber.* **1887**, *20*, 2942.
2. Japp, F. R.; Klingemann, F. *Ber.* **1887**, *20*, 3284.
3. Japp, F. R.; Klingemann, F. *Ber.* **1887**, *20*, 3398.
4. Japp, F. R.; Klingemann, F. *Ber.* **1888**, *21*, 549.
5. [R] Phillips, R. R. *Organic Reactions*; John Wiley and Sons, Inc.: Hoboken, NJ, 1959; pp 143.
6. Japp, F. R.; Klingemann, F. *J. Chem. Soc.* **1888**, *53*, 519.
7. Frank, R. L.; Phillips, R. R. *J. Am. Chem. Soc.* **1949**, 2804.
8. Manske, R. H. F.; Robinson, R. *J. Chem. Soc.* **1927**, 240.
9. Linstead, R. P.; Wang, A. B.-L. *J. Chem. Soc.* **1937**, 807.
10. Bessard, Y. *Org. Proc. Res. Dev.* **1998**, *2*, 214.
11. Khan, N. H.; Kidwai, A. R. *J. org. Chem.* **1973**, *38*, 822.

Jin Li

7.6 Krapcho Decarboxylation

7.6.1 Description

The Krapcho decarboxylation is the nucleophilic decarboxylation of malonate esters, β-ketoesters, α-cyanoesters, α-sulfonylesters, and related compounds. The reaction is done in dipolar aprotic solvents in the presence of salt and/or water at high temperatures.[1]

NaCN
DMSO, Δ

7.6.2 Historical Perspective

In 1967, Krapcho reported the utility of a method for decarboxylating geminal diesters using a salt such as NaCN dissolved in a polar aprotic solvent like dimethyl sulfoxide (DMSO). The reaction was discovered when an attempt to effect the following substitution reaction with **1** resulted in decarbomethoxylation to give **2**.[2] Shortly thereafter, a comprehensive study of decarbalkoxylations using mono- and di-substituted malonates was published.[3]

TsO, TsO, H, H, CO_2CH_3, CO_2CH_3 **1** — KCN, DMSO, 90 °C → NC, NC, H, H, CO_2CH_3, H **2**

Several reviews have been written which cover the history of the Krapcho reaction through 1982.[1,4] Further research in this area revealed the application of the decarboxylation method to compounds such as β-ketoesters, malonate esters, α-cyanoester, and α-sulfonylesters. The classical method for decarboxylation of these compounds usually involves acidic or basic hydrolysis, followed by thermal decarboxylation. Unfortunately, compounds containing acid or base sensitive functional groups are not compatible with these methods. Modern Krapcho conditions have replaced cyanide with less toxic halide anions. Additionally, several decarboxylations have occurred in the absence of salt.[4]

Many electron withdrawing functional groups (ketones, esters, nitriles, and sulfonyls), when positioned appropriately, enhance the efficiency of the decarboxylation process. Additionally, prior to decarboxylation, these groups lower the pK_a value of the central α-hydrogen, allowing deprotonation by weak and inexpensive bases to give relatively stable anions. These anions have found significant utility in several carbon–carbon bond forming processes. The need to remove the ester group following these transformations under mild, neutral conditions is largely responsible for the importance placed on the Krapcho reaction by modern synthetic chemists.

7.6.3 Mechanism

The mechanism of the Krapcho decarboxylation has been thoroughly examined in the literature.[4,5] Originating primarily in the Krapcho group, these studies reveal that the mechanism largely depends upon several factors. These include the absence or presence of a salt, the identity of the salt, and the structural features of the substrate to be decarboxylated. Based on the above, three general mechanisms have been observed. The first is the traditionally accepted mechanism for the Krapcho decarboxylation. As shown below for the malonate derivative, a nucleophilic ion, in a polar aprotic solvent, performs an S_N2 reaction on the methyl group of the ester. This produces a carboxylate ion which undergoes decarboxylation to produce an enolate with evolution of carbon dioxide. The concertedness of the decarboxylation step has been the subject of debate. Regardless, protonation of the enolate generates the ester. This mechanism is referred to as the $B_{AL}2$ mechanism in the literature and, generally speaking, is characteristic for decarboxylation of sterically hindered malonate derivatives.

RO, R', CH_3, :X⁻ → RO, R' → RO, R' + CO_2 —H_2O→ RO, R'

The other major mechanism results from an initial attack at the carbonyl to form a tetrahedral intermediate. Collapse of the tetrahedral intermediate results in expulsion of the enolate. The resulting acyl cyanide is rapidly hydrolyzed to give cyanide, carbon dioxide, and an alcohol. This mechanism predominates when cyanide salts are used and when less substitution is present at the central carbon of the malonate derivative.[6]

RO, R', R", ⁻CN → RO, R', CN, R" → RO, R' + NC(O)OR" —H_2O→ RO, R'

⁻CN + R"OH + CO_2 ←H_2O— NC(O)OR"

A third and final mechanism has been considered, whereby the reaction occurs via simple hydrolysis of the ester, resulting ultimately in thermal decarboxylation. This mechanistic pathway is favored in the absence of a salt, but cannot be ruled out as occurring even in the presence of a salt.[7]

7.6.4 Synthetic Utility

The Krapcho decarboxylation has found wide application in organic synthesis.[8–11] Its primary utility has been to convert malonate or acetoacetate derivatives to the corresponding carbonyls.[12–16] As malonates have found tremendous use in many areas of synthesis, the Krapcho has also found wide utility.

While investigating a novel rearrangement of isoxazoline-5-spirocycloalkane compounds, Brandi and co-workers utilized conditions which allowed the Krapcho decarboxylation to occur as part of a tandem process.[17] Following the cyclization of the nitrile oxide **3** with methylenecyclopropane (**4**), the isoxazoline **5** undergoes a thermal rearrangement to produce **8**. Mechanistically, homolytic cleavage of the N–O bond produces diradical **6**. Radical rearrangement provides **7** which recombines to form the six-membered ring **8**. Following acylation and tautomerization to give **9**, decarboxylation occurs to provide quinolizindione **10** in 31% overall yield for the three step process. Similar results were obtained in the synthesis of a related indolizindione.

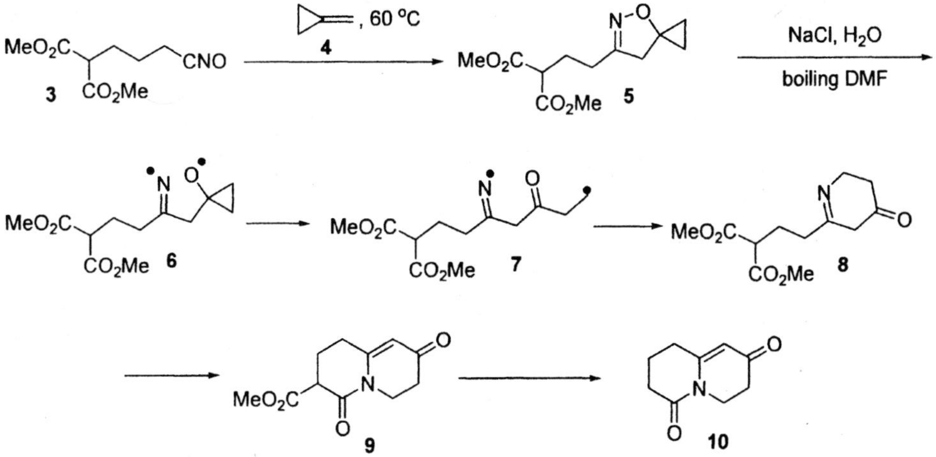

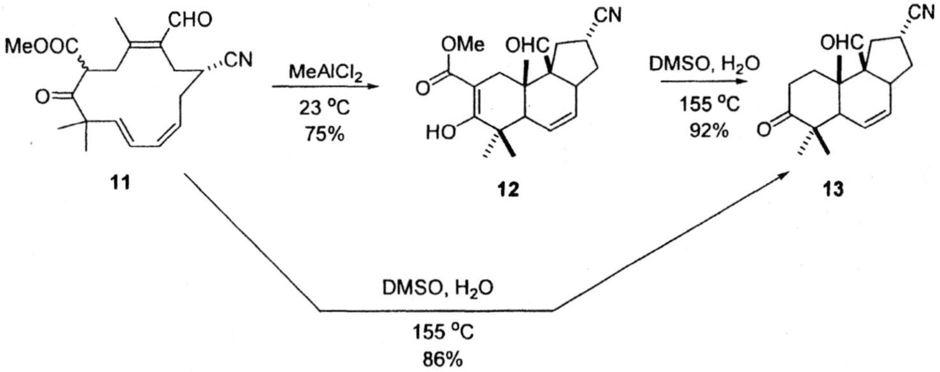

The Krapcho decarboxylation was also utilized as a beneficial side reaction in the key step in the Deslongchamps synthesis of (+)-maritimol.[18] In this asymmetric synthesis, the Lewis acid mediated intramolecular Diels–Alder reaction of **11** to produce **12** was followed by thermal decarboxylation to yield **13**. The thermal Diels–Alder reaction of **11** was then tested and found to proceed in slightly higher yield and resulted in decarboxylation occurring in the same pot. Remarkably, the complete diastereo- and enantiocontrol in this reaction is directed by the remote nitrile stereocenter. Additional examples exist in which the Krapcho decarboxylation occurs as the second step of a tandem process.[19]

An important utility of the Krapcho reaction is not necessarily the decarboxylation step itself. Rather, the fact that the decarboxylation can be made to occur allows several reactions that require malonates or their derivatives to find general synthetic utility. For example, elegant work in the area of rhodium carbenoid chemistry relies on diazomalonates to generate the carbenoid. As utilized by Wee,[20] diazomalonate **14** is treated with Rh_2OAc_4 to generate the carbenoid which inserts into the stereochemically defined tertiary C—H bond. The reaction proceeds exclusively with retention of configuration in forming the new quaternary carbon stereocenter. Decarboxylation of **15** under Krapcho's conditions provides lactone **16**, a key intermediate in the synthesis of (–)-eburnamonine.

14 → (Rh_2OAc_4, CH_2Cl_2, reflux; 90%) → **15**

15 → (10:1 v/v DMSO:H_2O, NaCl, 110 °C; 84%) → **16**

An additional example which further showcases the Krapcho utility in this regard comes from the Wu group in their efforts toward the synthesis of clavulactone, a compound possessing potent antitumor activity.[21] Seeking a general method for the synthesis of the dolabellane skeleton, troubles were encountered in the intramolecular radical conjugate addition. Principally, the *E*-isomer **17** (R_1 = H, R_2 = CO_2Et) provided the undesired cyclopentane $\mathbf{20_B}$ in excellent yield. It was postulated that the *Z*-isomer **18** would be selective for formation of $\mathbf{20_A}$. However, model studies demonstrated that isomerization to the *E*-olefin occurred under the reaction conditions. Thus, to solve this problem, **19** (R_1 = CO_2Et, R_2 = CO_2Et) was utilized, such that isomerization would not be an issue. As executed, a 2.4:1 ratio of $\mathbf{20_A}$:$\mathbf{20_B}$ was obtained. All that remained was the Krapcho decarboxylation to remove the ethoxycarbonyl providing the separable mixture $\mathbf{21_A}$ and $\mathbf{21_B}$. Though not tremendously efficient, the synthesis of $\mathbf{21_A}$ should prove general for the synthesis of most dolabellanes.

Bu_3SnH, AIBN, benzene, reflux

17 $R_1 = H$; $R_2 = CO_2Et$
18 $R_1 = CO_2Et$; $R_2 = H$
19 $R_1 = CO_2Et$; $R_2 = CO_2Et$

$R_1 = H$; $R_2 = CO_2Et$
20_B 88%
$R_1 = CO_2Et$; $R_2 = CO_2Et$,
$20_A:20_B = 2.4:1$

20_A + 20_B — DMSO, LiCl, H_2O, 190 °C → 21_A + 21_B

While synthesizing steroids containing a fluorine atom in the aromatic A-ring, the Krapcho decarboxylation was used to remove a methoxycarbonyl group whose purpose was to activate the methylene for use in an S_N2 alkylation reaction.[22] Thus, treatment of a mixture of **22** and **23** with anhydrous potassium carbonate resulted in efficient alkylation to produce a mixture of diastereomers of **24**. Decarboxylation was facile at 90 °C to produce **25** and **26**. Finally, a thermolysis reaction in 1,2,4-trichlorobenzene provided an impressive yield (79%) of cycloadducts **27** and **28** in a 53:47 ratio. This and other products from this study have the potential to present interesting biological applications, as other modified human steroids have shown such uses as radiolabeled compounds, chemotherapy agents, and possibly antiangiogenics.

22 + **23** — K_2CO_3, Acetone → **24**

NaCN, DMSO, 90 °C → **25** + **26**

200 °C, 16h

1,2,4-trichlorobenzene

27 + 28

The Krapcho has found significant use as a method which enables some interesting asymmetric conjugate addition chemistry. For example, diorganozinc reagents do not generally undergo conjugate addition reaction to acyclic unsaturated esters. The reactivity of unsaturated malonates is such that they are able to participate in these reactions. As demonstrated by Feringa and co-workers,[23] the asymmetric addition of dimethyl zinc to unsaturated malonate **29** proceeds in a high yield and with excellent enantioselectivity catalyzed by the phosphoramidite ligand **30**. The Krapcho decarboxylation is then utilized to provide methyl ester **32**. The three step homologation of **32** produces **33** setting the stage for a second conjugate addition reaction. Addition of the second methyl group proceeds with high selectivity, once again catalyzed by **30** to produce **34**. The process can be repeated further to produce the 1,3-dimethyl stereoarrays present in a variety of natural products, such as **35**.

29 → $Cu(OTf)_2$, Me_2Zn; **30**, toluene, –60 °C; 90%, 96% ee → **31**

LiCl, H_2O, DMF → **32** → **33**

$Cu(OTf)_2$, Me_2Zn; **30**, toluene, –60 °C; 100%, 97:3 dr → **34**

30

35

A second methodology relies on the use of unsaturated malonates as electrophiles in conjugate addition reactions, thus necessitating removal of the ancillary ester group. Enders and co-workers[24] have shown a general method for preparation of semialdehyde derivatives, such as **40**, which utilized the Krapcho decarboxylation reaction. In this case, conjugate addition of SAMP-hydrazone **36** proceeded with excellent selectivity to

37, resulting in the preparation of **38**. While one method for cleavage of the hydrazone auxiliary relies on ozonozlysis to yield the aldehyde, in this case, racemization at the newly formed stereocenter occurred on purification. Instead, the thioacetal **39** was procured with no epimerization and was followed by decarboxylation to produce **40** in high *ee*.

Ph Ph OMe N N H H **36** + CO_2Me CO_2Me **37** MgI_2, CH_2Cl_2 98%, 93:7 → **38**

HS SH (1.5 eq.) BF_3OEt_2 (2.5-5 eq.) 70% → **39** DMSO, H_2O, NaCl 150 °C 63–81% → **40**

Malonate and related activated methylene compounds have also been used as the nucleophile in conjugate addition/Michael reactions. Taylor and co-workers have developed a new methodology that utilizes (salen)aluminum complexes such as **43** as a catalyst to effect the enantioselective conjugate addition to α,β-unsaturated ketones by a variety of nucleophiles.[25] For example, nitriles, nitroalkanes, hydrazoic acids, and azides have found utility in this reaction. Additionally, cyanoacetate (**42**) has been demonstrated to undergo a highly enantioselective conjugate addition to **41**. The Krapcho decarboxylation is then necessary to produce cyanoketone **44**, an intermediate in the synthesis of enantioenriched 2,4-*cis*-disubstituted piperidine **45**.

41 1. NC CO_2CH_3 **42** (1.2 equiv) (*R*,*R*)-**43** (1.0 mol %) cyclohexane, 23 °C 2. DMSO/H_2O, 130 °C → **44** → **45**

(*R*,*R*)-[salen]Al
43 = [(*R*,*R*)-[salen]Al]$_2$O

An intramolecular conjugate addition of a β-ketoester **46** was utilized in the synthesis of racemic isoclovene.[26] Initially the crystalline tricylic compound **47** was synthesized by a facile intramolecular Michael addition using K_2CO_3 as the base. Removal of the β-keto ester was then completed under Krapcho conditions, producing the tricylic diketone **48**.

K_2CO_3, EtOH, 80%; NaCl, DMSO, 160 °C, 100%

46 **47** **48**

Activated methylene compounds such as dimethyl malonate have found substantial utility in palladium catalyzed allylic substitution reactions. Accordingly, the Krapcho decarboxylation is often used in conjunction with these reactions. As an example, the first total synthesis of enantiomerically pure (–)-wine lactone has utilized the sequence of reactions.[27] First, the allylic substitution reaction of 2-cyclohexen-1-yl acetate (**49**) with alkali sodium dimethylmalonate yielded **51** with high enantioselectivity, as a result of the use of chiral phosphine ligand **50**. The malonate was then subjected to Krapcho decarbomethoxylation using NaCl, H_2O, and DMSO at 160 °C to yield **52**. This reaction has been used similarly following the allylic substitution reaction with other malonate derivatives.[28–30]

$[C_3H_5PdCl]_2$, THF, $NaCH(COOCH_3)_2$, **50**; NaCl, H_2O, DMSO, 160 °C, 74%

49 **51** 91%, 95% *ee* **52**

In the synthesis of an intermediate to the antipsychotic drug ziprasidone, Krapcho conditions produced an unexpected side reaction.[31] In trying to decarboxylate compound **53** to produce **54**, by-products **55** and **56** were formed instead. Their formation was likely a result of the electron withdrawing substituents on the benzene ring. Although these products were unwanted and not part of the overall synthetic pathway, their formation is an interesting potential application of the Krapcho reaction when applied to compounds with electron withdrawing substituents, and could prove useful.

NaCl, H_2O, DMSO, 160 °C

53 **55** + **56**

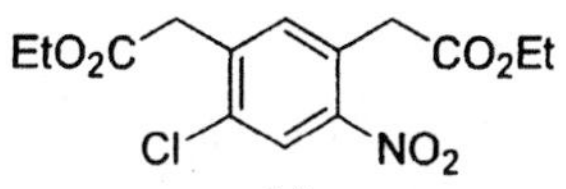

54

7.6.5 *Experimental*

57 → 58

Procedure for the Krapcho decarboxylation of 58[28]

A solution of compound **57** (90 mg, 0.24 mmol), sodium chloride (50 mg, 0.8 mol) and water (20 mg, 1 mmol) in DMSO (5 mL) was heated under reflux for 4 h. The reaction was cooled to ambient temperature before being diluted with ethyl acetate (50 mL). The organic phase was washed with water (3 × 20 mL) and brine (3 × 20 mL) before being dried over $MgSO_4$ and concentrated *in vacuo*. The crude product was purified with flash chromatography (ethyl acetate:light petroleum, 1:6) to yield the desired product **58** as a pale yellow solid (80%). Mp 81–82 °C. (Found: M^+, 309.1517. $C_{23}H_{19}N$ requires M^+, 309.1517.) $[\alpha]_D^{20}$-91.3 (*c* 0.46; $CDCl_3$), ν_{max} (Nujol) cm^{-1} 2252 (CN) and 1599 (C=C). δ_H (250 MHz; $CDCl_3$) 7.42-7.15 (15H, m, Ar-*H*), 6.29 (1H, d, *J*=10.4, Ph_2C=C*H*), 3.81 (1H, m, PhC*H*) and 2.71 (2H, d, *J*=6.9). δ_C (100.6 MHz; $CDCl_3$) 144.3, 141.4, 129.6, 128.5, 128.2, 127.7, 127.6, 127.4, and 127.0 (arom. *C* and *C*=C), 118.1 (*C*N), 41.5 (Ph*C*H) and 25.3 (*C*H_2CN).

7.6.6 *References*

1. Krapcho, A. P. *Synthesis* **1982**, 805–822 and 893–914.
2. Krapcho, A. P.; Mundy, B. P. *Tetrahedron Lett.* **1970**, 5437.
3. Krapcho, A. P.; Glynn, G. A.; Grenon, B. J. *Tetrahedron Lett.* **1967**, *3*, 215.
4. Krapcho, A. P.; Weimaster, J. F.; Eldridge, J. M.; Jahngen, Jr., E. G. E.; Lovey, A. J.; Stephens, W. P. *J. Org. Chem.* **1978**, *43*, 138.
5. Bernard, A. M.; Cerioni, G.; Piras, P. P. *Tetrahedron* **1990**, *46*, 3929.
6. Gilligan, P. J.; Krenitsky, P. J. *Tetrahedron Lett.* **1994**, *35*, 3441.
7. Liotta, C. L.; Cook, F. L. *Tetrahedron Lett.* **1974**, *13*, 1095.
8. Smith, III, A. B.; Kingery-Wood, J.; Leenay, T. L.; Nolen, E. G.; Sunazuka, T. *J. Am. Chem. Soc.* **1992**, *114*, 1438.
9. Hagiwara, H.; Uda, H. *J. Chem. Soc. Perkin Trans. I* **1991**, 1803.
10. Paquette, L. A.; Friedrich, D.; Pinard, E.; Williams, J. P.; St. Laurent, D.; Roden, B. A. *J. Am. Chem. Soc.* **1993**, *115*, 4377.
11. Dodd, D. S.; Oehlschlager, A. C. *J. Org. Chem.* **1992**, *57*, 2794.
12. Villhauer, E. B.; Anderson, R. C. *J. Org. Chem.* **1987**, *52*, 1186.
13. Lampe, J. W.; Hanna, R. G.; Piscitelli, T. A.; Chou, Y.-L.; Erhardt, P. W.; Lumma, Jr., W. C.; Greenberg, S. S.; Ingebretsen, W. R.; Marshall, D. C.; Wiggins, J. *J. Med. Chem.* **1990**, *33*, 1688.
14. Evans, D. A.; Scheidt, K. A.; Downey, C. W. *Org. Lett.* **2001**, *3*, 3009.
15. Aristoff, P. A.; Johnson, P. D.; Harrison, A. W. *J. Am. Chem. Soc.* **1985**, *107*, 7967.
16. Craig, D.; Henry, G. *Tetrahedron Lett.* **2005**, *46*, 2599.
17. Brandi, A.; Cordero, F. M.; De Sarlo, F.; Goti, A.; Guarna, A. *Synlett* **1993**, *1*, 1.
18. Toró, A.; Nowak, P.; Deslongchamps, P. *J. Am. Chem. Soc.* **2000**, *122*, 4526.
19. Henry, G. E.; Jacobs, H. *Tetrahedron* **2001**, *57*, 5335.
20. Wee, A. G. H.; Yu, Q. *Tetrahedron Lett.* **2000**, *41*, 587.
21. Zhu, Q.; Qiao, L.; Wu, Y.; Wu, Y.-L. *J. Org. Chem.* **2001**, *66*, 2692.

22. Maurin, P.; Ibrahim-Ouali, M.; Santelli, M. *Eur. J. Org. Chem.* **2002**, 151.
23. Schuppan, J.; Minnaard, A. J.; Feringa, B. L. *Chem Commun.* **2004**, 792.
24. Lassaletta, J. M.; Vásquez, J.; Prieto, A.; Fernández, R.; Raabe, G.; Enders, D. *J. Org. Chem.* **2003**, *68*, 2698.
25. Taylor, M. S.; Zalatan, D. N.; Lerchner, A. M.; Jacobsen, E. N. *J. Am. Chem. Soc.* **2005**, *127*, 1313.
26. Baraldi, P. G.; Barco, A.; Benetti, S.; Pollini, G. P.; Polo, E.; Simoni, D. *J. Org. Chem.* **1985**, *50*, 23.
27. Bergner, E. J.; Helmchen, G. *Eur. J. Org. Chem.* **2000**, 419.
28. Martin, C. J.; Rawson, D. J.; Williams, J. M. J. *Tetrahedron: Asymmetry* **1998**, *9*, 3723.
29. Bäckvall, J.-E.; Andersson, P. G.; Stone, G. B.; Gogoll, A. *J. Org. Chem.* **1991**, *56*, 2988.
30. Bäckvall, J.-E.; Nyström, J.-E.; Nordberg, R. E. *J. Am. Chem. Soc.* **1985**, *107*, 3676.
31. Gurjar, M. K.; Murugaiah, A. M. S.; Reddy, D. S. *Org. Process Rsch. & Devel.* **2003**, *7*, 309.

Richard J. Mullins, Michelle D. Hoffman, and Andrea L. Kelly

7.7 Nef Reaction

7.7.1 Description

The Nef reaction[1,2] is the conversion of nitroalkanes to ketones and aldehydes through treatment with base followed by acid.[3,4,5,6] For example, deprotonation of 1-nitrobutane (**1**) with aqueous NaOH followed by addition of excess aqueous sulfuric acid afforded butyraldehyde (**2**) in 85% yield (isolated as the oxime derivative).[7] These reactions proceed *via* intermediate nitronate anions, which are subsequently hydrolyzed to afford the carbonyl products. The overall transformation leads to formal polarity reversal of the carbon bearing the nitro group from a nucleophilic species to an electrophilic carbonyl carbon. Although the classical conditions for this process are quite harsh, a number of alternative procedures that employ mild reaction conditions have been developed.

NO_2 **1** → 1. 2.25 M NaOH; 2. 2 M H_2SO_4; 85% → O, H **2**

7.7.2 Historical Perspective

The first examples of the conversion of nitroalkanes to carbonyl compounds were described in 1893 by Konovaloff, who was examining the reactivity of nitroalkanes obtained from nitration of alkanes with nitric acid.[1] Konovaloff reported that treatment of 2-nitrohexane (**3**) with strong soda (NaOH) followed by reaction with Zn/HOAc afforded a mixture of methyl butyl ketone (**4**) and 2-aminohexane (**5**). Additionally, the reaction of the potassium salt of 2-phenylnitroethane with dilute aqueous acid provided mixtures of acetophenone and 2-phenylnitroethane.

NO_2 **3** → 1. NaOH; 2. Zn/HOAc → O **4** + NH_2 **5**

In independent studies published one year later, Nef reported that treatment of the sodium salts of nitroethane (**6**) or 2-nitropropane (**8**) with aqueous H_2SO_4 or HCl afforded acetaldehyde (**7**) or acetone (**9**), respectively.[2] In subsequent studies Nef demonstrated that this transformation had a sufficiently broad scope to be synthetically useful.

NO_2^- Na^+ **6** → aq H_2SO_4; 70% → O, H **7**

NO_2^- Na^+ **8** → aq H_2SO_4; 67% → O **9**

7.7.3 Mechanism

The mechanism of the classical Nef reaction is believed to involve deprotonation of the nitroalkane with base to provide a nitronate salt (**11**). Treatment of this species with weak acid typically leads to regeneration of the nitroalkane. However, under strongly acidic conditions this anion can undergo a hydrolysis reaction that is initiated by two sequential protonations of the oxygen atoms to afford **12** then **13**.[8] Attack of water on the electrophilic carbon atom of intermediate **13** provides **14**, which undergoes loss of a proton to give **15**. Dehydration of **15** to **16** followed by protonation generates **17**, which loses HNO to yield **18**. A final proton transfer step provides the carbonyl containing product (**19**).

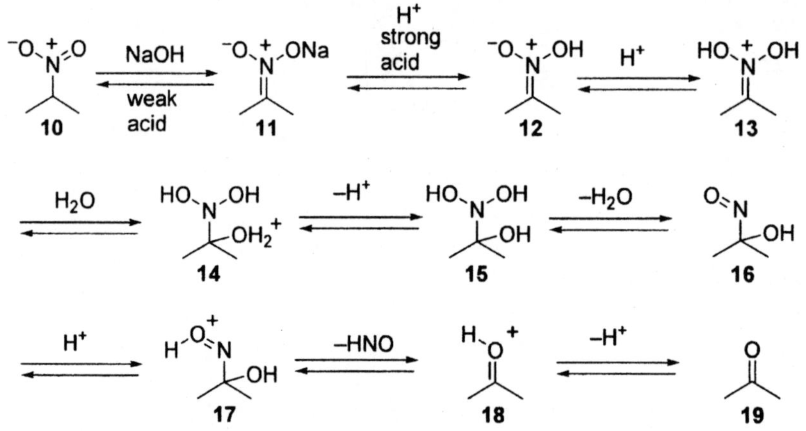

The mechanism of this transformation was elucidated through kinetic studies, and some intermediates in this process have been isolated (e.g., **12**).[5] Other mechanisms have been proposed,[9] and under certain circumstances dual pathways may operate simultaneously.[8b]

7.7.4 Variations and Improvements

Although the classical Nef reaction has proved quite useful for the conversion of nitroalkenes to carbonyl derivatives, the conditions originally developed for this transformation are quite harsh, and side reactions of functionalized molecules are frequently observed. In order to address this limitation, a considerable amount of effort has been dedicated toward the development of mild conditions that could be used in reactions of highly functionalized substrates that often serve as intermediates in complex molecule synthesis.

A number of methods to effect Nef-type reactions employ reducing agents in place of the acid/base combination. The most commonly used variant of this type was developed by McMurry in 1973 and involves treatment of nitroalkanes with aqueous $TiCl_3$.[10] In a representative example, these conditions were used for the conversion of **20** to **21** in 85% yield. These reactions likely proceed through generation of a Ti–O bond to afford **22**, followed by N–O bond cleavage to yield **23**. Tautomerization of the nitroso species (**23**) would provide oxime **24**, which is then hydrolyzed to afford **21**. Although these conditions are still somewhat acidic, several functional groups are tolerated including ketones, nitriles, and alkynes. Increased functional group tolerance is obtained through deprotonation of the substrates to afford nitronate anions, which can be reduced with $TiCl_3$ under buffered conditions at near-neutral pH. These modified conditions tolerate the presence of esters and ketals in addition to the functional groups noted above.

$TiCl_3$
H_2O/glyme
85%

NO_2 **20** **21**

$TiCl_3$

22 **23** **24**

Nitroalkanes have also been reduced to oximes with mixtures of $SnCl_2$•$2H_2O$, PhSH, and Et_3N.[11] As shown below, these conditions were used to achieve the transformation of **25** to **26** in 85% yield. The oxime products of these reactions can be converted to ketones under mild conditions *via* hydrolysis or oxidation. This method was employed in a key step of Trauner's recent synthesis of (–)-amathaspiramide F (see Section 7.7.6).[12]

1.5 equiv $SnCl_2$
3.0 equiv PhSH
3.0 equiv Et_3N
benzene, rt
85%

25 **26**

A number of other reagents have been used for the reductive conversion of nitroalkanes to carbonyl compounds or oximes,[3,4] including hexamethyl disilane,[13] Bu_3P/PhSSPh,[14] and catalytic Cu(acac).[15]

Nef reactions have also been effected through deprotonation of the nitroalkane followed by oxidation using reagents such as $KMnO_4$.[16,17] If conditions are carefully controlled several functional groups are tolerated including alcohols and alkenes. For example, treatment of **27** with methanolic KOH followed by addition of an aqueous solution of $KMnO_4$ and $MgSO_4$ afforded keto-alcohol **28** in 73% yield.[16b] Although aldehydes are obtained from primary nitroalkanes with ≤ 1 equiv of $KMnO_4$, use of excess oxidant provides carboxylic acid products. As shown below, a buffered solution of $KMnO_4$ was used to convert **29** to **30** in 75% yield.[18]

27 → 28: 1. KOH, MeOH; 2. $KMnO_4$, $MgSO_4$, H_2O; 73%

29 → 30: 1. NaOEt, EtOH; 2. $KMnO_4$, KOH, K_2HPO_4; 75%

Dimethyl dioxirane (DMDO) has also been employed as a mild and selective oxidant in Nef reactions.[19] For example, treatment of nitroalkane **31** with KO*t*-Bu followed by DMDO provided enone **32** in 72% yield.[20] The alkene isomerization that occurred in this case was desirable, as **32** represents the AB ring system of the natural product norzoanthamine. Many other oxidants have been used in Nef-type reactions,[3,4] including alkali metal percarbonates or mixtures of K_2CO_3 and H_2O_2.[21]

31 → 32: 1. KO*t*-Bu, THF; 2. DMDO, H_2O; 72%

Although early versions of the Nef reaction utilized saturated nitroalkanes as substrates, related transformations have also been carried out on nitroalkene substrates. These reactions are typically conducted under reducing conditions. For example, mixtures of NaH_2PO_2 and Raney-nickel have been employed to effect the conversion of nitroalkenes to aldehydes and ketones under very mild conditions that tolerate a range of functional groups including esters, acetals, and alcohols.[22] As shown below, **34** was prepared from **33** in 86% yield with this method.[22a] The generation of carbonyl derivatives from nitroalkenes has also been achieved with many other reagents.[3,4]

33 → 34: Ra-Ni, NaH_2PO_2, H_2O/EtOH; 86%

7.7.5 Synthetic Utility

The Nef reaction has been employed for over fifty years in carbohydrate synthesis. A two-step process referred to as the Snowden method has been widely used to effect the chain elongation of sugars *via* the nucleophilic addition of the nitromethane anion to an aldehyde group followed by Nef reaction to provide a new aldehyde.[6] For example, in early studies Sowden and Fischer described the conversion of **35** to **36** through addition of the nitromethane anion followed by deprotection and Nef reaction to afford **37** (isolated as its benzylphenyl hydrazone derivative).[23]

HO OH O
CH$_3$NO$_2$
NaOMe, MeOH
50%
HO OH OH
NO$_2$
Ph
35
Ph
36
1. H$_2$SO$_4$
2. NaOH
3. H$_2$SO$_4$
41%
HO OH OH
OH OH O
37

The Nef reaction has also found broad application to the synthesis of other natural products and biologically active molecules. A few representative recent examples of the use of this transformation in complex molecule synthesis are shown below.

As noted above, Trauner employed a Nef reaction as a key step in his recent synthesis of (–)-amathaspiramide F. Nitroalkane **38** was initially converted to oxime **39** in 85% yield through the use of $SnCl_2$/PhSH/Et_3N. This intermediate was then transformed to aminal **40** *via* oxidative cleavage of the oxime to provide an aldehyde, which underwent intramolecular trapping with the neighboring amide group.[12] Cleavage of the trifluoroacetamide protecting group in a subsequent step afforded the natural product.

Br MeO Br NO$_2$ N O NHMe F$_3$C O
38
SnCl$_2$, PhSH,
Et$_3$N, CH$_3$CN
85%
Br MeO Br N-OH N O NHMe F$_3$C O
39
IBX
THF, DMSO
65%
Br MeO Br OH NMe N O F$_3$C O
40

The Nef reaction has also been used in the construction of the 7-deoxy-ABC ring taxane skeleton. As shown below, conjugate reduction of nitroalkene **41** with $NaBH_4$ followed by oxidation of the resulting nitronate salt with K_2CO_3/H_2O_2 provided tetracycle **42** in 60% yield.[24] These reaction conditions are sufficiently mild to tolerate the presence of ketone and silyl ether functional groups in the molecule.

H O O$_2$N H O OTBS
41
1. NaBH$_4$, MeOH
2. K$_2$CO$_3$, H$_2$O$_2$
60%
H O O H O OTBS
42

Substrate **43** was converted to spiroacetal **45** using an alternative method to effect the Nef reaction of a nitroalkene under reducing conditions. Treatment of **43** with

$Pd(OH)_2/C$ in the presence of HCl and cyclohexene provided **45** in 64% yield. This reaction is believed to proceed *via* intermediate oxime **44**, which undergoes acid-mediated hydrolysis and ketalization to afford the product.[25] Use of other reducing agents, such as $TiCl_3/NH_4OAc/H_2O$, or hydrogenation with Pd/C and H_2 provided lower yields of the desired compound. The spiroketal **45** is a substructure of the γ-rubromycin family of natural products, which have shown activity against HIV reverse transcriptase and DNA helicase.

$Pd(OH)_2/C$
EtOH, HCl
Cyclohexene
64%

43 **44** **45**

In a recent asymmetric synthesis of cyclophellitol, Trost employed a Nef reaction for the conversion of nitrosulfone **46** to carboxylic acid **47**.[26] As shown below, **46** was treated with tetramethylguanidine to generate a nitronate salt, which was then oxidized with DMDO to afford the acid in 78% yield. The adjacent double bond was not epoxidized under these conditions.

1. tetramethylguanidine
2. DMDO, acetone
78%

46 **47**

7.7.6 *Experimental*

1. NaOMe, MeOH
2. $TiCl_3$, NH_4OAc, H_2O
90%

48 **49**

Levulinonitrile (49)[10]

A solution of **48** in methanol (0.5 M) was treated with 1.0 equiv NaOMe. To this mixture was added a buffered solution of $TiCl_3$ prepared by adding a solution of NH_4OAc (4.6 g, 60 mmol) in H_2O (15 mL) to a solution of 20% aqueous $TICl_3$ (100 mmol). The resulting mixture was stirred at room temperature until the starting nitroalkane had been completely consumed (45 min). The reaction mixture was then poured into ether and the layers were separated. The aqueous phase was extracted with ether and the combined organic layers were dried over anhydrous Na_2SO_4, filtered, and concentrated. The product was isolated in 90% yield by distillation.

1. KOH, MeOH
2. $KMnO_4$, $MgSO_4$
83%

NO_2 **50** → O **51**

Octanal (51)[16b]

Nitroalkane **50** (10.0 mmol) was dissolved in methanol (70 mL) and cooled to 0 °C. A solution of KOH (0.67 g, 10.1 mmol) in methanol (100 mL) was added dropwise over a period of 45 min, and the resulting mixture was stirred at 0 °C for an additional 15 min. A solution of $KMnO_4$ (1.06 g, 6.71 mmol) and $MgSO_4$ (0.89 g, 7.39 mmol) in water (100 mL) was then added dropwise with vigorous stirring at a rate such that the internal temperature of the reaction mixture did not exceed 0 ± 2 °C. When the addition was complete the mixture was allowed to stir for an additional 1 h at 0 °C, and was then filtered through a pad of celite. The solid material was washed with benzene (3 × 25 mL), triturated with 100 mL of benzene, and the combined benzene solutions were filtered through a new pad of celite. The aqueous and organic liquids were combined, diluted with brine (100 mL), and transferred to a separatory funnel. The layers were separated and the aqueous phase was extracted with benzene (4 × 100 mL). The combined organic phases were dried over anhydrous $MgSO_4$, filtered, and concentrated *in vacuo* to afford the title compound in 83 % yield.

7.7.7 *References*

1. (a) Konovaloff, M. *J. Russ. Chem. Soc.* **1893**, *25*, 472–500. (b) Konovaloff, M. *J. Russ. Chem. Soc.* **1893**, *25*, 509–546. (c) Konovaloff, M. *J. Chem. Soc. Abstracts* **1894**, *66*, 265. (d) Konovaloff, M. *J. Chem. Soc. Abstracts* **1894**, *66*, 277.
2. Nef. J. U. *Liebigs Ann.* **1894**, *280*, 263–291.
3. [R] Ballini, R.; Petrini, M. *Tetrahedron* **2004**, *60*, 1017–1047.
4. [R] Pinnick, H. W. *Org. React.* **1990**, *38*, 655–792.
5. [R] Noland, W. E. *Chem. Rev.* **1955**, *55*, 137–155.
6. [R] Petrus, L.; Petrusova, M.; Pham-Huu, D.-P.; Lattova, E.; Pribulova, B.; Turjan, J. *Monatsh. Chem.* **2002**, *133*, 383–392.
7. Johnson, K.; Degering, E. F. *J. Org. Chem.* **1943**, *8*, 10–11.
8. (a) Hawthorne, M. F. *J. Am. Chem. Soc.* **1957**, *79*, 2510–2515. (b) Sun, S. F.; Folliard, J. T. *Tetrahedron* **1971**, *27*, 323–330.
9. (a) Van Tamlen, E. E.; Thiede, R. J. *J. Am. Chem. Soc.* **1952**, *74*, 2615–2618. (b) Kornblum N. Brown, R. A. *J. Am. Chem. Soc.* **1965**, *87*, 1742–1747. (c) Cundall, R. B.; Locke, A. W. *J. Chem. Soc. B.* **1968**, 98–103.
10. McMurry, J. E.; Melton, J. *J. Org. Chem.* **1973**, *38*, 4367–4373.
11. (a) Bartra, M.; Romea, P.; Urpi, F.; Vilarrasa, J. *Tetrahedron* **1990**, *46*, 587–594. (b) Urpi, F.; Vilarrasa, J. *Tetrahedron Lett.* **1990**, *31*, 7499–7500.
12. Hughes, C. C.; Trauner, D. *Angew. Chem. Int. Ed.* **2002**, *41*, 4556–4559.
13. Hwu, J. R.; Tseng, W. N.; Himatkumar, V. P.; Wong, F. F.; Horng, D.-N.; Liaw, B. R.; Lin, L. C. *J. Org. Chem.* **1999**, *64*, 2211–2218.
14. Barton, D. H. R.; Motherwell, W. B.; Simon, E. S.; Zard, S. Z *J. Chem. Soc., Perkin Trans. 1* **1986**, 2243–2252.
15. Knifton, J. F. *J. Org. Chem.* **1973**, *38*, 3296–3301.
16. Freeman, F.; Yeramyan, A. *J. Org. Chem.* **1970**, *35*, 2061–2062.

17. (a) Shechter, H.; Williams, F. T., Jr. *J. Org. Chem.* **1962**, *27*, 3699–3701. (b) Steliou, K.; Poupart, M. –A. *J. Org. Chem.* **1985**, *50*, 4971–4973.
18. Seville-Stones, E. A.; Lindell, S. D. *Synlett* **1991**, 591–592.
19. Adam, W.; Makosza, M.; Saha-Moller, C. R.; Zhao, C.-G. *Synlett* **1998**, 1335–1336.
20. Williams, D. R.; Brugel, T. A. *Org. Lett.* **2000**, *2*, 1023–1026.
21. (a) Olah, G. A.; Arvanaghi, M.; Vankar, Y. D.; Surya Prakash, G. K. *Synthesis* **1980**, 662–663. (b) Narayana, C.; Reddy, N. K.; Kabalka, G. W. *Synth. Commun.* **1992**, *22*, 2587–2592.
22. (a) Monti, D.; Gramatica, P.; Speranza, G.; Manitto, P. *Tetrahedron Lett.* **1983**, *24*, 417–418. (b) Ballini, R.; Gil, M. V.; Fiorini, D.; Palmieri, A. *Synthesis*, **2003**, 665–667.
23. Sowden, J. C.; Fischer, H. O. L. *J. Am. Chem. Soc.* **1945**, *67*, 1713–1715.
24. Magnus, P.; Booth, J.; Diorazio, L.; Donohoe, T.; Lynch, V.; Magnus, N.; Mendoza, J., Pye, P.; Tarrant, J. *Tetrahedron* **1996**, *52*, 14103–14146.
25. Capecchi, T.; de Koning, C. B.; Michael, J. P. *J. Chem. Soc., Perkin Trans 1.* **2000**, 2681–2688.
26. Trost, B. M.; Patterson, D. E.; Hembre, E. J. *Chem. Eur. J.* **2001**, *7*, 3768–3775.

John P. Wolfe

7.8 Prins Reaction

7.8.1 *Description*

The Prins reaction involves the addition of an alkene **1** to an aldehyde **2** under acidic conditions. The major products of this reaction are often a 1,3-dioxane **3**, a diol **4**, or a homoallylic alcohol **5**.[1–9]

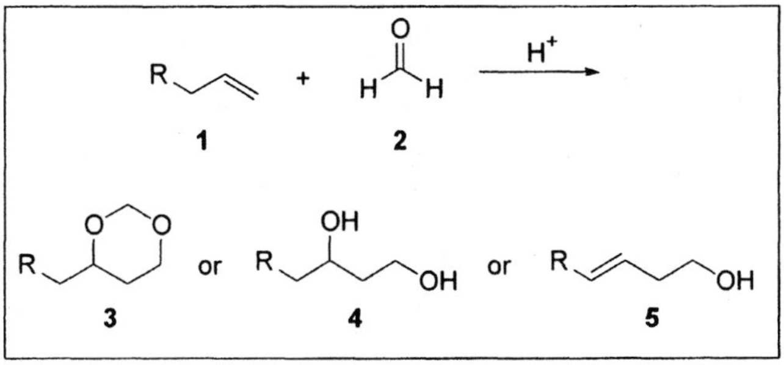

7.8.2 *Historical Perspective*

In 1919, in the journal *Chemisch Weekblad,* H. J. Prins reported the reaction of alkenes with formaldehyde in the presence of acid to give dioxanes, diols, and unsaturated alcohols.[3–5] The reaction was apparently discovered by Kriewitz in 1899,[1,2,6] but Prins carried out the most thorough initial investigation of the reaction and hence it carries his name.[6] In one example from this study, Prins found that styrene **6** reacted with formaldehyde **2** in the presence of sulfuric acid to give 1,3-dioxane **7**, allylic alcohol **8**, and diol **9**. When the reaction was carried out in glacial acetic acid, esters of this acid were the primary products.[3–5]

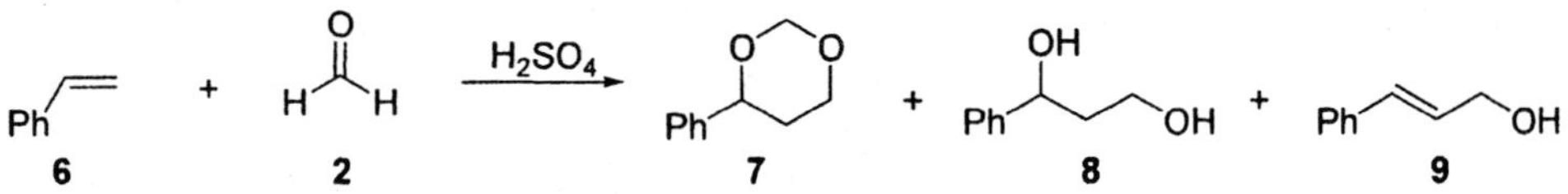

7.8.3 *Mechanism*

While several mechanisms for the Prins reaction have been proposed,[7,10] there is general agreement that the mechanism shown below is reasonable.[7] Attack of olefin **1** on protonated aldehyde **10** gives carbocation **11**. This carbocation can then be trapped with another molecule of aldehyde, generating adduct **12**. Cyclization of **12** affords dioxane **3**. Carbocation **11** can also be captured with water giving diol **4**. Finally, elimination of a proton from carbocation **11** affords homoallylic alcohol **5**. It is possible that the formation of **5** could also proceed *via* an ene-type mechanism.[8]

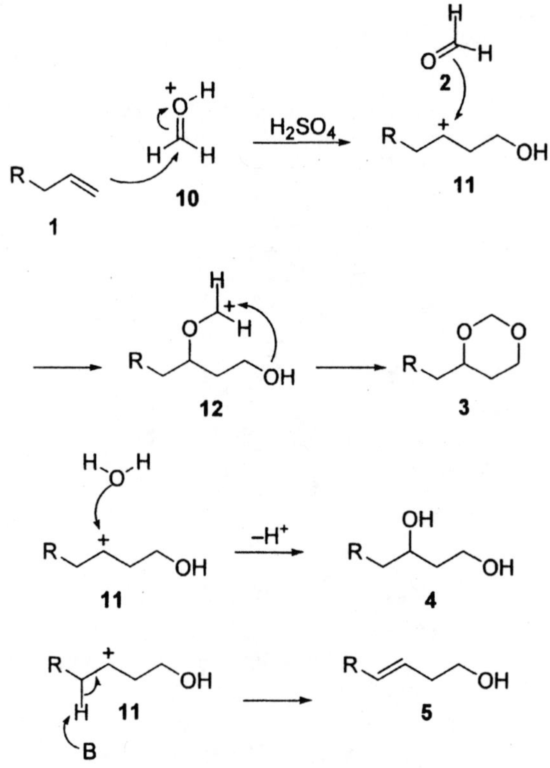

7.8.4 *Variations and Improvements*

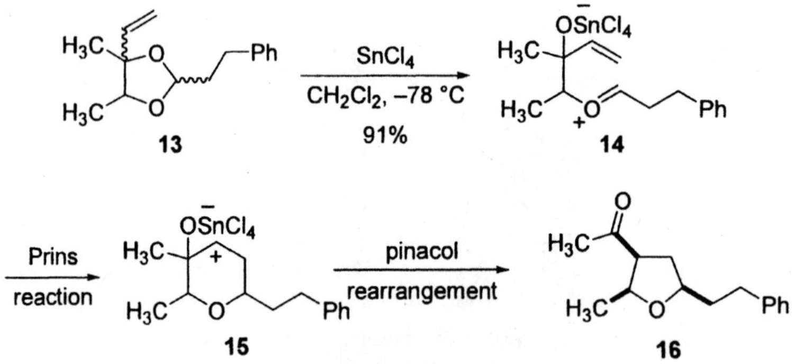

The most significant variation and improvement to the Prins reaction was developed by the Overman group.[11–15] Overman and co-workers found that they could control the fate of the carbocation intermediate formed during the Prins reaction by using that intermediate to trigger a pinacol rearrangement.[13] This led to the development of a sequential Prins cyclization–pinacol rearrangement reaction sequence as an extremely powerful method for the synthesis of substituted tetrahydrofurans. This sequence is designed to start with acetals such as **13**. Treatment of **13** with $SnCl_4$ generates oxonium ion **14**, which undergoes a Prins reaction to give carbocation **15**. Pinacol rearrangement of **15** affords tetrahydrofuran **16** in 91% yield as a single diastereomer.[14]

7.8.5 Synthetic Utility

An excellent synthetic example of the Prins reaction is found in MacMillan and Overman's total synthesis of (–)-7-deacetoxyalcyonin acetate (**21**), in which the core of the natural product is assembled *via* the Prins–pinacol reaction sequence.[16] Treatment of diol **17** with aldehyde **18** in the presence of BF_3•Et_2O presumably gives oxonium ion **19**, which then undergoes the Prins–pinacol reaction sequence to generate **20** in 79% yield, thereby establishing the entire core bicycle of **21** in one step. Intermediate **20** was then carried forward to complete the total synthesis of 7-deacetoxyalcyonin acetate (**21**).[16]

OH OH TMS **17** OHC OTIPS **18** BF_3•Et_2O, CH_2Cl_2 –55 → –20 °C 79% $\overset{-}{O}BF_3$ +O TMS OTIPS **19**

Prins-pinacol reaction sequence H CHO O OTIPS TMS **20** H H H H HO AcO O 17-deacetoxyalcyonin (**21**)

Another creative synthetic application of the Prins reaction was reported by Rychnovsky and co-workers.[17-20] Treatment of acetal **22** with BF_3•OEt_2 and HOAc effected a Prins desymmetrization reaction, affording pyran **23** in 42–51% yield after acylation. This intermediate was then used to complete a total synthesis of 17-deoxyroflamycoin **24**.[17]

O O OBn **22** 1) BF_3•OEt_2, HOAc cyclohexane, 23 °C 2) Ac_2O, DMAP, Et_3N 42-51% OAc OAc O OBn **23**

17-deoxyroflamycoin (**24**)

$$\textbf{25} \xrightarrow[\text{TfOH, } CH_2Cl_2,\ 87\%,\ dr > 30:1]{\textbf{26}} \textbf{27}$$

Finally, Huang and Panek utilized a Prins-type reaction between a crotylsilane and an aldehyde in their total synthesis of callipeltoside A (**28**).[21,22] In the presence of T_fOH, crotylsilane **25** and aldehyde **26** underwent mixed acetal formation and subsequent Prins cyclization to generate dihydropyran **27** in 87% yield as a single diastereomer. This intermediate was carried forward to complete a total synthesis of **28**.[22]

callipeltoside A (**28**)

7.8.6 *Experimental*

$$\textbf{13} \xrightarrow[CH_2Cl_2,\ -78\ °C,\ 91\%]{SnCl_4} \textbf{14}$$

Prins reaction → **15** → pinacol rearrangement → **16**

(±)-(2α,3α,5α)-1-(Tetrahydro-2-methyl-5-phenethylfuran-3-yl)ethanone (16)[14]
$SnCl_4$ (0.8 mL, 7 mmol) was added to a solution of the acetals **13** (478 mg, 3.37 mmol) and CH_2Cl_2 (10 mL) at –78 °C. The solution was warmed to –23 °C and maintained for 4 h. Saturated aqueous NaCl (10 mL) was added, and the mixture was extracted with CH_2Cl_2 (3 × 40 mL). The combined CH_2Cl_2 extracts were dried (Na_2SO_4) and concentrated, and the residue was purified by flash chromatography (1:2 EtOAc–hexane) to provide 681 mg (91%) of **16** as a colorless oil.

7.8.7 References

1. Kriewitz, O. *Ber.* **1899**, *32*, 57.
2. Kriewitz, O. *J. Chem. Soc.* **1899**, *76*, 298.
3. Prins, H. J. *Chem. Weekblad* **1919**, *16*, 64.
4. Prins, H. J. *Chem. Weekblad* **1919**, *16*, 1072.
5. Prins, H. J. *Chem. Weekblad* **1919**, *16*, 1510.
6. [R] Arundale, E.; Mikeska, L. A. *Chem. Rev.* **1952**, *51*, 505.
7. [R] Adams, D. R.; Bhatnagar, S. P. *Synthesis* **1977**, 661.
8. [R] Snider, B. B. *Comprehensive Organic Synthesis*; Trost, B. M. Ed.; Pergamon Press: Oxford, U. K., 1991; Vol. 2, pp. 527–561.
9. [R] Bode, S. E.; Wolberg, M.; Müller, M. *Synthesis* **2006**, 557.
10. Dolby, L. J. *J. Org. Chem.* **1962**, *27*, 2971 and references cited therein.
11. [R] Overman, L. E. *Acc. Chem. Res.* **1992**, *25*, 352.
12. [R] Overman, L. E. *Aldrichim. Acta* **1995**, *28*, 107.
13. Hopkins, M. H.; Overman, L. E. *J. Am. Chem. Soc.* **1987**, *109*, 4748.
14. Hopkins, M. H.; Overman, L. E.; Rishton, G. M. *J. Am. Chem. Soc.* **1991**, *113*, 5354.
15. Brown, M. J.; Harrison, T.; Herrinton, P. M.; Hopkins, M. H.; Hutchinson, K. D.; Mishra, P.; Overman, L. E. *J. Am. Chem. Soc.* **1991**, *113*, 5365.
16. MacMillan, D. W. C.; Overman, L. E. *J. Am. Chem. Soc.* **1995**, *117*, 10391.
17. Rychnovsky, S. D.; Yang, G.; Hu, Y.; Khire, U. R. *J. Org. Chem.* **1997**, *62*, 3022.
18. Rychnovsky, S. D.; Hu, Y.; Ellsworth, B. *Tetrahedron Lett.* **1998**, *39*, 7271.
19. Rychnovsky, S. D.; Thomas, C. R. *Org. Lett.* **2000**, *2*, 1217.
20. Jaber, J. J.; Mitsui, K.; Rychnovsky, S. D. *J. Org. Chem.* **2001**, *66*, 4679.
21. Huang, H.; Panek, J. S. *J. Am. Chem. Soc.* **2000**, *122*, 9836.
22. Huang, H.; Panek, J. S. *Org. Lett.* **2004**, *6*, 4383.

Dustin J. Mergott

7.9 Regitz Diazo Reactions

7.9.1 Description

The Regitz reaction involves the transfer of a diazo group from the tosyl azide or mesyl azide to active methylene compounds such as 1,3-diketones and their derivatives (**1**) in the presence of a base leading to 2-diazo-1,3-diketones (**2**).

TsN_3, Et_3N / MeCN; + $TsNH_2$

1 **2**

7.9.2 Historical Perspective

Regitz diazo transfer reactions have been reviewed previously.[1-3] The following two main routes have been known for the synthesis of diazo compounds: (1) diazotization of amines, oximes, nitrosoamines, and hydrazones; (2) transfer of the diazo function from tosyl or mesyl azides to active methylene compounds.

7.9.3 Mechanism

The possible mechanism for diazo transfer from *p*-toluenesulfonyl azide to active methylene compound **3** (flanked by carbonyl groups) is depicted below.[1,3] Deprotonation of α-keto ester **3** with NEt_3 leads to enolate **4** which attacks at the electrophilic N of the sulfonyl azide **5** to give intermediate tosyl derivative **6.** Proton transfer occurs within intermediate **6** followed by elimination of *p*-toluenesulfonamide, leading to diazo compound 7 and the by-product *p*-toluene sulfonamide **8.**[1,3]

NEt_3; enolate formation; **3** **4** **5** **6**

proton transfer; **6** **7** **8**

7.9.4 Variations and Improvements

The Regitz reaction provides a gateway for the synthesis of the following diazo derivatives: cyclopentadienes, cyclohexadienes, ketones, 1,2-, and 1,3-diketones and their derivatives, β-keto esters, α-iminoketones, α-ketohydrazones, amino-1,3-diketones, nitro-1,3-diketones and 1,3,5-triketones. Secondary reactions of diazo compounds lead to heterocycles such as 1,3,5-triazoles, 1,2,3-thiadiiazoles, and pyrazolines.

In most Regitz reactions, NMe_3 and NEt_3 are used as common bases. Modification of the Regitz reaction includes the use of KOEt/EtOH and phase transfer catalysts. Regitz reaction of sterically hindered ketones usually produces a low yield of diazo compounds and this difficulty is overcome by use of the more reactive 2,4,6-triisopropylbenzenesulfonyl azide in conjunction with a mixture of tetra-*n*-butylammonium bromide and 18-crown-6 as a catalyst and KOH as the base.[4a] Mesyl azide is a superior reagent for diazo transfer reactions and it is easily prepared in high yield from the inexpensive mesyl chloride and sodium azide in absolute MeOH. It is easily separated from the desired product upon washing the organic phase with 10% aq. NaOH solution. The diazo transfer works well for both β-ketoesters and formyl ketone. The use of *p*-carboxybenzenesulfonyl azide has been recommended because of its solubility in base, but its high cost makes mesyl azide the better choice.[5] The Regitz reaction proceeds moderately with 1,4-diazabicyclo[2.2.2]octane (DABCO) but an almost quantitative yield is achieved with 1,8-diaza-bicyclo[5.4.0]undec-7-ene (DBU) with a shorter reaction time.[4a]

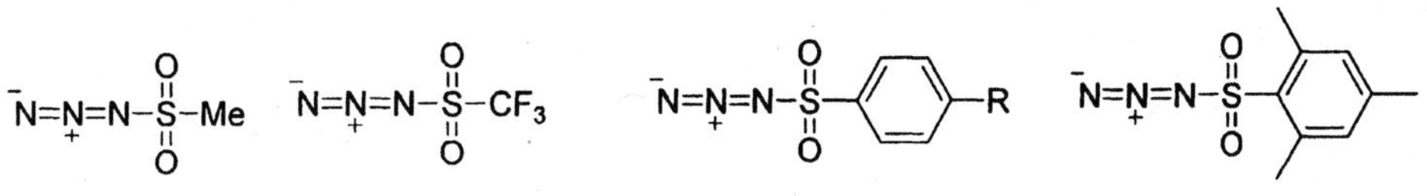

R = Me, CO_2H, NO_2

9 **10** **11** **12**

7.9.4.1 Diazo methylene compounds

Diazo group transfer to lithium cyclopentadienyl (**13**) with tosyl azide in ether affords diazocyclopentadiene (**14**).[6] Diazo transfer reactions from $TosN_3$ to anthrone (**15**, X = CO) and thioanthrone (**15**, X = SO_2) successfully furnished diazoanthrone[7] (**16**, X = CO) and diazothioanthrone[7] (**16**, X = SO_2).

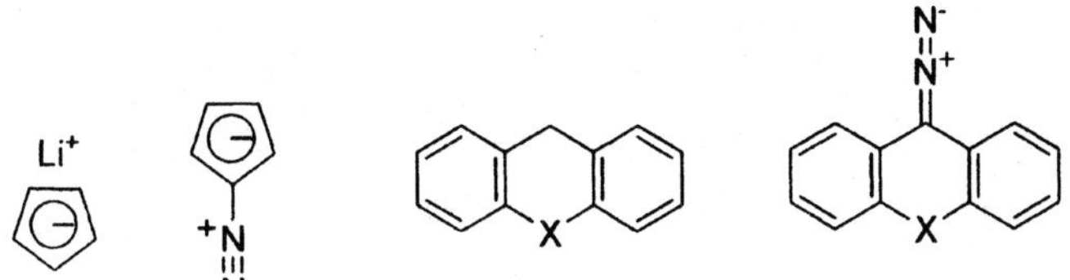

X = CO, SO_2

13 **14** **15** **16**

7.9.4.2 Diazo alkenes and acetylenes

Acetaldehyde and piperidine are allowed to condense in onepot to give imine **17**, which reacts with tosyl azide to yield diazomethane[8] (CH_2N_2) and tosyl derivative **18**. Mesyl sulfonyl azide reacts with defined orientation to ethoxy acetylene (**19**) to give 1,2,3-

triazole (**20**), which is either isolated as such or isomerizes to the open-chain diazo compound (**21**) or is present in solution in equilibrium.[9]

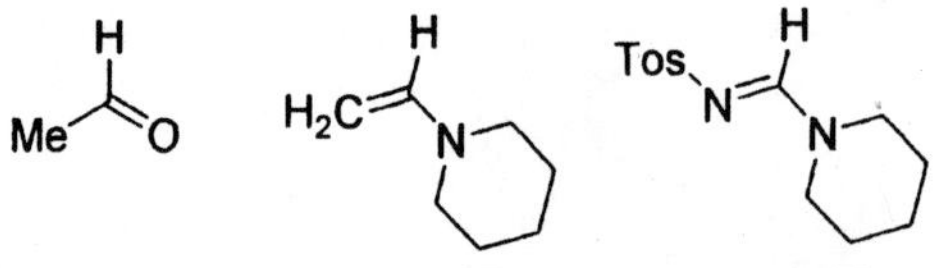

17 **18**

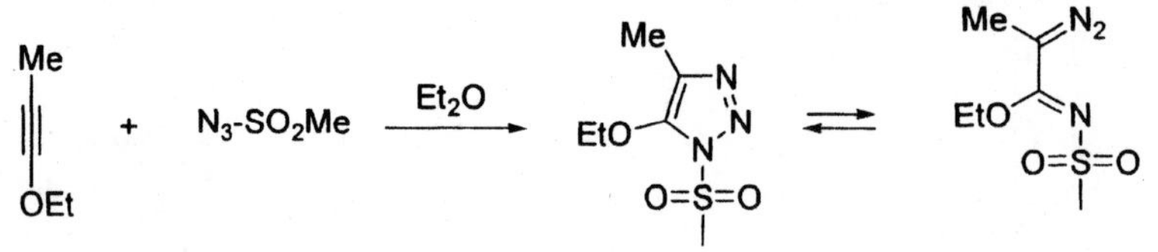

19 **20** **21**

7.9.4.3 *Diethyl diazo malonate*

Diethyl diazo malonate (**22**)[10] is prepared from diethyl malonate using TsN_3. The following diazo transfer agents are used to convert ethyl nitro acetate (**23**) to ethyl α-nitro-α-diazo acetate (**24**) in low yield: N_2O_5, tosyl azide, mesyl azide, 2-azido-3-ethylbenzothiazolium fluoroborate and *p*-carboxybenzenesulfonyl azide.[11] Trifluoromethanesulfonyl azide reacts smoothly with ethyl nitro acetate in acetonitrile upon addition of pyridine to generate ethyl α-nitro-α-diazo acetate (**26**) in 88% yield.[11] The diazo benzofuran derivative (**26**) is derived from substituted benzofuran acetate (**25**) using 4-azidosulfonylbenzoic acid and DBU as base in 77% yield.[12]

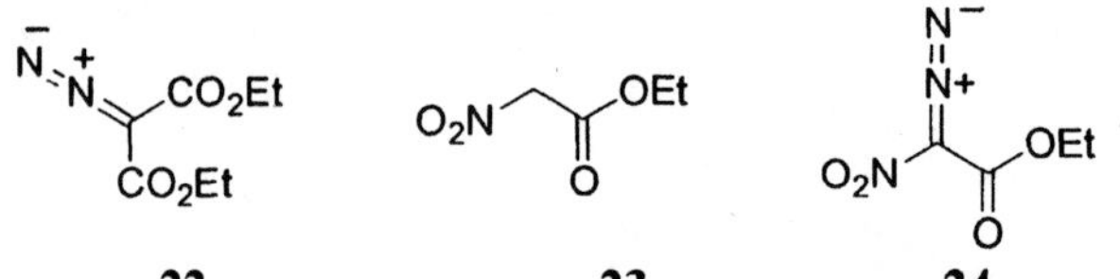

22 **23** **24**

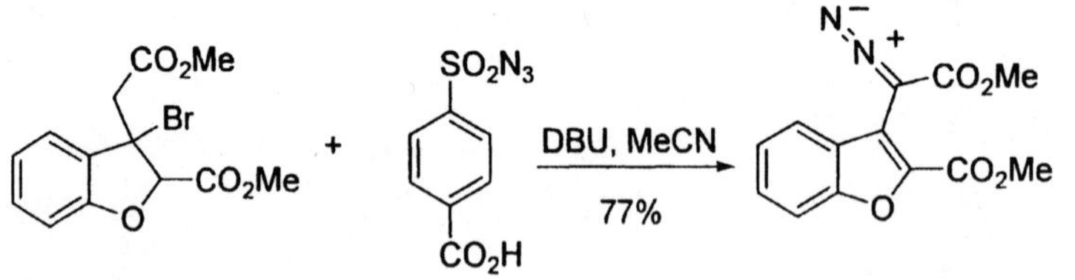

DBU, MeCN 77%

25 **26**

7.9.4.4 *α-Diazo sulfonyl compounds*

Ethyl 4-methylphenylsulfonylacetate (**27**) undergoes diazo transfer with TsN_3 only in the presence of NaOEt in $EtOH/Et_2O$ to give diazo ester (**28**), but the reaction then leads to the displacement of the ethoxy group to give the sodium salt of amide **29**.[13a] The cyclic β-oxosulfone (**30**) undergoes azo coupling in the reaction with tosyl azide; the product tautomeric hydrazone (**31**), upon warming undergoes decoupling, thus providing diazo sulfone (**32**).[13b]

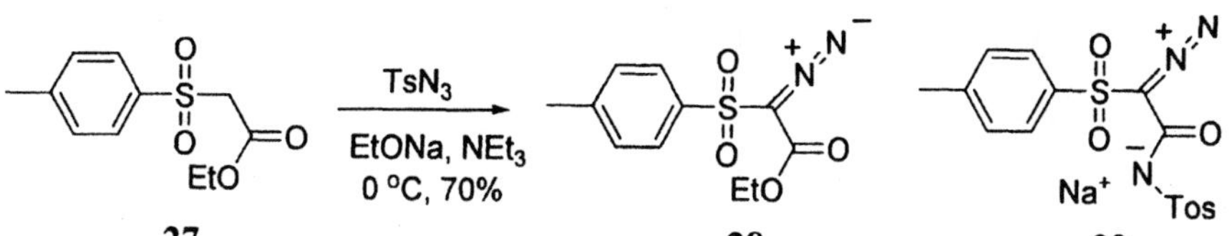

27 28 29

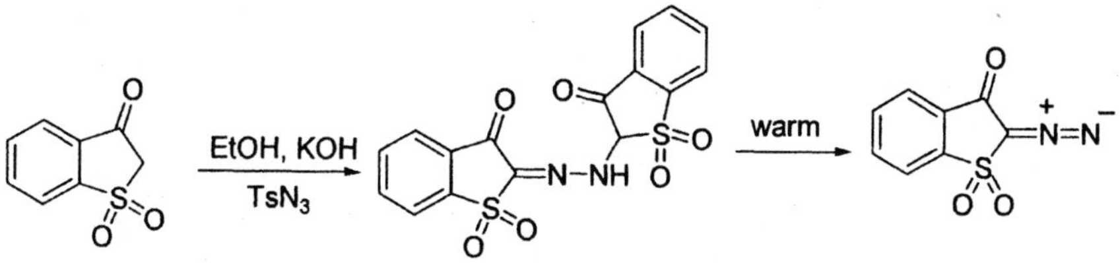

30 31 32

7.9.4.5 α-Diazo phosphono compounds

Diazo transfer from tosyl azide to diphenylphosphinoxide acetamide (**33**) is accomplished *via* 1,5 ring closure of **34** to afford 4-diphenylphosphinyl-5-hydroxy-1,2,3-triazole (**35**).[14] However, triazole **35** is further thermally isomerized in DMF back to diazo acetamide (**34**). Tosyl azide in the presence of piperidine converts phosphinoxide acetophenone (**36**) to diazo acetophenone **37**.[15]

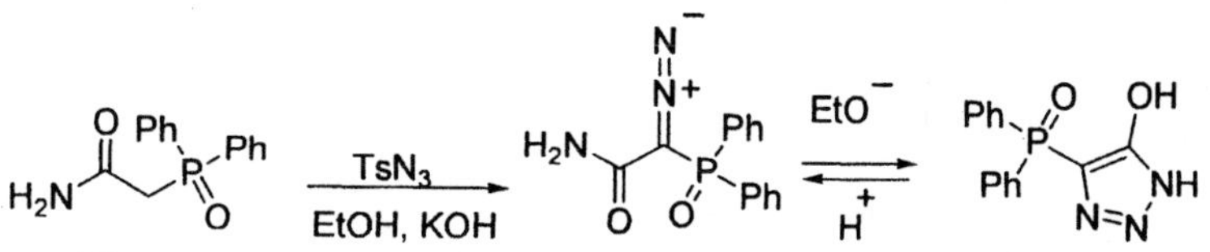

33 34 35

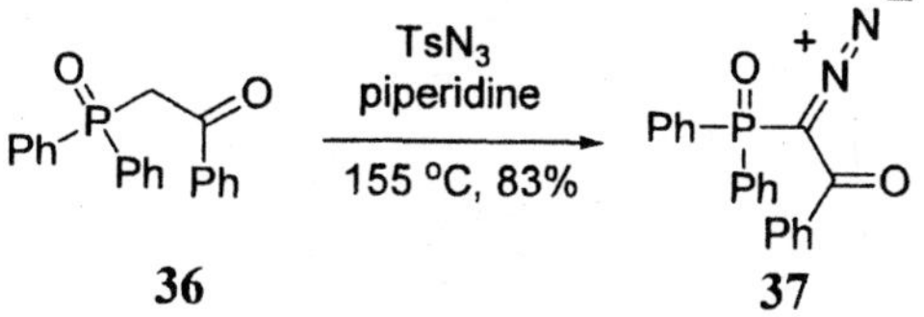

36 37

Diazo transfer to pyrazolidone (**39**) using azidinium salt (**38**) affords diazo pyrazolidone (**41**) and imine **40**; this method has proved superior to the aryl sulfonyl azide reactions in cases where azo coupling occurs.[16]

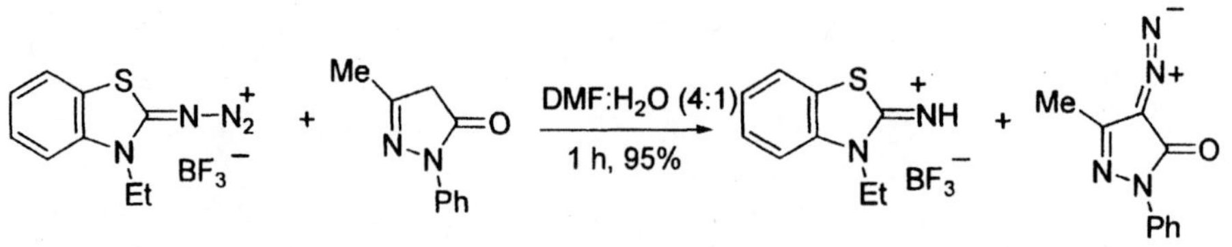

38 39 40 41

7.9.4.6 α-Diazo ketones

Mesyl azide[17] is used to transfer the diazo group to acyclic and cyclic ketones (**42, 44,** and **46**) to provide access to corresponding diazo ketones. α-Disubstituted ketone **44** is converted to the diazo ketone **45** *via* debenzoylation diazo transfer reaction using *p*-nitrobenzenesufonyl azide in the presence of DBU.[18]

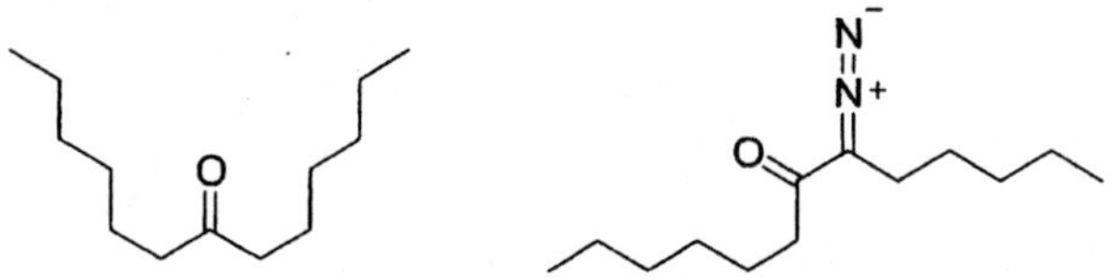

42 43

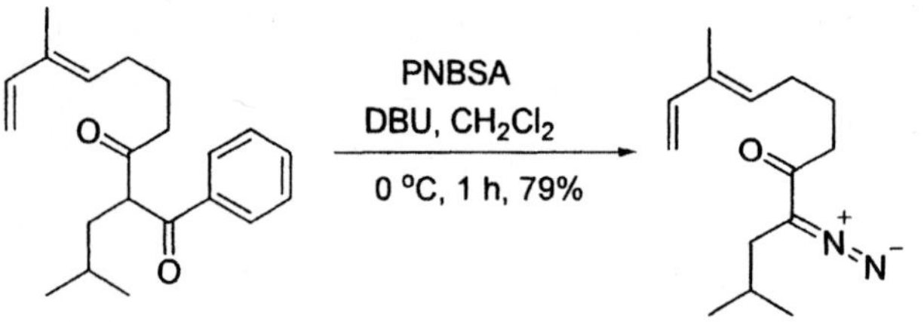

44 45

Proton removal from tetralone (**47**) is the limiting factor and the diazo transfer is carried out using the reactive azide triisopropylphenylsulfonyl azide (**46**) (TPPSA).[19]

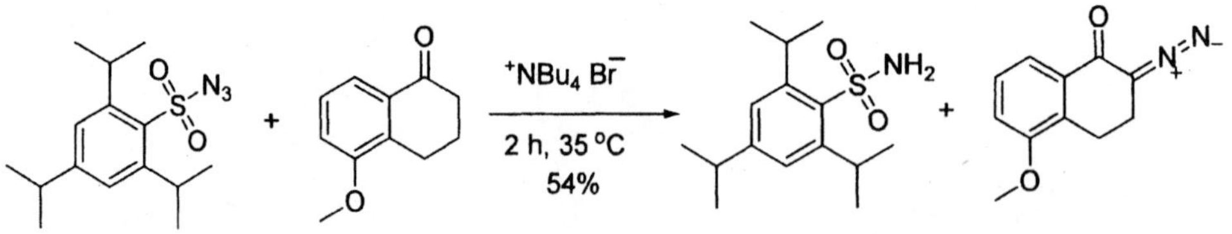

46 47 48 49

A direct diazo transfer reaction to bicyclo[3.2.1]oct-6-en-2-one[20] (**50**) is accomplished with 2,4,6-triisopropylbenzenesulfonyl azide and *t*-BuOK at –78 °C in THF.[20]

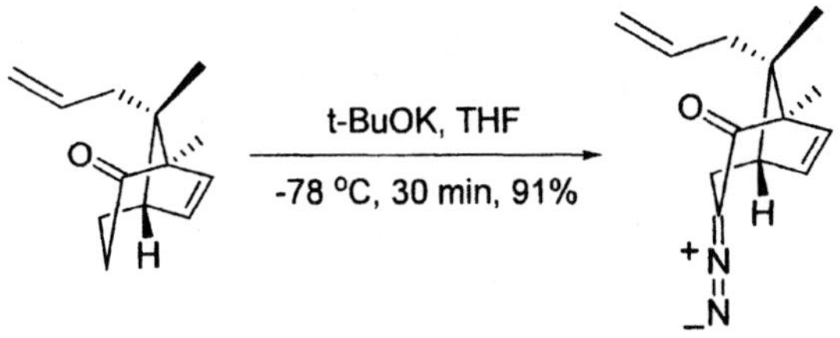

t-BuOK, THF

-78 °C, 30 min, 91%

50 51

The diazo group transfer reaction fails in nonactivated ketones because of insufficient proton labiality in the compound. This can be circumvented by introducing a formyl group *via* Claisen reaction to provide additional activation and this formyl group is eliminated in the course of the diazo transfer reaction.[4,5] The formylation followed by subsequent diazo transfer reaction is demonstrated with cyclopentanone (**52** → **53** → **54**)[4a,4b,4c] α,β-unsaturated ketone (**55** → **56** → **57**)[4d] and 4-*t*-Bu-cyclohexanone (**58** → **59** → **60**).[5,21]

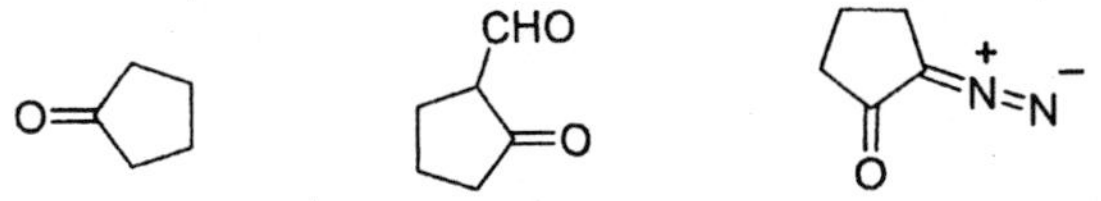

52 53 54

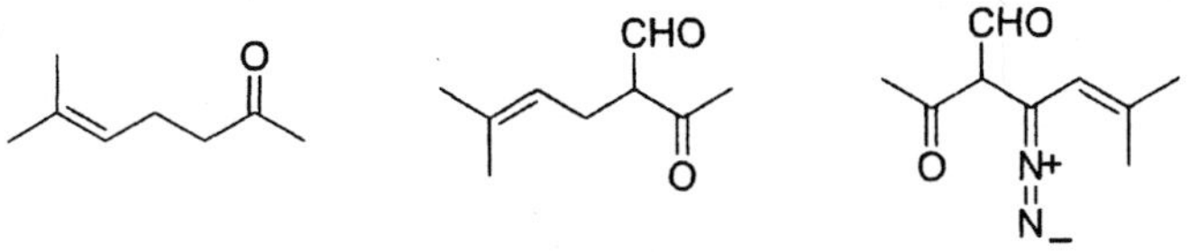

55 56 57

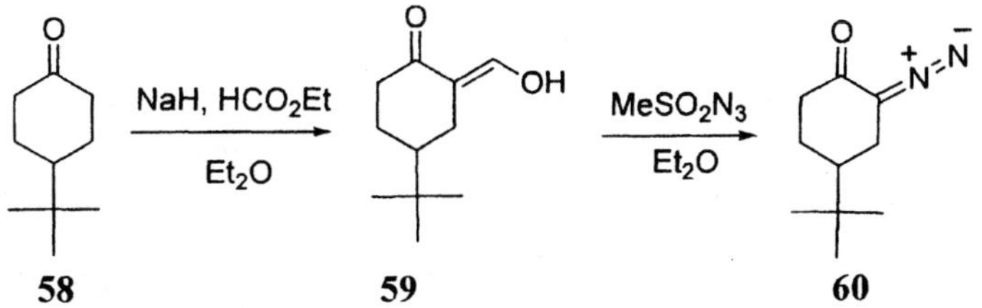

NaH, HCO_2Et, Et_2O → $MeSO_2N_3$, Et_2O

58 59 60

Diazo transfer to formyl methylene ketone **61** is accomplished using 4-carboxybenzenesulfonazide, and the resulting unsaturated alcohol is further oxidized by MnO_2 to unsaturated aldehyde **62**.[22]

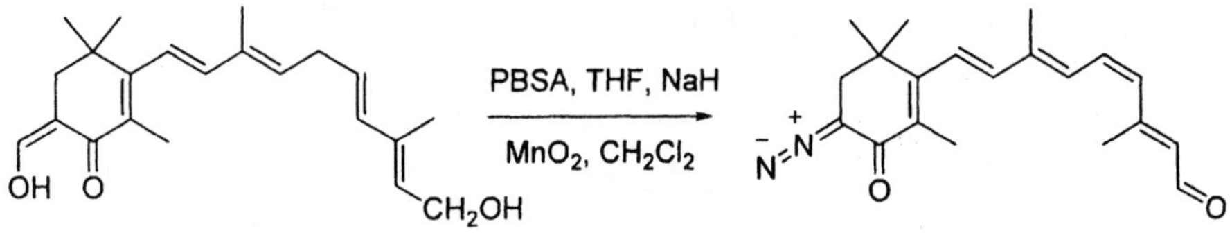

PBSA, THF, NaH

MnO_2, CH_2Cl_2

61 62

5-Diazojasmonate (**64**) is prepared from the 5-formyl derivative of **63** *via* a deformylation diazo group transfer reaction with the 4-carboxybenzenesulphonyl derivative.[23]

PBSA, NEt_3, MeCN
MeCN, 0 °C, 53%

63 **64**

Formylation followed by the deformylation diazo transfer reaction of 4-chromanone (**65**) produces 2-diazochromanone (**66**) in overall yield of 55–60%.[24] Diazo transfer from methanesulfonyl azide to dimethylbenzosuberone (**67**) is achieved *via* deformylation.[25]

65 **66**

HCO_2Et, NaOH, Et_2O
$MeSO_2N_3$, 3 h, 0 °C, r.t.
72%

67 **68**

C-Formylation or α-oximation of the highly hindered α-methylene group in the parent ketone is not possible. Direct transfer of the diazo group from tosyl azide under phase transfer conditions has been found unproductive.[19] A variety of highly hindered ketones are converted to corresponding diazo ketones with 2,4,6-triisopropylphenylsulfonyl azide under phase-transfer conditions.[19] Tetrabutylammonium bromide and 18-crown-6-ether are used as catalysts in these diazo transfer reactions. This method is superior to C-nitrosation followed by diazotization with chloramine. The diazo compounds **66** (55%) and **68** (72%) are obtained using this method.

The mixture of ketones (**69** and **70**) is transformed to the α-diazoketones (**71** and **72**) through reaction of their formyl derivatives using 4-carboxybenzenesulfonyl azide with potassium carbonate either in solution or in the solid phase.[26]

HCO$_2$Et, NaH, Et$_2$O; PBSA, K$_2$CO$_3$, MeCN, 1 h

69 + 70 → 71 + 72

Diazo transfer to cyclopropenyl ketone (**73**) is accomplished with diphenylphosphoro ester azide (DPPA).[27] The conversion of tricyclic ketone **75** to the α-formyl derivative followed by the reaction with tosyl azide yields the diazoketone **76** *via* deformylation.[28]

DPPA, THF, −78 °C, BuLi; 85%

73 → 74

1. HCO$_2$Et, NaH, THF, 14 h, rt; 2. TsN$_3$, NEt$_3$, 24 h; 60%

75 → 76

Treatment of hydroxymethylene bicyclic ketone (**77**) under standard conditions (TsN$_3$, NEt$_3$) produces the diazoketone (**78**).[29]

1. HCO$_2$Et, NaH, EtOH, 75%; 2. TsN$_3$, NEt$_3$, CH$_2$Cl$_2$, −10 °C, 41%

77 → 78

Similarly, 2-formyl cyclohexanone (**79**), 4-formyl seven membered heterocyclic ketones (**81**), and 5-formylbenzfuranone (**83**) are converted to the corresponding diazo ketones **80**, **82**, and **84** using *p*-tolylsulfonyl azide and *p*-benzoic acid sulfonylazide.[4a,30]

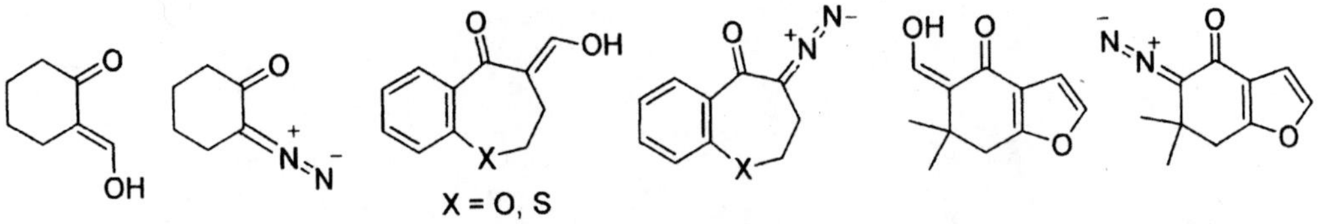

X = O, S

79 80 81 82 83 84

Tricyclodecanone (**85**) is formylated to **86,** then converted to the diazoketone (**87**) *via* deformylation diazo transfer reaction using *p*-toluene sulfonyl azide in the presence of NEt_3.[31]

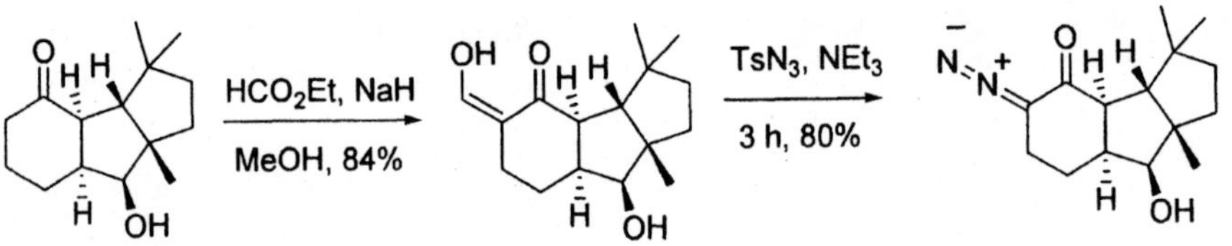

HCO₂Et, NaH / MeOH, 84%

TsN_3, NEt_3 / 3 h, 80%

85 86 87

α-Dinitroso amino ketone (**88**) is diazotized using chloramine to give diazoketone (**89**).[32a] α-Amidoketone is converted to diazoketone (**91**) *via* deamidation diazo transfer reaction in 56% yield.[32a,32b]

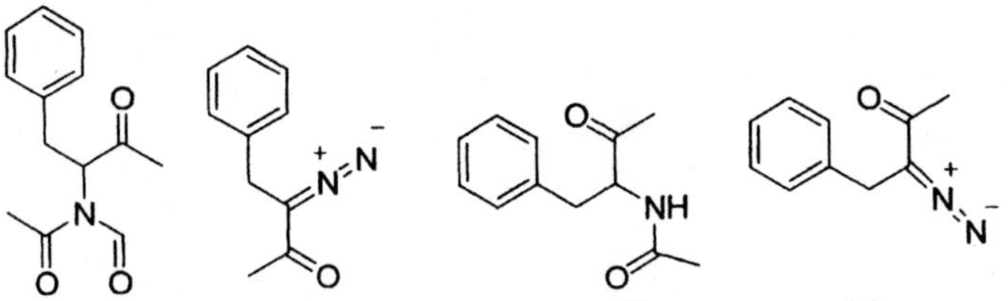

88 89 90 91

The *N*-fluorenyl diazoamide **93** could not be prepared using the tosyl azide; however, the diazo group transfer to diketone **92** is effected successfully using 2-azido-1-ethylpyridinium tetrafluoroborate to furnish diazo ketone **93.**[33]

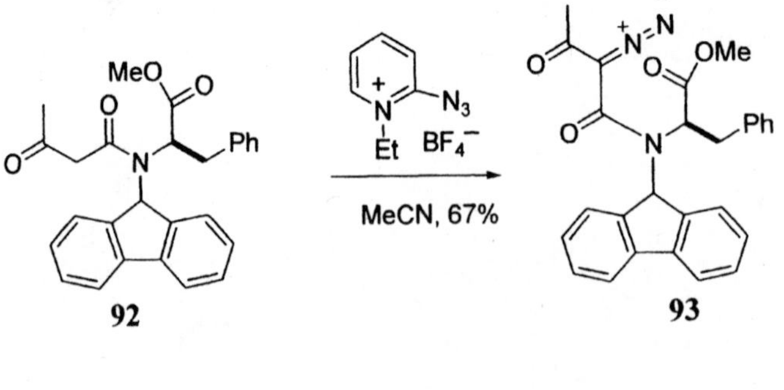

MeCN, 67%

92 93

7.9.4.7 1,2-Diketones to diazoketones

Diazo group transfer to 1,2-diketones *via* ketoxime and sulfonyl hydrazone is generally carried out in the presence of organic bases, and affords high yields of diazo compounds. Dichlorotricyclo[2.2.1.0^{2,6}]heptan-2,3-dione (**94**) is converted into its monotosylhydrazone **95** which is passed over basic alumina to afford yellow crystalline diazo ketone **96**.[34] The monohydrazone of 4,5-diketo-[9]paracyclophane (**97**) is prepared from the diketone and further oxidized with MnO_2 to give diazoketone **98**.[35]

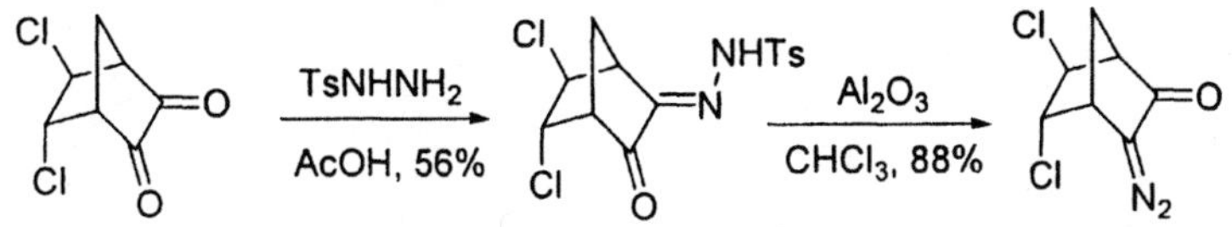

TsNHNH$_2$, AcOH, 56%; Al$_2$O$_3$, CHCl$_3$, 88%

94 **95** **96**

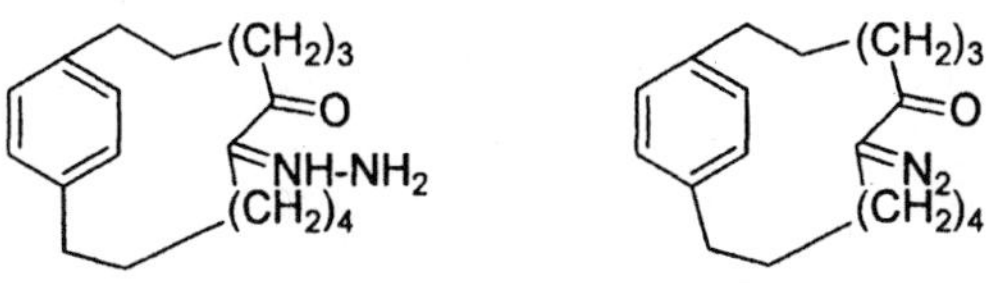

97 **98**

α-Diazo ketone **100** is synthesized *via* the base-catalyzed elimination of the *p*-toluenesulfinate ion from α-dicarbonyl mono-*p*-toluenesulfonylhydrazone (**99**).[36] The action of basic Al_2O_3 at 0 °C on substituted cyclobutanone tosylhydrazone (**101**) affords the diazoketone **102**.[37]

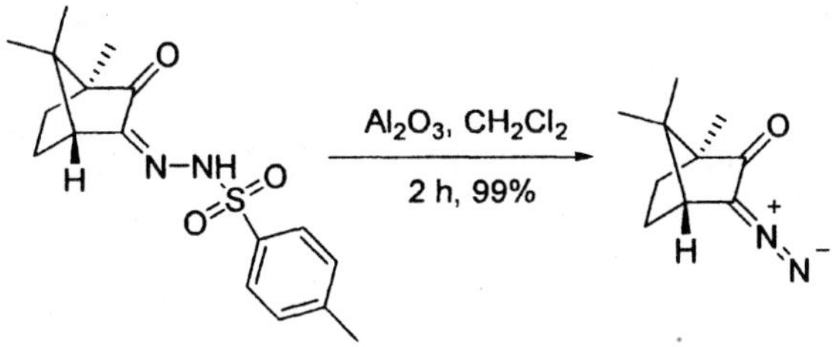

Al$_2$O$_3$, CH$_2$Cl$_2$, 2 h, 99%

99 **100**

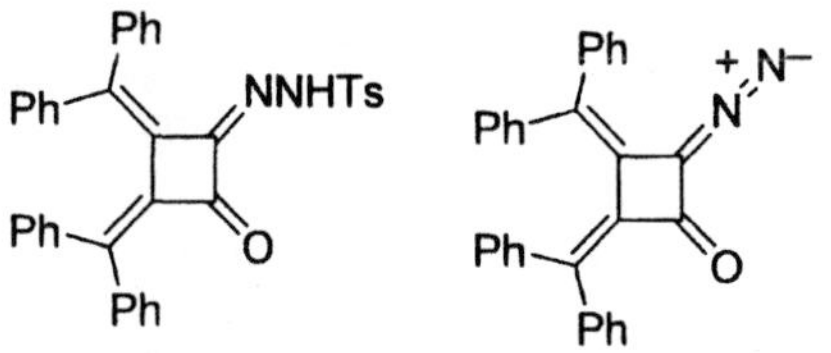

101 **102**

Diazo transfer to 2-(trimethylsilyl)cyclobutenocyclooctatetraenone (**103**) *via* formyl ketone was unsuccessful. Conversion to ketoxime (**104**) and thence to the diazo ketone **105** proceeds readily with NH_3 and NaOCl.[38] Similarly, indanoneoxime produces diazoketone **106** with NH_2Cl and KOH.[39] The hydroxymethylene-α-tetralone derivative (**107**) reacts with *p*-tolylsulfonyl azide to give diazo-α-tetralone (**108**) *via* deformylation.[40]

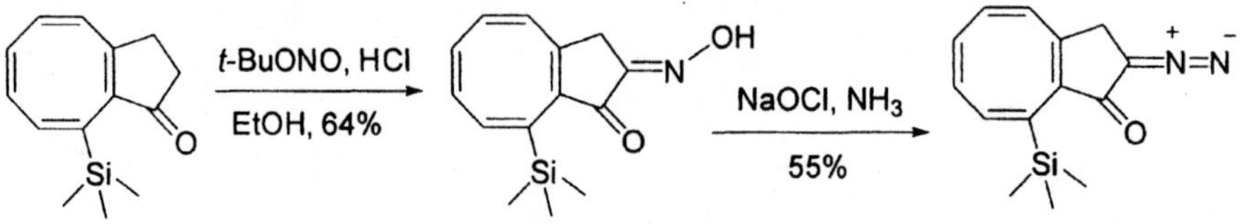

t-BuONO, HCl
EtOH, 64%
NaOCl, NH_3
55%

103 **104** **105**

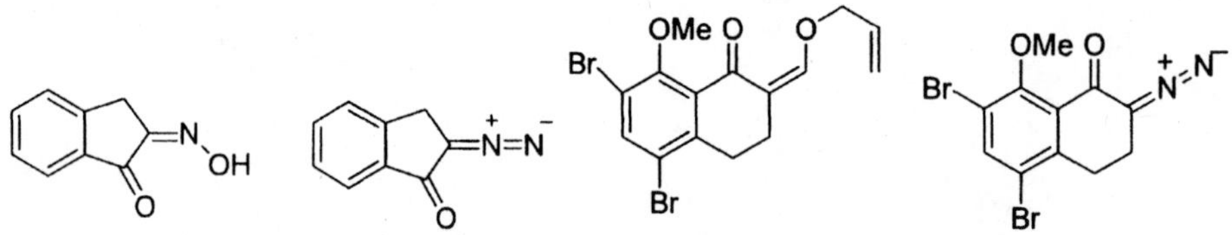

106 **107** **108**

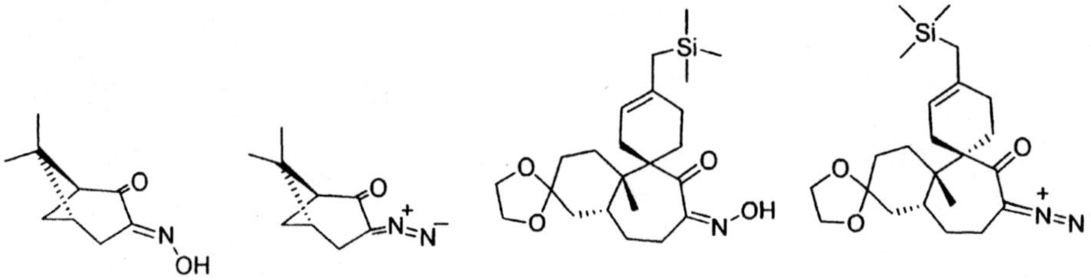

109 **110** **111** **112**

Ketoximes **109** and **111** are converted to the corresponding diazoketones **110** and **112** using chloramine (NH_2Cl).[41,42]

2-Diazo-A-norcholestan-1-one (**114**) is prepared from *syn*-2-oximino-A-norcholestan-1-one (**113**) with chloramine in aqueous ether in 63% overall yield.[20,43–45] Steroidal ketones **115** and **118** (D-ring and A-ring) are converted to the corresponding diazoketones **117** and **120** *via* ketoximes **116** and **119** using chloramine with *n*-butyl nitrite and *t*-BuOK in 55% and 78% yields, respectively.[46]

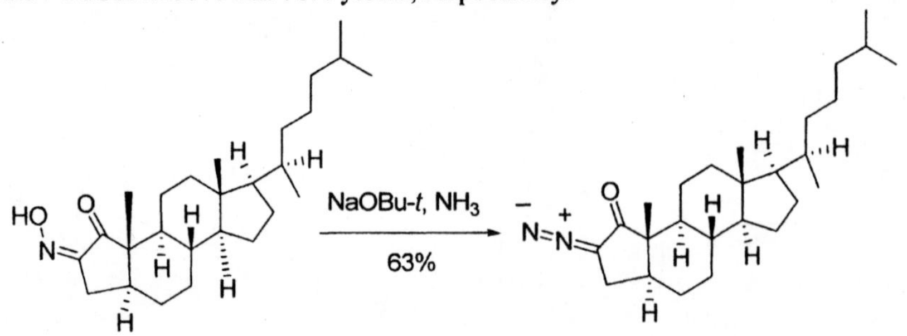

NaOBu-*t*, NH_3
63%

113 **114**

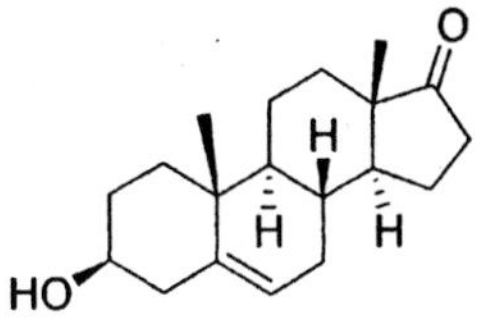

115 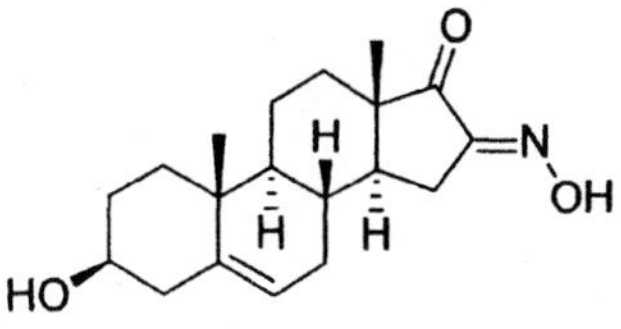 116 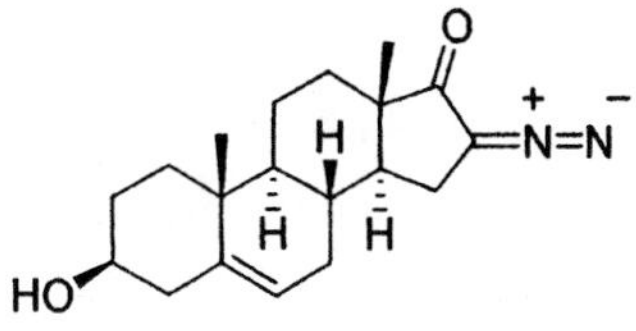 117

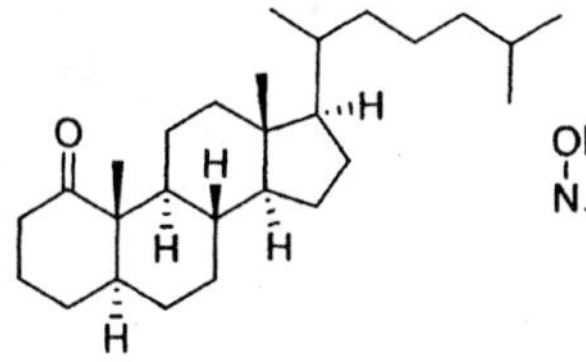

118 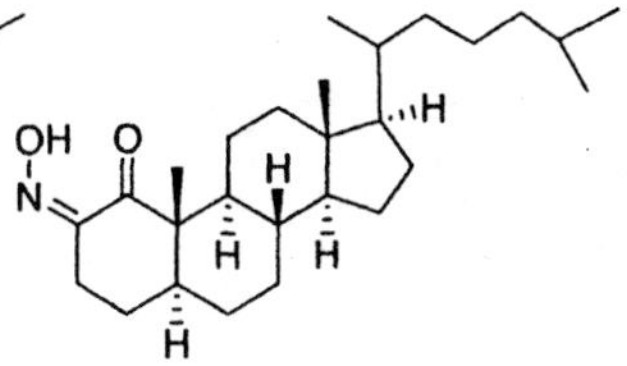 119 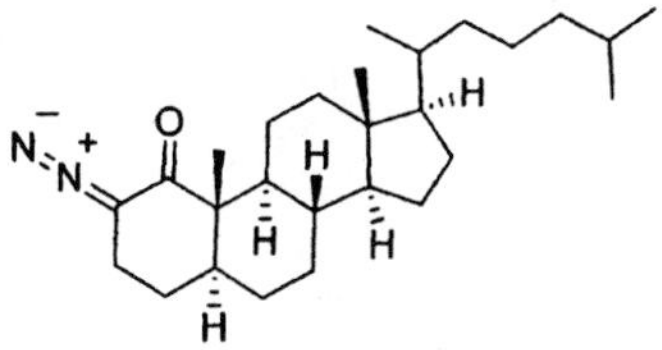 120

The mixture of eneaminoketones **121** is converted to the corresponding α-diazoketones **122** by alkylation with Bredereck's reagent followed by treatment with *p*-toluenesulfonyl azide.[47] The adenosine derivative (**123**) is treated with trifluoromethylsulfonyl azide to provide diazoketone **124** *via* deamination diazo transfer reaction in 72% yield.[48]

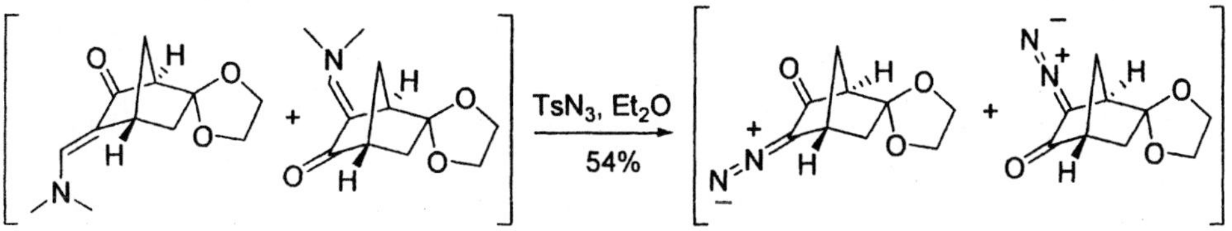

TsN_3, Et_2O 54%

121 122

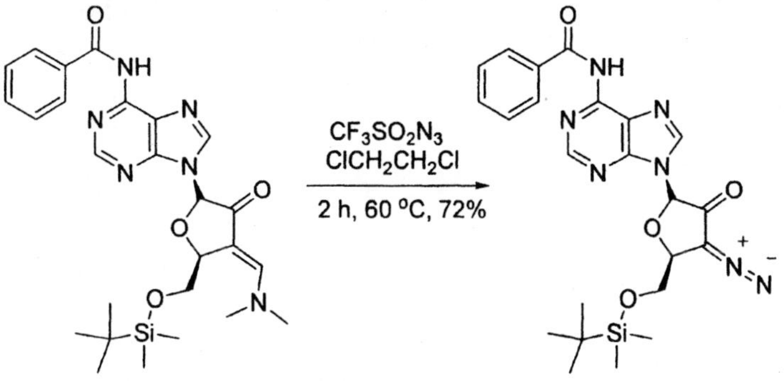

$CF_3SO_2N_3$ $ClCH_2CH_2Cl$ 2 h, 60 °C, 72%

123 124

7.9.4.8 Diazo 1,3-diketones

α-Diazo 1,3-diketones were prepared in excellent yield from 1,3-diketones with *p*-toluenesulfonyl azide and cesium carbonate[49]; the advantage is a milder, faster, easier product separation.[49] Diazo-1,3-diketones are prepared from 1,3-diketones using *p*-carboxybenzenesulfonyl azide and their yields are comparable with *p*-toluene sulfonyl azide.[30,50] Pentane-2,4-dione is conveniently converted to diazo compound **126** in high yield using azidotris-(diethylamino)phosphonium bromide (**125**) with a catalytic amount of base. Diazo transfer agent **125** is exceptionally safe to handle in the laboratory.[51] Diazo ketones are easily separated from the co-product hexaethylphosphoramide triamide hydrobromide (**127**).[51]

$(Et_2N)_3-P=N-N^+\equiv N$ Br^- **125**, *t*-BuOK, Et_2O

126 **127**

Cyclic ketones are converted smoothly to diazo ketones in the presence of HMPA. 4-Nitrophenyl azide (**128**) exhibits a diazo transfer reaction with malonates and cyclic 1,3-cyclohexanedione (**128**); in some cases diazo compound **129** further leads to fused triazole derivative (**130**) formation.[52]

NEt_3, HMPA, 55 °C, 1 h, 94%

128 **129** **130**

Treatment of 4-methylcyclohexane-1,3-dione (**131**) with tosyl azide in the presence of NEt_3 led to the corresponding diazo cyclohexane-1,3-dione (**132**).[53] Diazo transfer from tosyl azide successfully inserted the diazo group to cyclooctanedione (**133**) and provided **134** in 50% yield.[54]

131 **132** **133** **134**

Dimedone provides diazo dimedone (**135**) with *p*-benzoic acid sulfonylazide (84%) and with *p*-tolyl sulfonylazide (96%).[30] Weak nucleophilic base such as 1,8-diazabicyclo[5.4.0]undecen-7-ene (DBU) has been used as catalyst for the diazo transfer reactions of 1,3-diketones (**136 – 139**).[55] A highly efficient methodology in solid state has been developed for the synthesis of α-diazo carbonyl/sulfonyl compounds from 1,3-diketones using tosyl azide. This method avoids any aqueous workup and diazo compounds are obtained *via* a column filtration over silica gel.[30]

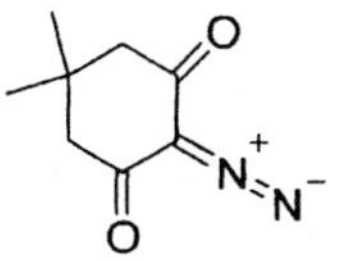

135 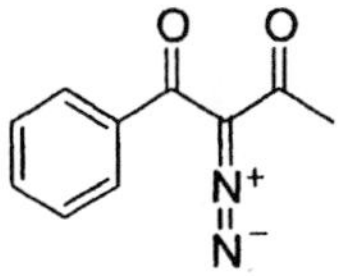 136 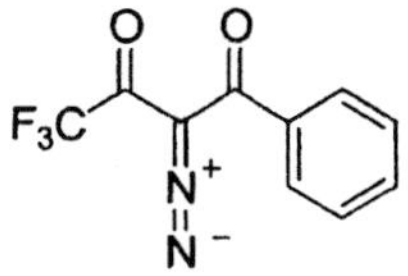 137

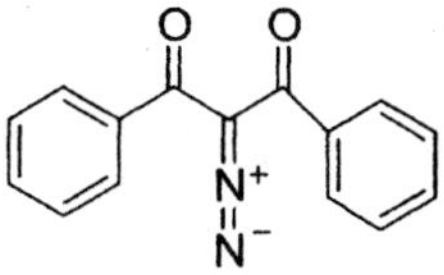

138 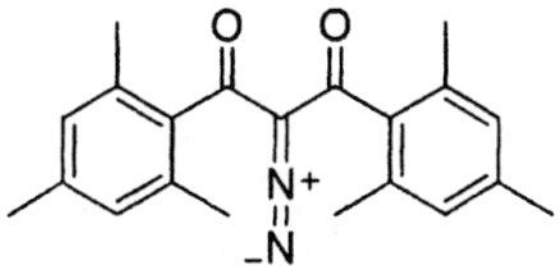 139

A mild method for the preparation of various 2-diazo-1,3-dicarbonyl compounds in the presence of solid acids such as clays in a heterogeneous manner provides good yield.[56] The following clays are natural, commercially available in bulk, and environmentally safe: smectite, atapulgite, and vermiculite. They are recovered and reused in diazo transfer reactions without losing their catalytic activities. Clays are filtered away from the product mix. By this method diazo dimedone (**135**) is prepared in 85% yield.[56–58] Diazo ketones **136** and **138** are also prepared from the corresponding ketones in 92% and 96% yields, using a triphase transfer system (tributylmethyl ammonium chloride immobilized on a polystyrene matrix in CH_2Cl_2).[59] Diazo ketone **136** is prepared using PS-SO_2N_3 and *p*-CBSA in quantitative yield. Unlike tosyl azide, polymer bound tosyl azide is safe and stable and improves the yield of the diazo 1,3-diketones.[30,60, 61]

Diazo 1,3-diketones **140** and **142** are prepared using TsN_3 in the presence of $CsCO_3$.[49,62] The advantages are mild reaction conditions, faster, excellent yields (91–99%), and easier product separation from the reaction mixture.[49,62]

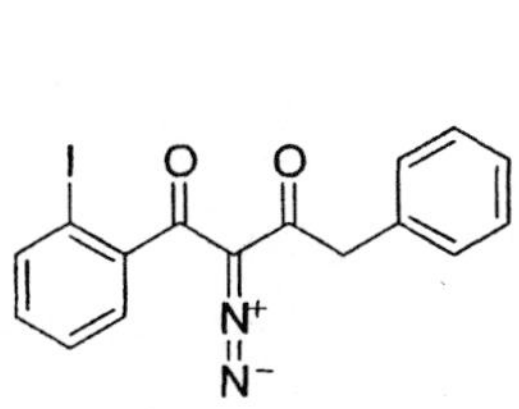

140 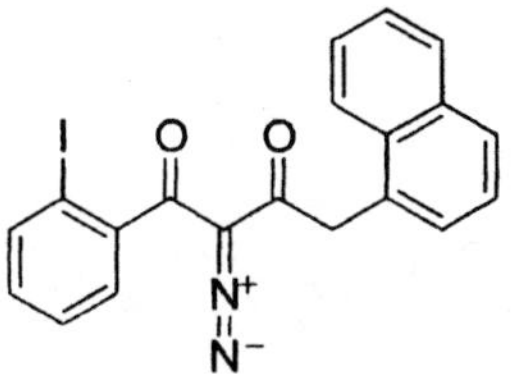 141 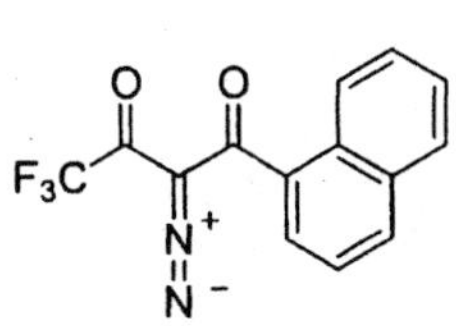 142

Substituted diazo indan-1,3-diones (**143,** R = H, 75%, R = Br 84%) and 1,3-diketone **144** are prepared using clays in superior yield. [57,58] Transfer of the diazo function to stereoisomeric *cis*- or *trans*-4,6-di-*t*-Bu-1,3-cyclohexanedione (**145**) is achieved using TsN_3 and KF as base; in the hindered ketone side reaction, formation of azine is avoided.[60]

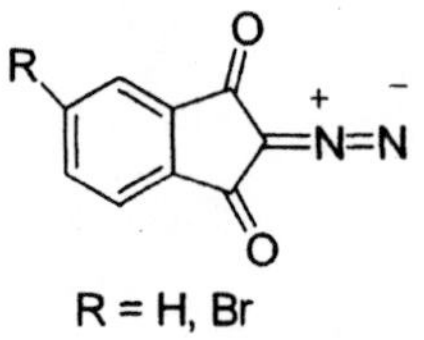

R = H, Br
143

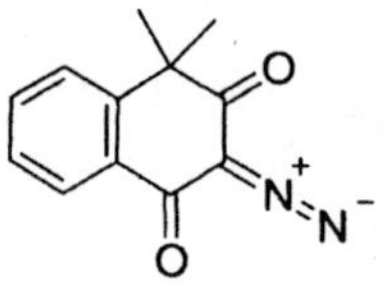

144

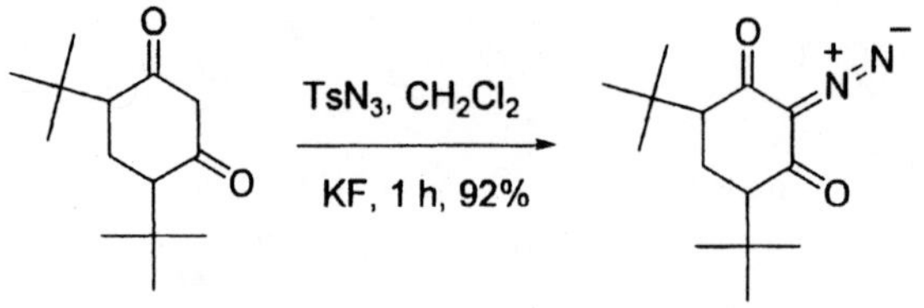

TsN_3, CH_2Cl_2
KF, 1 h, 92%

145 **146**

Polystyrene supported benzenesulfonyl azide ($PS\text{-}SO_2N_3$) offers several advantages over 4-carboxybenzenesulfonyl azide and 4-acetamidobenzenesulfonyl azide: $PS\text{-}SO_2N_3$ is thermally stable, and friction insensitive, with improved safety characteristics over other available reagents.[63] $PS\text{-}SO_2N_3$ allows rapid isolation of the α-diazo product with no aqueous workup. In most cases, there is no need for purification of the final product. The yields for the formation of various diazo 1,3-diketones are compared with diazo transfer reagents $PS\text{-}SO_2N_3$, *p*-CBSA, and tosyl azide.[64]

2-Nitro-1,3-diketone (**147**) is converted to 2-diazo-1,3-diketone (**148**) *via* reduction to amino followed by diazotization in 55% yield.[65] Diazo transfer reaction of 2-amino-1,3-diketone (**149**) to α-diazo-1,3-diketone (**150**) is achieved *via* diazotization in 91% yield.[66]

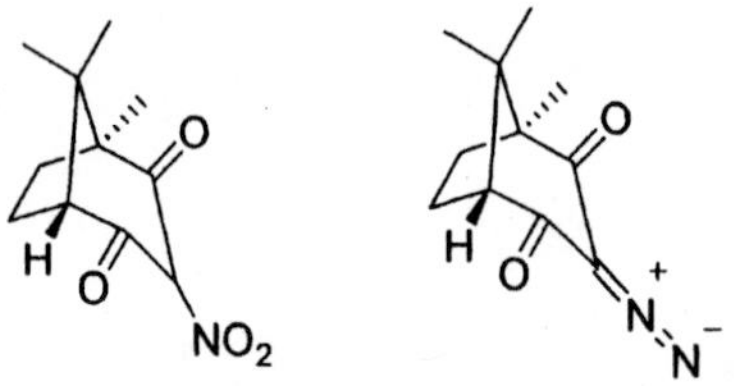

147 **148** 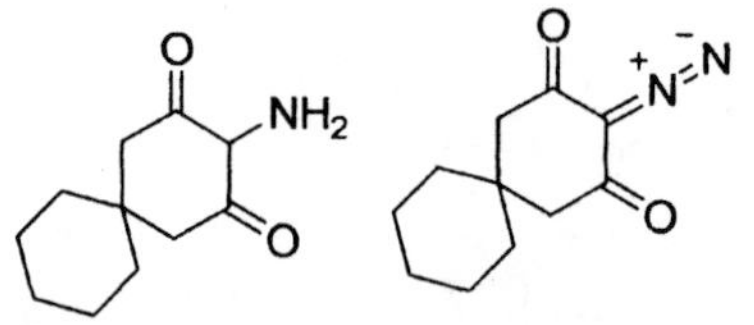 **149** **150**

The reaction of tosyl azide with indane-1,3-dione (143, R = H), carried out in HMPA or in diethyl ether in the presence of NEt_3, fails to afford the expected diazo transfer indan-1,3-dione.[67] However, the 2-phenylsulfanylindan-1,3-dione (**151**) in the presence of NEt_3 gives diazo diketone (**152**) in 36% yield.[49,67]

TsN$_3$, NEt$_3$, THF
50 °C, 36%

151 **152**

α-Acetyl and α-benzoyl substituted cyclic and acyclic 1,3-diketones are successfully converted to diazo 1,3-ketones using TsN_3 and *p*-nitrobenzenesulfonyl azide.[50,68] Acyl and benzoyl groups are eliminated during the course of diazo transfer reactions.

7.9.4.9 Diazo β-keto esters

The diazo keto ester **154** is prepared from the β-keto ester **153** using tosyl azide or *p*-NBSA.[69,70] The diazo ester (**156**) could not be prepared by diazo transfer with TsN_3/NEt_3, since elimination usually occurs to yield acrylate.[71] A diazo transfer agent, 1-ethyl-2-azidopyridinium tetrafluoroborate, successfully converted isoindole keto ester (**155**) to diazo keto ester (**156**).[71]

TsN$_3$, NEt$_3$
87%

153 **154**

NaOAc, EtOH, 42%

155 **156**

7.9.4.10 2,4-bis-Diazo-1,3,5-triketone

Pentane-1,3,5-trione (**157**) is converted into the bis-diazopentanetrione (**158**) using *p*-tolysulfonyl azide.[72] Bis-diazopentanetrione (**158**) is further oxidized with *t*-BuOCl and HCO_2H to pentaketone.[72]

2 TsN$_3$,
NEt$_3$, MeCN
88%

157 **158**

7.9.4.11 *Diazo oxazolidone*

The sodium enolate of *N*-acyloxazolidone (**159**) reacts with sodium hexamethyldisilazide and PNBSA and followed by treatment with TMSCl affords the diazo compound **160**.[73]

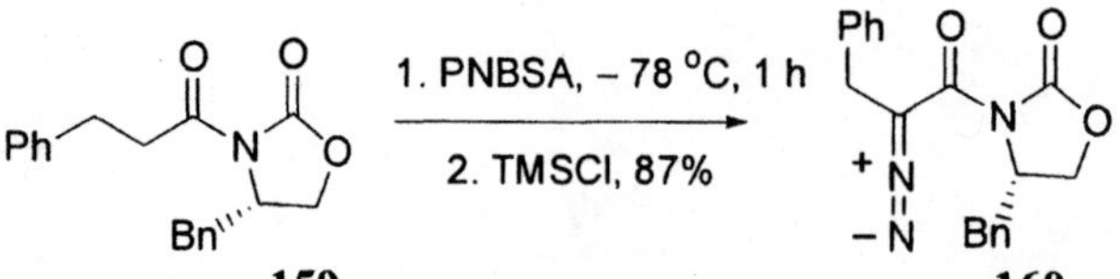

1. PNBSA, – 78 °C, 1 h
2. TMSCl, 87%

159 **160**

7.9.5. *Synthetic Utility*

Diazo compounds undergo substitution, addition, reduction, and cleavage reactions and the diazo group is retained with the final products. However, the diazo group is eliminated in the case of Wolff rearrangement, photolysis, and Rh-catalyzed C–H insertion reactions.

7.9.5.1 *Nitration*

Ethyl diazoacetate (**161**) is smoothly nitrated by N_2O_5.[74a] The reaction is initiated by electrophilic attack of a NO_2^+ at the carbon bearing the diazo function.[74a] Nitration of diazocyclopentadiene (**14**) with silver nitrate-benzyl chloride takes place in the C-1 and C-3 to afford the isomeric diazonitrocyclopentadienes (**163** and **164**) in the ratio 2:1.[75]

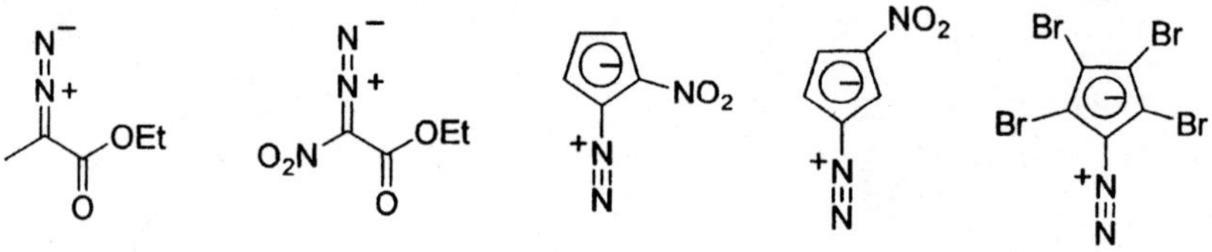

161 **162** **163** **164** **165**

7.9.5.2 *Halogenation*

Diazomethane itself undergoes transformation to chloro diazomethane at −100 °C, under the influence of *t*-BuOCl, but the product suffers decomposition with loss of N_2 at temperatures above −40 °C.[74b] Diazocyclopentadiene (**14**) is perbrominated to **165** using *N*-bromosuccinimide (NBS).[75]

7.9.5.3 *Metalation*

Diazomethane ($CH_2{=}N_2$) undergoes a hydrogen–lithium exchange with methyllithium in ether.[76] Triphenylmethylsodium (tritylsodium) has been used to introduce sodium into diazomethane.[77] Both diazomethyllithium and diazomethylsodium are only handled in the form of suspensions, since they are highly explosive in the dry state. Metalation of diazocyclopentadiene (**14**) by Hg(II)acetate followed by NaI treatment produces diazo bis-mercury iodide **166**, which is unstable and is further reacted with MeI to give 1-diazo-2,5-diiodocyclopentadiene (**167**).[75] The lithio derivative of ethyl diazo acetate (**168**) undergoes substitution reactions, for example, silylation to **169**.[78]

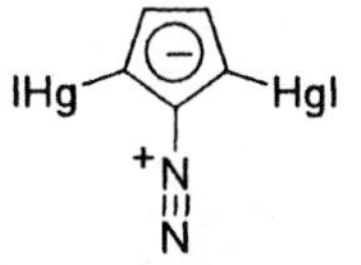

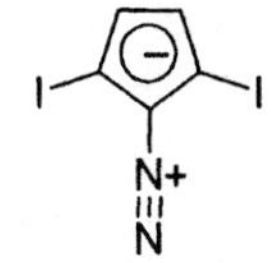

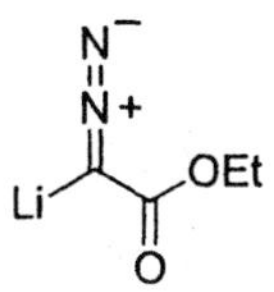

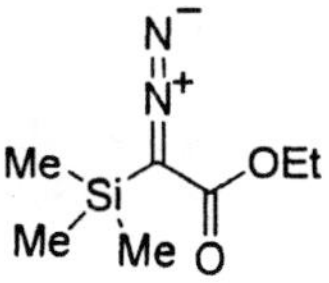

166 167 168 169

7.9.5.4 Aldol type addition

Aldol type addition of the ethyl diazo acetate **170** to indan-1,2-dione is particularly facile and catalyzed by the base.[79]

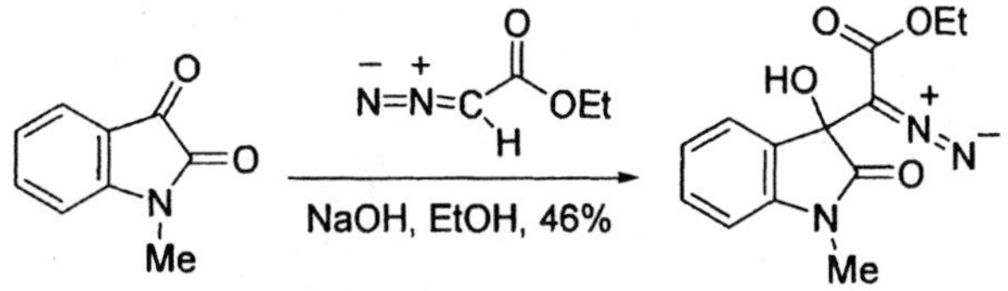

NaOH, EtOH, 46%

170 171

7.9.5.5 Addition to electron-rich alkenes

Electron-rich alkenes such as 1,1′,3,3′-tetraphenyl-2,2′-bisimidazolidinylidene **172** reacts with two equivalents of diazo **173** with cleavage to afford α-diazoaminal **174.**[80]

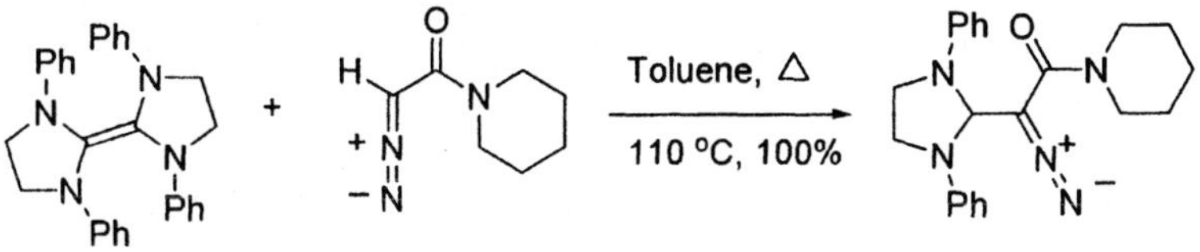

Toluene, Δ

110 °C, 100%

172 173 174

7.9.5.6 Acylation

Carboxylic acids are first treated with dicyclohexylcarbodiimide (DCC) to afford anhydrides which are allowed to react with the desired diazo alkane. This same sequence is also used to transform α-phthalimidocarboxylic acid **175** into diazoketone **176.**[81]

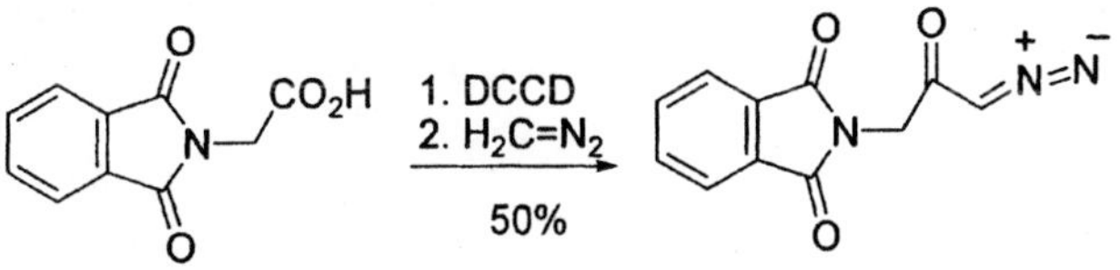

1. DCCD
2. $H_2C=N_2$

50%

175 176

7.9.5.7 Addition of diazo 1,3-ketones to α,β-unsaturated ketones

Addition of diazo-1,3-diketone **136** to cyclohexeneone affords the 1,4-addition product (**178**) in 3% yield and 1,2-addition product (**177**) in 56% yield. Similarly, addition of diazoketone **136** to α,β-unsaturated ketone **179** gives the 1,4-addition product (**181**) and 1,2-addition product (**180**).[82] In the absence of Lewis acid, Ti enolate with cyclohexenone gives a mixture of 1,2- and 1,4-addition products in low selectivity.[82] $TiCl_4$ activation gives the 1,4-addition product, whereas $BF_3 \cdot Et_2O$ activation slightly increases the amount of 1,2-addition when the enone is activated with $Ti(OPr)_4$, and the addition product is greatly increased.[82] The reaction of the Ti-enolate of α-diazo-β-ketone (**136**) with **179** $TiCl_4$ gives 1,2- and 1,4-addition products (**180** and **181**) in 70% yield, and with two equivalents of $TiCl_4$ leads to a major product **181** *via* 1,4-addition.[82] With $SnCl_4$, the selectivity for 1,4-addition is enhanced. On the other hand, with $BF_3 \cdot Et_2O$, the 1,2-addition product became predominant and with $Ti(Oi\text{-}Pr)_4$, further enhanced. Regiocontrol by Lewis acids has also been observed for cyclic ketones.[82]

$TiCl_4$, CH_2Cl_2, - 78 °C

136 **177** **178**

+ **136**, $TiCl_4$, CH_2Cl_2, - 78 °C

179 **180** **181**

7.9.5.8 Addition of diazo to sulfeneamine

The nucleophilic addition of Ti(IV) enolates of α-diazo-β-ketones **136** and **126** to *N*-tosylimine **182** and **184** is successfully promoted by the activation of $TiCl_4$ to give δ-(*N*-tosyl)amino substituted β-keto diazo carbonyl compounds **183** and **185.**[83]

$TiCl_4$, NEt_3, CH_2Cl_2, - 41 °C, 70%

136 **182** **183**

$TiCl_4$, NEt_3

CH_2Cl_2, - 41 °C

13%

126 **184** **185**

7.9.5.9 Addition of diazo to cyclic ketone

Nucleophilic addition of Ti(IV)-enolates of diazo ketone (**136**) to cycloheptanone (**186**) affords alcohol **187** which, on further dehydration with trifluoroacetic anhydride gives α,β-unsaturated carbonyl compounds (**188 & 189**) in 58% yield.[84]

$TiCl_4$, Ti (*i*-PrO_4)

THF, - 23 °C, 64%

136 **186** **187**

$(CF_3CO)_2O$, NEt_3

CH_2Cl_2, - 78 °C, 17 h

84%

188 **189**

7.9.5.10 Reduction of diazo-1,3-diketones to diazo hydroxyketones

Reduction of the carbonyl group in cyclic and acylic 2-diazo-1,3-diketones (**129, 135–137, 143, 146,** and **196**) with $NaBH_4$ in aqueous alcoholic solution, followed by hydrolysis of the reaction mixture over wet silica gel, affords the corresponding 3-hydroxy-2-diazoketones (**190–195** and **197**) in 58–87% yield.[85] Steric hindrance at α to the carbonyl decreases the yield of reduced products.[85]

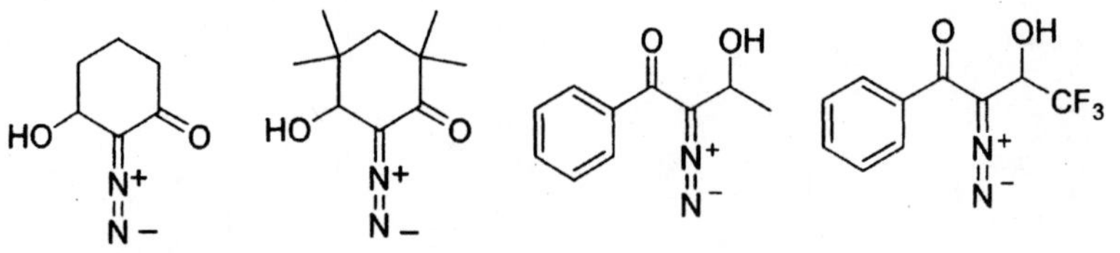

190 191 192 193

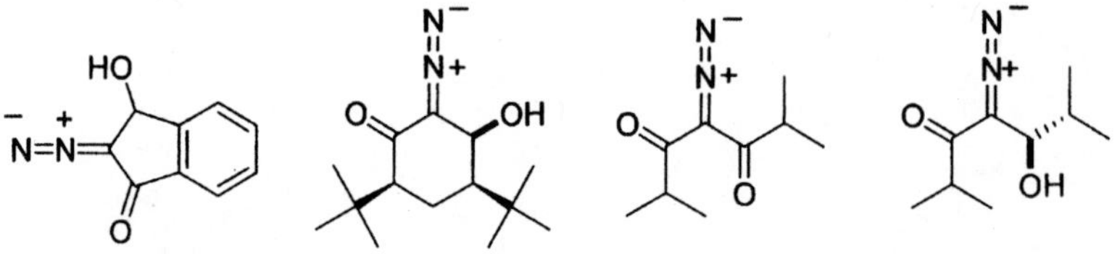

194 195 196 197

7.9.5.11 Silylation

The total synthesis of the isoprostane (±)-8-epi-PGF$_{2\alpha}$ ether is accomplished *via* cyclization of diazo keto silyl derivative **200**. Aldol condensation of diazoketone **198** with decadienal (**199**) in the presence of KHMDS followed by addition of LiBr provides an intermediate β-hydroxy adduct **201** which is not isolated but is immediately silylated to give silyl derivative **200.**[86] Diazo hydroxy ketones **201** and **203** are silylated with ClSi(Et)$_3$ and chloro dimethyl-*t*-tributylsilane to afford silylated diazoketones **202** and **204,** respectively.[86]

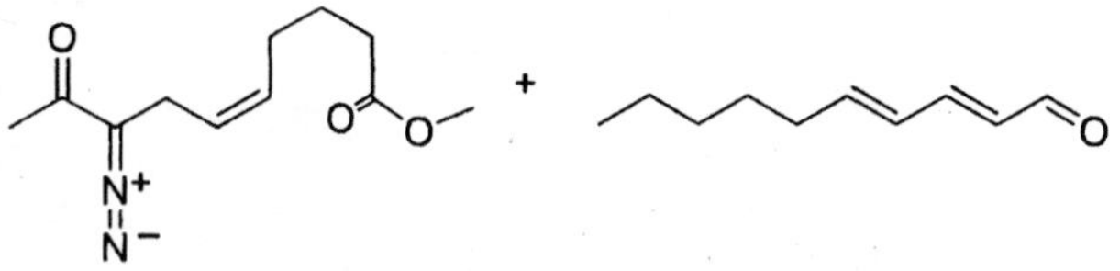

198 199

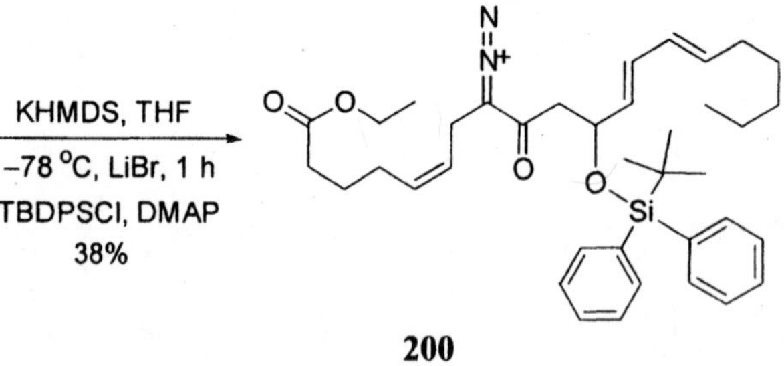

KHMDS, THF
−78 °C, LiBr, 1 h
TBDPSCl, DMAP
38%

200

201 202 203 204

7.9.5.12 Hydrogenation

Elimination of diazo group is observed during the catalytic hydrogenation of β-hydroxy diazoketone **205** to β-hydroxyketone **206** and it has the S configuration at C-4.[87]

H_2, Pd/C; EtOAc, 5h, 90%

205 206

7.9.5.13 Ring cleavage reactions

Sodium hydroxide cleavage of 2-diazo-1,3-dioxoindane (**146**, R = H) affords the expected ring cleavage product diazoketone **207** which cyclizes with concomitant loss of nitrogen upon acidification leading to isochroman-1,4-diketone (**208**).[88]

NaOH, 18 - 24 °C; 86%

207 208

7.9.5.14 Wolff rearrangement

Flash photolytic Wolff rearrangement of α-cyanodiazoacetophenone (**210**) produces α-cyano-α-phenylacetic acid (**211**) *via* hydration of the ketene intermediate.[89]

hν; H_2O, H^+

209 210 211

The irradiation of the diazo-1,3-ketone **132** in MeCN at 254 nm with methanol leads to the formation of cyclopentenone carboxylic esters (**212** and **213**) *via* Wolff rearrangement.[53] Photolysis of dichloro diazo ketone **96** in MeOH gives a mixture of methyl endo- and exo-*trans*-2,3-dichlorobicyclo[2.1.1]-hexane-5-carboxylate **214** in 85:15 ratio.[34]

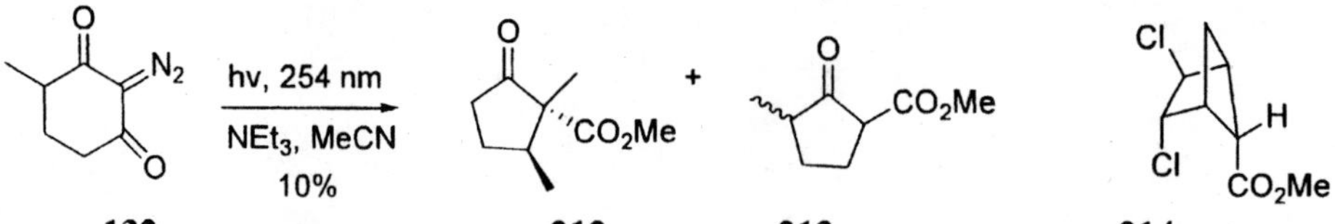

hv, 254 nm
NEt_3, MeCN
10%

132 **212** **213** **214**

A Wolff rearrangement of α-diazoketone (**51**) in the presence of water gives the ring contraction product, bicyclic[2.2.1]heptanecarboxylic acid (**215**).[20] Irradiation of the diazoketone mixture (**71** and **72**) in dioxane–water (2:1) afforded a mixture of *syn* and *anti* carboxylic acid (**216**).[26] Photolysis and Wolff rearrangement of diazoketone **76** at 150 °C in MeOH affords the methyl ester of tricyclic[3.2.1.$1^{2,4}$]nonane-*endo*-3-carboxylic acid in 51% yield, alkaline hydrolysis of which gives the *exo*-carboxylic acid (**217**).[28] Wolff rearrangement of cyclopentaannulene[8]diazoketone (**105**) affords the cyclobutaannulene[8] ester (**218**).[38]

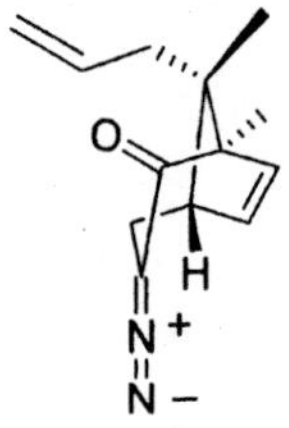

215 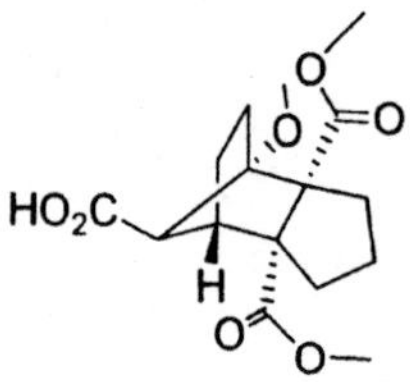 **216** 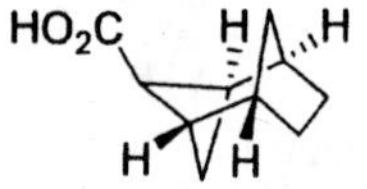 **217** 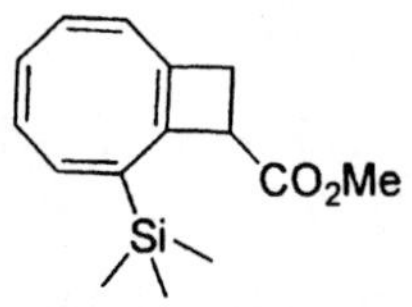 **218**

α-Diazoketone **219** is photolyzed in MeOH *via* Wolff rearrangement and provides the indene derivative (**220**).[18]

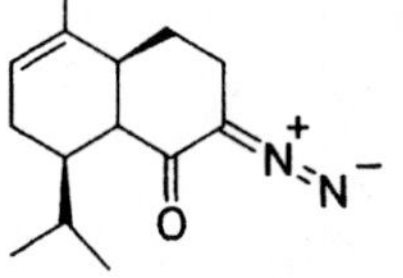

219 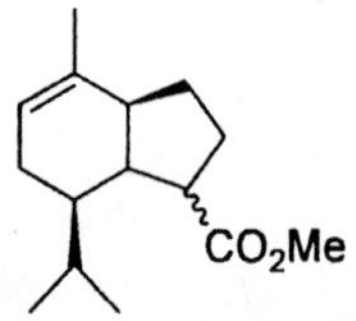 **220**

Irradiation of paracyclophane diazo ketone **98** affords the ring contraction acid **221** in 25% yield.[35] Photolysis of the α-diazobenzocycloheptanone (**68**) in methanol effected the Wolff rearrangement to give the tetralin ester (**222**).[25] Irradiation of 2-diazoindan-1-one in THF/H_2O produces benzcyclobutane carboxylic acid (**223**).[39]

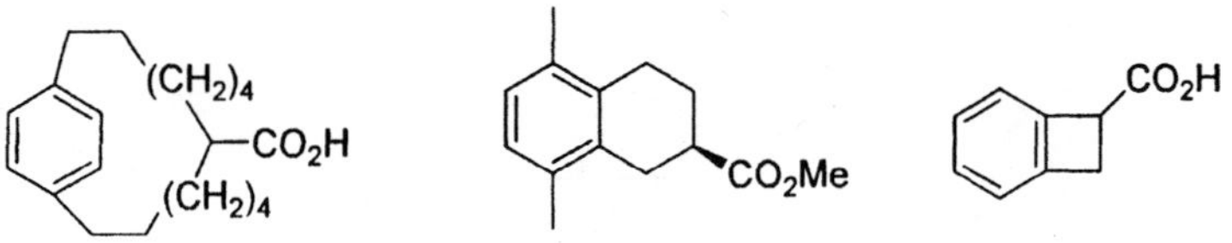
(CH2)4
CO2H
(CH2)4
CO2Me
CO2H

221 222 223

Irradiation of 2-diazo-3,4-bis(diphenylmethylene)-cyclobutanone (**102**) in the presence of alcohol and amine affords 1-alkoxycarbonyl- (**224**, R = OMe), and 1-carbamoyl-2,3-bis(diphenyl-methylene)cyclopropane (**224**, R = NHPh), respectively.[37]

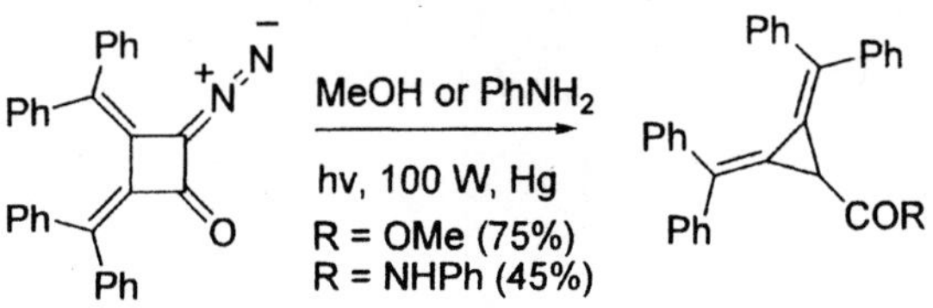
MeOH or PhNH2
hv, 100 W, Hg
R = OMe (75%)
R = NHPh (45%)
COR

102 **224**

Diazoketone **78**, under FVP conditions, undergoes a loss of nitrogen and Wolff rearrangement to the corresponding ketene (**225**) which subsequently undergoes loss of furan *via* retro Diels–Alder reaction and yields the desired propadienone (**226**).[29] Diazoketone **78** sublimes through a tube heated to 430 °C at 10^{-4} Torr and the pyrolyzates, furan and dimer of dimethylpropadienone are trapped at −196 °C.[29] Photolysis of diphenylphosphinylbenzoyldiazomethane (**227**) gives the phosphoric acid derivative **228**.[15]

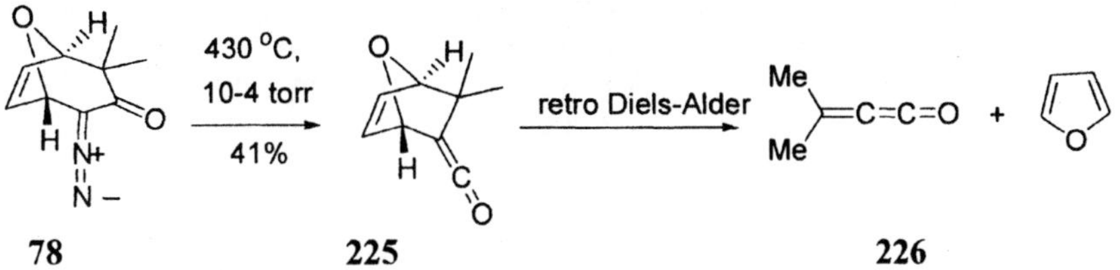
430 °C,
10-4 torr
41%
retro Diels-Alder
Me
Me
C=C=O
+

78 **225** **226**

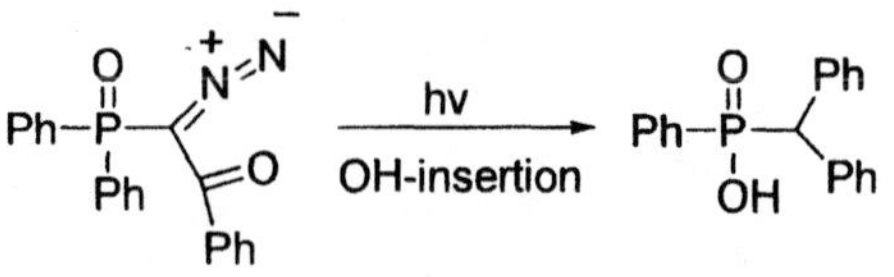
hv
OH-insertion

227 **228**

A number of modified steroids are known in which one of the rings is contracted by one carbon atom relative to the normal tetracyclic steroid nucleus *via* diazotization. Irradiation of diazoketones affords D-nor-, A-nor, and A-bis-nor-androstane carboxylic acids (**229**–**231**).[20,46]

hv, THF

aqu. $NaHCO_3$

44%

229

hv, THF

aqu. $NaHCO_3$

63%

230

hv, THF

aqu. $NaHCO_3$

32%

231

7.9.5.15 Elimination

3-Diazochromanone (**65**) undergos rapid elimination of the diazo group in the presence of BF_3-Et_2O to furnish chromone (**232**).[24] This method is complementary to procedures that are reported for the transformation of appropriately substituted chromanone to chromone.

CH_2Cl_2

$BF_3.Et_2O$

95%

65 232

7.9.5.16 Rh-mediated intramolecular C–H insertion

Rh-mediated intramolecular C–H insertion is effected by $Rh_2(OAc)_4$ to afford novel fused cyclic ketones.[70] The ring fused cyclopentenone derivative (**235**) is developed based on $Rh_2(NHCOMe)_4$-catalyzed diazo decomposition reaction of **233** and **234.**[84]

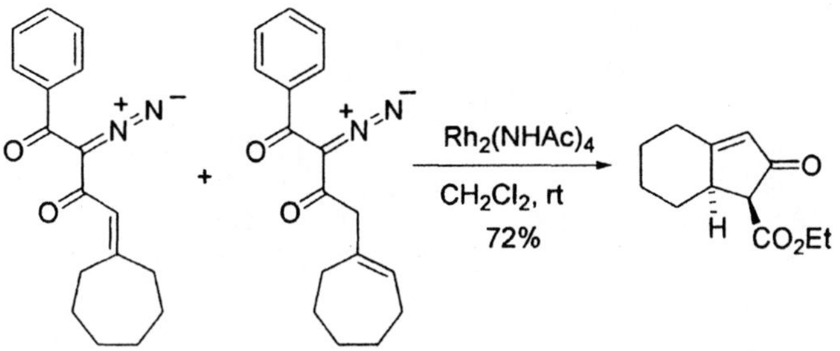

233 234 235

Diazo ester (**236**) is subjected to the action of $Rh_2(OAc)_4$ to yield the tetrahydroisoquinoline (**237**) as a single diastereoisomer.[71]

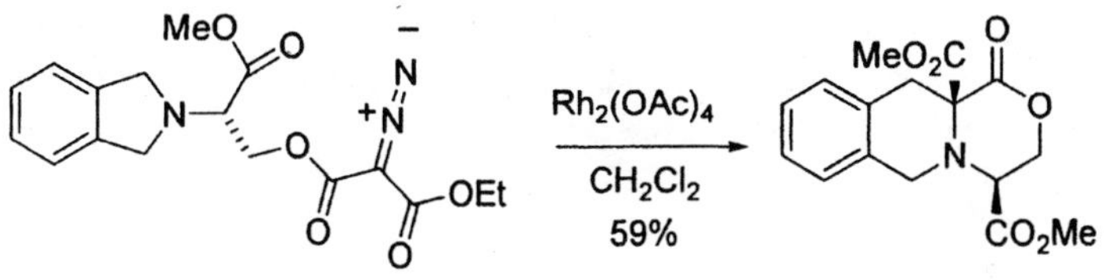

236 237

Rhodium(II) acetate catalyzes the decomposition of diazo diketone (**238**) to give nitroindole (**239**).[33]

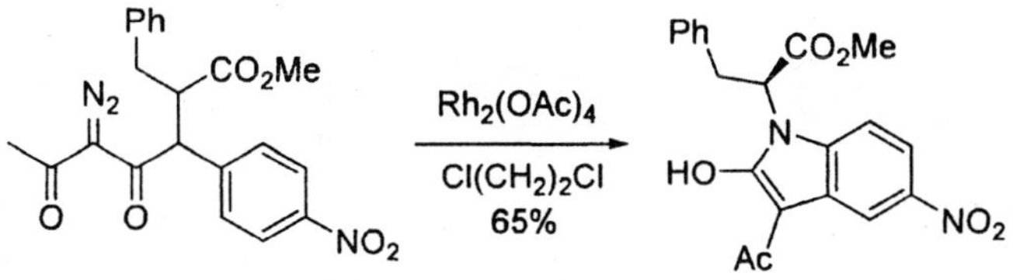

238 239

α-Diazo-β-keto ester (**154**) in the presence of a catalytic amount of $Rh_2(OAc)_4$ in CH_2Cl_2 at rt provides smooth conversion to the desired tricyclic ether (**240**).[69] Rh-mediated intramolecular C–H insertion gives fused cyclopentenone carboxylic acid.[90] Rhodium acetate catalyzed decomposition of 2-diazo-6-methylene-1,3-cyclooctanedione (**134**) gives propellane-2,8-dione (**241**) *via* transannular addition of a carbene to an exocyclic double bond.[54]

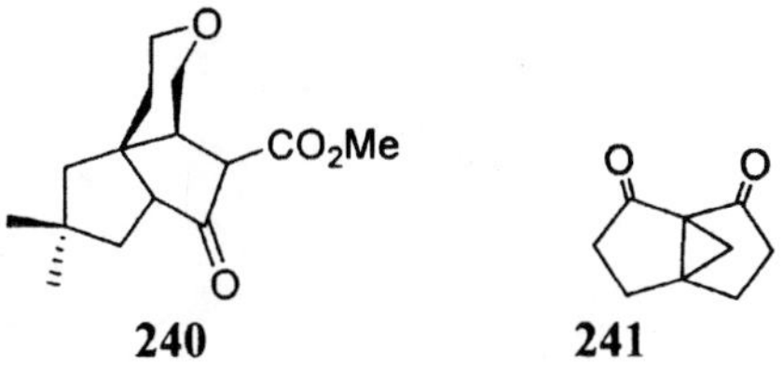

240 241

7.9.5.17 Synthesis of heterocycles

Synthesis of pyrrole derivatives

Although the chemistry of α-diazocarbonyl compounds have been extensively investigated the formation of pyrrole derivatives from α-diazo carbon bearing δ-(*N*-tosyl)amine group in unprecedented. The $Rh_2(OAc)_2$-catalyzed reaction of diazo 1,3-diketone **242** gives pyrrole derivatives (**243**) via intramolecular N–H insertion.[83]

$Rh_2(OAc)_4$, PhH, reflux, 75%

242 **243**

Diazo to pyrazole ring formation

Diazo transfer to cinnamyl chloride (**244**) using diazomethane, which in turn produced nitrosomethylurea, produces diazo acetyl pyrazoline (**245**).[91]

$[CH_2N_2]$, Ether

244 **245**

(*Z*)-6-Aryl-3-diazo-6-hydroxy-2,4-dioxohexenoic acid (**246**) reacts with aqueous NH_3 to give *Z*-6-aryl-3-diazo-6-hydroxy-2,4-dioxohexenamide (**248**) (28%) and 5-aryl-4-hydroxy-5-oxamamoyl-pyrazole (**247**, 53%).[92]

NH_4OH, 2 h, r.t., 41%

246 **247** **248**

Synthesis of 1,2,3-triazoles

Diazo transfer onto β-imino ketone (**249**) generally leads directly to 1,2,3-triazole (**250**). Benzoyl acetaldehyde anil produces 1-phenyl-4-benzoyl-1,2,3-triazole (**250**) in 90% yield.[1]

TsN_3, EtOH, KOH, 90%

249 **250**

Synthesis of triazolopyridine

Dehydrogenation of α-pyridinealdehyde hydrazone (**251**) yields not α-pyridyldiazomethane (**252**) but the cyclic isomer 1,2,3-triazolo[3,4-a]pyridine (**253**).[1] A similar ring closure is observed in diazo group transfers onto alkyl or aryl(2-pyridyl)methyl ketone (**254**) in MeOH/KOEt.[1] The diazo intermediate could not be isolated but 3-acyl[1,2,3]triazolo[3,4-a]pyridine (**255**) is obtained.[1]

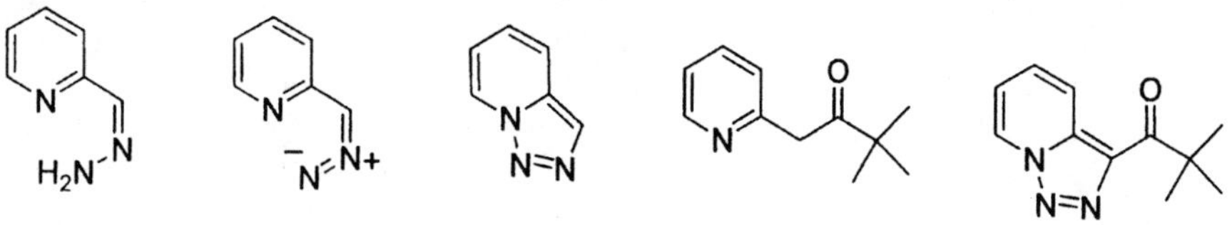

251 **252** **253** **254** **255**

Synthesis of 1,2,3-thiadiazoles

Diazo transfer from *p*-toluenesulfonyl azide onto acylthioacetamide (**256**) takes place in EtOH to give 90% of 4-acyl-5-phenylamino-1,2,3-thiadiazole (**258**).[1] The diazo intermediate (**257**) could not be isolated, since the cycloaddition is evidently faster than the diazo transfer reaction.[1] Thiadiazoles are also prepared by 1,3-dipolar cycloaddition of diazo carbonyl compounds to isothiocyanates in moderate yields.[1]

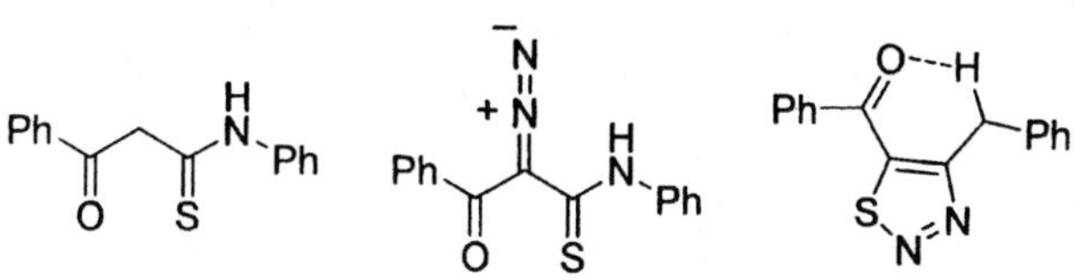

256 **257** **258**

7.9.6 Experimental

Diazo dihexyl ketone (43)[5]

A flame-dried, two-necked, 25-mL round-bottomed flask equipped with a septum and nitrogen purge was flushed with N_2 and charged with 144 mg (3.03 mmol) of 50% sodium hydride dispersion in mineral oil, one drop of absolute ethanol, and 2 mL of anhydrous ether. This mixture, while magnetically stirred, was cooled in an ice bath. Subsequently, 200 mg (1.01 mmol) of dihexyl ketone (**42**) and 222 mg (3.01 mmol) of ethyl formate in an additional 2 mL of ether were added dropwise. This reaction was stirred for 3 h in the ice/water bath and then overnight at room temperature. Mesyl azide (363 mg, 3.03 mmol) in 5 mL of ether was then added, and stirring was continued for an additional 2 h. The reaction was quenched with 1 mL of water. The organic layer was washed with 30 mL of 10% aqueous NaOH solution, and the aqueous layer was back extracted with three 30-mL portions of extraction solvent. The combined organic layers were dried over anhydrous $MgSO_4$ and concentrated *in vacuo*. The residue was chromatographed on 20 g of silica gel with 1% EtOAc/petroleum ether. The first 120 mL was discarded. The next 120 mL was concentrated *in vacuo* to give 160 mg (0.714 mmol, 71%) of diazo dihexyl ketone (**43**) as a yellow oil: R_f (10% EtOAc/hexane) 0.51.

1,4-Dimethyl-5,6,8,9-tetrahydro-6-diazo-7H-benzo[a]cyclohepten-7-one (68)[25]
An ice-cooled, magnetically stirred suspension of sodium hydride (0.04 g, 60% dispersion in mineral oil, washed with hexane) in anhydrous ether (2.0 cm^3) was treated with 1 drop of absolute ethanol and then with a mixture of the ketone **67** (0.188g, 1.0 mmol) and ethyl formate (0.074 g, 1 mmol) in ether (2.0 cm^3). Stirring was continued for 3 h at 0–8 °C and at room temperature overnight. At this point methanesulfonyl azide (0.121 g, 1.0 mmol) in ether (3.0 cm^3) was added and 2 h later the reaction was quenched with water (1.0 cm^3). The organic solution was separated, washed with 10% sodium hydroxide, and combined with back-extracts (3 × 10 cm^3 ether) of the alkaline solution, dried, and evaporated. The residue was chromatographed, using a 1:9 EtOAc–hexane mixture as eluent to afford the α-diazoketone **68** (0.154 g, 72%). υ max (film)/cm^1 2081, 1612.

2-Diazo-1-phenyl-butane-1,3-dione (136)[63]
Polystyrene benzenesulfonyl azide (500 mg, 0.75 mmol) was placed in a 5.0 mL disposable polypropylene/ polyethylene syringe and swollen with CH_2Cl_2. A mixture of 1-benzoylacetone (81.0 mg, 0.5 mmol) and Et_3N (0.21 mL, 1.5 mmol) in CH_2Cl_2 (2.0 mL) was drawn into the syringe containing the resin, placed on a LabQuake shaker, and rotated at room temperature. The reaction progress was monitored by TLC (1:1 Et_2O/heptane). After 4 h, the supernatant was collected, and the resin was washed with CH_2Cl_2 (3–5 mL). The washes were combined with the supernatant and concentrated to give a pale yellow solid that was reasonably pure by NMR. For the purpose of complete characterization, the solid was dissolved in a minimum volume of CH_2Cl_2 and passed through a short column of SiO_2 eluting with 3:1 heptane/EtO_2. Concentration of eluent provided an ivory solid (92.0 mg, 98%): mp 60–61 °C.

Indene methyl ester (220)[18]
A threaded Pyrex test tube, charged with a yellow solution of diazoketone **219** (10 mg, 0.049 mmol) in 15 mL of MeOH, was irradiated in a Rayonet apparatus for 16 h. The colorless solution was evaporated to a crude oil and chromatographed to yield methyl ester **220** (7.0 mg, 70%) as a clear oil which was a 4:1 mixture of diastereomers. TLC R_f (10% MTBE/petroleum ether) 0.81; IR (film) 1737 cm^{-1}.

Tricyclic α-keto ester (240)[69]
Cyclization of α-diazo ester **154** with rhodium acetate: Rhodium acetate (30 mg, 0.068 mmol) was added to a magnetically stirred solution of α-diazo ester **7** (1.20 g, 4.12 mmol) in dry methylene chloride (25 mL) under nitrogen at 25 °C. After 3 h, TLC showed complete conversion to a single, more polar, UV-active product. Methylene chloride was removed *in vacuo*. TLC mesh column chromatography on silica gel (35 g) with 85:15 petroleum ether/ethyl acetate yielded the tricyclic β-keto ester **240** as a colorless oil. Yield: 993 mg (91%). TLC R_f 0.21 on silica gel; 8:2 hexane/ethyl acetate. IR (CCl_4) 1760, 1735 cm^{-1}.

7.9.7 References

1 [R] Regitz, M. *Angew. Chem. Int. Ed. (Engl)* **1967**, 733.
2 [R] Regitz, M. *Synthesis* **1972**, 351
3 [R] Li, J. J. *Name Reactions*, 3rd ed., Springer-Verlag: New York, 2006, pp. 489–491.
4 (a) Rao Y. K.; Nagarajan, M. *Indian J. Chem.* **1986**, *25*, 735. (b) Dürr, H.; Hauck, G.; Brüeck, W.; Kober, H.; *Z. Naturforsch. B Anorg.Chem. Org. Chem.* **1981**, *36*, 1149. (c) Regitz, M.; Rüter, J. *Chem. Ber.* **1968**, *101*, 1263. (d) Regitz, M.; Menz, F.; Liedhegener, A. M. *Liebigs Ann. Chem.* **1970**, *739*, 174.
5 Taber, D. F.; Ruckle, R. E.; Hennessy, M. J. *J. Org. Chem.* **1986**, *51*, 4077.
6 Hünig, S.; Boes, O. *Justus Liebigs Ann. Chem.* **1953**, *579*, 28.
7 Regitz, M.; *Chem. Ber.* **1964**, *97*, 2742.
8 Regitz, M.; Himbert, G. *Liebigs Ann. Chem.* **1970**, *734*, 70.
9 Grünanger, P.; Finzi, P.V.; Scotti, C. *Chem. Ber.* **1965**, *98*, 623.
10 Regitz, M.; Liedhegener, A. *Chem. Ber.* **1966**, *99*, 3128.
11 Charette, A. B.; Wurz, R.P.; Ollevier, T. J. *Org. Chem.* **2000**, *65*, 9252.
12 Traulsen, T.; Friedrichsen, W. *J. Chem. Soc. Perkin Trans.* 1 **2000**, 1387.
13 (a) van Lewen, A. M.; Smid, P. N.; Strating, J. *Tetrahedron* **1965**, 337.
(b) Regitz, M. *Chem. Ber.* **1965**, *98*, 36
14 Regitz, M.; Anschütz, W. *Chem. Ber.* **1969**, *102*, 2216.
15 Regitz, M.; Anschütz, W.; Bartz, W.; Liedhegener, A. *Tetrahedron Lett.* **1968**, 3171.
16 Balli, H.; Gipp, R. *Liebigs Ann. Chem.* **1966**, *699*, 133
17 Boyer, J.H.; Mack, C. H.; Goebel, C. H.; Morgan, L. R., Jr. *J. Org. Chem.* **1959**, *23*, 1051.
18 Taber, D. F.; Kong, S.; Malcolm, S. C. *J. Org. Chem.* **1998**, *63*, 7953.
19 Lombardo, L.; Mander, L. N. *Synthesis* **1980**, **5**, 368.
20 Uyehara, T.; Takehara, N.; Ueno, M.; Sato, T. *Bull. Chem. Soc. Jpn.* **1995**, *68*, 2687.
21 Julia, S.; Linstrumelle, G. *Bull. Soc. Chim. Fr.* **1966**, 3490.
22 Borhan, B.; Kunz, R.; Wang, A. Y.; Nakanishi, K.; Bojkova, N.; Yoshihara, K. *J. Am. Chem. Soc.* **1997**, *119*, 5758.
23 Ward, J. L.; Gaskin, P.; Beale, M. H.; Sessions, R.; Koda, Y.; Wasternack, C. *Tetrahedron* **1997**, *53*, 8181.
24 Mandal, P.; Venkateswaran, R. V. *J. Chem. Res. (S)* **1998**, **88**.
25 Ho, T.-L.; Lin, Y.-J. *J. Chem. Soc. Perkin. Trans.* 1 **1999**, *9*, 1207.
26 Saha, G.; Ghosh, S. *Synth. Commun.* **1991**, *21*, 2129.
27 Meier, G.; Rang, H.; Emrich, R. *Liebigs Ann. Chem.* **1995**, *1*, 153.
28 Otterbach, A.; Musso, H. *Angew. Chem. Int. Ed. (Engl.)* **1987**, *26*, 554.
29 Brahms, J. C.; Dailey, W. P. *Tetrahedron Lett.* **1990**, *31*, 1381.
30 Ghosh, S.; Datta, I. *Synth. Commun.* **1991**, *21*, 191.
31 Ihara, M.; Suzuki, T.; Katogi, M.; Taniguchi, N.; Fukumoto, K. *J. Chem. Soc. Perkin Trans. 1* **1992**, 865.
32 (a) Franzen, V. *Justus Liebigs Ann. Chem.* **1957**, *602*, 199. (b) Reimlinger, H.; Skatteböl, L. *Chem. Ber.* **1960**, *93*, 2162.
33 Zaragoza, F. *Tetrahedron* **1995**, *51*, 8829.
34 Wiberg, K. B.; Ubersax, R. W. *J. Org. Chem.* **1972**, *37*, 3827.
35 Allinger, N. L.; Freiberg, L. A.; Hermann, R. B.; Miller, M. A. *J. Am. Chem. Soc.* **1963**, *85*, 1171.
36 Muchowski, J. M. *Tetrahedron Lett.* **1966**, 1773.
37 Ueda, K.; Toda, F. *Chem. Lett.* **1975**, 779.
38 Pirrung, M. C.; Krishnamurthy, N.; Nunn, D. S.; McPhail, A. T. *J. Am. Chem. Soc.* **1991**, *113*, 4910.
39 Horner, L.; Kirmse, W.; Muth, K. *Chem. Ber.* **1958**, *91*, 430
40 Pudleiner, H.; Laatsch, H. *Liebigs Ann. Chem.* **1990**, 423.
41 Ireland, R. E.; Dow, W. C.; Godfrey, J. D.; Thaisrivongs, S. *J. Org. Chem.* **1984**, *49*, 1001.
42 Meinwald, J.; Gassmann P. G. *J. Am. Chem. Soc.* **1960**, *82*, 2857.
43 Nace, H. R.; Nelander, D. H. *J. Org. Chem.* **1964**, *29*, 1677.
44 Cava, M. P.; Vogt, B. R. *Tetrahedron Lett.* **1964**, **2813**.
45 Cava, M. P.; Vogt, B. R. *J. Org. Chem.* **1965**, *30*, 3775.
46 Eadon, G.; Popov, S.; Djerassi, C. *J. Am. Chem. Soc.* **1972**, *94*, 1282.
47 Conti, P.; Kozikowski, A. P. *Tetrahedron Lett.* **2000**, *41*, 4053.
48 Norbeck, D. W.; Kramer, J. B. *J. Am. Chem. Soc.* **1988**, *11*, 7217.
49 Lee, J. C.; Yuk, J. Y. *Synth. Commun.* **1995**, *25*, 1511.
50 Wild, H. *J. Org. Chem.* **1994**, *59*, 2748.
51 McGuiness, M.; Shechter, H. *Tetrahedron. Lett.* **1990**, 31, 4987.
52 Benati, L.; Montevecchi, P. C.; Spagnolo, P. *Gazz. Chim. Ital.* **1992**, *122*, 249.
53 Cossy, J. *Tetrahedron. Lett.* **1989**, *30*, 4113.
54 Reingold, I. D.; Drake, J. *Tetrahedron Lett.* **1989**, *30*, 1921.
55 Kantin, G. P.; Nikolaev, V. A. *Russ. J. Org. Chem.* **2000**, *36*, 486.
56 de Rianelli, S. R.; de Souza, M. C. B. V.; Ferreira, V. F. *Synth. Commun.* **2004**, *34*, 951.
57 Regitz, M.; Schwall, H.; Heck, G.; Eistert, B.; Bock, G. *Liebigs Ann. Chem.* **1965**, *690*, 125.

58 Hark, R. R.; Hauze, D. B.; Petrovskaia, O.; Joullié, M. M. *Can. J. Chem.* **2001**, *79*, 1632.
59 Kumar, S. M. *Synth. Commun.* **1991**, *21*, 2121.
60 Popic, V. V.; Korneev, S. M.; Nikolaev, V. A.; Korobitsyna, I. K. *Synthesis* **1991**, 195.
61 Roush, W. R.; Feitler, D.; Rebek, J. *Tetrahedron Lett.* **1974**, 1391.
62 Yang, R.-Y.; Dai, L.-X. *Mendeleev Commun.* **1993**, *3*, 82.
63 Green, G. M.; Peet, N. P.; Metz, W. A. *J. Org. Chem.* **2001**, *66*, 2509.
64 Regitz, M.; Menz, F.; Ruter, J. *Tetrahedron Lett.* **1967**, 739.
65 Eistert, B.; Greiber, D.; Caspari, I. *Justus Liebigs Ann. Chem.* **1962**, *659*, 64.
66 Eistert, B.; Bock, G.; Kosch, E.; Spalink, F. *Chem. Ber.* **1960**, *93*, 1451.
67 Benati, L.; Calestani, G.; Montevecchi, P. C.; Spagnolo, P. *J. Chem. Soc., Perkin. Trans. 1* **1995**, *11*, 1381.
68 Taber, D. F.; Kanai, K. *Tetrahedron* **1998**, *54*, 11767.
69 Taber, D. F.; Schuchardt, J. L. *J. Am. Chem. Soc.* **1985**, *107*, 5289.
70 Nikolaev, V. A.; Kantin, G. P.; Utkin, P. Y. *Russ. J. Org. Chem.* **1994**, *30*, 1354.
71 Zaragoza, F. *Synlett* **1995**, 237.
72 Gleiter, R.; Krennrich, G.; Langer, M. *Angew. Chem. Int. Ed. (Engl)* **1986**, *25*, 999.
73 Evans, D. A.; Britton, T. C.; Ellman, J. A.; Dorow, R. L. *J. Am. Chem. Soc.* **1990**, *112*, 4011.
74 (a) Schöllkopf, U.; Schäfer, H. *Angew. Chem. Int. Ed. (Engl.)* **1966**, *77*, 379. (b) Closs, E. L.; Coyle, J. J. *J. Am. Chem. Soc.* **1965**, *87*, 4270.
75 Cram, D. J.; Patros, R. D. *J. Am. Chem. Soc.* **1963**, *85*, 1273.
76 Müller, E.; Ludsteck, D. *Chem. Ber.* **1954**, *87*, 1887.
77 Müller, E.; Disselhoff, H. *Liebigs Ann. Chem.* **1934**, *512*, 250.
78 Schollkopf, U.; Frasnelli, H., *Angew. Chem. Int. Ed. (Engl)* **1970**, *9*, 301.
79 Eistert B.; Borggrefe, G. *Liebigs Ann. Chem.* **1968**, *718*, 142.
80 Hoker, J.; Regitz, M.; Liedhegener, A. *Chem. Ber.* **1970**, *103*, 1486.
81 Penke, B.; Czombos, L.; Balaspiri, J.; Petres, K.; Kovacs, K. *Helv. Chim. Acta* **1970**, *53*, 1057.
82 Deng, G.; Tian, X.; Qu, Z.; Wang, J. *Angew. Chem. Int. Ed. (Engl.)* **2002**, *41*, 2773.
83 Deng, G.; Jiang, N.; Ma, Z.; Wang, J. *Synlett* **2002**, 1913.
84 Deng, G.; Tian, X.; Wang, J. *Tetrahedron Lett.* **2003**, *44*, 587.
85 Zhdanova, O. V.; Korneev, S. M.; Nikolaev, V. A. *Russ. J. Org. Chem.* **2004**, *40*, 316.
86 Taber, D. F.; Herr, R. J.; Gleave, D. M. *J. Org. Chem.* **1997**, *62*, 194.
87 Garcia, F. S.; Cebrian, G. M. P.; Lopez, A. H.; Herrera, F. J. L. *Tetrahedron* **1998**, *54*, 6867.
88 Holt, G.; Wall, D. K. *J. Chem. Soc. C.* **1966**, **857**.
89 Andraos, J.; Chiang, Y.; Kresge, A. J.; Pojarlieff, I. G.; Schepp, N. P.; Wirz, J. *J. Am Chem. Soc.* **1994**, *116*, 73.
90 Taber, D. F.; Schuchardt, J. L. *Tetrahedron* **1987**, *43*, 5677.
91 Moore, J. A. *J. Org. Chem.* **1955**, *20*, 1607
92 Zalesov, V. V.; Vyaznikova, N. G.; Andreichikov, Y. S. *Russ. J. Org. Chem.* **1995**, *31*, 1104.

Marudai Balasubramanian

7.10 Sommelet Reaction

7.10.1 Description

The Sommelet reaction, also known as the Sommelet oxidation, is the oxidation of benzyl halides (**1**) to the corresponding aldehydes (**3**) using hexamethylenetetramine (HMTA) followed by hydrolysis of the hexaminium salt **2**.[1–3] This reaction is closely related to the Delépine amine synthesis, the Duff aldehyde synthesis, and the Kröhnke reaction (see Section 7.10.3).

1 2 3

hexamethylenetetramine

7.10.2 Historical Perspective

In 1913, Marcel Sommelet (1877–1952) of the Faculté de Pharmacie of Paris University published his seminal article in *Comptes Rendus* with the title "*Decomposition of Alkyl Halide Addition Products of Hexamethylenetetramine*," where he delineated the formation of aldehydes by treatment of benzyl halides followed by hydrolysis.[4,5] In 1954, S. J. Angyal published a review in *Organic Reactions*, where he summarized the mechanism, scope, and limitations of the Sommelet reaction.[1] Angyal's review ignited a flurry of research and the Sommelet reaction became a well-known standard for the transformation of benzyl halides to the corresponding benzaldehydes.

7.10.3 Mechanism

At the time of its discovery, Sommelet recognized that the reaction took place in three stages:

a. Formation of a hexaminium salt **2** from benzyl halides **1** and HMTA;

1 2

b. Hydrolysis of this salt to an amine and its methylene derivative, that is, benzylamine $ArCH_2NH_2$ (**4**) and aldimine $ArCH_2NH{=}CH_2$ (**5**); both intermediates **4** and **5** were isolated from the reaction mixture;

c. Formation of an aldehyde (the Sommelet reaction proper).

However, he did not understand the nature of the third stage. He tentatively suggested that the reaction occurred by shift of the double bond in $ArCH_2NH{=}CH_2$ (**5**) to give $ArCH_2{=}NHCH_3$ (**6**), which in turn was hydrolyzed to benzaldehyde and methylamine. Quickly, Sommelet realized that the mechanism was not tenable and recognized the reaction was an *oxidation–reduction* process. He announced his views in two lectures,[6,7] but did not publish them and the true mechanism was largely unknown. In 1948, Shoppee's calculation indicated that the transformation of **5** to **6** was actually inherently impossible.[8] Angyal contributed significantly in deciphering the mechanism of the Sommelet reaction.[9] Based on the fact that $ArCH_2NH_2$ (**4**) and $ArCH_2NH{=}CH_2$ (**5**) are isolated intermediates, Angyal's mechanism may be summarized as follows:

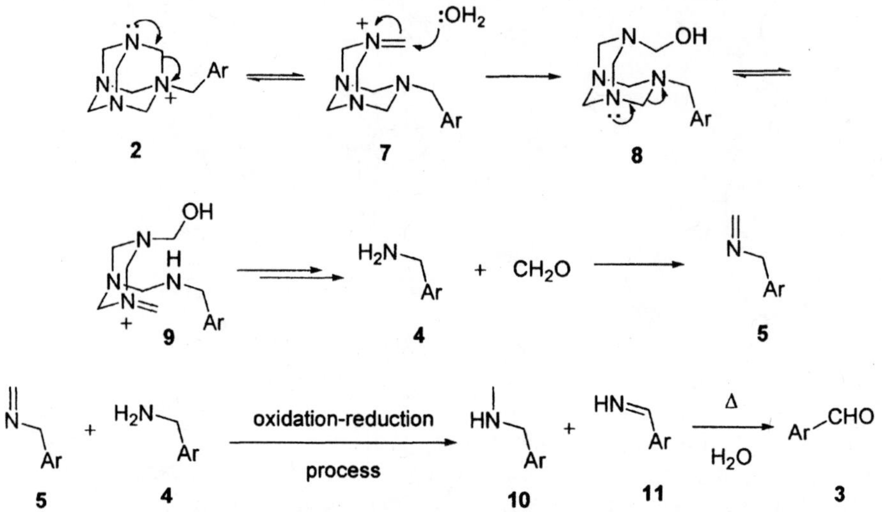

When benzyl halide **1** is treated with HMTA, the first step was the straightforward formation of hexaminium salt **2**. In the presence of water, an isomer of **2**, iminium **7**, is hydrolyzed to give hemiaminal **8**. Fragmentation of **8** affords iminium **9**, which collapses to benzylamine **4** and formaldehyde. Condensation of benzylamine **4** and formaldehyde provides another key intermediate, aldimine **5**. The crucial key *oxidation–reduction* process takes place between **4** and **5** with the exchange of two equivalents of hydride, giving rise to methylbenzylamine **10** and imine **11**, which is subsequently hydrolyzed to benzaldehyde **3**.

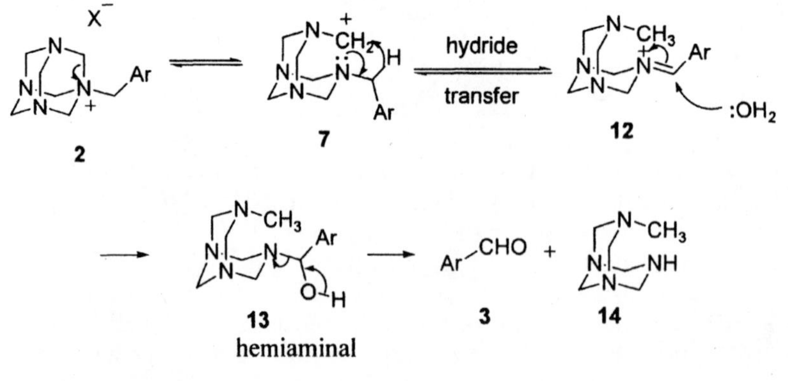

13
hemiaminal

While the aforementioned mechanism covers all the basis of experimental data, it is long and tedious. In 1961, Le Henaff simplified the mechanism to one that has been widely accepted by the chemistry community.[10–12] According to Le Henaff, hexaminium salt **2** isomerizes to iminium **7**, which undergoes a hydride transfer to provide iminium **12**. Hydrolysis of iminium **12** affords hemiaminal **13**, which then delivers benzaldehyde **3** and amine **14**.

7.10.4 Variations and Improvements

The Delépine reaction or the Delépine amine synthesis is the closest variation of the Sommelet reaction. The Delépine amine synthesis is the *acid hydrolysis* of the hexaminium salt **2** with strong acid in ethanol, giving rise to benzylamine **4** with the formaldehyde being removed as volatile formaldehyde diacetal.[13–15] In essence, it may be considered as "half" of the Sommelet reaction or the "interrupted" Sommelet reaction. The Delépine reaction works well for active halides such as benzyl, allyl halides, and α-halo-ketones.

Ar⁀X + hexamine (1) —Δ, $CHCl_3$→ 2 —3 M HCl, EtOH, 6 H_2O→ $ArCH_2NH_2$ (4)

The net outcome of the Sommelet reaction is the oxidation of the benzyl methylene position into the corresponding carbonyl, and in that sense, the Kröhnke reaction may be considered a close variant. The Kröhnke reaction transforms benzyl halides to benzaldehydes using pyridine and an aromatic nitroso compound **16**.[16,17] As shown below, quaternary pyridinium salt (**15**) is formed from addition of pyridine to benzyl halide **1**. Oxidation of **15** with *p*-nitrosodimethylaniline (**16**) affords nitrone **17**, which is then hydrolyzed to benzaldehyde **3** and hydroxylamine **18**.

Ar⁀X + pyridine → 15 —16, NaOH→ 17 —H_2O→ Ar-CHO (3) + 18

The third variation of the Sommelet reaction is the oxidation of benzyl halide **1** to the corresponding benzaldehyde **3** using **19**, the sodium salt of 2-nitropropane (**20**).[18,19]

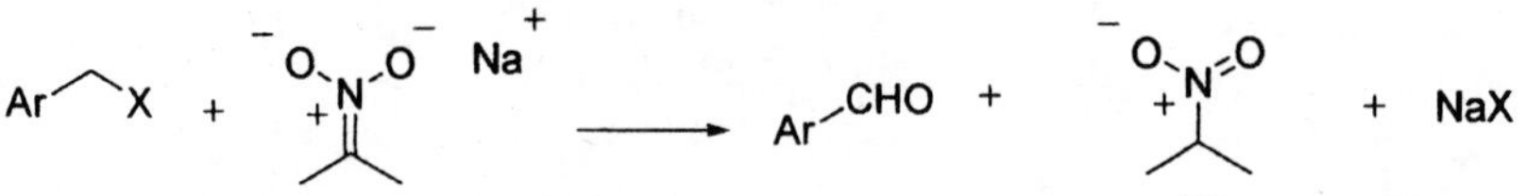

1 19 3 20

Submission of secondary amines such as **21** to Sommelet reaction conditions provided the corresponding two benzaldehydes such as **22** and **23**.[20] In addition, a variant of the Sommelet reaction transformed α,β-unsaturated imines into the corresponding α,β-unsaturated ketones.[21]

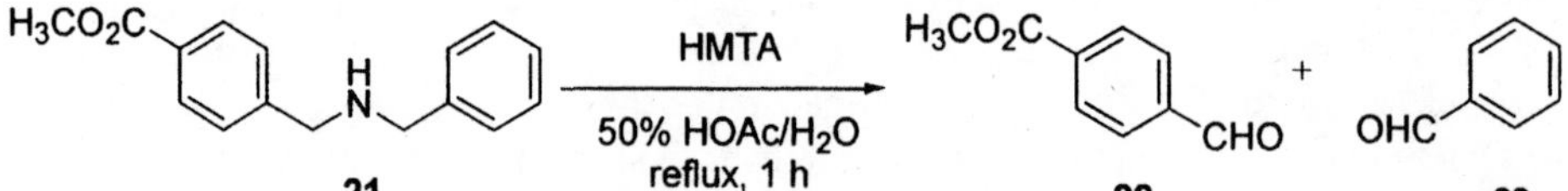

21 22 23

Finally, the Sommelet reaction may take place on the secondary benzyl halide such as **24a** to give the corresponding ketone **25**.[22] When the chlorine is replaced with an amine, analog **24b** also gave ketone **22** under Sommelet reaction conditions.

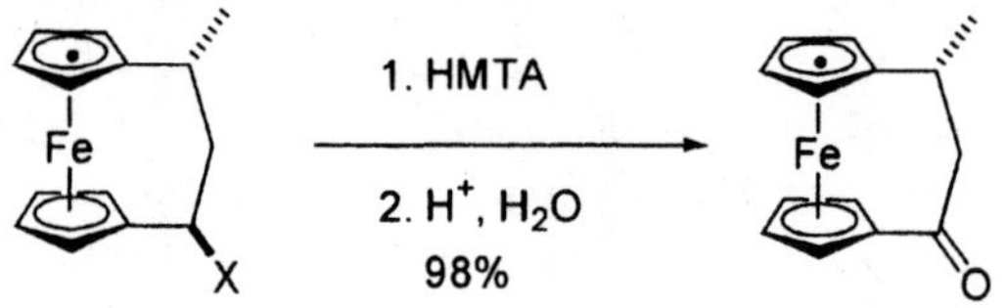

24 25

24a, X = Cl
24b, X = NH_2

7.10.5 Synthetic Utility

Although initial examples indicated that hindered substrates did not work well for the Sommelet reaction, Stokker discovered that some 2,6-di-substituted benzylamines underwent the amine-to-aldehyde conversion in 17–68% yields.[23] The best result was obtained with benzylamine **26**, which gave benzaldehyde **27** in 68% yield. In another report,[24] 2,6-difluorobenzyl chloride was converted to 2,6-difluorobenzylaldehyde in 62% yield although the steric hindrance of fluoride is insignificant in this case.

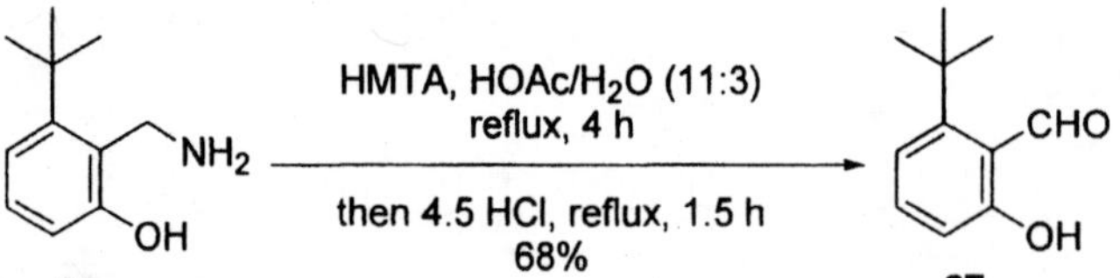

26 27

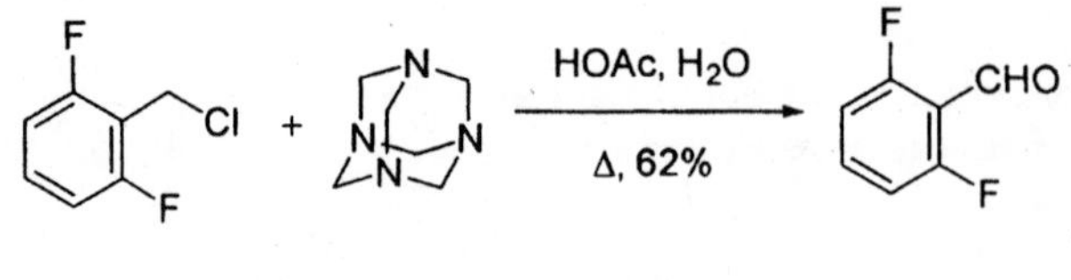

The scope of the Sommelet reaction has been explored. For instance, it was extended to make aromatic dialdehydes with yields ranging from 24% to 70%.[25] In addition, the Sommelet reaction has been extended to aliphatic allyl bromide although β-methylcrotonaldehyde (**29**) was obtained in only 35% from γ,γ-dimethylallyl bromide (**28**).[26]

HMTA, $CHCl_3$, then H_2O, reflux, H_2SO_4, 35%: **28** → **29**

The Sommelet reaction has been widely extended to heterocycles, which are important to medicinal chemists. With yields ranging from 50% to 57%, aldehydes of pyridine, isoquinoline, and thiazole were prepared from the corresponding heteroarylmethyl bromides.[27] 5-Benzyloxy-3-bromomethyl-benzo[*b*]thiophene (**30**) was converted to 3-carboxaldehyde **31** in 40% yield.[28] Thiazole **32** was the core structure in another Sommelet reaction to give **33** although in poor yield.[29] In contrast, a different thiazole substrate **34** gave aldehyde **35** in 49% yield.[30] Benzopyrans are also tolerated by the Sommelet reaction conditions.[31,32] For example, substrate **36** was converted to aldehyde **37** in 58% yield.[32]

30 → **31**: 1. HMTA, $CHCl_3$, reflux, 6 h; 2. 50% HOAc/H_2O, reflux, 3 h; 40%

32 → **33**: HMTA, $CHCl_3$, then H_2O, reflux; 18%

34 → **35**: HMTA, $CHCl_3$, then H_2O, reflux; 49%

36 → **37**: HMTA, 50% HOAc/H_2O, reflux, 0.5 h, 58%

As with heterocycles, the Sommelet reaction tolerates ester and phosphonates. For instance, benzyl bromide **38** was oxidized to benzaldehyde **39** in the presence of phosphonate functionality although the yield was not reported.[33]

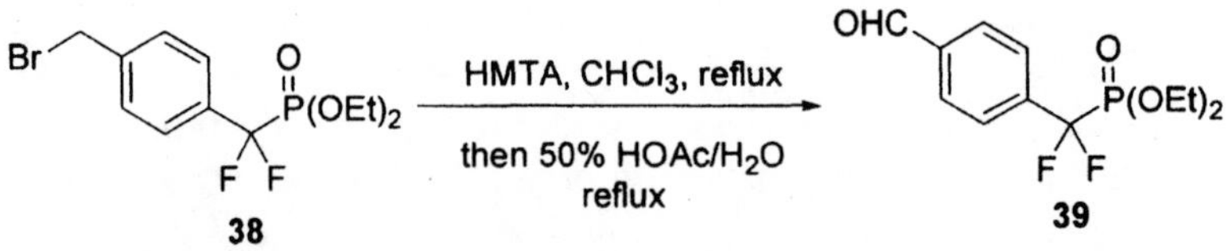

Finally, dibromomethyl derivatives also work well in the Sommelet reaction. For instance, bromination of 3,5-di-*tert*-butyltoluene (**40**) with *N*-bromosuccinamide (NBS) and benzoylperoxide (BPO) gave a mixture of monobromomethyl product (**41**) and dibromomethyl derivative (**42**) in 38% and 62% yield, respectively. Submission of both **41** and **42** to the Sommelet reaction conditions afforded the benzaldehyde **43** in 75% yield.[34]

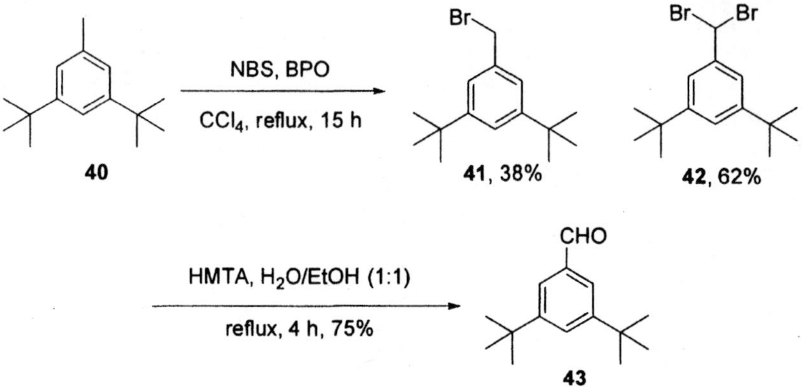

7.10.6 Experimental

According to Angyal's recommendation,[1] the optimal Sommelet reaction may be run without isolation of the hexaminium salt under such conditions: 0.1 mole of halomethyl compound is heated under reflux for one or two hours with 30 g of hexamine in 90 mL of 50% acetic acid, or with 16.5 g of hexamine in 90 mL of 60% ethanol.

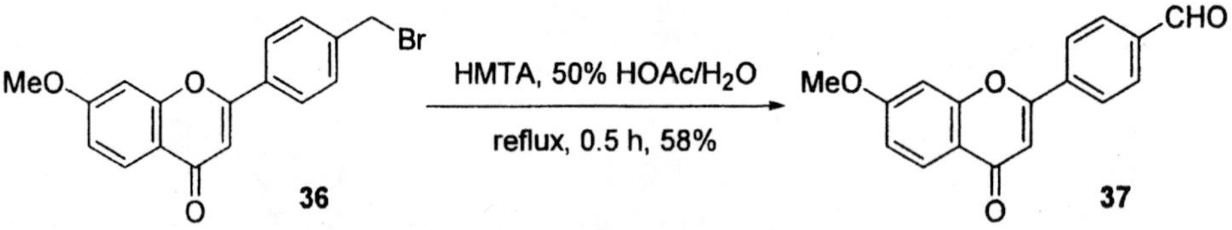

2-(4-Formylphenyl)-7-methoxy-4H-1-benzopyran-4-one (37)[32]

A mixture of **36** (0.85 g, 2.46 mmol) and HMTA (4 g, 28.6 mmol) in 30 mL of acetic acid (50%) was heated at reflux for 4 h. HCl (10 mL, 50%) was added and refluxing was continued for 0.5 h. The reaction mixture was diluted with water and the precipitate was collected. Recrystallization of the solid from EtOAc–*n*-hexane (50%) gave **37** as a white powder; yield (0.5 g, 58.4%): mp 183–184 °C.

7.10.7 References

1. [R] Angyal, S. J. The Sommelet reaction. *Org. React.* **1954**, *8*, 197–217.
2. [R] Blažević, N.; Kolbah, D.; Belin, B.; Šunjić, V.; Kajfež, F. *Synthesis* **1979**, 161–179.
3. [R] Kilenyi, S. N., in *Encyclopedia of reagents of Organic Synthesis*, ed. Paquette, L. A., Wiley: Hoboken, NJ, **1995**, *vol. 3*, p. 2666.
4. Sommelet, M. *Bull. Soc. Chim. France* [4], **1913**, *13*, 1085.
5. Sommelet, M. *Compt. Rend.* **1913**, *157*, 852.
6. Sommelet, M. *Bull. Soc. Chim. France* [4], **1915**, *17*, 82.
7. Sommelet, M. *Bull. Soc. Chim. France* [4], **1918**, *23*, 95.
8. Shoppee, C. W. *Nature* **1948**, *161*, 619.
9. Angyal, S. J.; Rassack, R. C. *Nature* **1948**, *161*, 723.
10. Le Henaff, P. *Compt. rend.* **1961**, *253*, 2706.
11. Le Henaff, P. *Ann. Chim.* **1962**, *7*, 367.
12. Schnekenburger J; Kaufmann R *Arch. Pharm.* **1971**, *304*, 259.
13. Delépine, M. *Bull. Soc. Chim.Paris* **1895**, *13*, 355; **1897**, *17*, 290.
14. Delépine, M. *Compt. Rend.* **1895**, *120*, 501; **1897**, *124*, 292.
15. [R] Blažzević, N.; Kolnah, D.; Belin, B.; Šunjić, V.; Kafjež, F. *Synthesis* **1979**, 161–178.
16. Kröhnke, F.; Böner, E. *Ber.* **1936**, *69*, 2006.
17. Kröhnke, F. *Ber.* **1938**, *71*, 2583.
18. Hass, H. B.; Bender, M. *J. Am. Chem. Soc.* **1949**, *71*, 1767.
19. Hass, H. B.; Bender, M. *Org. Synth.* **1950**, *30*, 99.
20. Snyder, H. R.; Demuth, John R. *J. Am. Chem. Soc.* **1956**, *78*, 1981.
21. Armesto, D.; Horspool, W. M.; Martin, J. A. F.; Perez-Ossorio, R. *Tetrahedron Lett.* **1985**, *26*, 5217.
22. Tebben, L.; Neumann, M.; Kehr, G.; Froehlich, R.; Erker, G.; Losi, S.; Zanello, P. *Dalton Transactions* **2006**, 1715.
23. Stokker, G. E.; Schultz, E. M. *Synth. Commun.* **1982**, *12*, 847.
24. Malykhin, E. V.; Steingarts, V. D. *J. Fluorine Chem.* **1998**, *91*, 19.
25. Wood, J. H.; Tung, C. C.; Perry, M. A.; Gibson, R. E. *J. Am. Chem. Soc.* **1950**, *72*, 2992.
26. Young, Wm. G.; Linden, Seymour L. *J. Am. Chem. Soc.* **1947**, *69*, 2912.
27. Angyal, S. J.; Penman, D. R.; Warwick, G. P. *J. Chem. Soc.* **1953**, 1740.
28. Campaigne, E.; Bosin, T.; Neiss, E. S. *J. Med. Chem.* **1967**, *10*, 270.
29. Simiti, I.; Hint, G. *Pharmazie* **1974**, 443.
30. Neville, M. C.; Verge, J. P. *J. Med. Chem.* **1977**, *20*, 946.
31. Atassi, G.; Briet, P. *Eur. J. Med. Chem.* **1985**, 393.
32. Göker, H.; Boykin, D. W.; Yildiz, S. *Bioorg. Med. Chem.* **2005**, *13*, 1707.
33. Holmes, C. P.; Li, X.; Pan, Y.; Xu, C.; Bhandari, A.; Moody, C. M.; Miguel, J. A.; Ferla, S. W.; De Francisco, M. N.; Frederick, B. T.; Zhou, S.; Macher, N.; Jang, L.; Irvine, J.D.; Grove, J. R. *Bioorg. Med. Chem. Lett.* **2005**, *15*, 4336.
34. Karamé, I.; Jahjah, M.; Messaoudi, A.; Tommasino, M. L.; Lemaire, M. *Tetrahedron: Asymmetry* **2004**, *15*, 1569.

Jie Jack Li

Appendixes

Appendix 1
Table of Contents for Volume 1: ***Name Reactions in Heterocyclic Chemistry***
Published in 2005

Appendix 2
Table of Contents for Volume 3: *Name Reactions for Chain Extension*
Due in 2009

Chapter 1 Organometallics

Chapter 3 Rearrangement

Section 3.1 Concerted rearrangement

3.1.1 Alder ene reaction
3.1.2 Claisen, Carrol rearrangement, Eschenmoser–Claisen, Johnson–Claisen and Ireland–Claisen rearrangements/Mislow–Evans rearrangement
3.1.3 Cope, oxy-Cope, anionic oxy-Cope rearrangements
3.1.4 Overman rearrangement
3.1.5 [1,2]-Wittig rearrangement
3.1.6 [2,3]-Wittig rearrangement

Section 3.2 Cationic rearrangement

3.2.1 Beckmann rearrangement
3.2.2 Demjanov rearrangement/Tiffeneau–Demjanov rearrangement
3.2.3 Meyer–Schuster rearrangement/Rupe rearrangement
3.2.4 Pinacol rearrangement
3.2.5 Pummerer rearrangement
3.2.6 Schmidt rearrangement
3.2.7 Wagner–Meerwein rearrangement

Section 3.3 Anionic rearrangement

3.3.1 Benzilic acid rearrangement
3.3.2 Brook rearrangement
3.3.3 Curtius rearrangement
3.3.4 Favorskii rearrangement rearrangement
3.3.5 Grob fragmentation
3.3.6 Neber rearrangement
3.3.7 Payne rearrangement
3.3.8 Smiles rearrangement

Section 3.4 Other rearrangement

3.4.1 Hofmann rearrangement
3.4.2 Lossen rearrangement
3.4.3 Stevens rearrangement
3.4.4 Wolff rearrangement

Chapter 4 Radical Chemistry

4.1 Barton–McCombie deoxygenation reaction
4.2 Barton nitrite photolysis
4.3 Sandmeyer reaction
4.4 Wohl–Ziegler reaction

Appendix 3
Table of Contents for Volume 4: *Name Reactions for Ring Formation*
Due in 2011

Chapter 5 Transformation of Carbocycles

Appendix 4
Table of Contents for Volume 5: ***Name Reactions in Heterocyclic Chemistry-2***
Due in 2013

10.3. Shaw pyrimidine synthesis

Chapter 11 Other Six-Membered Heterocycles

11.1 Allan–Robinson reaction
11.2 Baker–Venkataraman rearrangement
11.3 Bargellini reaction
11.4 Borsche cinnoline synthesis
11.5 Ferrario–Akermann thiocyclization
11.6 Gastaldi pyrazine synthesis
11.7 Gutknecht pyrazine synthesis
11.8 König benzoxazine synthesis
11.9 Lehmsted–Tanasrescu acridone synthesis
11.10 Niementowski quinazoline synthesis
11.11 Pechmann pyrone synthesis
11.12 Pechmann pyrazole synthesis
11.13 Robinson–Schöpf reaction
11.14 Simonis chromone cyclization
11.15 Wesseley–Moser rearrangement
11.16 Widman–Stoermer cinnoline synthesis
11.17 Wichterle reaction

Chapter 12 Potpourri

12.1 ANRORC mechanism
12.2 Boulton–Katritzky rearrangement
12.3 Chichibabin amination reaction
12.4 Dimroth rearrangement
12.5 Einhorn–Brunner triazole synthesis
12.6 Hantzsch synthesis of pyrroles, pyridines, and thiazoles
12.7 King–Ortoleva reaction
12.8 Traube purine synthesis

Subject Index

α-Alanine, Strecker amino acid synthesis, 492